Exclusively endorsed and approved by AQA

Teacher's Book

Geoff Carr • Ruth Miller

Series Editor: Lawrie Ryan

GCSE Biology

Published in 2006 by:
Nelson Thornes Ltd
Delta Place
27 Bath Road
CHELTENHAM
GL53 7TH
United Kingdom

06 07 08 09 10 / 10 9 8 7 6 5 4 3

A catalogue record for this book is available from the British Library

ISBN 0 7487 9642 8

Cover photographs: snail by Gerry Ellis/Digital Vision LC (NT); embryo by Biophoto/Science Photo Library; *E. coli* bacteria by Dr Gary Gaugler/Science Photo Library

Cover bubble illustration by Andy Parker
Illustrations by Bede Illustration

Page make-up by Wearset Ltd

Printed in Croatia by Zrinski

The following people have made an invaluable contribution to this book:

Pauline Anning, Jim Breithaupt, Nigel English, Ann Fullick, Patrick Fullick, Richard Gott, Keith Hirst, Paul Lister, Niva Miles, John Scottow, Glenn Toole.

GCSE Biology

Contents

GCSE Biology

Welcome to AQA Biology!

AQA Science for GCSE is the only series to be endorsed by AQA. The GCSE Biology Teacher Book is written by experienced science teachers and is designed to make planning the delivery of the specification easy – everything you need is right here! Information is placed around a reduced facsimile of the Student Book page, allowing you quick reference to features and content that will be used in the lesson.

LEARNING OBJECTIVES

These tell you what the students should know by the end of the lesson, linking directly to the Learning Objectives in the Student Book, but providing teachers with extra detail. The Learning Objectives for 'How Science Works' are all listed in Chapter 1. These should be integrated into the lessons chosen to teach various aspects of 'How Science Works' throughout the course.

LEARNING OUTCOMES

These tell you what the students should be able to do to show they have achieved the Learning Objectives. These are differentiated where appropriate to provide suitable expectations for all your students.

Teaching suggestions

Ideas on how to use features in the Student Book, suggestions for Gifted and Talented, Special Needs, ICT activities and different learning styles are all covered here, and more.

Practical support

For every practical in the Student Book you will find this corresponding feature which gives a list of equipment needed, safety references and further guidance to carry out the practical. A worksheet is provided on the e-Science CD ROM for each practical.

Activity notes

Each activity in the Student Book has background information notes on how to organise it effectively.

Icons

[icon] appears in the text where opportunities for investigational aspects of 'How Science Works' are signposted in the AQA specification.

? appears in the text where AQA have signposted opportunities to cover societal aspects of 'How Science Works' in the specification.

e-Science CD ROM

This contains a wide range of resources – animations, simulations, photopluses, Powerpoints, activity sheets, practical skill sheets, homework sheets – which are linked to Student Book pages and help deliver the activities suggested in the Teacher Book.

How science works

This is covered in the section at the beginning of the Student Book, in the main content, in the end of chapter spreads, and in the exam-style questions and 'How Science Works' questions. The corresponding teacher's notes give you detailed guidance on how to integrate 'How Science Works' fully into your lessons and activities.

Exam-Style Questions

There are multiple-choice questions for Biology 1a and 1b and structured questions for Biology 1, 2 and 3. They are ranked in order of difficulty for unit B1 and Higher Tier only questions are clearly marked for units B2 and B3. All questions in B1a and B1b are useful to complete, no matter which style of assessment is being taken, as they cover the same content. 'How Science Works' is integrated into some exam-style questions and there are separate 'How Science Works' questions to give additional practice in this area.

Lesson structure

This feature provides ideas for the experienced teacher, support for the newly qualified teacher and structure for cover lessons. Available for every double page lesson spread, it contains a variety of suggestions for how the spread could be taught, including starters and plenaries of varying lengths, as well as suggestions for the main part of the lesson.

Answers to questions

They're all here! All the questions in the Student Book are answered in the Teacher Book. Each answer is located in the corresponding feature in the Teacher Book. For example, answers to yellow in-text questions in the Student Book can be found in the yellow feature in the Teacher Book.

Key Stage 3 curriculum links:

This expands the 'What you already know' unit opener of the Student Book and gives QCA Scheme of Work references for relevant knowledge that may need revisiting before starting on the unit.

ACTIVITIES & EXTENSIONS

This highlights opportunities to extend a lesson or add activities, providing notes and tips on how to carry them out.

SPECIFICATION LINK-UP

This gives clear references to the AQA specification for the lesson, with additional notes and guidance where appropriate.

KEY POINTS

This feature gives ideas on how to consolidate the key points given in the Student Book, and how to use the key points as a basis for homework, revision or extension work.

Key Stage 3 Link-up

Sc1 Scientific enquiry

How Science Works does not relate directly to the individual statements in the Key Stage 3 Programme of Study. However, it builds on all of the knowledge, understanding and skills inherent in the statements.

It is expected that students will be familiar with:

- the need to work safely
- making a prediction
- controls
- the need for repetition of some results
- tabulating and analysing results
- making appropriate conclusions
- suggesting how they might improve methods.

RECAP ANSWERS

1 The hotter the water the more gas would be produced.

2 The temperature of the water.

3 The amount of gas produced.

4 E.g. the time the tubes were left.

5 E.g. the amount of pondweed used.

6

Temperature of water (°C)	Volume of gas produced (mm^3)
6	2
18	15
30	30
40	28

7 A suitable graph, with temperature on X axis and volume on Y axis. Points plotted correctly, axes labelled (including units) and line of best fit drawn.

8 E.g. as the temperature of the water increased, the volume of the gas produced increased, up to about 30°C to 40°C when it remained constant/started to reduce.

9 E.g. Kiah should have repeated her results.

Teaching suggestions

Finding out what they know

Students should begin to appreciate the 'thinking behind the doing' developed during KS3. It would be useful to illustrate this by a simple demonstration (e.g. *Elodea* bubbling oxygen) and posing questions that build into a flow diagram of the steps involved in a whole investigation. This could lead into the recap questions to ascertain each individual student's progress. Emphasis should be placed on an understanding of the following terms: prediction, independent, dependent and control variables and reliability.

The recap questions should identify each individual student's gaps in understanding. Therefore it is best carried out as an assessment. It might be appropriate to do questions 7, 8 and 9 for homework.

Activity notes

Collect newspaper articles and news items from the television to illustrate good and poor uses of science. There are some excellent television programmes featuring good and poor science. Have a competition to see who can bring in the poorest example of science used to sell a product – shampoo adverts are usually a good place to start!

H1 | How science works

This first chapter looks at 'How Science Works'. It is an important part of your GCSE because the ideas introduced here will crop up throughout your course. You will be expected to collect scientific evidence and to understand how we use evidence. These concepts will be assessed as the major part of your internal school assessment. You will take one or more 45-minute tests on data you have collected previously plus data supplied for you in the test. These are called Investigative Skills Assignments. The ideas in 'How Science Works' will also be assessed in your examinations.

What you already know

Here is a quick reminder of previous work with investigations that you will find useful in this chapter:

- You will have done some practical work and know how important it is to keep yourself and others safe.
- Before you start investigating you usually make a prediction, which you can test.
- Your prediction and plan will tell you what you are going to change and what you are going to measure.
- You will have thought about controls.
- You will have thought about repeating your readings.
- During your practical work you will have written down your results, often in a table.
- You will have plotted graphs of your results.
- You will have made conclusions to explain your results.
- You will have thought about how you could improve your results, if you did the work again.

RECAP QUESTIONS

Kiah wrote this account of a practical she did:

I wanted to find out how photosynthesis might be affected by temperature. I put five pieces of pondweed into five different test tubes and covered them with water. I put the test tubes into water baths at different temperatures and measured the gas coming off. I thought that the hotter the water the more gas would be produced.

One tube was left at room temperature which was 18°C and it made 15 mm^3 in five minutes. The second was in 6°C water and gave 2 mm^3. At 30°C there were 30 mm^3, at 40°C there were 28 mm^3 of gas. The test tubes were all left for thirty minutes.

1 What was Kiah's prediction?

2 What was the variable she chose to change? (We call this the independent variable.)

3 What was the variable she measured to judge the effect of varying the independent variable? (We call this the dependent variable. Its value *depends* on the value chosen for the independent variable.)

4 Write down one variable that Kiah controlled.

5 Write down a variable that Kiah did not say she had controlled.

6 Make a table of her results.

7 Draw a graph of the results.

8 Write a conclusion for Kiah.

9 How do you think Kiah could have improved her results?

Revealing to the students that they are using scientific thinking to solve problems during their everyday life can make their work in science more relevant. Other situations could illustrate this and should be discussed in groups or as a class.

For example:
'How can I make a really good cup of coffee? I know it takes about a spoonful of coffee, some hot water and some milk and sugar (knowledge). I have seen where these are kept and have seen what a good cup of coffee looks like' (observation).

SPECIFICATION LINK-UP

Section 10

This opening chapter of the course covers the complete specification for 'How science works – the procedural content'.

'How science works' is treated here as a separate chapter, it offers the opportunity to teach the 'thinking behind the doing' as a discrete set of procedural skills. However, it is of course an integral part of the way students will learn about science and those skills should be nurtured throughout the course.

It is anticipated that sections of this chapter will be taught as the opportunity presents itself during the teaching programme. The chapter should also be referred back to at appropriate times when these skills are required and in preparation for the internally assessed ISAs.

The thinking behind the doing

Science attempts to explain the world in which we live. It provides technologies that have had a great impact on our society and the environment. Scientists try to explain phenomena and solve problems using evidence. The data to be used as evidence must be reliable and valid, as only then can appropriate conclusions be made.

A scientifically literate citizen should, amongst other things, be equipped to question, and engage in debate on, the evidence used in decision-making.

The reliability of evidence refers to how much we trust the data. The validity of evidence depends on the reliability of the data as well as whether the research answers the question. If the data is not reliable, the research cannot be valid.

To ensure reliability and validity in evidence, scientists consider a range of ideas which relate to:

- *how we observe the world;*
- *designing investigations so that patterns and relationships between variables may be identified;*
- *making measurements by selecting and using instruments effectively;*
- *presenting and representing data;*
- *identifying patterns, relationships and making suitable conclusions.*

These ideas inform decisions and are central to science education. They constitute the 'thinking behind the doing' that is a necessary complement to the subject content of biology, chemistry and physics.

How science works

How science works for us

Science works for us all day, every day. You do not need to know how a mobile phone works to enjoy sending text messages. But, think about how you started to use your mobile phone or your television remote control. Did you work through pages of instructions? Probably not!

You knew that pressing the buttons would change something on the screen (***knowledge***). You played around with the buttons, to see what would happen (***observation***). You had a guess at what you thought might be happening (**prediction**) and then tested your idea (***experiment***).

If your prediction was correct you remembered that as a ***fact***. If you could repeat the operation and get the same result again then you were very pleased with yourself. You had shown that your results were **reliable**.

Working as a scientist you will have knowledge of the world around you and particularly about the subject you are working with. You will observe the world around you. An enquiring mind will then lead you to start asking questions about what you have observed.

Science moves forward by slow steady steps. When a genius such as Einstein comes along then it takes a giant leap. Those small steps build on knowledge and experience that we already have.

Each small step is important in its own way. It builds on the body of knowledge that we have. In 1675 a German chemist tried to extract gold from urine. He must have thought that there was a connection between the two colours. He was wrong, but after a long while, with an incredible stench coming from his laboratory, the urine began to glow.

He had discovered phosphorus. A Swedish scientist worked out how to manufacture phosphorus without the smell of urine. Phosphorus catches light easily. That is why most matches these days are manufactured in Sweden.

DID YOU KNOW?

The Greeks were arguably the first true scientists. They challenged traditional myths about life. They set forward ideas that they knew would be challenged. They were keen to argue the point and come to a reasoned conclusion.

Other cultures relied on long established myths and argument was seen as heresy.

Thinking scientifically

Figure 1 Tropical beach

ACTIVITY

Once you have got the idea of holidays out of your mind – look at the photograph in Figure 1 with your scientific brain.

Work in groups to ***observe*** the beach and the plants growing on it. Then you can start to think about why the plants can grow (***knowledge***) so close to the beach.

One idea could be that the seeds can float for a long while in the sea, without taking in any water.

You can use the following headings to discuss your investigation. One person should be writing your ideas down, so that you can discuss them with the rest of your class.

- What prediction can you make about the mass of the coconut seed and the time it spends in the sea water?
- What would be your independent variable?
- What would be your dependent variable?
- What would you have to control?
- Write a plan for your investigation.
- How could you make sure your results were reliable?

3

You can use your observations and your knowledge to make a prediction that mixing all of these together will produce a satisfactory result. You can then test your prediction and see what the results are. You check again the next day to see if you get the same results (reliability). Perhaps you have thought that you might want a bit more sugar, so you try adding an extra spoonful. Does it improve the results? This could lead to a discussion of the need for controls. Scientists work in exactly the same way – this is 'How science works'.

Collect newspaper articles and news items from the television to illustrate good and poor uses of science. There are some excellent television programmes illustrating good and poor science. Have a competition for who can bring in the poorest example of science used to sell products – shampoo adverts are a very good starter!

Checking for Misconceptions

Some common misconceptions that can be dealt with here and throughout the course are:

- The purpose of controls – some students believe that it is about making accurate measurements of the independent variable.
- The purpose of preliminary work – some believe that it is the first set of results.
- That the table of results is constructed after the practical work – students should be encouraged to produce the table before carrying out their work and complete it during their work.
- That precision is the number of places of decimals they can write down.
- That anomalies are identified after the analysis – they should preferably be identified during the practical work or, at the latest, before any calculation of a mean.
- They automatically extrapolate the graph to its origin.
- Lines of best fit must be straight lines
- Some will think you repeat readings to make sure your investigation is a fair test.

H2 Fundamental ideas about how science works

LEARNING OBJECTIVES

Students should learn:

- The difference between continuous, discrete, ordered and categoric variables.
- That evidence needs to be valid and reliable.
- That variables can be linked causally, by association or by chance.
- To distinguish between opinion based on scientific evidence and non-scientific ideas.

LEARNING OUTCOMES

Students should be able to:

- Recognise variables as continuous, ordered, discrete or categoric.
- Suggest how an investigation might demonstrate its reliability and validity.
- Know that variables can be linked causally, by association or by chance.
- Identify when an opinion does not have the support of valid and reliable science.

Teaching suggestions

- **Learning styles**

  Kinaesthetic: Handling cress seedlings to observe differences.

  Visual: Observing cress seedlings.

  Auditory: Listening to ideas of others on scientific opinions.

  Interpersonal: Discussing the variables associated with the cress seed growth.

  Intrapersonal: Considering the ethics of the thalidomide case and possibly the use of animals for testing human drugs.
- **Special needs.** Lists of possible variables could be made from which to select the most appropriate. Cloze statements can be used for essential notes.
- **Gifted and talented.** Discussion could range into the ethics of drug provision and the increased importance of having scientifically based opinions. It might develop into an appreciation of the limits of science in terms of ethical delivery of drugs. Decisions of this nature are not always as clear cut as scientists might want them to be. Some experts estimate the number of accidental deaths in the UK at 20 000 per year caused by prescribed medical drugs.

SPECIFICATION LINK-UP HOW SCIENCE WORKS

Section 10.2

Fundamental ideas

Evidence must be approached with a critical eye. It is necessary to look closely at how measurements have been made and what links have been established. Scientific evidence provides a powerful means of forming opinions. These ideas pervade all of 'How science works'.

Students should know and understand:

- *It is necessary to distinguish between opinion based on valid and reliable evidence and opinion based on non-scientific ideas (prejudice, whim or hearsay).*
- *Continuous variables where used (any numerical values) give more information than ordered variables, which are more informative than categoric variables. A variable may also be discrete, that is, restricted to whole numbers.*
- *Scientific investigations often seek to identify links between two or more variables. These links may be:*
  - *causal, in that a change in one variable causes a change in another*
  - *due to association – changes in one variable and a second variable are linked by a third variable*
  - *due to chance occurrence.*

Lesson structure

STARTER

Crazy science – Show a video clip of one of the science shows that are aimed at entertainment rather than education or an advert that proclaims a scientific opinion. This should lead into a discussion of how important it is to form opinions based on sound scientific evidence. (5–10 minutes)

Misconceptions! – Do a class survey of commonly held scientific misconceptions (this could be used for homework as a survey of friends and relatives – perhaps opinions could be related to age groups). E.g. 'If left long enough in a sealed container, meat will have maggots in it', 'The greenhouse effect and global warming are the same thing', 'Cold air cannot hold as much water vapour as warm air'. (10–15 minutes)

MAIN

- From a light-hearted look at entertainment science, bring the thalidomide example into contrast (if appropriate with video clips) and discuss how tragic situations can be created by forming opinions that are not supported by valid science.
- Show some cress seedlings grown in different light levels. Review some of the terminology from KS3. Discuss, in small groups, the different ways in which the independent and the dependent variables could be measured, identifying these in terms of continuous, ordered and categoric measurements.
- Discuss the usefulness in terms of forming opinions of each of the proposed measurements.
- Consider that this might be a commercial proposition and the students might be advising an investor in a company growing cress.
- Discuss how they could organise the investigation to demonstrate its validity and reliability to a potential investor.
- Discuss whether the relationship shows a causal, chance or by association linkage.

PLENARIES

Evidence for opinions – Bring together the main features of scientific evidence that would allow sound scientific opinions to be formed from an investigation. (5 minutes)

Analyse conclusions – Use an example of a poorly structured investigation and allow the students to critically analyse any conclusions drawn, e.g. data from an investigation into different forms of insulation, using calorimeters and cooling curves. (10 minutes)

Practical support

Equipment and materials required

Petri dishes growing cress to show differences in height and colour. There should be enough for small group work.

Answers to in-text questions

a) The original animal investigation did not include pregnant animals/was not carried out on human tissue and so was not valid, when the opinion was formed that it could be given to pregnant women.

b) Measure the heights of the seedlings. Continuous measurements (variables) are more powerful.

c) Control all (or as many as possible) of the other variables.

d) A simple causal relationship described.

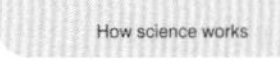

HOW SCIENCE WORKS

H2 Fundamental ideas about how science works

LEARNING OBJECTIVES

1. How do you spot when a person has an opinion that is not based on good science?
2. What is the importance of continuous, ordered and categoric variables?
3. What is meant by reliable evidence and valid evidence?
4. How can two sets of data be linked?

NEXT TIME YOU...

. . . read a newspaper article or watch the news on TV ask yourself if that research is valid and reliable. (See page 5.) Ask yourself if you can trust the opinion of that person.

Science is too important for us to get it wrong

Sometimes it is easy to spot when people try to use science poorly. Sometimes it can be funny. You might have seen adverts claiming to give your hair 'body' or sprays that give your feet 'lift'!

On the other hand, poor scientific practice can cost lives.

Some years ago a company sold the drug thalidomide to people as a sleeping pill. Research was carried out on animals to see if it was safe. The research did not include work on pregnant animals. The opinion of the people in charge was that the animal research showed the drug could be used safely with humans.

Then the drug was also found to help ease morning sickness in pregnant women. Unfortunately, doctors prescribed it to many women, resulting in thousands of babies being born with deformed limbs. It was far from safe.

These are very difficult decisions to make. You need to be absolutely certain of what the science is telling you.

a) Why was the opinion of the people in charge of developing thalidomide based on poor science?

Deciding on what to measure

You know that you have an independent and a dependent variable in an investigation. These variables can be one of four different types:

- A **categoric variable** is one that is best described by a label (usually a word). The colour of eyes is a categoric variable, e.g. blue or brown eyes.
- A **discrete variable** is one that you describe in whole numbers. The number of leaves on different plants is a discrete variable.
- An **ordered variable** is one where you can put the data into order, but not give it an actual number. The height of plants compared to each other is an ordered variable, e.g. the plants growing in the woodland are taller than those on the open field.
- A **continuous variable** is one that we measure. Therefore its value could be any number. Temperature (as measured by a thermometer or temperature sensor) is a continuous variable, e.g. 37.6°C, 45.2°C.

When designing your investigation you should always try to measure continuous data whenever you can. This is not always possible, so you should then try to use ordered data. If there is no other way to measure your variable then you have to use a label (categoric variable).

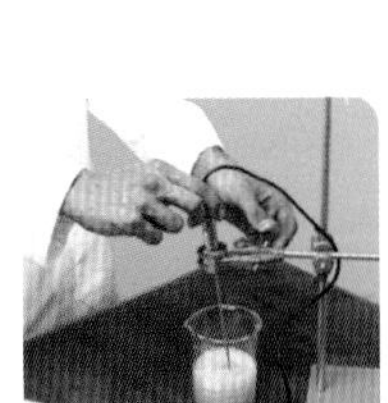

Figure 1 Student recording a range of temperatures

b) Imagine you were growing seedlings in different volumes of water. Would it be better to say that some were tall and some were short; or some were taller than others; or to measure the heights of all of the seedlings?

Making your investigation reliable and valid

When you are designing an investigation you must make sure that others can get the same results as you – this makes it **reliable**.

You must also make sure you are measuring the actual thing you want to measure. If you don't, your data can't be used to answer your original question. This seems very obvious but it is not always quite so easy. You need to make sure that you have ***controlled*** as many other variables as you can, so that no-one can say that your investigation is not **valid**. A valid investigation should be reliable *and* answer the original question.

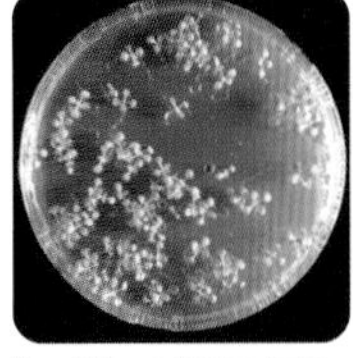

Figure 2 Cress seedlings growing in a Petri dish

c) State one way in which you can show that your results are valid.

How might an independent variable be linked to a dependent variable?

Variables can be linked together for one of three reasons:

- It could be because one variable has caused a change in the other, e.g. the more plants there are in a pond, the more oxygen there is in the water. This is a **causal link**.
- It could be because a third variable has caused changes in the two variables you have investigated, e.g. fields that have more grass also have more dandelions in them. There is an **association** between the two variables. This is caused by a third variable – how many sheep there are in the field!
- It could be due simply to **chance**, e.g. the type of weeds growing in different parts of your garden!

d) Describe a causal link that you have seen in biology.

SUMMARY QUESTIONS

1. Name each of the following types of variables described in a), b) and c).
   a) People were asked about how they felt inside a new shopping centre: 'warm', 'hot', 'quite warm', 'cold', 'freezing!'
   b) These people were asked as they entered the new shopping centre: 'Warmer than I did outside'; 'Colder than my shed!'
   c) These people had their body temperature measured using a clinical thermometer: 37.1°C; 37.3°C; 36.8°C; 37.0°C; 37.5°C.
2. A researcher claimed that the metal tungsten 'alters the growth of leukaemia cells' in laboratory tests. A newspaper wrote that they would 'wait until other scientists had reviewed the research before giving their opinion.' Why is this a good idea?

DID YOU KNOW?

Aristotle, a brilliant Greek scientist, once proclaimed that men had more teeth than women! Do you think that his data collection was reliable?

Figure 3 Sheep grazing in a field

KEY POINTS

1. Be on the lookout for non-scientific opinions.
2. Continuous data is more powerful than other types of data.
3. Check that evidence is reliable and valid.
4. Be aware that just because two variables are related it does not mean that there is a causal link between them.

SUMMARY ANSWERS

1. **a)** categoric
   b) ordered
   c) continuous
2. The investigation can be shown to be reliable if other scientists can repeat their investigations and get the same findings. Because it is reliable, opinions formed from it are more useful.

KEY POINTS

- Students should appreciate the need for sound science before opinions can be valued. They could challenge others in their group by presenting deliberately unfair tests and ask the others to spot the mistakes.
- Students should have notes that bring out the meaning of the key words, including continuous, discrete, categoric, ordered, valid and reliable.

H3 Starting an investigation

LEARNING OBJECTIVES

Students should learn:

- How scientific knowledge can be used to observe the world around them.
- How good observations can be used to make hypotheses.
- How hypotheses can generate predictions that can be tested.
- That investigations must produce valid results.

LEARNING OUTCOMES

Students should be able to:

- State that observation can be the starting point for an investigation.
- State that observation can generate hypotheses.
- Recall that hypotheses can generate predictions and investigations.
- Show that the design of an investigation must allow results to be valid.

Teaching suggestions

- **Gifted and talented.** It might be possible to discuss whether evolution is a fact or a theory. It is possible to view it as a theory which is well established or as a fact that can be explained by theories.
- **Learning styles**

  Visual: Observations made.

  Auditory: Listening to group discussions.

  Interpersonal: Discussing hypotheses and predictions.

  Intrapersonal: Answering question b).

  Kinaesthetic: Practical activities.
- **Teaching assistant.** The teaching assistant could be primed to ask appropriate questions and to prompt thought processes in line with the theme of the lesson.

Answers to in-text questions

a) No because we know that the glass is unlikely to be sensed by the plant, but light is.

c) e.g. Observation: the frog is green and so is the leaf. Hypothesis: the frog moves around until its skin is roughly the same colour as its surroundings.

d) No because his heart rate might increase because his hand is being held by his girlfriend and not just because of exercise. The results are not valid.

SPECIFICATION LINK-UP HOW SCIENCE WORKS

Section 10.3

Observation as a stimulus to investigation

Observation is the link between the real world and scientific ideas. When we observe objects, organisms or events we do so using existing knowledge. Observations may suggest hypotheses and lead to predictions that can be tested.

Students should know and understand:

- *Observing phenomena can lead to the start of an investigation, experiment or survey. Existing theories and models can be used creatively to suggest explanations for phenomena (hypotheses). Careful observation is necessary before deciding which are the most important variables. Hypotheses can then be used to make predictions that can be tested.*
- *Data from testing a prediction can support or refute the hypothesis or lead to a new hypothesis.*
- *If the theories and models we have available to us do not completely match our data or observations, then we need to check the validity of our observations or data, or amend the theories or models.*

Lesson structure

STARTER

Linking observation to knowledge – Discuss with students any unusual events they saw on the way to school. If possible take them into the school grounds to look and listen to events. Try to link their observations to their scientific knowledge. They are more likely to notice events that they can offer some scientific explanation for. This may well need prompting with some directed questions. Once students have got used to making observations, get them to start to ask questions about those observations. (10 minutes)

Demo observation – Begin the lesson with a demonstration – as simple as lighting a match or more involved such as a bell ringing in a bell jar, with air gradually being withdrawn. Students should, in silence and without further prompting, be asked to write down their observations. These should be collated and questions be derived from those observations. (10 minutes)

MAIN

- Work through the first section and leave them to discuss question a).
- If in the lab, allow students to participate in a 'scientific happening' of your choice, e.g. trying to ignite different foods. Preferably something that they have not met before, but which they will have some knowledge of. As an alternative, if possible take students into the school field where there will be many opportunities to observe plants (ideally the school pond), and questions should be around where certain plants/animals are to be found and their adaptations.
- If group needs some help at this point, they should try question b).
- In groups they should discuss possible explanations for one agreed observation. Encourage a degree of lateral thinking. You might need to pose the questions for some groups, e.g. why do some foods easily ignite and others do not? Or why does grass grow when it is cut but others plants are killed?
- Ask the group to select which of their explanations is the most likely, based on their own knowledge of science.
- Work these explanations into a hypothesis.
- Individually each student answers c). Gather in ideas and hypotheses.
- Students, working in groups, can now turn this into a prediction.
- They could suggest ways in which their prediction could be tested. Identify independent, dependent and control variables and the need to make sure that they are measuring what they intend to measure.
- Answer question d) as a class. For a touch of humour towards the end, you might get students to think about which variable you might be measuring.

PLENARIES

Poster – Students to design a poster that links Observation + knowledge – hypothesis – prediction – investigation. (10 minutes)

Discussion on evolution – The story of evolution could be used at many points in this chapter but is particularly useful here. Briefly evolution is of course closely related to the work of Charles Darwin (1809–1882). It was not a term that he used whilst gathering observations from around the world. The term 'evolution' had been used long before Darwin. Lamarck (1744–1829) used the term 'progression' in the way in which we would use evolution. His example of giraffes stretching their necks to reach the more succulent shoots and thus passing on these characteristics is very familiar. He also explained progression by an upward striving of organisms to higher levels of complexity. Darwin's grandfather, Erasmus, used the term 'evolution'. Malthus laid the foundation for the theory of evolution by natural selection by his observations and mathematical calculations that the supply of food could never keep up with the demand for food and so some organisms prospered whilst others struggled. This theory, linked to Darwin's observations, led to the theory of evolution by natural selection (1859). Those members of a species with an innate advantage would survive to pass on those characteristics, whilst others would perish. A.R. Wallace had also thought of the idea independently. Also independently, the Moravian monk, Mendel, proposed units of inheritance which he called 'factors' (1865). In 1869 Miescher, a Swiss scientist observed what he called nuclein, because he found the substance inside the nucleus of cells from the pus in bandages! Chromosomes were first observed in 1888 and, by the turn of the century, were thought to be involved in inheritance. Morgan, in the early years of the century, experimented with fruit flies. He zapped them with radiation, he put them in centrifuges, he heated them up and cooled them down, trying to induce mutations. By the 1940s it was widely agreed that genes existed on chromosomes. By 1953 the structure of DNA was revealed. Now we know most of the structure of human and other organisms' DNA.

SUMMARY ANSWERS

1 Observations when supported by scientific *knowledge* can be used to make a *hypothesis*. This can be the basis for a *prediction*. A prediction links an *independent* variable to a *dependent* variable. Other variables need to be *controlled*.

2 A hypothesis seeks to explain an observation – it is a good idea. A prediction tests the hypothesis in an investigation.

KEY POINTS

- Students should have notes relating observations to knowledge to hypothesis to prediction and to investigation. They should also have some notes on what it means to design validity into an investigation.

H3 Starting an investigation

LEARNING OBJECTIVES

1 How can you use your scientific knowledge to observe the world around you?
2 How can you use your observations to make a hypothesis?
3 How can you make predictions and start to design an investigation?

Figure 1 Plant showing positive phototropism

Observation

As humans we are sensitive to the world around us. We can use our many senses to detect what is happening. As scientists we use observations to ask questions. We can only ask useful questions if we know something about the observed event. We will not have all of the answers, but we know enough to start asking the correct questions.

If we observe that the weather has been hot today, we would not ask if it was due to global warming. If the weather was hotter than normal for several years then we could ask that question. We know that global warming takes many years to show its effect.

When you are designing an investigation you have to observe carefully which variables are likely to have an effect.

a) Would it be reasonable to ask if the plant in Figure 1 is 'growing towards the glass'? Explain your answer.

A farmer noticed that her corn was much smaller at the edge of the field than in the middle (observation). She noticed that the trees were quite large on that side of the field. She came up with the following ideas that might explain why this was happening:

- The trees at the edge of the field were blocking out the light.
- The trees were taking too many nutrients out of the soil.
- The leaves from the tree had covered the young corn plants in the spring.
- The trees had taken too much water out of the soil.
- The seeds at the edge of the field were genetically small plants.
- The drill had planted fewer seeds on that side of the field.
- The fertiliser spray had not reached the side of the field.
- The wind had been too strong over winter and had moved the roots of the plants.
- The plants at the edge of the field had a disease.

b) Discuss each of these ideas and use your knowledge of science to decide which four are the most likely to have caused the poor growth of the corn.

Observations, backed up by really creative thinking and good scientific knowledge can lead to a **hypothesis**.

DID YOU KNOW?

Some biologists think that we still have about one hundred million species of insects to discover – plenty to go for then! Of course, observing one is the easy part – knowing that it is undiscovered is the difficult bit!

What is a hypothesis?

A hypothesis is a 'great idea'. Why is it so great? – well because it is a great observation that has some really good science to try to explain it.

For example, you observe that small, thinly sliced chips cook faster than large, fat chips. Your hypothesis could be that the small chips cook faster because the heat from the oil has a shorter distance to travel before it gets to the potato in the centre of the chips.

c) Check out the photograph in Figure 2 and spot anything that you find interesting. Use your knowledge and some creative thought to suggest a hypothesis based on your observations.

Figure 2 A frog

When making hypotheses you can be very imaginative with your ideas. However, you should have some scientific reasoning behind those ideas so that they are not totally bizarre.

Remember, your explanation might not be correct, but you think it is. The only way you can check out your hypothesis is to make it into a prediction and then test it by carrying out an investigation.

Observation + knowledge → hypothesis → prediction → investigation

Starting to design a valid investigation

An investigation starts with a prediction. You, as the scientist, predict that there is a relationship between two variables.

- An **independent variable** is one that is changed or selected by you, the investigator.
- A **dependent variable** is measured for each change in your independent variable.
- All other variables become **control variables**, kept constant so that your investigation is a fair test.

If your measurements are going to be accepted by other people then they must be valid. Part of this is making sure that you are really measuring the effect of changing your chosen variable. For example, if other variables aren't controlled properly, they might be affecting the data collected.

d) Look at Figure 3. When investigating his heart rate before and after exercise, Darren got his girlfriend to measure his pulse. Would Darren's investigation be valid? Explain your answer.

Figure 3 Measuring a pulse

SUMMARY QUESTIONS

1 Copy and complete using the words below:

controlled dependent hypothesis independent knowledge prediction

Observations when supported by scientific …… can be used to make a …… . This can be the basis for a …… . A prediction links an …… variable to a …… variable. Other variables need to be …… .

2 Explain the difference between a hypothesis and a prediction.

KEY POINTS

1 Observation is often the starting point for an investigation.
2 Hypotheses can lead to predictions and investigations.
3 You must design investigations that produce valid results if you are to be believed.

H4 Building an investigation

LEARNING OBJECTIVES

Students should learn:

- How to design a fair test.
- The purpose of a trial run.
- How to ensure accuracy and precision.

LEARNING OUTCOMES

Students should be able to:

- Design a fair test and understand the use of control groups.
- Know how to manage fieldwork investigations.
- Use trial runs to design valid investigations.
- Design accuracy into an investigation.
- Design precision into an investigation.

Teaching suggestions

- **Learning styles**

  Kinaesthetic: Carrying out experiment.

  Visual: Reading instruments.

  Auditory: Listening to the outcomes of experiments from different groups.

  Interpersonal: Discussing quality of results.

  Intrapersonal: Considering fair tests in relation to athletics.

- **Teaching assistant.** The teaching assistant could support pupils who have physical or coordination difficulties with practical work. The purpose of the experiment will need reinforcing for lower abilities.

Answers to in-text questions

a) Any variables associated with the weather, the soil, the genetic variability of the seeds, human error in planting.

b) Drawing of thermometer scale with 'true value' shown with a set of results closely clustered around this true value.

SPECIFICATION LINK-UP HOW SCIENCE WORKS

Section 10.4

Designing an investigation

An investigation is an attempt to determine whether or not there is a relationship between variables. Therefore it is necessary to identify and understand the variables in an investigation. The design of an investigation should be scrutinised when evaluating the validity of the evidence it has produced.

Students should know and understand:

- *An independent variable is one that is changed or selected by the investigator. The dependent variable is measured for each change in the independent variable.*
- *Any measurement must be valid in that it measures only the appropriate variable.*

Fair test

- *It is important to isolate the effects of the independent variable on the dependent variable. This may be achieved more easily in a laboratory environment than in the field where it is harder to control all variables.*
- *A fair test is one in which only the independent variable affects the dependent variable, as all other variables are kept the same.*
- *In field investigations it is necessary to ensure that variables that change their value do so in the same way for all measurements of the dependent variable.*
- *When using large-scale survey results, it is necessary to select data from conditions that are similar.*
- *Control groups are often used in biological and medical research to ensure that observed effects are due to changes in the independent variable alone.*

Choosing values of a variable

- *Care is needed in selecting values of variables to be recorded in an investigation. A trial run will help to identify appropriate values to be recorded, such as the number of repeated readings needed and their range and interval.*

Accuracy and precision

- *Readings might be repeated to improve the reliability of the data. An accurate measurement is one that is close to the true value.*
- *The design of an investigation must provide data with sufficient accuracy.*
- *The design of an investigation must provide data with sufficient precision to form a valid conclusion.*

Lesson structure

STARTER

Head start – Start, for example, with a video clip of a 100 m race. This has to be a fair test. How is this achieved? Then show the mass start of the London marathon and ask if this is a fair test. Then move on to ask why there is no official world record for a marathon. Instead they have world best times. This could lead discussion into how difficult it is to control all of the variables in the field. You could go back to suggest that athletes can break the 100 m world record and for this not to be recognised because of a helping wind. (10 minutes)

That's not fair! – Challenge students with a test you set up in an 'unfair' way. You can differentiate by making some errors obvious and some more subtle. Students can observe then generate lists of mistakes in small groups. Ask each group to give one error from their list and record what should have been done to ensure fair testing until all suggestions have been considered. (10 minutes)

MAIN

- Move into group discussion of in-text question a). Other examples of field testing might include setting up wind farms, testing durability of paints or corrosion of metals at sea.

- Group discussions on how and why we need to produce survey data. Use a topical issue here. It might be appropriate to see how it should *NOT* be done by using a vox pop clip from a news programme.
- Students will be familiar with the idea of a placebo, but possibly not with how it is used to set up a control group. This might need explanation.
- Consider the case of whether it is possible to tell the difference in taste if the milk is put in before or after the tea. R.A. Fisher tested this, using a doubleblind test and went on to devise 'Statistical Methods for Research Workers'.
- It is important that students appreciate the difference between accuracy and precision. They could be given a metre rule and asked to use it to assess reaction times. They should be asked to think carefully about how they can get accurate results. To increase accuracy they should be told to repeat their readings. They should be told that the most precise readings will be taken by those that use their equipment the most carefully.
- On completion each group submits its method for obtaining accurate results. The group with the most accurate method is declared the winner.
- Repeat the class competition, but this time consider the range of their repeat measurements, i.e. judge their precision.
- Use the data gathered to award a prize to the group showing the greatest precision.
- Find the maximum range for the whole class – Who got the highest reading/who got the lowest? Can we explain why? Gather suggestions.

PLENARIES

Prize giving! – The winning group should explain why they think they have the most accurate results.

Precision and accuracy – Discuss in-text question b) in small groups and then gather feedback. (10 minutes)

The winning group should explain how they know that they have the results that show the greatest precision.

SUMMARY ANSWERS

1 Trial runs give you a good idea of whether you have the correct *conditions*; whether you have chosen the correct *range*; whether you have enough *readings*; if you need to do *repeat* readings

2 Any example that demonstrates understanding of the two terms.

3
- Keep all variables other than the independent and the dependent variables controlled.
- Attempt to keep all of the other variables varying to the same extent.
- Use a sample that is treated in the same way except for the independent variable.

KEY POINTS

- The students must be able to appreciate the difference between accuracy and precision. Ask students for their own definitions without the textbook.
- They sometimes confuse accuracy with the sensitivity of an instrument. Also, in calculations, some students will write irrelevant places of decimals. Ask students to comment on an example.

H4 Building an investigation

LEARNING OBJECTIVES

1 How do you design a fair test?
2 How do you make sure that you choose the best values for your variables?
3 How do you ensure accuracy and precision?

Figure 1 Corn being harvested

Fair testing

A **fair test** is one in which only the independent variable affects the dependent variable. All other variables are controlled, keeping them constant if possible.

This is easy to set up in the laboratory, but almost impossible in fieldwork. Plants and animals do not live in environments that are simple and easy to control. They live complex lives with lots of variables changing constantly.

So how can we set up fieldwork investigations? The best you can do is to make sure that all of the many variables change in much the same way, except for the one you are investigating. Then at least the plants get the same weather, even if it is constantly changing.

a) Imagine you were testing how close together you could plant corn to get the most cobs. You would plant five different plots, with different numbers of plants in each plot. List some of the variables that you could not control.

If you are investigating two variables in a large population then you will need to do a survey. Again it is impossible to control all of the variables.

Imagine you were investigating the effect of diet on diabetes. You would have to choose people of the same age and same family history to test. The larger the sample size you test, the more reliable your results will be.

Control groups are used in investigations to try to make sure that you are measuring the variable that you intend to measure. When investigating the effects of a new drug, the control group will be given a placebo.

The control group think they are taking a drug but the placebo does not contain the drug. This way you can control the variable of 'thinking that the drug is working' and separate out the effect of the actual drug.

Choosing values of a variable

Trial runs will tell you a lot about how your early thoughts are going to work out.

Do you have the correct conditions?
A photosynthesis investigation that produces tiny amounts of oxygen might not have enough:

- light, • pondweed, • carbon dioxide, or
- the temperature might not be high enough.

Have you chosen a sensible range?
If there is enough oxygen produced, but the results are all very similar:

- you might not have chosen a wide enough range of light intensities.

Have you got enough readings that are close together?
If the results are very different from each other:

- you might not see a pattern if you have large gaps between readings over the important part of the range.

Accuracy

Accurate results are very close to the ***true value***.

Your investigation should provide data that is accurate enough to answer your original question.

However, it is not always possible to know what that true value is.

How do you get accurate data?
- You can repeat your results and your mean is more likely to be accurate.
- Try repeating your measurements with a different instrument and see if you get the same readings.
- Use high quality instruments that measure accurately.
- The more carefully you use the measuring instruments, the more accuracy you will get.

Precision and reliability

If your repeated results are closely grouped together then you have precision and you have improved the reliability of your data.

Your investigation must provide data with sufficient precision. It's no use measuring a person's reaction time using the seconds hand on clock! If there are big differences within sets of repeat readings, you will not be able to make a valid conclusion. You won't be able to trust your data!

How do you get precise and reliable data?
- You have to use measuring instruments with sufficiently small scale divisions.
- You have to repeat your tests as often as necessary.
- You have to repeat your tests in exactly the same way each time.

A word of caution!

Be careful though – just because your results show precision does not mean your results are accurate. Look at the box opposite.

b) Draw a thermometer scale showing 4 results that are both accurate and precise.

The difference between accurate and precise results

Imagine measuring the temperature after a set time when a food is burned and used to heat a fixed volume of water. Two students repeated this experiment, four times each. Their results are marked on the thermometer scales below:

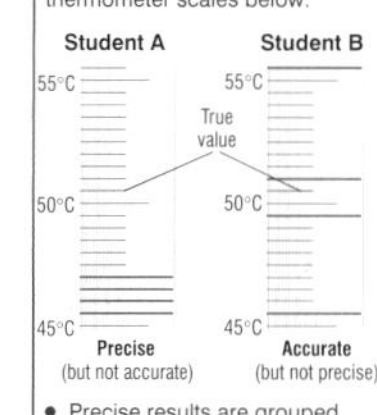

- Precise results are grouped closely together.
- Accurate results will have a mean (average) close to the true value.

SUMMARY QUESTIONS

1 Copy and complete using the following terms:

range **repeat** **conditions** **readings**

Trial runs give you a good idea of whether you have the correct ……; whether you have chosen the correct ……; whether you have enough ……; if you need to do …… readings.

2 Use an example to explain how results can be accurate, but not precise.

3 Briefly describe how you would go about setting up a fair test in a laboratory investigation. Give your answer as general advice.

KEY POINTS

1 Care must be taken to ensure fair testing – as far as is possible.
2 You can use a trial run to make sure that you choose the best values for your variables.
3 Careful use of the correct equipment can improve accuracy.
4 If you repeat your results carefully they are likely to become more reliable.

H5 Making measurements

LEARNING OBJECTIVES

Students should learn:

- That they can expect results to vary.
- That instruments vary in their accuracy.
- That instruments vary in their sensitivity.
- That human error can affect results.
- What to do with anomalies.

LEARNING OUTCOMES

Students should be able to:

- Differentiate between results that vary and anomalies.
- Explain why it is important to use equipment properly.
- Explain that instruments vary in their accuracy and sensitivity.
- Explain the difference between random and systematic errors.

Teaching suggestions

- **Special needs.** Students will need support when interpreting data on body temperature and identifying evidence for random and systematic error.
- **Gifted and talented.** Demonstrate a different experiment in which there is a built-in systematic error, e.g. aerobic yeast respiration and carbon dioxide production.
- **Learning styles**

  Kinaesthetic: Taking measurements.

  Visual: Observing systematic error.

  Auditory: Listening to explanations of differences between random and systematic errors.

  Interpersonal: Group discussions to determine how to accurately measure a person's height.

  Intrapersonal: Answering questions d) and e).
- **ICT link-up.** Using data logging to exemplify minor changes in dependent variables provides a good opportunity to include ICT in the lesson.

SPECIFICATION LINK-UP HOW SCIENCE WORKS

Section 10.5

Making measurements

When making measurements we must consider such issues as inherent variation due to variables that have not been controlled, human error and the characteristics of the instruments used. Evidence should be evaluated with the reliability and validity of the measurements that have been made in mind.

A single measurement

- *There will always be some variation in the actual value of a variable no matter how hard we try to repeat an event.*
- *When selecting an instrument, it is necessary to consider the accuracy inherent in the instrument and the way it has to be used.*
- *The sensitivity of an instrument refers to the smallest change in a value that can be detected.*
- *Even when an instrument is used correctly, human error may occur which could produce random differences in repeated readings or a systematic shift from the true value which could, for instance, occur due to incorrect use or poor calibration.*
- *Random error can result from inconsistent application of a technique. Systematic error can result from consistent misapplication of a technique.*
- *Any anomalous values should be examined to try to identify the cause and, if a product of a poor measurement, ignored.*

Lesson structure

STARTER

Demonstration – Demonstrate different ways of measuring the temperature of a water bath. Use a range of thermometers including digital ones. Discuss the relative merits of using each of these devices for different purposes. Discuss the specification of the measuring instrument – its percentage accuracy, its useful range and its sensitivity. (10 minutes)

MAIN

- In small groups, devise the most accurate way to measure a person's height. They can have any equipment they need. Students will need to think about what a person's height includes, e.g. hair flat or not, shoes off or not. They might suggest a board placed horizontally on the head, using a spirit level, removing the person being measured and then using the laser/sonic measurer placed on the ground. Take care to follow manufacturer's instructions when using the laser.
- Choose a person to try out this technique. Stress that we do not have a true answer. We do not know the person's true height. We trust the instrument and the technique that is most likely to give us the most accurate results.
- In groups, answer question a).
- Individually answer questions b) and c).
- Demonstrate an experiment in which there is a built-in systematic error, e.g. weighing some chemicals using a filter paper without using the tare or, for the really able, measuring the effect of temperature on the rate of carbon dioxide production by yeast, whilst the yeast is growing.
- Demonstrate an experiment in which there is a random error, e.g. weighing potato chips before and after an osmosis investigation without standardising the drying technique.
- Point out the difference between systematic errors and random errors. Also, how you might tell from results which type of error it is.
- Ask students to complete questions d) end e) on their own.
- Encourage students to identify anomalies whilst carrying out the investigation so that they have an opportunity to check and replace them.

PLENARIES

Check list – Draw up a check list for an investigation so that every possible source of error is considered. (10–15 minutes)

Human vs computer – Class discussion of data logging compared to humans when collecting data. Stress the importance of data logging in gathering data over extended or very short periods of time. (See 'Activity & extension ideas' box.)

ACTIVITY & EXTENSION

Use data logging to illustrate how detailed measurements taken frequently can show variation in results that would not have been seen by other methods. It can increase considerably the accuracy of readings that can be taken where other readings might not be possible to take accurately. It can also illustrate a surfeit of data that might detract from the data required.

Practical support

Equipment and materials needed

A laser/sonic measure and the instructions to illustrate its usage and sensitivity. It would be a good idea to have these printed for the students. Also tape, metre rule and 30 cm rule.

Other equipment is dependent on the type of demonstration used to show systematic and random error.

HOW SCIENCE WORKS

H5 Making measurements

LEARNING OBJECTIVES

1. Why do results always vary?
2. How do you choose instruments that will give you accurate results?
3. What do we mean by the sensitivity of an instrument?
4. How does human error affect results and what do you do with anomalies?

Using instruments

Do not panic! You cannot expect perfect results.

Try measuring the temperature of a beaker of water using a digital thermometer. Do you always get the same result? Probably not. So can we say that any measurement is absolutely correct?

In any experiment there will be doubts about actual measurements.

a) Look at Figure 1. Suppose, like this student, you tested the rate at which oxygen was produced using an enzyme. It is unlikely that you would get two readings exactly the same. Discuss all the possible reasons why.

When you choose an instrument you need to know that it will give you the accuracy that you want. That is, it will give you a true reading.

If you have used an electric water bath, would you trust the temperature on the dial? How do you know it is the true temperature? You could use a very expensive thermometer to calibrate your water bath. The expensive thermometer is more likely to show the true temperature. But can you really be sure it is accurate?

You also need to be able to use an instrument properly.

b) In Figure 1 the student is reading the amount of gas in the measuring cylinder. Why is the student unlikely to get a true reading?

Figure 1 Student testing the rate at which oxygen is produced using an enzyme

When you choose an instrument you have to decide how accurate you need it to be. Instruments that measure the same thing can have different sensitivities. The **sensitivity** of an instrument refers to the smallest change in a value that can be detected.

Choosing the wrong scale can cause you to miss important data or make silly conclusions. For example, 'The cells were all the same size – they were less than 1 millimetre.'

c) Match the following types of measuring instrument to their best use:

Used to measure	Sensitivity of measuring instrument
Size of a cell	millimetres
Human height	metres
Length of a running race to test fitness	micrometres
Growth of seedlings	centimetres

FOUL FACTS

William Buckland was, like the rest of his family, known for being brilliant but slightly odd. On a visit to Italy they were told by a guide that the spot where a saint was martyred produced blood each morning. Kneeling down to observe more closely, he tasted the 'blood'. He proclaimed that it was not blood, it was bat's urine!

10

How science works

Errors

Even when an instrument is used correctly, the results can still show differences.

Results may differ because of **random error**. This is most likely to be due to a poor measurement being made. It could be due to not carrying out the method consistently.

The error might be a **systematic error**. This means that the method was carried out consistently but an error was being repeated.

Check out these two sets of data that were taken from the investigation that Mark did. He tested 5 different volumes of enzyme. The third line is the amount of oxygen that was expected from calculations:

Amount of enzyme used (cm^3)	1	2	3	4	5
Oxygen produced (cm^3)	3.2	8.9	9.5	12.7	15.9
	3.1	6.4	9.7	12.5	16.1
Calculated oxygen production (cm^3)	4.2	8.4	12.5	16.6	20.7

d) Discuss whether there is any evidence for random error in these results.
e) Discuss whether there is any evidence for systematic error in these results.

Anomalies

Anomalous results are clearly out of line. They are not those that are due to the natural variation you get from any measurement. These should be looked at carefully. There might be a very interesting reason why they are so different. If they are simply due to a random error, then they should be discarded (rejected).

If anomalies can be identified while you are doing an investigation, then it is best to repeat that part of the investigation.

If you find anomalies after you have finished collecting data for an investigation, then they must be discarded.

DID YOU KNOW?

Sir Alexander Fleming was showing his research assistant some plates on which he had grown bacteria. He noticed an anomaly. There was some mould growing on one of the plates and around it there were no bacteria. He investigated further and grew the mould, identifying it as *Penicillium rubrum*.

He persuaded an assistant to taste it and he said it tasted like Stilton cheese! He later injected the assistant with it – and he didn't die!

Only because Fleming checked out his anomaly did it lead to the discovery of penicillin. Oh, and Fleming also let his nose dribble onto one plate and he discovered lysozyme!!

SUMMARY QUESTIONS

1 Copy and complete using the words below:

accurate discarded random sensitivity systematic use variation

There will always be some …… in results. You should always choose the best instruments that you can to get the most …… results. You must know how to …… the instrument properly. The …… of an instrument refers to the smallest change that can be detected. There are two types of error – …… and …… . Anomalies due to random error should be …… .

2 Which of the following will lead to a systematic error and which to a random error?

a) Using a weighing machine, which has something stuck to the pan on the top.
b) Forgetting to re-zero the weighing machine.

KEY POINTS

1. Results will nearly always vary.
2. Better instruments give more accurate results.
3. Sensitivity of an instrument refers to the smallest change that it can detect.
4. Human error can produce random and systematic errors.
5. We examine anomalies; they might give us some interesting ideas. If they are due to a random error, we repeat the measurements. If there is no time to repeat them, we discard them.

11

SUMMARY ANSWERS

1 There will always be some *variation* in results. You should always choose the best instruments that you can to get the most *accurate* results. You must know how to *use* the instrument properly. The *sensitivity* of an instrument refers to the smallest change that can be detected. There are two types of error, *random* and *systematic*. Anomalies due to random error should be *discarded*.

2 i) systematic
ii) random.

KEY POINTS

- The students need to demonstrate an appreciation that accuracy and precision of measurement is a key feature in a successful investigation.

Answers to in-text questions

a) Generally a failure to control variables, e.g. temperature of the reactants might have changed.

b) Student is, e.g., standing too far away and at the wrong angle to accurately read the measuring cylinder.

c)

Used to measure	Sensitivity of measuring instrument
Size of a cell	micrometres
Human height	centimetres
Length of a running race to test fitness	metres
Growth of seedlings	millimetres

d) First attempt for 2 cm^3 of enzyme is the random error.

e) Average results are close to individual results, which are consistently different to the calculated oxygen production.

H6 Presenting data

LEARNING OBJECTIVES

Students should learn:

- What is meant by the range and the mean of a set of data.
- How to use tables of data.
- How to display data.

LEARNING OUTCOMES

Students should be able to:

- Express accurately the range and mean of a set of data.
- Distinguish between the uses of bar charts and line graphs.
- Draw line graphs accurately.

Teaching suggestions

- **Gifted and talented.** These students could be handling two dependent variables in the table and graph, e.g. cooling and weight loss of a beaker of water with time, with repeat readings included.
- **Learning styles**

  Kinaesthetic: Practical activities.

  Visual: Making observations and presenting data.

  Auditory: Listening to group discussions.

  Interpersonal: Participating in group discussions.

  Intrapersonal: Considering how to produce their own table and graph.
- **ICT link-up.** Students could use a set of data with, e.g., Excel to present the data in different ways, such as pie charts, line graphs, bar charts, etc. Allow them to decide on the most appropriate form. Care needs to be given to 'smoothing' which does not always produce a line of best fit.

SPECIFICATION LINK-UP HOW SCIENCE WORKS

Section 10.6

Presenting data

To explain the relationship between two or more variables, data may be presented in such a way as to make the patterns more evident. There is a link between the type of graph used and the variable they represent. The choice of graphical representation depends upon the type of variable they represent.

Students should know and understand:

- *The range of the data refers to the maximum and minimum values.*
- *The mean (or average) of the data refers to the sum of all the measurements divided by the number of measurements taken.*
- *Tables are an effective means of displaying data but are limited in how they portray the design of an investigation.*
- *Bar charts can be used to display data in which the independent variable is categoric and the dependent variable continuous.*
- *Line graphs can be used to display data in which both the independent and dependent variables are continuous.*

Lesson structure

STARTER

Excel – Prepare some data from a typical investigation that the students may have recently completed. Use all of the many ways of presenting the data in Excel to display it. Allow students to discuss and reach conclusions as to which is the best method. (10 minutes)

Newspapers– Choose data from the press – particularly useful are market trends where they do not use (0,0). This exaggerates changes. This could relate to the use of data logging which can exaggerate normal variation into major trends. (5 minutes)

MAIN

- Choose an appropriate topic to either demonstrate or allow small groups to gather data. E.g., cooling of water against time; using food labels to determine saturated oil content of different foods. Any topic that will allow rapid gathering of data. Be aware that some data will lead to a bar chart, this might be more appropriate to groups struggling to draw line graphs.
- Students should be told what their task is and therefore know how to construct an appropriate table. This should be done individually prior to collecting the data. Refer to the first paragraph under 'Table'.
- Group discussion on the best form of table.
- Carry out data gathering, putting data directly into table. Refer to second paragraph under 'Tables'.
- Individuals produce their own graphs. Refer to the 'Next time you . . .' box.
- Graphs could be exchanged and marked by others in the group, using the criteria in the paragraph mentioned above.

PLENARIES

Which type of graph? – Give students different headings from a variety of tables and ask them how best to show the results graphically. This could be done as a whole class with individuals showing answers as the teacher reveals each table heading. Each student can draw a large letter 'L' (for the line graph) on one side of the paper and 'B' (for bar chart) on the other, ready to show their answers (5 minutes)

Key words – Students should be given key words to prepare posters for the lab. Key words should be taken from the summary questions in the first six sections. (10 minutes)

Crossword – The students should complete a crossword based on the previous five lessons. (15 minutes)

ACTIVITY & EXTENSION IDEAS

Lower-attaining students should start with bar charts and move on to line graphs.

More-able students ought to be practising the skills learned earlier whilst gathering their data. They could also be given more difficult contexts that are more likely to produce anomalies. They could, for example, be given a context that produces both random and systematic errors.

Answers to in-text question

a) Dependent variable 4–28 mm.

HOW SCIENCE WORKS

H6 Presenting data

LEARNING OBJECTIVES

1. What do we mean by the 'range' and the 'mean' of the data?
2. How do you use tables of results?
3. How do you display your data?

For this section you will be working with data from this investigation:

Mel spread some bacteria onto a dish containing nutrient jelly. She also placed some discs onto the jelly. The discs contained different concentrations of an antibiotic. The dish was sealed and then left for a couple of days.

Then she measured the diameter of the clear part around each disc. The clear part is where the bacteria have not been able to grow. The bacteria grew all over the rest of the dish.

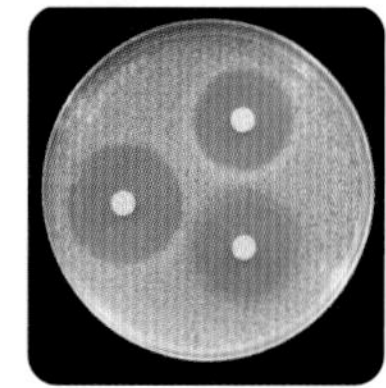

Figure 1 Petri dish with discs showing growth inhibition

Tables

Tables are really good for getting your results down quickly and clearly. You should design your table ***before*** you start your investigation.

Your table should be constructed to fit in all the data to be collected. It should be fully labelled, including units.

In some investigations, particularly fieldwork, it is useful to have an extra column for any notes you might want to make as you work.

While filling in your table of results you should be constantly looking for anomalies.

- Check to see if a repeat is sufficiently close to the first reading.
- Check to see if the pattern you are getting as you change the independent variable is what you expected.

Remember a result that looks anomalous should be checked out to see if it really is a poor reading or if it might suggest a different hypothesis.

Planning your table

Mel knew the values for her independent variable. We always put these in the first column of a table. The dependent variable goes in the second column. Mel will find its values as she carries out the investigation.

So she could plan a table like this:

Concentration of antibiotic (μg/ml)	Size of clear zone (mm)
4	
8	
16	
32	
64	

Or like this:

Concentration of antibiotic (μg/ml)	4	8	16	32	64
Size of clear zone (mm)					

All she had to do in the investigation was to write the correct numbers in the second column to complete the top table.

Mel's results are shown in the alternative format in the table below:

Concentration of antibiotic (μg/ml)	4	8	16	32	64
Size of clear zone (mm)	4	16	22	26	28

The range of the data

Pick out the maximum and the minimum values and you have the range. You should always quote these two numbers when asked for a range. For example, the range is between .... (the lowest value) and .... (the highest value) – and don't forget to include the units!

a) What is the range for the dependent variable in Mel's set of data?

The mean of the data

Often you have to find the mean of each repeated set of measurements.

You add up the measurements in the set and divide by how many there are. Miss out any anomalies you find.

The repeat values and mean can be recorded as shown below:

Concentration of antibiotic (μg/ml)	Size of clear zone (mm)			
	1st test	2nd test	3rd test	Mean

Displaying your results

Bar charts

If you have a categoric or an ordered independent variable and a continuous dependent variable then you should use a bar chart.

Line graphs

If you have a continuous independent and a continuous dependent variable then a line graph should be used.

Scatter graphs (or scattergrams)

Scatter graphs are used in much the same way as line graphs, but you might not expect to be able to draw such a clear line of best fit. For example, if you wanted to see if lung capacity was related to how long you could hold your breath, you would draw a scatter graph with your results.

SUMMARY QUESTIONS

1. Copy and complete using the words below:

   categoric continuous mean range

   The maximum and minimum values show the ...... of the data. The sum of all the values divided by the total number of the values gives the .... Bar charts are used when you have a ...... independent variable and a continuous dependent variable.
   Line graphs are used when you have ...... independent and dependent variables.
2. Draw a graph of Mel's results from the top of this page.

NEXT TIME YOU...

... make a table for your results remember to include:
- headings,
- units,
- a title.

... draw a line graph remember to include:
- the independent variable on the *x*-axis,
- the dependent variable on the *y*-axis,
- a line of best fit,
- labels, units and a title.

GET IT RIGHT!

Marks are often dropped in the exam by candidates plotting points incorrectly. Also use a **line of best fit** where appropriate – don't just join the points 'dot-to-dot'!

KEY POINTS

1. The range states the maximum and the minimum values.
2. The mean is the sum of the values divided by how many values there are.
3. Tables are best used during an investigation to record results.
4. Bar charts are used when you have a categoric or an ordered independent variable and a continuous dependent variable.
5. Line graphs are used to display data that are continuous.

SUMMARY ANSWERS

1. The maximum and minimum values show the *range* of the data. The sum of all the values divided by the total number of the values gives the *mean*. Bar charts are used when you have a *categoric* independent and a *continuous* dependent variable. Line graphs are used when you have *continuous* independent and dependent variables.
2. Graph drawn with concentration on the horizontal (x) axis and size of clear zone on the vertical (y) axis.
   Axes labelled with units and points plotted correctly.
   Line of best fit should be drawn.

KEY POINTS

- The students should be able to produce their own correctly labelled table and graph.

H7 Using data to draw conclusions

LEARNING OBJECTIVES

Students should learn:

- How to use charts and graphs to identify patterns.
- How to draw accurate conclusions from relationships.
- How to improve the reliability of an investigation.

LEARNING OUTCOMES

Students should be able to:

- Express accurately the range and mean of a set of data.
- Distinguish between the uses of bar charts and line graphs.
- Draw line graphs accurately.

Teaching suggestions

- **Special needs.** Provide a flow diagram so that students can see the process as they are going through it.
- **Gifted and talented.** Could take the original investigation and design out some of the flaws, producing an investigation with improved validity and reliability.

  Summary question 2 could be examined in some detail and the work researched on the web.
- **Learning styles**

  Kinaesthetic: Drawing graphs.

  Visual: Observing tests.

  Auditory: Class and small group discussion.

  Interpersonal: Class and small group discussion.

  Intrapersonal: Evaluating the reliability and validity of their own tests.
- **ICT link-up.** ICT can be used to quickly create data for graphs in the lesson and this will generate a good volume of data for discussion. Some ICT support is needed for rapid production of graphs.

SPECIFICATION LINK-UP HOW SCIENCE WORKS

Section 10.7

Identifying patterns and relationships in data

The patterns and relationships observed in data represent the behaviour of the variables in an investigation. However, it is necessary to look at patterns and relationships between variables with the limitations of the data in mind.

Students should know and understand:

- *Patterns in tables and graphs can be used to identify anomalous data that require further consideration.*
- *A line of best fit can be used to illustrate the underlying relationship between variables.*
- *The relationships that exist between variables can be linear (positive or negative), directly proportional, predictable curves, complex curves and relationships not easily represented by a mathematical relationship.*

Evaluation

- *In evaluating a whole investigation, the reliability and validity of the data obtained must be considered. Reliability and validity of an investigation can be increased by looking at data obtained from secondary sources, through using an alternative method as a check and by requiring that the results are reproducible by others.*
- *Conclusions must be limited by the data available and not to go beyond them.*

Lesson structure

STARTER

Starter graphs – Prepare a series of graphs that illustrate the various types of relationships in the specification. Each graph should have fully labelled axes. Students should, in groups, agree statements that describe the patterns in the graphs. Gather feedback from groups and discuss. (10 minutes)

MAIN

- Using the graphs from the previous lesson, students should be taught how to produce lines of best fit. (Develop own data if this is not available or is in the form of a bar graph.) Students could work individually with help from the first section of H7 'Using data to draw conclusions'.
- They should identify the pattern in their graph.
- They now need to consider the reliability and validity of their results. They may need their understanding of reliability and validity reinforced. How this is achieved will depend on the investigation chosen in H6 'Presenting data'. Questions can be posed to reinforce their understanding of both terms. If the investigation was not carefully controlled, then it is likely to be unreliable and invalid, thus posing many opportunities for discussion. There is also an opportunity to reinforce other ideas such as random and systematic errors.
- If the previous activity is unlikely to yield these opportunities, then a brief demonstration of a test, e.g. finding the energy transfer when burning crisps of different mass, could be used. Students should observe the teacher and make notes as the tests are carried out. They should be as critical as they can be, and in small groups discuss their individual findings. One or two students could be recording the results and two more plotting the graph as the teacher does the tests. These could be processed immediately onto the screen.
- Return to the original prediction. Look at the graph of the results. Ask how much confidence the group has in the results.
- Review the links that are possible between two sets of data. Ask them to decide which one their tests might support.
- Now the word 'conclusion' should be introduced and a conclusion made . . . if possible! It is sometimes useful to make a conclusion that is 'subject to . . . e.g. the reliability being demonstrated'.

PLENARIES

Flow diagram – When pulling the lesson together it will be important to emphasise the process involved – graph – line of best fit – pattern – question the reliability and validity – consider the links that are possible – make a conclusion – summarise evaluation. This could be illustrated with a flow diagram generated by class discussion. (5 minutes)

Key words – Students should be given key words to complete the posters for the lab. (10 minutes)

ACTIVITY & EXTENSION

Students should be able to transfer these skills to examine the work of scientists and to become critical of the work of others. Collecting scientific findings from the press and subjecting them to the same critical appraisal is an important exercise. They could be encouraged to collect these or be given photocopies of topical issues suitable for such appraisal.

H7 Using data to draw conclusions

LEARNING OBJECTIVES

1. How do we best use charts and graphs to identify patterns?
2. What are the possible relationships we can identify from charts and graphs?
3. How do we draw conclusions from relationships?
4. How can we improve the reliability of our investigation?

Identifying patterns and relationships

Now that you have a bar chart or a graph of your results you can begin to look for patterns. You must have an open mind at this point.

Firstly, there could still be some anomalous results. You might not have picked these out earlier. How do you spot an anomaly? It must be a significant distance away from the pattern, not just within normal variation.

A line of best fit will help to identify any anomalies at this stage. Ask yourself – do the anomalies represent something important or were they just a mistake?

Secondly, remember a line of best fit can be a straight line or it can be a curve – you have to decide from your results.

The line of best fit will also lead you into thinking what the relationship is between your two variables. You need to consider whether your graph shows a **linear** relationship. This simply means can you be confident about drawing a straight line of best fit on your graph? If the answer is yes – then is this line positive or negative?

a) Say whether graphs (i) and (ii) in Figure 1 show a positive or a negative linear relationship.

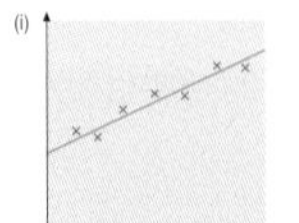

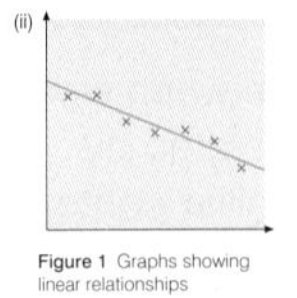

Figure 1 Graphs showing linear relationships

Look at the graph in Figure 2. It shows a positive linear relationship. It also goes through the origin (0,0). We call this a **directly proportional** relationship.

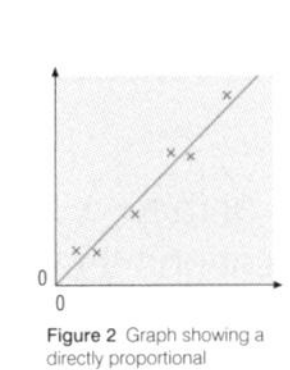

Figure 2 Graph showing a directly proportional relationship

Your results might also show a curved line of best fit. These can be predictable, complex or very complex! Look at Figure 3 below.

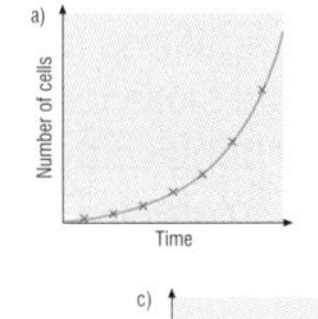

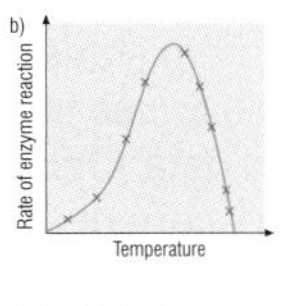

c) Air temperature / Time

Figure 3 a) Graph showing predictable results. b) Graph showing complex results. c) Graph showing very complex results.

Drawing conclusions

Your graphs are designed to show the relationship between your two chosen variables. You need to consider what the relationship means for your conclusion.

There are three possible links between variables. (See page 5.) They can be:

- causal,
- due to association, or
- due to chance.

You must decide which is the most likely. Remember a positive relationship does not always mean a causal link between the two variables.

Poor science can often happen if a wrong decision is made here. Newspapers have said that living near electricity sub-stations can cause cancer. All that scientists would say is that there is possibly an association. Getting the correct conclusion is very important.

You will have made a prediction. This could be supported by your results. It might not be supported or it could be partly supported. Your results might suggest some other hypothesis to you.

Your conclusion must go no further than the evidence that you have. A grass snake is not poisonous, but this does not mean that all snakes are safe to handle!

Evaluation

If you are still uncertain about a conclusion, it might be down to the reliability and the validity of the results. You could check these by:

- looking for other similar work on the Internet or from others in your class,
- getting somebody else to re-do your investigation, or
- trying an alternative method to see if you get the same results.

SUMMARY QUESTIONS

1. Copy and complete using the words below:

   anomalous complex directly negative positive

   Lines of best fit can be used to identify ...... results. Linear relationships can be ...... or .... If a graph goes through the origin then the relationship could be ...... proportional. Often a line of best fit is a curve which can be predictable or ....
2. Nasma found a newspaper article about slimming. (See above.) Discuss the type of experiment and the data you would expect to see to support this conclusion.

DID YOU KNOW?

Thomas Huxley, a famous biologist, said 'The great tragedy of Science – the slaying of a beautiful hypothesis by an ugly fact.'

NEXT TIME YOU...

... read scientific claims, think carefully about the evidence that should be there to back up the claim.

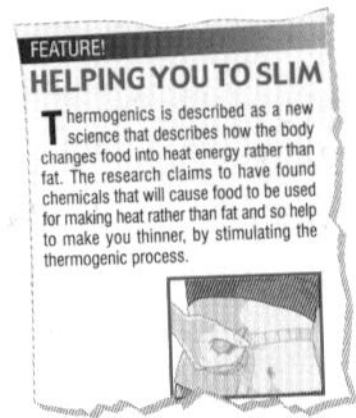
FEATURE!

HELPING YOU TO SLIM

Thermogenics is described as a new science that describes how the body changes food into heat energy rather than fat. The research claims to have found chemicals that will cause food to be used for making heat rather than fat and so help to make you thinner, by stimulating the thermogenic process.

KEY POINTS

1. Drawing lines of best fit help us to study the relationship between variables.
2. The possible relationships are linear, positive and negative; directly proportional; predictable and complex curves.
3. Conclusions must go no further than the data available.
4. The reliability and validity of data can be checked by looking at other similar work done by others, perhaps on the Internet. It can also be checked by using a different method or by others checking your method.

SUMMARY ANSWERS

1. Lines of best fit can be used to identify *anomalous* results. Linear relationships can be *positive* or *negative*. If a graph goes through the origin then the relationship could be *directly* proportional. Often a line of best fit is a curve which can be predictable or *complex*.
2. Two large groups of people – chosen to be of similar, e.g., weight, body shape, age, sex, etc. Two groups chosen who will be on the same diet and exercise routines for an appropriate length of time. Idea of control group with placebo. (More-able students might be able to appreciate the idea of a double-blind test.) Measurement of thermogenic process? Might expect increased rate of sweating or some measurement of increased production of heat! Certainly some measurement of the loss of weight between the two groups. Significant difference in the mean differences between the two groups.

Answers to in-text questions

a) i) is positive linear.
ii) is negative linear.

KEY POINTS

- The students should appreciate the process involved from the production of a graph to the conclusion and evaluation.

H8 Scientific evidence and society

LEARNING OBJECTIVES

Students should learn:

- That science must be presented in a way that takes into account the reliability and the validity of the evidence.
- That science should be presented without bias from the experimenter.
- That evidence must be checked to appreciate whether there is any political influence.
- That the status of the experimenter can influence the weight attached to a scientific report.

LEARNING OUTCOMES

Students should be able to:

- Make judgements about the reliability and the validity of scientific evidence.
- Identify when scientific evidence might have been influenced by bias or political influence.
- To judge scientific evidence on its merits, taking into account the weight given to it by the status of the experimenter.

Teaching suggestions

- **Special needs.** Provide a diagram with the key points in order of discussion so that students can see the process as they are going through it.
- **Gifted and talented.** Students could attend a local public enquiry or even the local town council as it discusses local issues with a scientific context.
- **Learning styles**

  Kinaesthetic: Role play.

  Visual: Researching data.

  Auditory: Class and small group discussion on possible bias in newspaper reporting of scientific issues.

  Interpersonal: Class and small group discussion on GM foods.

  Intrapersonal: Consideration of the influences on research to personal life.
- **ICT link-up.** The Internet exercise on researching the issue of GM foods in the main part of the lesson is a good ICT activity. This could considerably increase the volume and improve the presentation of data for discussion. Some students may require support for downloading data from the Internet.

SPECIFICATION LINK-UP HOW SCIENCE WORKS

Section 10.8

Societal aspects of scientific evidence

A judgement or decision relating to social-scientific issues may not be based on evidence alone, as other societal factors may be relevant.

Students should know and understand:

- *The credibility of the evidence is increased if a balanced account of the data is used rather than a selection from it which supports a particular pre-determined stance.*
- *Evidence must be scrutinised for any potential bias of the experimenter, such as funding sources or allegiances.*
- *Evidence can be accorded undue weight, or dismissed too lightly, simply because of its political significance. If the consequences of the evidence might provoke public or political disquiet, the evidence may be downplayed.*
- *The status of the experimenter may influence the weight placed on evidence; for instance, academic or professional status, experience and authority. It is more likely that the advice of an eminent scientist will be sought to help provide a solution to a problem than a scientist with less experience.*

Lesson structure

STARTER

Ask a scientist – It is necessary at this point to make a seamless join between work which has mostly been derived from student investigations to work generated by scientists. Students must be able to use their critical skills derived in familiar contexts and apply them to second-hand data. One way to achieve this would be to bring in newspaper cuttings on a topic of current scientific interest. They should be aware that some newspaper reporters will 'cherry-pick' sections of reports to support sensational claims that will make good headlines.

Students could be prompted by the key word posters to question some of the assumptions being made. This could be presented as a 'wish-list' of questions they would like to put to the scientists who conducted the research and to the newspaper reporter. (10 minutes)

Researching scientific evidence – With access to the Internet, students could be given a topic to research. They should use a search engine and identify the sources of information from, say, the first six web pages. They could then discuss the relative merits of these sources in terms of potential for bias.

MAIN

- The following points are best made using topics that are of immediate importance to your students. The examples used are only illustrative. Some forward planning is required to ensure that there is a plentiful supply of newspaper articles, both local and national, to support the lesson. These could be displayed and/or retained in a portfolio for reference.
- Working in pairs, students should answer question a). They should write a few sentences about the headline and what it means to them. Follow this with a class discussion, building up many more questions that need to be answered.
- It might be possible to follow this article on the Internet to find out what is unsafe about GM foods. It should lead to a balanced discussion of the possible benefits and hazards of GM foods.
- Use the next section to illustrate the possibility of bias in reporting science. Again use small group discussions followed by whole class plenary.

- If you have access to the Internet for the whole class, then it is worth pursuing the issue of GM foods in relation to their political significance. Pose the question: 'What would happen to the economy of this country if we decided to refuse to grow all GM products?' Would different people come together to suppress information that might support GM products? Should they be allowed to suppress scientific evidence? Stress that there is a balanced argument here and that some genetic modification already exists in our environment, yet people have a fear of GM. Why do they have that fear? Should scientists have the task of reducing that fear to proper proportions? There is much to discuss.
- Small groups can imagine that they are preparing a case against the use of pesticides on crops close to their village. They could be given data that relates to toxicity levels and spread using different spraying techniques. Up-to-date data can be obtained from the web, e.g. from the DEFRA website. Students could be given the data as if it were information provided at a public enquiry. They should be asked to prepare a case that questions, e.g., the reliability and the validity of the data.

ACTIVITY & EXTENSION

Students could role play a public enquiry. They could be given roles and asked to prepare a case for homework. The data should be available to them so that they all know the arguments before preparing their case. Possible link here with the English department. This activity could be allocated as a homework exercise

PLENARIES

Group report – Groups should report their findings on GM foods to the class. (10 minutes)

Scientific data posters – Groups could prepare posters that use scientific data to present their case for or against any of the developments discussed. (10–15 minutes)

KEY POINTS

- The students should show in discussions that they can apply the skills they have developed in their own investigative work and that those skills are applicable to scientific evidence generated by professional scientists.

H8 Scientific evidence and society

LEARNING OBJECTIVES

1. How can science encourage people to have faith in its research?
2. How might bias affect people's judgement of science?
3. Can politics influence judgements about science?
4. Do you have to be a professor to be believed?

Now you have reached a conclusion about a piece of scientific research. So what is next? If it is pure research then your fellow scientists will want to look at it very carefully. If it affects the lives of ordinary people then society will also want to examine it closely.

DID YOU KNOW?

During the middle part of the last century, Stalin had an iron grip on the agriculture in what was the USSR. The effect was to starve to death up to 8 million people. Biological research was to be the scapegoat and many scientists had their work expunged. Vavilov was on a field trip, he had just discovered a new species of wheat, when he was hunted down, put into prison and died of starvation two years later.

You can help your cause by giving a balanced account of what you have found out. It is much the same as any argument you might have. If you make ridiculous claims then nobody will believe anything you have to say.

Be open and honest. If you only tell part of the story then someone will want to know why! Equally, if somebody is only telling you part of the truth, you cannot be confident with anything they say.

a) A disinfectant claims that it kills 99.9% of germs on surfaces that you come in contact with every day. What is missing? Is it important?

You must be on the lookout for people who might be biased when representing scientific evidence. Some scientists are paid by companies to do research. When you are told that a certain product is harmless, just check out who is telling you.

b) Bottles of perfume spray contain this advice 'This finished product has not been tested on animals.' Why might you mistrust this statement?

Suppose you wanted to know about how to slim. Who would you be more likely to believe? Would it be a scientist working for 'Slimkwik', or an independent scientist? Sometimes the differences are not quite so obvious.

We also have to be very careful in reaching judgements according to who is presenting scientific evidence to us. For example, if the evidence might provoke public or political problems, then it might be played down.

Equally others might want to exaggerate the findings. They might make more of the results than the evidence suggests. Take as an example the data available on animal research. Animal liberation followers may well present the ***same*** evidence completely differently to pharmaceutical companies wishing to develop new drugs.

c) Check out some websites on smoking and lung cancer. Do a balanced review looking at tobacco manufacturers as well as anti-smoking lobbies such as ASH. You might also check out government websites.

The status of the experimenter may place more weight on evidence. Suppose a lawyer wants to convince a jury that a particular piece of scientific evidence is valid. The lawyer will choose the most eminent scientist in that field. Cot deaths are a particularly difficult problem for the police. If the medical evidence suggests that the baby might have been murdered then the prosecution and the defence get the most eminent scientists to argue the reliability and validity of the evidence. Who does the jury believe?

EXPERT WITNESS IN COT DEATH COURT CASE MISLED THE JURY

A child abuse expert was struck off as a doctor today for giving seriously misleading evidence in a court case. The court case led to a woman being wrongly convicted of murdering her two children.

DID YOU KNOW?

A scientist who rejected the idea of a causal link between smoking and lung cancer was later found to be being paid by a tobacco company.

SUMMARY QUESTIONS

1. Copy and complete using the words below:

   status balanced bias political

   Evidence from scientific investigations should be given in a …… way. It must be checked for any …… from the experimenter.
   Evidence can be given too little or too much weight if it is of …… significance.
   The …… of the experimenter is likely to influence people in their judgement of the evidence.
2. Collect some newspaper articles to show how scientific evidence is used. Discuss in groups whether these articles are honest and fair representations of the science. Consider whether they carry any bias.
3. This is the opening paragraph from a review of GM foods.

   The UK government has been promoting ... a review of the science of GM, led by Sir David King (the Government's Chief Scientific Adviser) working with Professor Howard Dalton (the Chief Scientific Adviser to the Secretary of State for the Environment, Food and Rural Affairs), with independent advice from the Food Standards Agency.

   Discuss this paragraph and decide which parts of it make you want to believe the evidence they might give. Then consider which parts make you mistrust any conclusions they might reach.

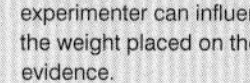

KEY POINTS

1. Scientific evidence must be presented in a balanced way that points out clearly how reliable and valid the evidence is.
2. The evidence must not contain any bias from the experimenter.
3. The evidence must be checked to appreciate whether there has been any political influence.
4. The status of the experimenter can influence the weight placed on the evidence.

SUMMARY ANSWERS

1. Evidence from scientific investigations should be given in a *balanced* way. It must be checked for any *bias* from the experimenter. Evidence can be given too little or too much weight if it is of *political* significance. The *status* of the experimenter is likely to influence people in their judgement of the evidence.
2. Identification of any bias in reports.
3. Discussion.

Answers to in-text questions

a) E.g. what are the 0.1% of bacteria that are not killed?; how dangerous are they?; how quickly do they reproduce?; are they more likely to reproduce in the absence of other competitor bacteria?; what are the surfaces that we come in contact with?; what evidence is there?; who did the research?

b) It might not be safe for humans to use. The constituents of the perfume might have been tested on animals before being made into the final product.

c) Identification of any political bias; this could be from companies and individuals as well as governments.

H9 How is science used for everybody's benefit?

Note that if there are particular difficulties with teaching this context of science then equivalent work can be located in the Chemistry and Physics books.

LEARNING OBJECTIVES

Students should learn:

- That scientific enquiry can result in technological developments.
- That scientific and technological developments can be exploited by different people in different ways.
- That scientific and technological developments can raise ethical, social, economic and environmental issues.
- That different decisions concerning these issues are made by different groups of people.
- That there are many scientific questions unanswered and some questions that can never be answered by science.

LEARNING OUTCOMES

Students should be able to:

- Recognise links between science and technology.
- Recognise when people exploit scientific and technological developments.
- Recognise ethical, social, economic and environmental issues raised by scientific and technological developments.
- Show how scientific and technological developments raise different issues for different groups of people.
- Discuss how scientific questions remain unanswered and recognise the limitations of science for answering some questions.

SPECIFICATION LINK-UP HOW SCIENCE WORKS

Section 10.8 continued

- *Scientific knowledge gained through investigations can be the basis for technological developments.*
- *Scientific and technological developments offer different opportunities for exploitation, to different groups of people.*
- *The uses of science and technology developments can raise ethical, social, economic and environmental issues.*
- *Decisions are made by individuals and by society on issues relating to science and technology.*

Section 10.9

Limitations of scientific evidence

Science can help us in many ways but it cannot supply all the answers.

We are still finding out about things and developing our scientific knowledge. There are some questions that we cannot answer, maybe because we do not have enough reliable and valid evidence.

And there are some questions that science cannot answer at all. These tend to be questions where beliefs and opinions are important or where we cannot collect reliable and valid scientific evidence.

Lesson structure

STARTER

Technological development Part 1 – Choose a technological development where students can appreciate the science that underpins its operation. Your choice will be largely based on the students' level of scientific knowledge. Examples might include: antiseptics, antibiotics, sewage treatment, fish farming. They might research the science for homework. Asking students to offer any scientific knowledge that they have related to that technological development can generate discussions. The knowledge can then be evaluated in terms of its importance to the technological development. (10 minutes)

MAIN

- Divide the students into small groups and give each group a different technological development (photographs would be appropriate). They should be asked to discuss the different uses to which they have been put. They could offer other ideas for which they might not have been used. They do not have to stay with that actual example but can drift into other developments.
- A plenary session in which one of the group reports their findings to the others. Discussions could then include whether these uses were of general benefit or not.
- If any groups come up with novel ideas as to how to make better use of some of the technology then you could spin the idea of economic development.
- Do any of the ideas raise environmental issues?
- Do any of the ideas raise ethical issues?
- Do any of the ideas raise social issues?
- At this point it might be appropriate to cut to the example of contraceptives in the student book. Students could read through this on their own or it could be read and explained as a class or in small groups with support. It might be appropriate to intervene at points to make sure the relevance of each of the learning objectives is appreciated.
- How the discussion is handled will depend on how the reading was organised. It would be useful to allow small group discussion for questions 3 and 4.

PLENARIES

Technological development Part 2 – Take another technological development, such as biological washing powders and use it as an example to help the group discuss all of the issues raised by the lesson. (5 minutes)

SUMMARY ANSWERS

1 That diosgenin, a precursor of progesterone, was available in yam plants. That progesterone was a hormone involved in human fertility.

2 Some people misinterpreted the science and used yam as a contraceptive. Pharmaceutical companies manufactured it. Women took it to avoid pregnancy. Other researchers produced different pills, some of which were less dangerous. The male pill was developed. The morning after pill was developed. Some institutions wanted to ban it.

3 **a)** (i) E.g. it encouraged sex outside marriage, (ii) it put birth control under women's control, (iii) extra funding was required from the taxpayer, (iv) pollution in rivers causing fish to become sterile.

b) (i) and (ii) decided by individuals, (iii) and (iv) decided by the government. The latter on the basis that it might be possible – but expensive – to remove the hormones before they enter the river.

ACTIVITY & EXTENSION

You could visit a local museum to see how local industry might have used science for its technologies.

HOW SCIENCE WORKS

H9 How is science used for everybody's benefit?

LEARNING OBJECTIVES

1 How does science link to technology?
2 How is science used and abused?
3 How are decisions made about science?
4 What are the limitations of science?

The development of oral contraceptives shows how science can be used for technological development. There were many unscientific ways in which women would attempt contraception.

In Europe, women wore the foot of a weasel around their neck to prevent them becoming pregnant. In North Africa the flower silphium was thought to be an oral contraceptive. It became very expensive to buy and eventually was used so much that it became extinct.

These ideas might have had some basis, because some plants do contain human sex hormones. Some plants such as black kohosh, are used today as relief from problems related to the menopause.

DID YOU KNOW?
The yam has been used as a medicine for at least 2000 years in China, Japan and South-East Asia.

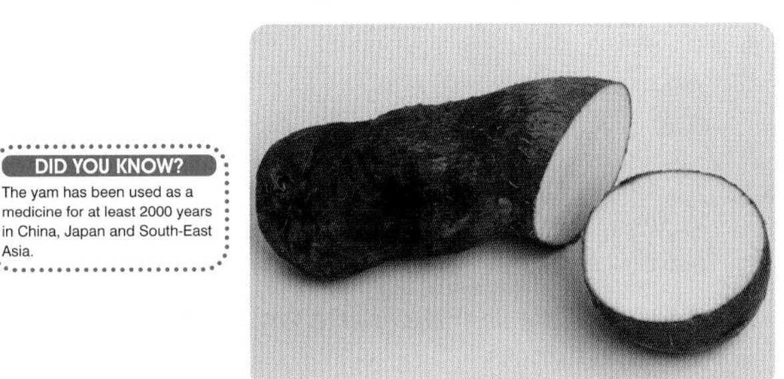

Figure 1 A yam plant

Figure 2 Frank Colton

Yam was used as a pain relief. A Japanese scientist found diosgenin in yam plants. Diosgenin was used as the starting point for an investigation that led to the development of the hormone progesterone. This led to the development of the first contraceptive pill. It was some unscientific thinking that encouraged people to eat yam plants as a natural contraceptive – they aren't!

Frank Colton developed Enovid, one of the first oral contraceptives, in 1960. By 1961 the birth control pill was available to 'everyone'. Some doctors were in a dilemma for social as well as medical reasons. They would not prescribe it to unmarried women, because it 'encourages sex outside marriage'.

The UK Government said it couldn't afford the cost. They said the pill could have long-term effects. A woman's body was likened to a clock; 'Whilst it was running well it should be left alone,' said Sir Charles Dodds, a leading expert on drugs. The pill allowed women to take control of their own fertility.

18

How science works

It was known that there was a risk of heart disease and stroke. However, new developments have reduced the dose and therefore this risk.

Today there are hundreds of different contraceptive pills. There are even male contraceptive pills. The morning-after pill has raised problems for some people who consider it is a form of abortion.

Some of these hormones are now in such a high concentration in river water that it affects the ability of some fish to reproduce.

Figure 3 Contraceptive pills

The contraceptive pill still raises social, ethical, economic and even environmental issues.

There are many questions left for science to answer. For example, how to develop an oral contraceptive that is 100% safe for all people. However, science cannot answer questions about whether or not we should use contraception.

SUMMARY QUESTIONS

Use the account of the development of contraceptive pills to answer these questions.

1 What scientific knowledge was available to Frank Colton that enabled him to develop Enovid?
2 How did different groups of people react to the development of Enovid?
3 a) Identify some of these issues raised by the development of contraceptive pills: i) ethical, ii) social, iii) economic, iv) environmental.
b) Which of these issues are decided by individuals and which by society?

KEY POINTS

1 Scientific knowledge can be used to develop technologies.
2 People can exploit scientific and technological developments to suit their own purposes.
3 The uses of science and technology can raise ethical, social, economic and environmental issues.
4 These issues are decided upon by individuals and by society.
5 There are many questions left for science to answer. But science cannot answer questions that start with 'Should we .... ?'

19

Teaching suggestions

- **Special needs.** Small groups will need sympathetic support to work their way through these ideas. It might be more appropriate to separate the lesson in two, dealing with the science behind the technologies separately from the issues raised.
- **Gifted and talented.** Students could research their own scientific story and respond to the same issues, e.g., the story of the development of statins. The science started with Dr Endo and Dr Kuroda in Japan. They reasoned that microorganisms such as Penicillium may well make inhibitors of an enzyme that helps to manufacture cholesterol. They would do this to defend against other organisms because other organisms would need this enzyme to manufacture cholesterol for their cell walls. They identified and separated the inhibitor from Penicillium citrinum. Merck and Co. developed the first commercial statin from the mould Aspergillus terreus.
- **Learning styles**

  Kinaesthetic: Examining the technology.

  Visual: Observing equipment.

  Auditory: Class and small group discussion on the uses of science and technological developments.

  Interpersonal: Class and small group discussion.

  Intrapersonal: Reflecting on the issues related to contraception.

KEY POINTS

- The students should show in discussions that they appreciate the link between science and technology.
- Students should be more aware that science and technology raises different issues for different people.

SUMMARY ANSWERS

1 Could be some differences which would be fine, e.g. prediction; design; safety; controls; method; table; results; repeat; graph; conclusion; improve.

2 **a)** Scientific opinion is based on reliable and valid evidence, an opinion might not be.
b) Continuous variable because it is more powerful than an ordered or a categoric variable.
c) A causal link is where only one independent variable has an effect on one dependent variable, an association has a third variable involved.

3 **a)** A hypothesis is an idea that fits an observation and the scientific knowledge that is available.
b) Increasing concentrations of sulphur dioxide reduces the growth rate of lichens.
c) A prediction can be tested.
d) The hypothesis could be supported or refuted or it might cause you to change your hypothesis.
e) The theory on which you based the hypothesis might have to be changed.

4 **a)** When all variables but the one being used are kept constant.
b) Important variables are constantly changing, e.g. weather.
c) You would set up the investigation so that all of the changing variables affected the plants in the same way.
d) If you have the correct conditions; if you have the correct range; if you have the correct interval readings; if you need to repeat your readings.

5 E.g. were the readings taken at the exact times stated? Could the bubbles be accurately counted? Were the bubbles all the same size? Were the correct concentrations measured accurately? Was time given to allow the yeast to start respiring rapidly?

6 **a)** Take the highest and the lowest reading.
b) The sum of all the readings.
c) When you have an ordered or categoric independent variable and a continuous dependent variable.
d) When you have a continuous independent variable and a continuous dependent variable.

7 **a)** Examined to see if it is an error, if so, repeat it. If identified from the graph, it should be ignored. Be aware that it could lead to something really interesting and lead to a new hypothesis.
b) Identify a pattern.
c) That it does not go further than the data, the reliability and the validity allow.
d) By repeating results, by getting others to repeat your result by checking other equivalent data.

8 **a)** The science is more likely to be accepted.
b) They might be biased due to who is funding the research or because they are employed by a biased organisation. There might be political influences, the public might be too alarmed by any conclusions.

9 **a)** For many scientific developments there is a practical outcome which can be used – a technological development. Many technological developments allow further progress in science.
b) Society.

10 **a)** That stomata are very small holes on the surface of leaves.
b) The size of the stomata affected the rate of diffusion of carbon dioxide.
c) The smaller the stomata, the greater the rate of diffusion of carbon dioxide.
d) Diameter of the hole.
e) Volume of carbon dioxide diffusing per hour.
f) 2 mm–22.7 mm
g) To ensure that temperature changes did not affect the rate of diffusion.
h) Yes, because it was a practical solution to a difficult problem *or* no, it did not represent the actual sizes of stomata.
i) Before repeat readings should have been taken compare results with those of other groups/compare results with those from secondary sources e.g. research on the Internet.
j) Yes, because there are differences between the readings that form a pattern. However, without repeat readings it is not certain that readings of less than 6.0 mm might not show any real differences.
k) Suitable graph with diameter of the hole on the X axis and volume on the Y axis.
l) Directly proportional.
m) That increased size of the hole increases the rate of diffusion through the hole.
n) No.

HOW SCIENCE WORKS: H1 – H9

SUMMARY QUESTIONS

1 Fit these words into order. They should be in the order in which you might use them in an investigation.

design; prediction; conclusion; method; repeat; controls; graph; results; table; improve; safety

2 a) How would you tell the difference between an opinion that was scientific and a prejudiced opinion?

b) Suppose your were describing the height of plants for some fieldwork. What type of variable would you choose and why?

c) Explain the difference between a causal link between two variables and one which is due to association.

3 You might have observed that lichens do not grow where there is air pollution. You ask the question why. You use some theory to try to answer the question.

a) Explain what you understand by a hypothesis.

b) Sulfur dioxide in the air forms acids that attack the lichens. This is a hypothesis. Develop this into a prediction.

c) Explain why a prediction is more useful than a hypothesis.

d) Suppose you have tested your prediction and have some data. What might this do for your hypothesis?

e) Suppose the data does not support the hypothesis. What should you do to the theory that gave you the hypothesis?

4 a) What do you understand by a fair test?

b) Explain why setting up a fair test in fieldwork is difficult.

c) Describe how you can make your results valid in fieldwork.

d) Suppose you were carrying out an investigation into how pulse rates vary with exercise. You woulc need to carry out a trial. Describe what a trial wou tell you about how to plan your method.

5 Suppose you were watching a friend carry out an investigation measuring the carbon dioxide produced by yeast cells. You have to mark your friend on how accurately she is making her measurements. Make a list of points that you would be looking for.

6 a) How do you decide on the range of a set of data?

b) How do you calculate the mean?

c) When should you use a bar chart?

d) When should you use a line graph?

7 a) What should happen to anomalous results?

b) What does a line of best fit allow you to do?

c) When making a conclusion, what must you take into consideration?

d) How can you check on the reliability of your results

8 a) Why is it important when reporting science to 'tell the truth, the whole truth and nothing but the truth'

How science works

b) Why might some people be tempted not to be completely fair when reporting their opinions on scientific data?

a) 'Science can advance technology and technology can advance science.'
What do you think is meant by this statement?

b) Who answers the questions that start with 'Should we . . . '?

You can see from this electron micrograph below that stomata are very small holes in the leaves of plants. They allow carbon dioxide to diffuse into the leaf cells for photosynthesis. The size of the hole is controlled by guard cells. It was suggested that the size of the hole might affect the rate at which carbon dioxide diffused through the hole.

Electron micrograph of stomatal guard cells

Stomata are very small holes (when fully open they are 10–20 μm in diameter). The question was: Are small holes better than large holes? This would seem reasonable as plants have very small stomata. The hypothesis was that small holes would allow more carbon dioxide to pass through than large holes.

It was decided to use much larger holes than the stomata because it would be easier to get accurate measurements. The investigation was carried out and the results were as follows.

Diameter of hole (mm)	Volume of CO_2 diffusing per hour (cm^3)
22.7	0.24
12.1	0.10
6.0	0.06
3.2	0.04
2.0	0.02

a) What was the observation on which this investigation was based?
b) What was the original hypothesis?
c) What was the likely prediction?
d) What was the independent variable?
e) What was the dependent variable?
f) What is the range for the diameter of the hole?
g) Why was the temperature kept the same during the investigation?
h) Was this a sensible range of size of holes to use? Explain your answer.
i) How could the investigation have its reliability improved?
j) Was the sensitivity of the instrument measuring volumes of CO_2 satisfactory? Provide some evidence for your answer from the data in the table.
k) Draw a graph of the results in the table above.
l) Describe the pattern in these results.
m) What conclusion can you make?
n) Does your conclusion support the prediction?

How science works teaching suggestions

- **Literacy guidance.** The externally set test for every ISA has a question in which the scoring of marks is in part dependent on skills such as presenting information, developing an argument and drawing a conclusion.
- Where students are asked for an explanation, they have the opportunity to write answers in continuous prose and practise their literary skills. Look for good grammar, clear expression and the correct spelling of scientific terms. Question 5, for example, would encourage the development of an argument. Question 6 – how to present data and question 7c) – drawing a conclusion. Questions relating to terms used in How Science Works should allow students to express their understanding of those terms, e.g. question 3a).
- **Higher- and lower-level answers.** Clear understanding is needed for the answer to question 3a) and higher-attaining students would be expected to include reference to the hypothesis matching observations to accepted theory. Lower-attaining candidates should be able to say that a hypothesis is a 'good idea'. The most demanding question is question 9.
- **Special needs.** Students may be able to cope with question 1 if provided with the words on flash cards and asked to assemble them in the best order, and with question 4a). Lower-ability students would be generally better served by considering these questions in class where they can have access to support, texts and group discussion. Some of the questions could be altered and put into a context taken from the text.
- **How and when to use these questions**
  - The questions are page referenced and most could be used as summary questions for homework or for discussion and plenary sessions in the lesson.
  - Question 1 is a summary of the design of an investigation and, together with a clear diagram, would make an excellent Revision card.
  - Questions 2, 3, 5 and 8 could be used for class discussion.
  - Question 10 should be prepared for homework and discussed in small groups. It brings together many of the skills that have/will have been learned throughout the course.

B1a | Human biology

Key Stage 3 curriculum links

The following link to **'What you already know'**:

- The need for a balanced diet; the principles of digestion, including the role of enzymes; food is used as a fuel during respiration to maintain the body's activity and as a raw material for growth and repair.
- The physical and emotional changes that take place during adolescence; the human reproductive system, including the menstrual cycle and fertilisation.
- The abuse of alcohol, solvents and other drugs affects health.
- The growth and reproduction of bacteria and the replication of viruses can affect health and how the body's natural defences may be enhanced by immunisation and medicines.

QCA Scheme of work
7B Reproduction
8A Food and digestion
8C Microbes and disease
9B Fit and healthy

BIOLOGY

B1a | Human biology

What you already know

Here is a quick reminder of previous work that you will find useful in this unit:

- When you reach puberty your body changes so that you can reproduce.
- Girls have a regular menstrual cycle when their body prepares for pregnancy. Boys begin making sperm which can fertilise an egg.
- Eating a balanced diet is an important part of keeping healthy.
- A balanced diet will include carbohydrates, proteins, fats, minerals, vitamins, fibre and water.
- The food you eat is used as a fuel during respiration. It provides energy for the cells of your body. Your food also gives you the raw materials you need to grow and to repair worn out cells.
- Your food needs breaking down (digesting) before it is any use in your body.
- Diseases caused by bacteria and viruses can affect your health.
- Your body can defend itself against disease-causing microorganisms. However sometimes you give it a helping hand by taking medicines or being immunised.
- Both legal and illegal drugs can damage your health if you abuse them. It is against the law to use illegal drugs.

RECAP QUESTIONS

1 What are the main food groups you need to eat to have a balanced diet?

2 How does your body use the food that you eat?

3 What are the main changes which take place in:
a) boys and
b) girls
when they go through puberty?

4 How often is an egg produced in the menstrual cycle?

5 a) Which types of microorganism can cause disease?
b) What is an *infectious* disease?
c) Give the names of three infectious diseases.

6 How does your body defend itself against disease?

7 Explain how medicines and immunisation can help to keep you healthy.

22

RECAP ANSWERS

1 Carbohydrates, proteins, fats, minerals, vitamins, fibre and water.

2 Food is digested (broken down) and used for fuel, growth and repair.

3 **a)** Boys get taller; the voice breaks; sex organs grow; testes start producing sperm; pubic hair grows; face and body hair grow.
b) Girls get taller; develop breasts; body shape changes – fat on thighs and buttocks; periods start (menstrual cycles begin, start producing eggs); pubic hair grows.

4 Once a month.

5 **a)** Bacteria, viruses, (fungi, protists).
b) A disease that is caused by microorganisms and can be passed on from one person to another.
c) Any infectious disease, e.g. measles, influenza, chicken pox.

6 The body defends itself by stopping microorganisms getting in, making scabs over cuts, producing mucus to trap microbes, producing white blood cells.

7 Antibiotics kill bacteria. Immunisation gives your body a chance to meet dangerous microorganisms in a safe way; so if you meet the real thing your body can destroy the bacteria or viruses before they make you ill.

Activity notes

- The poster activity suggested can be done on separate pieces of paper (suggested size: A4) so that the posters can be displayed for class appraisal. Displaying individual posters encourages students to use a simple design, which will have more of an impact and also gives opportunities for peer appraisal.
- The emphasis on the language and images used might depend on the locality of the school. Are there specific groups of people who need to be targeted? Is English the first language of all the residents? Where should the posters be displayed?
- Students could be encouraged to use a variety of media, Internet searches, cuttings from magazines and newspapers, photographs, etc. so that those who do not consider themselves to be 'artistic' do not feel at a disadvantage.
- There are extensions to this activity, including the production of a series of posters, as it could become obvious that trying to get a single message across is difficult and it could be better to aim posters at different groups in society.
- Would video clips be the answer to get your message across more effectively? Such things are now quite common in shopping centres, supermarkets, post offices and other locations. The advantages of such a campaign could be discussed, particularly with respect to people who cannot read or for whom English is not their first language.

SPECIFICATION LINK-UP

Unit Biology B1a

Human biology

How do human bodies respond to changes inside them and to their environment?

The nervous system and hormones enable us to respond to external changes. They also help us to control conditions inside our bodies. The menstrual cycle is controlled by hormones.

What can we do to keep our bodies healthy?

A combination of a balanced diet and regular exercise are needed to keep the body healthy.

How do we use/abuse medical and recreational drugs?

Drugs affect our body chemistry. Medical drugs are developed to relieve illness or disease. Drugs may also be used recreationally as people like the effect on the body, e.g. alcohol and tobacco. People cannot make sensible decisions about drugs unless they know their full effects.

What causes infectious diseases and how can our bodies defend themselves against them?

Our bodies provide an excellent environment for many microbes which can make us ill once they are inside. Our bodies need to stop most microbes getting in and deal with any microbes which do get in.

Human biology

Making connections

ACTIVITY

Many people think that doctors are only there to make them better when they feel ill. As you can see, most doctors do much more than this!

Design a poster to encourage people to use their GP to help them get healthy and stay healthy.

Your poster is going to be put up in all sorts of places – schools, libraries, shopping centres – so make sure it is clear, easy to read and gets the message across.

Chapters in this unit

Co-ordination and control — Healthy eating — Drug abuse — Controlling infectious disease

23

Teaching suggestions

The unit opener sets the scene for the first four topics of the specification with a brief outline of some of the issues raised.

- 'What you already know' and the 'Recap questions' can be used together as a starter and take about 20 minutes, allowing students to jot down the answers and for these to be checked. Alternatively, the 'Recap questions' could be used as a quiz with teams of students competing against each other.
- **Misconceptions from KS3** – The answers given by the students to the 'Recap questions' will indicate topics with which they are unfamiliar or have forgotten from KS3. Some general areas that might need clarification are:
  - The concept of a 'balanced diet': not only does a diet need all the components listed, but needs the correct balance of energy-giving and growth-promoting foods.
  - Not all bacteria and viruses cause diseases. It is worth giving some examples of bacteria that are useful, e.g. decomposers, bacteria that bring about fermentations (yoghurt production).
  - Some other microorganisms cause diseases, e.g. fungi (athlete's foot and ringworm) and protists (malaria and amoebic dysentery).
  - An 'infectious' disease is one that can be passed from one person to another, unlike scurvy or beri-beri which are 'deficiency' diseases, or cystic fibrosis or arthritis.
- **Making connections** – The picture references could promote some class discussion about the message conveyed by the first picture. It could be relevant that only one of the pictures shows a person being prescribed medicine; the others involve prevention or investigation. The picture references link in to topics covered in subsequent spreads. For example:
  - Fig. 1 links with 4.6 'Immunity'.
  - Fig. 2 links with 1.4 'The artificial control of fertility'.
  - Fig. 4 links with 3.3 'Alcohol – the acceptable drug?' and 3.4 'Smoking and health'.
  - Fig. 5 links with 4.3 'Using drugs to treat disease'.
  - Fig. 6 links with 2.1 'Diet and exercise' and 2.2 'Weight problems' and could be referred to when introducing the topics covered in those spreads.
- The illustrations could also be used as a starter to introduce the unit. Students could be asked to look at them and then suggest which topic in the unit they were linked with, giving them a foretaste of what is to come and to see a pathway through the unit.
- A class discussion about the activity could form the main part of an introductory lesson on the unit. The posters could be group efforts or individual ones, depending on the time available. Initial discussions could include references to size, the use of colour, how to get across the information. Do you use words only, pictures only or a combination of both?
- There are some things that GPs do that have not been shown in the picture references, so it could be useful to try to include some of these e.g. holiday jabs and regular screening procedures.
- Some students could come up with the idea that there should be a series of posters, as it is difficult to get too much information on one poster. An example of this is the poster campaign to get older people to have flu jabs in the autumn.
- Some of the suggestions in the 'Activity notes' could be used in plenary sessions, particularly the appraisal and discussion of locations.

Chapters in this unit

- Co-ordination and control
- Healthy eating
- Drug abuse
- Controlling infectious disease

B1a 1.1 Responding to change

LEARNING OBJECTIVES

Students should learn that:

- The nervous system enables humans to react to their surroundings by means of cells called receptors, which detect changes in the environment.
- There are receptors sensitive to touch, light, sound, changes in position, chemicals, pressure, pain and temperature changes and information from these receptors is co-ordinated in the brain.
- Some processes in the body are co-ordinated by hormones, which are chemical substances secreted from glands and transported to their target organs in the bloodstream.

LEARNING OUTCOMES

Most students should be able to:

- Identify the sense organs involved in responding to light, sound, changes of position, chemicals, touch, pressure and temperature in humans.
- Describe how the nervous system works and the difference between the functions of a sensory and a motor neurone.
- Describe how hormones are involved in control and co-ordination.

Some students should also be able to:

- Explain the differences between hormonal and nervous control.

Answers to in-text questions

a) Co-ordination and control; awareness of surroundings.

b) Chemical.

c) **i)** Ears. **ii)** Skin. **iii)** Nose.

d) A neurone is a single nerve cell; a nerve is a lot of neurones bundled together.

e) A sensory neurone carries information from sensory receptors to the CNS; motor neurone carries impulses from the CNS to the effector organs.

SPECIFICATION LINK-UP B1a.11.1

- *The nervous system enables humans to react to their surroundings and coordinate their behaviour.*
- *Receptors detect stimuli which include light, sound, changes in position, chemicals, touch, pressure, pain and temperature. (The structure and functions of sense organs such as the eye and the ear are not required.)*
- *Information from receptors passes along cells (neurones) in nerves to the brain. The brain coordinates the response.*
- *Many processes in the body are coordinated by chemical substances called hormones. Hormones are secreted by glands and are transported to their target organs by the bloodstream.*

Lesson structure

STARTER

'Senses working overtime' – Play and discuss this song by XTC; this is a song about the senses and can be sampled at online music shops or the official XTC site at www.xtcidearecords.co.uk. (5 minutes)

Circus of activities – Place 'feely' bags around the room involving senses and containing mystery objects, e.g. a sniff test with different essences in film containers, guess the sound on a Walkman, mystery object photos or down a microscope, identifying mystery fruit (care with hygiene). This starter could be as short as 5–10 minutes if only one or two things used, but it could be expanded to be part of the main lesson.

Name the hormones – Using Figure 2 on page 24 of the Student Book, give students 5 minutes to write down the names of the hormones secreted by the glands shown on the diagram. Check the answers, awarding bonus points to students who can name more than one hormone for each gland. (10 minutes)

MAIN

The main part of the lesson could be a practical session involving the testing of sensitivity using the following:

- **Identification of the density of nerve endings** – We will investigate sensitivity of different areas of the skin. Working in pairs, one student is blindfolded or told to look in a different direction, while another student touches them on the back of the hand with either one or two pieces of blunt wire about 1 cm apart mounted in a cork. The blindfolded student has to say whether it was one point or two points that touched them. In addition to the back of the hand, other areas of the body, such as the upper arm or the back of the leg, could be investigated. In this way, a comparison could be made about the sensitivity of different areas of the body.
- 'How Science Works' concepts can be introduced here: the accuracy of the measurements, the calculation of means and the control of variables. If class results are collated, some indication of the variability can be discussed.

PLENARIES

What is the significance of the results? – What do the results of the experiments tell us? Is it touch, pressure or pain receptors that are being stimulated? Could this type of experiment be modified for other receptors such as hot or cold receptors? Suggest students link up the density of the receptors to the areas tested. (10 minutes)

Does the colour smell right? – A variation of the above could involve the conflicting information from odd combinations of colour and smell, e.g. pink food smelling of peppermint, blue food smelling of strawberry. This could trigger a discussion of how senses are used and that information comes from more than one sense organ. (10 minutes)

B1a 1.1 Responding to change

LEARNING OBJECTIVES

1 How is your body controlled?
2 What is the difference between your nervous system and your hormones?
3 How do you respond to changes in your surroundings?

You need to know what is going on in the world around you. Your ***nervous system*** makes this possible. It enables you to react to your surroundings and co-ordinate your behaviour.

Your nervous system carries electrical signals (***impulses***) which travel fast – from 1 to 120 metres per second. This means you can react to changes in your surroundings very quickly indeed.

Figure 1 Your body is made up of millions of cells which have to work together. Whatever you do with your body – whether it's winning a race or playing on the computer – your movements need to be co-ordinated. The conditions inside your body must also be controlled.

a) What is the main job of the nervous system?

Hormones are chemical substances. They control many of the processes going on inside your body. Special ***glands*** make and release (***secrete***) these hormones into your body. Then the hormones are carried around your body to their target organs in the bloodstream. They can act very quickly, but often their effects are quite slow and long lasting.

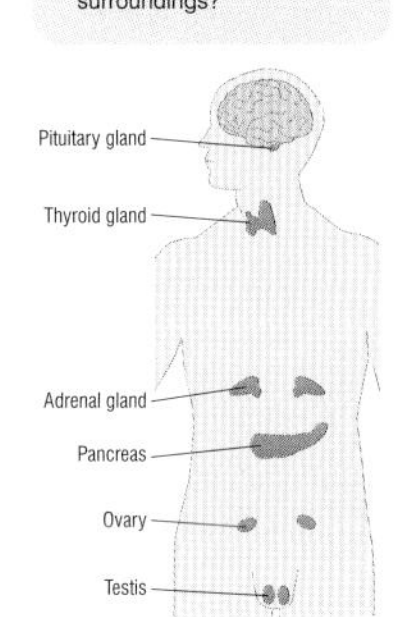

Figure 2 Hormones act as chemical messengers. They are made in glands in one part of your body but having an effect somewhere else entirely

b) What type of messengers are hormones?

The nervous system

Like all living things, you need to avoid danger, find food and – eventually – find a mate! This is where your nervous system comes into its own. Your body is particularly sensitive to changes in the world around you. Any changes (known as **stimuli**) are picked up by cells called ***receptors***.

These receptors are usually found clustered together in special **sense organs**, such as your eyes and your skin. You have many different types of sensory receptors (see Figure 3).

c) Where would you find receptors which would respond to i) a loud noise, ii) touching a hot oven, iii) a strong perfume?

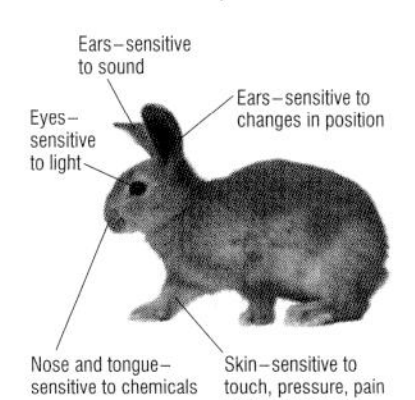

Figure 3 Look at this rabbit. Being able to detect changes in the environment is important. It can often be a matter of life and death.

How your nervous system works

Once a sensory receptor detects a stimulus, the information (sent as an electrical impulse) passes along special cells called **neurones**. These are usually found in bundles of hundreds or even thousands of neurones known as ***nerves***.

The impulse travels along the neurone until it reaches the **central nervous system** or **CNS**. The CNS is made up of the brain and the spinal cord. The cells which carry impulses from your sense organs to your central nervous system are called ***sensory neurones***.

d) What is the difference between a neurone and a nerve?

Your brain gets huge amounts of information from all the sensory receptors in your body. It co-ordinates the information and sends impulses out along special cells. These cells carry information from the CNS to the rest of your body. The cells are called ***motor neurones***. They carry impulses to make the right bits of your body – the **effector organs** – respond.

Effector organs are muscles or glands. Your muscles respond to the arrival of impulses by contracting. Your glands respond by releasing (**secreting**) chemical substances.

The way your nervous system works can be summed up as:

receptor → sensory neurone → co-ordinator → motor neurone → effector (CNS)

e) What is the difference between a sensory neurone and a motor neurone?

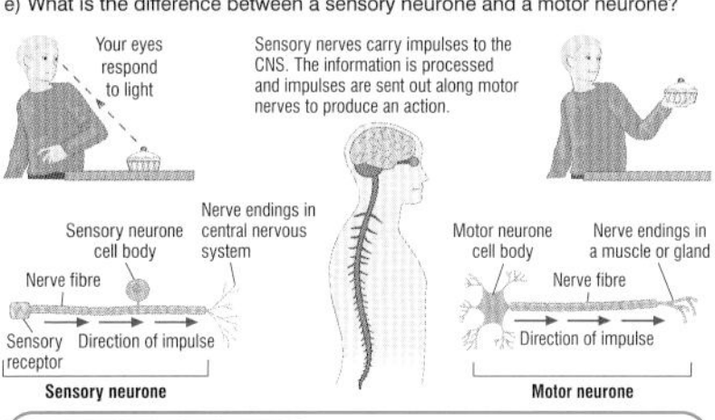

Figure 4 The rapid responses of our nervous system allow us to respond to our surroundings quickly – and in the right way!

DID YOU KNOW?

Some male moths are so sensitive to chemicals that they can detect the scent of a female several kilometres away. What's more they can follow the scent trail and find her!

GET IT RIGHT!

Be careful to use the terms **neurone** and **nerve** correctly. Talk about **impulses** (*not* **messages**) travelling along a neurone.

SUMMARY QUESTIONS

1 Copy and complete using the words below:

blood chemical electrical glands nervous

Your ...... system carries fast ...... impulses. Your hormones are ...... messengers secreted by special ...... and carried around the body in the ....

2 Make a table to show the different types of sense receptors. For each one, give an example of the sort of things it responds to, e.g. touch receptors respond to an insect crawling on your skin.

3 Explain i) what happens in your nervous system when you see a piece of chocolate, pick it up and eat it, ii) the differences between hormonal and nervous control in your body.

KEY POINTS

1 Hormones, secreted by special glands, are chemicals that help control and co-ordinate processes in your body.
2 The nervous system uses electrical impulses to enable you to react to your surroundings and co-ordinate what you do.
3 Cells called receptors detect stimuli (changes in the environment).
4 Impulses from receptors pass along sensory neurones to the brain. Impulses are sent from the brain to the effector organs along motor neurones.

Teaching suggestions

- **Special needs.** Provide cards with the separate parts of a nervous pathway, such as 'receptor', 'sensory neurone' etc., on them and ask students to place in the correct order in the pathway.
- **Learning styles**

  Interpersonal: Discussing and evaluating colours and smells.

  Intrapersonal: Writing a report of the experiment.

  Kinaesthetic: Carrying out practical work on the density of touch receptors in the skin.

  Visual: Making observations in the practical investigation.

  Auditory: Explaining results to others.
- **Homework.** Writing up the investigations could give students some practice in the use of continuous prose. Drawing graphs and summarising class results could also be set.

Practical support

Identifying of the density of nerve endings

Equipment and materials required

Small pieces of blunt wire (unbent paperclip, blunt tapestry needles) mounted in pieces of cork; if two wires are used they should be about 1 cm apart.

ACTIVITY & EXTENSION IDEAS

- **How fast is a nerve impulse?** If you stub your toe, there is a very short interval of time between the action of stubbing and the sensation of pain. Discuss this and suggest to students that they try to work out how the time interval between being touched on the toe and feeling the sensation of touch could be measured. Can this measurement be used to work out how fast the nerve impluse is transmitted? Hint: how might measuring the length of the leg help in the calculations.
- **Differences between nervous control and hormonal control.** A useful table of the differences between nervous and hormonal control could be built up using the information in the Student Book spread

SUMMARY ANSWERS

1 Nervous, electrical, chemical, glands, blood.

2 Table showing receptors for sight, sound, position, smell with student example for each one.

3 **i)** Eye → brain → hand; tongue → brain → mouth/jaw.
ii) Hormonal – by chemicals. Nervous – by electrical impulses.

GET IT RIGHT!

Students do need to understand the difference between nerve and neurone. A clear definition of each could be made into a revision card. It could be helpful to students to have a good understanding of the different types of neurone and know that nerves can contain either sensory or motor neurones or a mixture of both.

KEY POINTS

The key points given here would make excellent revision cards.

B1a 1.2 Reflex actions

LEARNING OBJECTIVES

Students should learn that:

- Reflex actions are automatic and rapid responses to stimuli.
- Simple reflex actions involve receptors, sensory neurones, motor neurones and relay neurones, together with synapses and effectors.
- Reflex actions take care of basic functions, such as breathing, and help to avoid danger or harm to the body.

LEARNING OUTCOMES

Most students should be able to:

- Explain what is meant by the term 'reflex action'.
- Describe the roles of receptors, sensory, relay and motor neurones, synapses and effectors in a reflex action.
- Explain why reflex actions are so important.

Some students should also be able to:

- Analyse a specific reflex action in terms of stimulus → receptor → co-ordinator → effector → response.
- Explain how a synapse works.

Teaching suggestions

- **Gifted and talented.** These students could investigate the work of Pavlov and his dogs in the context of the reflex action.
- **Special needs.** These students could be given cards with the words needed to complete question 1 of the Summary Questions on them and asked to place them in the correct places on a large copy of the passage.
- **Learning styles**

  Interpersonal: Discussion of the results of the practical exercise.

  Intrapersonal: Involvement in the knee jerk or student reflexes demonstrations.

  Kinaesthetic: Practical work in the stick-drop experiment.

  Visual: Observing the reflex actions.

  Auditory: Listening to the ideas put forward by other students.

SPECIFICATION LINK-UP B1a.11.1

- *Reflex actions are automatic and rapid. They often involve sensory, relay and motor neurones.*
- *The role of receptors, sensory neurones, motor neurones, relay neurones, synapses and effectors in simple reflex actions.*

Lesson structure

STARTER

- **Knee jerk reflex** – Ask for a volunteer, or select a suitable student, from the class and demonstrate the knee jerk reflex on them. If appropriate, allow students to work in pairs and try it out on each other (caution needed here!). Discuss what is happening. (5–10 minutes)
- **Eye reflexes** – Working in pairs, the alteration in pupil size when the eyes are opened in bright light can be easily observed by students. Discuss the value of this reflex in protecting the eyes. (5–10 minutes)
- **Quick quiz** – Prepare a short list of questions (definitions, spellings, etc.) linked to the previous spread about the nervous pathway, so that students are prepared for the special features of the reflex pathway. (10 minutes)

MAIN

- **The stick-drop test** for measuring reaction time – Working in pairs, one student holds a metre rule vertically at the zero end, between the thumb and forefinger of another student, so that the 50 cm mark is level with the top of the forefinger. Without warning, the first student drops the rule and the second student attempts to catch it between the thumb and forefinger, noting the distance on the ruler just above the forefinger. Repeat several times, so that a mean can be calculated. Then change around so that everyone gets a turn. Write a report of the experiment.
- 'How Science Works' concepts can be introduced here: the accuracy of the measurements, the calculation of means and the control of variables. If class results are collated, some indication of the variability can be discussed.
- Show Animation B1a 1.2 'Reflex actions' from the CD ROM for GCSE Biology.

PLENARIES

Summary – What do tests such as the stick-drop test tell us? Ask students to identify the parts of the body involved. What senses are being used? Can we train ourselves to react more quickly? Does practice make perfect? (5–10 minutes)

Demonstration of knee-jerk reflex and **blinking reflexes** – Discuss how this links in with the diagram of a reflex pathway in the Student Book. (5–10 minutes)

Pass the zap – Provide some volunteer students with A4 sheets on which the names of parts of the reflex pathway have been printed. The students should then arrange themselves in the correct order. They could be timed as a challenge. Using a lightning-shaped zap, ask the students to talk through their bit of the process as the impulse (zap) gets passed to them. (5–10 minutes, but could take longer if more than one group have a go)

Can we alter reflex actions? – Students to think of situations where it is possible to alter the automatic response (not dropping a hot object, deliberately breathing more slowly, etc.). Are there some reflex actions over which we have no control? Discuss the situations and build up a list. (10 minutes)

ACTIVITY & EXTENSION IDEAS

- The practical on reaction times could be extended to see if times are improved by practice.
- Research on the effects of alcohol on reaction times; this could link up with a later lesson in this unit.
- The stick-drop practical lends itself to further work, such as investigating whether reaction times alter with age, time of day or intake of caffeine. Results can be tabulated and the class results compared. Boys can be compared with girls, and distribution curves could be drawn.

Practical support

The stick-drop test

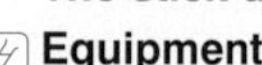

Equipment and materials required

A metre rule, access to computers for the interactive reaction timers. It could be helpful to have pre-printed sheets on which to record reaction times, so that it is easy for students to gather class results.

Pass the zap

Equipment and materials required

A4 sheets with receptor, sensory neurone, synapse, relay neurone, synapse, motor neurone, effector, gland, muscle printed on them and a zap in the shape of a flash of lightning.

BIOLOGY CO-ORDINATION AND CONTROL

B1a 1.2 Reflex actions

LEARNING OBJECTIVES

1 What is a reflex?
2 Why are reflexes so important?

Your nervous system lets you take in information about the world around you and respond in the right way. However some of your responses are so fast that they happen without giving you time to think.

When you touch something hot, or sharp, you pull your hand back before you feel the pain. If something comes near your face, you blink. Automatic responses like these are known as **reflexes**.

What are reflexes for?

Reflexes are very important both for human beings and for other animals. They help you to avoid danger or harm because they happen so fast. There are also lots of reflexes which take care of your basic body functions. These functions include breathing and moving the food through your gut.

It would make life very difficult if you had to think consciously about those things all the time – and could be fatal if you forgot to breathe!

a) Why are reflexes important?

How do reflexes work?

Reflex actions involve just three types of neurone. These are:

- sensory neurones,
- motor neurones, and
- relay neurones which simply connect a sensory neurone and a motor neurone. We find relay neurones in the CNS, often in the spinal cord.

An electrical impulse passes from the sensory receptor along the sensory neurone to the CNS. It then passes along a relay neurone (usually in the spinal cord) and straight back along a motor neurone. From there the impulse arrives at the effector organ (usually a muscle in a reflex). We call this a *reflex arc*.

The key point in a reflex arc is that the impulse bypasses the conscious areas of your brain. The result is that the time between the stimulus and the reflex action is as short as possible. When you put your hand on something hot, you have moved your hand away before you feel the pain!

b) Why is it important that the impulses in a reflex arc do not go to the conscious brain?

How synapses work

Your nerves are not joined up directly to each other. There are junctions between them called **synapses**. The electrical impulses travelling along your neurones have to cross these synapses but they cannot leap the gap. Look at Figure 1 to see what happens next.

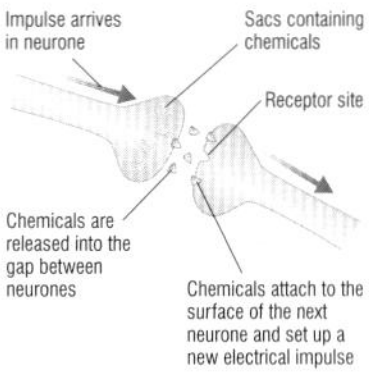

Impulse arrives in neurone
Sacs containing chemicals
Receptor site
Chemicals are released into the gap between neurones
Chemicals attach to the surface of the next neurone and set up a new electrical impulse

Figure 1 When an impulse arrives at the junction between two neurones, chemicals are released which cross the synapse and arrive at **receptor sites** on the next neurone. This starts up an electrical impulse in the next neurone.

26

Human biology

The reflex arc in detail

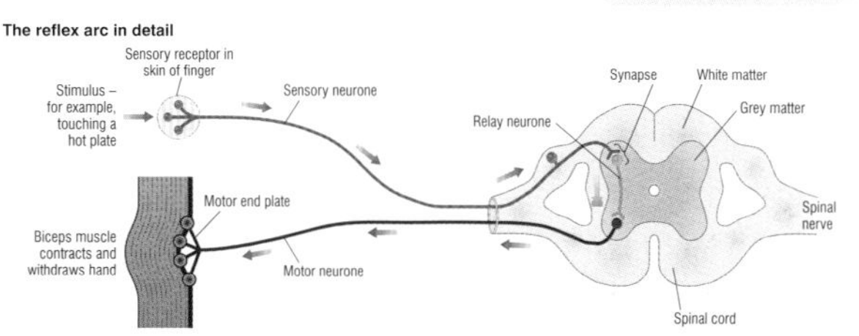

Figure 2 The reflex action which moves your hand away from something hot can save you from a nasty burn!

Look at Figure 2. It shows what would happen if someone touched a hot object.

When they touch it, a receptor in their skin is stimulated. An electrical message passes along a sensory neurone to the central nervous system – in this case the spinal cord.

When an impulse from the sensory neurone arrives in the synapse with a relay neurone, a chemical message is released. This crosses the synapse to the relay neurone and sets off an electrical impulse that travels along the relay neurone.

When the impulse reaches the synapse between the relay neurone and a motor neurone returning to the arm, another chemical message is released.

This crosses the synapse and starts an electrical impulse travelling down the motor neurone. When the impulse reaches the organ (effector), it is stimulated to respond. In this example the impulses arrive in the muscles of the arm, causing them to contract and move the hand rapidly away from the source of pain.

Most reflex actions can be shown as follows:

stimulus → receptor → co-ordinator → effector → response

This is not very different from a normal conscious action. However, in a reflex action the co-ordinator is a relay neurone either in the spinal cord or in the unconscious areas of the brain. The whole reflex is very fast indeed.

SUMMARY QUESTIONS

1 Copy and complete using the words below:

conscious motor reflex relay response sensory stimulus

In a ...... arc the electrical impulse bypasses the ...... areas of your brain. The time between the ...... and the ...... is as short as possible. Only ...... neurones, ...neurones and ...... neurones are involved.

2 Explain why some actions, such as breathing and swallowing, are reflex actions, while others such as speaking and eating are under your conscious control.

3 Draw a flow chart to explain what happens when you step on a pin. Make sure you include an explanation of how a synapse works.

DID YOU KNOW?

Newborn babies have a number of special reflexes, which disappear as they grow. If something touches the palm of the hand of a newborn baby it will grip on tightly by reflex. In theory the baby would hang from a washing-line! Doctors check for these reflexes to show that a new baby is fit and well.

Figure 3 A baby's gripping reflex is very strong

KEY POINTS

1 Some responses to stimuli are automatic and rapid and are called reflex actions.
2 Reflex actions run everyday bodily functions and help you to avoid danger.

27

SUMMARY ANSWERS

1 Reflex, conscious, stimulus, response, sensory, relay, motor.

2 Reflex actions that need to operate automatically, even when you are asleep, cannot rely on conscious thought processes, unlike speaking and eating.

3 Stimulus → sensory neurone → synapse → chemical message → relay neurone → chemical message → motor neurone → muscles in leg lift the foot.

Answers to in-text questions

a) To protect the body and take care of basic body functions, e.g. breathing.
b) To speed up reaction time.

KEY POINTS

The points made here are very general and students should be advised to make themselves familiar with specific examples of reflex pathways. If revision cards are made, then a diagram of a reflex pathway could be included and annotated with the names and functions of the components.

B1a 1.3 The menstrual cycle

LEARNING OBJECTIVES

Students should learn that:

- The menstrual cycle is controlled by several hormones secreted by the pituitary gland and the ovaries.
- FSH (follicle stimulating hormone), secreted by the pituitary gland, causes an egg to mature in one of the ovaries and also stimulates the secretion of oestrogen by the ovaries.

LEARNING OUTCOMES

Most students should be able to:

- Name the hormones that control the menstrual cycle and state where each is produced.
- Describe the functions of each hormone and relate each hormone to the different phases of the cycle.

Some students should also be able to:

- Explain the relationships between the different hormones and how they interact in the cycle.

SPECIFICATION LINK-UP

B1a.11.1

- *Hormones regulate the functions of many organs and cells. For example, the monthly release of an egg from a woman's ovaries and the changes in the thickness of the lining of her womb are controlled by hormones secreted by the pituitary gland and by the ovaries.*
- *Several hormones are involved in the menstrual cycle of a woman. Those hormones involved in promoting the release of an egg include:*
  - *FSH, which is secreted by the pituitary gland and causes an egg to mature in one of the ovaries, and also stimulates the ovaries to produce hormones including oestrogen.*
  - *Oestrogen, which is secreted by the ovaries and inhibits the further production of FSH as well as stimulating the pituitary gland to produce a hormone called LH.*

Lesson structure

STARTER

'What do we know about hormones?' – Recap knowledge of hormones from B1a 1.1, using sheets of A3 paper and working in groups of three. The sheets from each group can be shown and compared. (5 minutes)

The female reproductive system – Give students unlabelled diagrams of side view and front view (should be known from KS3), or project an image and ask for labels which can be filled in. If a large unlabelled poster is available, a set of labels could be made which the students attempt to pin in the right place. If unlabelled diagrams are used, check spellings of the different parts! (10 minutes)

MAIN

This lesson does have to focus on the menstrual cycle, so the following suggestions describe some ways in which the information can be put across.

- **PowerPoint® presentation introducing the vocabulary and linking the actions of the hormones** – A series of PowerPoint® diagrams to illustrate the stages could be prepared. Firstly, show the pituitary gland and the female reproductive system; secondly, show secretion of FSH from the pituitary affecting the ovaries in two ways ('stimulation of oestrogen production' and 'stimulation of egg development'); thirdly show oestrogen production linked to the uterus, labelled 'developing lining' and two links back to the pituitary gland, one labelled 'negative feedback – inhibits FSH production' and the other labelled 'stimulates LH production'; lastly, link from pituitary to ovaries labelled 'ovulation triggered'.
- The PowerPoint® diagrams can be used in conjunction with a human torso model if available, so that the location of the pituitary gland and the female reproductive systems can be seen easily. This reinforces some of the properties of hormones, as the students could be asked to consider how the hormones get from one place to another.
- **A video of ovulation** – This would be useful if available. The series *'The Human Body'* (BBC) shows ovulation *in situ* in detail.
- **Demonstration of negative feedback** – This can be achieved using an electric heater, such as a low voltage coil of resistance wire in a series circuit with a bimetal strip arranged so that when it heats up it switches off, and when it cools down it switches back on again. Failing that, a discussion of how the thermostat on a central heating system or on an immersion heater works would be of benefit. This then needs to be translated into the events of the cycle.
- **Using Figure 3 in the Student Book** – Explaining it using a series of PowerPoint® diagrams could be helpful.

PLENARIES

'What do we know about hormones?' – Add more links to the sheets if this was used as a starter. (5 minutes)

True or false? – Present the students with a series of statements about the hormones and the cycle, some of which are true and others not. Check answers at end. This offers an opportunity to make clear any points about the cycle that students do not understand. (10 minutes)

Answers to in-text questions

a) Hormones made in the pituitary gland in the brain and in the ovary.
b) To support the developing baby.
c) FSH and LH.
d) Oestrogen.

ACTIVITY & EXTENSION

- Students could look at slides under the microscope, or projected slides, of mammalian ovaries to show follicles at various stages of development.
- They could look at preserved specimens of ovaries if available (possibly from hens).
- Start a class discussion on mammals: 'Do other mammals have an equivalent of the menstrual cycle? Compare the monthly cycle of human females with the breeding seasons of other mammals.'
- Ask: 'Do males have hormones equivalent to FSH, LH and oestrogen?'
- There are some very good pre-birth web sites (some have a distinct pro-life angle).
- View and discuss *'Window on Life'* (*Sunday Times* free CDs).

DID YOU KNOW?

It might be interesting to get the students to calculate roughly how many of the eggs a baby girl is born with actually develop and are released between puberty and when ovulation ceases at the menopause.

Teaching suggestions

- **Special needs.** Give students broken sentences to sequence, describing the role of oestrogen and the outline only of the menstrual cycle.
- **Gifted and talented.** These students could be encouraged to think of other examples of negative feedback in everyday life and in biological situations. Extra bonus marks if they can think of other body systems where negative feedback operates!
- **Learning styles**

  Kinaesthetic: Pinning labels in the correct places.

  Intrapersonal: Reviewing previous knowledge.

  Interpersonal: Discussing the route of the hormones.

  Visual: Following the sequence of diagrams of hormone action.

  Auditory: Discussing negative feedback.

BIOLOGY | CO-ORDINATION AND CONTROL

B1a 1.3 The menstrual cycle

LEARNING OBJECTIVES

1. How is the menstrual cycle controlled?
2. When is a woman most likely to conceive?

Hormones control the functions of many of your body organs. They also control the activities of your individual cells. A woman's **menstrual cycle** is a good example of how this control works.

Hormones made in a woman's brain and in her ovaries control her menstrual cycle. The levels of the different hormones rise and fall in a regular pattern. This affects the way her body works.

Figure 1 The bodies of young boys and girls work in very similar ways. But once the sex hormones kick in during puberty, some big differences appear in the shape of their bodies and how they work. This shows you the power of the hormones!

What is the menstrual cycle?

The average length of the menstrual cycle is about 28 days. Each month the lining of the womb thickens ready to support a developing baby. At the same time an egg starts maturing in the ovary.

About 14 days after the egg starts maturing it is released from the ovary. This is known as **ovulation**. The lining of the womb stays thick for several days after the egg has been released.

If the egg is fertilised by a sperm, then pregnancy takes place. The lining of the womb provides protection and food for the developing embryo. If the egg is not fertilised, the lining of the womb and the dead egg are shed from the body. This is the monthly bleed or ***period***.

All of these changes are brought about by hormones. These are made and released by the **pituitary gland** (a pea sized gland in the brain) and the **ovaries**.

a) What controls the menstrual cycle?
b) Why does the lining of the womb build up each month?

How the menstrual cycle works

Once a month, a surge of hormones from the pituitary gland in the brain starts eggs maturing in the ovaries. The hormones also stimulate the ovaries to produce the female sex hormone ***oestrogen***.

- **FSH:** secreted by the pituitary gland. It makes eggs mature in the ovaries. *FSH* also stimulates the ovaries to produce ***oestrogen***.
- **Oestrogen:** made and secreted by the ovaries. It stimulates the lining of the womb to build up ready for pregnancy. It also stimulates the pituitary gland to make another hormone known as ***LH***.
- **LH:** secreted by the pituitary gland. It stimulates the release of a mature egg from one of the ovaries in the middle of the menstrual cycle.

c) Which hormones are made in the pituitary gland?
d) Which hormone is made by the ovary?

Pituitary gland
Fallopian tube
Ovary
Uterus

Figure 2 Hormones from the pituitary and the ovaries work together to control a woman's fertility

28

Human biology

The hormones produced by the pituitary gland and the ovary act together to control what happens in the menstrual cycle. As the oestrogen levels rise they inhibit (slow down) the production of FSH and encourage the production of LH by the pituitary. When LH levels reach a peak in the middle of the cycle, they stimulate the release of a mature egg.

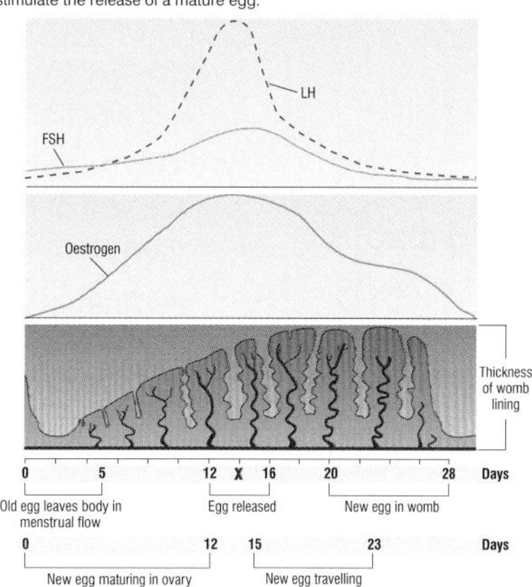

Figure 3 The changing levels of the female sex hormones control the different stages of the menstrual cycle

DID YOU KNOW?

A baby girl is born with ovaries full of immature eggs, but they do nothing until she has gone through the changes of **puberty**.

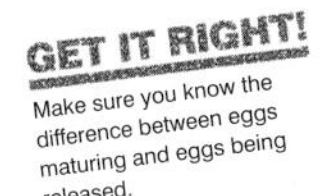

GET IT RIGHT!
Make sure you know the difference between eggs maturing and eggs being released.

SUMMARY QUESTIONS

1. Copy and complete using the list below:

   28 hormones FSH LH menstrual oestrogen ovary

   During the ...... cycle a mature egg is released from the ...... about every ...... days. The cycle is controlled by several ...... including ......, ... and ....

2. Look at Figure 3.

   a) On which day is the woman most likely to get pregnant?
   b) On which days is she having a menstrual period?
   c) On which day is the level of LH highest?
   d) Which hormone controls the build up of the lining of the womb?

3. Produce a leaflet to explain the events of the menstrual cycle to women who are hoping to start a family. You will need to explain the graphs at the top of this page and show when they are most likely to get pregnant.

KEY POINTS

1. Hormones control the release of an egg from the ovary and the build up of the lining of the womb in the menstrual cycle.
2. The main hormones involved are FSH and LH from the pituitary gland and oestrogen from the ovary.

Practical support

Demonstration of negative feedback

Equipment and materials required

Variable power supply (low voltage), resistance wire coil, metallic strip, connecting wires.

SUMMARY ANSWERS

1. Menstrual; ovary; 28; hormones; FSH, LH and oestrogen.
2. **a)** Day 14. **b)** Days 1–5. **c)** Day 14. **d)** Oestrogen.
3. Leaflet.

GET IT RIGHT!

Students can be confused by the development of an immature egg while it is in the ovary and the release of the egg. If the events are linked to the hormones involved, then it can become clearer.

KEY POINTS

The PowerPoint® presentations on the events of the menstrual cycle should reinforce the key points for this spread.

B1a 1.4 The artificial control of fertility

LEARNING OBJECTIVES

Students should learn that:

- Oral contraceptives containing hormones can prevent pregnancy by inhibiting FSH production so that no eggs mature.
- FSH can be used as a 'fertility drug', stimulating the production of mature eggs in women whose own FSH levels are too low.

LEARNING OUTCOMES

Most students should be able to:

- Explain how oral contraceptives inhibit FSH production and prevent pregnancy.
- Describe how treatment with FSH can help a woman produce mature eggs if her own FSH production is too low.
- Describe how FSH is used in IVF treatments.

Some students should also be able to:

- Discuss issues arising from the use of hormones to control fertility artificially.

Teaching suggestions

- **Special needs.** These students could make a poster showing the stages of a course of IVF treatment.
- **Gifted and talented.** Mature eggs from a woman undergoing fertility treatment can be stored. Students could investigate the exact conditions needed for storage. Why are such conditions necessary? Is there any chance of deterioration if the eggs are kept in storage for long periods of time?
- **Learning styles**

  Intrapersonal: Writing a paragraph about how it feels to be born as a result of a new treatment.

  Interpersonal: Discussion on alternatives to the pill.

  Kinaesthetic: Building up a diagram of the action of the contraceptive pill or the sequence of IVF treatment.

  Visual: Viewing video or animation.

  Auditory: Listening to the opinions of others.
- **Homework.** Students to make their own revision cards of the sequence of IVF treatment and/or how the contraceptive pill prevents pregnancy.

SPECIFICATION LINK-UP B1a.11.1

- *The uses of hormones in controlling fertility include:*
  - *giving oral contraceptives which contain hormones to inhibit FSH production so that no eggs mature*
  - *giving FSH as a 'fertility drug' to a woman whose own FSH is too low to stimulate eggs to mature.*

Lesson structure

STARTER

'What do I know about contraception?' – The students could be shown a packet of contraceptive pills, or a photo from the Internet, and then asked to write a paragraph entitled: 'What do I know about contraception?' This could lead to a discussion on what the pills contain and how they work. (10 minutes)

Review of hormones in the menstrual cycle – Show the students a diagram of the events in the menstrual cycle (suggest the bottom half of Figure 3 on page 29 – leave out the hormone graphs) and ask them to indicate the hormones involved at each stage. Award bonus points for any student who can draw in the correct curves for the different hormones. This will remind them of the natural sequence. (10 minutes).

MAIN

This lesson needs to focus on the two main issues: the use of oral contraceptives and the use of FSH as a 'fertility drug'. Both these issues could trigger discussions, so allow time for questions and for students to express their own opinions.

- **Animation on oral contraceptives** – Use Animation B1a 1.3 'The menstrual cycle' on the GCSE Biology CD ROM to illustrate the normal events of the menstrual cycle and then show how the hormones in the contraceptive pill affect the sequence of events. The students could discuss exactly what the pill does and its effect on the secretion of other hormones involved in the cycle.
- The discussion could be extended to include the consequences of failing to take the pill regularly. What happens if the level of artificial hormones drops suddenly?
- **A video of IVF treatment** – Use free download – IVF orientation video from San Diego Fertility Center.

PLENARIES

- **Louise Brown – the first 'test tube' baby** – Louise Brown was the first baby to be born as a result of IVF in 1978. Tell the students the story and ask them to write a short paragraph about how it might feel to be the first person to be born as the result of a new treatment. Select some students to read their accounts. (10 minutes)
- **Injections and patches** – The hormones used in contraceptive pills can be given as injections or as patches that stick to the skin. Discuss the advantages and disadvantages of the use of these alternatives. (5–10 minutes)

ACTIVITY & EXTENSION IDEAS

- **Alternatives to the contraceptive pill?** The contraceptive pill is not the only way to avoid pregnancy? It is not suitable for everyone and there may be medical reasons why it is not appropriate to prescribe it. Ask students to build up a list of other methods of contraception and discuss the advantages and disadvantages. This could lead to some discussion on sexually transmitted diseases and the incidence of HIV/AIDS.
- **How would a contraceptive pill for males work?** There has been some research on this, but students could be encouraged to work out what would need to happen.
- **What happens to the spare embryos?** Students could carry out a web search for information on this topic. Use web sites of newspapers, TV channels and the British Fertility Society to find stories to discuss.

DID YOU KNOW?

In addition to the one given, the following might be of interest:

- When commercial production of the contraceptive pill first started, oestrogen in the air caused a lot of the male workers in the factory to start developing breasts.

BIOLOGY CO-ORDINATION AND CONTROL

B1a 1.4 The artificial control of fertility

LEARNING OBJECTIVES

1. How can hormones be used to stop pregnancy?
2. How can hormones help to solve the problems of infertility?

For centuries people have tried to control when they have children. They have used substances from camel dung to vinegar to try and stop people having babies. Other people have carved fertility figures, made sacrifices and swallowed horrible herbs to try and have a child.

But it is only in the last fifty years or so that scientists have really been able to help couples control their own fertility, if they choose to do so.

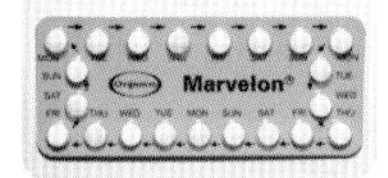

Figure 1 The contraceptive pill contains a mixture of hormones which effectively trick the body into thinking it is already pregnant, so no more eggs are released

Contraceptive chemicals

In the 21st century it is possible to choose when to have children – and when not to have them. One of the most important and widely used ways of controlling fertility is to use ***oral contraceptives*** (the ***contraceptive pill***).

The pill contains female hormones, particularly oestrogen. The hormones affect your ovaries, preventing the release of any eggs. The pill inhibits (stops) the production of FSH so no eggs mature in the ovaries. Without mature eggs, you can't get pregnant.

Anyone who uses the pill as a contraceptive has to take it very regularly. If they forget to take it, the artificial hormone levels drop. Then their body's own hormones can take over very quickly. This can lead to the unexpected release of an egg – and an unexpected baby!

Fertility treatments

In the UK as many as one couple in six have problems having a family when they want one. There are many reasons for this infertility. It may be linked to a lack of female hormones. Some women want children but simply do not make enough FSH to stimulate the eggs in their ovaries. Fortunately artificial FSH can be used as a fertility drug. It stimulates the eggs in the ovary to mature and also triggers oestrogen production.

Figure 2 Most people who take fertility drugs end up with one or two babies. But the Walton family in the UK had six baby girls who all survived and are now young adults in their own right!

30

Human biology

Fertility drugs are also used when a couple is trying to have a baby by IVF (*in vitro* fertilisation). If your fallopian tubes are damaged, eggs cannot reach your womb so you cannot get pregnant naturally.

Fortunately doctors can now help. They remove eggs from the ovary and fertilise them with sperm outside the body. Then they place the tiny developing embryos back into the uterus of the mother, bypassing the faulty tubes.

To produce as many ripe eggs as possible for IVF, the woman is given fertility drugs as part of her treatment. IVF is expensive and not always successful.

Advantages and disadvantages

The use of hormones to control fertility has been a major scientific breakthrough. But like most things there are pros and cons!

In the developed world, using the pill has helped make families much smaller than they used to be. There is less poverty because people have fewer mouths to feed.

The pill has also helped to control population growth in countries such as China, where they find it difficult to feed all their people. In many other countries of the developing world the pill is not available because of a lack of money, education and doctors.

The pill can cause health problems so a doctor always oversees its use.

The use of fertility drugs can also have some health risks for the mother and it can be expensive for society. A large multiple birth can be tragic for the parents if some or all of the babies die. It also costs hospitals a lot of money to keep very small premature babies alive.

Controlling fertility artificially also raises many ethical issues for society and individuals. For example, some religious groups think that preventing conception is denying life and ban the use of the pill.

The mature eggs produced by a woman using fertility drugs may be stored, or fertilised and stored, until she wants to get pregnant later. But what happens if the woman dies, or does not want the eggs or embryos any more?

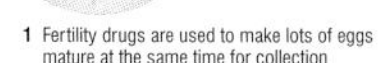

1 Fertility drugs are used to make lots of eggs mature at the same time for collection

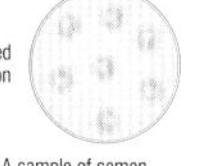
2 The eggs are collected and placed in a special solution in a petri dish

3 A sample of semen is collected

4 The eggs and sperm are mixed in the petri dish

5 The eggs are checked to make sure they have been fertilised and the early embryos are developing properly

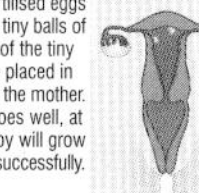
6 When the fertilised eggs have formed tiny balls of cells, 1 or 2 of the tiny embryos are placed in the uterus of the mother. Then, if all goes well, at least one baby will grow and develop successfully.

Figure 3 New reproductive technology using hormones and IVF has helped thousands of infertile couples to have babies

DID YOU KNOW?

In the early days of using fertility drugs there were big problems with the doses used. In 1971 an Italian doctor removed fifteen four-month-old fetuses (ten girls and five boys) from the womb of a 35-year-old woman after treatment with fertility drugs. Not one of them survived.

SUMMARY QUESTIONS

1. Define the following terms: oral contraceptive, fallopian tube, fertility drug, *in vitro* fertilisation.
2. Explain how artificial female hormones can be used to:
   a) prevent unwanted pregnancies,
   b) help people overcome infertility.
3. What, in your opinion, are the main advantages and disadvantages of using artificial hormones to control female fertility?

KEY POINTS

1. Hormones can be used to control fertility.
2. Oral contraceptives contain hormones, which stop FSH production so no eggs mature.
3. FSH can be used as a fertility drug for women, to stimulate eggs to mature in their ovaries. These eggs may be used in IVF treatments.

31

SUMMARY ANSWERS

1. Oral contraceptive – A pill taken by mouth that prevents pregnancy.
   Fallopian tube – The tube between the ovary and the uterus.
   Fertility drug - Artificial hormone (FSH) to help eggs mature in ovary and to trigger oestrogen production.
   In vitro fertilisation – Fertilisation that takes place outside the body.
2. **a)** Stops eggs maturing in the ovaries.
   b) Stimulate eggs to mature in the ovary and triggers oestrogen production.
3. Advantages and disadvantages described.

KEY POINTS

There is a Homework activity suggested which covers the Key Points for this spread.

B1a 1.5 Controlling conditions

LEARNING OBJECTIVES

Students should learn that:

- The nervous system and hormones help us to control conditions inside the body.
- Internal conditions such as temperature, blood sugar levels and the balance of water and ions are controlled.
- It is important to control the internal environment.

LEARNING OUTCOMES

Most students should be able to:

- Describe how the temperature, blood sugar levels and balance of water and ions are controlled.
- Explain why it is important to control the internal environment.

Some students should also be able to:

- Evaluate the claims of the manufacturers of sports drinks.

Teaching suggestions

- **Special needs.** These students could be given a pre-printed copy of the summary question 1 and slips with the words on to fill in.
- **Gifted and talented.** These students could carry out a detailed analysis of the contents of sports drinks. The results could be distributed to other members of the class or made into a poster on a notice board in the laboratory.
- **Learning styles**

  Interpersonal: Discussion of the questions in the Starter.

  Intrapersonal: Reflecting on the data obtained from investigation on sports drinks.

  Kinaesthetic: Practical work on temperature measurements.

  Visual: Reading measurements from thermometers and presenting the data.

  Auditory: Taking part in discussions on the benefits of sports drinks.
- **Homework.** Summary questions are quite detailed and could be set as a homework exercise.

SPECIFICATION LINK-UP B1a.11.1

- *Internal conditions which are controlled include:*
  - *the water content of the body – water leaves the body via the lungs when we breathe out and via the skin when we sweat, and excess water is lost via the kidneys in the urine*
  - *the ion content of the body – ions are lost via the skin when we sweat and excess ions are lost via the kidneys in the urine*
  - *temperature – to maintain the temperature at which the enzymes work best*
  - *blood sugar levels – to provide the cells with a constant supply of energy.*

Students should use their skills, knowledge and understanding of 'How Science Works':

- *to evaluate the claims of manufacturers about sports drinks.*

Lesson structure

STARTER

Wipeout exercise on board – Put key words from the student spread on the board (suggest 'internal environment', 'homeostasis', 'ions', 'hormone', 'enzymes' and 'pancreas'). Read out definitions or descriptions of what the words mean. These words are wiped out when the students recognise them. This has the advantage of identifying difficult concepts. (10–15 minutes)

'What happens if . . .?' – Put a series of questions on the board starting with the words 'What happens if . . .?' (your body loses too much water, you lose too much salt, you eat a bag of sweets, you feel hot). Students jot down responses and then check answers. (10–15 minutes)

Why do we . . .?' – Similar to above, but ask questions such as 'Why do we sweat?', 'Why do we urinate less frequently when the weather is warm?' and others. (10–15 minutes)

MAIN

Each of the following suggestions could occupy the main part of the lesson, but the first two could be put together if time permitted. The control of blood sugar does highlight the differences between the nervous system and hormone action.

- **Measuring body temperature of class** – Using forehead thermometers if available. Using clinical thermometers could be too time-consuming and impractical. It is possible to obtain a mean value and also interesting to plot the variation, so that students understand that there is not a 'fixed' body temperature.
- Some of the concepts in 'How Science Works' can be introduced into this activity: the accuracy of the measurements; the mean and range of a set of data; how the data can be displayed.
- **Linking change in body temperature (skin temperature) with increase in exercise** – Students can record skin temperature before undergoing a period of exercise (such as running on the spot, step jumps, anything that can be arranged without too much disruption!) for a set period of time (1–3 minutes of exercise should suffice). Skin temperature after the exercise should be recorded. Again, class results could be collated: the differences are relevant here. A discussion about why the changes occur would be relevant.
- More 'How Science Works' concepts can be introduced here: the relationship between different variables; the concept of the *change* in temperature being important and not the actual values. Time should be allowed for the gathering together of results.
- **Control of blood sugar levels** – In this suggested activity, a comparison can be made between the action of hormones and the nervous system. Introduce with a discussion of the 'sugar rush' from eating several jelly babies (or other sweets) in a row. Observations could be made and the discussion widened to include younger siblings capacity for sweets. It would be useful here to use a torso model (or the diagram on page 24 in the Student Book) to show location of the pancreas.

ACTIVITY & EXTENSION IDEAS

- **More on blood sugar levels.** As a follow-up to the suggestion on 'Control of blood sugar levels', the process of what happens to the sugar can be presented or volunteered by the students to build up a flow chart so that the way in which the jelly babies affect the production of a hormone in the pancreas can be shown.
- **How good are sports drinks?** Students are expected to be able to evaluate the claims made by the manufacturers of sports drinks. This can be carried out by individual students sampling the drinks or by means of a class project. For the evaluation to be reliable, the students should all carry out the same exercise and then compare the effects of using a sports drink with just drinking water.
- **Hypothermia – what are the signs?** Find out about hypothermia and how it affects older people. What should they do to avoid it?

PLENARIES

Water balance – Write on board 'hot day', 'running a marathon' and 'lazy day at home'. Ask students to suggest how their fluid intake and urine output would vary under these different circumstances and why. (5–10 minutes)

Sports drinks – are they a waste of money? – Have a can of a sports drink to show the students and write up on the board a list of its constituents. Ask students whether they use such drinks and what is the purpose of them. Do they work? (This can be used as an introduction to an Extension or Activity, or be used to introduce a Homework project.) (10 minutes)

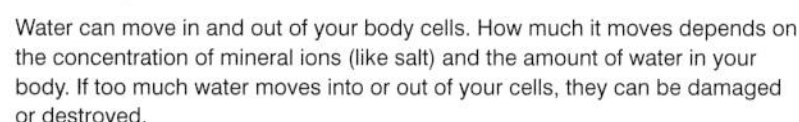

B1a 1.5 Controlling conditions

LEARNING OBJECTIVES

1. How are conditions inside your body controlled?
2. Why is it so important to control your internal environment?

The conditions inside your body are known as its ***internal environment***. Your organs cannot work properly if this keeps changing. Many of the processes which go on inside your body aim to keep everything as constant as possible. This balancing act is called **homeostasis**.

It involves your nervous system, your hormone systems and many of your body organs.

a) Why is homeostasis important?

Controlling water and ions

Water can move in and out of your body cells. How much it moves depends on the concentration of mineral ions (like salt) and the amount of water in your body. If too much water moves into or out of your cells, they can be damaged or destroyed.

You take water and minerals into your body as you eat and drink. You lose water as you breathe out, and in your sweat. You lose salt in your sweat as well. You also lose water and salt in your ***urine***, which is made in your ***kidneys***.

Your kidneys can change the amount of salt and water lost in your urine, depending on your body conditions. They play an important part in controlling the balance of water and mineral ions in your body. The concentration of the urine produced by your kidneys is controlled by a combination of nerves and hormones.

b) What do your kidneys control?

Figure 1 Running a marathon affects your internal environment

PRACTICAL

Helping your body out

When you do a lot of exercise you lose a lot of salt and water from your body as you sweat. It is important to keep your cells hydrated so your body can work properly.

There are lots of special 'sports drinks' you can buy. They claim to rehydrate your body fast, supply you with energy and replace the salt you have lost. Some people think that plain water is just as good! Your kidneys control your internal environment very effectively unless you are exercising really hard for a long time. However, the manufacturers of sports drinks have scientific evidence to back up their claims. You can investigate these claims, and discover just what the drinks contain. See if they help you to perform better!

- How will you carry out your investigation?

Figure 2 A real help in sport – or a good way of making money? Sports drinks are becoming more and more popular, but do most of us really need them?

NEXT TIME YOU...

... drink a lot of water all in one go, watch how often you need to go to the toilet afterwards! Your kidneys will remove the extra water from your blood and you will produce lots of pale urine.

Controlling temperature

It is vital that your deep core body temperature is kept at 37°C. At this temperature your **enzymes** work best. At only a few degrees above or below normal body temperature the reactions in your cells stop and you will die.

Your body controls your temperature in several ways. For example, you can sweat to cool down and shiver to warm up. You can also change your clothing or turn up the heating! Your nervous system is very important in coordinating the way your body responds to changes in temperature.

c) What is the ideal body temperature?

Controlling blood sugar

When you digest a meal, lots of sugar (glucose) passes into your blood. Left alone, the blood glucose levels would keep changing. The levels would be very high straight after a meal, but very low again a few hours later. This would cause chaos in your body. However, the concentration of glucose in your blood is kept constant by hormones made in your ***pancreas***.

d) What would happen to your blood sugar level if you ate a packet of sweets?

Figure 3 Sweets like this are almost all sugar. When you eat them your body has to deal with the effect on your blood.

SUMMARY QUESTIONS

1. Copy and complete using the words below:

   body constant homeostasis hormones internal environment nervous system

   Your ...... cannot work properly if your .... keeps changing. ...... describes the processes which keep everything as ...... as possible. This control of conditions involves your .... and your ....
2. Why is it important to control:
   a) water levels in the body
   b) the body temperature
   c) sugar (glucose) levels in the blood?
3. Look at the marathon runners in Figure 1 on page 32. List the ways in which the running is affecting their:
   a) water balance, b) ion balance, c) temperature.
   d) It is much harder to run a marathon in a costume than in running clothes. Explain why this is.
4. Outline in a flow chart how you could evaluate the claims of the manufacturers of sports drinks.

FOUL FACTS

Once your body temperature drops below 35°C you are at risk of dying from hypothermia. Several hundred old people die from the effects of cold each year. So do a number of young people who get lost on mountains or try to walk home in the snow. Once you lose the ability to control your body temperature, cold kills!

KEY POINTS

1. Humans need to maintain a constant internal environment, controlling levels of water, ions, blood sugar and temperature.
2. Homeostasis is the result of the coordination of your nervous system, your hormones and your body organs

SUMMARY ANSWERS

1. Body, internal environment, homeostasis, constant, nervous system, hormones.
2. a) to stop water moving in or out of cells, damaging and destroying them; b) because the enzymes work best at 37°C; c) because blood sugar which is too high or too low causes problems in the body.
3. a) losing water through sweating; b) losing salt through sweating; c) temperature going up with exercising; d) sweating cools you down and helps to keep the body temperature down – costume makes you sweat more (as you get hotter) which means you lose more water – but also makes it harder for sweat to evaporate (so you don't cool so effectively), also costume heavy so harder work to run.
4. Select sample of people to test → carry out test, e.g. running a set distance → do task with sports drink and with water taken → measure performance with and without sports drink → check conclusions with research done by others (e.g. Internet search).

Answers to in-text questions

a) It is important because cells of the body need a constant environment in which to work properly.

b) The kidneys control the balance of water and mineral ions in the blood.

c) 37°C.

d) It would go up.

KEY POINTS

These are important and form a definition of homeostasis. The points are supported by the Lesson structure. Students, working in pairs, could write two revision questions each then let their partners attempt the questions.

B1a 1.6 Controlling fertility

SPECIFICATION LINK-UP

B1a.11.1

Substantive content that can be revisited in this spread:

- *The uses of hormones in controlling fertility include:*
  - *giving oral contraceptives which contain hormones to inhibit FSH production so that no eggs mature*
  - *giving FSH as a 'fertility drug' to a woman whose own level of FSH is too low to stimulate eggs to mature.*

Students should use their skills, knowledge and understanding of 'How Science Works':

- *to evaluate the benefits of, and the problems that may arise from, the use of hormones to control fertility, including IVF.*

BIOLOGY CO-ORDINATION AND CONTROL

B1a 1.6 Controlling fertility

The contraceptive question

About 3.5 million women in the UK rely on the contraceptive pill to help them plan their families. This month's post-bag is full of queries about this widely-used form of contraception. We're letting our experts get to grips with your FAQs!

Question:

'My daughter is getting married soon and she has gone on the pill. How does it work, and can she really rely on it?' — **Mary**

Our expert's reply:

The pill contains female hormones, which are very similar to the ones in your daughter's body that control her menstrual cycle. The pill stops the production of a hormone called FSH made in your daughter's brain. FSH stimulates the eggs in your daughter's ovaries to mature each month. No FSH means no mature eggs – and no pregnancy.

Your daughter and her husband can certainly rely on the pill to stop her getting pregnant – but only if your daughter remembers to take her pill regularly! If the artificial hormone levels drop, her own body hormones can take over and an egg can be released – which can lead to an unplanned pregnancy!

A tiny pill containing a mixture of female hormones can trick your body and prevent it from releasing any eggs – which makes sure you can't get pregnant!

Question:

'What are the risks of using the pill?' — **Geeta**

Our expert's reply:

The big advantage of the pill is that it prevents pregnancy – and half a million women die each year around the world giving birth. But no medical treatment is risk-free.

The pill can have some mild side effects like headaches and slight weight gain. But in a few people it can cause blood clots, high blood pressure, heart attacks and strokes.

For most people the benefits of the pill outweigh the risks – but only a doctor can prescribe the pill. This ensures your health is looked after carefully and the risks kept as small as possible.

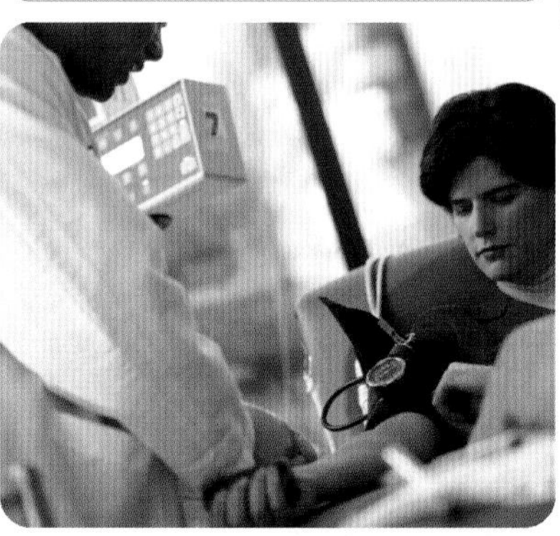

Anyone in the UK who takes the contraceptive pill will have a regular check-up of their blood pressure and general health to make sure they get the benefits of the pill with as little risk as possible

Question:

'If the pill is so good, why doesn't everyone use it so we can really control the world population?' — **Sarah**

Our expert's reply:

The pill is playing an important part in helping to control population growth in many countries, including China. It has also helped people out of the poverty trap in parts of the developing world – smaller families mean more food to go around.

But in some countries the pill is not available because there isn't enough money, education or doctors for it to be used properly. Also, there are some religious groups who feel that preventing conception is denying life, and so they ban the use of the pill for their believers. You're right – it's a difficult issue!

34

Teaching suggestions

This topic follows on from 1.3 'The menstrual cycle', and 1.4 'The artificial control of fertility'. There are some sensitive issues raised, involving religious beliefs and practices. Some of the teaching suggestions made here have already been made in previous spreads, but there is scope for much discussion and debate before encouraging students to attempt either of the activities proposed in the Activity box on page 35 of the Student Book.

The contraceptive question – There are several issues here that could be researched and debated in class discussions. With a topic like this, where there are different ethical and religious beliefs, it could be valuable to allow all views to be aired (if not already covered in recent PSHE). Apart from a generalised discussion about the uses of contraception, it could be useful to select more specific areas:

- How exactly does the contraceptive pill work? Some research on the different types of pill and the combination of hormones they contain could be informative.
- What are the risks of taking the pill? The side effects could be discussed and then a balance sheet drawn up of the pros and cons.
- Can the contraceptive pill help to solve world food problems? A properly organised debate could be held. Students can decide to make a case in favour (i.e. to propose the motion that 'The contraceptive pill can solve world food problems') or a case against. For homework, students could compose a short speech in favour of the motion or against it. Proposers and seconders for each side of the argument can be chosen to present their arguments, followed by a debate involving the whole class.
- 'A contraceptive pill for men?' Research has been done on a male contraceptive pill, but little is heard of this. Ask: 'What would such a pill need to do?' Information is available from BUPA (www.bupa.co.uk) and BBC (www.bbc.co.uk) web sites. It is well worth some research and a discussion, particularly as most of this issues spread is about women!

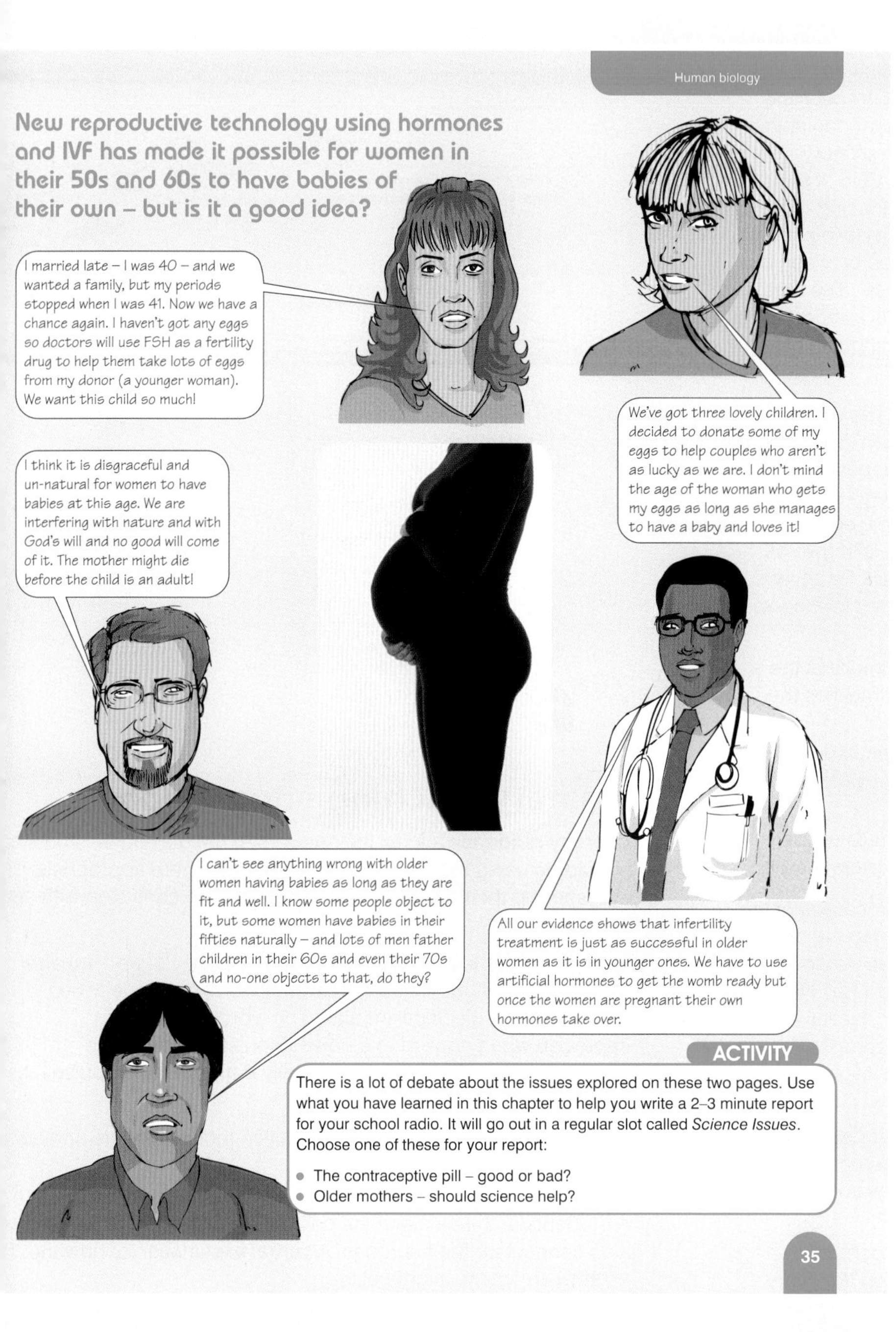

Human biology
New reproductive technology using hormones and IVF has made it possible for women in their 50s and 60s to have babies of their own – but is it a good idea?
I married late – I was 40 – and we wanted a family, but my periods stopped when I was 41. Now we have a chance again. I haven't got any eggs so doctors will use FSH as a fertility drug to help them take lots of eggs from my donor (a younger woman). We want this child so much!
We've got three lovely children. I decided to donate some of my eggs to help couples who aren't as lucky as we are. I don't mind the age of the woman who gets my eggs as long as she manages to have a baby and loves it!
I think it is disgraceful and un-natural for women to have babies at this age. We are interfering with nature and with God's will and no good will come of it. The mother might die before the child is an adult!
I can't see anything wrong with older women having babies as long as they are fit and well. I know some people object to it, but some women have babies in their fifties naturally – and lots of men father children in their 60s and even their 70s and no-one objects to that, do they?
All our evidence shows that infertility treatment is just as successful in older women as it is in younger ones. We have to use artificial hormones to get the womb ready but once the women are pregnant their own hormones take over.
ACTIVITY
There is a lot of debate about the issues explored on these two pages. Use what you have learned in this chapter to help you write a 2–3 minute report for your school radio. It will go out in a regular slot called Science Issues. Choose one of these for your report:
The contraceptive pill – good or bad?
Older mothers – should science help?
35

ACTIVITY & EXTENSION IDEAS

Teenage pregnancies – An extension to any discussion on contraception would be to consider the statistics on the number of teenage pregnancies. Britain has a higher incidence of teenage pregnancies than many other countries. Ask: 'Why is this? What part could more information about contraception play in reducing the number?'

- ? The teaching suggestions made above could help students discuss some of the topics involved and to decide on which issue they would like to do their report. The time of the report is quite short, so the importance of getting the message across concisely needs to be stressed. This is good practice in writing concisely and explaining scientific concepts clearly.

Old wives' tales – An interesting slant on the topic of contraception would be from the historical viewpoint. Again, some research into different methods and when they were practised would be needed. The scientific background to these methods of contraception could be discussed and compared with the way in which the pill works. Old wives' tales could also lead into a discussion about when conception is likely to occur and thus into the topic of the fertility of older women. The BBC have a web site with some information on teenage myths about contraception. Search for 'contraception at www.bbc.co.uk.

The use of fertility drugs in older women – Some background information/statistics about the age of the menopause, the ages at which women are having their first child, the number of surrogate births, etc., could be gathered. Such information could expand the ideas in the speech bubbles in the Student Book. You could divide the class into groups and give them 15 minutes to prepare a short comment on one of the images, using extra information and their own ideas. Each one has moral issues as well as some scientific considerations. Again, this could help with the suggested activity.

? **Should the NHS fund fertility treatment for older women?** – Gather up ideas as to how much the treatment costs and then weigh this up against treating some other conditions. This could be a fairly open-ended discussion.

New research – You could start a discussion on new methods of contraception using hormones and the possibility of contraceptive implants.

A comparison of the use of FSH to inhibit contraception and to stimulate egg production – There is a possibility of confusion, so the differences should be made absolutely clear with OHP or other presentations, particularly for lower attaining students.

SUMMARY ANSWERS

1 a) F **b)** C **c)** D **d)** B **e)** A **f)** E

2 a) It enables you to react to your surroundings and to co-ordinate your behaviour.

b) i) Eye. **ii)** Ear. **iii)** Skin. **iv)** Skin.

c) Diagram of reflex arc. The explanation needs to include the following points: reference to three types of neurone – a sensory neurone, a motor neurone and a relay neurone. The relay neurone is found in the CNS, often in the spinal cord. An electrical impulse passes from the sensory receptor, along the sensory neurone to the CNS. It then passes to a relay neurone and straight back along a motor neurone to the effector organ (usually a muscle in a reflex). This is known as the 'reflex arc'. The junction between one neurone and the next is known as a 'synapse'. The time between the stimulus and the reflex action is as short as is possible. It allows you to react to danger without thinking about it.

3 a) It is the monthly cycle of fertility in women. The average length of the menstrual cycle is about 28 days. Each month the lining of the womb thickens ready to support a developing baby, and at the same time an egg starts maturing in the ovary. About 14 days after the egg starts maturing, it is released from the ovary. This is known as 'ovulation'. The lining of the womb stays thick for several days after the egg has been released before it is shed as the monthly 'period'.

b) i) FSH is made by the pituitary gland; it stimulates the maturation of eggs in the ovaries and stimulates the ovaries to produce oestrogen.

ii) LH is made in the pituitary gland and stimulates the release of a mature egg from one of the ovaries in the middle of the menstrual cycle.

iii) Oestrogen is made in the ovary and stimulates the lining of the womb to build up ready for pregnancy.

4 a) Hormones are chemicals that control the processes of the body. Hormones are released from glands into the blood. Some hormones act quickly, but many act more slowly over a longer period of time than nervous control. Nervous control can be very fast, especially reflexes. It involves the transmission of electrical impulses along neurones. Transmission from one neurone to the next involves chemical substances.

b) A synapse is the junction between two neurones. It enables the impulses to be transmitted from one neurone to the next. It controls the transmission of impulses and ensures that the impulses travel in the right direction.

c) Oral contraceptives (the contraceptive pill) contain hormones, including oestrogen, which act on the ovary preventing the release of any eggs. The production of FSH is inhibited so that no eggs mature in the ovaries. If there are no mature eggs, then you cannot get pregnant. If a woman does not make enough FSH to stimulate eggs to mature, she may be given FSH as a 'fertility' drug. As well as stimulating the development of eggs, FSH triggers oestrogen production.

5 a) You would get hot from Sun and exercise, but the body temperature needs to be kept stable. Heat is lost by sweating, but there will be water and mineral ions lost in the sweat. In order to keep a stable balance of mineral ions and water, the kidney would reabsorb some so that you would produce very little urine. You would feel thirsty which encourages you to drink as soon as possible to replace the water lost from the body.

b) Look for careful thought, planning, awareness of risk, understanding of possibilities and limitations etc.

CO-ORDINATION AND CONTROL: B1a 1.1 – B1a 1.6

SUMMARY QUESTIONS

1 Match up the following parts of sentences:

a) Many processes in the body	A ...... effector organs.
b) The nervous system allows you	B ...... secreted by glands.
c) The cells which are sensitive to light	C ...... to react to your surroundings and co-ordinate your behaviour.
d) Hormones are chemical substances	D ...... are found in the eyes.
e) Muscles and glands are known as	E ...... are known as nerves.
f) Bundles of neurones	F ...... are controlled by hormones.

2 a) What is the job of your nervous system?

b) Where in your body would you find nervous receptors which respond to:
i) light?
ii) sound?
iii) heat?
iv) touch?

c) Draw a simple diagram of a reflex arc. Explain carefully how a reflex arc works and why it allows you to respond quickly to danger.

3 a) What is the menstrual cycle?

b) What is the role of:
i) FSH
ii) LH
iii) oestrogen
in the menstrual cycle?

4 a) Explain carefully the difference between nervous and hormone control of your body.

b) What is a synapse and why are they important in your nervous system?

c) How can hormones be used to control the fertility of a woman?

5 It is very important to keep the conditions inside the body stable.

a) Taking part in school sports on a hot day without a drink bottle for the afternoon would be difficult for your body. Explain how your body would keep the internal environment as stable as possible.

b) Plan an investigation to see whether sports drinks or water are most effective in helping you perform well when you are exercising.

36

EXAM-STYLE QUESTIONS

1 Oral contraceptives can stop someone becoming pregnant by . . .

A preventing the ovaries from releasing an egg.
B killing eggs that have been released from the ovari
C killing sperms before they can reach an egg.
D preventing a fertilised egg from implanting in the uterus lining.

2 A person puts their foot on a sharp object. They automatically lift their foot. The structures involved i this reflex action are shown below:

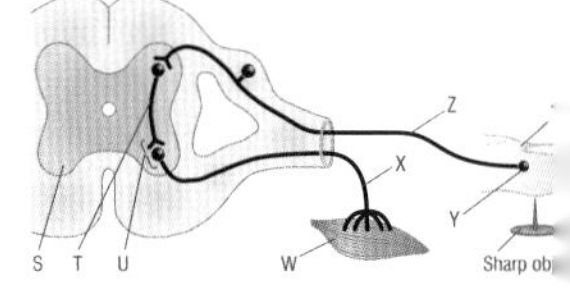

(a) The relay neurone is labelled with the letter
A T **B** U
C X **D** Z

(b) The motor neurone is labelled with the letter
A T **B** X
C Y **D** Z

(c) A synapse is labelled with the letter
A S **B** U
C W **D** Y

(d) The structure labelled W is known as
A a synapse **B** a receptor
C a coordinator **D** an effector organ

(e) In this reflex action the correct path taken by an impulse is . . .
A sensory neurone → receptor → coordinator → motor neurone → effector.
B effector → coordinator → receptor → sensory neurone → motor neurone.
C receptor → sensory neurone → coordinator → motor neurone → effector.
D coordinator → receptor → sensory neurone → effector → motor neurone.

Summary teaching suggestions

- **Literacy guidance** – In all the answers to the questions, students should be using the correct scientific terms where appropriate and spelling them correctly, so that there is no confusion with other terms.
- **Misconceptions** – When answering questions 2c) and 4a) and b), check that students do not use the word 'message' when referring to the electrical impulse.The word 'message' is imprecise when applied to a nerve impulse: students are expected to use the more scientifically correct term 'impulse'.
- **Special needs**
  - In question 1, students could be given the statements printed out and asked to match them.
  - In question 2c), the diagram of the reflex arc could be given to students and they make the connections by filling in the arrows to connect up the neurones. A verbal explanation of how the reflex arc works might help.
  - In question 3b), the roles and the hormones could be written on cards and matched.
- **Learning styles**
  Visual: Learning the description of the reflex arc, the role of the hormones in the menstrual cycle and the description of the synapse can be reinforced using diagrams.
- **When to use these questions?**
  - These questions can be used as an end-of-chapter test or quiz.
  - They are also a useful Revision exercise.
  - Question 1 is quite general and covers the topic of 'control'. It could be used at the end of teaching 1.1.
  - Question 2 could be used after teaching 1.1 and 1.2.
  - Question 3 could be used as a summary to 1.3.

The graph below shows the concentrations of three hormones involved in the menstrual cycle:

Release of egg from an ovary

Hormone concentration in blood | Follicle Stimulating Hormone (FSH) | LH

7 14 21 28
Time/days

Hormone concentration in blood | Oestrogen

7 14 21 28
Time/days

(a) Both FSH and LH are produced in the same gland. What is the name of this gland? (1)

(b) Where in the body is the hormone oestrogen produced? (1)

(c) What is the name given to the release of an egg from the ovary? (1)

(d) The lining of the uterus thickens from around day 5 until some days after the egg has been released. Suggest two purposes for this thickened lining? (2)

(e) Use the information in the graph as well as your knowledge to explain how the concentration of oestrogen affects and controls the release of an egg during the menstrual cycle. (4)

4 Some women are unable to have children naturally. The hormone FSH can sometimes be used to help these women have children.

(a) (i) Harriet does not produce eggs at ovulation. Explain how FSH could be the cause of the problem. (1)

(ii) Explain how FSH could help Harriet to produce and release an egg. (4)

(b) Sharon has had an infection of her Fallopian tubes that has left them blocked. Although she still produces eggs, they are unable to pass down the Fallopian tubes. Describe a method by which Sharon and her partner John could still have children. Include in your account the role of FSH. (4)

HOW SCIENCE WORKS QUESTIONS

The girls in the class set a challenge to the boys. The girls suggested that they had much better control over their nervous reactions than the boys did. The boys accepted the challenge and agreed to the investigation.

The equipment was set up as you can see in this picture.

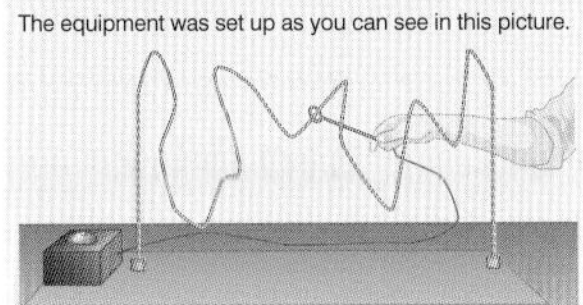

Five girls and five boys took it in turns to move the metal ring along the wire. If anyone touched the ring onto the wire the circuit would be completed and the bell would ring. The teacher counted the number of times the bell rang for each student.

The results are in this table:

Girls' names	Number of touches	Boys' names	Number of touches
Alexandra	6	Arthur	7
Farzana	0	Barnaby	2
Kerry	4	Zahir	4
Summer	3	Jameel	1
Annabel	8	Terry	5

a) In this investigation, which was the dependent variable? (1)

b) Suggest one variable that had not been controlled. (1)

c) Why did the group decide to use the teacher to record the results? (1)

d) Calculate the average for:
i) the boys ii) the girls. (2)

e) Which of the following words would you use to describe the independent variable?
i) continuous ii) categoric
iii) ordered iv) discrete. (1)

f) How might you present these results?
i) bar chart ii) line graph
iii) scattergraph iv) use a line of best fit. (1)

g) Do you think that the girls' prediction is supported by the data collected? Explain your answer. (2)

EXAM-STYLE ANSWERS

1 **A** *(1 mark)*

2 a) **A**

b) **B**

c) **B**

d) **D**

e) **C** *(1 mark each)*

3 a) Pituitary (gland) *(1 mark)*

b) Ovaries *(1 mark)*

c) Ovulation *(1 mark)*

d) It provides food *(1 mark)* and protection *(1 mark)* for the fertilised egg.

e) At the start of each cycle FSH is secreted. The graph shows the level of FSH rising from days 1–5. *(1 mark)*
FSH stimulates the production of oestrogen and helps an egg to mature in one of the ovaries. Graph shows the oestrogen level rising on days 5–12. *(1 mark)*
High oestrogen levels slow down the production of FSH and stimulate the production of LH. Graph shows LH level rising and FSH level falling as oestrogen levels rise. *(1 mark)*
High LH levels stimulate the release of an egg from one of the ovaries. Graph shows egg is released on day 13 when LH level is at its highest. *(1 mark)*

4 a) (i) Harriet may not be producing enough FSH to stimulate an egg to mature in her ovary each month. *(1 mark)*

(ii) Giving Harriet injections of FSH should raise its level to the point where eggs mature in her ovaries. *(1 mark)*
FSH also stimulates the ovaries to produce oestrogen. *(1 mark)*
Oestrogen stimulates the pituitary gland to produce LH. *(1 mark)*
LH stimulates the release of a mature egg from one of the ovaries. *(1 mark)*

b) Eggs can be removed from Sharon's ovaries using a fine needle. *(1 mark)*
These eggs can be fertilised by John's sperm outside the body (in vitro fertilisation). *(1 mark)*
The developing embryos can be implanted into Sharon's uterus/womb where they may develop into a fetus. *(1 mark)*
There is always a risk that some eggs may not fertilise or the embryos may not develop. To help ensure that there are a sufficient number of eggs, Sharon is given an injection of FSH before the eggs are removed so that a large number mature at once and hence many eggs are available for in vitro fertilisation. *(1 mark)*

Exam teaching suggestions

Questions 1 and 2 (multiple choice) are a useful way to test recall and understanding of the whole class at the end of a lesson. They are quick to mark with no interpretation of answers needed. Students can therefore mark their own, or they can mark each others, with no margin for disputes.

In question 3e), it is important that, in answering this, students 'use the information in the graph' as well as their own knowledge. They must therefore support each statement they make by referring to some aspect of the graph. The mark should be withheld if a statement is not supported by evidence up to a maximum of 2 marks.

In question 4b), the last mark should help to differentiate foundation and higher-level students. It not only requires more detail but also explains the role of FSH. Although the question specifically asks for such an explanation, it is surprising how many students fail to do so.

Question 4 could be used to revise some facts about reproductive hormones and as an introduction to a discussion on the moral, ethical and economic aspects of IVF and other fertility treatments.

HOW SCIENCE WORKS ANSWERS

a) The dependent variable is the number of touches.

b) Examples of variables that were not controlled include how long it took to do the test, which hand was used, whether the hand could be supported, how long you were allowed to touch the wire for one ring of the bell.

c) Because the teacher was unbiased

d) (i) The average number of touches for the boys was 3.8 touches.
(ii) The average number of touches for girls was 4.2 touches, both could reasonably be rounded off to 4 touches.

e) (ii) Categoric describes the independent variable.

f) (i) Bar chart, because the independent variable is categoric.

g) No. No real difference between the boys and the girls. Bigger difference between individuals than between the sexes.

How science works teaching suggestions

- **Higher- and lower-level questions.** Questions f) and g) are higher level and questions a) and b) are lower level. The answers for these questions have been provided at these levels.
- **Gifted and talented.** There is the possibility of developing a technique that controls all of the variables listed in the answer to question b). This could prompt some discussion on the controlling and 'taking account of' variables that might be needed in a behaviour-type investigation. There are other variables, such as individual skill levels and previous experience, that might be considered.
- **How and when to use these questions.** The questions could develop from a simple reaction timing investigation involving the whole class. Group results should be considered as a way of reducing the influence of certain uncontrollable variables.

B1a 2.1 Diet and exercise

LEARNING OBJECTIVES

Students should learn that:

- A healthy diet contains the right balance of the different foods you need and provides the right amount of energy.
- The metabolic rate is the rate at which the chemical reactions in the body are carried out.
- The less exercise you take the less food you need; people who exercise regularly are usually fitter than those who take little exercise.

LEARNING OUTCOMES

Most students should be able to:

- Describe the constituents of a healthy diet.
- Define metabolic rate and explain how it can vary according to the amount of activity carried out and the proportion of muscle to fat in the body.
- Describe the relationships between food intake, exercise and fitness.

Some students should also be able to:

- Explain all the interactions between food intake, exercise, fitness, metabolic rate, gender, genetic factors etc. which affect body mass

Teaching suggestions

- **Learning styles**

  Interpersonal: Discussing the different food groupings.

  Intrapersonal: Making deductions about energy use.

  Kinaesthetic: Measuring the energy content of food and the role-playing exercise.

  Visual: Reading the temperature values on thermometers.

  Auditory: Listening to the feedback from other students.

SPECIFICATION LINK-UP B1a.11.2

- *A healthy diet contains the right balance of the different foods you need and the right amount of energy. A person is malnourished if their diet is not balanced. This may lead to a person being too fat or too thin. It many also lead to deficiency diseases.*
- *The rate at which all chemical reactions in the cells of the body are carried out (the metabolic rate) varies with the amount of activity you do and the proportion of muscle to fat in your body. It may be affected by inherited factors.*
- *The less exercise you take and the warmer it is, the less food you need. People who exercise regularly are usually fitter than people who take little exercise. If you exercise your metabolic rate stays high for some time after you have finished.*

Lesson structure

STARTER

Sorting out food groups – Prepare six A4 sheets, each with the name of a major food group written in large letters on it and make a separate list of foods of all types. Give one A4 sheet to each of the first six students who come in through the door. As the other students enter, assign them a food from your list ('You are a tomato'. 'You are a pint of milk.', etc.), making sure you have all food types covered. Ask the students to move to the food group that they feel they belong to, adding that they may well be able to fit into two groups. Go through a group at a time finding out which food groups are where and discussing any anomalies. (10–15 minutes)

Do you have a healthy diet? – Each student to write out what they ate for their previous evening meal, assigning the items to the correct food groups. Compare with the other students in small groups. Which food groups were eaten? Were any missing? (5–10 minutes)

MAIN

- **Measuring energy in foods** – This practical is based on the burning 'peanut' experiment. Test tubes are set up, each containing a measured volume of cold water and a thermometer used to record the initial temperature. A piece of chosen food is placed on the end of a 20 cm length of wire, or a mounted needle, and ignited. As soon as it is alight, it is held as close under the test tube of water as possible. When it has finished burning, record the highest temperature reached in the water in the test tube. Suitable foods for testing are Bite-sized Shredded Wheat, corn snacks such as Wotsits and dry bread. If using sweets, beware of falling hot sugar, and fatty foods have a tendency to spit. Students can work in groups or individually, and results pooled. It is advisable to avoid using peanuts due to allergies.
- This activity provides plenty of scope for the introduction of concepts covered in 'How Science Works'. The accuracy of the measurements, the quantities of food used, the control of variables and evaluation of the results can all be discussed.
- **How much energy do I use when . . .?** – This practical is based on the fact that 10 J of energy is required to raise a 1 kg mass a distance of 1 m. Ask students to raise a 100 g mass up into the air for 1 m and then tell them they have done 1 J of work. You can vary this with different masses depending on availability.
- It is possible to extend this practical by calculating the energy used/work done when carrying out activities such as climbing stairs or stepping up on to an object. The mass of the student should be measured and the height of the object or staircase determined. If several volunteers are used, the work done can be calculated and then this value can be used to work out the quantity of sugar they would need to eat to replace the energy. [100 g sugar contains 1630 kJ of energy, so not much!]

PLENARIES

Role playing exercise – Ask students to take roles of nutritional advisors and people with different energy needs, such as a pregnant woman, top athlete, body builder, etc. (10 minutes)

Matching diets to people – Write 'Energy intake (in kJ)' down one side of the board, and different occupations and ages (such as adolescent, male manual worker, female secretary) down the other side. Students are asked to match the energy intake with occupation. (5 minutes)

Practical support

Measuring energy in foods

Equipment and materials required

Each group will need: boiling tubes/test tubes, mounted needles or 20 cm lengths of wire, a Bunsen burner, a test tube rack, a heat-proof mat, a metal-jawed clamp, a thermometer, a range of foods cut into cubes or small pieces (exclude peanuts due to allergic reactions).

Safety: See CLEAPSS Laboratory Handbook/CD ROM section 9.4.2.

How much energy do I use when . . .?

Equipment and materials required

A supply of weights ranging from 100 g to 1 kg would be useful, together with a metre rule. Access to a staircase is needed for the main experiment, together with a measuring tape and scales to weigh the student volunteers.

ACTIVITY & EXTENSION

As an extension activity, students could be asked to investigate the different types of 'fitness' equipment available at their local gym or from articles in magazines. Ask them to evaluate these against the exercise they get from games' periods in school and ordinary activities. Why do they think these machines have been devised and who benefits from them?

BIOLOGY — HEALTHY EATING

B1a 2.1 Diet and exercise

LEARNING OBJECTIVES

1. What does a healthy diet contain?
2. Why can some people eat lots of food without getting fat?
3. How does an athlete's diet differ from yours?

Figure 1 Everyone needs a source of energy to survive – and your energy source is your food. Whatever food you eat – whether you prefer sushi, dahl, or roast chicken – most people eat a varied diet that includes everything you need to keep your body healthy.

What makes a healthy diet?

A healthy diet contains:

- carbohydrates,
- proteins,
- fats,
- vitamins,
- minerals,
- fibre and
- water

and the energy you need to live, all in the right amounts!

If your diet isn't balanced, you will end up ***malnourished***. If you don't take in enough vitamins and minerals, you will end up with deficiency diseases like scurvy. (Scurvy is caused by a lack of vitamin C.)

Fortunately, in countries like the UK, most of us take in all the minerals and vitamins we need from the food we eat. However, our diet can easily be less well balanced in terms of the energy we take in. If we take in too much energy we get fat – but if we don't eat enough we get too thin.

It isn't always easy to get it right because different people need different amounts of energy.

a) Why do you need to eat food?

How much energy do you need?

The amount of energy you need to live depends on lots of different things. Some of these things you can change and some you can't.

If you are male, you will need to take in more energy than a female of the same age – unless she is pregnant.

If you are a teenager, you will need more energy than if you are in your 70s – and there isn't much you can do about it!

b) Why does a pregnant woman need more energy than a woman who isn't pregnant?

SCIENCE @ WORK

Fitness instructors at your local leisure centre or gym can measure the proportion of your body which is made up of fat. They can advise you on the right food to eat and the exercise you need to take to become thinner, fitter – or both!

The amount of exercise you do affects the amount of energy you use up. If you do very little exercise, then you don't need much food. The more you exercise, the more food you need to take in. Your food supplies energy to your muscles as they work.

People who exercise regularly are usually much fitter than people who take little exercise. They make bigger muscles – and muscle tissue burns up much more energy than fat. But exercise doesn't always mean time spent training or 'working out' in the gym. Walking to school, running around the house and garden looking after small children or having a physically active job all count as exercise too.

38

Human biology

c) Why do athletes need to eat more food than the average person?

Figure 2 Athletes who spend a lot of time training and playing a sport will have a great deal of muscle tissue on their bodies – up to 40% of their body mass. So they have to eat a lot of food to supply the energy they need.

The temperature where you live affects your energy needs as well. The warmer it is, the less energy you need. This is because you have to use less energy keeping your body temperature at a steady level. So you need to take in less food!

Figure 3 If you live somewhere really cold, you need lots of high-energy fats in your diet. You need the energy to keep warm!

The metabolic rate

Imagine two people who are very similar in age, sex and size. However, they may still need quite different amounts of energy in their diet. This is because the rate at which all the chemical reactions in the cells of the body take place (the **metabolic rate**) varies from person to person.

The proportion of muscle to fat in your body affects your metabolic rate. Men generally have a higher proportion of muscle to fat than women, so they have a higher metabolic rate. You can change the proportion of muscle to fat in your body by exercising and building up more muscle.

Your metabolic rate is also affected by the amount of activity you do. Exercise increases your metabolic rate for a time even after you stop exercising.

Finally, scientists think that your basic metabolic rate may be affected by factors you inherit from your parents.

SUMMARY QUESTIONS

1. What do we mean by 'a balanced diet'?
2. a) Why does an old person need less energy in their diet than a teenager?
   b) Why does a top footballer need more energy in their diet than you do? Where does the energy in the diet come from?
3. a) What is meant by the 'metabolic rate'?
   b) Explain why some people put on weight more easily than others.

DID YOU KNOW?

Between 60–75% of your daily energy needs are used up in the basic reactions needed to keep you alive. About 10% is needed to digest your food – and only the final 15–30% is affected by your physical activity!

Figure 4 When a hedgehog like this wakes up after a long winter sleep, it is vital that it gets moving and feeding again as quickly as possible. Hibernating animals store special brown fat, which has a very high metabolic rate. When they use it at the end of hibernation it produces heat rapidly – and helps them survive.

GET IT RIGHT!

Metabolic rate is not the same as heart rate or breathing rate – make sure you know the difference.

KEY POINTS

1. Most people eat a varied diet, which includes everything needed to keep the body healthy.
2. Different people need different amounts of energy.
3. The metabolic rate varies from person to person.
4. The less exercise you take, the less food you need.

39

SUMMARY ANSWERS

1. A diet that contains the right amount of carbohydrates, proteins, fats, vitamins, minerals, fibre and water, and the right amount of energy.
2. a) An old person does not move around so much, uses less energy and is not growing.
   b) A top footballer probably has more muscle, which uses a lot of energy. The energy comes from fats and carbohydrates.
3. a) 'Metabolic rate' is the rate at which all the chemical reactions in the cells of the body are carried out.
   b) Some people have a slower metabolic rate, some take less exercise, some eat more and do not use up all the energy they take in as food so they store the excess as fat.

Answers to in-text questions

a) To supply the body cells with energy.

b) Because a pregnant woman has to provide energy for a growing baby as well as herself.

c) Athletes have a lot of muscle tissue and muscle tissue burns up a lot of energy.

GET IT RIGHT!

As well as knowing that metabolic rate is not the same as heart rate or breathing rate, students need to be able to define metabolic rate.

KEY POINTS

If the key points are made into revision cards, then add some facts to these generalisations. An example would be to add to number 3 key point a selection of occupations and relative energy requirements.

B1a 2.2 Weight problems

LEARNING OBJECTIVES

Students should learn that:

- In the developed world, excess food and lack of exercise lead to high levels of obesity and associated health problems.
- Arthritis, diabetes, high blood pressure and heart disease are more common in overweight people than in thinner people.
- In the developing world, some people suffer from health problems linked to lack of food.

LEARNING OUTCOMES

Most students should be able to:

- Describe the problems associated with excess food in the diet and how these may be overcome by diet and exercise.
- Describe the health problems linked to lack of food.

Some students should also be able to:

- Evaluate, when supplied with relevant information, the claims made by different slimming programmes.

Teaching suggestions

- **Gifted and talented.** Higher attaining students could speculate on societal factors that might correlate with body mass index.
- **Special needs.** Students may be able to play the starter 'The food groups beetle game' suggested earlier, with some help.
- **Learning styles**

  Interpersonal: Reporting on exercise during the week.

  Intrapersonal: Caption writing for photographs.

  Kinaesthetic: Measurements of BMI.

  Visual: Presenting statistics on obesity.

  Auditory: Listening to other students reporting on their findings.
- **Homework.** Write a letter of advice to a person who suffers from either obesity or anorexia, trying to be supportive as well as helpful.

SPECIFICATION LINK-UP B1a.11.2

- *In the developed world too much food and too little exercise are leading to high levels of obesity and the diseases linked to excess weight:*
  - *arthritis (worn joints)*
  - *diabetes (high blood sugar)*
  - *high blood pressure*
  - *heart disease.*
- *Some people in the developing world suffer from health problems linked to lack of food. These include:*
  - *reduced resistance to infection*
  - *irregular periods (in women).*

Students should use their skills, knowledge and understanding of 'How Science Works':

- *to evaluate the claims made by slimming programmes.*

Lesson structure

Weight problems are widespread and attention is drawn to the potential sensitivity of these topics. It would be wise to warn students in advance in order to prevent nasty comments.

STARTER

The food groups beetle game – Recap the 'Diet and exercise' lesson by playing this game. Students can be supplied with an outline of a beetle with no legs but the letters C, P, F, V, M, F around the thorax (or they could draw their own). Either issue dice and instructions (1 = carbohydrate, 2 = proteins, 3 = fats, etc.), or read out names of foods and students label the legs with the name of the food as they play. A timer can be used on this if you can digitally project (search for 'free countdown timer' in a search engine). (10 minutes)

Caption writing – Find pictures of a Sumo wrestler and female body builder on the web (www.images.google.com) and ask students to write a caption/text for each one. Students can be selected to read out their efforts or write them on the board. (5 minutes).

Class survey – Survey the class as to how much exercise they take per week (over and above what is compulsory). It is possible to make some generalisations about the effect of exercise on body mass. (10 minutes)

MAIN

- **Measurement of BMI** – The concept of BMI can be introduced and then students can measure their own mass and height and use the formula to calculate their own BMI. The actual activity should be optional for reasons of sensitivity: a sheet of data for fictitious characters with heights and body masses could be supplied. The actual formula is not difficult to use, but a BMI calculator can be found at www.bbc.co.uk. Search the web for 'BMI calculators' to find graphs and normal ranges. There are also graphs for displaying the data and turning the BMI into a descriptor.
- If a set of fictitious characters is used, then you can ask the students to do the calculations and classify them into the correct categories. Those that come in the 'obese' and 'overweight' groups could then be recommended a slimming programme. Due to the sensitive nature of this, it is better to do it with fictitious characters and not the class members.
- Search the web for 'obesity statistics' about various groups of people (different age groups; different ethnic groups; different countries). Students can discuss these statistics and suggest how they can be displayed to have an impact. It could be useful here to distinguish between overweight, moderately obese and clinically obese, using BMI values.

- **Are 'slimming', 'low fat' or 'diet' foods worth buying if you want to lose weight?** – A comparison of such foods with 'normal' brands can be made by checking their fat and energy content and other constituents from their labels. This would work quite well with different brands of yoghurt or cereal bars. Things to remember are differences in size, differences in mass and differences in contents as well as differences in price. Students could be encouraged to suggest/bring in their own for a general class comparison. This is a useful exercise in evaluation (more 'How Science Works'), as the students may find it interesting to work out fat content per gram and energy content per gram, in order to make their investigation reliable.

PLENARIES

'What advice would you give Homer Simpson on how and why to lose weight?' – Search the web for an image or cartoon clip of Homer Simpson or Peter Griffin of Family Guy. Use either as a stimulus and award a doughnut as a prize for the best advice! (5–10 minutes)

The science behind the slimming diet – Compare different slimming programmes/techniques, such as Weight Watchers, Atkins Diet, Glycaemic Index. What is the scientific basis of each? (10 minutes)

ACTIVITY & EXTENSION IDEAS

Ask students how they could persuade their local doctor that they needed to employ a nutritionist or a dietician in their practice?

What is the ideal body shape? – Start a discussion on role models and the pressure to be thin. Discuss the 'Real women' campaigns (Body Shop and Marie Claire) showing models representative of the population. Is the ideal body shape the same throughout the world? Some pictures of people of different shapes and sizes from around the world could help here. In order not to be sexist in a mixed school, include men and women, especially as men can suffer from weight problems as well.

BIOLOGY — HEALTHY EATING

B1a 2.2 Weight problems

LEARNING OBJECTIVES

1. What health problems are linked to being overweight?
2. Why is it unhealthy to be too thin?

Human beings come in all sorts of shapes and sizes. Most people look about right but there will always be extremes. Some people are very overweight and others appear unnaturally thin. Scientists and doctors don't just measure what you weigh. They look at your ***body/mass index*** or ***BMI***. This compares your weight to your height in a simple formula:

$$BMI = \frac{weight}{(height)^2}$$

Most people have a BMI in the range 20–30. But if you have a BMI of below 18.5, or above 35, then you may have some real health problems.

a) What does your body/mass index measure?

Figure 1 In spite of some of the media hype, most people are not obese – but the amount of weight people carry certainly varies a great deal!

Obesity

If you take in more energy than you use, the excess is stored as fat. You need some body fat to cushion your internal organs. Your fat also acts as an energy store for when you don't feel like eating. But if someone eats a lot more food than they need, over a long period of time, they could end up ***obese***.

Carrying too much weight is often inconvenient and uncomfortable. Far worse, it can lead to serious health problems. Obese people are more likely to suffer from **arthritis** (worn joints), **diabetes** (high blood sugar levels which are hard to control), ***high blood pressure*** and ***heart disease***. They are more likely to die young than slimmer people.

b) What health problems are linked to obesity?

Losing weight

Many people want to lose weight. This might be for their health or just to look better. You gain fat by taking in more energy than you need, so there are three main ways you can lose it.

- You can reduce the amount of energy you take in by cutting back the amount of food you eat – particularly energy-rich foods like biscuits, crisps and chips.
- You can increase the amount of energy you use up by taking more exercise.
- And the best way to lose weight is to do both – reduce your energy intake and exercise more!

Many people find it easier to lose weight by attending slimming groups. At these weekly meetings they get lots of advice, plus support from other slimmers. All the different slimming programmes involve eating less food and/or taking more exercise!

Increasing your exercise levels can be an important part of losing weight and getting fitter. However, you need to take care. If you suddenly start working out hard in the gym, or taking other vigorous exercise, you can cause other health problems.

SCIENCE @ WORK

A nutritionist uses scientific knowledge about food, diet and health to help other people understand the importance of food and develop well-being. They are important in the food industry (for people *and* animals), in government departments, in hospitals and education in the UK and abroad.

40

Different slimming programmes approach weight loss in different ways. Many simply give advice on healthy living. They advise lots of fruit and vegetables, not too much fat or too many calories and plenty of exercise. Some are more extreme suggesting you cut out almost all of the fat or the carbohydrates from your diet.

Others claim that 'slimming teas' or 'herbal pills' will enable you to eat what you like and still lose weight. What sort of evidence would you look for to decide which approaches worked best?

c) What must you do to lose weight?

Starvation

In some parts of the world obesity is rare, because the biggest problem is lack of food. Civil wars, droughts and pests can destroy local crops so people cannot get enough to eat. Starvation leads to a number of symptoms including:

- You become very thin and your muscles waste away.
- Your immune system can't work properly so you pick up infections.
- If you are female, your periods will become irregular or stop altogether.

These symptoms are also sometimes seen in the developed world in people suffering from the mental disorder called **anorexia** (loss of appetite) **nervosa**.

d) What are the main symptoms of starvation?

Figure 2 Hundreds of thousands of people around the world suffer the symptoms of malnutrition and starvation. There is simply not enough food for them to eat.

SUMMARY QUESTIONS

1. Copy and complete using the words below:

   energy fat less more obese

   If you take in more ...... than you use, the excess is stored as .... If you eat too much over a long period of time, you will eventually become .... To lose weight you need to eat ...... and exercise ....
2. Plan a simple information sheet about the dangers of being overweight and how to lose weight sensibly.
3. Research the claims of two slimming programmes. Compare and evaluate the claims they make.

Human biology

DID YOU KNOW?

The heaviest man ever recorded was Jon Brower Minnoch (USA, 1941–1983). He was 185 cm (6 ft 1 in) tall and was overweight all his life. At his heaviest he weighed 635 kg (100 stone). The heaviest recorded woman was another American, Rosie Bradford, who weighed 544 kg (85 stone) in 1987.

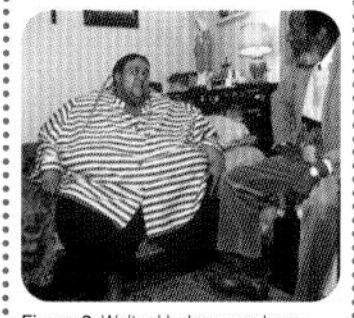

Figure 3 Walter Hudson was by no means the fattest man in the world – but he made a living out of the media interest in his enormous weight until he died of flu in his early forties.

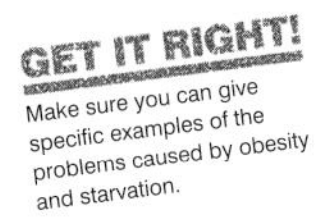

GET IT RIGHT!

Make sure you can give specific examples of the problems caused by obesity and starvation.

KEY POINTS

1. If you take in more energy than you use, you will store the excess as fat.
2. Obese people have more health problems than people of normal weight.
3. People who do not have enough to eat can develop serious health problems.

41

SUMMARY ANSWERS

1. Energy, fat, obese, less, more.
2. Information sheet.
3. Results of research. Look for scientific reasoning to back up the comparisons made.

Answers to in-text questions

a) It compares your weight and height (w/h^2).

b) Arthritis, diabetes, high blood pressure, heart disease plus any other correct answers such as breathlessness.

c) Reduce your food (energy) intake, increase your exercise (energy output) or both.

d) Loss of weight, reduced resistance to infection/poor immune system, irregular periods/periods stop (for women).

KEY POINTS

These key points can be reinforced by the letter writing exercise suggested as a homework activity. If revision cards are made, it would be good to expand the statements by adding examples, such as listing some of the health problems.

B1a 2.3 Fast food

LEARNING OBJECTIVES

Students should learn that:

- Cholesterol is carried around the body by two types of lipoproteins – LDLs (bad) and HDLs (good)
- The amount of cholesterol, made by the liver and present in the blood, depends on the diet and inherited factors.
- Saturated fats increase blood cholesterol levels and high levels are linked with an increased risk of disease of the heart and blood vessels.
- In addition to high levels of fat, some processed foods may contain a high proportion of salt which may lead to increased blood pressure.
- Mono and polyunsaturated fats can lower blood cholesterol and help balance LDLs and HDLs.

LEARNING OUTCOMES

Most students should be able to:

- Explain what cholesterol is.
- Describe the effects of different types of fat on blood cholesterol levels.
- Describe the effects that too much salt in the diet may have on some people.

Some students should also be able to:

- Evaluate information on the effects of cholesterol and salt on our health.
- Explain the importance of the LDL/HDL balance in the blood and how this is affected by diet.

Teaching suggestions

- **Special needs.** In order to explain the different types of fat, students could be shown a shirt 'saturated' in water, a 'monorail' (the monorail put through Springfield in the Simpsons) and 'Polyfilla'.
- **Gifted and talented.** Students could investigate the chemical structure of cholesterol.
- **Learning styles**
  Interpersonal: Discussing the documentary shown.
  Intrapersonal: Considering how our lifestyles have changed.
  Kinaesthetic: Testing foods for fats.
  Visual: Predicting lifestyle changes for Inuits.
  Auditory: Explaining the terms used.

SPECIFICATION LINK-UP B1a.11.2

- *Cholesterol is a substance made by the liver and found in the blood. The amount of cholesterol produced by the liver depends on a combination of diet and inherited factors. High levels of cholesterol in the blood increase the risk of disease of the heart and blood vessels.*
- *Cholesterol is carried around the body in two types of lipoproteins. Low-density lipoproteins (LDLs) are 'bad' and can cause heart disease. High-density lipoproteins (HDLs) are 'good' cholesterol. The balance of these is very important to good heart health.*
- *Saturated fats increase blood cholesterol levels. Mono-unsaturated and polyunsaturated fats may help both to reduce blood cholesterol levels and to improve the balance between LDLs and HDLs.*
- *Too much salt in the diet can lead to increased blood pressure for about 30% of the population.*
- *Processed food often contains a high proportion of fat and/or salt.*

Students should use their skills, knowledge and understanding of 'How Science Works':

? • *to evaluate information about the effect of food on health.*

Lesson structure

STARTER

A diet of fast food for a month – Start a discussion of the film documentary 'Super-size me', where a reporter, Morgan Spurlock, ate nothing but fast food for a month. (5–10 minutes)

What's in a burger and chips? – Bring in a fast food meal and show it to the class. Ask: 'What is in it? What problems might you get if you ate lots of these in the short term? What happens if it becomes a way of life (i.e. long term)?' (5 minutes)

'Shall we get a takeaway and a video?' – Why is fast food so popular? Discuss and make a list of how and why eating habits have changed in the last 30 years. (10 minutes)

MAIN

- **Practical on testing for fats in burger and chips** – Or test any other fast foods, such as crisps, pizza, KFC – the students could bring in some small samples of their own. Grind up portions in a pestle and mortar with a little water. Allow to settle and then decant or filter off some of the liquid. Add to half a small test tube of ethanol and shake vigorously. (Caution with eyes and naked flames.) If there is fat present, a creamy emulsion is obtained. Demonstrate what an emulsion looks like by shaking up some cooking oil in a gas jar three-quarters full of water (or, on a smaller scale in a test tube and then each group could do their own).
- An alternative to the emulsion test is to wipe pieces of fatty food on to greaseproof paper. A translucent mark is left if there is fat present. This might be quicker and less messy to do than the emulsion test if large numbers of food items are to be tested.
- **'Good to eat Fred the Red'** – This is an interactive food and nutrition programme to download from the Science Year CD web site (www.sycd.co.uk), from Manchester United Football Club or on the Science Year CDs. If computers are available, the students can work through it themselves, or it can be projected, or accessed as homework.

PLENARIES

Explain the terms – Write or tack up key words from this topic on the board and pick/invite two students to come to the front and explain one each. They remove the word they have explained, if they are judged to have been successful in explaining it to the rest of the class. They can then choose the next pair and the key words to be explained. If stuck, a student can choose someone to help them. (5–10 minutes)

Why don't Inuits have high cholesterol levels? – Inuit tribesmen [show pictures] traditionally eat large amounts of fat in the form of seal and whale blubber. They do not have high average cholesterol readings. Ask: 'Why might this be? What factors of their lifestyle and living conditions could account for this? Write down recommendations to an Inuit who is giving up the traditional lifestyle for a sedentary life.' (10–15 minutes)

ACTIVITY & EXTENSION

- Search the web for 'video heart surgery' to show the layers of fat around the heart.
- Use a digital sphygmomanometer (these devices are inexpensive and sold in high streets) to record blood pressures: it shows how they can vary and also provides a stimulus to students thinking of medical careers. Some 'How Science Works' concepts can be introduced here: taking a range of measurements, working out mean values and considering a range of values in a population.
- Ask the school kitchen for nutritional data on the fat and salt content of some sample school meals. This might need to be negotiated first! Alternatively, it could be useful to invite the person in charge of your school canteen in for a discussion about the nutritional guidelines they work to in producing school meals.
- Moly-models could be used to demonstrate the differences in the structure of saturated and unsaturated fats. Compare fats and oils from plant and animal sources and what they are used for.

NEXT TIME YOU . . .

. . . go food shopping, compare the fat content of different brands of foods, such as yoghurts. A burger and fries has an obvious fat content, but many foods have 'hidden fat', added to make it tastier.

GET IT RIGHT!

It might be helpful to make sure that students are clear about the different types of fat in the diet and their effects on the cholesterol in the blood.

BIOLOGY HEALTHY EATING

B1a 2.3 Fast food

LEARNING OBJECTIVES

1. What is cholesterol?
2. Why do your cholesterol levels matter?
3. Is too much salt bad for us?

People eat fast processed food because it is quick and easy and fits in with their busy lives. But it often contains a lot of fat and salt. These make the food taste good. However, there are some real concerns about the effect that too much fat and salt in your diet can have on your health.

a) What substances do you often find in fast foods?

Figure 1 Fast food tastes good. But it has had lots of things added to make it easy to cook and eat. These often include fat and salt.

Cholesterol

Fat is an energy-rich food. So too much fat in your diet can easily make you overweight. But that isn't the only problem with fatty food. The amount and type of fat you eat also seems to affect the levels of **cholesterol** in your blood.

Cholesterol is a substance which you make in your liver. It gets carried around your body in your blood. You need it to make the membranes of your body cells, your sex hormones and the hormones that help your body deal with stress. Without cholesterol, you wouldn't survive. Yet people often talk about cholesterol as if it is a bad thing. Why?

High levels of cholesterol in your blood seem to increase your risk of getting heart disease or diseased blood vessels. The cholesterol builds up in your blood vessels and can even block them. Heart disease is one of the main causes of death in the UK and USA, so no wonder doctors are worried.

b) Why do you need cholesterol in your body?

Controlling cholesterol

The amount of cholesterol you have in your blood depends on two things:

- The way your liver works, which is something you inherit from your parents and cannot change.
- The amount of fat in your diet.

Some people have livers that can deal with almost any amount of fat. Their blood cholesterol seems to stay within healthy levels. But for many people the level of cholesterol in their blood is linked to the amount and type of fat they eat.

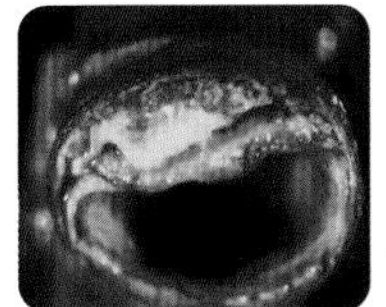

Figure 2 When you get cholesterol building up in the wrong place – like the arteries leading to your heart – it can be very serious indeed

It isn't just the overall level of cholesterol in your blood which affects your risk of developing heart disease. Cholesterol is carried around your body by two types of ***lipoproteins***:

- **Low-density lipoproteins (LDLs)** are known as 'bad' cholesterol. Raised levels of LDLs increase your risk of heart problems.
- **High-density lipoproteins (HDLs)** are known as 'good' cholesterol and they reduce your risk of heart disease.

The balance of LDLs and HDLs in your blood is very important for a healthy heart.

42

Human biology

There are three main types of fats in the food you eat and they seem to have different effects on your cholesterol:

- **Saturated fats** increase (raise) blood cholesterol levels. You find them in animal fats like meat, butter and cheese.
- **Mono-unsaturated fats** seem to have two useful effects. They may reduce your overall blood cholesterol levels and improve the balance between LDLs and HDLs in your blood. You find them in foods like olive oil, olives, peanuts and lots of margarines.
- **Polyunsaturated fats** seem to be even better at reducing your blood cholesterol levels and balancing LDLs and HDLs than mono-unsaturates. You find them in foods including corn oil, sunflower oil, many margarines and oily fish.

c) What is the big difference between saturated fats and the other types of fats?
d) Why are raised blood cholesterol levels a worry?

NEXT TIME YOU...

. . . eat a burger and fries, think about all the fat you are taking on board. Will your liver be able to deal with it, or are your blood cholesterol levels about to go up?

What about salt?

Like fat, salt is vital in your diet. Without it, your nervous system would not work and the chemistry of all your cells would be in chaos. But for about a third of you (30% of the population), too much salt in your diet can lead to high blood pressure. This can damage your heart and kidneys and increase your risk of a stroke.

Many people eat too much salt each day without knowing it. That's because many processed, 'fast' foods contain large amounts of salt. But you can control your salt intake by doing your own cooking – or by reading the labels very carefully when you buy ready-made food!

Oil/fat	% saturated fat	% polyunsaturated fat	% mono-unsaturated fat
Butter	66	4	30
Corn oil	13	62	25
Olive oil	14	12	74
Sunflower oil	11	69	20

Figure 3 Once you start to look at the different fats in the food you are buying, shopping can get very complicated!

SUMMARY QUESTIONS

1 Copy and complete using the words below:

salt heart salt blood pressure fat cholesterol

Fast food can contain too much ...... and .... Raised ...... in the blood can lead to ...... disease, while too much ...... can give some people high ....

2 Look at Figure 3 and use it to help you answer these questions:

a) Which fat or oil has the highest percentage of mono-unsaturates?
b) Which fat or oil has the highest percentage of polyunsaturates?
c) Which fat or oil has the highest percentage of saturated fats?
d) Decide which of these fats or oils would be the best to use for a healthy heart, and which would be the worst. Explain your answer carefully, including the balance of LDLs and HDLs in your blood.

3 Many people want to lower the amount of salt in fast foods.

a) Explain why salt is important in your diet.
b) Why are people worried about high salt levels in food?
c) Would lowering the salt levels in processed foods make everyone healthier? Explain your answer.

KEY POINTS

1. Fast food often contains high proportions of fat and/or salt.
2. Cholesterol is made in the liver and found in the blood. High cholesterol levels have been linked to heart disease.
3. The level and type of cholesterol in your blood is influenced by the type of fat you eat.
4. Too much salt in the diet can lead to raised blood pressure in about a third of the population.

43

SUMMARY ANSWERS

1 Fat, salt, cholesterol, heart, salt, blood pressure.

2 a) Olive.

b) Sunflower.

c) Butter.

d) Sunflower oil would be best – it has the most poly-unsaturates which are best at lowering cholesterol levels and has the best total of poly- and mono-unsaturates, thereby improving the balance of LDLs and HDLs in your blood. Butter is worst – it has the highest saturated fat level but it does have some mono-unsaturates as well.

3 a) Salt is needed for the cells and the nervous system.

b) Salt can cause raised blood pressure in some people.

c) No, because raised salt levels only affect about one person in three.

Answers to in-text questions

a) Salt and fat.

b) To make cell membranes, sex hormones and stress hormones.

c) Saturated fats cause raised cholesterol levels; the others lower cholesterol levels.

d) They are linked to heart disease and blood vessel disease.

KEY POINTS

- Key points are reinforced by the plenary 'Explain the terms'.
- If key points are made into revision cards, then add some extra information as examples. The types of fat in the diet could be named. These additions will help students make the links between fats and cholesterol.

B1a 2.4 Health issues

SPECIFICATION LINK-UP

B1a.11.2/3

Students should use their skills, knowledge and understanding of 'How Science Works':

- *to evaluate information about the effect of food on health (11.2)*
- *to evaluate claims made by slimming programmes (11.2)*
- *to evaluate the effect of statins on cardio-vascular disease (11.3).*

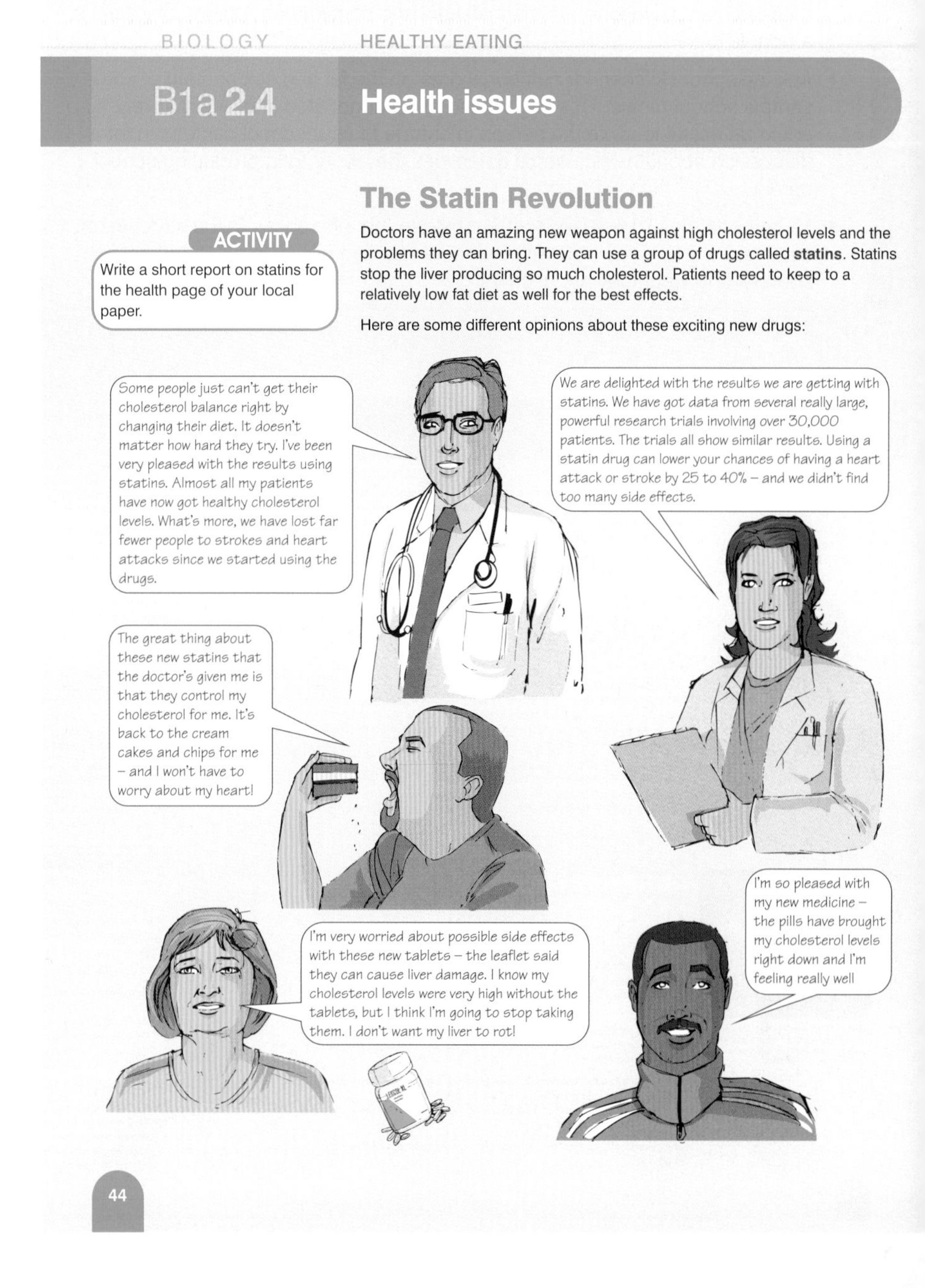

BIOLOGY HEALTHY EATING

B1a 2.4 Health issues

The Statin Revolution

Doctors have an amazing new weapon against high cholesterol levels and the problems they can bring. They can use a group of drugs called **statins**. Statins stop the liver producing so much cholesterol. Patients need to keep to a relatively low fat diet as well for the best effects.

Here are some different opinions about these exciting new drugs:

ACTIVITY

Write a short report on statins for the health page of your local paper.

Teaching suggestions

The Statin Revolution – There is a great deal of stimulus material in this section. To enable students to understand the issues raised by the use of statins, then they might need to do some prior research or be supplied with answers to the following questions, which could be given on a fact sheet:

- What sort of chemicals are statins? Where do they come from? [Some students may already know (from publicity about Flora and Benecol) that they are derived from plant compounds.]
- What is considered to be a 'normal' or 'healthy' level of cholesterol in the blood?
- How are drugs researched? Does a trial involving 30 000 patients seem enough when the adult population is in millions? A discussion about how drugs are tested and trials done might be beneficial here.
- What is meant by side effects? Reference could be made to the informative leaflets contained in any medicines.

With this information, the class could discuss each of the talking heads from the student book in turn, and then consider whether statins are the answer to lowering cholesterol levels in the whole population. Ask: 'Does more information need to be available?' The approach presented here is to consider statins as medicine (pills to be prescribed), but should food manufacturers include statins in processed foods such as spreads and yogurt drinks? Ask: 'What are the dangers of buying the processed foods containing statins?'

The report for the health page of the local paper should present a balanced view and contain some science, remembering that the readers of the paper might not necessarily have studied Biology.

Scientists wear blinkers! – This interesting article highlights the different approaches to the results of scientific research and also that there is not necessarily one simple, straightforward answer to conditions associated with diet and disease. As with the statin story, some background information could be beneficial to an understanding of the article.

NEWS

Scientists wear blinkers – and we pay the price!

For many years now scientists and doctors have been telling us that we are at risk from heart disease because we eat too much animal fat and our blood cholesterol is too high. But a lot of people still die of heart disease. Now it seems that vitamins might be just as important to our hearts as fat. What's more, this idea was first discovered years ago – so why didn't we find out sooner?

Thirty years ago, Kilmer McCully was a young researcher at Harvard University in the USA. He discovered a possible link between an amino acid called homocysteine and changes in the blood supply to the heart which can lead to heart attacks. High levels of homocysteine are linked to low levels of B vitamins in the diet – and these B vitamins are often missing in processed foods!

Changing your diet or taking a cheap supplement of B vitamins lets your body remove the homocysteine and prevents the damage to your heart.

Unfortunately McCully did his research at the same time as many top scientists were supporting the link between fats and heart disease. Time and money had been spent developing anti-cholesterol drugs and low-fat foods. No-one wanted to hear about McCully's cheap and simple solution. He lost his funding at Harvard and his ideas were quashed.

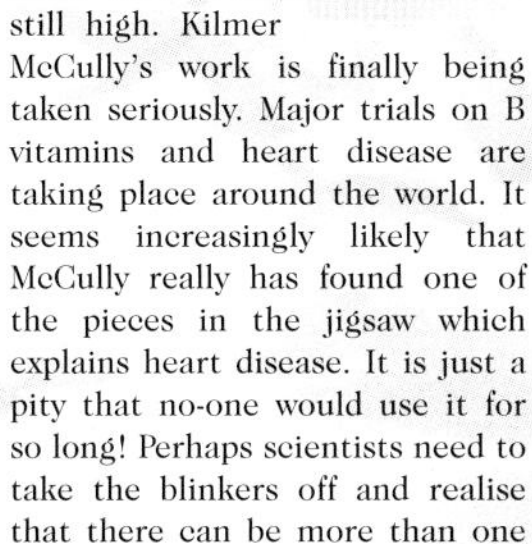

Thirty years on – and in spite of the fact that we have all cut back on our fat levels and taken our anti-cholesterol medicines, deaths from heart disease are still high. Kilmer McCully's work is finally being taken seriously. Major trials on B vitamins and heart disease are taking place around the world. It seems increasingly likely that McCully really has found one of the pieces in the jigsaw which explains heart disease. It is just a pity that no-one would use it for so long! Perhaps scientists need to take the blinkers off and realise that there can be more than one solution to a problem!

ACTIVITY

Write a letter:
Either from the young Kilmer McCully to a friend explaining what you have discovered about a link between B vitamins and heart disease and what it might mean for patients;

Or from a senior scientist who has been working on treatments for high cholesterol and heart disease to one of his colleagues about McCully's work and how you feel about it.

Menu 1	Menu 2
Turkey twizzlers	Char-grilled chicken
Chicken nuggets	Spaghetti Bolognese
Pizza	Fish with pesto topping
Chips	Baked potato
Spaghetti hoops	Fresh fruit
Iced bun	Yoghurt
Doughnut	

ACTIVITY

Plan an assembly for the year 7 pupils in your school on the importance of a healthy diet. It should include help with the food they should choose in the school canteen for lunch.

- The higher attaining students could research the B vitamins. Ask: 'What are they needed for in the body? In what foods do they occur? What quantities of B vitamins are needed for this treatment to be effective?' Research on the Internet might provide the answer.
- Ask: 'What quantities of B vitamins are present in processed foods?' Find out the RDA (Recommended Daily Amount) of the B vitamins and compare these with the content of some staple foods. (Bender and Bender's Food Tables have values.)

With the additional information, it could be of value to discuss the two approaches and evaluate the advantages and disadvantages of each type of treatment. The cost to the NHS, the awareness of the general public and the advantages of educating people to eat sensibly could all form part of the debate.

Which one would you choose? – This suggested activity encourages students to try to put into practice the science they have learned. Year 7 students will not have encountered the 'Food and Digestion' topic at Key Stage 3, so the planning of the assembly would need to take this into account.

ACTIVITY & EXTENSION IDEAS

- ? The recommended activities linked with the statin story and the link between B vitamins and cardiovascular disease could be used for homework and enable the students to have some practice in writing in continuous prose.
- Students could find out about members of their family or anyone they know who has been prescribed statins.
- More information about the different types of fat in foods could be researched. Most food labels will list the saturated and unsaturated fat content.
- ? Diets in other parts of the world differ from ours. For example, much has been written about the benefits of a Mediterranean-type diet including red wine and olive oil. Research into different diets and the incidence of cardiovascular disease could highlight the benefits of eating fewer fats or different types of fat.
- ? Sample menus could be designed for different diets, such as the Atkins diet, giving reasons for the inclusion of the different foods. It could be better to suggest that a menu is designed for a 24 hour period or longer, as it is not always relevant to base a diet on one meal.
- ? If menus from different restaurants can be obtained, students could choose what to eat if they are following a particular slimming diet. For each course, they should give reasons for their choice. This could also be done using menus from Indian or Chinese takeaways, and also from burger chains and other fast food outlets.
- ? Have the fast-food chains done enough to make their food healthier? Ask students to collect literature from the fast food outlets (McDonald's, Burger King, KFC) about efforts they have made to cut down on saturated fats and salt content of their food.

SUMMARY ANSWERS

1 **a) i)** A balanced diet is when carbohydrates, proteins, fats, vitamins, minerals, fibre and water are in the right amounts for health.

ii) Metabolic rate is the rate at which all the chemical reactions in the cells of the body take place.

b) The amount of exercise you do affects the amount of energy you use up. The more you exercise, the more food you need to take in to supply energy to your muscles as they work. Athletes are very fit and have big muscles. Muscle tissue burns up much more energy than fat, so athletes need a lot of food to supply the energy needed by their bodies when they exercise.

2 **a)** Obesity is being very overweight with a BMI of over 30.

b) Obese people are more likely to suffer from diseases such as arthritis, diabetes, high blood pressure and heart disease and to die young.

c) Eat less food overall, change of diet to include more fruit and vegetables, eat less fatty food and take more exercise.

d) Following a slimming diet involves reducing the amount of energy you take in by eating less and increasing the amount of energy you use by taking more exercise. A slimming diet is designed to help you lose a certain amount of weight and stay healthy. There is no element of choice (normally) in starvation. The body breaks down muscles which become wasted. The immune system is damaged and women become infertile as their periods stop.

3 **a)** A high fat diet is often linked to raised levels of blood cholesterol: it depends on the type of fats in the diet. High levels of saturated fats in the diet seem to have a bad effect on cholesterol levels in some people. The mono- and poly-unsaturated fats are considered to be linked to healthy levels of cholesterol.

b) Raised cholesterol levels are linked to heart disease and diseases of the blood vessels.

c) Fast foods may be high in saturated fats and salt. The high salt content can cause a rise in blood pressure in some people. The high saturated fat content can lead to obesity and other problems. Occasionally eating fast food is no problem, but some people may eat too much. Although some people realise the dangers, the food tastes good, it is relatively cheap and easy to obtain, needing no preparation.

d) The answer to this question should be presented in the form of a report, and credit would be given for relevant scientific points (balanced diet, lots of fruit and vegetables, fat and salt content, enough energy content) combined with clear presentation. It should be remembered that the listeners would not necessarily have the detailed scientific background, but would be concerned about the health of their children.

4 Look for a sympathetic approach which recognises the importance of self-image to young people and the risks of eating disorders, malnutrition and obesity if the diet isn't balanced.

a) Points such as: posture, be realistic about weight, balanced diet with three meals a day, plenty of fresh fruit and veg will keep weight in the ideal range, keep skin and hair looking good, give plenty of energy etc. Avoid snacking by having good lunch. If really hungry, eat fruit etc.

b) Need plenty of calories to support growth and sport. calcium rich food for bone growth, lots of protein to help muscles, fats and carbohydrates for energy. Eat regularly and build up stamina levels with regular exercise etc.

HEALTHY EATING: B1a 2.1 – B1a 2.4

SUMMARY QUESTIONS

1 a) Define the following:
i) Balanced diet.
ii) Metabolic rate.
b) A top athlete needs to eat a lot of food each day. This includes protein and carbohydrate. Explain how they can eat so much without putting on weight.

2 a) What is obesity?
b) Why is obesity a threat to your health?
c) Suggest some ways in which an obese person might lose weight.
d) How do following a slimming diet and suffering from starvation differ?

3 a) What is the link between a high-fat diet, raised blood cholesterol and the LDL/HDL balance in your blood?
b) Why are doctors concerned if a patient has raised cholesterol levels or a high ratio of LDLs to HDLs?
c) Fast food is often linked to an unhealthy lifestyle. Explain the problems with fast foods – and why people still eat them.
d) Recently there has been a lot of media interest in school dinners. People think they contain far too many 'fast foods' and not enough fresh produce, fruit and vegetables.
Plan a short report for your local radio station on why healthy school meals are important for the future health of the children who eat them.

4 Here are two young people who have written to a lifestyle magazine problem page for advice about their diet and lifestyle. Produce an 'answer page' for the next edition of the magazine.
a) Melanie: *I'm 16 and I worry about my weight a lot. I'm not really overweight but I want to be thinner. I've tried to diet but I just feel so tired when I do – and then I buy chocolate bars on the way home from school when my friends can't see me! What can I do?*
b) Jaz: *I'm nearly 17 and I've grown so fast in the last year that I look like a stick! So my clothes look pretty silly. I'm also really good at football, but I don't seem as strong as I was and my legs get really tired by the end of a match. I want to build up a bit more muscle and stamina – but I don't just want to eat so much I end up getting really heavy. What can I do about it?*

EXAM-STYLE QUESTIONS

1 The table is about some conditions that affect the bo[…] as a result of certain diets.
Match descriptions **A**, **B**, **C** and **D** with the words 1 to […] in the table.
A an imbalance of nutrients in the diet
B a severe shortage of food in the diet
C a psychological disorder leading to a dangerously low body mass
D a body/mass index above 30

	Condition
1	Obesity
2	Anorexia
3	Malnutrition
4	Starvation

2 The body/mass index (BMI) compares body mass to height. The BMI is calculated using the following formula:

$$BMI = \frac{\text{body mass in kg}}{(\text{height in metres})^2}$$

(a) A woman has a body mass of 60 kg and a height o[…] 1.6 metres. Her BMI is equal to …
A 18.7 **B** 23.4 **C** 29.2 **D** 37.5

The graph shows the percentage of a population […] people in different BMI groups who suffer from a[…] form of arthritis.

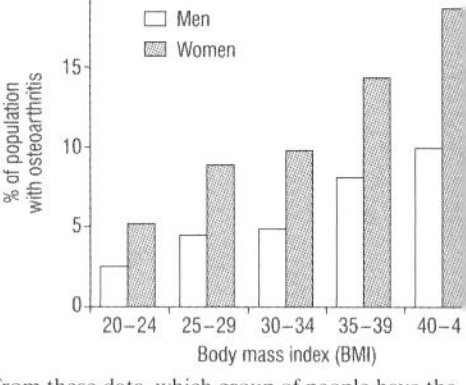

(b) From these data, which group of people have the highest percentage of osteoarthritis?
A Men with a BMI of 30–34.
B Men with a BMI of 35–39.
C Women with a BMI of 35–39.
D Women with a BMI of 40–44.

(c) The data were collected by measuring the BMI of […] large sample of people. What is the main advanta[…] of using a large sample of people?
A The data obtained are more reliable.
B The mean BMI can be calculated.
C The results obtained are fairer.
D People of many ages are included.

Summary teaching suggestions

- **Literacy guidance** – Several of these questions require explanations in continuous prose. These can be used to stress the importance of good sentence construction, correct grammar and spelling.
- **Misconceptions** – Students should be aware of the difference between what is considered *fact* (a definition of obesity) and what is considered a *possibility* (the link between a high fat diet and cholesterol). Also, not all obese people suffer from diabetes or heart disease.
- **Special needs**
Students could answer question 3d) by designing a series of healthy school meals. They could draw up menus or cut out and paste pictures on to cards.
- **Learning styles**
Interpersonal: Some of these questions deal with explanations and complex answers that could be debated and discussed before students write their answers. Such discussions help students to sort out their ideas and appreciate the depth of the answer required.
Auditory: A variety of answers to question 3d) could be read aloud to the class. This gives students the opportunity to judge good answers and also to take on board the ideas of others.
- **When to use the questions?**
  - Parts of question 1 relate to B1a 2.1 'Diet and exercise' and could be used as a homework exercise after studying the topic.
  - If students have difficulty in writing definitions, such as required in question 1a), they could be encouraged to make revision cards of correct answers.
  - Question 2 relates to B1a 2.2 'Weight problems' and all the answers required need to be written in continuous prose.
  - Question 3c) could be used as a follow-up to B1a 2.3 'Fast food'.

(d) What conclusion can be drawn from the data?
  A Women with a BMI of 35–39 are the group most likely to suffer from osteoarthritis.
  B The higher the BMI in both men and women the greater the risk of suffering from osteoarthritis.
  C The lower the BMI in men, the more likely they are to suffer from osteoarthritis.
  D Slimming will prevent osteoarthritis. (1)

(a) State the seven components that make up a healthy diet. (2)
An investigation was carried out over four different periods to find the energy intake of 14- and 15-year-old girls and boys. The results are shown in the table below.

Period of time	Average energy intake/kJ per day	
	Boys	Girls
1930s	12873	11088
1960s	11739	9534
1970s	10962	8484
1980s	10478	8316

(b) Calculate the percentage decrease in energy intake for girls between the 1930s and the 1980s. Show your working. (1)
(c) Explain why the intake of energy for both boys and girls decreased between the 1930s and the 1980s. (2)
(d) If the same study had been carried out with groups of 70-year-olds, how might the results have been different? (1)
(e) Suggest a reason why girls need to take in less energy than boys of the same age. (2)
(f) What would be the best way to display the data from the table above? Explain your choice. (2)
(g) Calculate the mean of the average energy intake of boys from the 1930s to the 1980s. (1)
(h) Who has the larger range of average energy intake between the 1930s and 1980s – boys or girls? Show your working out. (1)

What a person eats can affect their health. Explain how doing each of the following might help to keep a person healthy:
(a) Reducing the amount of saturated fat that is eaten. (3)
(b) Eating less salt. (2)

Human biology

HOW SCIENCE WORKS QUESTIONS

A class of students were asked to test some fruit juices for their vitamin C content. They were given the apparatus set up as in the diagram below. They had to put the sample of fruit juice into the test tube and add the dye (DCPIP) from the burette drop by drop until the mixture retained the blue colour of the dye.

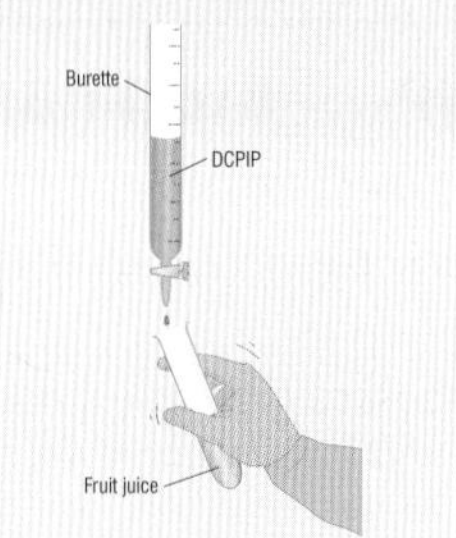

a) Name a control variable they would have to use. (1)
b) Describe how they should use the burette to get accurate results. (5)
The class results were as follows:

Juice	Volume of DCPIP added/cm³				
Orange	21.1	26.2	24.8	25.5	25.7
Apple	20.9	19.7	21.3	20.5	21.0

c) Calculate the mean for the amount of DCPIP added by the class to the apple juice. (1)
d) It was suggested that the first result for the orange juice was an anomaly. Why was this thought to be an anomaly? (1)
e) Calculate the average for the amount of DCPIP added to the orange juice. (1)
f) What conclusion can you make from these results? (1)
g) One student commented that when carrying out the titration with the orange juice it was quite difficult to tell when the DCPIP stayed blue. She thought this was due to the colour of the orange juice. How does this idea affect your conclusion? (2)

EXAM-STYLE ANSWERS

1 **A** **3**
B **4**
C **2**
D **1** *(1 mark each)*

2 a) **B**
b) **D**
c) **A**
d) **B** *(1 mark each)*

3 a) Carbohydrates
Proteins
Fats
Vitamins
Minerals
Fibre
Water *(7 correct = 2 marks, 5 or 6 correct = 1 mark, 0–4 correct = no marks)*

b) Girls' energy intake during 1930s = 11 088 kJ
Girls' energy intake during 1980s = 8316 kJ
Difference = 2772 kJ

$$\% \text{ decrease} = \frac{2772 \times 100}{11088} = 25\%$$

(With correct working shown 1 mark)
(Without working shown or incorrect working no mark)

c) *Either* Less exercise is taken *(1 mark)*
because there are more labour-saving devices. *(1 mark)*
Or Less energy is needed to keep warm *(1 mark)*
because there is central heating. *(1 mark)*
Any valid reason why teenagers in the 1980s take less exercise or are better at keeping warm should be allowed.

d) The energy consumption would have been less/reduced. *(1 mark)*
A higher-level student might point out that it is true for both sexes on all four occasions.

e) Girls/females/women have a lower proportion of muscle to fat than boys/males/men. *(1 mark)*
They therefore have a lower metabolic rate and hence require less energy. *(1 mark)*
For both marks, the link must be made between girls having relatively less muscle than boys and the fact that this means they need less energy.
A higher-level student might also state that muscle is an active tissue with a much higher metabolic rate than fat.

f) As a bar chart. *(1 mark)*
Because the 'period of time'/independent variable is a categoric variable. *(1 mark)*

g) 11 513 kJ (12 873 + 11 739 + 10 962 + 10 478 = 46 052
46 052 ÷ 4 = 11 513) *(1 mark)*

h) Girls – Range for boys = 12 873 − 10 478 = 2395 kJ
Range for girls = 11 088 − 8316 = 2772 kJ *(1 mark)*

4 a) • Saturated fat increases blood cholesterol levels.
• Especially low-density lipoproteins (LDLs).
Higher levels of LDLs increase the risk of heart disease.
(1 mark for each of the above.)

b) • Too much salt in the diet can increase blood pressure.
• Increased blood pressure can cause heart damage/cause kidney damage/increase the risk of stroke.
(1 mark for each of the above)

HOW SCIENCE WORKS ANSWERS

a) An example of a control variable that could be used is the same amount of fruit juice.
b) Use the burette to get accurate results by reading the bottom of the meniscus; add drops slowly; swirling the mixture and watching for the colour change; read burette level again.
c) Mean for the amount of DCPIP added by the class to the apple juice is 20.7 cm³ or 20.68 cm³.
d) 3.7 cm³ difference to next reading, much more than any other difference.
e) The mean for the amount of DCPIP added to the orange juice is 25.55 cm³ – leaving out the anomaly.
f) That this sample of orange juice contains more vitamin C than the apple juice sample.
g) It could be that more DCPIP was added than was needed and so the vitamin C in the orange juice might have been over-estimated.

How science works teaching suggestions

- **Literacy guidance**
  - Key terms that should be clearly understood are: mean, anomaly, conclusion
  - Questions expecting longer answers, where students can practise their literacy skills: are b) and g).
- **Higher- and lower-level questions.** Question g) is higher level and question a) is lower level. Answers for both these questions have been provided at these levels.
- **Gifted and talented.** There is a possibility of bringing out the difference between an anomaly and expected variation.

B1a 3.1 Drugs

LEARNING OBJECTIVES

Students should learn that:

- Drugs may cause harm by changing the chemical processes in the body, affecting behaviour and causing damage to major organs, such as the brain, lungs and liver.
- People using drugs may become dependent on them and unable to manage without them (addiction).
- Cannabis may cause psychological problems and the use of hard drugs such as cocaine and heroin can seriously damage health.

LEARNING OUTCOMES

Most students should be able to:

- Give a definition of the term 'drug'.
- Explain what is meant by addiction.
- Describe some of the problems caused by drug abuse.

Some students should also be able to:

- Understand that the impact of drug use varies from individual to individual.

Teaching suggestions

- **Gifted and talented.** These students could investigate further the nature of drugs, how they affect nerve transmission and whether they act in the brain or on the nervous system generally.
- **Learning styles**

  Interpersonal: Discussing and contributing to contents of the medicine cabinet.

  Intrapersonal: Feedback to class on reaction times.

  Kinaesthetic: Investigating the effects of caffeine.

  Visual: Making measurements of reaction times.

  Auditory: Listening to other students' opinions on social problems.
- **ICT link-up.** Reaction timers (search the web) can be used in the experiments on reaction times.

SPECIFICATION LINK-UP B1a.11.3

- *Drugs can be beneficial but may harm the body.*
- *Many drugs derived from natural substances have been known to indigenous peoples for many years.*
- *Drugs change the chemical processes in people's bodies so that they may become dependent or addicted to them and suffer withdrawal symptoms without them. Heroin and cocaine are very addictive.*

Lesson structure

STARTER

Are there drugs in your medicine cabinet? – Write up a list of some of the contents of a First Aid kit or a typical family medicine cabinet [paracetamol, calamine, anti-histamine cream, antiseptic or brand names such as Calpol, Nurofen, Dettol] and ask the students to identify which they think are drugs, or contain drugs, and which are other substances. Check some of the proprietary pain-relievers for substances such as caffeine, as well as the more obvious ones. (5–10 minutes)

Quick quiz on different categories of drugs – Hand out lists of different drugs to the students as they enter the classroom and ask them to put M (medicinal), R (recreational) or I (illegal) alongside each one. Check through the list, pointing out that there is some overlap between the different categories. (5–10 minutes)

What are the side effects? – Provide copies of an information leaflet from a packet of over-the-counter drugs, such as a pain-reliever. Discuss the possible side effects of the drug and balance this against the benefits of taking the drug. Why do you think all this information is necessary? (10–15 minutes)

MAIN

- **Investigation of the effect of caffeine on reaction times** – Caffeine has a mild stimulatory effect increasing alertness. Using the stick-drop method of testing reaction times, the effect of caffeine can be measured. Students can volunteer to drink measured amounts of coffee, with a known/controlled caffeine content, and have their reaction times measured before and after drinking the coffee. A period of time (about 10 minutes) has to be allowed for the coffee to be absorbed before the second test is carried out. Remind the students that it is the difference between the two times which is significant.
- This experiment can be used to introduce 'How Science Works' concepts as practice for internal assessment – predictions can be made, measurements repeated and controlled conditions are easy to ensure. However, generalising from data collected from one individual or a small group brings in the variety of factors that make humans different, and the need for large sample sizes in investigations where all variables cannot be controlled. Care needs to be taken with drinking in a laboratory (could be done outside).
- **The effect of caffeine on memory** – If the stick-drop test has already been done with the class, students could be given a series of numbers or words to remember and then asked to write them down before and after drinking a set amount of coffee. The investigation can be set up as above, with the usual controls, repetitions and calculations of means. This could be a more appropriate experiment to use as it introduces the possibility of using words, numbers or pictures, rather than the reflex response of catching the ruler. More 'How Science Works' concepts are also introduced here.
- **Where does the drug have an effect?** – Many drugs act directly on the synapse and affect the rate of transmission of impulses across the junction of two neurones. Review the synapse with a PowerPoint® presentation of the events occurring and the effect of a stimulant (such as caffeine) compared with that of large doses of nicotine, which blocks transmission. Opiates act on the brain reducing the response to painful stimuli; you can link this in to the use of opiates in pain control. This can be used in conjunction with either of the experiments suggested if time permits. Alternatively, use the Animation Bla 1.2 'Reflex actions' available on the GCSE Biology CD ROM.

PLENARIES

Discussion on social problems linked with drug abuse – Ask students to compile a list of five important social problems associated with the use of drugs, including tobacco and alcohol. Compare lists and decide which ones top the list. (10 minutes)

How old do I have to be before I can buy a drink or a packet of cigarettes? – Do students know the rules? Why do they think these rules are needed? Do they work? (10 minutes)

ACTIVITY & EXTENSION

- More work can be done with the results of the stick-drop experiment. The distance dropped can be converted into a time: it is the square root of 2 × the distance dropped in metres divided by 9.81 (g the acceleration due to gravity).
- The experiment can compare the effect of caffeine on males and females. Is there a difference? If there is, can it be explained? Also, the effects of different quantities may differ between different sexes and age groups.

Practical support

Investigation of the effect of caffeine on reaction times

Equipment and materials required

Metre rulers, cups of coffee (not to be consumed in the laboratory).

BIOLOGY

DRUG ABUSE

B1a 3.1 Drugs

LEARNING OBJECTIVES

1. What is a drug?
2. What is addiction?
3. What problems do drug addiction cause?

Figure 1 Millions of pounds worth of illegal drugs are brought into the UK every year. It is a constant battle for the police to find and destroy drugs like these.

A drug is a substance that alters the way in which your body works. It can affect your mind, your body or both. In every society there are certain drugs which people use for medicine, and other drugs which they use for pleasure.

Many of the drugs that are used both for medicine and for pleasure come originally from natural substances, often plants. Many of them have been known to and used by indigenous peoples for many years. Usually some of the drugs that are used for pleasure are socially acceptable, while others are illegal.

a) What do we mean by 'indigenous peoples'?

Drugs are everywhere in our society. People drink coffee and tea, smoke cigarettes and have a beer, an alcopop or a glass of wine. They think nothing of it. Yet all of these things contain drugs – caffeine, nicotine and alcohol (the chemical ethanol). These drugs are all legal.

Other drugs, such as cocaine, ecstasy and heroin are illegal. Which drugs are legal and which are not varies from country to country. Alcohol is legal in the UK as long as you are over 18, but it is illegal in many Arab states. Heroin is illegal almost everywhere.

b) Give an example of one drug which is legal and one which is illegal in the UK.

Because drugs affect the chemistry of your body, they can cause great harm. This is even true of drugs we use as medicines. However, because medical drugs make you better, it is usually worth taking the risk.

But legal drugs, such as alcohol and tobacco, and illegal substances, such as solvents, cannabis and cocaine, can cause terrible damage to your body. Yet they offer no long-term benefits to you at all.

What is addiction?

Some drugs change the chemical processes in your body so that you may become addicted to them. You can become dependent on them. If you are addicted to a drug, you cannot manage properly without it.

Once addicted, you generally need more and more of the drug to keep you feeling normal. When addicts try to stop using drugs they usually feel very unwell. They often have aches and pains, sweating, shaking, headaches and cravings for their drug. We call these ***withdrawal symptoms***.

c) What do we mean by 'addiction'?

DID YOU KNOW?

In 2003, 1042 men and boys and 346 women and girls died as a result of using illegal drugs in the UK. 766 of these deaths were due to heroin or methadone, the chemical used to help people stop taking heroin.

48

Human biology

The problems of drug abuse

People take drugs for a reason. Drugs can make you feel very good about yourself. They can make you feel happy and they can make you feel as if your problems no longer matter. Unfortunately, because most recreational drugs are addictive, they can soon become a problem themselves.

No drugs are without a risk. Cannabis is often thought of as a relatively 'soft' – and therefore safe – drug. But evidence is growing which shows that it can cause serious psychological problems to develop in some people.

Hard drugs, such as cocaine and heroin, are extremely addictive. Using them often leads to very severe health problems. Some of these come from the drugs themselves, and some come from the lifestyle which often goes with drugs. Because they are illegal, they are expensive. Young people often end up turning to crime to pay for their drug habit. They don't eat properly or look after themselves. They can also end up with serious illnesses, such as hepatitis, STDs and HIV/AIDS.

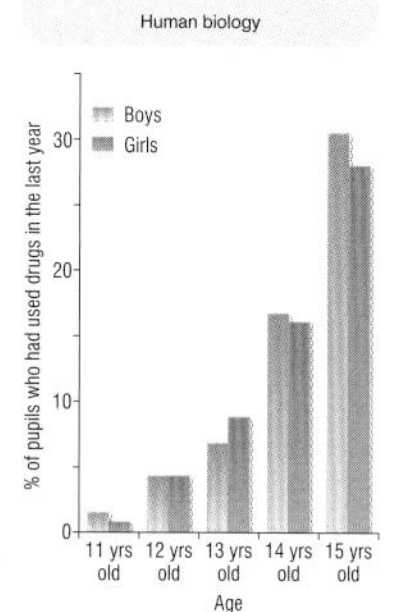

Figure 2 Most of the young people who have used drugs have smoked cannabis – but the number of 15-year-old pupils who have tried drugs is causing a lot of concern

DON'T LET DRUG DEALERS CHANGE
THE FACE OF YOUR NEIGHBOURHOOD.

Figure 3 Drugs can seem appealing, exciting and fun when you first take them. Many people use them for a while and then leave them behind. But the risks of addiction are high, and no-one can predict who drugs will affect most.

SUMMARY QUESTIONS

1 Copy and complete using the words below:

mind cocaine ecstasy legal alcohol drug body

A …… alters the way in which your body works. It can affect the ……the …… or both. Some drugs are …… e.g. caffeine and …… . Other drugs, such as ……, …… and heroin are illegal.

2 a) Why do people often need more and more of a drug?
b) What happens if you stop taking a drug when you are addicted to it?

3 a) Look at Figure 2. Explain what this tells you about drug-taking in young people.
b) Why do people take drugs?
c) Explain some of the problems linked with using cannabis, cocaine and heroin.
d) Why do you think young people continue to take these drugs when they are well aware of the dangers?
e) Why does the impact of drug use vary from person to person?

KEY POINTS

1. Drugs change the chemical processes in your body, so you may become addicted to them.
2. Alcohol, tobacco and illegal drugs may harm your body.
3. Smoking cannabis may cause psychological problems.
4. Hard drugs, such as cocaine and heroin, are very addictive and can cause serious health problems.

49

SUMMARY ANSWERS

1 Drug, mind, body, legal, alcohol, cocaine, ecstasy.

2 **a)** Because they become addicted and so they cannot function properly without it.
b) You suffer withdrawal symptoms/feel ill.

3 **a)** With increase in age, more and more young people take drugs. By the age of 15, almost a third (30%) of young people have tried drugs.
b) Drugs make them feel good; they feel they can cope with problems and fit in with the crowd.
c) Psychological damage; damage to the brain, liver, lungs and heart; serious addiction; death.
d) Young people become addicted; they daren't be different from their mates; etc.
e) Different people may be more prone to addiction and/or psychological problems.

Answers to in-text questions

a) People who have traditionally inhabited a region since ancient times.

b) Any appropriate answers, such as alcohol and tobacco are legal, but cocaine and heroin are not.

c) 'Addiction' means that you cannot function properly without the drug and suffer withdrawal symptoms if you are without it.

KEY POINTS

The teaching of the key points can be reinforced by using the summary questions as a homework exercise. These are fairly general points which are expanded in subsequent spreads.

B1a 3.2 Legal and illegal drugs

LEARNING OBJECTIVES

Students should learn that:

- Many recreational drugs affect the brain and the nervous system
- Drugs may be used recreationally and some are more harmful than others.
- Some drugs are illegal and others legal; the overall impact of legal drugs on health is greater than that of illegal drugs.

LEARNING OUTCOMES

Most students should be able to:

- List some common legal and illegal drugs.
- Explain why some people use illegal drugs for recreation.
- Explain why the use of legal drugs has a greater impact on health than the use of illegal drugs.

Some students should also be able to:

- Evaluate the different types of drugs and their effects.
- Evaluate the impact of different types of drugs on people.

Teaching suggestions

- **Gifted and talented.** These students could find out about the drugs used for doping horses and dogs. Are they the same as those used by athletes?
- **Special needs.** These students could be given the names of the different recreational drugs and asked to place them in the correct categories under the headings 'Legal' and 'Illegal'.
- **Learning styles**

  Interpersonal: Working in groups in the practicals.

  Intrapersonal: Reflecting on the issues involved in taking illegal drugs.

  Kinaesthetic: Practical work on reaction times.

  Visual: Looking at the PowerPoint® presentation.

  Auditory: Taking part in the debate about which drugs are most dangerous.
- **Homework.** Write up the accounts of the experiments and present the data in suitable tables.

SPECIFICATION LINK-UP B1a.11.3

- ? *Some people use drugs recreationally. Some of these recreational drugs are more harmful than others. Some of these drugs are legal, some illegal.*
- *The overall impact of legal drugs on health is much greater than the impact of illegal drugs, because far more people use them.*

Students should use their skills, knowledge and understanding of 'How Science Works':

- *to evaluate the different types of drugs and why some people use illegal drugs for recreation.*

Lesson structure

STARTER

Legal or illegal? – Quick quiz on which drugs are legal and which are illegal. This could be done using 'flash cards' with the names of different drugs on them. Hold these up in front of the class and ask the students to indicate which are legal (hold up an 'L') and which illegal (hold up an 'I') or write 'L' or 'I' on 'show me' boards. (5–10 minutes)

How much caffeine is there in . . .? – Prepare a number of samples of everyday beverages (tea, coffee, chocolate, decaff coffee, sports drink, cola etc.) and ask students to place them in order according to how much caffeine they contain. Discuss the results, pointing out that some people may be taking in more caffeine than they think. (10 minutes)

MAIN

- **Effect of caffeine on reaction times** – It would be possible to use one of the experiments described on the previous spread. There is an opportunity here to try a different approach: for example, if the stick drop test was used, then try the memory test or time reactions to a visual or auditory stimulus.
- **Effect of caffeine on heart rate** – Changes in heart rate (measured by measuring the pulse rate) can be recorded before and after caffeine intake. Is there any difference? What variables need to be controlled?
- 'How Science Works' concepts can be introduced. Techniques will have been learnt from the previous experiments and students will be familiar with handling the variables. They can evaluate the reliability and validity of their investigation.
- There is an opportunity to extend one of the practicals previously described by using one of the extension ideas from the previous spread, e.g. using different quantities of caffeine.
- **Drugs in sport** – PowerPoint® presentation on the use of drugs in sport, to include drugs which build up muscle, ways of making the body produce more blood, speeding up reactions and making competitors more alert. The presentation could include some case histories, frequency and details of testing for a range of different sports.
- ? The presentation could be accompanied by articles from newspapers relating to the illegal use of performance-enhancing drugs. Students could discuss the issues and express their opinions. What about the use of caffeine as a stimulant? Should everyone be tested rather than random tests?

PLENARIES

Collation of results – It would be worthwhile gathering together the results of the experiments done in the first two spreads of this chapter. The results could be discussed and some possible conclusions drawn up. (10 minutes)

What have we learnt about the effects of drugs? – Give students a list of the drugs mentioned in this spread and ask which ones speed up the activity of the brain, making people feel more alert, and which slow down brain activity, making people feel calmer. (5–10 minutes)

ACTIVITY & EXTENSION IDEAS

- **Debate the issue.** Which are more dangerous – legal or illegal drugs? This question could be debated by the class. Each student could prepare a short case for and against and then the motion. Draw lots to decide which students speak.

BIOLOGY DRUG ABUSE

B1a 3.2 Legal and illegal drugs

LEARNING OBJECTIVES

1. How do drugs like caffeine and heroin affect your nervous system?
2. Why do people use illegal drugs?
3. Which are more dangerous – legal or illegal drugs?

What is the most widely used drug in the world? It is probably one that most of you will have used at least once today, yet no-one really thinks about. The caffeine in your cup of tea, mug of coffee or can of cola is a drug.

Many people find it hard to get going in the morning without their first mug of coffee – they are probably addicted to the drug! Caffeine is a mild stimulant. It stimulates your brain and increases your heart rate and blood pressure.

a) What makes a can of cola a drug?

Figure 1 Even everyday drugs, like the caffeine we take for granted, have an effect on your nervous system and brain

How do drugs affect you?

Many of the drugs used for medical treatments have little or no effect on your nervous system. However, all of the drugs which people use for pleasure (see table on page 51) affect the way your brain and nervous system work. It is these changes which people enjoy when they use the drugs. The same changes can cause addiction, so your body doesn't work properly without the drug.

Some drugs like caffeine, nicotine and cocaine speed up the activity of your brain. They make you feel more alert and energetic.

Others like alcohol and cannabis slow down the responses of your brain, making you feel calm and better able to cope. Heroin actually stops impulses travelling in your nervous system, so you don't feel any pain or discomfort. Other drugs like cannabis produce vivid waking dreams. You see or hear things which are not really there.

Why do people use drugs?

People use these drugs for a variety of reasons. For example, people feel that caffeine, nicotine and alcohol help them cope with everyday life. Few people who use these drugs would think of themselves as addicts, yet the chemicals can have a big impact on your brain (see Figure 2).

As for the other recreational drugs – people who try them may be looking for excitement or escape. They might want to be part of the crowd or just want to see what happens. Yet because many of these drugs are addictive, you don't have to try them many times before your body starts to demand a regular fix!

Figure 2 NASA scientists have shown that common house spiders spin their webs very differently when given some of the commonly used drugs shown in the table on page 51. The effect of caffeine on the nervous system of a spider is particularly dramatic!

50

Human biology

Some of these recreational drugs are more harmful than others. Most media reports on the dangers of drugs use focus on illegal drugs (see table below). But in fact the impact of legal drugs on health is much greater than the impact of illegal drugs. That's because far more people take them. Millions of people in the UK smoke – but only a few thousand take heroin.

b) Why do legal drugs cause many more health problems than illegal drugs?

Legal recreational drugs	Illegal recreational drugs
Ethanol (alcoholic drinks)	cannabis
Nicotine (cigarette smoke)	cocaine
Caffeine (coffee, tea, cola etc)	heroin
	ecstasy
	LSD

Drugs in sport

The world of sport has a major problem with the illegal use of drugs. In theory competition in sport is to find the best natural athlete. The only difference between the competitors should be their natural ability and the amount they train. However, there are many drugs which can enhance your performance in sport – and sadly some athletes use them and cheat. Drugs can build up your muscle mass, make you body produce more blood, make you more alert and speed up your reactions.

The sports authorities produce new tests for drugs and run random drugs tests to try and identify the cheats. Athletes are banned from competing if they are discovered using illegal drugs. But competitors are always looking for new ways to get ahead, so the illegal use of drugs in sport continues.

SCIENCE @ WORK

Many people are involved in developing new drugs tests, carrying out random tests on athletes and analysing the results. You need science qualifications to be involved!

FOUL FACTS

There are drugs that can be given to horses and dogs to make them perform better – or worse – than expected. Some people use these to make a lot of money through betting on a result they already know!

Figure 3 At the 2000 Summer Olympic Games in Sydney, Australia, the Romanian gymnast Andrea Raducan won a gold medal. It was taken away when she tested positive for a banned stimulant.

SUMMARY QUESTIONS

1. Copy and complete using the words below:

   brain health illegal legal recreation

   Drugs which people use for ...... all affect the ...... and nervous system. Some of these drugs are legal but some of them are ...... More people suffer ...... problems caused by the ...... drugs than illegal ones.

2. a) What are the main reasons for using illegal drugs?
   b) Plan a TV advert against the use of illegal substances in sport to be shown in the run-up to the 2012 Olympics in the UK.

3. Compare the impact of legal and illegal drugs on individuals and on society.

KEY POINTS

1. Many recreational drugs affect the brain and nervous system.
2. Some recreational drugs are legal and others are illegal.
3. The overall impact of legal drugs on health is much greater than illegal drugs because more people use them.

51

SUMMARY ANSWERS

1. Recreation, brain, illegal, health, legal.
2. a) Help cope with life, to calm down, to enjoy the effects, addiction, etc.
   b) Outline of advert.
3. Look for recognition that legal drugs cost society more because of the huge numbers of people adversely affected by them.

Answers to in-text questions

a) It contains caffeine.

b) Because many more people take (and abuse) legal drugs.

KEY POINTS

These key points can be emphasised by the addition of specific examples. For example, it is useful for students to know how specific drugs affect the brain and nervous system and which drugs are legal and which illegal.

B1a 3.3 Alcohol – the acceptable drug?

LEARNING OBJECTIVES

Students should learn that:

- Alcohol affects the body by slowing down reactions.
- It may lead to lack of self-control, unconsciousness or coma.
- It may cause damage to body organs, particularly the liver and brain.

LEARNING OUTCOMES

Most students should be able to:

- Describe how alcohol affects body organs such as the liver and the brain.
- Explain what is meant by the term 'binge drinking'.
- Recognise some of the effects of drinking on society.

Some students should also be able to:

- Analyse data showing the impact of drink on behaviour.
- Evaluate the impact of alcohol as a legal drug on society.

Teaching suggestions

- **Gifted and talented.** Students could compose letters of complaint from body parts to the individual about what the alcohol he/she drinks is doing to them.
- **Special needs.** Students could play a dominoes-style game matching the effects of alcohol with associated problems (e.g. 'damages liver' goes with 'cirrhosis', 'slows reaction time' goes with 'dangerous to drive', etc.).
- **Learning styles**

  Interpersonal: Collaborating and discussing while compiling lists.

  Intrapersonal: Deducing how much has been drunk at a party.

  Kinaesthetic: Investigating the alcoholic content of drinks.

  Visual: Seeing the effects of drinking alcohol.

  Auditory: Explaining the alcohol content of drinks and the legal limits.
- **Homework.** Write a letter to Homer Simpson showing concern over the amount of 'Duff' beer he drinks.

SPECIFICATION LINK-UP B1a.11.3

- *Alcohol affects the nervous system by slowing down reactions and helps people relax, but too much may lead to lack of self-control, unconsciousness or even coma, eventually damaging the liver and brain.*

Lesson structure

STARTER

Drinking and driving – The Department for Transport has a road safety web site entitled *Think! Campaign*, which has downloadable videos and also posters, both of which could be used for a discussion starter. See www.dft.gov.uk. (10–15 minutes)

What is in the bottle? – Prepare a large bottle (Winchester or similar from the Chemistry department) with labels such as 'Toxic', 'If taken in large quantities causes paralysis', and other dire warnings. Students are asked to guess what it is. Have alcohol of some sort available for students to sniff. (Care needed here in view of solvent abuse!) (5 minutes)

I've only had a couple of drinks . . . – Get students to compile a list of ways in which they think they could tell if someone had drunk too much. Some care needed here as this could be a sensitive issue; it could lead to stories of how drinking too much has affected them or someone they know. (10 minutes)

MAIN

- **Data analysis** – Provide sheets of data about drink-related problems for the students to break down and analyse. Excel spreadsheets are available from government sources (National Statistics Online – see www.statistics.gov.uk).
- **Effects of drinking alcohol on the body** – Show a video of a volunteer taking a reaction test before and after drinking alcohol. Does it slow them down? Do they appear to be clumsy, vague or muddled?
- Combine the video with slides or a PowerPoint® presentation on alcohol (download from ASE Science Year CD at www.sycd.co.uk). You can consider the physical effects, such as brain damage, cirrhosis of the liver, accidents and lack of inhibition. In addition, a flow chart of how the alcohol is absorbed and where it goes could be built up using an OHP presentation.
- **Alcohol content of drinks** – Into some empty beer, strong beer/cider, wine, fortified wine and spirit bottles pour water equivalent to the alcohol content of the beverage. For example, 4% of the volume of the beer bottle, 12% of the volume of the wine bottle and so on. Follow this up with a demonstration of quantities of each which represent one unit. A common misconception is that a pint of beer is equivalent to a glass of wine or a measure of spirits – it is not: it is twice as much.
- Search the web for pictures of people at a party and, for each person, list the quantity and type of alcoholic beverage they have consumed. Ask the students to calculate how much (how many units) of alcohol each person has consumed. Assuming that alcohol is broken down in the liver at the rate of one unit per hour, calculate how long it will be before each person is safe to drive.
- This would be a good point to discuss how quickly alcohol gets into the blood and what the legal limits are. The detection of the presence of alcohol in the breath, the blood and the urine (things that can be tested) can be described and explained.

ACTIVITY & EXTENSION

- To support the 'Foul fact' information, show footage of Jimi Hendrix (search the web for Jimi Hendrix video). Play very loudly and stop it abruptly. Discuss how he died [inhalation of vomit] and how to avoid it happening to you. Show what to do, with a trousered volunteer on a clean floor, if you find someone unconscious from drink at a party.
- Show a video of giraffes or elephants after eating fermented fruit. Why don't they get alcohol problems (apart from short term ones!)?
- Invite a guest speaker from Alcoholics Anonymous, or a similar organisation, to give a talk about the problems of alcohol addiction.
- There is some evidence that certain groups of people lack the enzymes needed to break down alcohol in the liver. What effect would alcoholic drinks have on these people?

Answers to in-text questions

a) Alcohol affects the body and is addictive.

b) Unable to judge distances accurately or responding too slowly in an emergency when driving.

c) The brain and the liver.

BIOLOGY DRUG ABUSE

B1a 3.3 Alcohol – the acceptable drug?

LEARNING OBJECTIVES

1 How does alcohol affect your body?
2 What is binge drinking?

Figure 1 You can buy alcoholic drinks like these legally in the UK once you are 18. But you have to be over 21 in many places in the USA and they are completely illegal in some other countries.

For many people **alcohol** is part of their social life. They like to share a drink with friends or enjoy a glass of wine with a meal. They probably don't think of themselves as drug users.

In small amounts, alcohol makes people feel relaxed and cheerful. It makes you less inhibited. So shy people can feel more confident when they've had an alcoholic drink.

But alcohol has a powerful effect on your body. It is very addictive and it is also very poisonous. Just imagine if alcohol was discovered today. It would almost certainly be illegal and thought of as a very dangerous drug.

In fact, alcohol is one of the most widely used drugs in the UK. Although some religions ban the use of alcohol, it is accepted all over the world. Perhaps this is because alcohol has been around for thousands of years. We also see that many important and famous people like a drink!

a) Why is alcohol described as a drug?

How does alcohol affect your body?

Alcohol is poisonous. However, your liver can usually break it down. Your liver gets rid of the alcohol before it causes permanent damage and death.

When you have an alcoholic drink, the alcohol passes through the wall of your gut and goes into your bloodstream. From your blood, the alcohol passes easily into nearly every tissue of your body. It gets into your nervous system and brain. This slows down your reactions. It can make you lose your self-control.

When you have had too much to drink, you lack judgement. You can end up making stupid or dangerous decisions. Some people end up making mistakes they regret for the rest of their lives.

If you drink large amounts of alcohol, like a whole bottle of spirits, your liver simply cannot cope. You would suffer from alcohol poisoning. This can quickly lead to unconsciousness, coma and death.

b) Give an example of a poor decision that someone under the influence of alcohol might make.

Some people drink heavily for many years, becoming **alcoholics**. They are addicted to the drug. Their liver and brain suffer long-term damage and eventually the drink may kill them.

They may develop **cirrhosis of the liver**. This disease destroys your liver tissue. They can also get ***liver cancer***, which spreads quickly and can be fatal. In some heavy drinkers their brain is so damaged (it becomes soft and pulpy) that it can't work any longer. This causes death.

Short bouts of very heavy drinking can cause the same symptoms to develop quite quickly.

c) Which organs are most affected by heavy drinking?

DID YOU KNOW?

It takes your liver about one hour to break down the alcohol in a glass of wine or half a pint of beer (one unit of alcohol).

52

Human biology

The effects of drinking on society

Alcohol can also put you at risk because of the way you behave under the influence of the drug. Because alcohol slows down your reactions, you are much more likely to have an accident. This is very dangerous if you drive after drinking. Alcohol is a factor in about 20% of all fatal road accidents in the UK.

Alcohol abuse affects personal lives as well. Domestic violence is often linked to patterns of heavy drinking. Many crimes take place when people are under the influence of alcohol, often mixed with other drugs.

Binge drinking is a recent problem. This often involves young people. They go out and get very drunk several nights a week. They become violent and abusive, damage property and put their own health at risk.

Alcohol related crime in the UK costs us around 20 billion pounds a year and causes great unhappiness. Add to this the medical costs of alcohol abuse and you see that we all pay a high price for this socially acceptable drug!

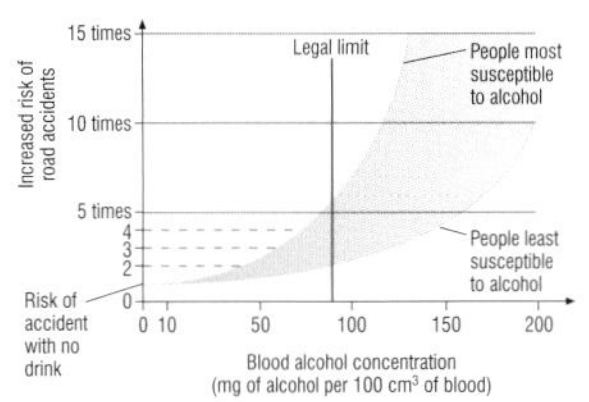

Figure 3 Alcohol affects your brain and your reactions – it is not surprising that if you drink alcohol and then drive a car you are much more likely to have an accident

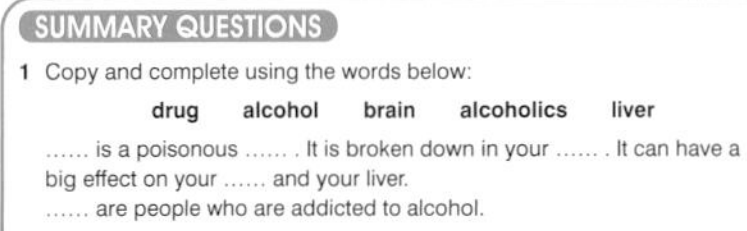

SUMMARY QUESTIONS

1 Copy and complete using the words below:

drug alcohol brain alcoholics liver

...... is a poisonous .... It is broken down in your .... It can have a big effect on your ...... and your liver.
...... are people who are addicted to alcohol.

2 a) How does alcohol reach your brain after you have had a drink?
b) What effect does alcohol have on your brain?

3 Look at Figure 3.
a) What is the approximate legal limit of alcohol that you are allowed in your blood before you drive a car?
b) Young people are often easily affected by alcohol. If a young person drinks enough to have 125 mg of alcohol per 100 cm³ of their blood and then drives a car, how will this affect their risk of having an accident?
c) Police advise you not to drink alcohol at all if you are going to drive a car. Based on the evidence of the graph, explain why this is good advice?
d) Summarise the effects of alcohol on society and discuss banning its use.

FOUL FACTS

There are around 28 000 alcohol related deaths a year in England alone. Young people can die of acute alcohol poisoning – or they may just become unconscious and choke to death on their own vomit while they sleep.

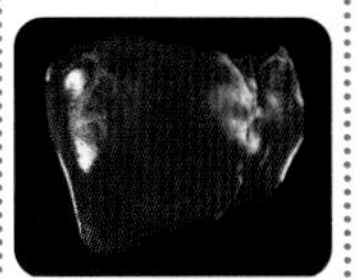

A healthy liver

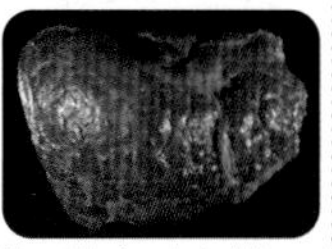

Diseased liver from a heavy drinker with cirrhosis

Figure 2 Your liver deals with all the poisons you put into your body. But if you drink too much alcohol, your liver may not be able to cope. The difference between the healthy liver and the liver with cirrhosis shows just why people are warned against heavy drinking!

KEY POINTS

1 Alcohol affects your nervous system by slowing down your reactions.
2 Alcohol can lead to loss of self-control, unconsciousness, coma and death.
3 Alcohol can cause damage to your liver and brain.

PLENARIES

Cost/benefit sheet – Ask the students to draw up a sheet showing the pros and cons of drinking alcohol. (10 minutes)

Loop game – Prepare a set of cards with questions and answers on them of alcohol facts and figures, but the answer on each card does not match its question. Give out to students and start off with a student asking a question. The student who has the answer on their card then asks the next question. This should work out that the student who started the game will have the answer to the last question. More information can be found on Science Year CD web site. Time it for a competition next lesson. (5 minutes)

Recommended limits – Introduce the idea of the weekly recommended maximum of 14 units of alcohol for females and 21 for males. Why is there a difference? This could lead to a short discussion of the problems of binge drinking. (10–15 minutes)

SUMMARY ANSWERS

1 Alcohol, drug, liver, brain, alcoholics.

2 a) It is absorbed through the gut wall into the blood stream and carried around to all the tissues of the body (including the brain) in the blood.

b) One is about 15 times more likely to have an accident.

3 a) 85–90 mg of alcohol per 100 cm³ of blood.

b) The risk increases by more than a factor of 10.

c) Almost everyone is more likely to have an accident, even after a small amount of alcohol. They are more susceptible/five times more at risk of an accident by the time they are at the legal limit; even the least susceptible have slightly raised risk. Most people do not know how easily they are affected. It is safest if no-one drinks and drives.

d) Look for arguments for and against the banning of alcohol, i.e. a balanced argument.

KEY POINTS

These are reinforced by playing the 'loop game' and drawing up a cost/benefit sheet as suggested in the plenaries.

B1a 3.4 Smoking and health

LEARNING OBJECTIVES

Students should learn that:

- Tobacco smoke contains nicotine, which is addictive, and carcinogens.
- The carbon monoxide in tobacco smoke reduces the oxygen-carrying capacity of the blood.

LEARNING OUTCOMES

Students should be able to:

- Describe how tobacco smoke affects the lungs.
- Link the substances in tobacco smoke with diseases of the lungs.
- Explain the effect of carbon monoxide on the oxygen-carrying capacity of the blood.
- Explain the risks of smoking during pregnancy.

Some students should also be able to:

- Analyse data on the health risks of smoking.
- Evaluate different methods of giving up smoking.

Teaching suggestions

- **Gifted and talented.** Students could research the constituents of cigarette smoke and make a bar chart to show the proportions of the main ones, such as nicotine, tar and carbon monoxide.
- **Special needs.** Students could be given a word search with key words to find.
- **Learning styles**

  Interpersonal: Discussing pro- and anti-smoking.

  Intrapersonal: Interpreting pictures of the lungs.

  Kinaesthetic: Investigating the tar content of cigarettes.

  Visual: Following the sequence of the effect of smoke on the respiratory tract.

  Auditory: Defining passive smoking.

Answers to in-text questions

a) Approximately 4000 chemicals.

b) Nicotine.

c) **i)** Pink. **ii)** Grey.

d) Black tar from cigarette smoke builds up in the lungs of the smoker turning them from pink to grey.

SPECIFICATION LINK-UP B1a.11.3

- *Nicotine is the addictive substance in tobacco smoke. Tobacco smoke contains carcinogens.*
- *Tobacco smoke also contains carbon monoxide which reduces the oxygen-carrying capacity of the blood. In pregnant women this can deprive a fetus of oxygen and lead to a low birth mass.*

Lesson structure

STARTER

Anti-smoking or pro-smoking? – Write, or project, on to the board a number of statements about smoking. There should be a mixture of anti-smoking and pro-smoking statements. Ask students to allocate each statement a mark from 1 (strongly agree) to 5 (strongly disagree). Using numbered lines on the floor or stickers on the wall or board, students could line up to form human bar charts for each statement. This quickly shows the group response to the statements. (10 minutes)

What I think of smoking, in three words only – Choose one student to begin with their three-word statement about smoking; that student then chooses the next one and so on until the range of opinions runs out. (5–10 minutes)

MAIN

The suggestions for the main part of the lesson depend very much on what has already been presented to the students in other lessons such as PSHE.

- **The tar content of a cigarette** – This is a practical demonstration to show the tar content in a cigarette and can be done by attaching a lighted cigarette to a 'smoking machine' (drawing air through and catching the tar on a filter). This should be done in a fume cupboard so that no smoke affects students. Compare the tar deposits from a filter-tipped cigarette and a cigarette without a filter tip. Start a discussion on the benefits of filter tips. Widen the discussion to compare pipe smoking and cigars.
- Show students pictures of lungs from smokers and lungs from non-smokers and ask them to discuss the relationship between the pictures and the results of the demonstration of the tar content of the cigarettes.
- **Carbon monoxide in cigarette smoke** – Introduce the students to carbon monoxide detectors and consider why and where they should be placed in the home. It would be good to have some examples to show or to discuss/demonstrate the way in which they work (commercially available pellets which discolour when in contact with CO). How sensitive are they? Does a detector react to the quantity of carbon monoxide in a single cigarette?
- **The effects of cigarette smoke on the respiratory tract** – Use a PowerPoint® presentation on the way in which the cilia in the respiratory tract work. This should include the nose, trachea, etc. Discuss how they work normally and then discuss what happens if they are paralysed. (See www.sycd.co.uk for presentation.)
- Show videos of the effects of smoking-related conditions, such as emphysema and bronchitis. Search the web for 'smoking emphysema video'. These could be linked with statistics on the number of working days lost through illness with these conditions, the cost to employers and the cost to the NHS.

PLENARIES

'Diamond nines' cards – Write out nine statements regarding smoking and health on to cards. The students have to arrange these cards in order, with the most important statement at the top of the diamond down to the least important statement at the bottom. They must be prepared to justify their positioning. (10 minutes)

Human bar chart activity – Re-run the starter to establish if any changes of opinion have occurred. (5 minutes)

Passive smoking: what is it and how dangerous is it? – Ask students to define 'passive smoking' and then to assess the risks of tar and carbon monoxide from other people's cigarettes on their own health. (5–10 minutes)

ACTIVITY & EXTENSION

- Reference is made in 'Next time you . . .' to the 4000 chemicals in cigarette smoke. What are these chemicals? Students could carry out some research of their own using the Internet.
- Evaluate some of the advertisements that have been shown to discourage people from smoking (thinking particularly of the one showing the fatty deposits, or the ones showing distressed sufferers or the children of smokers). [Care is needed here if any students are directly affected.]
- Show students what one pound/half a kilogram of tar looks like. It should be possible to work out the thickness of the layer coating the air sacs if you divide by the area of the lungs!
- Smokers who give up smoking often comment on how much better their food tastes. Design an experiment to find out if smoking affects the taste buds or the sense of smell. Remember to use controls and to take into consideration age and number of cigarettes smoked per day.

GET IT RIGHT!

The names of the conditions caused by smoking should be known and preferably spelt correctly. The description of the effect of cigarette smoke on the cilia (small hairs) is also important and would be required to gain high marks.

KEY POINTS

Students could use ICT to design posters on the key points that could be displayed around the school as part of an anti-smoking campaign.

BIOLOGY DRUG ABUSE

B1a 3.4 Smoking and health

LEARNING OBJECTIVES

1. How does smoking tobacco affect your health?
2. How do pregnant women put their babies at risk by smoking?

Smoking is big business. There are 1.1 billion smokers worldwide, smoking around 6000 billion cigarettes each year!

As a cigarette burns it produces a smoke cocktail of around 4000 chemicals. If you smoke, you breathe this cocktail straight into your lungs. You absorb some of the chemicals into your blood stream. This carries them around your body to your brain.

a) How many chemicals do you inhale in cigarette smoke?

Nicotine is the addictive drug in ***tobacco smoke***. It makes people feel calm, contented and able to cope. But you gradually need more and more of the drug to get the same effect. So the number of cigarettes you smoke each day tends to increase over the years. What's more, you feel awful if you don't get your regular dose of nicotine.

b) What is the drug in tobacco smoke?

Figure 1 Cigarette smoking increases your risk of developing many serious and fatal diseases. Every packet of cigarettes sold in the UK has to carry a clear health warning. Yet people still buy them in their millions!

Smoking related diseases

If you are a non-smoker, small hairs in your breathing system are constantly moving mucus away from your lungs. The mucus traps dirt, dust and bacteria from the air you breathe in. The hairs make sure you get rid of it all.

If you smoke, each cigarette anaesthetises these hairs. They stop working for a time, allowing dirt down into your lungs. This makes you much more likely to suffer from colds and other infections. The mucus also builds up and causes coughing (smoker's cough!).

Tar is a sticky black chemical in tobacco smoke that builds up in your lungs, turning them from pink to grey. It makes smokers much more likely to develop bronchitis. The build-up of tar in your lungs can also lead to the delicate air sacs in the lungs breaking down. We call this ***emphysema***. It makes the lungs much less efficient. Your breathing becomes difficult and you can't get enough oxygen.

c) What colour are the lungs of i) a non-smoker? ii) a smoker?
d) What causes the difference in colour in c)ii)?

NEXT TIME YOU...

. . . see someone lighting up a cigarette, think of the 4000 chemicals they are drawing down into their lungs – and the damage that some of them can do!

Tar is also a major **carcinogen** (a cancer causing substance). A build-up of tar can cause lung cancer. We can cure this cancer if it is caught early enough. However, it often grows in the lungs with no obvious symptoms. By the time doctors diagnose it, it has spread to other parts of the body and has become fatal. There are many other carcinogens in tobacco smoke as well.

Many of the chemicals in cigarette smoke are carried right round your body in the blood. Some of them affect your heart and blood vessels. Smoking raises your blood pressure and makes it more likely that your blood vessels will become blocked. This can cause heart attacks, strokes and thrombosis.

FOUL FACTS

Smoking 20 to 60 cigarettes a day will coat your lungs in 1 to 1.5 pounds of tar every year!

54

Human biology

Many people want to give up smoking but because nicotine is so addictive, it isn't easy. There are many different ways of giving up. Some are more effective than others. Some are much more expensive than others. The most important thing is always how much you want to give up. Each smoker has to find the method that suits them best and helps them to become a non-smoker!

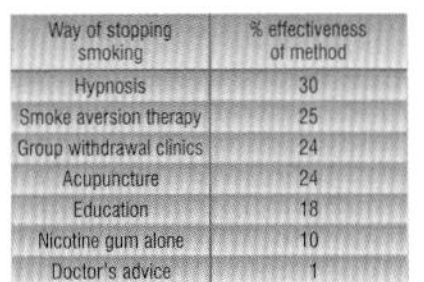

Way of stopping smoking	% effectiveness of method
Hypnosis	30
Smoke aversion therapy	25
Group withdrawal clinics	24
Acupuncture	24
Education	18
Nicotine gum alone	10
Doctor's advice	1

Figure 2 Many people try to give up smoking, but because nicotine is so addictive it isn't easy. There are lots of different methods which can help you – some seem to be more effective than others!

GET IT RIGHT!

Be specific when you describe the effects of smoking on the body.

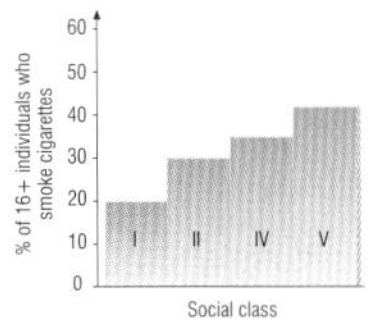

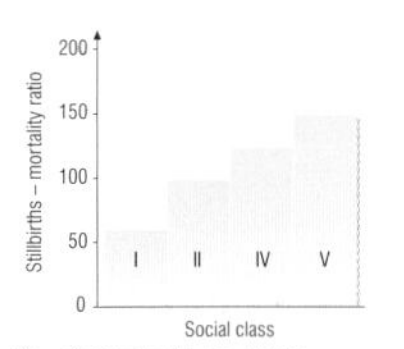

Figure 3 Data like this show that the people who smoke more are more likely to have stillborn babies. The same pattern is there for babies who are born weighing less than they should.

Smoking and pregnancy

Carbon monoxide is a very poisonous gas found in cigarette smoke. It is picked up by your red blood cells. This reduces the amount of oxygen carried in your blood. After smoking a cigarette, up to 10% of a smoker's blood will be carrying carbon monoxide rather than oxygen! This is one reason why smokers often get breathless going up the stairs!

Carbon monoxide in cigarette smoke affects pregnant women in particular. During pregnancy a woman needs oxygen, not just for her own cells but for her developing fetus as well.

If she smokes, the amount of oxygen in her blood will be lower than normal. This means her fetus will be deprived of oxygen. Then it may not grow as well as it should.

Mums who smoke when they are pregnant have a much higher risk of having:

- a premature birth (baby born too early so may struggle to survive),
- a baby with a low birth mass (so it is more at risk of developing problems),
- a stillbirth (where the baby is born dead).

SUMMARY QUESTIONS

1. Define the following words: a) **nicotine**; b) **tar**; c) **carbon monoxide**; d) **carcinogenic**
2. Smokers are more likely to get infections of their breathing tubes and lungs than non-smokers. Explain why this is the case.
3. Explain how cigarette smoking is linked to an increased risk of getting:
   a) lung cancer b) emphysema c) heart disease.
4. a) Look at Figure 3. What is the evidence that smoking during pregnancy can be dangerous for the developing baby?
   b) Explain how smoking causes these problems during pregnancy.
   c) How do scientists try to make sure that the correlations they spot in their data are as reliable as possible?
   d) Carry out some research to find two ways of helping smokers kick the habit. Compare and contrast each method.

KEY POINTS

1. Tobacco smoke contains substances which can help to cause lung cancer, lung diseases such as emphysema and bronchitis, and diseases of the heart and blood vessels.
2. Tobacco smoke contains carbon monoxide, which reduces the oxygen-carrying capacity of the blood. In pregnant women this can deprive a fetus of oxygen and lead to a low birth mass or death.

SUMMARY ANSWERS

1. **Nicotine:** the drug found in tobacco smoke.
   Tar: sticky black (carcinogenic) chemical found in tobacco smoke.
   Carbon monoxide: poisonous gas found in tobacco smoke.
   Carcinogenic: cancer-causing.
2. Small hairs in your breathing system are constantly moving mucus away from your lungs. The mucus traps dirt, dust and bacteria from the air you breathe in, and the hairs make sure you get rid of it all. If you smoke, these hairs are anaesthetised by each cigarette. They stop working for a time, allowing dirt down into the lungs so you are much more likely to suffer from colds and other infections.
3. **a)** Tar coats the lungs; tar is carcinogenic so it causes lung cancer to develop.
   b) Tar coats the lungs, so it damages the delicate structure of the air sacs which then cannot work properly.
   c) Chemicals carried in the blood cause raised blood pressure and an increased chance of blood vessels blocking up, so there is an increased risk of heart attack.
4. **a)** The graph shows the greater the number of cigarettes smoked by the mother, the higher the chance that the baby will be born dead.
   b) If a woman smokes, carbon monoxide from the smoke is picked up by her red blood cells, reducing the oxygen-carrying capacity of her blood by up to 10%, so the amount of oxygen in her blood will be lower than normal. This means that her fetus will be deprived of oxygen, so that it may not grow as well as it should.
   c) They collect data from large samples of people to increase reliability.
   d) Comparison of methods, e.g. nicotine patches/hypnosis.

B1a 3.5 Lung cancer and smoking

SPECIFICATION LINK-UP

B1a.11.3

Students should use their skills, knowledge and understanding of 'How Science Works':

- *to evaluate how the link between smoking tobacco and lung cancer gradually became accepted*
- *to evaluate different ways of trying to stop smoking.*

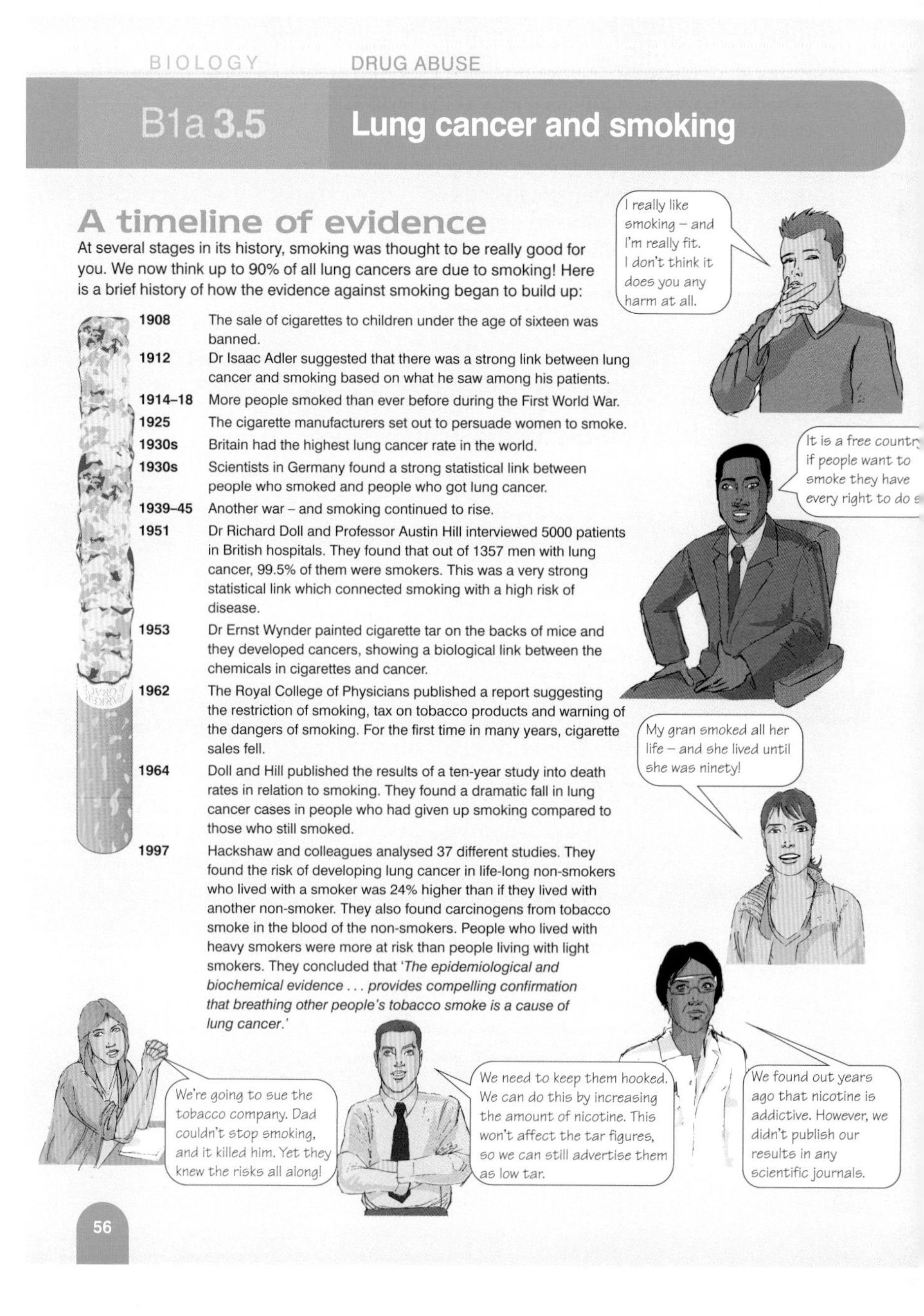

BIOLOGY DRUG ABUSE

B1a 3.5 Lung cancer and smoking

A timeline of evidence

At several stages in its history, smoking was thought to be really good for you. We now think up to 90% of all lung cancers are due to smoking! Here is a brief history of how the evidence against smoking began to build up:

1908	The sale of cigarettes to children under the age of sixteen was banned.
1912	Dr Isaac Adler suggested that there was a strong link between lung cancer and smoking based on what he saw among his patients.
1914–18	More people smoked than ever before during the First World War.
1925	The cigarette manufacturers set out to persuade women to smoke.
1930s	Britain had the highest lung cancer rate in the world.
1930s	Scientists in Germany found a strong statistical link between people who smoked and people who got lung cancer.
1939–45	Another war – and smoking continued to rise.
1951	Dr Richard Doll and Professor Austin Hill interviewed 5000 patients in British hospitals. They found that out of 1357 men with lung cancer, 99.5% of them were smokers. This was a very strong statistical link which connected smoking with a high risk of disease.
1953	Dr Ernst Wynder painted cigarette tar on the backs of mice and they developed cancers, showing a biological link between the chemicals in cigarettes and cancer.
1962	The Royal College of Physicians published a report suggesting the restriction of smoking, tax on tobacco products and warning of the dangers of smoking. For the first time in many years, cigarette sales fell.
1964	Doll and Hill published the results of a ten-year study into death rates in relation to smoking. They found a dramatic fall in lung cancer cases in people who had given up smoking compared to those who still smoked.
1997	Hackshaw and colleagues analysed 37 different studies. They found the risk of developing lung cancer in life-long non-smokers who lived with a smoker was 24% higher than if they lived with another non-smoker. They also found carcinogens from tobacco smoke in the blood of the non-smokers. People who lived with heavy smokers were more at risk than people living with light smokers. They concluded that '*The epidemiological and biochemical evidence . . . provides compelling confirmation that breathing other people's tobacco smoke is a cause of lung cancer.*'

56

Teaching suggestions

- **How social attitudes to smoking have changed** – The series of pictures and the timeline through the twentieth century show a variety of different attitudes to smoking. Students could do more research on some other topics. For example, did smoking tobacco originate amongst the American Indian tribes? How did Raleigh come across tobacco? Smoking was encouraged during WW1 – during both WW1 and WW2 troops received a ration of cigarettes. Soldiers who smoked were considered manly. Macho image persisted until the middle of the twentieth century. For support material, search for smoking at www.britishpathe.com. Students could compile a display of different attitudes, making posters about the ones shown in the student book and thinking of others. Include a discussion of the data shown on page 57 in the Student Book.
- **Cigarette advertisements** – If possible, display a range of different advertisements for cigarettes (this could be a PowerPoint® presentation – search the web for 'cigarette advertising'). It would be good to have a selection which includes some from the early and middle part of the twentieth century, showing women as well as men. These could be put up as posters or projected. For each one, the students should decide who the adverts are aimed at and what they are trying to say about smoking? What are the images that come across? It could be interesting to get them to play an 'Associations' game as each one is presented to them. What is the first thing that comes into their minds? This could be contrasted with the anti-smoking campaigns that are run.

An anti-smoking campaign in your school – A review of social attitudes and the images put across in posters or advertisements for cigarettes could help give the students some ideas on how to plan an anti-smoking campaign. A successful campaign needs more than one approach and also needs to be co-ordinated. This activity can involve all the students in the class, different groups being responsible for different aspects: one group could compile the

A change of image?

Tobacco arrived in Britain in the 16th century, when it was seen as a new and exciting thing. Since then it has been seen as a foul habit, the height of fashion and a calming influence during the wars. Now in the developed world, scientific evidence shows that smoking is a serious health risk. But in the developing world – and for many young people even in the UK – smoking is still seen as a glamorous and desirable habit . . .

ACTIVITY

Discuss the evidence shown here in small groups.

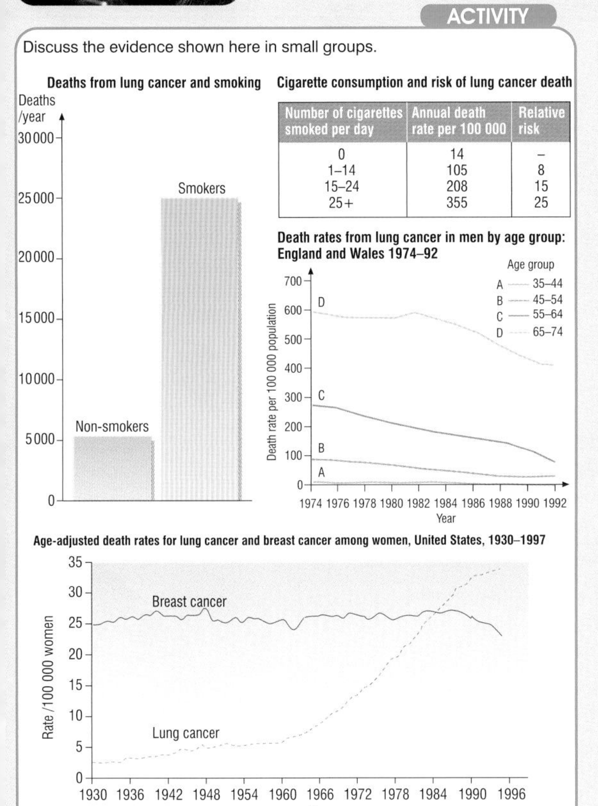

Cigarette consumption and risk of lung cancer death

Number of cigarettes smoked per day	Annual death rate per 100 000	Relative risk
0	14	–
1–14	105	8
15–24	208	15
25+	355	25

Evidence like this gradually builds up a compelling picture showing that smoking is a major risk factor in the development of lung cancer. Similar evidence can be put together linking smoking with cancers of the throat and tongue and with heart disease.

ACTIVITY

In 21st-century Britain most people accept that there is a link between smoking and lung cancer, heart disease and other health problems – but it doesn't stop them smoking! There has also been a lot in the press about ***passive smoking*** – breathing in other people's cigarette smoke. In some countries smoking has been banned in almost all public places.

Look at the evidence published by Hackshaw and colleagues in the *British Medical Journal* in 1997 along with all the other evidence here. Use it to carry out one of these two tasks:

Either: Plan an anti-smoking campaign in your school. Target teachers and pupils alike. Plan an article for the school magazine and a presentation which can be used in Citizenship lessons with year 9 pupils.

Or: Plan a campaign to have smoking banned in all public places in your local community – shops, pubs, cafes and restaurants, bars and bowling alleys. Think carefully about how to get the issues across to the general public. You might need posters, leaflets and/or a speech to make at a public meeting.

Whichever task you chose, use plenty of scientific evidence to help make people take notice!

ACTIVITY & EXTENSION IDEAS

- Asking around among the students will usually produce a broken clay pipe they have found in their garden. If they look at the bowl size, they will notice that the earlier the date of the pipe, the smaller the size of the bowl. Tobacco was a luxury item. It could then be interesting to compare clay pipes with other smoking devices, such as 'hookahs' (a Middle-eastern/Asian smoking device) and the American Indian practice of passing the peace pipe around.
- Many of the activities and related suggestions could be set as homework projects and provide opportunities for students to write in continuous prose. Another way of finding out opinions is to ask students to design questionnaires, which can be evaluated by their peers – there is an art in designing a good questionnaire so that it yields the information you need!
- It is important for the association between the smoking-related diseases to be understood and the evidence known.

article for the school magazine, another group could design a presentation for use with Year 9 students (along the lines of 'Don't even try it . . .') and a further group could consider a poster campaign.

- **No smoking in public places** – This activity could run along similar lines to the above. Students could start by finding out where smoking is allowed in the public places in the local community. It might also be interesting to investigate the attitudes of the organisations that control the smoking arrangements in the public areas. Once the background information has been gathered, it should be easier to think of ways of getting the message across.
- **Survey of smoking habits** – Students could do some research of their own on the smoking habits of their family and friends. How many have never smoked? How many have smoked and given up? How many smoke and would like to give up?
- **How easy is it to give up smoking?** – This could link up with an earlier spread on keeping healthy and what doctors can do for you. Some examples of anti-smoking devices and methods could be displayed in the laboratory. Students could research the different methods of giving up: nicotine patches; chewing gum; gradually cutting down. Find out from family and friends how they did it. How many times do people try and fail? When you stop smoking, when do you miss it most?
- **A total ban on smoking in public places?** – Some countries, such as the Republic of Ireland, have a total ban on smoking in public areas. Such a ban has been discussed for the UK, but there is not agreement on a total ban. Discuss the implications of a total ban and how it would affect places such as restaurants and pubs.

B1a 3.6 Does cannabis lead to hard drugs?

SPECIFICATION LINK-UP

B1a.11.3

Students should use their skills, knowledge and understanding of 'How Science Works':

- *to evaluate claims made about the effect of cannabis on health and the link between cannabis and addiction to hard drugs.*

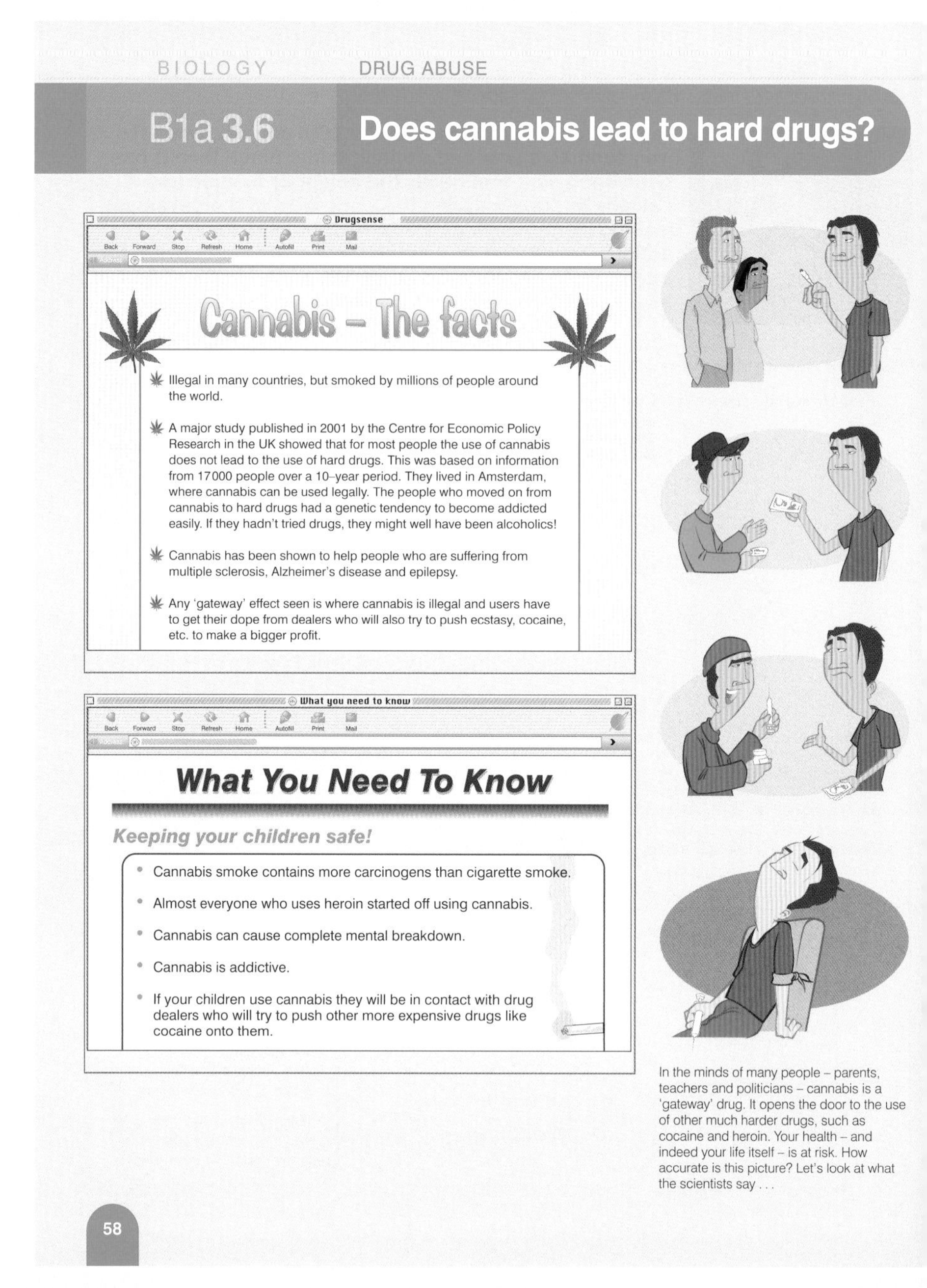

BIOLOGY DRUG ABUSE

B1a 3.6 Does cannabis lead to hard drugs?

- Illegal in many countries, but smoked by millions of people around the world.
- A major study published in 2001 by the Centre for Economic Policy Research in the UK showed that for most people the use of cannabis does not lead to the use of hard drugs. This was based on information from 17000 people over a 10-year period. They lived in Amsterdam, where cannabis can be used legally. The people who moved on from cannabis to hard drugs had a genetic tendency to become addicted easily. If they hadn't tried drugs, they might well have been alcoholics!
- Cannabis has been shown to help people who are suffering from multiple sclerosis, Alzheimer's disease and epilepsy.
- Any 'gateway' effect seen is where cannabis is illegal and users have to get their dope from dealers who will also try to push ecstasy, cocaine, etc. to make a bigger profit.

- Cannabis smoke contains more carcinogens than cigarette smoke.
- Almost everyone who uses heroin started off using cannabis.
- Cannabis can cause complete mental breakdown.
- Cannabis is addictive.
- If your children use cannabis they will be in contact with drug dealers who will try to push other more expensive drugs like cocaine onto them.

In the minds of many people – parents, teachers and politicians – cannabis is a 'gateway' drug. It opens the door to the use of other much harder drugs, such as cocaine and heroin. Your health – and indeed your life itself – is at risk. How accurate is this picture? Let's look at what the scientists say . . .

58

Teaching suggestions

- In any classroom sessions on this topic, it is vital to have some facts and figures to hand, so that discussions are meaningful and do not end up in generalisations or get too personal. Students need to have accurate information in order to be able to form their opinions and to be able to present a balanced argument or viewpoint.
- The activity described in the Student Book is for the class to hold a debate on the motion: 'We believe that cannabis should not be made a legal drug'. Each student is asked to prepare a speech for the motion and a speech against the motion. The writing of both speeches is important, so that students get some practice in putting forward both sides of an argument. The students are recommended to include references to different aspects of the topic and so need to collect information from a variety of sources. A display of useful references, newspaper articles, booklets and other material could be provided in the classroom or the students directed to find their own. The research and writing of the speeches could be set as homework tasks, so the students would be well-prepared for the debate in advance of the lesson.
- **The facts about cannabis and hard drugs** – As preparation for the debate, it could be useful to review what the students already know about cannabis and hard drugs. This could be a question and answer session, or a 'True' or 'False' session, and will indicate how much accurate (and inaccurate) information they have been given from other sources or picked up from their peers. The chemical nature of the various drugs could be researched, their medicinal use understood and the precise effects that they may have can be described or shown in videos or photographs. If forewarned, students could carry out their own research prior to the lesson as a homework exercise.
- **Facts and figures** – It is possible to obtain up-to-date statistics about drug use in the UK and other countries, amongst different age groups etc. These statistics could form part of a wall display. It could be interesting to carry out a review of national/local newspapers for a week or more, cutting out all the articles that are drug-related. How large a part is played by drug use in incidents? There have been several newspaper articles in which footballers and other sportspersons have been accused of taking drugs.

Scientist A: In some people cannabis use can trigger mental illness, which can be very serious and permanent. The people who are affected have usually been ill or have a family history of mental illness. In fact it seems as if these people are particularly attracted to drug use.

There have been some major studies in New Zealand involving many thousands of young people. These show a definite link between using cannabis and both depression and schizophrenia. In fact one research team calculated that if we stopped people using cannabis in the UK, the number of people with schizophrenia would drop by 13%.

Scientist B: There is very little evidence based on serious scientific research to say that cannabis use affects the long- or short-term memory of users or that cannabis has many of the physically damaging effects often reported in the popular press.

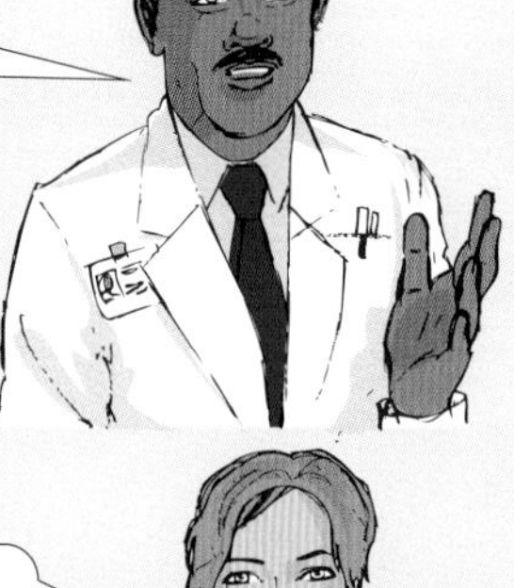

Scientist C: Almost all heroin users were originally cannabis users. This sounds as if cannabis use leads to heroin use. But think again! This is not a case of 'cause and effect' as we call it in science. Almost all cannabis users were originally smokers – but we don't claim that smoking cigarettes leads to cannabis use! In fact the vast majority of smokers do not go on to use cannabis – and the vast majority of cannabis users do not move on to hard drugs like heroin.

Scientist D: The number of people who are damaged or killed as a result of the use of illegal drugs each year is a tiny fraction of the people affected by the legal drugs alcohol and tobacco.

A lot of scientific research has been done into the effects of cannabis on our health, and on the links between cannabis use and addiction to hard drugs.

Unfortunately many of the studies have been quite small. They have not used large sample sizes, so the evidence is not reliable.

Most of the bigger studies show that the effects of cannabis use are not as serious as was thought. They also show that cannabis acts as a 'gateway' to other drugs. That's ***not*** because it makes people want a stronger drug but because it puts them in touch with illegal dealers.

The UK Government downgraded cannabis in 2004, although it is still an illegal drug. More states in the USA are looking at decriminalising cannabis use as the evidence grows steadily that it is less dangerous than alcohol. Some scientists support the moves, while others feel that cannabis should remain illegal.

ACTIVITY

You are going to set up a classroom debate. The subject is: *'We believe that cannabis should not be made a legal drug.'*
You are going to prepare ***two*** short speeches – one ***for*** the idea of legalising cannabis and one ***against***.
You can use the information on these pages and also look elsewhere for information – in books and leaflets from PSHE, in the media and on the Internet.
In both of your speeches you must base your arguments on scientific evidence as well as considering the social, moral and ethical implications of any change in the law. You have to be prepared to argue your case (both 'for' and 'against') and answer any questions – so do your research well!

- **Role play** – Once the students have some facts and figures, as well as their opinions, a series of role play exercises could be set up. The trigger for these could be based on the material given in the text of the student book. Some examples could include a worried parent ('Is my child smoking cannabis?'), a multiple sclerosis sufferer ('Cannabis helps relieve my symptoms – why must I feel like a criminal?'), a student ('It's legal in Amsterdam, so why is it not legal here?') or a concerned friend ('My best friend has been behaving oddly and I think he/she has been taking drugs').

ACTIVITY & EXTENSION IDEAS

Learn to say 'No' – An extension of the 'Keeping your children safe!' information on the Whatyouneedtoknow.org web site would be to design a poster campaign for the school, warning of the dangers of cannabis. Students could be encouraged to use information from both web sites to make their posters accurate and avoid generalisations. As well as designing posters, students could be encouraged to think about where these posters need to be sited around the school for the maximum effect.

- It has been suggested that smoking cannabis is more dangerous than smoking cigarettes, as there are more carcinogens in cannabis. Students could research the contents of smoke from cannabis and make a comparison of the two. If some research was done about the components of cigarette smoke for the previous spread, this can be displayed in the laboratory and start off the comparison.
- A comparison of the statistics of the numbers of people affected by alcohol or tobacco and those affected by cannabis could be made. If it is true that more people are damaged by alcohol and tobacco than by cannabis, why are these drugs not banned? Invite a former cannabis user/drug user in to give a talk about the effects and temptations of trying hard drugs after smoking cannabis.

SUMMARY ANSWERS

1 **a)** A drug is a substance that alters the way in which the body works. It can affect your mind or your body or both.

b) They are carried all around your body. Many of them are poisons. They can affect lots of organs in the body and cause damage anywhere.

c) Any two appropriate examples of each, e.g. tobacco and alcohol, and heroin and cannabis.

d) Drugs change the chemical processes in your body so that you may become dependent on them. If you are addicted to a drug, you cannot manage properly without it.

2 **a)** Alcohol passes through the wall of your gut, into your bloodstream and then easily into nearly every tissue of your body including the nervous system and the brain. It slows down your reactions and can make you lose your self-control. When you have had too much to drink, you lack judgement and often make stupid or dangerous decisions.

b) In small quantities, alcohol makes people feel relaxed and cheerful. You feel less inhibited, making you feel more confident if you are shy. It is used for this reason on social occasions. Some people like the taste of alcoholic drinks.

c) Alcohol is broken down slowly in the liver. If you drink large quantities, your liver cannot cope and you suffer alcohol poisoning, which can lead to unconsciousness, coma and death. People who drink heavily for many years can cause damage to the liver and brain. They may develop cirrhosis of the liver (when the liver tissue is destroyed) or liver cancer that spreads fast and can be fatal. In heavy drinkers, the brain may be damaged (becoming soft and pulpy) so that it cannot work any longer and this causes death.

d) The answer to this more open-ended question should/could include reference to the following areas: been used for centuries, hard to get support for making it illegal, would make a lot of people criminals, socially acceptable, influential people use alcohol, brings in a lot of money through taxes.

3 **a)** Tobacco, 4000, absorbed/passed, nicotine, addictive, blood, tar, lungs.

b) i) If you are a non-smoker, small hairs in your breathing system are constantly moving mucus away from your lungs. The mucus traps, dirt, dust and bacteria from the air you breathe in, and the hairs make sure you get rid of it all. If you smoke, these hairs are anaesthetised by each cigarette and so stop working for a time. This allows dirt down into the lungs. The mucus builds up and causes coughing (smoker's cough).

ii) Tar, a sticky black chemical in tobacco smoke, is not absorbed into your blood stream, but it builds up on the surface of the lung tissue, turning it from pink to grey. The build-up of tar leads to the tissues of the delicate air sacs breaking down (emphysema). Non-smokers do not have this tar, so are much less likely to get emphysema.

iii) Tar is also a major carcinogen. Many cases of lung cancer are linked to the build-up of tar in the lungs, so smokers are at greater risk than non-smokers.

iv) Cigarette smoke contains the poisonous gas carbon monoxide, which can be picked up by the red blood cells. It reduces the oxygen-carrying capacity of the blood and leads to a shortage of oxygen. During pregnancy, the developing baby gets oxygen from the mother's blood. If the mother smokes, the quantity of oxygen in her blood will be reduced and the fetus will be deprived of oxygen. This could result in the baby not growing as well as it should. As a result, the baby is at risk of being premature, with a low birth mass or born dead.

4 **a)** 83%

b) Bar chart with labelled axes.

DRUG ABUSE: B1a 3.1 – B1a 3.6

SUMMARY QUESTIONS

1 a) What is a drug?
b) Why can drugs be so harmful?
c) Give two examples of legal drugs and two examples of illegal drugs.
d) What is addiction to a drug?

2 a) How does alcohol affect your body?
b) Why do so many people use alcohol?
c) Alcohol can cause serious damage to your body, both if you take a single big overdose or if you drink too much over many years.
Explain what happens to your body in both cases.
d) Alcohol costs our society millions of pounds in health care and in sorting out the social problems that it causes. It is a very dangerous drug.
Explain why you think it is still legal and easy to get hold of.

3 a) Copy and complete:
Every cigarette contains ...... leaves which burn to produce around ...... chemicals which are breathed into your lungs. Some of those chemicals are ...... into the blood stream to be carried around your body and to your brain.
...... is the......drug found in tobacco smoke. It is absorbed into your .... On the other hand ......stays in your ...... where it can cause cancer.
b) Explain the following facts about smokers.
i) They are more likely than non-smokers to cough.
ii) They are more likely than non-smokers to suffer from lung diseases like emphysema.
iii) They are more likely than non-smokers to suffer from lung cancer.
iv) They are more likely than non-smokers to have a baby which is born dead or has a low birth mass.

4 Use the data on page 57 to help you answer the following questions:
a) What percentage of the people who die of lung cancer are smokers?
b) Draw a bar chart to show the effect of the number of cigarettes you smoke on your relative risk of dying.
c) i) What happened to the death rates from cancer from 1974–1992 in men in England and Wales?
ii) What do you think happened to the numbers of men smoking over the same time period? Explain your answer.
d) The numbers of women smoking increased steadily from the 1950s. How does the evidence suggest that smoking is linked to lung cancer but not to breast cancer.

EXAM-STYLE QUESTIONS

1 The table below contains statements about certain dru[gs]. Match the words **A**, **B**, **C** and **D** with the statements **1** t[o] **4** in the table.

A Alcohol **B** Nicotine
C Cannabis **D** Heroin

	Statement
1	May cause cirrhosis of the liver.
2	May cause psychological problems.
3	Is an illegal drug.
4	Acts as a stimulant.

2 Which of the following substances in tobacco smoke is addictive?

A Tar **B** Nicotine
C Carbon monoxide **D** Carbon dioxide

3 Which of the following is **not** a disease that can be caused by smoking cigarettes?

A Liver damage **B** Bronchitis
C Emphysema **D** Lung cancer

4 Which of the options **A**, **B**, **C** or **D** in the following statement is **not** correct?
Women who smoke during pregnancy have a higher ri[sk] of having a:

A longer pregnancy.
B premature birth.
C baby of lower than average birth mass.
D baby that is dead at birth.

5 The graph shows the number of cigarettes smoked per male per year in the UK from 1910–1980. The dotted line shows the number of male deaths from lung cance[r] over the same period.

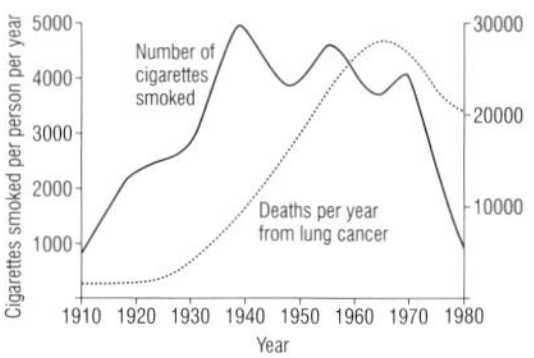

c) i) Overall they fell in all age groups affected.

ii) The number of smokers fell because statistics show that most people who die of lung cancer are smokers.

d) Deaths from breast cancer have shown no significant increase after the 1950s whereas deaths from lung cancer have increased dramatically.

Summary teaching suggestions

- **Special needs**
  - Students could be given the words needed on cards and then asked to put them into the passage in question 3 in the correct places.
- **Gifted and talented**
  - Once the students have answered the questions about alcohol and smoking, they could do some further research and find statistics to back up their answers: all kinds of resources could be used from libraries, Internet, to their local health centre or organisations such as Alcoholics Anonymous and anti-smoking campaigns.
- **When to use these questions?**
  - Some of the questions here require one-word answers (questions 1c) and 3a)), but most need answering in greater depth.
  - The questions are grouped and can be used at the end of the relevant topics, such as drugs, alcohol and smoking, as homework exercises.
  - Although it is beneficial for students to gain practice in writing full answers in continuous prose, if time is short then there is value in assessing the marks allocated and writing down the main points that they would include in an answer. For example, in question 3b) iv), the answer needs to include reference to carbon monoxide, red blood cells, the carriage of oxygen, how the baby gets oxygen, effects of carbon monoxide on red blood cells, shortage of oxygen to baby, baby does not grow. A list of the important points makes a good revision resource.

(a) Name the independent variable shown on the graph. (1)

(b) Most graphs display data collected on one dependent variable.
   (i) Why is this graph unusual? (1)
   (ii) Explain why the data has been presented like this. (1)

(c) Use the information in the graph to show that there is a link between cigarette smoking and lung cancer. (2)

(d) Assuming smoking does cause lung cancer, give two reasons why there is a time lag of around 30 years between a high level of cigarette smoking and a high rate of death from lung cancer. (2)

(e) What is the effect on the body of the carbon monoxide in tobacco smoke? (1)

The table shows the results of a survey carried out in 1995 to find out how much alcohol youngsters drink in a week.

Amount drunk* (units of alcohol)	Aged 11 (%)	Aged 12 (%)	Aged 13 (%)	Aged 14 (%)	Aged 15 (%)
None	75.7	64.1	56.7	44.7	31.4
1–6	20.1	29.0	29.7	34.2	38.8
7–10	1.7	3.5	5.4	8.8	13.3
11–14	1.6	1.4	3.7	5.2	5.4
15–20	0.7	1.0	2.1	3.8	4.6
21+	0.3	1.0	2.4	3.3	7.5

(* One unit = half a pint of beer/one glass of wine/one measure of spirit).

(a) Calculate the percentage of 14-year-olds who drink more than six units of alcohol in a week. (1)

(b) In a school with 240 pupils aged 15, how many of these pupils does the data suggest drink more than 21 units per week? (1)

(c) Calculate the mean percentage of 11 to 15 year olds who drink 1–6 units of alcohol in a week. (1)

(d) What is the range of the proportion of youngsters aged 11–15 who drink 15–20 units of alcohol per week? (1)

(e) Explain why driving a car under the influence of alcohol can be dangerous. (3)

(f) Suggest two other effects on society of alcohol abuse apart from drink-driving. (2)

HOW SCIENCE WORKS QUESTIONS

Some students decided to test whether drinking coffee could affect heart rate. They asked the class to help them with their investigation. They divided the class into two groups. Both groups had their pulses taken. They gave one group a drink of coffee. They waited for 10 minutes and then took their pulses again. They then followed the same procedure with the second group.

a) What do you think the second group were asked to drink? (1)

b) State a control variable that should have been used. (1)

c) Explain why it would have been a good idea not to tell the two groups exactly what they were drinking. (1)

d) Which of the following best describes the type of dependent variable being measured?
i) continuous
ii) discrete
iii) categoric
iv) ordered. (1)

e) Study this table of results that they produced.

Group	Increase in pulse rates (beats per min.)
With caffeine	12, 15, 13, 10, 15, 16, 10, 15, 16, 21, 14, 13, 16
Without caffeine	4, 3, 4, 5, 7, 5, 7, 4, 2, 6, 5, 4, 7

Can you detect any evidence for systematic error in these results? If so, describe this evidence. (2)

f) Is there any evidence for a random error in these results? If so, describe this evidence. (2)

g) What is the range for the increase in pulse rates without caffeine? (1)

h) What is the mean (or average) increase in pulse rate:
i) with caffeine
ii) without caffeine. (2)

i) What conclusion, if any, does the data collected suggest? (1)

j) How could you make your data more reliable? (1)

EXAM-STYLE ANSWERS

1 A 1

B 4

C 2

D 3 *(1 mark each)*

2 B *(1 mark)*

3 A *(1 mark)*

4 A *(1 mark)*

5 a) The year. *(1 mark)*

b) i) There are two dependent variables. *(1 mark)*

ii) So that the relationship between the number of cigarettes smoked and the number of deaths from lung cancer can be displayed. *(1 mark)*

c) As the number of cigarettes smoked increased, so the number of deaths from lung cancer also increased, but many years later. *(1 mark)*

As the number of cigarettes smoked decreased, so the number of deaths from lung cancer also decreased, but many years later. *(1 mark)*

d) Because cigarette smoking has a cumulative effect, it takes a number of years of smoking for harmful chemicals (carcinogens) in tobacco to build up to a level where they cause cancer. *(1 mark)*

Lung cancer is a progressive disease and it is often many years before its symptoms are noticed and many more years before it causes death. *(1 mark)*

e) Tobacco smoke reduces the oxygen-carrying capacity of the blood (by combining more readily with haemoglobin than oxygen does). *(1 mark)*

The information in brackets is not essential to obtain the mark but is the sort of detail that a higher-level candidate might be expected to provide.

6 a) $8.8 + 5.2 + 3.8 + 3.3 = 21.1\%$ *(1 mark)*

The question says 'more than' and so candidates are expected to add the four larger categories of units drunk (7–10, 11–14, 15–20 and 21+) for 14 year olds. Common errors are adding in the 1–6 category as well or using the wrong age group. Students are not asked to show their working and so just the answer '21.1%' will suffice.

b) $\frac{7.5}{100} \times 240 = 18$ *(1 mark)*

Again, no working is required and so just the figure 18 is enough to gain the mark.

c) 30.4% *(1 mark)*

d) 0.7% to 4.6% *(1 mark)*

e) Alcohol enters the nervous system and brain *(1 mark)*
slowing reactions/impairing judgement *(1 mark)*
therefore increasing the time it takes to stop the car in an emergency. *(1 mark)*

For the last mark students must show why it is 'dangerous'. They must therefore link slowed reactions with some kind of risk to themselves or others.

f)
- Antisocial behaviour (e.g. abusiveness, vandalism, rowdiness). *(1 mark)*
- Domestic (and other) violence. *(1 mark)*
- Absenteeism. *(1 mark)*
- Increased criminal activity (e.g. robbery). *(1 mark)*

(Maximum 2 marks)

HOW SCIENCE WORKS ANSWERS

a) Second group were asked to drink coffee without caffeine.

b) Examples of control variables that could have been used:
drink the same amount of coffee
wait for the same amount of time
ensure that the students rest in the same conditions between drinking and pulse measurement.

c) They could be expecting their pulse to rise and as a result it might have done.

d) The type of dependent variable being measured is discrete.

e) The students drinking decaffeinated coffee should not have an increase in pulse rate so there could be systematic error.

f) The differences in the individual results are due to uncontrolled variables but the increase of 21 bpm in the 'with caffeine' group could be random error.

g) The range for the increase in pulse rates without caffeine is 2 bpm to 7 bpm.
(i) Taking caffeine increases pulse rate.
(ii) Increase the sample size of students in each group.

h) (i) 14 beats per min.
(ii) 5 beats per min.

How science works teaching suggestions

- **Higher- and lower-level questions.** Question f) is a higher-level question and the answer provided above is also at this level. Question b) is lower level and the answer provided is also lower level.
- **Gifted and talented.** Possibility of developing the idea of control groups, particularly in medical research. Some might be able to appreciate the use of double-blind trials.
- **Misconceptions.** The difference between random and systematic error. Stress to students that systematic error will produce a consistency of error, i.e. either always too high or always too low (usually because a technique is being applied incorrectly or an instrument is faulty, e.g. not zeroed or calibrated accurately).

B1a 4.1 Pathogens

LEARNING OBJECTIVES

Students should learn:

- The differences between bacteria and viruses.
- How pathogens cause infectious diseases.
- How Semmelweiss tried to control the spread of infectious disease caused by microorganisms.

LEARNING OUTCOMES

Most students should be able to:

- Define the term 'pathogen' and describe how bacteria and viruses differ from other organisms.
- Explain how pathogens cause disease.
- Describe the contribution made by Semmelweiss to the control of the spread of infection in hospitals.

Some students should also be able to:

- Understand the process by which Semmelweiss came to his conclusions.
- Explain why Semmelweiss' ideas were not immediately adopted.

Teaching suggestions

- **Special needs.** Students could be given the programme 'Wordshark 3', putting in the key vocabulary and playing the games generated. Alternatively, a pairs cards activity with the key words from the spread could be played like the game 'Fish'.
- **Gifted and talented.** Students could research the life and times of Semmelweiss. How did his work rank alongside the contributions made by Lister and Pasteur?
- **Learning styles**

  Interpersonal: Discussing and collaborating in compiling list of diseases.

  Intrapersonal: Understanding Semmelweiss's attempts to control infections in hospitals.

  Kinaesthetic: Practical work associated with the investigations.

  Visual: Viewing presentation on bacteria and viruses.

  Auditory: Explaining differences in the colours of the milk samples.
- **Homework.** Use Question 3 of the Summary Questions as a Homework exercise.

SPECIFICATION LINK-UP B1a.11.4

- *Microorganisms that cause infectious disease are called pathogens.*
- *Bacteria and viruses may reproduce rapidly inside the body producing poisons (toxins), which make us feel ill. Viruses damage cells in which they reproduce.*

Students should use their skills, knowledge and understanding of 'How Science Works':

- *to evaluate the contribution of Semmelweiss in controlling infection to solving modern problems with the spread of infection in hospitals.*

Lesson structure

STARTER

Diseases we have in our group – Give each small group of students a sheet of A3 paper and get them to write on the names of any diseases they have had. Which bench had the most? What were they caused by? Were the diseases infectious or not? [Note: some sensitivity needed here with regard to things students do not wish to discuss.] This could be extended to 'My family'. (10 minutes)

Bush tucker challenge – eat some bacteria! – Provide some small pieces of blue cheese, yoghurt and 'helpful bacteria' culture drinks for the students to sample (under hygienic conditions in food technology room and check for allergies). Alternatively, just allow students to inspect and smell the foods. Discuss the usefulness of bacteria, illustrating not all bacteria are 'baddies'. (5–10 minutes)

MAIN

- **Microorganisms and disease** – Search image banks on the web for 'bacteria and virus' and use them to show the differences between bacteria and viruses. There are some very good images of different bacteria and viruses (good e.m. pictures available) and such a presentation could include references to size, what can be seen with the naked eye, with a microscope and with the electron microscope. Include some examples of other pathogenic microorganisms, such as fungi and protists.
- **Experiment to demonstrate the presence of microorganisms in the air** – This experiment is similar to one carried out by Louis Pasteur. It can be done as a demonstration and set up a few days before the lesson. Make up some nutrient broth in a test tube (using a broth tablet and 10 cm^3 of distilled water). Boil the broth to sterilise it and then pour half of it into each of two test tubes. One test tube should then have a cotton wool plug through which a straight piece of glass tubing is inserted so that it does not reach the top of the liquid. The other test tube should also have a cotton wool plug, but the piece of glass tubing is longer, bent into an S-shape and inserted so that there is a straight piece going through the cotton wool and the S-shape arranged outside. Both test tubes should then be sterilised by heating them in a pressure cooker for 15 minutes and allowed to cool (or boiled over a Bunsen burner for about a minute – care needed). Look at the tubes and their contents at intervals over the next few days. Ask: 'Which tube goes cloudy first? Why?' Discuss what is happening in both sets of apparatus.
- Some sterile agar plates, which have previously been partially sealed, exposed to the air around the laboratory and incubated, can be fully sealed and examined for signs of bacterial colonies and compared with the results of the broth experiment.
- 'How Science Works' concepts can be introduced here. For example, the experiment illustrates the need for controls, replication of results for reliability and evaluation of the method used.

PLENARIES

Crossword – Use a crossword of key words and definitions in the spread. This could begin in the lesson and students could write more clues for homework. (5 minutes)

How do microbes get in and what is their fate? – Watch a clip from *The Simpsons* 'Marge in chains' from the video 'Crime and Punishment', showing microbes getting in to various characters and the immune system attacking them, followed by a discussion. (5–10 minutes)

ACTIVITY & EXTENSION

- Introduce the bio-hazard warning symbol.
- A number of good web sites are available, including Michigan State University 'Microbe Zoo' to view on-line. (This can also be purchased.)
- **Was Semmelweiss right?** The benefits of washing hands can be demonstrated by touching the surface of a sterile agar plate with unwashed fingers, replacing the lid and sealing the plate. Wash hands thoroughly, dry them and then touch the surface of a similar sterile agar plate, replacing the lid and sealing as before. Label both plates. Incubate at 25°C and observe what grows. Warning: These plates should on no account be incubated at a higher temperature or opened, and should be disposed of in a safe manner. Alternatively, this could be set up as a demonstation.
- Some Fungi and Protoctista are also pathogenic, causing diseases such as athlete's foot and ringworm (Fungi) and malaria (Protoctista); students could investigate how these diseases are spread. Are they infectious?

Practical support

Microbes in air demonstration
(See main lesson notes)

Equipment and materials required

Nutrient broth, 3 test tubes, distilled water, water bath, 250 cm^3 glass beaker, tripod and gauze, Bunsen burner, heat proof mat, cotton wool, pressure cooker/autoclave, straight glass tube, S-shaped glass tube, prepared agar plates (sealed).

KEY POINTS

A plenary involving the key points has been suggested. Students are provided with the beginnings of a crossword and they are to write clues for homework.

BIOLOGY — CONTROLLING INFECTIOUS DISEASE

B1a 4.1 Pathogens

LEARNING OBJECTIVES

1. What are the differences between bacteria and viruses?
2. How do pathogens cause disease?
3. How did Ignaz Semmelweiss change the way we look at disease?

Infectious diseases are found all over the world, in every country. The diseases range from relatively mild ones, such as the common cold and tonsillitis, through to known killers, such as tetanus, influenza and HIV/AIDS.

An infectious disease is caused by a **microorganism** entering and attacking your body. People can pass these microorganisms from one person to another. This is what we mean by ***infectious***.

Microorganisms which cause disease are called **pathogens**. Common pathogens are bacteria and viruses.

a) What causes infectious diseases?

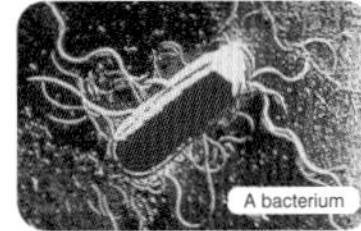

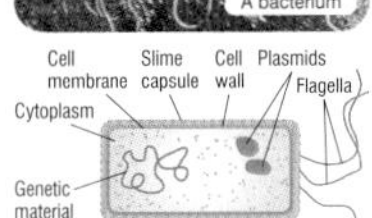

Figure 1 Bacteria come in a variety of shapes and sizes, which help us to identify them under the microscope, but they all have the same basic structure

The differences between bacteria and viruses

Bacteria are single celled living organisms that are much smaller than animal and plant cells.

A bacterium is a single cell. It is made up of cytoplasm surrounded by a membrane and a cell wall. Inside the bacterial cell is the genetic material. Unlike animal and plant cells, this genetic material is not contained in a nucleus.

Although some bacteria cause disease, many are harmless and some are really useful to us. We use them to make food like yoghurt and cheese, in sewage treatment and to make medicines.

b) How are bacteria different from animal and plant cells?

Viruses are even smaller than bacteria. They usually have regular shapes. A virus is made up of a protein coat surrounding simple genetic material. They do not carry out any of the functions of normal living organisms except reproduction. But they can only reproduce by taking over another living cell. As far as we know, all naturally occurring viruses cause disease.

c) Give one way in which viruses differ from bacteria?

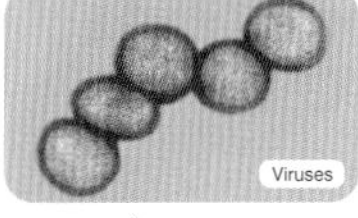

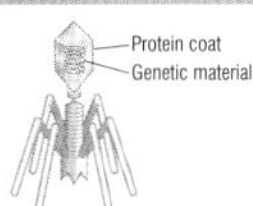

Figure 2 Viruses are really tiny with a very simple structure. Scientists are still arguing about whether they are actually living organisms or not.

How pathogens cause disease

Bacteria and viruses cause disease because once they are inside the body they reproduce rapidly. Bacteria simply split in two. They often produce toxins (poisons) which affect your body. Sometimes they directly damage your cells. Viruses take over the cells of your body as they reproduce, damaging and destroying them. They very rarely produce toxins.

Common disease symptoms are a high temperature, headaches and rashes. These are caused by the damage and toxins produced by the pathogens. The symptoms also appear as a result of the way your body responds to the damage and toxins.

d) How do pathogens make you feel ill?

62

Human biology

The work of Ignaz Semmelweiss

When Ignaz Phillipp Semmelweiss was a doctor in the mid-1850s, many women who gave birth in hospital died a few days later. They died from childbed fever, but no-one knew what caused it.

Semmelweiss realised that his medical students were going straight from dissecting a dead body to delivering a baby without washing their hands. He wondered if they were carrying the cause of disease from the corpses to their patients.

Then another doctor cut himself while working on a body and died from symptoms which were identical to childbed fever. Now Semmelweiss was sure the fever was caused by an infectious agent.

He insisted that his medical students wash their hands before delivering babies. Immediately, fewer mothers died.

Getting his ideas accepted

Semmelweiss presented his findings to other doctors. He thought his evidence would prove to them that childbed fever was spread by doctors. But his ideas were mocked.

Many doctors thought that childbed fever was God's punishment to women. They didn't want to accept the idea that the disease was caused by something invisible passed from patient to patient. Also it was hard for doctors to admit that they might have spread the disease and killed their patients instead of curing them.

What's more, hand-washing seemed a strange idea at the time. There was no indoor plumbing, the water was cold, and the chemicals used eventually damaged the skin of your hands. It is difficult for us to imagine just how difficult hand-washing must have seemed in the 19th century! It took years for Semmelweiss's ideas to be accepted.

In hospitals today, bacteria such as MRSA, which are resistant to antibiotics, are causing lots of problems (see page 68). Getting doctors, nurses and visitors to wash their hands more often is seen as part of the solution – just as it was in Semmelweiss's time!

GET IT RIGHT!

Make sure you know the differences between bacteria and viruses.

Figure 3 Ignaz Semmelweiss – his battle to persuade medical staff to wash their hands to prevent infections is still going on today!

DID YOU KNOW?

Semmelweiss couldn't bear to think of the thousands of women who died because other doctors ignored his findings. By the 1860s he suffered a major breakdown and in 1868, aged only 47, he died – from an infection picked up from a patient during an operation!

SUMMARY QUESTIONS

1. Copy and complete using the words below:

   toxins viruses microorganisms reproduce pathogens damage symptoms bacteria

   The ...... which cause infectious diseases are known as .... Once ...... and ...... get inside your body they ...... rapidly. They ...... your tissues and may produce ...... which cause the ...... of disease.
2. Bacteria and viruses can both cause disease. Make a table which shows how bacteria and viruses are different, and how they are similar.
3. Give five examples of the way we now accept the germ theory of disease in our everyday lives, e.g. washing your hands after using the toilet.
4. Write a letter by Ignaz Semmelweiss to a friend explaining how you formed your ideas and the struggle to get them accepted.

KEY POINTS

1. Infectious diseases are caused by microorganisms such as bacteria and viruses.
2. Microorganisms which cause disease are called pathogens.
3. Bacteria and viruses reproduce rapidly inside your body. They may produce toxins which make you feel ill.
4. Viruses use and damage your cells as they reproduce. This can also make you feel ill.

63

SUMMARY ANSWERS

1. Microorganisms, pathogens, bacteria, viruses, reproduce, damage, toxins, symptoms.
2.

Bacteria	Viruses
Smaller than animal and plant cells	Smaller than bacteria
Have cell wall, cytoplasm, genetic material	Have protein coat and genetic material
No nucleus	No nucleus
Cause infectious disease	Cause infectious disease
Divide by splitting in two	Divide by taking over your cells
Often produce toxins	Rarely produce toxins
Sometimes damage cells	Always damage cells

3. Any sensible suggestions such as wiping work surfaces, cleaning toilets, using tissues to blow nose, etc.
4. Students should show in their letter the main points made on page 63 in the Student Book, including an appreciation of why the new ideas met resistance.

Answers to in-text questions

a) Pathogens/microorganisms/bacteria and viruses.

b) Bacteria are smaller; the genetic material is not in a nucleus; cell wall material is different from plant cells; they have a cell wall (animal cells do not have cell walls).

c) Viruses are much smaller/they have very regular shapes/they only have a protein coat and genetic material/they have no cytoplasm or other contents.

d) Pathogens reproduce rapidly inside your body; they damage your cells; they produce toxins that make you feel ill. Your body reacts to pathogens and the damage they cause/toxins they make, which also makes you feel ill.

B1a 4.2 Defence mechanisms

LEARNING OBJECTIVES

Students should learn:

- How pathogens get into the body and are spread from person to person.
- That the body has different ways of preventing the entry of pathogens.
- That white blood cells help to defend the body against pathogens that do gain entry.

LEARNING OUTCOMES

Most students should be able to:

- Explain the ways in which infectious diseases are spread.
- Describe the ways in which the body prevents the entry of pathogens.
- Describe the functions of the white blood cells within the body.

Some students should also be able to:

- Explain in detail the role of the white blood cells.

Teaching suggestions

- **Special needs.** Students could be provided with a series of pictures showing a bacterium being engulfed by a white blood cell and asked to put them in order.
- **Gifted and talented.** Students should be capable of extending the concept map, putting in further links and connections.
- **Learning styles**

  Interpersonal: Discussing defence strategies.

  Intrapersonal: Interpreting demonstration of clotting.

  Kinaesthetic: Role play at 'numbskulls party'; playing dice game.

  Visual: Matching disease with entry into the body; concept map.

  Auditory: Responding to questions and explanations.

SPECIFICATION LINK-UP B1a.11.4

- *The body has different ways of protecting itself against pathogens. White blood cells help to defend against pathogens:*
  - *by ingesting pathogens*
  - *by producing antibodies which destroy particular bacteria or viruses*
  - *by producing antitoxins which counteract the toxins (poisons) released by pathogens.*

Lesson structure

STARTER

How does it get in? – Give the students a picture of four doors labelled 'Droplets', 'Direct contact', 'Food and drink' and 'Breaks in the skin' and a list of diseases underneath (flu, TB, impetigo, herpes, salmonellosis, AIDS, hepatitis). Ask the students to join the disease with the way it enters the body, leaving undone any they do not know and completing these as the lesson proceeds. Check and discuss. (5–10 minutes)

Defence strategies – Show a picture of a castle and list the possible ways in and the possible defences. Show a picture of the human body alongside it and draw up a similar list. (10–15 minutes)

Traffic light cards – Give the students red (not sure at all), green (entirely sure) and amber (partly sure) cards, and ask them to hold these up in response to questions about how certain organisms get into the body and how they are fought off. Alternatively, the responses could be made into key word cards. (10 minutes)

MAIN

The lesson suggestions here concentrate on students gaining an understanding of the ways in which the blood is involved in defending the body against pathogens. The activities vary in length; a shorter one could be paired with a longer one, e.g. show a video clip and play the dice game.

- Search the web for 'sneeze video' to show a clip of a sneeze in slow motion and/or video footage, or show Animation B1a 4.2 'How white blood cells protect us from disease' from the GCSE Biology CD ROM.
- **To demonstrate clotting** – Before the lesson make a video using a digital video camera showing a finger being pricked with a sterile lancet, a drop of blood being forced out and then the tip of a needle being drawn through the blood until it starts to pick up threads of fibrin. These can very soon be drawn out from the blood. All materials used should be disposed of hygienically and this should not be done during the lesson.
- **Bacteria versus white blood cells** – Play a dice game in pairs. One student is allocated a bacterial disease. The student must state which way the disease is going to try to get into the body (the other student must check to see if this is appropriate). The first student must throw a six before entry can be gained. Once inside, they start to produce toxins, one for every point on the dice. They take turns with their opponent, who represents a defending white blood cell producing anti-toxins. When the opponent throws the dice, the points represent anti-toxins, which counteract the toxins produced by the bacterium. A running score should be kept until the white blood cell throws a six, which represents an antibody and kills the bacterium to win the game. If the running score of toxins reaches 10, the white cell dies and the bacterium wins.
- The game can be continued by playing against more partners. If, as a white cell, you have produced an antibody against a specific bacterium before, you can kill the bacterium with any even number, not just a six.

PLENARIES

Fill in the missing concepts – Students to complete a pre-prepared concept map, which has the connections made and labelled already; they fill in the concepts. (10 minutes)

Overcrowded refugee camp – Pin up or project a picture of an overcrowded refugee camp. Ask the students to describe why the people shown are in danger from infectious diseases. (5–10 minutes)

Numbskulls party – This is a role-play exercise involving bacteria meeting inside a body, introducing themselves to each other, finding out what each other does, discussing how they got there, all done *à la* cocktail party chitchat. (10–15 minutes)

ACTIVITY & EXTENSION

- Link the blood clotting video to haemophilia and discuss clotting times.
- Also link blood clotting to the self-sealing fuel tanks on fighter aircraft and some racing cars.
- Show a jar of pickled onions and discuss why they have not gone rotten. Link to the bactericidal effect of stomach acid.

BIOLOGY CONTROLLING INFECTIOUS DISEASE

B1a 4.2 Defence mechanisms

LEARNING OBJECTIVES

1. How does your body stop pathogens getting in?
2. How do white blood cells protect us from disease?

There are a number of ways in which we can spread pathogens from one person to another. The more pathogens that get into your body, the more likely it is that you will get an infectious disease.

Droplet infection: When you cough, sneeze or talk you expel tiny droplets full of pathogens from your breathing system. Other people breathe in the droplets, along with the pathogens they contain. So they pick up the infection, e.g. 'flu (influenza), tuberculosis or the common cold. (See Figure 1.)

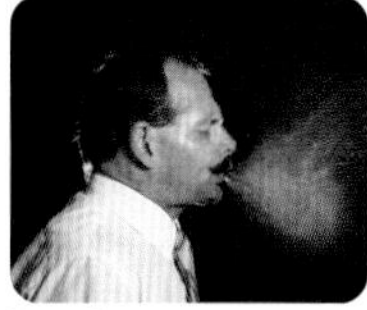

Figure 1 Droplets carrying millions of pathogens fly out of your mouth and nose at up to 100 miles an hour when you sneeze!

Direct contact: Some diseases are spread by direct contact of the skin, e.g. impetigo and some sexually transmitted diseases like genital herpes.

Contaminated food and drink: Eating raw or undercooked food, or drinking water containing sewage can spread disease, e.g. diarrhoea or salmonellosis. You get these by taking large numbers of microorganisms straight into your gut.

Through a break in your skin: Pathogens can enter your body through cuts, scratches and needle punctures, e.g. HIV/AIDS or hepatitis.

When people live in crowded conditions, with no sewage treatment, infectious diseases can spread very rapidly.

a) What are the four main ways in which infectious diseases are spread?

Preventing microbes getting into your body

Each day you come across millions of disease-causing microorganisms. Fortunately your body has a number of ways of stopping these pathogens getting inside.

Your skin covers your body and acts as a barrier. It prevents bacteria and viruses from reaching the vulnerable tissues underneath.

If you damage or cut your skin in any way you bleed. Platelets in your blood quickly help to form a clot which dries into a scab. The scab forms a seal over the cut, stopping pathogens getting in through the wound. (See Figure 2.)

Figure 2 When you get a cut, the platelets in your blood set up a chain of events to form a clot which dries to a scab. This stops pathogens from getting into your body. It also stops you bleeding to death as well!

Your breathing system could be a weak link in your body defences. That's because every time you breathe you draw air loaded with pathogens right inside your body. However, your breathing organs produce a sticky liquid, called mucus, which covers the lining of your lungs and tubes. It traps the pathogens. The mucus is then moved out of your body or swallowed down into your gut. Then the acid in your stomach destroys the microorganisms. In the same way, the stomach acid destroys most of the pathogens you take in through your mouth.

b) What are the three main ways in which your body prevents pathogens from getting in?

How white blood cells protect you from disease

In spite of your various defence mechanisms, some pathogens still manage to get inside your body. Once there, they will meet your second line of defence – the ***white blood cells*** of your **immune system**. The white blood cells help to defend your body against pathogens in several ways:

Role of white blood cell	How it protects you against disease
Ingesting microorganisms	Some white blood cells ingest (take in) pathogens, destroying them and preventing them from causing disease.
Producing antibodies Antibody, Antigen, Bacterium, White blood cell, Antibody attached to bacterium	Some white blood cells produce special chemicals called **antibodies**. These target particular bacteria or viruses and destroy them. You need a unique antibody for each type of bacterium or virus. Once your white blood cells have produced antibodies against a particular pathogen, they can produce them again very rapidly if that pathogen invades again.
Producing antitoxins Antitoxin molecule, Toxin and antitoxin joined together, Toxin molecule, Bacterium	Some white blood cells produce antitoxins. These counteract the toxins (poisons) released by pathogens.

Figure 3 Ways in which your white blood cells destroy pathogens and protect you against disease

FOUL FACTS

Have you ever wondered why the mucus produced from your nose turns green when you have a cold? Some white blood cells containing green coloured enzymes are secreted in your mucus to destroy the cold viruses – and any bacteria which decide to infect the mucus at the same time. The dead white blood cells along with the dead bacteria and viruses are removed in the mucus, making it look green.

GET IT RIGHT!

Avoid using words like 'battle' and 'fight' when you explain how antibodies work. Such words suggest that the white blood cells think about what they are doing.

SUMMARY QUESTIONS

1 Explain how diseases are spread by:

a) droplet infection
b) direct contact
c) contaminated food and drink
d) through a cut in the skin.

2 Certain diseases mean you cannot fight infections very well. Explain why the following symptoms would make you less able to cope with pathogens.

a) Your blood won't clot properly.
b) The number of white cells in your blood falls.
c) Your skin is damaged to expose a large area of raw tissue underneath.

3 Here are four common things we do. Explain carefully how each one helps to prevent the spread of disease.

a) Washing your hands before preparing a salad.
b) Throwing away tissues after you have blown your nose.
c) Making sure that sewage does not get into drinking water.
d) Putting your hand in front of your mouth when you cough or sneeze.

4 Explain in detail how the white blood cells in your body work.

KEY POINTS

1. Your body has several methods of defending itself against the entry of pathogens using the skin, the mucus of the breathing system and the clotting of the blood.
2. Your white blood cells help to defend you against pathogens by ingesting them, making antibodies and making antitoxins.

SUMMARY ANSWERS

1 **a)** Droplets full of pathogens pass into the air to be breathed in by someone else.
b) Pathogens on skin passed to someone else's skin on contact.
c) Pathogens taken in on food or in drink.
d) Pathogens can get through the barrier of the skin to the tissue underneath.

2 **a)** You cannot stop pathogens getting into cuts.
b) You have not got enough white blood cells to ingest pathogens or to produce antibodies/antitoxins so pathogens can take hold.
c) There is too big an area for a clot to form over all of it fast enough, so it provides entry points for pathogens.

3 **a)** Prevents pathogens getting from your hands to the food.
b) Removes pathogens from where they might come into contact with other people or get on your hands.
c) Prevents pathogens from the gut being taken in with drinking water.
d) Prevents pathogens in droplets getting into the air that other people are breathing.

4 Explanation to include the ingestion of microorganisms, the production of antibodies and antigens.

Answers to in-text questions

a) Droplet infection; direct contact; contaminated food and drink; through a break in the skin.

b) Skin acts as a barrier; breathing organs produce mucus to trap pathogens; blood uses platelets to produce clots to seal wounds.

KEY POINTS

Filling in the missing concepts will help students to remember the key points.

B1a 4.3 Using drugs to treat disease

LEARNING OBJECTIVES

Students should learn that:

- Medicines, such as painkillers, relieve symptoms but do not kill pathogens.
- Antibiotics help to cure bacterial diseases by killing infective bacteria inside the body.
- Antibiotics cannot kill viral pathogens which live and reproduce inside cells.

LEARNING OUTCOMES

Most students should be able to:

- Explain what is meant by the term 'medicine'.
- Describe how antibiotics can be used to treat bacterial infections.
- Explain why antibiotics are not used to treat diseases caused by viruses.

Some students should also be able to:

- Explain the difficulty of developing antiviral drugs.

Teaching suggestions

- **Special needs.** Students could be shown pictures of people with various complaints and asked to decide which medicines they should be given.
- **Gifted and talented.** Students could be given references to find out how mutations come about as a preparation for the next topic and link to work on DNA and base sequences. The frequency of mutations is also an interesting topic.
- **Learning styles**

  Interpersonal: Discussing and evaluating old remedies.

  Intrapersonal: Interpretating the process of clotting.

  Kinaesthetic: Practical work on sensitivity of bacteria to antibiotics.

  Visual: Following sequence of the discovery of penicillin.

  Auditory: Listening to 'horrible history' and discussion.

SPECIFICATION LINK-UP B1a.11.4

- *Some medicines, including painkillers, help to relieve the symptoms of disease, but do not kill the pathogens.*
- *Antibiotics, including penicillin, are medicines that help to cure bacterial disease by killing infective bacteria inside the body. Antibiotics cannot be used to kill viral pathogens, which live and reproduce inside cells. It is difficult to develop drugs which kill viruses without also damaging the body's tissues.*

Lesson structure

STARTER

What medicine do I need? – Either pretend to feel unwell yourself or pick someone from the class, wrap them up in a scarf and give them a hot water bottle. Produce a bottle of over-the-counter cough medicine, a box of aspirins or paracetamol, some throat sweets and a bottle of prescription antibiotics. Discuss what should be given to the 'patient' and why. (5–10 minutes)

Horrible history! – Read a description of someone dying of an infection in the past, before the days of penicillin. For example, Lord Caernarvon following the discovery of Tutankhamun's tomb. Ask: 'Was the curse really an ancient biological hazard warning?' As an alternative or an addition, there are pictures of people dying of tetanus that could be projected. Discuss. (10 minutes)

Old remedies, did they work? – Ask: 'Is there any truth in old wive's tales and ancient remedies for healing?' Show a piece of mouldy bread in a sealed plastic bag, a jar of honey, a bottle of vinegar, a soldering iron (for cauterising), some wood ash and some cobwebs. Discuss the scientific background to these remedies and consider what was available to people before there were antibiotics. (10 minutes)

MAIN

There are several important issues in this spread and some interesting ideas for practical work. The class could be split, some doing the sensitivity of bacteria exercise and others investigating the mouthwashes, etc. Comparisons could be made as each group reports to the other.

- **The discovery of penicillin** – Search the web to find pictures about Alexander Fleming and his work.
- **Sensitivity of bacteria to antibiotics** – This experiment can either be set up as a demonstration or carried out by the students in groups. Agar plates are inoculated with a suitable bacterium, such as *Bacillus subtilis*, have antibiotic-impregnated discs placed on them and incubated for 24 hours. The antibiotics diffuse from the discs into the agar and inhibit the growth of bacteria around them, resulting in clear zones in the agar. The diameter of these clear zones can be measured. The experiment can be carried out using Oxoid multodiscs (available from suppliers), with several antibiotics; or different concentrations of one antibiotic could be used.
- If this is to be carried out by the students, then all the usual precautions need to be taken. The agar plates could be set up for them and sealed, but if facilities allow it is more instructive if they do it themselves following all the safety measures involved with the handling of sterile equipment and bacteria. (Teacher should be trained in aseptic techniques.)
- Predictions can be made and this could form part of a coursework assignment. It involves the introduction of many 'How Science Works' concepts. [Note: There are restrictions on the use of bacteria in schools, so guidelines would need to be consulted and all suitable precautions taken if this is to be used as a class experiment.]
- As an addition, or alternative, to the experiment above, the antimicrobial properties of mouthwashes, toothpastes and cleaning products could be investigated in a similar way. Agar plates are inoculated with a harmless bacterium and have wells cut into them by removing cylinders of agar with a cork borer. Solutions of the antimicrobial substances being tested could be placed in the wells and the plates incubated as before. The relative effects can be judged by the clear areas around the wells. (See CLEAPSS Handbook/CD ROM Section 15.2.)

- A comparison can be made with the antibiotics, but it should be pointed out that the substances being tested here are for use outside the body, whereas antibiotics are designed to work inside the body.

PLENARIES

Finish your medicine! – Draw out a flow chart with decision boxes to show the consequences of not finishing a course of antibiotics, or of using antibiotics too widely. (5–10 minutes)

Should antibiotics be used for . . . ? – Provide a list of statements about the use of antibiotics and ask students to say whether or not each is a good idea, with reasons to back up their decision. Statements could include:

'Chickens raised in barns are given antibiotics in their food.'
'If you have a cold you should go to the doctor for some antibiotics.'
'In some countries, antibiotics can be bought over the counter.'
'Milking cows may have tubes of antibiotics placed in their udders.'
'Some chopping boards have antibiotics/antibacterial substances built into them.'
(5–10 minutes)

Quick quiz – Use a quiz on the meanings of key terms used in the topic. (10 minutes)

ACTIVITY & EXTENSION IDEAS

- Remind students that aspirin has a role to play in the treatment of heart and other diseases as well as being a painkiller. It is also an anti-inflammatory which can be tolerated well by people with arthritis and other muscle conditions.
- 'What about people who are allergic to penicillin?' Introduce the idea that there are different antibiotics, perhaps mentioning narrow-spectrum and broad-spectrum antibiotics. Ask students if they have been prescribed antibiotics other than penicillin.

KEY POINTS

These can be emphasised by having a quiz on the key terms and points in this spread as suggested as a plenary.

BIOLOGY CONTROLLING INFECTIOUS DISEASE

B1a 4.3 Using drugs to treat disease

LEARNING OBJECTIVES
1 What is a medicine?
2 Why can't we use antibiotics to treat diseases caused by viruses?

When you have an infectious disease, you often take medicines which contain useful drugs. Often the medicine has no effect at all on the pathogen that is causing the problems. It just eases the symptoms and makes you feel better.

For example, drugs like aspirin and paracetamol are very useful as painkillers. When you have a cold they will help relieve your headache and sore throat. On the other hand, they will have no effect on the virus which has invaded your tissues and made you feel ill!

Many of the medicines you can buy at a chemist's are like this. They are symptom relievers rather than pathogen killers. They do not make you better any faster. You have to wait for your immune system to overcome the invading microorganisms.

a) Why don't medicines like aspirin actually cure your illness?

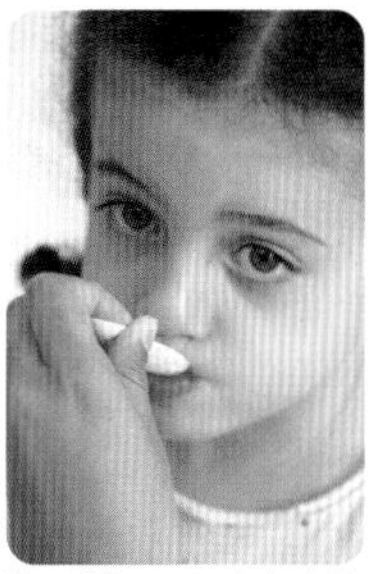

Figure 1 Taking paracetamol will make this child feel better, but she will not actually get well any faster as a result.

Antibiotics

While drugs which make us feel better are useful, what we really need are drugs that can *cure* us. We use antiseptics and disinfectants to kill bacteria outside the body. But they are far too poisonous to use inside you. They would kill you and your pathogens at the same time!

The drugs which have really changed the way we treat infectious diseases are **antibiotics**. These are medicines which can kill disease-causing bacteria inside your body.

b) What is an antibiotic?

Alexander Fleming was a scientist who studied bacteria and was keen to find ways of killing them. In 1928, he was growing lots of bacteria on agar plates. Alexander was a rather sloppy scientist, and his lab was quite untidy. So he often left the lids off his plates for a long time. He also often forgot about cultures he had set up!

When he came back from a holiday, Fleming noticed that lots of his culture plates had mould growing on them. Just before he put the plates in the washing up bowl, he noticed a clear ring in the jelly around some of the spots of mould. Something had killed the bacteria covering the jelly.

There and then Fleming saw how important this was. He worked hard on the mould, extracting a 'juice' which he called **penicillin**. But he couldn't get much penicillin from the mould. He couldn't make it keep, even in the fridge. So he couldn't prove it would actually kill bacteria and make people better. By 1934 he gave up on penicillin and went on to do different work!

About 10 years after penicillin was first discovered, Ernst Chain and Howard Florey set about trying to use it on people. Eventually they managed to make penicillin on an industrial scale. The process was able to produce enough penicillin to supply the demands of the Second World War. We have used it as a medicine ever since.

DID YOU KNOW?
A patient called Albert Alexander was dying of a blood infection when Florey and Chain gave him some of their experimental penicillin for five days. The effect was almost miraculous and Albert recovered. But then the penicillin ran out. Florey and Chain even tried to collect spare penicillin from Albert's urine, but it was no good. The infection came back and sadly Albert died.

66

Human biology

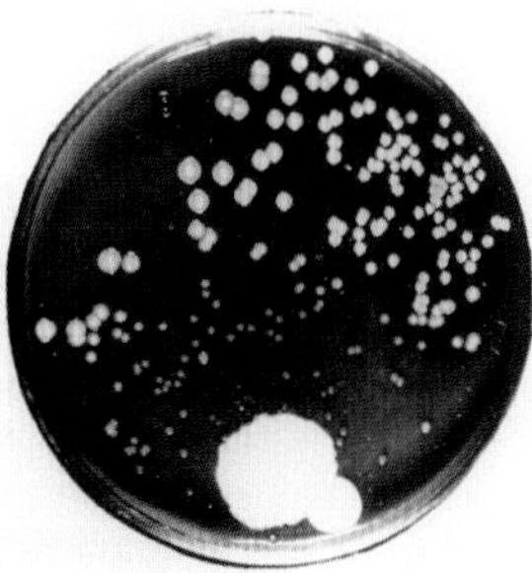

Figure 2 Alexander Fleming was on the lookout for something that would kill bacteria. As a result of him noticing the effect of this mould on his cultures, millions of lives have been saved around the world.

c) Who was the first person to discover penicillin?

How antibiotics work

Antibiotics, such as penicillin, work by killing the bacteria which cause disease while they are inside your body. They damage the bacterial cells without harming your own cells. They have had an enormous effect, because we can now cure bacterial diseases such as plague and TB. These same diseases killed millions of people in years gone by.

Unfortunately antibiotics have not been a complete answer to the problem of infectious diseases. Antibiotics have no effect on diseases caused by viruses. What's more, developing drugs which do have an effect on viral diseases is proving very difficult indeed.

The problem with viral pathogens is that they reproduce inside the cells of your body. It is extremely difficult to develop drugs which kill the viruses without damaging the cells and tissues of your body at the same time.

d) How do antibiotics work?

Figure 3 Penicillin was the first antibiotic. Now we have many different ones which kill different types of bacteria. In spite of this, scientists are always on the look out for new antibiotics to keep us ahead in the battle against the pathogens.

SUMMARY QUESTIONS
1 What is the main difference between drugs like paracetamol and drugs such as penicillin?
2 a) How did Alexander Fleming discover penicillin?
b) Why was it so difficult to make a medicine out of penicillin?
c) Who developed the industrial process which made it possible to mass-produce penicillin?
3 Explain why it is so much more difficult to develop medicines against viruses than it has been to develop anti-bacterial drugs.

KEY POINTS
1 Some medicines relieve the symptoms of disease but do not kill the pathogens which cause it.
2 Antibiotics cure bacterial diseases by killing the bacteria inside your body.
3 Antibiotics do not destroy viruses because viruses reproduce inside the cells. It is difficult to develop drugs that can destroy viruses without damaging your body cells.

67

SUMMARY ANSWERS

1 Paracetamol relieves symptoms/makes you feel better, whereas antibiotics kill the bacteria and actually make you better.

2 **a)** He noticed a clear area around mould growing on bacterial plates.
b) It was difficult to get much penicillin out of the mould and it does not keep easily.
c) Florey and Chain.

3 Viral pathogens reproduce inside your cells, so it is very difficult to develop a drug that destroys them without destroying your cells as well.

Answers to in-text questions

a) Because they do not kill the pathogens that are making you ill.
b) A drug that kills pathogenic bacteria in your body.
c) Alexander Fleming.
d) They kill the bacteria that make you ill inside your body.

B1a 4.4 Changing pathogens

LEARNING OBJECTIVES

Students should learn that:

- Bacteria develop resistance to antibiotics as a result of natural selection.
- Over-use of antibiotics should be avoided in order to prevent more resistant strains of bacteria arising.
- Bacteria and viruses can mutate causing new strains of disease to appear.

LEARNING OUTCOMES

Most students should be able to:

- Explain how antibiotic resistant strains of bacteria arise.
- Describe how mutations of bacteria and viruses give rise to new strains of disease.
- Describe how new strains of disease can spread rapidly causing epidemics and pandemics.

Some students should also be able to:

- Explain in detail how bacteria become increasingly resistant to antibiotics.
- Evaluate the problems of preventing the spread of a new disease such as a mutated form of bird flu.

Teaching suggestions

- **Special needs.** These students could make a poster with reasons why you should finish your course of antibiotics or they could sort the list of infectious diseases into those that are caused by bacteria and those caused by a virus.
- **Gifted and talented.** MRSA is not the only problem or cause of infection in hospitals. These students could research the incidence of other infections which can spread in a hospital environment.
- **Learning styles**

  Interpersonal: Working with others on script-writing activity.

  Intrapersonal: Taking part in Loop game.

  Kinaesthetic: Making a poster.

  Visual: PowerPoint® presentations on mutation and antibiotic resistance.

  Auditory: Taking part in class discussions.

SPECIFICATION LINK-UP B1a.11.4

- *Many strains of bacteria, including MRSA, have developed resistance to antibiotics as a result of natural selection. To prevent further resistance arising it is important to avoid over-use of antibiotics.*

Students should use their skills, knowledge and understanding of 'How Science Works' to:

- *evaluate the consequences of mutations of bacteria and viruses in relation to epidemics and pandemics, e.g. bird influenza.*

Lesson structure

STARTER

Lets get it clear! – This is a good opportunity to remind students of which diseases are caused by bacteria and which by viruses. Draw up a list of ailments on the board and ask students to put a 'B' by those caused by bacteria and 'V' for those caused by a virus. (5–10 minutes)

Refer back – One of the Plenaries from the previous lesson could be used here if not used before. (See page 67 in this book).

MAIN

It is important that students understand how antibiotic resistance arises and how mutations occur. It would be possible to combine two or more of these suggestions if time permits.

- **What is a mutation and how does it occur?** – Show a PowerPoint® presentation or a video on mutations and how they occur. Provide students with a worksheet which they can complete as the presentation proceeds. It could be worth pointing out that mutations occur under natural circumstances all the time, but that the mutation rate in microorganisms appears to be greater as they reproduce more rapidly than other organisms.
- **Antibiotic resistance** – A presentation on this topic could follow the suggestion made above. Link in with the practical work suggested in the Main lesson notes in this book on page 66 on the sensitivity of bacteria to different antibiotics.
- **The MRSA story** – Provide groups of students with reference material, such as suitable web sites, newspaper and magazine articles and information given to hospitals, and suggest that they write a script for a radio or TV programme about MRSA. The emphasis is to be on the facts rather than on sensational reporting.
- **Pandemics and epidemics – what they are and how to survive them** – This could take the form of a general discussion of the difference between the two terms, with examples. Using the information in the Student Book, students could draw up a list of how diseases spread from country to country. Alongside each method of spread, suggestions for a control could be made. Finally, the students could decide on how they personally would recommend precautions that they and their families could take to avoid exposure to the disease and thus survive.

PLENARIES

Get your 'flu jab! – Ask students why it is important that people aged 65 and over should be vaccinated against influenza every year. Who else qualifies for the 'flu jab? Why? Should it be given to everyone? (5–10 minutes)

Loop game – Prepare cards with the key words and terms and their definitions to play this game. (5–10 minutes)

ACTIVITY & EXTENSION IDEAS

- **Survival Poster.** Students to design a poster setting out how people can avoid/reduce their chances of catching 'flu or other infectious diseases. The poster could be displayed in schools, surgeries and public places.
- **Research previous 'flu pandemics.** Students could find it interesting to find out more about major outbreaks of influenza. Some of their grandparents might remember the outbreak in the 1950s. History web sites could provide information on the rapidity with which the disease spread and how long it lasted. An interesting comparison could be made between the times involved previously and the predicted time scales for any future outbreaks.
- **Anti-viral medicines and vaccines.** Class discussion on the use of anti-viral medicines to bridge the gap between the outbreak of the disease and the development of a suitable vaccine. Why does it take so long for the vaccine to be developed? How do anti-viral medicines work?

Answers to in-text questions

a) To prevent more antibiotic resistant strains appearing.

b) Bacteria.

B1a 4.4 Changing pathogens

LEARNING OBJECTIVES

1 What is antibiotic resistance?
2 Why is mutation such a problem?

If you are given an antibiotic and use it properly, the bacteria which have made you ill are almost all killed. The ones that are left are the ones which have a natural **mutation** which means they are not affected by the antibiotic. They are resistant to it – but your body finishes them off!

Antibiotic-resistant bacteria

If antibiotics are used too often, or you don't take the full course of medicine prescribed by your doctor, more of the resistant bacteria survive. If they go on to make someone else ill, they will not be killed by the original antibiotic. They are ***resistant*** to that antibiotic. This resistance is the result of a process of **natural selection**. (See Figure 1.)

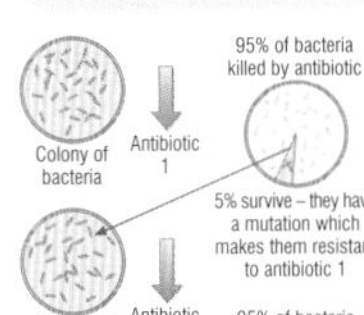

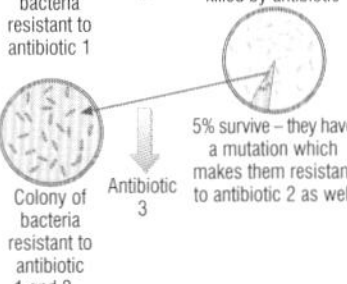

Figure 1 Bacteria can develop resistance to many different antibiotics in a process of natural selection

As more types of bacteria become resistant to more antibiotics, so diseases caused by bacteria are becoming harder and harder to treat.

To prevent more resistant strains of bacteria appearing it is important not to over-use antibiotics. It's best to only use them when you really need them. It is also very important that people finish their course of medicine every time.

a) Why is it important not to use antibiotics too frequently?

The MRSA story

Hospitals treat many patients with infectious diseases. They use large amounts of antibiotics. As a result of natural selection, hospitals often contain a number of bacteria which are not affected by most of the commonly used antibiotics. This is what has happened with ***MRSA*** (the bacterium ***methicillin resistant Staphylococcus aureus***).

In hospitals, where doctors and nurses move from patient to patient, these antibiotic-resistant bacteria are spread easily. MRSA alone now causes around a thousand deaths every year in hospital patients who, although ill, might otherwise have recovered.

How can we control the spread of these antibiotic-resistant bacteria in hospitals? There are a number of simple steps which can have a big effect on the spread of microorganisms such as MRSA. We have known some of them since the time of Semmelweiss, but they sometimes get forgotten!

- Doctors, nurses and other medical staff wash their hands between patients.
- Visitors wash their hands as they come into and leave the hospital.
- Look after patients infected with the bacteria in isolation from other patients.
- Keep hospitals clean – high standards of hygiene.
- Medical staff wear either disposable clothing or clothing which is regularly sterilised.

b) Is MRSA a bacterium or a virus?

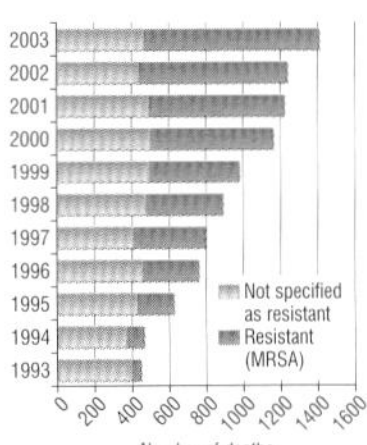

Source: National Statistics Office

Figure 2 The growing impact of MRSA in our hospitals can be seen from this data

Mutation and pandemics

Another problem caused by the mutation of bacteria and particularly viruses is that new forms of diseases can appear. Because no-one is immune to them, they can cause widespread illness. A good example of this in action is influenza, commonly known as 'flu.

'Flu is caused by a virus which mutates easily, so every year new strains appear which can fool your immune system. These new strains are usually quite similar to the old 'flu but just occasionally a very different form of 'flu virus appears.

These usually cause a 'flu ***epidemic*** (in one country) or even a ***pandemic*** (across several countries). In 1918–19 a new strain of 'flu emerged which spread rapidly around the world and killed between 20 and 40 million people.

Animals such as chickens and pigs get 'flu-like diseases too. Sometimes an animal virus will mutate so it can infect people. For example, the 1918–19 'flu pandemic may have been linked to bird 'flu.

Many scientists think that a new and serious form of human 'flu is likely to be linked to one of the bird influenzas which keep appearing in China, Thailand and Asia.

We are trying to prevent a new pandemic using research to discover what makes certain types of 'flu so dangerous. The World Health Organisation is monitoring all 'flu outbreaks. Scientists are also trying to produce different ***vaccines*** to protect us against new and dangerous forms of the disease.

There is no 'antibiotic for viruses' but there are several drugs which can reduce the length of time you suffer with 'flu. This lessens the chance of it spreading.

Finally, many countries have plans to restrict travel and even put people in isolation if there is a 'flu outbreak. This worked very well in 2003 when SARS, a 'flu-like illness, appeared for the first time in China.

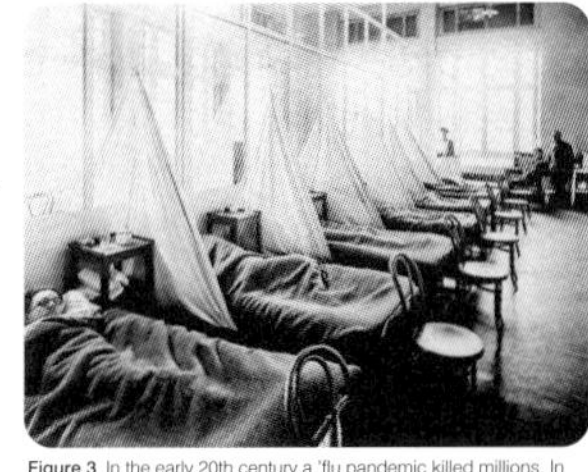

Figure 3 In the early 20th century a 'flu pandemic killed millions. In the 21st century a new form of 'flu could sweep the world even more quickly because of the way we all travel around on aircraft for business and holidays.

GET IT RIGHT!

Remember bacteria don't 'want' to develop resistance – they just do!
Make sure you understand how bacteria become resistant to antibiotics through natural selection.

SUMMARY QUESTIONS

1 Copy and complete using the words below:

antibiotics bacterium better disease mutation mutate resistant virus

If bacteria change or ...... they may become ...... to ...... This means the medicine no longer makes you .... A ...... in a ...... or ......can also lead to a new form of ....

2 Make a flow chart to show how bacteria develop resistance to antibiotics.

3 A new strain of bird 'flu has been discovered in Asia and it is in the national news. Write a piece for the science pages of your local paper. Explain what people can do to protect themselves and what problems they might face preventing the new disease spreading.

KEY POINTS

1 Many types of bacteria have developed antibiotic resistance as a result of natural selection. To prevent the problem getting worse we must not over-use antibiotics.
2 If bacteria or viruses mutate, new strains of a disease can appear which spread rapidly to cause epidemics and pandemics.

SUMMARY ANSWERS

1 Mutate, resistant, antibiotics, better, mutation, virus, bacterium, disease.

2 Colony of bacteria treated with antibiotic 1 → 5% have mutation and survive → the surviving bacteria are treated with antibiotic 2 → 5% have a mutation and are resistant to antibiotic 1 and 2 → etc.

3 Look for an appreciation that there are no antibiotics for viruses and the problems of isolating infected people.

KEY POINTS

The Key Points are covered well by the content of the lesson. Revision cards using the key words and definitions would be useful.

B1a 4.5 Developing new medicines

LEARNING OBJECTIVES

Students should learn:

- That new medical treatments and drugs need to be extensively tested and trialled before being used.
- About possible consequences if drugs are not tested thoroughly.

LEARNING OUTCOMES

Most students should be able to:

- Describe and explain the reasons for testing new drugs.
- Explain the dangers of using drugs that have not been thoroughly tested.

Some students should also be able to:

- Explain the main stages in testing drugs.
- Demonstrate understanding of the flaws in the original development of thalidomide.
- Evaluate the benefits and dangers of using thalidomide to treat diseases such as leprosy and cancer.

Teaching suggestions

- **Special needs.** Lead a session with students on household safety in handling drugs and medicines. Concentrate on things like taking the whole prescribed course, keeping medicines away from children, taking care with the right dose and times at which medicines taken, and discarding out-of-date drugs.
- **Gifted and talented.** Students could be encouraged to explore the medical issues involved, such as the dilemma that doctor's have in prescribing expensive treatments in the light of their budgets. Ask: 'Who gets them? Does it depend on age? Does it depend on post code?'
- **Homework.** Ask students to write a short report on the advantages and disadvantages of testing drugs on animals.
- **Learning styles**

  Interpersonal: Discussing the features of a good medicine.

  Intrapersonal: Considering whether you would volunteer for drug trials.

  Kinaesthetic: Researching the drug thalidomide.

  Visual: Following the sequence on drug testing.

  Auditory: Explaining the meanings of words.

SPECIFICATION LINK-UP B1a.11.3

- *When new medical drugs are devised, they have to be extensively tested and trialled before being used. Drugs are tested in the laboratory to find out if they are toxic. They are then trialled on human volunteers to discover any side effects.*
- *Thalidomide is a drug that was developed as a sleeping pill. It was also found to be effective in relieving morning sickness in pregnant women. However, it had not been tested for this use. Unfortunately, many babies born to mothers who took the drug were born with severe limb abnormalities. The drug was then banned, but more recently is being used successfully to treat leprosy.*

Lesson structure

STARTER

A brand new medicine: is it OK to take it? – Show students a pill or medicine bottle and tell them it is a brand new medicine. Get them to write down what they think happens from the initial idea or discovery, up to the time it is obtainable from the chemist. Build up a sequence on the board from the ideas that the class come up with. (10 minutes)

What do the words mean? – Write the words 'safe', 'effective', 'stable', 'incorporated' and 'excreted' on the board and discuss with the class how these words can be defined in terms of drugs. (10 minutes)

What do we expect of a good medicine? – Discuss what features are important, apart from the medicine relieving the symptoms or curing the disease. (5 minutes)

MAIN

- **The thalidomide story: what went wrong?** – Build up a more complete picture of the drug thalidomide by extending the information given in the Student Book. A video, or projected pictures, and a commentary could be used. More information about the development of the drug, the consequences of its use as a treatment for morning sickness during pregnancy, and the current possibilities of its use as a treatment for leprosy and AIDS can be obtained from the Internet. There are several web sites providing information, e.g. www.britishpathe.com.
- **Drug testing** – Produce a PowerPoint® presentation on good medicines and the stages of drug testing. Produce worksheets to accompany the presentation and allow opportunities for the students to discuss points and complete their sheets as you progress through the presentation. You might include consideration of the number of people involved in research and testing, the time scale and the size of any trials carried out, thereby including important elements of 'How Science Works'.
- **Would you volunteer for drug testing?** – Ask: 'If so, under what circumstances? If you would not, give your reasons and suggest other ways in which drugs should be trialled and tested.' This discussion can start in groups and then widen to include the whole class. This activity could follow the drug testing presentation suggested above.

PLENARIES

Drug testing policy – Suggest to the students that they are the Minister for Health and they have to write a statement about their policy on drug testing. A PhotoPLUS resource B1a 4.5 'Drug testing' is available on the GCSE Biology CD ROM. Give them 10 minutes and then choose some to read out. (15 minutes)

Get the sequence right – Make sets of cards, each with a sentence on regarding the process of drug testing. Students, working in small groups, are to put them into the correct order. This could be a competition to see which group can do it in the shortest time. If you have the sets made up in different colours, then they are easier to sort out. Keep in separate bags. (5–10 minutes)

ACTIVITY & EXTENSION IDEAS

- ? Discuss the problems associated with the testing of drugs on animals. Look at the activities of animal rights' supporters and try to get across the difference between anti-vivisectionist groups and those people concerned with animal welfare (often confused). Are people justified in making the protests they do? How would drugs be tested if animals were not used?
- ? Discuss the finances of the drugs industry. The costs of development have to be recovered, but is it right that diseases for which there are drugs available are not treated in developing countries because they cannot afford them?
- ? Another aspect of the finances could be the basis of a class discussion. Who would you target if you wanted to maximise your profits? Are there any groups of the population who have lots of drug needs, but which there would be little point in targeting? What could be done to address this problem?
- ? Provide each student with an A4 sheet divided into four sections: one section for the feelings of the doctor who prescribed thalidomide, one for the feelings of the person who was affected, one for the feelings of the parent and one for the feelings of the drug company. When completed, this could be discussed in class or used in a role play exercise.
- Find out what the term 'placebo' means and suggest why some people are given a placebo when new drugs are being trialled (How Science Works).
- There has been a great deal of controversy about the cost of drugs. Discuss whether cheaper drugs to treat HIV and AIDS should be made available to developing countries.
- A PhotoPLUS resource, B1a 4.4 'New drugs' could be used here if not shown previously.

BIOLOGY CONTROLLING INFECTIOUS DISEASE

B1a 4.5 Developing new medicines

LEARNING OBJECTIVES

1 What are the stages in testing and trialling a new drug?
2 Why is testing new drugs so important?

We are developing new medicines all the time, as scientists and doctors try to find ways of curing more diseases. Every new medical treatment has to be extensively tested and trialled. This process makes sure that it works well and is as safe as possible.

A good medicine is:

- **Effective** – it must prevent or cure the disease it is aimed at, or at least make you feel better.
- **Safe** – the drug must not be toxic (poisonous) and there must be no unacceptable side effects.
- **Stable** – you need to be able to use the medicine under normal conditions and store it for some time.
- **Successfully taken into and removed from your body** – a medicine is no use unless it can reach its target in your body. Then your body must be able to remove the medicine once it has done its work.

Figure 1 No matter how many medicines we have, there is always room for more as we tackle new diseases!

Developing and testing a new drug

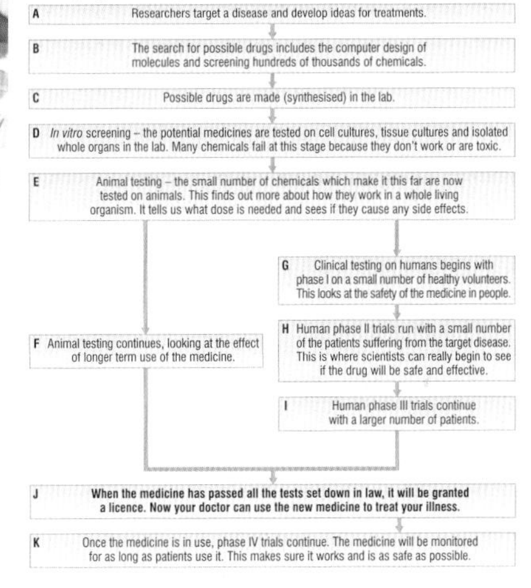

A Researchers target a disease and develop ideas for treatments.
B The search for possible drugs includes the computer design of molecules and screening hundreds of thousands of chemicals.
C Possible drugs are made (synthesised) in the lab.
D In vitro screening – the potential medicines are tested on cell cultures, tissue cultures and isolated whole organs in the lab. Many chemicals fail at this stage because they don't work or are toxic.
E Animal testing – the small number of chemicals which make it this far are now tested on animals. This finds out more about how they work in a whole living organism. It tells us what dose is needed and sees if they cause any side effects.
F Animal testing continues, looking at the effect of longer term use of the medicine.
G Clinical testing on humans begins with phase I on a small number of healthy volunteers. This looks at the safety of the medicine in people.
H Human phase II trials run with a small number of the patients suffering from the target disease. This is where scientists can really begin to see if the drug will be safe and effective.
I Human phase III trials continue with a larger number of patients.
J When the medicine has passed all the tests set down in law, it will be granted a licence. Now your doctor can use the new medicine to treat your illness.
K Once the medicine is in use, phase IV trials continue. The medicine will be monitored for as long as patients use it. This makes sure it works and is as safe as possible.

Figure 2 It takes a long time and a lot of work to develop a successful new drug

GET IT RIGHT!

You won't have to recall all the steps in testing a new drug. However, you will need to know that they are extensively tested and trialled before the public can use them.

70

Human biology

When scientists research a new medicine they have to make sure that all these conditions are met. This is why it takes a very long time. It can take up to 12 years to bring a new medicine into your doctor's surgery. It can also cost a lot of money, up to about £350 million!

a) What are the important properties of a good new medicine?

Testing drugs

We test new medicines in the laboratory. This is to find out if they are toxic and if they seem to do their job. We also trial new medicines on human volunteers. This is to discover if they have any side effects.

Take a look at all the stages a new drug has to go through – no wonder it is a slow process! (See Figure 2.)

Why do we test new medicines so thoroughly?

Thalidomide is a medicine which was developed in the 1950s as a sleeping pill. This was before we had agreed standards for studying the effects of new medicines. In particular, the specific animal tests on pregnant animals which are now known to be essential were not carried out.

Then it was discovered that thalidomide stopped sickness in pregnancy. Because thalidomide seemed very safe for adults, it was assumed that it was also safe for unborn children. Doctors gave it to pregnant women to relieve their morning sickness.

Tragically, thalidomide was ***not*** safe for developing fetuses. It affected many of the women who took the drug in the early stages of pregnancy. They went on to give birth to babies with severe limb deformities.

The thalidomide tragedy led to a new law which set standards for the testing of all new medicines. Since the Medicines Act 1968, new medicines ***must*** be tested on animals to see if they have an effect on developing fetuses.

There is another twist in the thalidomide story. Although thalidomide is never given to anyone who is or might become pregnant, doctors are finding more and more uses for the drug. They can use it to treat leprosy and autoimmune diseases (where the body attacks itself). There have also been some very exciting results using thalidomide to treat certain types of cancer.

b) Why was thalidomide prescribed to pregnant women?

SUMMARY QUESTIONS

1 Copy and complete using the words below:

effective trialled safe medicine stable tested

Every new ...... has to be extensively ...... and ...... before you can use it to make sure that it works well. A good medicine can be taken into and removed from your body, and it is ......, ... and ....

2 a) Testing a new medicine costs a lot of money and can take up to 12 years. Explain the main stages in testing new drugs.
b) What were the flaws in the original development of Thalidomide?
c) Comment on the benefits and drawbacks of using Thalidomide to treat leprosy and cancer.

SCIENCE @ WORK

No new drug could be developed without the skills of a huge team of people – scientists specialising in biology, chemistry and pharmacology, project managers, technicians, vets ...... and almost all of them have science qualifications!

Figure 3 This man has limb deformities because his mother took thalidomide during her pregnancy. He was just one of thousands of people affected by the thalidomide tragedy, many of whom have gone on to live full and active lives.

KEY POINTS

1 When we develop new medicines they have to be tested and trialled extensively before we can use them.
2 Drugs are tested to see if they work well. We also make sure they are not toxic and have no unacceptable side effects.
3 Thalidomide was developed as a sleeping pill and was found to prevent morning sickness in early pregnancy. It had not been fully tested and it caused birth defects. Thalidomide is now used to treat leprosy and other diseases.

71

SUMMARY ANSWERS

1 Medicine tested/trialled, trialled/tested, safe, effective, stable.

2 **a)** Refer to flow diagram on page 70 in Student Book.
b) The drug was not tested on pregnant animals.
c) It treats the diseases but extreme care needs to be taken that women of child-bearing age do not become pregnant whilst taking thalidomide. Therefore they are not prescribed the drug.

Answers to in-text questions

a) Effective, safe, stable, successfully absorbed and excreted from our bodies.

b) To help stop morning sickness.

KEY POINTS

The details of the thalidomide story have been made very clear in this spread and in the suggested activities which include PhotoPLUS resources from the GCSE Biology CD.

B1a 4.6 Immunity

LEARNING OBJECTIVES

Students should learn:

- How the immune system works.
- How vaccination can protect you against bacterial and viral diseases.

LEARNING OUTCOMES

Most students should be able to:

- Describe how the immune system responds to pathogens in the body.
- Explain how vaccines work.
- List some of the advantages and disadvantages of being vaccinated against a particular disease.

Some students should also be able to:

- Evaluate the advantages and disadvantages of being vaccinated against a particular disease.

Teaching suggestions

- **Special needs.** Students can be given sheets with key words in one column with an empty box next to each one. Write the definitions in another column with a letter in a box next to each one. Students to match the words with the definitions and put the letter corresponding to the correct definition in the box next to each word.
- **Gifted and talented.** Students could be given more raw data to analyse, making predictions and extrapolating trend lines.
- **Learning styles**

  Interpersonal: Showing and discussing vaccination scars.

  Intrapersonal: Evaluating the advantages and disadvantages of immunisation.

  Kinaesthetic: Role playing in MMR debate; making a poster.

  Visual: Presenting data on deaths from infectious diseases.

  Auditory: Hearing and explanation of key words.

SPECIFICATION LINK-UP B1a.11.4

- *People can be immunised against a disease by introducing small quantities of dead or inactive forms of the pathogen into the body (vaccination). Vaccines stimulate the white blood cells to produce antibodies that destroy the pathogens. This makes the person immune to future infections by the microorganism, because the body can respond by rapidly making the correct antibody, in the same way as if the person had previously had the disease. An example is the MMR vaccine used to protect children against measles, mumps and rubella.*

Students should use their skills, knowledge and understanding of 'How Science Works':

- *to evaluate the advantages and disadvantages of being vaccinated against a particular disease.*

Lesson structure

STARTER

The work of Edward Jenner – Find a web page about Edward Jenner and vaccination to discuss. (10–15 minutes)

Show injection scars on shoulders: who has any? – How many students have parents with smallpox vaccination scars? Discuss the vaccinations they have had. Has anyone had special vaccinations in order to visit certain countries? Discuss what might be in the injections. (10 minutes)

MAIN

- **Vaccination** – Introduce the key words: 'antigen, immunity, immunisation and vaccination'. Distinguish 'antigen' from similar words like 'antibody, antibiotic and antitoxin', by establishing clear definitions (use for revision cards). Draw out the links between the words, building up the connections in the context of defence against disease.
- Link with a presentation on what happens when you have your jabs and the importance of the second dose and the boosters. This can be illustrated by using a graph. Why do we need to keep up with tetanus jabs?
- If you have it, show extracts of the documentary film on the MMR dilemma – *Does the MMR jab cause autism?* (Horizon, BBC May 2005) followed by discussion and, if time, the suggested continuation below.
- Read discussions about MMR and autism at www.bbc.co.uk. Print the various points made by the public, cut them out and categorise the opinions expressed.
- Using a computer, design a poster or a leaflet persuading parents to have their children vaccinated.
- **Deaths from infectious diseases** – Use graphs to show how many people used to die of infectious/contagious diseases in the past. Information can be obtained from the Internet (try BBC, Wellcome Museum, etc. – either for statistics for individual diseases such as cholera and TB, or for more general information). As a practical exercise to emphasise and visualise the numbers involved, use grains of rice, one for each person who dies. By weighing and calculation, you can work out how many grains per gram and therefore how heavy the piles of rice for each year should be.
- It could be interesting to compare deaths from diseases such as TB in different countries, or to compare the decline in deaths from diseases such as smallpox with the increase in deaths from AIDS, pointing out that there are always some infectious diseases about!

PLENARIES

Key words challenge – Return to the key words on the board and challenge students to make a sentence containing any two of the words. This can be a competition. (5–10 minutes)

Role play exercise – Following the viewing of the video on MMR, ask one student to play Dr Wakefield (see lesson 4.7), one to play a worried parent, one to play someone from the medical authority insisting there is no link to MMR, others to support. (10–15 minutes)

ACTIVITY & EXTENSION IDEAS

- Extend the 'grains of rice' idea from the main section to make a display around the school highlighting how many people in the world die from preventable diseases now. This can be done for different countries, for deaths from specific diseases or as part of a wider campaign to draw attention to poverty in developing countries (campaigns such as *Make Poverty History*).
- If the key words exercise was not done, then ask the students: 'Why do we need booster doses of vaccines?' Show a graph of what happens after a single dose of vaccine, followed by a second dose. Discuss in relation to different diseases, such as polio, diphtheria and some of the less well-known ones, such as yellow fever and cholera.
- What are the consequences of totally eradicating infectious diseases such as smallpox? Discuss this in relation to a possible terrorist attack involving the release of diseases into highly populated areas.
- If students are interested in the past, there are web sites that give details of the causes of death in different parts of the country: data collected from records of death certificates. (See www.statistics.gov.uk.) They could also research parish registers. The history of the Great Plague of 1665: its spread to Eyam and the consequences are well-documented and accessible via the Internet.

Answers to in-text questions

a) Your immune system recognises that the antigens on the microorganisms that get into your system are different from your own cells. Your white blood cells then make antibodies to destroy the pathogens. Once your white blood cells have learnt the right antibody needed to tackle a particular pathogen, they can make that antibody very quickly if you meet the pathogen again, and so you are immune to that disease.

b) They trigger your natural immune response to disease, so your white cells make antibodies. Then if you meet the live pathogen, you can make the right antibodies really rapidly before you get ill.

c) Bacterial: tetanus, diphtheria or any other sensible choice. Viral: measles, mumps, rubella, polio or any other sensible choice.

B1a 4.6 Immunity

LEARNING OBJECTIVES

1. How does your immune system work?
2. How does vaccination protect you against disease?

Every cell has unique proteins on its surface called **antigens**. The antigens on the microorganisms which get into your body are different to the ones on your own cells. Your immune system recognises they are different.

Your white blood cells then make antibodies to attack the antigens. This destroys the pathogens. (See page 65.)

Your white blood cells seem to 'remember' the right antibody needed to tackle a particular pathogen. If you meet that pathogen again, they can make the same antibody very quickly. So you become ***immune*** to that disease.

The first time you meet a new pathogen you get ill. That's because there is a delay while your body sorts out the right antibody needed. The next time, you completely destroy the invaders before they have time to make you feel unwell.

a) How does your immune system work?

Figure 1 No-one likes having a vaccination very much – but they save millions of lives!

Vaccination

Some pathogens can make you seriously ill very quickly. In fact you can die before your body manages to make the right antibodies. Fortunately, you can be protected against many of these diseases by **immunisation** (also known as ***vaccination***).

Immunisation involves giving you a ***vaccine***. A vaccine is usually made of a dead or weakened form of the disease-causing microorganism. It works by triggering your natural immune response to invading pathogens.

A small amount of dead or inactive pathogen is introduced into your body. This gives your white blood cells the chance to develop the right antibodies against the pathogen ***without*** you getting ill.

Then if you meet the live pathogens, your white blood cells can respond rapidly. They can make the right antibodies just as if you had already had the disease. This is how vaccination protects you against disease.

b) How do vaccines work?

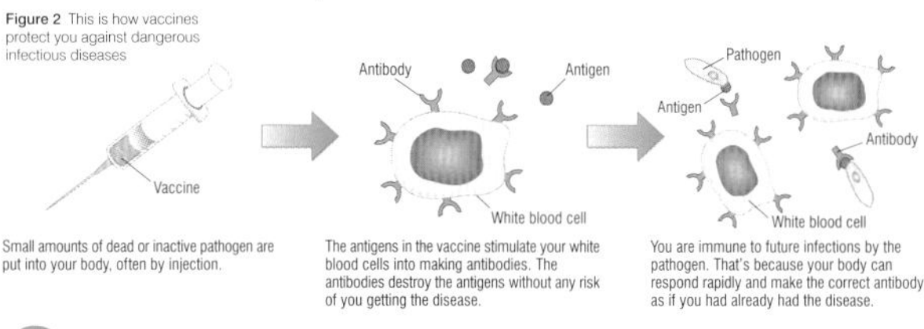

Vaccine
Antibody
Antigen
White blood cell
Pathogen
Antigen
Antibody
White blood cell
Small amounts of dead or inactive pathogen are put into your body, often by injection.
The antigens in the vaccine stimulate your white blood cells into making antibodies. The antibodies destroy the antigens without any risk of you getting the disease.
You are immune to future infections by the pathogen. That's because your body can respond rapidly and make the correct antibody as if you had already had the disease.

Figure 2 This is how vaccines protect you against dangerous infectious diseases

We use vaccines to protect us against both bacterial diseases (e.g. tetanus and diphtheria) and viral diseases (e.g. polio, measles, mumps and rubella). Vaccines have saved millions of lives around the world. One disease – smallpox – has been completely wiped out by vaccinations. It also looks as if polio will disappear as well in the next few years.

c) Give an example of one bacterial and one viral disease which you can be immunised against.

The vaccine debate

No medicines are completely risk free. Very rarely, a child will react badly to a vaccine with tragic results. Making the decision to have your baby immunised can be difficult.

Society as a whole needs as many people as possible to be immunised against as many diseases as possible. This keeps the pool of infection in the population as low as we can get it. On the other hand, by taking your healthy child along for a vaccination, you know there is a remote chance that something will go wrong.

Because vaccines are so successful, we never see the terrible diseases they protect us against. We forget that 60 years ago in the UK thousands of children died every year from infectious diseases. Many more were left permanently damaged. So parents today are often aware of the very small risks from vaccination – but sometimes forget about the terrible dangers of the diseases we vaccinate against.

If you are a parent it can be difficult to find unbiased advice to help you make a decision. The media emphasise scare stories which make good headlines. The pharmaceutical companies want to sell vaccines. Doctors and health visitors can weigh up all the information, but they have vaccination targets set by the government.

Most people agree that vaccination is a good thing both for you and for society. The great majority of parents still choose to protect their children – but it is not always an easy choice. (See pages 74 and 75.)

DID YOU KNOW...

... in the first 10 years of the 20th century, nearly 50% of ***all*** deaths in people aged up to 44 years old were caused by infectious diseases? The development of antibiotics and vaccines means that now only 0.5% of all deaths in the same age group are due to infectious disease!

High levels of antibodies do not stay in your blood forever – immunity is the ability of your white blood cells to produce them quickly if you are re-infected by a disease.

SUMMARY QUESTIONS

1. Copy and complete using the words below:

   antibodies pathogen immunised dead immune inactive white blood

   People can be ...... against a disease by introducing small quantities of ...... or ...... forms of a ...... into your body. They stimulate the .... cells to produce ...... to destroy the pathogen. This makes you ...... to the disease in future.
2. Explain carefully, using diagrams if they help you:
   a) how the immune system of your body works,
   b) how vaccines use your natural immune system to protect you against serious diseases.
3. Make a table to show the advantages and disadvantages of giving your child the MMR vaccine. What would you choose to do?

KEY POINTS

1. You can be immunised against a disease by introducing small amounts of dead or inactive pathogens into your body.
2. Your white blood cells produce antibodies to destroy the pathogens. Then your body will respond rapidly to future infections by the same pathogen, by making the correct antibody. You become immune to the disease.
3. We can use vaccination to protect against both bacterial and viral pathogens.

SUMMARY ANSWERS

1. Immunised, dead, inactive, pathogen, white blood, antibodies, immune.
2. **a)** Every cell has unique proteins on its surface called 'antigens'. Your immune system recognises that the antigens on the microorganisms that get into your system are different from the ones on your own cells. Your white blood cells then make antibodies to destroy the antigens/pathogens. Once your white blood cells have learnt the right antibody needed to tackle a particular pathogen, they can make that antibody very quickly if you meet the pathogen again, and so you are immune to that disease.
   b) A small quantity of dead or inactive pathogen is introduced into your body. This gives your white blood cells the chance to develop the right antibodies against the pathogen without you getting ill. Then if you meet the live pathogens, your body can respond rapidly, making the right antibodies just as if you had already had the disease.
3. **For:** protects against measles and mumps that can kill them or cause permanent damage; protects against rubella damaging unborn babies; keeps the level of diseases very low in the whole of society. **Against:** very small risk of damage to the child.

KEY POINTS

These have been well covered in the lesson suggestions in the main part of the lesson and in the plenary 'Key words challenge'. The summary questions could be set as a homework exercise.

B1a 4.7 How *should* we deal with disease?

SPECIFICATION LINK-UP

B1a.11.4

Substantive content that can be re-visited in this spread:

- *People can be immunised against a disease by introducing small quantities of dead or inactive forms of the pathogen into the body (vaccination). An example is the MMR vaccine used to protect against measles, mumps and rubella.*

Students should use their skills, knowledge and understanding of 'How Science Works' to:

- *explain how the treatment of disease has changed as a result of increased understanding of the action of antibiotics and immunity*
- *evaluate the advantages and disadvantages of being vaccinated against a particular disease.*

BIOLOGY CONTROLLING INFECTIOUS DISEASE

B1a 4.7 How *should* we deal with disease?

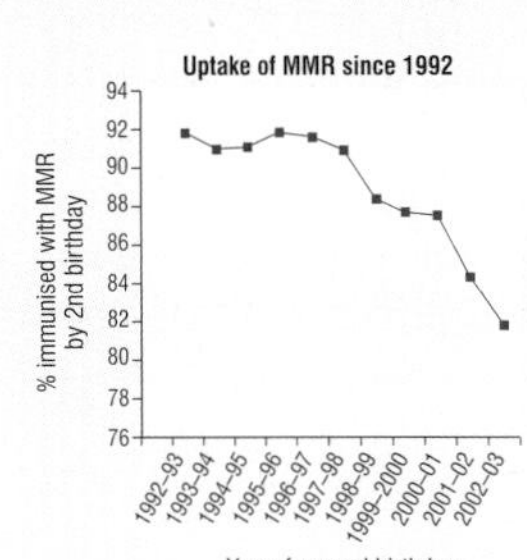

(Source: MMR Decision Making Study, Durham University)

Media reporting of the 'controversial MMR vaccine' helped to feed people's fears about the safety of their children. In fact, as far as the great majority of scientists and doctors were concerned, there was *no* controversy. All the other evidence before and since has shown no link between MMR and autism.

ACTIVITY

Find out as much as you can about the MMR vaccine and the way the media reported the situation (www.thevaccinesite.org is a good starting point).
Either Design a web site for parents which answers the sort of questions they might ask about MMR. Make it user-friendly – the sort of thing health workers could use to reassure couple [B] about the importance of the MMR vaccine.
Or Produce a PowerPoint® presentation on the importance of responsible media reporting of science and medicine, using the MMR case as your main example.

The MMR dilemma

MMR is a vaccine which protects you against measles, mumps and rubella. Measles and mumps can both cause serious problems such as brain damage and death. Rubella can damage unborn babies. Once the MMR vaccine was introduced into the UK, cases of these diseases fell rapidly until no more children died of measles or mumps.

In February 1998, Dr Andrew Wakefield and his team published a paper in *The Lancet* (a medical journal). His research suggested that the MMR vaccine might be linked to the development of autism in some children. (Autistic children tend to be very withdrawn from people and cannot cope easily with normal life.)

The story got a lot of media coverage. People became very worried and uptake of the MMR vaccine started to fall. By 2000, the number of measles cases in Ireland had gone up hugely. For the first time in years, two babies died of the disease. By 2001, uptake of the MMR vaccine in England had also dropped – from 92% to 75%. These levels are not enough to keep people safe.

It turned out that the research had been done on a tiny group of twelve children. The scientist had been paid £55 000 by the parents of some of the children to prepare evidence against the vaccine for a court case. What's more, Dr Wakefield had developed some measles treatments which would not have been used if parents were confident in MMR. No-one knew this when he published his results.

This research has now been completely discredited. Many studies on thousands of children have shown no link between MMR and autism. Sadly, children have been damaged and some have died from measles because of the poor research and irresponsible media reporting that managed to scare so many parents unnecessarily.

I'm sure doctors wouldn't recommend the vaccine if they didn't think it was safe. We trust our GP, and we don't want to risk Kirsty having any of these dreadful diseases so she's going to have the MMR.

There's no smoke without fire. I'm sure these scientists couldn't publish their research if it wasn't true. We just daren't risk Cameron becoming autistic so we're not giving him the MMR.

74

Teaching suggestions

The following suggestions can be set as homework exercises or form the basis of class discussions prior to carrying out one of the suggested activities. The 'I did things *my* way . . . ' suggestion gives practice in writing prose. A selection could be chosen to be read to the class and followed up by contrasting the testing of new treatments nowadays, and whether any in the class would volunteer (see lesson 4.5). Information about MMR can be found at www.sciencemuseum.org.uk as well as at www.who.org.

- **Killer 'flu: what can we do?** – Research the epidemic of 1918 and why it caused so much devastation. Remember that it occurred immediately after WW1, ask: 'Do you think that this had an effect on the numbers of people affected?' Students can find out what they can about other major 'flu epidemics, with respect to what strain caused them and how far apart these epidemics have been. They should consider the chances of another major epidemic occurring soon and think about why some people get the 'flu and others do not. Draw up a list of suggestions of the precautions you would need to take in the event of such a major epidemic occurring in the near future.
- **Understanding the 'flu vaccine** – Find pictures of the virus that causes influenza and project them. Discuss the structure of the virus, the way in which it replicates and what a mutation might do to its structure. Help students understand how the virus mutates and that new vaccines need to be developed every year. They could assess the success of the vaccination of elderly and vulnerable people. Ask: 'How does this benefit the NHS? Should everyone be offered the vaccine?'
- **'I did things *my* way . . .'** Ask the students to choose one of the following, but to do some further research so that the accounts they write are as accurate as possible:
  - Imagine that you are one of the people used in Lady Mary Wortley Montagu's experiments with smallpox. Describe what your feelings were when being given the smallpox and how you were treated.

I did things my way . . .

Each of these scientists made a great breakthrough in the treatment of disease – but the methods they used would not be acceptable today!

Edward Jenner (1794): I was sure it was the pus from cowpox which was protecting the milkmaids. I knew from smallpox! I proved my point by scratching the pus from some cowpox spots into the skin of young James Phipps, who soon went down with cowpox. Two months later I scratched lots of pus from a smallpox blister into his arm – and the young lad showed no signs of illness at all!

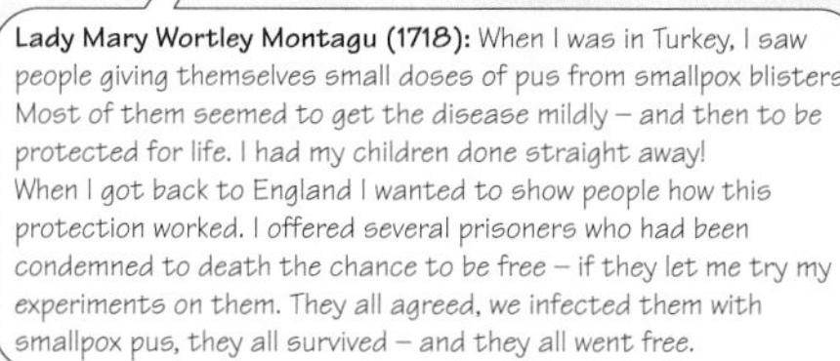

Lady Mary Wortley Montagu (1718): When I was in Turkey, I saw people giving themselves small doses of pus from smallpox blisters. Most of them seemed to get the disease mildly – and then to be protected for life. I had my children done straight away!
When I got back to England I wanted to show people how this protection worked. I offered several prisoners who had been condemned to death the chance to be free – if they let me try my experiments on them. They all agreed, we infected them with smallpox pus, they all survived – and they all went free.

Louis Pasteur (1885): I had been working on my vaccine against rabies for years when I came up with a new idea, using the spinal cords of rabbits which had died of rabies. I had not had the chance to test my vaccine properly when a little boy was brought to me. He'd been badly bitten by a dog with rabies two days earlier. Two doctors agreed he was sure to die so – with the permission of his family – I gave him thirteen injections of my vaccine. He went home – and he never developed rabies. It was a triumph but I told no-one except my dear wife. I needed more evidence!

ACTIVITY

It worked – but was it right?

Some of the early workers on vaccines had some amazing successes – but the methods they used would not be acceptable today. Compare their methods with what you found out about modern medicines testing on pages 70 and 71.

a) Take each scientist in turn and explain how they carried out their work and why it would not be allowed today.
b) Do you think that the methods they used can be justified because of the great advances in medicine they brought about?
c) Draw up a schedule for how a new vaccine might be developed today.

ACTIVITIES & EXTENSIONS IDEAS

- **Killer 'flu** – The activities suggested in the student book link with the teaching suggestions given above. 'Make up a slogan' could be run as a competition in the class and any posters designed could be displayed around the laboratory.
- **It worked – but was it right?** – The activity suggested in the student book is a useful homework project and brings together many of the concepts covered in this chapter.
- **Compile a list of diseases** – The list should show diseases that affect large numbers of people in developing countries and for which there are no simple, cheap treatments available at the moment.
- **Is malnutrition a disease?** – Lack of the necessary vitamins in the diet results in deficiency diseases, which could affect a person's immune system and their ability to recover from an infectious condition. This topic could be discussed with reference to conditions in refugee camps.
- **Do we rely too much on drugs and medicines?** – Is it too easy to take a pill for the slightest ache or pain? Should drugs like statins and other former prescription drugs be available over-the-counter? Does this lead to self-diagnosis rather than making an appointment to see the doctor?

– Young James Phipps kept a diary (called a journal in those days, probably), which has just been discovered. Unfortunately, it fell apart as soon as it was found. What do you think he wrote in his diary during the time that Edward Jenner was using him in his experiment?

– Imagine a conversation between Louis Pasteur and his wife about his work. She has complained about the time he is spending on his work, and he is trying to tell her what he has achieved with the rabies vaccine and why he has kept quiet about it.

- **The MMR dilemma.** If not already used in the previous spread, show students of the documentary film *Does the MMR jab cause autism?* (Horizon, BBC May 2005). The class could discuss the film and the subsequent investigation which discredited Dr Wakefield's results. This provides an excellent opportunity to discuss 'How Science Works' regarding issues of reliability and validity of data. This could be followed by consideration of the consequences of not giving the MMR jab or of giving separate vaccinations for the diseases.
- **Measles, mumps and rubella – what are the risks?** Because so many children have been vaccinated, these diseases are relatively rare in the UK. Find out what the symptoms and effects are. Why can they become life-threatening? This activity can be extended to other infectious diseases for which we have vaccinations, such as diphtheria and polio. Some of this information could be used in the activities, designed to persuade parents that their children should be vaccinated, suggested in the Student Book.

SUMMARY ANSWERS

1 a) i) A disease which is caused by a microorganism and which can be passed from one person to another.

ii) A very small organism which can only be seen using a microscope, e.g. bacteria, virus.

iii) A microorganism that causes disease.

iv) A poison made by a bacterium or virus.

b) **Bacterium:** single cell with cytoplasm, membrane, cell wall, genetic material but no nucleus; variety of different shapes and sizes (rod, round, comma, spiral); some cause disease, but many harmless and some really useful.
Virus: smaller than bacteria, regular shapes, protein coat surrounding simple genetic material; do not carry out any functions except reproduction; can only reproduce by taking over another living cell; all naturally occurring viruses cause disease.

c) Bacteria and viruses cause disease because once inside the body they reproduce rapidly. Bacteria split in two. They often produce toxins that affect your body, and sometimes they directly damage your cells. Viruses take over the cells of the body as they reproduce, damaging and destroying them and they very rarely produce toxins. The disease is caused by the damage and toxins produced by the pathogens and also by the way your body responds to the damage and the toxins.

2 The answer to this question must cover how diseases are spread: droplet infection, direct contact, contaminated food and drink, and through breaks in the skin. Any sensible precautions, such as washing hands, getting rid of used tissues, hand in front of mouth when coughing, etc. Credit should be given for clarity of explanations and suitability for target audience.

3 a) Medicine which makes you feel better and relieves symptoms, e.g. painkiller; medicine which makes you better destroys pathogens causing disease, e.g. antibiotics.

b) An antibiotic is a chemical/medicine which can kill disease-causing bacteria inside your body.

c) Not all antibiotics kill all bacteria; bacteria can develop antibiotic resistance.

d) They have developed resistance to different antibiotics through a process of natural selection. They are then spread from patient to patient on the hands of doctors and nurses, or on the objects around in a hospital.

e) Use antibiotics carefully, only when they are needed, and make sure that people always finish the course.

4 a) Safety issues: not poisonous; no serious side effects and the medicine must work.

b) The number of stages of testing that is required legally means it takes a lot of time and a lot of people to develop one new drug.

c) Credit for arguments and ideas about relevant points – could use example of thalidomide or Pasteur, Jenner, etc.

5 a) Student diagram. [Mark diagram for accurate points.]

b) Vaccination has virtually wiped out these diseases. Vaccines work by triggering the natural immune response to invading pathogens.

CONTROLLING INFECTIOUS DISEASE: B1a 4.1 – B1a 4.7

SUMMARY QUESTIONS

1 a) Define the following terms:
i) infectious disease
ii) microorganism
iii) pathogen
iv) toxin.

b) What are the main differences between bacteria and viruses?

c) How do tiny organisms like bacteria and viruses make a large person like you ill?

2 There is going to be a campaign to try and stop the spread of colds in Year 7 of your school. There is going to be a poster and a simple PowerPoint® presentation.
Make a list of all the important things that the Year 7 children need to know about how diseases are spread. Also cover how to reduce the spread of infectious diseases from one person to another.

3 a) What is the difference between a medicine which makes you feel better and a medicine which actually makes you better?

b) What is an antibiotic?

c) What are the limitations of antibiotics?

d) Where have antibiotic-resistant bacteria (like MRSA found in hospitals) actually come from?

e) What can we do to prevent the problem of antibiotic resistance getting worse?

4 a) Why do new medicines need to be tested and trialled before doctors can use them to treat their patients?

b) Why is the development of a new medicine so expensive?

c) Do you think it would ever be acceptable to use a new medicine before all the trials had been completed?

5 a) Draw a labelled flow diagram to show how your immune system works.

b) Explain why thousands of children in the UK no longer die of diphtheria or become paralysed by polio.

c) Make a table to show the risks and benefits of having children vaccinated against serious diseases.

d) The media like a good story. Many people read the papers and watch television.
Explain why it is important that stories about medical issues like vaccination should be reported very carefully in the media.

EXAM-STYLE QUESTIONS

1 The table below contains statements about chemicals that act against pathogens.
Match the list of chemicals **A**, **B**, **C** and **D** with the statements **1** to **4** in the table.
A Antibodies **B** Antitoxins
C Antibiotics **D** Antiseptics

	Statement
1	Counteract poisons released by pathogens
2	Produced by the white blood cells to destroy particular bacteria or viruses
3	Kill infective bacteria outside the body
4	Drugs that kill infective bacteria inside the body

2 Which one of the following is **not** used by white blood cells to protect us against pathogens?
A Production of antibodies to destroy bacteria
B Sealing of wounds to prevent infection
C Production of antitoxins to counteract poisons released by pathogens
D Ingestion of pathogens.

3 MRSA is a bacterium that kills around a thousand hospital patients each year. Which one of the followin has led to these MRSA infections?
A Ineffective vaccines
B Overcrowding in hospitals
C Antibiotic resistance
D Shortage of hospital equipment

4 Which of the following would **not** help control the spread of MRSA?
A Medical staff having regular health checks
B Medical staff washing their hands between seeing patients
C Visitors washing their hands as they enter and leave the hospital
D Cleaning hospitals with antiseptics

5 New drugs are thoroughly trialled and tested to ensur they have certain features that make them good medicines. When tested on human volunteers, it is no possible to keep every control variable constant. So h can researchers make their data reliable?
A Use only animals instead of humans in tests.
B Only test people over the age of 65.

c)

Risks of having vaccine	Benefits of having vaccine
Very tiny risk of reaction to vaccine and child suffering permanent damage	Children protected against relatively high risk of getting a dangerous disease if no-one is vaccinated
	Society protected against the risk of the disease killing children
	Financial benefit to society and individuals, as so many serious and fatal diseases no longer need hospital treatment etc.

d) Many people get their ideas about science from what they see in the media. If medical issues are reported in a biased or inaccurate way people make decisions based on poor information, e.g. children having died because the levels of MMR uptake fell after the media reported the work of Dr. Andrew Wakefield.

Summary teaching suggestions

- **Special needs**
  - In question 1, where there are definitions, students could be given the words on cards and the definitions separately and asked to match the word with the definition.
  - In question 2, the students could be given a series of cards or pictures of the way in which diseases are spread, and asked to select ones which they would use and then make a poster.
- **Learning styles**
  Interpersonal: Questions 1c), 3e), 4 and 5d) all lend themselves to a class discussion. Following this, the students could then write their answers.

C Use as large a sample of people as possible.
D Calculate the mean age of the people tested. (1)

Measles is an extremely infectious disease that is caused by a virus. Measles can cause brain damage and death. To try to prevent epidemics of measles, the MMR vaccine was developed.

(a) Which other two diseases does the MMR vaccine protect against? (2)

(b) The MMR vaccine contains the virus that causes measles. Why does this virus not cause measles when the MMR vaccine is injected into a child? (1)

(c) Name the type of chemical released by certain white blood cells to destroy viruses. (1)

(d) Explain how vaccinating a child with the MMR vaccine makes them immune to measles. (5)

(e) A child that has not been immunised with the MMR vaccine develops measles. Suggest a reason why antibiotics will not cure the child of measles. (1)

(f) In the UK, by 2001, the uptake of the MMR vaccine had fallen from 92% to 75%. Suggest one reason for this decrease. (1)

(a) A health centre gives the following advice about antibiotics to its patients. In each case suggest **one** reason why the advice is given.
- (i) Do not take an antibiotic for a viral infection like a cold or the flu.
- (ii) Do not take an antibiotic that has been prescribed for someone else.
- (iii) Ask whether an antibiotic is likely to help your illness or whether there are alternatives.
- (iv) Do not stop taking the antibiotic once you feel better – always complete the course of drugs. (4)

(b) Describe how antibiotic resistance arises. (5)

HOW SCIENCE WORKS QUESTIONS

Look at the newspaper advert below and answer the following questions:

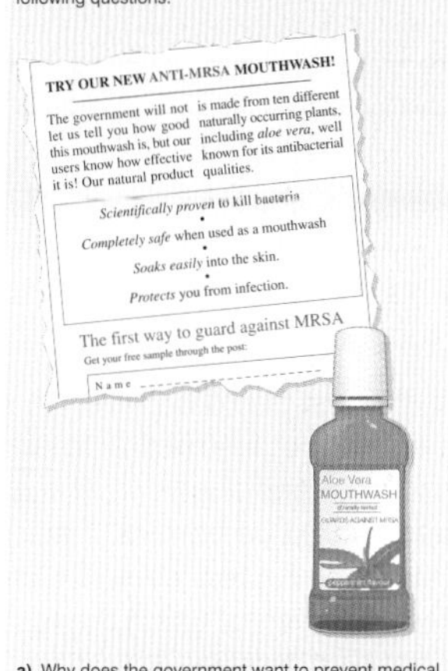

TRY OUR NEW ANTI-MRSA MOUTHWASH!
The government will not let us tell you how good this mouthwash is, but our users know how effective it is! Our natural product is made from ten different naturally occurring plants, including aloe vera, well known for its antibacterial qualities.
Scientifically proven to kill bacteria
Completely safe when used as a mouthwash
Soaks easily into the skin.
Protects you from infection.
The first way to guard against MRSA
Get your free sample through the post:
Name
Aloe Vera
MOUTHWASH

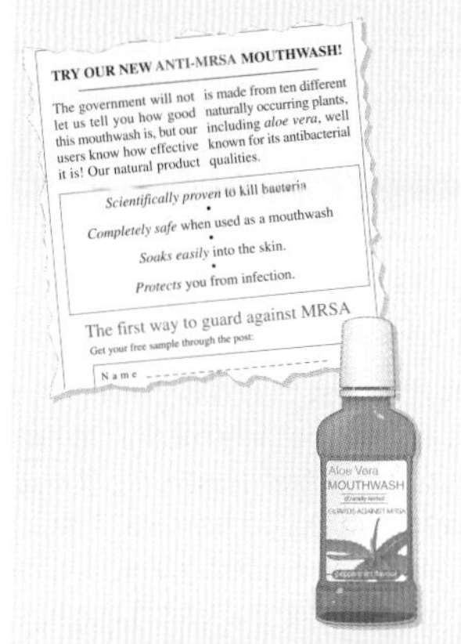

a) Why does the government want to prevent medical claims for this mouthwash? (1)

b) Describe the tests that you think should have been carried out to make the claim that it is 'Scientifically proven to kill bacteria'. (4)

c) Why might you mistrust any claims made by this company? (2)

d) Mr Skeptic commented that 'it must be crazy to think that a plant could kill something as difficult as MRSA'. What do you think? (1)

e) Describe the sort of investigation needed to be able to say that the mouthwash is 'completely safe'. (3)

Intrapersonal: Students' answers to many of the questions depend on them considering the issues and coming to their own opinions.
Visual: Diagrams are helpful in answering questions 1b) and 5a).

Homework

Question 5d) could be used for homework. Students could be directed to find and comment on a story about a medical issue from either a newspaper article or a news item on the TV.

When to use these questions?

- Many of these questions can be used to start a class discussion linked with a particular spread.
- Each question has links with a particular spread and would provide an excellent stimulus for a revision session on that topic.

EXAM-STYLE ANSWERS

1 A 2
B 1
C 4
D 3 *(1 mark each)*

2 B *(1 mark)*

3 C *(1 mark)*

4 A *(1 mark)*

5 C *(1 mark)*

6 a) Mumps *(1 mark)*
Rubella (German measles) *(1 mark)*

b) The virus has been weakened/made inactive. *(1 mark)*

c) Antibodies. *(1 mark)*

d)
- The injection of MMR vaccine introduces an inactive form of the measles virus into the body.
- White blood cells respond to antigens on the surface of the measles virus by developing antibodies.
- That are specific to the measles virus.
- Should living measles virus enter the body at some later date, white blood cells will rapidly produce these antibodies again.
- The antibodies which destroy the measles viruses that have infected the body before they can multiply and cause harm.

(1 mark for each)

e) Viruses live inside body cells and antibiotics are not able to reach them and so cannot destroy them. *(1 mark)*

f) Parents were afraid of side-effects from the MMR vaccine (as a result of research, now discredited, suggesting a link between the MMR vaccine and autism). *(1 mark)*

7 a) (i) Viruses live inside cells where antibiotics cannot reach them and hence any antibiotic taken would not be effective. *(1 mark)*

(ii) Different antibiotics are designed to work in different ways depending on the type of disease. Someone else's antibiotic might not be effective against your disease. *(1 mark)*

(iii) To prevent more resistant strains of bacteria developing, it is important only to use antibiotics when they are essential. *(1 mark)*

(iv) The length of the course of drugs is calculated to kill all the pathogens it is aimed at. Stopping the course early allows the most resistant bacteria to survive and multiply. *(1 mark)*

b)
- Within any population of bacteria there are many different types that have evolved over many years by natural mutation.
- An antibiotic will kill almost all these bacteria except the few that have a natural mutation that makes them resistant to the antibiotic.
- Under normal circumstances the surviving bacteria possess no advantage and so the body's immune system destroys them.
- However, if the antibiotic is present there is a greater chance that resistant bacteria will survive while the non-resistant ones are killed.
- In time, and with continued use of the antibiotic, all the non-resistant bacteria are destroyed and only the resistant strains survive.

(1 mark for each)

HOW SCIENCE WORKS ANSWERS

a) The Government wants to prevent medical claims for this mouthwash because it has not been tested scientifically to be effective against MRSA bacteria.

b) The tests that should have been carried out to make the claim that it is 'Scientifically proven to kill bacteria' are as follows:
- controlled investigation
- bacteria grown
- mouthwash added
- bacteria killed.

c) The claims might be mistrusted because:
- they are not telling the whole truth
- they might be biased
- the advert is misleading as it refers to 'bacteria' as if they were just one type.

d) Many plants are effective at killing bacteria.
- MRSA might be killed by plant extracts.

e) The sort of investigations needed to be able to say that the mouthwash is 'completely safe' included:
- could be tested on animals
- could be tested on human cells grown in a laboratory
- could be tested on the human population and any adverse reactions recorded.

B1a Examination-Style Questions

Examiner's comments

This examination paper is intended to provide an end of unit B1b test with questions in the style of those that students are likely to encounter in their actual examination.

It would be most useful as a class exercise completed under examination conditions, especially after a homework has been set to revise the whole unit. In this way it would introduce students to working without assistance and under time constraints. One problem with this approach is that it might reinforce wrong answers, especially in the minds of those that do not revise thoroughly. It is therefore important that, when marking the papers, the responses are analysed and a list of the common mistakes and wrong answers is drawn up. This allows the essential follow-up session to focus on these points rather than giving equal emphasis to all answers. It is important that all students leave with a complete set of correct answers to avoid wrong answers being reinforced. Whether this is achieved by the students correcting their own scripts as the questions are gone over, or by issuing a full printed set of correct answers, is a matter of teacher judgement.

If preferred, the test could be given to students to do as a homework and under instruction to use textbooks and their own notes to research their answers. The expectation would be that most answers would be correct. If students do not have access to textbooks at home the test could be carried out as a class exercise. The purpose of using the test in this manner is to reinforce what has been taught rather than to test what has been learnt. Even when used in this way, it is still important that the students' responses are checked thoroughly as some students will be convinced that they know a correct answer and so not bother to check its accuracy. If these errors are not rectified misconceptions can be reinforced in their minds.

Answers to Questions

Biology A

1 **A** **4**

B **2**

C **1**

D **3** *(1 mark each)*

2 (a) **B** *(1 mark)*

Common error is:
Option A (fewer individuals smoked). The crucial word is 'individuals'. The table states that the figures are '**percentages**' of the population. We would need to know the total population size for both years before we could conclude that more **individuals** smoked in 1978.

(b) **A** *(1 mark)*

All options seem to attract a number of students.
Option B – blood tests might not detect an occasional smoker because nicotine does not remain in the blood indefinitely.
Option C – even if tobacconists were willing to divulge information on their customers, those buying the cigarettes are not necessarily the ones smoking them.
Option D – students are attracted by words such as 'researcher' and 'recorded' because they sound scientific. As smokers do not smoke continuously, a proportion of those observed not smoking will in fact be smokers.

continues opposite ›

EXAMINATION-STYLE QUESTIONS

Biology A
In question **1** match the letters with the numbers.
Use **each** answer only **once**.

1 The menstrual cycle in women is controlled by hormones.
Match words **A**, **B**, **C** and **D** with the spaces in **1** to **4** in the sentences.

A LH **B** Pituitary gland
C FSH **D** Oestrogen

Each month, the hormone ……**1**…… is produced from the ……**2**……. and causes an egg to mature in a woman's ovaries. The ovaries, in turn, produce a hormone called ……**3**…… that stimulates the production of another hormone called ……**4**…… which causes the mature egg to be released. *(4 marks)*

See pages 28–9

GET IT RIG
Start with what you k
In question 1 you have
match four letters with
numbers. Read the **wh**
question first and then
up the pairs that you a
certain about. If you c
match all four, all well
good. If however, you
only sure about three
them, do not worry as
last one can be paire
process of eliminatio
the best examination
technique (far better
all four) but preferab
leaving blank spaces

Question 2
In **each** part choose only **one** answer.

2 Cigarette smoking has been shown to increase the chances of early death due to diseases such as lung cancer, bronchitis and emphysema.
The table below shows the percentage of the adult UK population by age group who were smokers. The figures were taken at five different times over a 25-year period.

	Age 16–19	Age 20–24	Age 25–34	Age 35–49	Age 50–59	Age 60+
1978	34	44	45	45	45	30
1988	28	37	36	36	33	23
1998	31	40	35	30	27	16
2000	29	35	35	29	27	16
2003	26	36	34	30	25	16

Figures show the percentage UK population who were smokers.

(a) Which of the following statements is the only one supported by the data in the table?
A Fewer individuals smoked in 2003 that in 1978.
B A greater percentage of the population smoked in 1978 than in 2003.
C In 1978 a greater percentage of 20–24-year olds smoked than 25–34-year olds.
D From 16 years old, the proportion of the population who smoke declines steadily with age. *(1 mark)*

(b) What would be the most reliable way of collecting the data in the table?
A Samples of people were asked about their smoking habits.
B Samples of people had their blood tested for nicotine.
C Tobacconists were asked about the people who buy cigarettes.
D Researchers recorded the number of people smoking in various places. *(1 mark)*

See page 5

GET IT RIGHT!

Advice is given in the Student Book on answering questions involving matching four words with numbers that replace words in a sentence. Students should be trained to read all four sentences first before trying to match any of the words. They should then match the pair that they are most confident is correct followed by the one they are next most certain about. Even if they don't know the last pairing it can still be correctly deduced by a process of elimination. To convince students of the value of this approach, try constructing similar questions but include one matching pair that is clearly not within their breadth of knowledge. By elimination they can even gain a mark for something they never knew! One caveat however; it is important to impress upon them that this is only a last resort and that the most certain way of gaining marks is to know and understand the biology, thereby leaving nothing to chance.

Advice is also given on using mark allocations as a guide to the detail needed in an answer. Students need to realise that examiners do not read an answer and then give an impression mark. Instead they have very precise mark schemes with a specific number of points, each of which carries 1 mark. If there are 5 marks, almost invariably means there are 5 marking points.

BUMP UP THE GRADE

To say 'read the question carefully' seems blindingly obvious. Yet, however many times one repeats it, candidates still fail to obtain marks as a result of not reading questions thoroughly. Often candidates do not spend sufficient time re-reading the question and trying to understand what the examiner wants by way of a response. Rather, they see a key word or words and then simply write what they know about that topic. It is not just in the rubric of a question where careless reading costs marks but also in the reading of tables and graphs. There are a number of occasions in these end of unit questions that illustrate how careful reading could earn extra marks and so bump up the grade. It is often helpful to get students to look at some examination papers and to state what they think the examiner wants in particular questions. They often read more accurately when doing this without the pressure that they feel in an examination. The next stage is then to persuade them (using the examples of where they lost marks in this end of unit examination) that taking time and thinking what is required is an effective means of improving their grade.

ology B

In 2003 a health survey was carried out on over 8000 men and women over 16 to find out their blood cholesterol levels. From this survey a table of results was made. It shows . . .

1 The mean level of cholesterol for each group.

2 The percentage of the group with a cholesterol level above the recommended healthy limit of 5.0 mmol/dm³.

	Age 25–34 %	Age 35–44 %	Age 45–54 %	Age 55–64 %	Age 65–74 %
Men					
Group size	718	789	675	585	401
Mean ($mmol/dm^3$)	5.3	5.8	5.9	5.8	5.5
% 5.0 and above	59.8	76.9	81.0	79.7	67.4
Women					
Group size	717	794	674	603	455
Mean ($mmol/dm^3$)	5.0	5.4	5.8	6.3	6.2
% 5.0 and above	54.9	69.3	79.3	83.7	77.1

(a) The sample groups were of different sizes. Suggest which one provides the most reliable result and why. *(2 marks)*

(b) To remain healthy a maximum level of 5.0 $mmol/dm^3$ for blood cholesterol is often recommended.

(i) Which group is most at risk from ill-health due to their cholesterol levels? *(1 mark)*

(ii) Which group is the second most at risk from ill-health due to their cholesterol levels? *(1 mark)*

(iii) What in particular is the health risk from having a high blood cholesterol level? *(2 marks)*

(c) In which organ of the body is cholesterol made? *(1 mark)*

(d) What a person eats affects how much cholesterol the body makes. What other factor also affects how much is made? *(1 mark)*

(e) The amount and type of fat in the diet affects the level of cholesterol in the blood. For each of the following, state whether their presence in the diet increases, decreases or has little effect on the blood cholesterol level.

(i) saturated fats

(ii) mono-unsaturated fats

(iii) polyunsaturated fats *(3 marks)*

GET IT RIGHT!

Check the mark allowance. The number of marks allowed for a question gives you valuable information on the extent and detail of an answer. For example, in question 1 part (b) (iii), there are 2 marks compared to only 1 mark for the other sections of part (b). This suggests that two health risks are needed. (Think about which parts of the body are affected.)

See page 8

See pages 42–3

› *continues from previous page*

Biology B

1 (a) The sample for women aged 35–44. *(1 mark)*

Because this is the largest sample (numerically). *(1 mark)*

No mark for just 'aged 35–44' – 'women' must be stated.

(b) (i) Women aged 55–64 *(1 mark)*

(ii) Men aged 45–54 *(1 mark)*

(iii) *For 2 marks*
Either cardiovascular disease *or* disease of the heart and blood vessels.
For 1 mark
Either heart disease *or* disease of the blood vessels.

(c) The liver *(1 mark)*

(d) Inherited/genetic factors *(1 mark)*

(e) (i) Increases

(ii) Little effect

(iii) Decreases *(1 mark each)*

B1b | Evolution and environment

Key Stage 3 curriculum links

The following link to **'What you already know'**:

- Fertilisation in humans and flowering plants is the fusion of a male and female cell.
- Environmental and inherited causes of variation within a species; selective breeding can lead to new varieties.
- Ways in which living things and the environment can be protected and the importance of sustainable development. Some organisms are adapted to survive daily and seasonal changes in their habitats. Predation and competition for resources affects the sizes of populations.
- Food webs composed of several food chains. Toxic materials can accumulate in food chains.

QCA Scheme of work
7B Reproduction
7C Environment and feeding relationships
7D Variation and classification
8D Ecological relationships
9A Inheritance and selection
9D Plants for food

RECAP ANSWERS

1 **a)** Information from two different parents combines in different ways so no two offspring are the same as either their parents or their siblings. This means every member of a species is slightly different from the others – and some are very different.

b) Things like the amount of food available to animals, or the amount of light or minerals available to plants, make a big difference as to how well they grow and so they can be genetically quite similar but look very different.

2 The animals or plants that are best suited to the conditions will be most likely to survive and breed.

3 You would begin by choosing two dogs (one male and one female) with the curliest tails you can find, and allowing them to breed. Some of the puppies should have even curlier tails than either of their parents. Choose the puppy in the litter that has the curliest tail and use that puppy to breed with another curly-tailed dog until you get puppies with the very curly tail that you want. [Look for an understanding of the principle of selecting parents and then offspring with the desired feature.]

4 There are many things that could be mentioned here, so any sensible and relevant points well-explained could be given credit.

5 **a)** A group of organisms of the same species living in the same habitat at the same time.

b) Animals compete for food, space to live, and mates. Their population numbers can also be affected by the number of predators and disease.

c) Bacteria compete for nutrients (food), oxygen and space. [Any other sensible suggestions can be accepted.]

6 **a)** Accept any suitable examples, but each must begin with a plant, e.g. grass, cow, human.

b) The toxin builds up in the tissues of the animals at each trophic level in the food chain. In the lower levels of the chain, the concentration of the toxin is so low that it does not cause harm. As the food chain gets longer, the concentrations increase, as the members of the higher levels eat more of the organisms in the level beneath them until they reach the top carnivores, which are poisoned and die.

Activity notes

The activity suggested in the student text follows on from the relationship between the fig wasp and its adaptations. As this activity requires some research, the students need to be forewarned; it could be appropriate to set the research as a homework exercise. Some hints could be given:

- Many plants depend on one type of insect for pollination and this is shown in adaptations of the shape of the flower and the arrangement of the floral parts, their scent or their colour.
- Most parasites have special adaptations to ensure that they pass from one specific host to another.

It would be good to ensure that there are plant and animal examples chosen.

BIOLOGY

B1b | Evolution and environment

What you already know

Here is a quick reminder of previous work that you will find useful in this unit:

- In sexual reproduction, fertilisation happens when a male and a female sex cell join together.
- There are variations between members of the same species. Because of these variations, some individuals are more successful than others. In harsh conditions, the fittest survive.
- In sexual reproduction, information from two parents is mixed to make a new plan for the offspring. This leads to variation between members of a species.
- Variation between organisms of the same species has ***environmental*** as well as ***inherited causes***.
- You can produce animals and plants with the features you want by a process called **selective breeding**.
- There are ways in which we can protect living organisms and the environment they live in.
- Sustainable development (when we replace the plants and animals we use) is becoming more and more important.
- Competition for the resources available is one thing which affects the size of populations of animals and plants.
- Toxic (poisonous) materials that we produce can build up in food chains and cause big problems.

Our environment is precious and needs protecting

RECAP QUESTIONS

1 a) How does sexual reproduction lead to variation within a species?
b) How can the environment cause variation in a species?
2 What do we mean by 'the fittest survive'?
3 Imagine that you want to produce a type of dog with a very curly tail using selective breeding. Explain how you would do this. You can use diagrams if you think they will be helpful.
4 Think of as many ways as you can in which we might protect the environment and the living things in it. Explain why it is important to do this.
5 a) What do we mean by a 'population' of animals or plants?
b) What sort of things do animals compete for? List as many things as you can which might affect the numbers of animals in a population.
c) There is competition in a culture of bacteria. What sort of things might bacteria compete for?
6 a) Give two examples of food chains.
b) Give an example of the way toxins can build up in a food chain.
c) Explain how it is the carnivores at the end of the chain which are most likely to be affected.

80

Teaching suggestions

- 'Recap questions' can be carried out using a 'Blockbusters' game.
- Some of the 'Recap questions' can be used to trigger class discussion of principles: for example, question 3 can be set up on the board with pictures of dogs with curly tails and possible offspring. Then the students can be asked to choose the puppies and explain why.
- **Selective breeding** – Search the web for pictures to show selective breeding of horses, cattle, plants; bread wheat is a well-documented example of selective breeding of an important crop and there are other examples in cattle. When students understand the principles of choosing characteristics, ask them to suggest how they would set about breeding a champion racehorse. They need to make a list of the desirable

SPECIFICATION LINK-UP

Unit Biology B1b

Evolution and environment

What determines where particular species live and how many of them there are?

Animals and plants are well-adapted to survive in their normal environment. Their population depends on many factors including competition for the things they need, being eaten for food and being infected by disease.

Why are individuals of the same species different from each other? What new methods do we have for producing plants and animals with the characteristics we prefer?

There are not only differences between different species of plants and animals but also between individuals of the same species. These differences are due partly to the information in the cells they have inherited from their parents and partly to the different environments in which the individuals live and grow. Non-sexual reproduction can be used to produce individuals exactly like their parents. Scientists can now add, remove or change genes to produce the plants and animals they want.

Why have some species of plants and animals died out? How do new species of plants and animals develop?

Changes in the environment of plants and animals may cause them to die out. Particular genes, or accidental changes in the genes of plants or animals may give them characteristics which enable them to survive better. Over time this may result in entirely new species.

How do humans affect the environment?

Humans often upset the balance of different populations in natural ecosystems, or change the environment so that some species find it difficult to survive. With so many people in the world, there is a serious danger of causing permanent damage, not just to local environments but also to the global environment.

Making connections

There are about 700 different species of fig trees. Each one has its own species of pollinating wasps, without which it will die! The fig flowers of the trees are specially adapted so that they attract the right wasps.

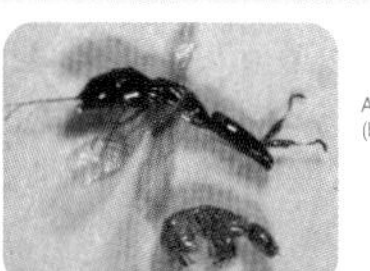

A female (top) and a male (bottom) fig wasp

Male fig wasps vary. Some species can fly but others are adapted to live in a fig fruit all their life. If they are lucky, a female wasp will arrive in the flower. Then the male will fertilise her. After this, he digs an escape tunnel for the female through the fruit and dies himself! The male wasp has special adaptations, such as a loss of his wings and very small eyes. These adaptations help him move around inside the fig fruit to find a female.

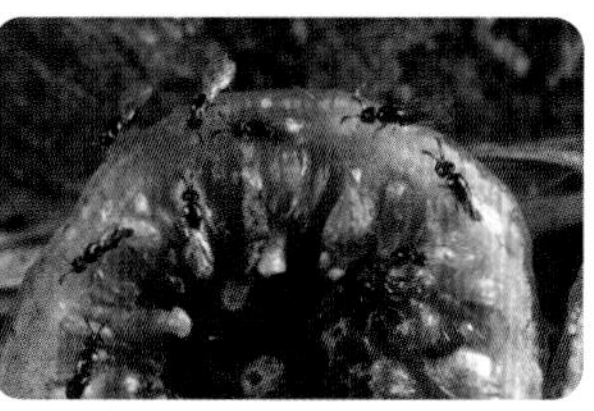

Female fig wasps have specially shaped heads for getting into fig flowers. They also have ovipositors. These allow them to place their eggs deep in the flowers of the fig tree.

Dr James Cook

If a fig tree cannot attract the right species of wasp, it will never be able to reproduce. In fact in some areas the trees are in danger of extinction because the wasp populations are being wiped out. Dr James Cook and his team at Imperial College, London are looking at the adaptations of the different wasps and their genetic material. They are trying to work out the relationships between all the different species.

ACTIVITY

Fig wasps are very strange animals. They have lots of adaptations which help them to reproduce sexually in just one species of tree.
Make a list of as many different types of animals or plants as you can that have strange ways of reproducing. Use a large piece of paper or a white board to record your ideas. You can include animals or plants which depend on just one other type of organism to be successful in life!

Chapters in this unit

characteristics, deciding which ones are hereditary and which could be due to the environment. Alternatively, they could consider the breeding of cattle for meat or for milk. Would they look for different characteristics and, if so, why?

- **Protecting the environment** – Question 4 of the 'Recap questions' is very open-ended and could be an exercise involving the whole class. A list could be built up and the importance discussed. At the end of the discussion, it would be appropriate to choose the five most important ways and then set the students the task of explaining the importance for homework.
- **Competition for the longest food chain** – Students asked to work out the longest food chain they can think of, i.e. the one with the most (sensible) links.
- **Pass the toxins game** – Using Post-it notes or similar stickers, use 16 or more with simple plants drawn on them; 8 with caterpillar drawings, 4 with small bird drawings and 1 with a bird of prey drawn on it. Then give one sticker to each student to put on to their shoulder as they enter the room. Choose a farmer and give him/her stickers with skull and crossbones on them (or red dots if you are pushed for time). The farmer then sticks one of these on each of the 'plant' students, who then have to stay still ('plants' do not move). The 'caterpillars' can then 'eat' the 'plants' but they collect the poison stickers as well. The small birds catch and eat the caterpillars, collecting any stickers that the caterpillars have on them. The bird of prey eats the small birds and thereby collects all the poison stickers and dies.
- Show the video on fig tree wasps at www.figweb.org – *Pollination Ecology in Tropical Figs – a Case of Mutualism* (by Georges Michaloud, Cameraman Alain Devez, produced by Service du Film de Recherche Scientifique (SFRS) 92170 VANV; currently available on streaming video). [It lasts 26 minutes.] Alternatively, a more recent TV film is 'Queen of trees' from Granada by Mark Deeble and Victoria Stone. Have some fresh or dried figs as prizes for those students who can best describe the relationship between the wasps and the trees. It would be good to show them some fresh figs anyway! (Do not eat in class.)

Special needs

The 'Recap questions' could be simplified and given on a pre-printed sheet. Students could also benefit from discussion of some of the issues raised by the questions as suggested earlier.

Misconceptions from KS3

The questions based on 'What you already know' could reveal some gaps in knowledge or misunderstandings:

- The differences between sexual and non-sexual (asexual) reproduction need to be made clear. It is worth emphasising that many plants can do both, but non-sexual reproduction is rare in animals.
- The definition of a 'population' might need clarifying.
- Not all variations between organisms are inherited. For example, the children of bodybuilders do not have well-developed muscles.

Chapters in this unit

B1b 5.1 Adaptation in animals

LEARNING OBJECTIVES

Students should learn that:

- Animals are adapted for survival in their particular habitat.
- There is a relationship between body size and surface area : volume ratio.
- Hair and body fat can provide insulation.

LEARNING OUTCOMES

Most students should be able to:

- Define the term 'adaptation'.
- Describe how animals are adapted to survive in cold climates.
- Describe how animals are adapted to life in a dry climate.

Some students should also be able to:

- Suggest how organisms are adapted to the conditions in which they live, when provided with appropriate information.

Teaching suggestions

- **Special needs.** Students could have a floor dominoes session with matching animals and their adaptations.
- **Gifted and talented.** Students could be asked to design an experiment to investigate whether people who regularly swim in the sea have a different surface area to volume ratio than those who only swim in heated pools.
- **Learning styles**

  Interpersonal: Discussing the effects of insulation.

  Intrapersonal: Making deductions about the practical investigations.

  Kinaesthetic: Practical work on SA/V and heat loss.

  Visual: Observing animal adaptations.

  Auditory: Describing adaptations of animals.
- **ICT link-up.** The use of data loggers in the practical work and the use of computers to produce graphs of the results link up to ICT.

SPECIFICATION LINK-UP B1b.11.5

- *Organisms have features (adaptations), which enable them to survive in the conditions in which they normally live.*
- *Animals may be adapted for survival in the conditions where they normally live, e.g. deserts, the Arctic.*
- *Animals may be adapted (e.g. with spines, poisons and warning colours to deter predators) to cope with specific features of their environment.*

Lesson structure

STARTER

Can you tell where I live from what I look like? – Bring in some live or stuffed animals or alternatively project some good pictures. Then discuss their adaptations, drawing some conclusions about the conditions in the habitats in which they might be found. The points to get across are that the adaptations are physical features that you can touch or see, but that there are also behavioural adaptations that are also important. (10 minutes)

Temperature regulation! – Get a student to dress up in a fur hat, coat and gloves (or dress up yourself). Contrast with pictures of Newcastle United football supporters taking their shirts off in the snow at matches. Discuss effects on temperature regulation. (5 minutes)

Surface area: volume ratio demonstration with building blocks – Use 1 block, 8 blocks built into a cube, 27 blocks built into a cube, and so on. Work out the surface area : volume ratio for each cube and plot a graph of length of side (number of blocks) against SA/V ratio. This will show that as the dimensions of the cube increase, there is a smaller surface area per unit of volume. (10–15 minutes)

MAIN

- Search the web for video of arctic animals (see Discovery Channel at www.education.discovery.com). Emphasise the size and shape of the bodies of the animals.
- **Surface area : volume ratios and heat loss** – A practical session to introduce the concept of surface area : volume ratios having an effect on heat loss. There are several ways of doing this. The simplest way is to give students cups and saucers and digital thermometers. Pour the same volume of hot (about 60°C) water into each cup and measure the temperature drop. Alternatively, pour 1 litre of hot water into a litre beaker and divide another litre of water equally between ten 100 ml beakers. Monitor the temperature. Data loggers can be used here, either by the students or if used as a demonstration to project a temperature graph as it forms.
- It is also possible to use different sizes of flasks, allowing students to carry out their own temperature readings and plot their own graphs. This could be used for coursework as predictions can be made, readings carried out and repeated, and conclusions drawn. This presents an excellent opportunity to teach various concepts important to 'How Science Works'.
- **Effect of insulation on heat loss** – In order to demonstrate that the thickness of an insulating coat will affect the temperature loss, two conical flasks of the same volume can be filled with hot water. One flask is left uncovered and the other surrounded by an insulating layer of cotton wool, or other material. The temperature drop can be recorded as before.
- This experiment could be done as a demonstration or by groups of students. It could be done at the same time as the previous experiment. This investigation is also useful for teaching and assessing investigative aspects of 'How Science Works', as it involves taking measurements, plotting graphs and drawing conclusions.

PLENARIES

Mix and match adaptations and functions – Put a list of adaptations on the board alongside a list of functions. Ask students to come and link an adaptation to a function. (5 minutes)

Comparing SA/V ratios – Provide students with some data about the sizes of different animals and get them to work out the SA/V ratios. For example, a comparison between a mouse and an elephant would be suitable. You can simplify the data by supplying them with the dimensions of a box into which the animal would fit. (10 minutes)

ACTIVITY & EXTENSION IDEAS

- Demonstrate the cooling effect of sweating by wiping the backs of the hands of volunteer students with cotton wool soaked in ethanol and asking them how it feels. To show that it is the evaporation of alcohol that is doing the cooling, students could be given a test tube of ethanol in a rack with a digital thermometer in it. The temperature can be read and recorded. The thermometer should then be repeatedly dipped into the ethanol and waved around in the air (gently and carefully) and the lowest temperature reached recorded. You could have a competition to see who can record the lowest temperature.
- So far, only the adaptations shown by mammals have been mentioned, so it could be useful to find out how animals such as invertebrates, amphibians, reptiles etc. cope with cold and very dry climates.
- The thickness of the insulation in the experiment could be varied and could be the basis of a different prediction. Ask what type of variable is the 'thickness of the insulation', i.e. 'discrete' if they count the number of layers of insulation, but 'continuous' if they measure the thickness. Can they think of an investigation of a categoric variable? [For example, the type of insulation used.] (These types of variable relate to 'How Science Works'.)

DID YOU KNOW?

In addition to the information given, polar bear hair is not white but colourless. It is hollow and transparent to allow the light to fall on to its skin, which is black to absorb the rays. It has been said that polar bears hide their noses with their paws when they are hunting to prevent their prey from spotting them.

B1b 5.1 Adaptation in animals

LEARNING OBJECTIVES

1. How can hair help animals survive in very cold climates?
2. What are the advantages – and disadvantages – of lots of body fat?
3. What is your surface area : volume ratio?

The variety of conditions on the surface of the Earth is huge. If you are a living organism, you could find yourself living in the dry heat of a desert or in wastelands of ice and snow. Fortunately, living organisms have special features (known as **adaptations**). These make it possible for them to survive in their particular habitat – however extreme the conditions might be!

Animals in cold climates

To survive in a cold environment you must be able to keep yourself warm. Arctic animals are adapted to reduce the heat they lose from their bodies as much as possible. You lose body heat through your body surface (mainly your skin). The amount of heat you lose is closely linked to your surface area : volume (SA/V) ratio.

Look at Figure 2. This explains why so many Arctic mammals, such as seals, walruses, whales, and polar bears, are relatively large. It keeps their surface area : volume ratio as small as possible and so helps them hold on to their body heat.

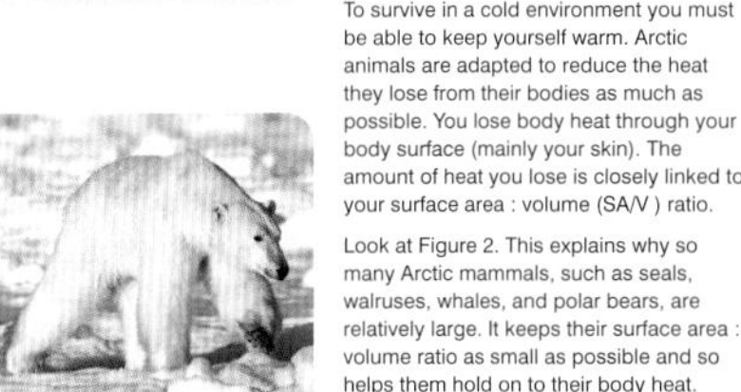

Figure 1 The Arctic is a cold and bleak environment. However, the animals which live there are well adapted for survival. Notice the large size, small ears, thick coat and white camouflage of this polar bear.

1 cm
1 cm
1 cm
sa : vol ratio = 6 : 1

3 cm
3 cm
3 cm
sa : vol ratio = 54 : 27 = 2 : 1

Figure 2 The ratio of surface area to volume falls as objects get bigger. You can see this clearly in the diagram. This is very important when you look at the adaptations of animals which live in cold climates.

a) Why are so many Arctic animals large?

Animals in very cold climates often have other adaptations too. The surface area of the thin skinned areas of their bodies – like their ears – is usually very small. This reduces their heat loss – look at the ears of the polar bear in Figure 1.

Many Arctic mammals have plenty of insulation, both inside and out. Blubber – a thick layer of fat that builds up under the skin – and a thick fur coat on the outside will insulate an animal very effectively. They really reduce the amount of heat lost through their skin.

The fat layer also provides a food supply. Animals often build up their blubber in the summer. Then they can live off their body fat through the winter when there is almost no food.

b) List three ways in which Arctic animals keep warm in winter.

Camouflage is important both to predators (so their prey doesn't see them coming) and to prey (so they can't be seen). The colours which would camouflage an Arctic animal in summer against plants would stand out against the snow in winter. Many Arctic animals, including the Arctic fox, the Arctic hare and the stoat, exchange the greys and browns of their summer coats for pure white in the winter.

DID YOU KNOW...

. . . that polar bears stay white all year round? They don't need any camouflage for two reasons. Firstly polar bears don't have any predators on the land – who would dare to attack a polar bear? Secondly they hunt their prey in the sea among the ice all year round. So the white colour makes them less visible to the seals they hunt.

Surviving in dry climates

Dry climates are often also hot climates – like deserts! Deserts are very difficult places for animals to live. There is scorching heat during the day, followed by bitter cold at night, while water is in short supply.

The biggest challenges if you live in a desert are:

- coping with the lack of water, and
- stopping your body temperature from getting too high.

Many desert animals are adapted to need little or no drink. They get the water they need from the food they eat.

Mammals keep their body temperature the same all the time. So as the environment gets hotter, they have to find ways of keeping cool. Most mammals sweat to help them cool down. But this means they lose water, which is not easy to replace in the desert.

c) Why do mammals try to lose heat without sweating in hot, dry conditions?

Desert animals have other adaptations for cooling down. They are often most active in the early morning and late evening, when the temperature is comfortable. During the cold nights and the heat of the day they rest. You find them in burrows well below the surface, where the temperature doesn't change much.

Many desert animals are quite small, so their surface area is large compared to their volume. This helps them to lose heat through their skin. They often have large, thin ears as well to increase their surface area for losing heat.

Another adaptation of many desert animals is that they don't have much fur. Any fur they do have is fine and silky. They also have relatively little body fat stored under the skin. Both of these features make it easier for them to lose heat through the surface of the skin. The animals keep warm during the cold nights by retreating into their burrows.

Figure 3 Animals like this fennec fox have many adaptations to help them cope with the hot dry conditions. How many can you spot?

Figure 4 An elephant is pretty big but it lives in hot, dry climates. Its huge wrinkled skin would cover an animal which was much bigger still. The wrinkles increase the surface area to aid heat loss.

SUMMARY QUESTIONS

1. a) List the main problems which face animals living in cold conditions like the Arctic.
   b) List the main problems which face animals living in the desert.
2. Give three ways in which animals that stay in the Arctic throughout the winter keep warm. Explain how these adaptations work.
3. Give three ways in which animals which live in a desert manage to keep cool without sweating so they don't lose water.
4. Explain why being quite large helps many Arctic animals to keep warm.

GET IT RIGHT!

Remember, the ***larger*** the animal, the ***smaller*** the surface area : volume (SA/V) ratio.
Animals often have ***increased*** surface areas in ***hot*** climates, and ***decreased*** surface areas in ***cold*** climates.

FOUL FACTS

Animals from the deep oceans are adapted to cope with enormous pressure, no light and very cold water. But if these deep-water organisms are brought up to the surface too quickly, they explode because of the rapid change in pressure.

KEY POINTS

1. All living things have adaptations which help them to survive in the conditions where they live.
2. Animals which are adapted for cold environments are often large, with a small surface area: volume (SA/V) ratio. They have thick insulating layers of fat and fur.
3. Changing coat colour in the different seasons gives animals year-round camouflage.
4. Adaptations for hot, dry environments include a large SA/V ratio, thin fur, little body fat and behaviour patterns that avoid the heat of the day.

SUMMARY ANSWERS

1. **a)** It is very cold, so there is a problem in keeping warm and finding enough food.
   b) It is very hot, so the main problems are keeping the body cool and finding enough water.
2. Animals who stay in the Arctic throughout the winter have small ears (to reduce the surface area of thin-skinned tissue for heat loss), thick fur (provides an insulating layer to help prevent heat loss), and a layer of fat/blubber (further insulation against heat loss).
3. Animals living in the desert keep cool without sweating by avoiding the heat of the day and by having large SA/V ratio to increase heat loss (e.g. large ears, baggy skin, little fur, thin and silky fur etc.).
4. The larger the animal, the smaller the surface area : volume ratio, which means the proportion of heat lost will be less. Bigger animals keep warm more easily because less heat is lost.

Answers to in-text questions

a) It keeps the surface area : volume ratio as small as possible and so helps them to reduce heat loss.

b) Arctic animals have small ears, thick fur, and a layer of fat/blubber.

c) Because sweating results in loss of water from the body. There is not much water in the desert, so they cannot rely on finding more to drink.

KEY POINTS

The summary questions are based on the key points and could be used as a homework task. In addition, it could be helpful to make a table contrasting adaptations for cold conditions with those for hot, dry conditions.

B1b 5.2 Adaptation in plants

LEARNING OBJECTIVES

Students should learn:

- How plants are adapted to live in dry conditions.
- How changes in the surface area of plants affect the rate at which water is lost.
- That plants living in dry conditions may store water in their tissues.

LEARNING OUTCOMES

Most students should be able to:

- Describe the adaptations shown by plants that live in dry environments.

Some students should also be able to:

- Explain how these adaptations reduce the quantity of water lost by the plant.
- Explain the importance of water-storage tissues in desert plants.

Teaching suggestions

- **Special needs.** Students could measure the leaf areas of two contrasting plants by wax rubbing over large squared graph paper. Ask them to predict from this which plant will need more water.
- **Gifted and talented.** Students could try to work out a method for estimating the total leaf surface area on a tree.
- **Learning styles**

  Interpersonal: Discussing the location of stomata.

  Intrapersonal: Writing a report of adaptations shown by plants.

  Kinaesthetic: Practical work on leaf size and transpiration.

  Visual: Observing differences in grass leaves.

  Auditory: Listening to descriptions of how stomata function.

Answers to in-text questions

a) For photosynthesis and to keep the plant upright.

b) By evaporation through the stomata.

c) Because water is lost through the surface of the leaf, so if the surface area is smaller, there will be less water lost.

d) Leaves, stem and/or roots.

SPECIFICATION LINK-UP B1b.11.5

- *Plants may be adapted for survival in the conditions where they normally live, e.g. deserts, the Arctic.*
- *Plants may be adapted to cope with specific features of their environment, e.g. thorns, poisons and warning colours.*

Lesson structure

STARTER

Losing water – Choose two identical soft-leaved plants (tomato plants or whatever is available) and two similar cactus plants. Leave one of each to dry out, so that the soft-leaved plant is wilted, but water the other two thoroughly. Present these to the students explaining how they have been treated. Ask the students what the differences are. Ask: 'Why are the differences not as great between the two cactus plants as they are between the two soft-leaved plants?' (5–10 minutes)

Differences in grasses – As each student enters the room, give them a blade of marram grass and a blade of grass from the playing field or a lawn. Set them a time limit of 3 minutes to write down as many differences between the two grasses as they can. Follow this up with a list on the board and a discussion about why there are these differences. If they are not familiar with marram grass, they could be asked to suggest where the grass came from. (10 minutes)

What is a succulent? – Introduce the term 'succulent'. Ask: 'What does it mean? How are succulents different from cacti?' Lead a discussion on their distribution, adaptations and uses (you could show them a bottle of tequila and/or some aloe vera). (5–10 minutes)

MAIN

- Search the web to show pictures of adaptations of plants to arid conditions. See Animation B1b 5.2 'How stomata work' on the GCSE Biology CD ROM. It would be useful to have a number of succulents and cactus plants available for students to be able to feel the texture and examine the structures in detail. Ask students to write a report of adaptations.
- **Leaf surface area and transpiration** – This is a practical investigation into how variation in leaf size affects transpiration. Using pieces of blotting paper of known surface area, make up some 'leaves' of different sizes. Attach a piece of string to one end of the blotting paper and pass the string down a drinking straw, so that the 'leaf' can be supported in a boiling tube (or small measuring cylinder). The blotting paper could have a thin card backing to give it strength. If a known volume of water is placed in each boiling tube (or the 'leaves' are placed in small measuring cylinders) the volume of water lost can be calculated in cm^3/hour against area in cm^2. The results can be shown graphically.
- This exercise can be set up as a demonstration where groups of students are given a 'leaf' of a different size, the measurements taken and then collected together and discussed. Many 'How Science Works' concepts are introduced here.
- If you have a potometer (or a barometric pressure sensor) you can connect it to a leafy shoot. The rate of water uptake (equivalent to the rate of water loss) can be measured for the intact shoot. When several readings have been taken and a mean rate calculated, several leaves can be removed and the readings repeated. This can be done again removing more leaves. The effect will depend on the type of shoot chosen – one with soft leaves is better than something like laurel or rhododendron.
- The surface area of the leaves can be measured as they are removed, by drawing around them on squared paper, cutting round the outlines and weighing them. If you know the mass of a known area of the squared paper, then it is possible to calculate the area of each leaf. (Dividing the mass of the leaf by the mass of 1 cm^2 will give the area of the leaf.)
- **Practical on distribution of stomata** – Transparent nail varnish can be used to make stomatal peels of leaves. These can be mounted on microscope slides and viewed under the low power lens. This is an activity that can be demonstrated or done by the students. Ensure laboratory is well ventilated.
- Different types of leaf, e.g. *Tradescantia*, can be chosen and peels made of upper and lower surfaces. Ask: 'Where are most stomata found?'

PLENARIES

Where do I come from? – Show a series of pictures of plants and ask students to write down the environment from which they come. (5 minutes)

Game of snap – Students play with cards showing features and functions of plants in arid conditions. (10 minutes)

Demonstration of way in which stomata function – Blow up a long balloon which has had a piece of Sellotape stuck down one side so that it bends when inflated, to represent a guard cell. Match with a pre-prepared one to show the stoma between the two guard cells. Show how the stoma gets smaller when the balloons are deflated. (10 minutes)

ACTIVITY & EXTENSION IDEAS

- **Demonstration of expanding stem** – Fold a piece of green card into corrugations and Sellotape the ends together. Self-shading can be demonstrated and also, by pulling it wide and closing it up, the ability of barrel cacti to expand when water is plentiful.
- Ask students to make thumbnail sketches predicting what graphs would look like for the following:
  - cuticle thickness vs. average temperature
  - surface area vs. average humidity
  - yearly rainfall vs. water storage capacity.

  (These relate to 'How Science Works' – relationships between variables.)
- Not all plants showing adaptations to prevent loss of water live in hot, dry conditions. Ask students to consider other environments in which water may be unavailable to plants. Show them pictures of conifers and salt-marsh plants (many of which are succulent).
- How are plants, such as cacti, able to make enough food by photosynthesis if their leaves are reduced to spines?
- Look at flat-bladed cacti such as prickly pear (*Platyopuntia*). Shine a light on it and move the light around to model the apparent movement of the Sun during the day. Which orientation would be best for the cactus? Would the orientation change in different hemispheres?

BIOLOGY — ADAPTATION FOR SURVIVAL

B1b 5.2 Adaptation in plants

LEARNING OBJECTIVES

1. How are plants adapted to live in dry conditions?
2. How do plants store water?

There are some places where plants simply cannot grow. In deep oceans no light penetrates, and no plants can grow. In the icy wastes of the Antarctic, no plants grow.

Almost everywhere else, including the hot, dry areas of the world, we can find plants growing. Without them there would be no food for the animals. But plants need water both for photosynthesis and to keep their tissues upright. If a plant does not get the water it needs, it wilts and eventually dies.

a) Why do plants need water?

Plants take in water through their roots in the soil. It moves up through the plant and is lost through the leaves in the **transpiration stream**. Plants lose water all the time through their leaves. There are small openings called **stomata** in the leaves of a plant. These open to allow gases in and out for photosynthesis and respiration. But at the same time water is lost by evaporation.

The rate at which a plant loses water is linked to the conditions it is growing in. When it is hot and dry, photosynthesis and respiration take place quickly. As a result, plants also lose water very fast. So how do plants that live in dry conditions cope? Most of them either reduce their surface area so they lose less water or they store water in their tissues. Some do both!

b) How do plants lose water from their leaves?

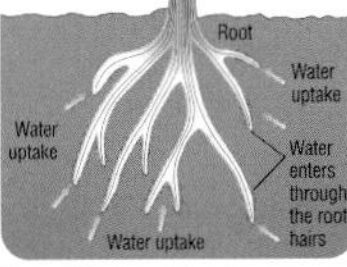

Figure 1 The transpiration stream means plants are losing water all the time by evaporation from their leaves. When the conditions are hot and dry, they lose water very quickly.

Changing surface area

When it comes to stopping water loss through the leaves, the surface area : volume ratio (see page 82) is very important to plants. There are a few desert plants which have broad leaves with a large surface area. These leaves collect the dew that forms in the cold evenings. They then funnel the water towards their shallow roots

Some plants in dry environments have curled leaves. This reduces the surface area of the leaf. It also traps a layer of moist air around the leaf which really cuts back the amount of water they lose by evaporation.

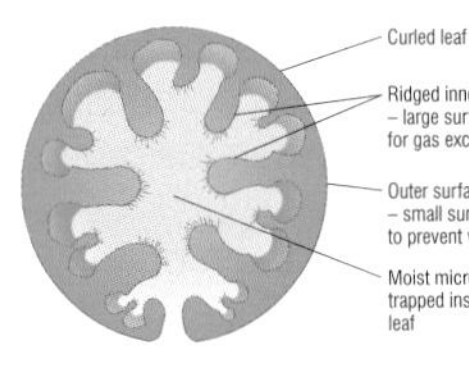

Figure 2 Plants that live on sand dunes near the sea have to survive very dry conditions. This marram grass, which you can find all around the British coast, has tightly curled leaves, which reduce the surface area available for water loss.

84

Evolution and environment

However, most plants which live in dry conditions have reduced the surface area of their leaves. This cuts down the area from which water can be lost. Some desert plants have small fleshy leaves with a thick cuticle to keep water loss down. The cuticle is a waxy covering on the leaf that stops water evaporating away.

The best-known desert plants are the cacti. Their leaves have been reduced to spines with a very small surface area indeed. This means the cactus only loses a tiny amount of water – and the spines put animals off eating the cactus as well!

c) Why do plants often reduce the surface area of their leaves to help them prevent water loss?

GET IT RIGHT!

Remember that plants need their stomata open for photosynthesis and respiration. This is why they lose water by evaporation from their leaves.

Storing water

Plants can also cope with dry conditions by storing water in their tissues. When there is plenty of water available after a period of rain, the plant stores it. Plants which store water in their fleshy leaves, stems or roots are known as **succulents**.

Cacti don't just rely on their spiny leaves to help them survive in dry conditions. They are succulents as well. The fat green body of a cactus is its stem, which is full of water-storing tissue. All these adaptations make cacti the most successful plants in a hot dry climate.

d) In which parts can a plant store its water?

DID YOU KNOW?

An apple tree in the UK can lose a whole bath of water from its leaves every day. A large saguaro cactus in the desert loses less than one glass of water in the same amount of time!

Keep away!

One of the biggest problems for plants is being eaten by animals. Plants have a wide variety of adaptations designed to deal with this. Vicious thorns, unpleasant tastes and poisonous chemicals can all put animals off!

We have made use of some of these adaptations. For example, we use the bitter chemical in the bark of the cinchona tree to make quinine. This helps relieve the symptoms of malaria. What's more, the poison digitalis from foxgloves is used as a very effective heart medicine.

Figure 3 Cacti are well adapted to survive in desert conditions

SUMMARY QUESTIONS

1. Copy and complete using the words below:

   adaptations desert plants spiny stem water

   Cacti are …… which live in the …… . They have two main …… to help them survive. Their leaves have become …… and they store …… in their …… .
2. a) Explain why plants lose water through their leaves all the time.
   b) Why does this make living in a dry place such a problem?
3. Explain three adaptations which help plants living in dry conditions to reduce water loss from their leaves.

KEY POINTS

1. Plants lose water all the time by evaporation from their leaves.
2. Plants which live in dry places have adaptations which help to reduce water loss. These adaptations may often include reduced surface area of their leaves and/or water-storage tissues.

85

SUMMARY ANSWERS

1. Plants, desert, adaptations, spiny, water, stems.
2. a) Water is lost by transpiration. The stomata are open for gaseous exchange in photosynthesis and respiration and water is lost by evaporation at the same time.
   b) Dry places are often hot, so photosynthesis and respiration occur at a faster rate. The stomata are open more, so there is more evaporation. If the air is dry, evaporation occurs at a faster rate.
3. Small leaves; curled leaves – reduce surface area: Thick cuticle; also reduces rate of evaporation.

KEY POINTS

Students need to understand why plants lose water by transpiration. They should be able to list some structural adaptations that prevent water loss in dry conditions. These features could be added to the general point about changes to surface area and to water storage, thus reinforcing the key points.

B1b 5.3 Competition in animals

LEARNING OBJECTIVES

Students should learn:

- What is meant by competition and why animals compete with each other.
- What qualities are needed in order to become a successful competitor.

LEARNING OUTCOMES

Most students should be able to:

- Explain how competition is necessary for survival.
- Describe those characteristics which make an animal a successful competitor.
- Suggest the factors for which an animal is competing in a given habitat.

Some students should also be able to:

- Explain why certain characteristics make an animal a successful competitor.

Teaching suggestions

- **Special needs.** Students could be shown some plastic animals or pictures and asked to pick out for each one an adaptation which makes it a good competitor.
- **Gifted and talented.** Introduce the concept of the ecological niche. The students could research interesting or unusual examples of ecological niches.
- **Learning styles**

  Interpersonal: Discussing and working in pairs for the 'Competition for grass' starter.

  Intrapersonal: Interpreting courtship rituals.

  Kinaesthetic: Investigating a tree for food web; using apparatus.

  Visual: Obtaining information in the card game; making observations.

  Auditory: Listening to the presentation on inter/intraspecific competition.

SPECIFICATION LINK-UP B1b.11.5

- *Animals often compete with each other for food, mates and territory.*

Students should use their skills, knowledge and understanding of 'How Science Works':

- *to suggest factors for which organisms are competing in a given habitat.*

Lesson structure

STARTER

Survive! – Prepare small laminated cards, each having one of the following words written on it: 'food', 'shelter', 'water' or 'mate'. Place similar cards around the room, enough so that most but not all students can collect a complete set of all four cards. Tell the students to move around the room collecting cards according to the rules: 'no swapping or taking from others by force and you can only hold one of each kind'. Students who collect a full set can go to the front (or sit down) – they have survived. Those who do not manage to collect the set do not survive and have to show how they died (hunger, no mates etc.). As an extension, you could allow swapping around the remainder to illustrate social effects (or altruism?) e.g. mothers feeding offspring but starving themselves. (10–15 minutes)

Competition for grass? – With the students working in pairs, ask them to list as many animals as they can that eat grass. Give them a time limit of 2 minutes. Students claiming to have the greatest number are asked to read out their list so that the rest of the class can agree or disagree. This leads to a discussion on competition. (5 minutes)

What are we competing for? – With the students working in groups, supply them with A3 sheets of paper and ask them to list as many types of competition from everyday life as they can in 3 minutes. Swap sheets over with other groups to share ideas. This leads into a discussion on what wild animals would compete for. (10 minutes)

MAIN

- **Courtship displays and mating behaviour** – Find video or images of courtship displays (e.g. peacocks) or competition between males (e.g. sea lions). References to the breeding plumage of different birds, behaviour of stags in the rutting season etc. are other examples available. Search at Discovery Channel www.education.discovery.com; BBC nature www.bbc.co.uk; Google www.video.google.com. Peacocks' feathers and deer antlers could be passed around. This emphasis on courtship could lead into a discussion of passing on the genes.
- Introduce the ideas of 'interspecific' and 'intraspecific' competition**:** the words do not necessarily have to be used, only the idea that two types of animal eating the same food in the same habitat will be in competition with each other. There are some data available for the variation in numbers of two closely related species of flour beetle (*Tribolium*) living in the same culture, and for two species of *Paramecium*: only one species will survive. You could also link this with the introduction of species, such as the grey squirrel, into this country and the rabbit (or more recently the camel) into Australia. For competition amongst members of the same species, there are data showing that when the density of limpets on a rocky shore increases, their length and biomass decrease.
- These examples can be presented to the students as OHPs, and for each one discuss what they are competing for and why one wins and the other loses.
- **Food web** – Depending on the season, investigate a tree or a clump of plants in the school grounds to show the relationships between the different animals that feed on the plant. Using pooters and sweep nets, the small animals can be trapped and identified and a food web built up. There will be caterpillars and beetles eating the leaves; greenfly feeding directly on the plant sap; butterflies and moths feeding on the flowers; and other invertebrates feeding on the bark. Each group of animals can find plenty of food without being in competition with another species.

PLENARIES

- **Discuss a TV competition show** – Discuss, for example, *Big Brother* or *The Weakest Link*. What parallels can be drawn between what goes on in such a show and competition in nature? (10 minutes)
- **Survive!** – Repeat the starter, but this time the students can challenge each other to 'mock battles' (or answer a question related to competition) to get their cards. (5–10 minutes)

ACTIVITY & EXTENSION

- Research the story of the way in which rabbits overran some parts of Australia and how the problem was solved; or research proposals to cull wild camels.
- **Camouflage effectiveness** – Find some good pictures of camouflaged animals and ask the students to time how long it takes them to identify the animals. There are some good examples, e.g. flatfish, amphibians, snakes etc. Does camouflage play a role in competition between animals?
- **Competition in birds** – Birds compete all the time for food, mates and territory. Ask students to research the different ways in which birds compete using colour, food preferences and song. Good examples to get the students started are robins and blackbirds.

Answers to in-text questions

a) There is only a limited amount of food, water and living space in an area.

b) Any sensible choices here. For example, for a plant-eater: eating a wide range of plants, sensitive hearing to hear predators. For a carnivore: ability to run fast; good eyesight; sharp teeth etc.

BIOLOGY ADAPTATION FOR SURVIVAL

B1b 5.3 Competition in animals

LEARNING OBJECTIVES

1. What is competition?
2. What makes an animal a good competitor?

Animals and plants grow alongside lots of other living things, some from the same species and others completely different. In any area there will only be a limited amount of food, water and space, and a limited number of mates. As a result, living organisms spend their time competing for the things they need.

The best adapted organisms are most likely to be the winners of the ***competition*** for resources. They will be most likely to survive and produce healthy offspring.

a) Why do living organisms compete?

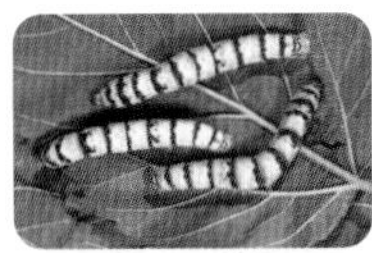

Figure 1 Some herbivores, like these silk worms eating their mulberry leaves and the panda with its bamboo, only feed on one particular plant. They are very open to competition from other animals or to a disease that damages their food plant.

What do animals compete for?

Animals compete for many things, including:

- water
- territory
- mates

Competition for food is very common. Herbivores (animals which eat only plants) sometimes feed on many types of plant, and sometimes on only one or two different sorts. Many different species of herbivores will all eat the same plants. Just think how many types of animals eat grass!

The animals which eat a wide range of plants are most likely to be successful. If you are a picky eater, you risk dying out if anything happens to your only food source. An animal with wider tastes will just eat something else for a while!

Competition is common among carnivores (animals which eat only meat). They compete for prey. Small mammals like mice are eaten by animals like foxes, owls, hawks and domestic cats. The different types of animals all hunt the same mice. So the animals which are best adapted to the area will be most successful.

Carnivores have to compete with other members of their own species for their prey as well as with members of different species. Successful predators are adapted to have long legs for running fast and sharp eyes to spot prey. These features will be passed on to their offspring.

Animals often avoid direct competition with members of other species when they can. It is the competition between members of the same species which is most intense!

Prey animals compete with each other too – to be the one which ***isn't*** caught! Adaptations like camouflage colouring, so you don't get seen, and good hearing, so you pick up a predator approaching, are important for success.

b) Give one adaptation which would be useful to a plant-eater and one which would be helpful to a carnivore.

Competition for mates can be fierce. In many species the male animal puts a lot of effort into impressing the females. The males compete in different ways to win the privilege of mating with her.

DID YOU KNOW?

Some animals have warning colours which let predators know they are poisonous or taste nasty. Poison arrow frogs – which give us the curare used in medicine – have very bright warning colours. Other frogs which aren't poisonous at all mimic the warning colours so that predators leave them alone.

86

Evolution and environment

In some species – like deer and lions – the males fight between themselves. Then the winner gets the females.

Many male animals display to the female to get her attention. Some birds have spectacular adaptations to help them stand out. Male peacocks have the most amazing tail feathers. They use them for displaying to other males (to warn them off) and to females (to attract them).

Figure 2 The spectacular display of a male peacock certainly attracts the attention of the females. And unlike animals which fight for their mates, the peacock doesn't risk getting hurt when he tries to win over the females.

What makes a successful competitor?

A successful competitor is an animal which is adapted to be better at finding food or a mate than the other members of its own species. It also needs to be better at finding food and water than the members of other local species. What is more, it must also breed successfully.

Many animals are successful because they avoid competition with other species as much as possible. They feed in a way that no other local animals do, or they eat a type of food that other animals avoid. For example, one plant can feed many animals without direct competition. While caterpillars eat the leaves, greenfly drink the sap, butterflies suck nectar from the flowers and beetles feed on pollen.

It is much harder to avoid competition within the same species, but many animals try to do just that. They may set up and defend a **territory** – an area where they live and feed. This is a common way of making sure that they will be able to find enough food for themselves and for their young when they breed.

FOUL FACTS

Different types of African dung beetles avoid competition with each other by attacking the same pile of dung at different times of day and in different ways. The most active beetles work in the heat of the day and make balls of dung which they roll away. The quieter tunnellers and the beetles that actually live in the dung heaps work as dusk is falling.

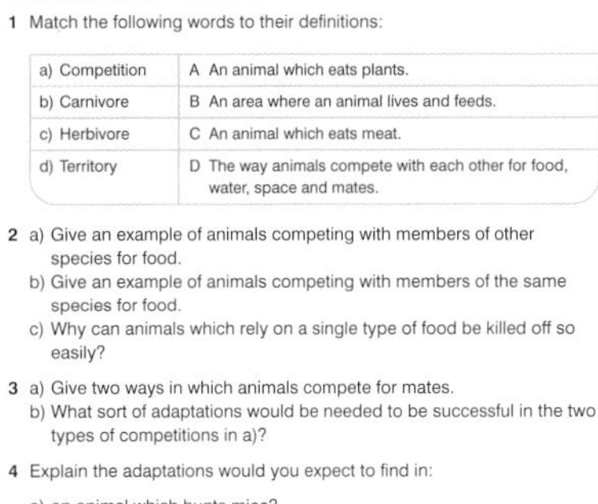

SUMMARY QUESTIONS

1 Match the following words to their definitions:

a) Competition	A An animal which eats plants.
b) Carnivore	B An area where an animal lives and feeds.
c) Herbivore	C An animal which eats meat.
d) Territory	D The way animals compete with each other for food, water, space and mates.

2 a) Give an example of animals competing with members of other species for food.
b) Give an example of animals competing with members of the same species for food.
c) Why can animals which rely on a single type of food be killed off so easily?

3 a) Give two ways in which animals compete for mates.
b) What sort of adaptations would be needed to be successful in the two types of competitions in a)?

4 Explain the adaptations would you expect to find in:

a) an animal which hunts mice?
b) an animal which eats grass?
c) a fish which feeds on other fish?
d) an animal which feeds on the tender leaves at the top of trees?

GET IT RIGHT!

Learn to look at an animal and spot the adaptations which make it a successful competitor!

KEY POINTS

1. Animals often compete with each other for food and territories.
2. Animals compete for mates.
3. Animals have adaptations which make them good competitors.

87

SUMMARY ANSWERS

1 **a)** D **b)** C **c)** A **d)** B

2 **a)** Any suitable examples, such as lions, cheetahs and leopards etc.

b) Any suitable examples, such as rabbits, limpets on a sea shore.

c) If anything happens to their food supply, such as another animal eating it, fire or disease, then they will starve.

3 **a)** Fighting or displaying.

b) Fighting: strength, antlers, teeth, etc. Displaying: spectacular appearance, colours, part of body to display (e.g. peacock's tail).

4 **a)** Quite small, moves stealthily, sharp teeth, good eyesight and hearing; claws, hunts at time mice are active.

b) Special teeth to grind grass, ability to run fast away from predators, good all-round eyesight, good hearing to detect predators, etc.

c) Ability to breathe and swim in water, swim fast, teeth that prevent the escape of the smaller fish.

d) Teeth and gut adapted to eating plants, ability to reach the top of trees (long neck or good at climbing), ability to grip on to branches, possibly use tail for balance.

KEY POINTS

- Use question 4 of the 'Summary questions' to help students remember these points. It would be useful to add some examples of adaptations to any revision cards made of the key points.
- A display of pictures of adaptations around the laboratory could also be useful.

B1b 5.4 Competition in plants

LEARNING OBJECTIVES

Students should learn that:

- Plants compete with each other for water and nutrients.
- Plants compete for light.

LEARNING OUTCOMES

Most students should be able to:

- Explain why plants need light.
- Explain why plants need minerals.
- Suggest the factors for which plants are competing in a given habitat.

Some students should also be able to:

- Suggest why some plants are better competitors in a given habitat.
- Evaluate the strategies used by plants to make them successful competitors, e.g. seed dispersal mechanisms.

Teaching suggestions

- **Special needs.** Supply students with a collection of seeds/fruits and ask them to match a dispersal/distribution method to each one. Sycamore, burdock, dandelion, strawberry, nuts and tomatoes are good examples.
- **Gifted and talented.** Students could be asked to find the best wing surface area to weight ratio for sycamore seeds by making small models from paper and paperclips. Provide them with a template for the wing, digital balances and litterpickers for dropping from a chosen height.
- **ICT link-up**
  - Data loggers are used for registering light intensity.
  - Computer programs can be used to plot graphs, and some of the higher attaining students might like to try to use a statistics package to see if results are significant.
- **Learning styles**

  Interpersonal: Reporting the results of experiments.

  Intrapersonal: Deducing methods of seed dispersal.

  Kinaesthetic: Practical work on investigating competition in plants.

  Visual: Making observations of competition beneath the tree.

  Auditory: Discussing results.

SPECIFICATION LINK-UP B1b.11.5

- *Plants often compete with each other for light and for water and nutrients from the soil.*

Students should use their skills, knowledge and understanding of 'How Science Works':

- *to suggest the factors for the distribution of (animals or) plants in a particular habitat.*

Lesson structure

STARTER

Competition under a tree – In summer, or fine weather, take students outside and look under a tree. What are the plants growing under the tree competing for? Ask a volunteer to record the suggestions and write a list to look at more closely back in the laboratory. (10 minutes)

How do coconuts disperse their seeds? – Have a coconut complete with husk and show a clip from the opening scenes of *Monty Python 'The Holy Grail'*, where the King is trying to explain why they are imitating the sound of horses' hooves with coconut shells. Show the students the coconut, float it in a bucket of water and lead into a discussion of seed distribution techniques. (5–10 minutes)

Seed dispersal – Have a range of seeds and fruits around the room, labelled with numbers. As the students enter the room, hand each one a list of numbers and get them to write the method of dispersal against each number using W (for wind), A (for animal) or E (for explosive). Allow 10 minutes, then check lists. (10–15 minutes)

MAIN

- **Investigation into competition in plants** – A spacing trial can be set up using radishes in late spring to early autumn. (See 'Practical support'.)
- 'How Science Works' concepts, such as experimental design, predicted outcomes, recording measurements and drawing conclusions can be demonstrated here.
- **Competition between weeds and crop plants** – Fill a number of identical small pots with compost and sow radishes and 'weed' (any other seeds such as marigolds etc.) seeds at different densities e.g. one radish to ten weeds, five radishes to five weeds, etc. Water the pots regularly and keep all other conditions (light, temperature) the same. Harvest the radishes at the appropriate time and compare the mass of radishes harvested at the different densities. The weeds could also be harvested and their wet mass determined, so that this can be compared with the mass of the radishes. Lead a discussion of the results prior to the students writing up a report.
- Experiments involving the growth of plants need time to yield results, so this should be planned ahead. Other crops and weeds could be used and the plants could be harvested as soon as there is sufficient growth (you do not need to have actual radishes before the effect of the competition is noticeable).
- **Do plants shade out the competition?** – Measure the surface area of nettle leaves growing in shady conditions and compare with the surface area of nettle leaves growing in brightly lit conditions. Squared paper can be used to measure leaf surface area and light meters or data loggers used to record light intensity.
- This is a good opportunity to teach many 'How Science Works' concepts. Predictions can be made and the results plotted as light intensity against surface area. Variable warning! There are some complex variables here: it is best to stick to the light intensity and pick leaves at the same height above the ground.

PLENARIES

Matching seeds to strategies – Use the Java-based matching game software to create an exercise for the whiteboard. Locate the free software by searching for Hot Potato, Jmatch.

Explosive fruits and seeds – Show videos, or pictures, of explosive fruits such as lupins or exploding cucumbers. (10 minutes)

ACTIVITY & EXTENSION IDEAS

- The dried seed pods of the common weed thale cress (*Arabidopsis thaliana*) will explode nicely when touched if caught at the right time. It is very common with a fast life cycle (six weeks). [It is also the first plant to have its entire DNA sequence worked out.]
- Use data loggers to investigate light intensity in different sites in the school garden or on school grounds. Correlate light intensity with the type of vegetation present.
- Investigate the distribution of sycamore seedlings under a sycamore tree. Sycamore seeds drop to the ground beneath the tree and will germinate and grow. The ones which drop furthest away will survive longest and grow larger. Students can measure the height of seedlings at measured distances away from the tree and comment on size and distribution. This works well if you have access all round the tree, so that students can work in groups and aspect (N, S, E or W) can also be taken into consideration. [This could be tried for other trees such as oak and ash.]
- Have a demonstration of seeds germinated and grown in light and some in the dark. This shows 'etiolation' (the seeds germinated in the dark will grow into yellow, straggly seedlings) and indicates that chlorophyll does not develop unless the seedlings are exposed to light. Cress is very good for this.

Practical support

Investigation into competition in plants

Equipment and materials required

Balance for weighing seedlings; for each group or demonstration: radish seeds, potting compost, small trays.

Details

Plant seeds into small trays of moist potting compost at increasing distances apart on both *x*- and *y*-axes of the trays. The distances should be clearly marked along the sides of the trays, so that the experiment can be replicated. The trays should be watered regularly and kept in the same conditions of light and temperature. Weigh the seedlings, or plants, when grown to find the ideal spacing. This can be carried out on a larger scale with plants in a school garden or with rapid-cycling brassicas under a light bank at any time of year.

B1b 5.4 Competition in plants

LEARNING OBJECTIVES

1. What do plants compete for?
2. Why do plants need light?

Plants might look like peaceful organisms, growing silently in your local park. But the world of plants is full of cut-throat competition. They compete with each other for light, for water and for nutrients (minerals) from the soil.

They need light for photosynthesis, to make food using energy from the Sun. They need water for photosynthesis and to keep their tissues rigid and supported. And plants need minerals so they can make all the chemicals they need in their cells.

a) What do plants compete with each other for?

EXPERIMENTAL DATA

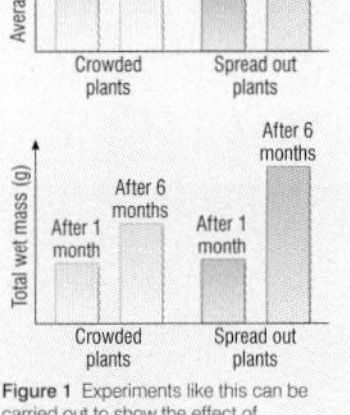

Figure 1 Experiments like this can be carried out to show the effect of competition on plants. All the conditions – light level, the amount of water and minerals available and the temperature were kept exactly the same for both sets of plants. The differences in their growth were the result of overcrowding and competition for resources in one of the groups.

Why do plants compete?

Just like animals, plants are in competition both with other species of plants and with their own species. Big, tall plants like trees take up a lot of water and minerals from the soil and prevent light from reaching the plants beneath them. So the plants around them need adaptations to help them to survive.

If a plant sheds its seeds and they land nearby, the parent plant will be in direct competition with its own seedlings. Because the parent plant is large and settled, it will take most of the water, minerals and light. So the plant will deprive its own offspring of everything they need to grow successfully!

If the seeds from a plant all land close together – even if they are a long way from their parent – they will then compete with each other as they grow. So many plants have special adaptations which help them to spread their seeds over a wide area.

b) Why is it important that seeds are spread as far as possible from the parent plant?

Coping with competition

When plants are growing close to other species they often have adaptations which help them to avoid competition.

Small plants found in woodlands often grow and flower very early in the year. Although it is cold, plenty of light gets through the bare branches of the trees. The dormant trees take very little water out of the soil. The leaves shed the previous autumn have rotted down to provide minerals in the soil.

Plants like snowdrops, anemones and bluebells are all adapted to take advantage of these things. They flower, set seeds and die back again before the trees are in full leaf.

Another way plants compete successfully is by having different types of roots. Some plants have shallow roots taking water and minerals from near the surface of the soil. Others have long, deep roots, which go far underground. Both compete successfully for what they need without affecting the other.

If one plant is growing in the shade of another, it may grow taller to reach the light. It may also grow leaves with a bigger surface area to take advantage of all the light it does get.

c) How can short roots help a plant to compete successfully?

Spreading the seeds

To compete successfully, a plant has to avoid competition with its own seedlings. Usually, the most important adaptation for success is the way they shed their seeds.

Many plants use the wind to help them. Some produce seeds which are so small that they are carried easily by air currents. Many others produce fruits with special adaptations for flight to carry their seeds as far away as possible. Examples include the parachutes of the dandelion 'clock' and the winged seeds of the sycamore.

d) How do the fluffy parachutes of dandelion seeds help the seeds spread out?

Some plants use mini-explosions to spread their seeds. The pods dry out, twist and pop, flinging the seeds out and away.

Juicy berries, fruits and nuts are produced by plants to tempt animals to eat them. Once the fruit gets into the animal's gut, the tough seeds travel right through. They are deposited with the waste material in their own little pile of fertiliser, often miles from where they were eaten!

Fruits which are sticky or covered in hooks get caught up in the fur or feathers of a passing animal. They are carried around until they fall off hours or even days later.

Sometimes the seeds of different plants land on the soil and start to grow together. The plants which grow fastest will compete successfully against the slower-growing plants. For example:

- the plants which get their roots into the soil first will get most of the available water and minerals;
- the plants which open their leaves fastest will be able to photosynthesise and grow faster still, depriving the competition of light.

Plants compete at all levels, from spreading their seeds to the height they grow and how early they flower each year. The winners of the competitions are the ones we see. The losers just don't make it!

FOUL FACT

The roots of some desert plants produce a chemical which inhibits (prevents) seeds from germinating. They murder the competition before it has a chance to get growing!

Figure 2 Coconuts will float for weeks or even months on ocean currents which can carry them hundreds of miles from their parents – and any other coconuts!

SUMMARY QUESTIONS

1. a) Give two ways in which plants can overcome the problems of growing in the shade of another plant.
   b) Explain how a primrose plant manages to grow and flower successfully in spite of living under a large oak tree.
2. a) Why is it so important that plants spread their seeds successfully?
   b) Give three examples of successful adaptations for spreading seeds.
3. The dandelion is a successful weed. Carry out some research and evaluate the adaptations that make it a better competitor than other plants on a school field.

KEY POINTS

1. Plants often compete with each other for light, for water and for nutrients (minerals) from the soil.
2. Plants have many adaptations, which make them good competitors.

Answers to in-text questions

a) Light, water and minerals/nutrients.
b) So that there is no competition between the parent and the offspring.
c) The plant roots can take in water and minerals near the surface of the soil, while other plants with deeper roots take water from lower down in the soil.
d) The fluffy parachutes help the seeds to float in the air so that they can be blown as far as possible from the parent plant.

KEY POINTS

Students will remember the key points more easily if each factor is linked with a strategy. For example, competition for water could involve having deep roots or very shallow spread out roots. Examples of advantages are always required in the more demanding examination questions.

SUMMARY ANSWERS

1. a) May grow taller, may have deeper/shallower roots, flower at different time of year.
   b) Primroses grow in deciduous woodlands and they produce flowers before the oak trees leaves have grown to full size, so they are not shaded.
2. a) To avoid competition between the seedlings and the parent plants and to avoid competition between the seedlings as far as possible.
   b) Primrose tolerates low temperatures and can photosynthesise well in low light levels and low temperatures. It has wide spread shallow roots which reduces competition with tree roots for minerals and water. So primrose leaves grow up early in the spring and the plant flowers and sets seeds before the main oak leaf canopy develops. Then leaves grow very big to capture as much light as they can before dying back ready to overwinter and produce many new leaves in the new growing season.
3. E.g. deep taproot (difficult to remove, can regenerate well if severed); low rosette of leaves (avoids blades of lawnmowers and grazing animals) long flowering period, produces large numbers of seeds, very effective wind dispersal of seed over a large area.

DID YOU KNOW?

Tumbleweeds, found on the plains and deserts of North America, use the whole plant to scatter their seeds. When the seeds are ripe, the plants break off at the roots and are blown away, often travelling miles across the plains, scattering seeds as they go.

B1b 5.5 How do *you* survive?

SPECIFICATION LINK-UP

B1b.11.5

Students should use their skills, knowledge and understanding of 'How Science Works':

- *to suggest how organisms are adapted to the conditions in which they live.*

Substantive content that can be revisited in this spread:

- *organisms have features (adaptations) which enable them to survive in the conditions in which they normally live.*

BIOLOGY ADAPTATION FOR SURVIVAL

B1b 5.5 How do *you* survive?

The most amazing plants in the world?

Most plants die without water – but not the resurrection plants. They can survive massive water losses. They don't prevent water loss or store water – they have adapted to cope with water loss when it happens. These amazing plants can lose up to 95% of their water content without suffering permanent damage.

When conditions get dry, the plants lose more and more water until all that is left are the small, shrivelled remains. The plant looks dead. It can last like this for weeks – but within about 24 hours of watering the tissues fill up with water again (rehydrate). The plant looks as good as new!

Dr Peter Scott and his team at the University of Sussex are trying to find out just how this survival mechanism works, because resurrection plants aren't just a fascinating fact. All over the world crops fail every year because conditions are too dry, and so millions of people don't get enough food. If scientists can find a way to produce 'resurrection crops', then starvation might become a thing of the past.

ACTIVITIES

a) How might 'resurrection crops' prevent starvation in the world?

b) You want to get money for some research into the adaptations of a very unusual animal or plant (you can make one up!). Write a brief application for funding for your project. Using the example of the resurrection plants, explain how information about an unusual adaptation might lead to great benefits for people. Use this to back up your claim for money!

The difference 24 hours and some water can make to a resurrection plant!

The fastest predator in the world?

It takes you about 650 milliseconds to react to a crisis. But the star-nosed mole takes only 230 milliseconds from the moment it first touches its prey to gulping it down. That's faster than the human eye can see!

What makes this even more amazing is that star-nosed moles live underground and are almost totally blind. Their main sense organ is a crown of fleshy tendrils around the nose – incredibly sensitive to touch and smell but very odd to look at!

It seems likely that they have adapted to react so quickly because they can't see what is going on. They need to grab their prey as soon as possible after they touch it. If they don't it might move away or try to avoid them. Then they wouldn't know where it had gone.

When you've got ultra-sensitive tendrils which can try out 13 possible targets every second, who needs eyes?!

The star-nosed mole

Teaching suggestions

- **The fastest predator in the world?** – After some discussion of the rapid reaction time of the star-nosed mole, get the students to test their own reaction times. The best way of doing this is to use an Internet-based reaction timer to see who can react fastest in the class. (Search the web for 'reaction timer'.) Alternatively, record a time for everyone and calculate the mean. The times can be compared with the star-nosed mole. A reaction timer which records the time taken to respond to a sound would be good, because the students could be blindfolded as an even better comparison with the star-nosed mole. What sense is most highly developed in humans?

- **Death by infection** – A fully-grown komodo dragon weighs about 160 lb (73 kg) and it can eat 80% of its body weight in one go. Ask: 'How many quarter-pounder beefburgers could it eat at one go?' This is an entertaining exercise which could finish a discussion about the animal. In addition, students might like to consider the dangers of living at the same time as *Megalonia*, a larger relative of the komodo dragon that was around some 20 000 years ago. *Megalonia* weighed in at about 1000 lb (454 kg). If it, too, could eat 80% of its body weight in one go, how many students from the class could it consume?

- **A carnivorous plant** – Venus fly trap plants are readily available from many garden centres and nurseries ('Homebase' often have them for kids to grow). A practical exercise can be carried out using a paintbrush to trigger the mechanism. The students can find out how many times the trigger hairs have to be brushed before the mechanism starts to operate. On a longer term, they can find out how many times the trap mechanism can be triggered before it turns black and dies.

Death by infection

The Komodo dragon – largest reptile in the world

The Komodo dragon is the largest reptile in the world. A big male can be over three metres long! They live in Indonesia and their colour varies depending on which island they make their home. They have long, forked tongues which give them an excellent sense of smell. They can smell rotting meat five miles away!

The Komodo dragon eats carrion (dead animals) but it is also a predator. But the dragons are reptiles. They cannot run fast for long or pounce on their prey, yet they can kill a huge water buffalo. How do they do it?

The dragons have 52 sharp teeth, but it is not the sharpness which makes them deadly, it is the bacteria which grow on them. A dragon will lie in wait for a water buffalo, and then rush out and grab one of the hind legs. It tears the leg, and the 15 different species of bacteria growing on its teeth get straight into the buffalo's blood stream. Within a couple of days, the buffalo will be dead from the lethal infection. The dragon just follows after its prey until the buffalo dies.

The dragon then eats almost 80% of its own weight in buffalo meat before resting quietly for a very long time!

A carnivorous plant

The Venus flytrap is a plant that grows on bogs. Bogs are wet and their peaty soil has very few minerals in it. This makes them a difficult place for plants to live.

Venus flytraps have special 'traps' which contain a sweet-smelling nectar. They sit wide open, displaying their red insides. Insects are attracted to the colour and the smell.

Inside the trap are many small, sensitive hairs. As the insect moves about to find the nectar, it will brush against these hairs. Once the hairs have been touched, the trap is triggered. It shuts, trapping the insect inside.
Special enzymes then digest the insect inside the trap.

The Venus flytrap – an insect-eating plant!

The Venus flytrap uses the minerals from the digested bodies of its victims in place of the minerals it can't get from the bog soil. After the insect has been digested, the trap reopens ready for its next victim.

ACTIVITIES

c) Look carefully at the information you have been given about resurrection plants, star-nosed moles, Komodo dragons and Venus flytraps. For each one make a list of the adaptations which help them to survive.
d) There are so many different living organisms, each with their own adaptations. Choose three organisms that you know something about – or find out about three organisms which interest you. Make your own fact file on their adaptations and how these adaptations help them to compete successfully. Choose one organism that has an adaptation which has been used by people in some way. Try to include at least one plant!

- **Insectivorous plants** – There are a number of insectivorous plants with different mechanisms for trapping their prey. A good example is the pitcher plant (*Sarracenia*) which digests insects that fall into it. They have a slippery back surface and are full of digestive enzymes that break down the bodies of the unfortunate victims. Some good sources of information are Kew Gardens and the *Dorling Kindersley Plant Book* (edited by Janet Marinelli, 2004).
- **The most amazing plants in the world?** – Find out more about the work on resurrection plants carried out by Dr Peter Scott and his team at the University of Sussex. This could be an activity for higher attaining students and lead nicely into the next chapter. Ask: 'What kinds of crop plants would be the most suitable?' Students to think about the types of crop grown in areas that are prone to drought and the staple foods of the people who live in those regions. The suggested activity in the student book could be set for higher attaining students with a hint about genetic engineering and the isolation of 'survival' genes.

Special needs

Students could be given pictures of three animals, such as the giraffe, the penguin and the ivy (or any suitable well-adapted organisms), and asked to describe or annotate those features of each one that make them successful in their particular environments.

Fact file activities

- A suggestion for this activity would be to find out more about some of the plant species which have become problems because they compete so well that they are becoming more abundant than our native species. These include
  - The water hyacinth (*Eichornia crassipes*), which is considered to be one of the worst weeds in the world.
  - Japanese knotweed (*Fallopia japonica*), which is now a serious pest of the countryside and is almost impossible to control.
  - Milfoil (*Myriophyllum aquaticum*), which is used in aquaria and ends up in freshwater systems where it competes with native plants.

  More examples can be found in the *Dorling Kindersley Plant Book* in the chapter on invasive plants.
- Some suggestions of highly adapted animals are:
  - The vicuna, which lives at altitudes above 5000 m.
  - The camels: single-humped dromedaries or twin-humped Bactrian camels are adapted to life in the desert.
  - Other desert animals, such as kangaroo rats, prairie dogs, bighorn sheep and American desert pigs.
  - Springtails, which are insects that can live in Antarctica at temperatures as low as –38°C.
  - Social insects, such as ants, termites, wasps and bees, which are highly organised.

  More examples can be found in most animal reference books. Some bird groups are worth investigating; for example, the humming birds and their relationships with specific plants, the flightless birds, cuckoos and the ruddy duck.
- There are numerous videos and other resources to back up these activities. A search of the BBC Natural History web site could prove fruitful.

SUMMARY ANSWERS

1 **a)** They are cold blooded – temperatures in the Arctic are too low for reactions of the body to work and so for them to survive.

b) Problems – overheating in day, too cold at night, water loss. How they cope with problems: bask in sun in morning to warm up, hide in burrows or shade of rocks to avoid heat of day and cold of night, reduce water loss by behaviour and do not sweat.

c) Large surface area : volume ratio allows them to lose heat effectively.

2 **a)** Lots of water loss through the leaves, not much water taken up by roots.

b) Most of the water is lost through the leaves, so less leaf surface area means less water loss.

c) Spines, rolled leaves.

d) Water storage in stem, roots or leaves, thick waxy cuticle, ability to withstand dehydration.

e) They have several different adaptations to enable them to withstand water loss/little water available (spines, water storage in stem etc.).

3 **a)** It makes sure that there is plenty of food for the animals and their young.

b) Pandas feed almost exclusively on bamboo, so if it dies out they have no food and will die out as well. For other animals bamboo is only part of the diet, so they can simply eat other plants.

4 Because they are competing for exactly the same things.

5 **a)** In the first month, the crowded seedlings grow taller than the spread out seedlings. Crowded seedlings shade each other, so each seedling grows taller to avoid the shade. Spread seedlings do not have that pressure. But over six months, the crowded seedlings do not get as much light as they shade each other. They photosynthesise less so they cannot grow as tall as the spread out seedlings, which can make as much food as possible.

b) i) They relied mainly on the food stored in the seed. The crowded ones were taller but the spread ones had thicker stems and bigger leaves.

ii) As in a), the spread out seedlings get the full effect of the light and grow as well as possible, making lots of new plant (and so wet) mass. The crowded plants each get less light, therefore less photosynthesis and less wet mass.

c) To make their data more reliable (and give greater validity to any conclusions drawn).

d) i) Light level, the volume of water, minerals available and temperature.

ii) To make the test as fair as possible.

Summary teaching suggestions

- **Special needs**
  Students could be given pictures of animals and plants from cold regions and animals and plants from desert regions and asked to place them under an appropriate heading: 'Arctic' or 'Desert'. This activity covers the content of question 1 and question 2.
- **Gifted and talented**
  In question 5, students could comment on the results as shown on the bar chart. Ask: 'Is there a correlation between total wet mass and average height? What factors would have to be controlled in taking such measurements?'
- **Misconceptions**
  Not all adaptations are structural, so students could answer question 1b) by referring to the behaviour of desert animals.

ADAPTATION FOR SURVIVAL: B1b 5.1 – B1b 5.5

SUMMARY QUESTIONS

1 Cold-blooded animals like reptiles and snakes take their body temperature from their surroundings and cannot move until they are warm.
  a) Why do you think that there are no reptiles and snakes in the Arctic?
  b) What problems do reptiles face in desert conditions and how do they cope with them?
  c) Most desert animals are quite small. How does this help them survive in the heat?

2 a) What are the main problems for plants living in a hot, dry climate?
  b) Why does reducing the surface area of their leaves help plants to reduce water loss?
  c) Describe two ways in which plants can reduce the surface area of their leaves.
  d) How else are some plants adapted to cope with hot, dry conditions?
  e) Why are cacti such perfect desert plants?

3 a) How does marking out and defending a territory help an animal to compete successfully?
  b) Bamboo plants all tend to flower and die at the same time. Why is this such bad news for pandas, but doesn't affect most other animals?

4 Why is competition between animals of the same species so much more intense than the competition between different species?

5 Use Figure 1 on page 88 to answer these questions.
  a) Describe what happens to the height of both sets of seedlings over the first six months and explain why the changes take place.
  b) The total wet mass of the seedlings after one month was the same whether or not they were crowded. After six months there was a big difference.
    i) Why do you think both types of seedling had the same mass after one month?
    ii) Explain why the seedlings that were more spread out each had more wet mass after six months?
  c) When scientists carry out experiments such as the one described on page 88, they try to use large sample sizes. Why?
  d) i) Name the control variable mentioned in the caption to Figure 1.
    ii) Why were these variables kept constant?

EXAM-STYLE QUESTIONS

1 Which of the following adaptations would **not** reduce the rate of transpiration in a plant?
  A Fleshy succulent leaves B Thick waxy cuticle
  C Few stomata D Reduced leaf area

2 Which of the following will **not** help an arctic mamma such as a polar bear, to survive cold temperatures?
  A Thick fur coat
  B Thick layer of fat beneath the skin
  C Small extremities such as ears
  D A large surface area : volume ratio

3 To investigate competition between species, a series o 20 enclosures were set up to look at the effect of kangaroo rats on other, smaller rodents. Kangaroo ra were removed from half of the enclosures. Over three years the numbers of the smaller rodents were counte in both sets of enclosures.
The results are shown on the graph.

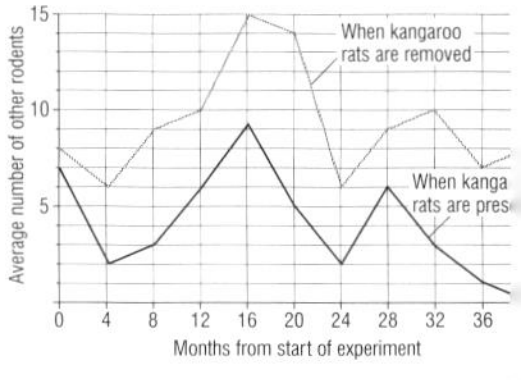

(a) In the enclosure with kangaroo rats, how many ot rodents, on average, were present after 12 months
  A 4 B 6 C 8 D 10

(b) After how many months was there the biggest difference between the average number of other rodents in the two sets of enclosures?
  A 16 B 20 C 32 D 40

(c) The conclusion that can be drawn from the experiment is that:
  A kangaroo rats eat other rodents.
  B kangaroo rats compete with other rodents for fo
  C kangaroo rats compete with other rodents and limit their population size.
  D kangaroo rats make it more difficult for other rodents to breed.

- **Learning styles**
  Interpersonal: Several parts of questions, such as questions 1c), 2e) and 4, lend themselves to class discussion before writing the answers.
  Visual: Annotated diagrams are a good way of learning answers to question 2d) and e). It would be an alternative way of asking the questions.
  Auditory: The questions can be put verbally to the students who can then respond.
- **When to use these questions?**
  - Each question relates to a different lesson, so question 1 could be used after 5.1 etc. as a plenary or a homework exercise, if the summary questions are used in the lesson.
  - For revision sessions.
  - Some of the questions and answers could be turned into revision cards. For example, using question 2, students could make a revision card with the problems identified in a) and the adaptations needed to answer c) and d).

EXAM-STYLE ANSWERS

1 **A** *(1 mark)*

2 **D** *(1 mark)*

3 a) **B** *(1 mark)* b) **B** *(1 mark)* c) **C** *(1 mark)* d) **A** *(1 mark)*

4 a)
- The arctic fox is a carnivore and therefore needs to camouflage itself so that its prey is not frightened away before it can attack.
- In winter the ground is covered in snow and so a white coat makes it almost invisible to its prey.
- In summer, with the snow melted, a brown coat is better camouflage against the brown of the soil/tundra.

(1 mark for each point)

b) Polar bears live in regions where there is ice and snow all year round. Remaining white camouflages them from the seals and fish that they eat. *(1 mark)*

5 a) i) Time of day. *(1 mark)*
ii) Continuous. *(1 mark)*

Evolution and environment

(d) How might the reliability of the results of this experiment be improved?
- A By increasing the number of enclosures of each type.
- B By using the same number of enclosures but increasing the number of kangaroo rats and other rodents in each type.
- C By using just one enclosure of each type rather than ten of each.
- D By having a third set of ten enclosures with only kangaroo rats present. (1)

Animals that live in the arctic have a range of adaptations that allow them to survive.

(a) Explain why the coat of an arctic fox is brown in summer and white in winter. (3)

(b) If the arctic fox adapts by changing its coat colour between summer and winter, why then does the polar bear remain white throughout the year? (1)

The gemsbok is a large herbivore living in dry desert regions of South Africa. It feeds on grasses that are adapted to the dry conditions by obtaining moisture from the air as it cools at night.
The table below shows the water content of these grasses and the feeding activity of the gemsbok over a 24-hour period.

Time of day	% water content of grasses	% of gemsboks feeding
03.00	18	40
06.00	23	60
09.00	25	20
12.00	8	17
15.00	6	16
18.00	5	19
21.00	7	30
24.00	14	50

(a) (i) Name the independent variable investigated. (1)
(ii) Is this a categoric, ordered, discrete or continuous variable? (1)

(b) How does the water content of the grasses change throughout the 24 hour period? (1)

(c) Between which recorded times are more than 30% of the gemsboks feeding? (1)

(d) Suggest three reasons why the gemsboks benefit from feeding at this time. (3)

HOW SCIENCE WORKS QUESTIONS

Maize is a very important crop plant. Amongst many other foods, it is made into cornflakes. It is also grown for animal feed. The most important part of the plant is the cob which fetches the most money. In an experiment to find the best growing conditions, three plots of land were used. The young maize plants were grown in different densities in the three plots.

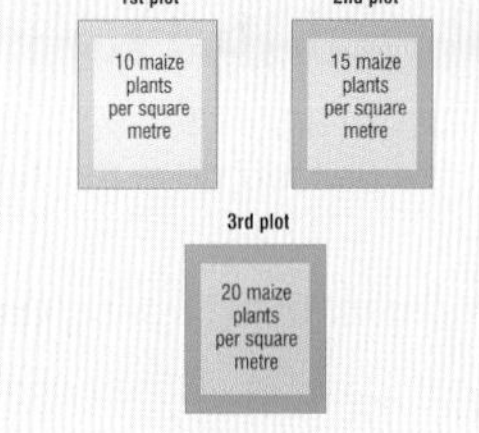

The results were as follows:

	Planting density (plants/m^2)		
	10	15	20
Dry mass of shoots (kg/m^2)	9.7	11.6	13.5
Dry mass of cobs (kg/m^2)	6.1	4.4	2.8

a) What was the independent variable in this investigation? (1)

b) Draw a graph to show the effect of the planting density on the mass of the cobs grown. (3)

c) What is the pattern shown in your graph? (1)

d) This was a fieldwork investigation. What would the experimenter have taken into account when choosing the location of the three plots? (2)

e) Did the experimenter choose enough plots? Explain your answer. (1)

f) What is the relationship between the mass of cobs and the mass of shoots at different planting densities? (1)

g) The experimenter concluded that the best density for planting the maize is ten plants per m^2. Do you agree with this as a conclusion? Explain your answer. (2)

b) It rises during the night (some time after 1800 hours) to a maximum of 25% at 0900 hours and then falls (more rapidly at first) to a minimum of 5% at 1800 hours. *(1 mark)*
Some accurate reference to actual figures in the table is necessary to obtain the mark.

c) Between 2400 hours and 0600 hours. *(1 mark)*
The important words here are 'more than'. Candidates who ignore these words will include the figure of 30% and therefore give a response of 2100 hours to 0900 hours.

d)
- The water content of the grasses that it eats are high over this period.
- It is night and the gemsboks are therefore less easily seen by predators.
- It is cooler and therefore they are less likely to have to sweat and so this helps them conserve precious water.

(1 mark for each point)

Exam teaching suggestions

As always, the multiple choice questions (numbers 1–3) provide an easy-to-mark test for use at the end of a topic. While the first two test recall, question 3 is altogether different. It presents students with a situation that will be entirely new to them. Parts a) and b) require them to interpret the graph and extract some data from it while parts c) and d) test their understanding of experimental technique. The important thing is for students to read the information provided and study the graph thoroughly in order to gain an understanding of what is taking place before even looking at the questions.

Equally they must not be phased by being presented by an unknown situation – it is the application of their knowledge rather than its recall that is being tested. All this makes this question more suited to higher-tier students.

For question 3c), students frequently choose option B (kangaroo rats compete with other rodents for food). While there is no doubt that kangaroo rats limit the size of the other rodents' population due to competition, there is no evidence from the graph that this competition was for food rather than some other factor.

In question 3d), A is correct because it gives a larger sample size. Students often opt for option B as this also gives a larger sample size. However, increasing the number of animals in the existing enclosures also increases the density of the populations within these enclosures. This might introduce other complicating factors such as interspecific competition and an increased risk of disease. Option C reduces the sample size making it **less** valid. Option D has no bearing on the experiment that is investigating the effects of kangaroo rats on other rodents and not the reverse.

Question 4 allows students to write prose in an organised fashion and so demonstrate their ability to put forward a coherent explanation.

Question 5 is useful for students of all abilities as it provides for different levels of answer. For example, in part d) most students should be able to give one reason but only the most able are likely to provide all three.

HOW SCIENCE WORKS ANSWERS

a) The independent variable in this investigation was planting density.

b) Graph with correctly labelled axes and points plotted accurately. Density on the X axis and dry mass on the Y axis.

c) The pattern should show increasing density of planting reducing dry mass of cobs.

d) The three plots should have the same type of soil, the same amount of water and similar weather patterns.

e) Three plots was too small to be certain of a pattern, five would have been better.

f) As the dry mass of cobs gets less per m^2, the dry mass of the shoots increases.

g) It seems from this investigation to be correct, but does the pattern continue? If so, then planting fewer plants could give an even higher yield. The experimenter should investigate lower densities of planting.

How science works teaching suggestions

- **Literacy guidance**
  - Key terms that should be clearly understood are: independent variable, pattern, fieldwork investigation and conclusion.
  - Questions expecting longer answers, where students can practise their literacy skills are: d), f) and g).
- **Higher- and lower-level questions.** Questions d), e) and g) are higher level and the answers provided above are at this level. Question and answer for question a) is lower level.
- **Gifted and talented.** Possibility of developing arguments around the difficulty of choosing suitable areas of land on which to carry out field trials. They might be introduced to the idea of economic value in the different parts of the crop, and perhaps the expense of removing any waste.
- **How and when to use these questions.** When students are confident enough to understand the requirements for field trial investigations. To develop an appreciation of the limitations of some evidence derived scientifically.
- **Homework.** Students living in rural areas could research the optimum growing distance for commercial crops or simply in a vegetable garden. They might consider the biological reasons behind these recommendations.
- **Special needs.** Some help will be needed by some students to interpret units used in the table and the use of the word 'density' in this context.
- **Misconceptions.** That patterns can always be extrapolated. That extrapolation is not reasonable! To address this choose an absurd example that is obviously invalid to extrapolate. For example, the pulse rate against time spent running for a marathon runner in the first minute of the race. Try extrapolating for over 2 hours!

B1b 6.1 Inheritance

LEARNING OBJECTIVES

Students should learn that:

- Genetic information is passed from parent to offspring in the gametes.
- The genetic information is located on the chromosomes in the nucleus of a cell.

LEARNING OUTCOMES

Most students should be able to:

- Explain that genetic information is present in the gametes from each parent.
- Describe the relationship between chromosomes and genes.
- State clearly where the chromosomes are located in the cells.
- Explain that each gene affects a different characteristic.

Some students should also be able to:

- Understand that genetic information is carried on the DNA molecule.
- Recognise that genes are found in pairs on paired chromosomes.

Teaching suggestions

- **Special needs.** Students could be given a list of simpler animals for the 'Offspring naming race' starter.
- **Gifted and talented**
  Discuss complementary base pairing (A-T and G-C).
- **Homework.** An interesting homework exercise would be to ask students to research their own families for inherited characteristics. There are some obvious ones, such as tongue rolling, straight thumbs vs. bendy thumbs, dimples vs. no dimples, ear lobes vs. no ear lobes. They might be able to produce a family tree or pedigree.
- **Learning styles**
  Interpersonal: Discussing the results of human karyotypes.
  Intrapersonal: Deducing the colours of flowers from the conkers.
  Kinaesthetic: Manipulating chromosomes in the human karyotype.
  Visual: Viewing the clips on the structure of cells and division.
  Auditory: Discussing naming offspring in the starter.
- **ICT link-up.** Try finding and using animations of the process of mitosis.

SPECIFICATION LINK-UP B1b.11.6

- *The information that results in plants and animals having similar characteristics to their parents is carried by genes which are passed on in the sex cells (gametes) from which the offspring develop.*
- *Different genes control the development of different characteristics.*
- *The nucleus of a cell contains chromosomes. Chromosomes carry genes that control the characteristics of the body.*

Lesson structure

STARTER

Offspring naming race – Show the students a list of names of ten types of animal and ask them to name their offspring. Include some difficult ones in the list. In a discussion at the end, show photographs and talk over why they resemble each other to establish current knowledge. (5–10 minutes)

What colour will the flowers be? – Show pictures of horse chestnut trees with two colours of flowers, red and white. Hand round some conkers and ask the students what will decide whether they produce red or white flowers when grown. Lead into a discussion. (5 minutes)

Where are my genes? – Remind students of the structure of cells by projecting, or drawing, a generalised cell and getting them to name the parts: cytoplasm, cell membrane, and nucleus. Project an image of a sperm and an egg cell and ask students where they think the genetic material is. Search the web for 'drosophilia chromosome'. (5–10 minutes)

MAIN

- **Organisation of genetic material** – Following on from the 'Where are my genes?' starter, show the students a series of pictures or a video clip of cells undergoing mitotic cell division. Ask: 'What are the rods that can be seen? What would we see if we could look closer?'
- This leads into a discussion of how the genetic material is organised and how it is being shared out equally between the daughter cells. Show a picture of a 'giant' plastic chromosome from a *Drosophila* salivary gland and then build up a picture of a chromosome, so that the relationship between chromosomes and genes is clear.
- Show the students a **human karyotype**, emphasising that there are 23 pairs of chromosomes. One of each pair comes from the father and the other from the mother. If the karyotypes are large enough, it is possible for the students to work in groups and do a cut and stick pairing exercise. In order to help students, some of the more difficult ones could be done for them and they finish it off with a small number of obviously different chromosomes.
- It would be interesting to give the students a mixture of male and female karyotypes and see if they can discover which they have been given, by identifying the 'odd' male chromosome. In a karyotype of a Down's syndrome person, it should be possible to show trisomy 22 or the other variation [not that it is the chromosomes that are important here, but it is a good way of showing that some conditions can show up in a karyotype].
- **Pink flowers or white flowers?** – For a long-term investigation, students could take the conkers used in the 'What colour will the flowers be?' starter, plant them and label them for future groups to find the answer. A time capsule description of the experiment (the best selected by the class) should be placed in a plastic bag, put inside a cigar tube and glued shut. Instructions could be written on a laminated tag and attached to the trees. (Before considering whether or not to do this exercise, think about whether you want the students to carry out one of the cloning exercises, which also involves growing plants – you may run out of pots or growing space!)

PLENARIES

Clues for key words – Give the students the key words: 'cell, nucleus, chromosome, gene'. Ask students to identify and then place them in order of size. (5 minutes)

'Window on Life' – Show the visual summary *Window on Life* (*Sunday Times* CD). (5 minutes)

Karyotypes – If the human karyotype exercise was used above, discuss the value of such karyotypes. Ask: 'Who would use them? What conditions could they show?' This could lead into a discussion about the value of amniocentesis and the inheritance of chromosome abnormalities. (5–10 minutes)

ACTIVITY & EXTENSION

- Show a three-dimensional model of DNA to illustrate the structure.
- The story of Mendel and his peas, and descriptions of some of the experiments using *Drosophila* provide some background to inheritance.
- Students can make their own models of DNA from kits (some are available from the Science Museum catalogue) – check web site.

BIOLOGY

VARIATION

B1b 6.1 Inheritance

LEARNING OBJECTIVES

1 How do parents pass on genetic information to their offspring?
2 Where is the genetic information found?

Young animals and plants resemble their parents. Horses have foals and people have babies. Chestnut trees produce conkers which grow into little chestnut trees. Many of the smallest organisms that live in the world around us are actually identical to their parents. So what makes us the way we are?

Figure 1 This picture shows a mother pig and her offspring. They aren't exactly the same as each other, but they are obviously related!

Why do we resemble our parents?

Most families have characteristics which we can see clearly from generation to generation. People find it funny and interesting when one member of a family looks very much like another. Characteristics like nose shape, eye colour and dimples are ***inherited*** (passed on to you from your parents).

Your resemblance to your parents is the result of ***genetic information*** passed on to you in the sex cells (**gametes**) from which you developed. This genetic information determines what you will be like.

a) Why do you look like your parents?

Genes and chromosomes

The genetic information which is passed from generation to generation during reproduction is carried in the nucleus of your cells. Almost all of the cells of your body contain a nucleus. And it contains all the plans for making and organising a new cell. What's more, the nucleus contains the blueprint for a whole new you!

Imagine the plans for building a car. They would cover many sheets of paper! Yet in every living organism, the nucleus of the cells contains the information to build a whole new animal, plant, bacterium or fungus. A human being is far more complicated than a car. So where does all the information fit in?

b) Where is the genetic information stored?

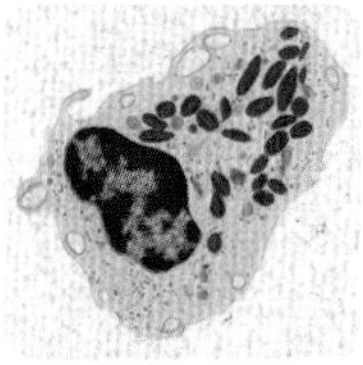

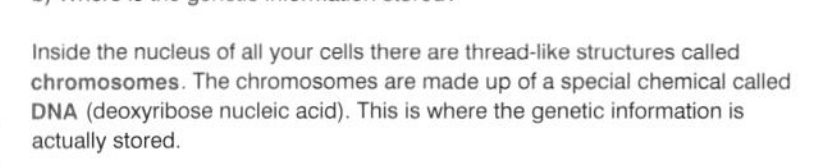

Inside the nucleus of all your cells there are thread-like structures called **chromosomes**. The chromosomes are made up of a special chemical called **DNA** (deoxyribose nucleic acid). This is where the genetic information is actually stored.

DNA is a long molecule made up of two strands which are twisted together to make a spiral. This is known as a double helix – imagine a ladder that has been twisted round!

Each different type of organism has a different number of chromosomes in their body cells. Humans have 46 chromosomes while turkeys have 82! You inherit half your chromosomes from your mother and half from your father, so chromosomes come in pairs. You have 23 pairs of chromosomes in all your normal body cells.

Each of your chromosomes contains thousands of **genes** joined together. These are the units of inheritance.

Figure 2 This micrograph shows a highly magnified human cell. In fact the nucleus of the cell would only measure about 0.005 mm! All the instructions for making you and keeping you going are inside this microscopic package. It seems amazing that they work!

94

Evolution and environment

Each gene is a small section of the long DNA molecule. Genes control what an organism is like – its size, its shape and its colour. Each gene affects a different characteristic about you.

Your chromosomes are organised so that both of the chromosomes in a pair carry genes controlling the same things in the same place. This means your genes also come in pairs, one from your father and one from your mother.

c) Where would you find your genes?

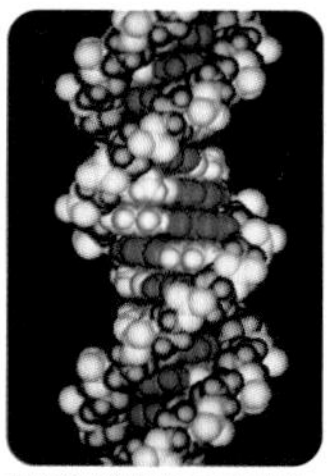

Some of your characteristics are decided by a single pair of genes. For example, there is one pair of genes which decides whether or not you will have dimples when you smile! However most of your characteristics are the result of several different genes working together. For example, your hair and eye colour are both the result of several different genes.

Figure 3 DNA! This huge molecule is actually made up of lots of smaller molecules joined together. Each gene is a small section of the big DNA strand.

Science pioneers: Cracking the code

For a very long time no-one knew how inheritance worked. By the 1940s most scientists thought that DNA was probably the molecule which carried inherited information from one generation to the next.

In the 1950s James Watson (a young American) and Francis Crick (from the UK) were working on the DNA problem at Cambridge. They took all the information they could find on DNA – including X-ray pictures of the molecule taken by another team, Maurice Wilkins and Rosalind Franklin in London.

Watson and Crick tried to build a model of the DNA molecule that would explain everything they knew. When they finally realised that the bases always paired up in the same way they had cracked the code. The now famous DNA double helix was seen for the first time.

SUMMARY QUESTIONS

1 Copy and complete using the words below:

chromosomes DNA genes genetic information gametes nucleus resemble

Offspring ...... their parents because of .... passed on to them in the ...... (sex cells) from which they developed. The information is contained in the ......, made of a chemical called ...... found in the ...... of the cell. The information is carried in small units called ....

2 a) What is the basic unit of inheritance?
b) Offspring inherit information from their parents, but do not look exactly like them – why not?

3 a) Which molecule carries genetic information?
b) Why do chromosomes come in pairs?
c) Why do genes come in pairs?
d) How many genes do scientists think human beings have?

DID YOU KNOW...

. . . that scientists are still not sure exactly how many genes you have? The best estimates put it at between 20 000 to 25 000 – not that many to make a whole human being!

GET IT RIGHT!

Make sure you are clear about the difference between a cell, the nucleus, chromosomes, genes and characteristics.

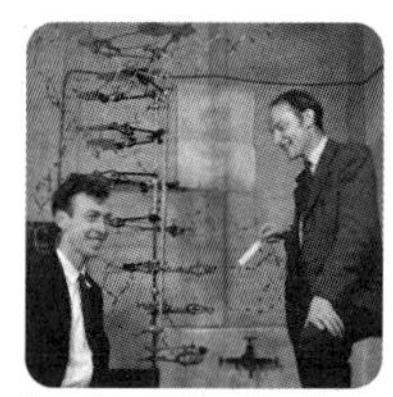

Figure 4 Francis Crick and James Watson – the two men who first showed the world how DNA works. Watson, Crick and Wilkins all received the Nobel Prize for their work. Rosalind Franklin died of cancer before the prizes were awarded.

KEY POINTS

1 Young animals and plants have similar characteristics to their parents. That's because of genetic information passed on to them in the sex cells from which they developed.
2 The nucleus of your cells contains chromosomes. Chromosomes carry the genes that control the characteristics of your body.

95

SUMMARY ANSWERS

1 Resemble genetic information, gametes, chromosomes, DNA, nucleus, genes.

2 **a)** The gene.
b) Offspring inherit information from both parents and so end up with a combination of characteristics, some from father and some from mother.

3 **a)** DNA.
b) You inherit one from each parent.
c) Genes are carried on the chromosomes, so because chromosomes come in pairs, so do the genes – one from each parent.
d) 20 000–25 000.

Answers to in-text questions

a) Because you have inherited genetic information from your parents. This determines what you will look like.

b) In the nucleus of your cells.

c) On your chromosomes in the nucleus of your cells.

DID YOU KNOW?

We do know that we share more than 95% of our DNA with chimpanzees, so we obviously have some genes in common with them!

KEY POINTS

Use the plenary 'Clues for key words' to help students learn key points.

B1b 6.2 Types of reproduction

LEARNING OBJECTIVES

Students should learn:

- The differences between sexual and asexual reproduction.
- That the offspring produced by asexual reproduction are very similar and show little variation.
- That the offspring produced by sexual reproduction differ slightly from each other and from their parents.

LEARNING OUTCOMES

Most students should be able to:

- Describe why asexual reproduction produces identical offspring.
- Describe how variety is achieved in individuals produced by sexual reproduction.

Some students should also be able to:

- Explain the genetic differences between sexually and asexually produced offspring.

Teaching suggestions

- **Gifted and talented.** Students could research and explain the types of nuclear division involved in both types of reproduction.
- **ICT link-up.** There are some interactive programs on the inheritance of different genes and how variation is achieved.
- **Learning styles**

  Interpersonal: Discussing the inheritance of different features.

  Intrapersonal: Making deductions about the inheritance of dimples.

  Kinaesthetic: Practical exercise on the inheritance of different alleles.

  Visual: Naming fruits and vegetables.

  Auditory: Brainstorming discussion on why some plants reproduce sexually and asexually.

KEY POINTS

The students could build up a flow diagram, to show the location and nature of the genes in cells, starting with cell, then nucleus, chromosomes, DNA, gene. They could show the difference between gametes and body cells, i.e. single chromosomes in gametes and pairs in body cells.

SPECIFICATION LINK-UP B1b.11.6

- *There are two forms of reproduction:*
  - *sexual reproduction – the joining (fusion) of male and female gametes. The mixture of the genetic information from two parents leads to variety in the offspring*
  - *asexual reproduction – no fusion of gametes and only one individual is needed as the parent. There is no mixing of genetic information and so no variation in the offspring. These genetically identical individuals are known as clones.*

Lesson structure

STARTER

The name game – Produce a variety of organs of vegetative reproduction, such as bulbs (onion), corms (crocus), rhizome (iris), stem tuber (potato), root tuber (dahlia) and get students to name them. (5 minutes)

Fruit or vegetable? – Have a range of fruit and vegetables available, to include onions, potatoes, yams, root ginger, peas in pod, runner beans, and tomatoes. Ask students to say whether they are produced as the result of sexual or asexual reproduction. (5 minutes)

Who has a dominant characteristic? – Write the list of human genes with two possible alleles given on page 97 of the student book. Ask the students to look at each other and decide which version of each characteristic they have. By adding up the numbers for each one, it is possible to decide on which are the dominant characteristics (alleles). (10–15 minutes)

MAIN

- **OHP presentation of variety in sexual reproduction** – Using the example in the student book for the inheritance of dimples, show the following possibilities:
  - Both parents with both alleles for dimples.
  - Both parents with both alleles for no dimples.
  - One parent with both alleles for dimples, the other parent with both alleles for no dimples.
  - One parent with one allele for dimples and one allele for no dimples, the other parent with both alleles for no dimples.
  - One parent with both alleles for dimples and the other parent with one allele for dimples and one allele for no dimples.
  - Both parents with one allele for dimples and the other allele for no dimples.
- Provide worksheets for the students and explain about the relationship between chromosomes, genes and alleles. Remind them that dimples dominates over no dimples. After showing each possibility, get them to decide whether or not the offspring will have dimples. They have to justify their answers. After they have had time to decide and write their answers, discuss in class, or proceed to the next activity to back up the theoretical predictions.
- **Practical to demonstrate how sexual reproduction produces variety** – For this experiment, you will need two sets (about 50 in each set) of different coloured beads for each group of students. If beads are difficult, it is possible to use haricot beans dyed different colours. Each group of students will need two beakers into which 50 beads can be mixed. The different possibilities listed above can be tested, but the significant ones showing variety are if both parents are heterozygous, one parent heterozygous and one homozygous recessive or one parent heterozygous and one homozygous dominant. Into one beaker put the 'alleles' of one parent and into the second beaker the 'alleles' of the other. If the parent is heterozygous, mix 25 beads of one colour with 25 of another colour. With closed eyes, the student should take one bead from each beaker and record what colours they have selected. Beads should continue to be chosen and recorded until all the beads in the beakers have been used up. Simple arithmetic will show what the ratios/numbers of the offspring having a particular characteristic are.
- This can be extended by choosing to investigate two pairs of alleles. Ask: 'What are the chances of having dangly ear lobes and dimples if one of your parents has dangly ear lobes and the other has dimples?' You will need four colours, but keep the alleles of one gene separate from the alleles of the other gene: the students need to choose a pair of beads for the dangly earlobes and a pair for the dimples.

ACTIVITY & EXTENSION

- Suggest to students that they investigate family photographs and family members for similarities and differences. Sometimes it is possible that a characteristic 'skips' a generation, e.g. 'he has his grandfather's nose'. Students could write a short paragraph about any inherited tendencies in their family. If this is a sensitive issue, then supply pictures of the Hapsburgs or other dynasty where there are obvious family characteristics.
- Students could be asked to find out other examples of single genes controlling characteristics and link this in with some inherited genetic disorders, such as cystic fibrosis and Huntington's disease.
- Those students who keep pets, such as budgerigars or mice, could investigate some of their characteristics. Coat colour in mice is a good example.

SUMMARY ANSWERS

1 **a)** No fusion of gametes, only one parent, no variety.
b) Two parents, fusion of gametes, variety.
c) Sex cell.
d) The differences between individuals as a result of their genetic material.

2 **a)** Sexual reproduction depends on the gametes from two individuals meeting.
b) Sexual reproduction introduces variety, which makes a species more likely to survive.

3 **a)** They have the best of both worlds – safe reproduction through bulbs and variety from seeds.
b) There will be genetic variety in the sexually produced offspring as they inherit characteristics from both parents whereas there is no genetic variety resulting from asexual reproduction as this only involves one parent plant.

BIOLOGY VARIATION

B1b 6.2 Types of reproduction

LEARNING OBJECTIVES

1 Why does asexual reproduction result in offspring that are identical to their parents?
2 How does sexual reproduction produce variety?

Figure 1 A mass of daffodils like this can contain hundreds of identical flowers. This is because they come from bulbs which reproduce asexually.

Reproduction is very important to living things. It is during reproduction that genetic information is passed on from parents to their offspring. There are two very different ways of reproducing – ***asexual reproduction*** and ***sexual reproduction***.

Asexual reproduction

Asexual reproduction only involves one parent. The process produces more organisms completely identical to itself. There is no joining of special sex cells and there is no variety in the offspring.

Asexual reproduction gives rise to offspring known as **clones**. Their genetic material is identical both to the parent and to each other. Although there is no variety, asexual reproduction is very safe – you don't have to worry about finding a partner!

a) Why is there no variety in offspring from asexual reproduction?

Asexual reproduction is very common in the smallest animals and plants and in bacteria. However, many bigger plants like daffodils, strawberries and brambles do it too. Bulbs, corms, tubers (like potatoes), runners and suckers are all ways in which plants reproduce asexually. Asexual reproduction also takes place all the time in your own body, as cells divide to grow and to replace worn-out tissues.

Sexual reproduction

The other way of passing information from parents to their offspring is through sexual reproduction. Sexual reproduction involves the joining of a male sex cell and a female sex cell from two parents. These two special cells (gametes), one from each parent, join together to form a new individual.

If you are the result of sexual reproduction, you will inherit genetic information from both parents. You will have some characteristics from both of your parents, but won't be exactly like either of them. This introduces variety. In plants the gametes involved in sexual reproduction are found within ovules and pollen. In animals they are called ova (eggs) and sperm.

Sexual reproduction is more risky than asexual reproduction, because it relies on the sex cells from two individuals meeting. In spite of this, variety is so important to survival that sexual reproduction is seen in species of organisms ranging from bacteria to people!

b) How does sexual reproduction cause variety in the offspring?

Variation

Sexual reproduction involves the joining of different genetic information. This results in offspring which show much more variation than the offspring from asexual reproduction. This is a great advantage in making sure the species survives. That's because the more variety there is in a group of individuals, the more likely it is that at least a few of them will have the ability to survive difficult conditions.

GET IT RIGHT!

Asexual reproduction results in identical genetic information being passed on. Sexual reproduction makes sure the genetic information is mixed so there is variety in the offspring.

96

Evolution and environment

If we take a closer look at how sexual reproduction works, it becomes clear how variation appears in the offspring.

Each pair of genes affects a different characteristic about you. However the genes in a pair can come in different forms. These different versions of the same gene are called **alleles**. Most things about you are controlled by lots of different pairs of genes. Luckily some of your characteristics are controlled by one gene with just two possible alleles. For example, there are genes which decide whether:

- your earlobes are attached closely to the side of your head or hang freely,
- your thumb is straight or curved,
- you have dimples when you smile,
- you have hair on the second segment of your ring finger.

We can use these genes to help us understand how inheritance works.

c) Why is variety important?

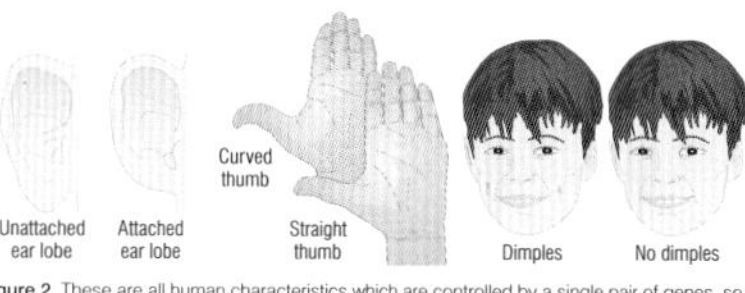

Figure 2 These are all human characteristics which are controlled by a single pair of genes, so they can be very useful in helping us to understand how sexual reproduction introduces variety and how inheritance works.

The gene which controls dimples has two possible forms – an allele for dimples, and an allele for no dimples. The gene for dangly earlobes also has two possible alleles – one for dangly earlobes and one for earlobes which are attached. Some features have lots of different possible alleles.

You will get a random mixture of thousands of alleles from your parents – which is why you don't look exactly like either of them!

NEXT TIME YOU...

... look at your finger tips, take a close look at your fingerprints. They are one of your few characteristics that aren't controlled completely by your genes! Even identical twins (who have identical genetic information) have different fingerprint patterns. Something that goes on before you are even born must also be involved in making your unique fingerprint patterns.

Figure 3 Although these young people have some family likenesses, the variety which results from the mixing of their parents' genetic information can clearly be seen!

SUMMARY QUESTIONS

1 Define the following words:
a) asexual reproduction b) sexual reproduction
c) gamete d) variety.

2 a) Why is sexual reproduction more risky for individuals than asexual reproduction?
b) What is the big advantage of sexual reproduction over asexual reproduction?

3 A daffodil reproduces asexually using bulbs and sexually using flowers.
a) How does this help to make them very successful plants?
b) Explain the genetic differences between a daffodil's sexually and asexually produced offspring.

KEY POINTS

1 In asexual reproduction there is no joining of gametes and only one parent. There is no genetic variety in the offspring.
2 In sexual reproduction male and female gametes join. The mixture of genetic information from two parents leads to genetic variety.

PLENARIES

What are the benefits of asexual vs. sexual reproduction? – This question produces a quick balance sheet of ideas with suggestions from the students. (5 minutes)

Why do some plants do both? – This is a brainstorming session on why a number of plants have both strategies. Ask: 'Has it got something to do with the differences between plants and animals?' (5–10 minutes)

More than one gene involved? – The human characteristics described in the student book are examples of variation due to a single gene, but these are the exceptions rather than the rule. Ask: 'What other characteristics are inherited? How are hair colour and height affected by inheritance?' (5–10 minutes)

Answers to in-text questions

a) Offspring only have one parent; there is no joining or fusion of sex cells from two parents; so identical genetic information is passed on.

b) There is genetic information from both parents, so there is a mixture of characteristics from both.

c) There is a great advantage in making sure that the species survives. The more variety there is in a group of individuals, the more likely it is that at least a few of them will have the ability to survive difficult conditions.

B1b 6.3 Cloning

LEARNING OBJECTIVES

Students should learn that:

- A clone is genetically identical to its parent.
- Taking cuttings is a rapid and cheap method of obtaining new plants.
- Modern cloning techniques, such as tissue culture and embryo transplants, use small groups of cells to produce many identical offspring.

LEARNING OUTCOMES

Most students should be able to:

- Define a clone.
- Explain the importance to gardeners and plant growers of cloning plants.
- Describe cloning by tissue culture in plants.
- Describe the process of embryo transplanting in animals.

Some students should also be able to:

- Interpret information about the advantages and disadvantages of different cloning techniques.
- Make informed judgements on the economic and ethical issues concerning embryo cloning.

Teaching suggestions

- **Special needs.** Students could be shown a clip from *Star Wars 2 – Attack of the clones* showing the clone army (or search the web for a trailer). Ask them to write, or record, instructions to the aliens on how to build a clone army.
- **Gifted and talented.** Students could research tissue culture as practised in horticultural research and development establishments. Provide the students with a list of words associated with the topic (laminar flow cabinet, callus, micropropagation, plantlets, etc.) for them to find the meanings of.
- **Learning styles**
  Interpersonal: Discussing and evaluating cloning.
  Intrapersonal: Understanding cloning techniques.
  Kinaesthetic: Practical work on growing plants from cuttings.
  Visual: Obtaining information from photographs.
  Auditory: Discussion of issues on cloning after watching *Window on Life* (*Sunday Times* CD).

SPECIFICATION LINK-UP B1b.11.6

- *In asexual reproduction, there is no mixing of genetic information, and so no variation in the offspring. These genetically identical individuals are known as clones.*
- *New plants can be produced quickly and cheaply by taking cuttings from older plants. These new plants are genetically identical to the parent plant.*
- *Modern cloning techniques include:*
  - *tissue culture – using small groups of cells from part of a plant*
  - *embryo transplants – splitting apart cells from a developing animal embryo before they become specialised, then transplanting the identical embryos into host mothers.*

Students should use their skills, knowledge and understanding of 'How Science Works':

- ? *to make informed judgements about economic, social and ethical issues concerning cloning . . .*

Lesson structure

STARTER

Photograph of 'Minime' and Dolly – Show the photograph from the Austin Powers films. Ask: 'How was he made?' Show a photograph of Dolly the Sheep for a real-life version and ask: 'How was she made?' (10 minutes)

Photographs of identical twins – Search the web for a photograph of identical twins (from within the school, if possible, with permission). Ask: 'How did they get to be identical?' Write down bullet points of stages. (5–10 minutes)

MAIN

This topic lends itself to practical work on cloning plants. The ideal time to do this is spring or early summer, but provided cuttings are kept in suitable conditions it can be done at any time of year. With both experiments, students could be given a sheet of instructions to carry out.

- **Growing potatoes** – This practical is best carried out during April–May, when pots can be left outside or in glasshouses. You will need large potatoes that have obvious 'eyes', and preferably the beginnings of shoots (chits). Give each group of students one potato; tell them to cut it into sections, each with at least one 'eye'. The sections should be allowed to dry off and then placed in pots of compost. The pots should be stored in a frost-free area, kept watered and supplied with nutrients. They can be re-potted as necessary, noting developments (digital photos can help here), until harvesting when the flowers have died off. A small prize can be awarded to the student whose pot yields the greatest mass at harvest. All students can take their potatoes home at the end of the experiment.
- If this is run as a competition, then the pots need to be kept in the same place, given the same quantities of water and nutrients and harvested at the same time, introducing the students to the idea of controlling those variables we can, and making the test as fair as possible. Be prepared to allow some time for checking progress and measuring during the course of the investigation. This activity can be used to teach aspects of 'How Science Works'.
- **Geranium cuttings** – This exercise is similar, but can be done at any time of year using a propagator. Take cuttings from geraniums, or zonal pelargoniums, by using 10 cm growing tips cut straight across just beneath a node. Remove most of the leaves, leaving only the top two or three, which can be reduced in size to cut down on water loss. Fill 10 cm pots to the brim with cutting compost and use a dibber to make a hole in the compost. Place the cutting in the hole so that the base of the stem is about halfway down the depth of the compost. Firm in, water and label. Pots can be placed in a propagator or covered with a polythene bag over the top, secured round the pot with an elastic band and left on a windowsill. It is not usually necessary to use hormone rooting compounds, but the cut tips can be dipped in some if you want to explain their use (see 'Activity and extension ideas'). Students can take a series of digital photographs for a PowerPoint® presentation of their plant's development.
- Search the web for 'cloning video'. Prepare a sheet of questions based on the content of the film as an aid to focussing attention. You may need to stop the film from time to time to talk over points or to allow for completion of notes.

PLENARIES

Cloning cattle – Give students an empty flow diagram for cloning cattle and the associated labels, but in the wrong order. Ask them to sort them out and complete. They could put in additional links or information boxes for an extension. (10 minutes)

'Window on Life' – If you have not used a video as the main part of the lesson, watch the summary of cloning in the animations section of *Window on Life* (*Sunday Times* CD) and talk through the issues. (10 minutes)

Crossword – Use a crossword to summarise the learning points for this section. Ask students to write more clues for homework or as an extension. (10 minutes)

ACTIVITY & EXTENSION

- If you do the geranium cuttings exercise, then it could be interesting to compare some cuttings grown without using rooting hormone with some that you have dipped in rooting hormone. Also, allow some cuttings to root in water (use small conical flasks), so that the developing root systems can be observed. Pots of transparent gel are also commercially available.
- Try growing cuttings from a variety of different plants. If your school has an Open Day/fete/Bring and Buy sale, these could be sold to bring in money for a charity or for a school project.
- **Look at the mythological creature 'Hydra'** – Ask: 'How might this creature have grown new heads?' Students could compare it with the cnidarian *Hydra* and the work currently going on to identify how this polyp is able to grow copies of itself from its body.

BIOLOGY — VARIATION

B1b 6.3 Cloning

LEARNING OBJECTIVES

1. What is a clone?
2. Why do we want to create clones?

A clone is an individual which has been produced asexually from its parent. It is therefore genetically identical to the parent. Many plants reproduce naturally by cloning, and this has been used by farmers and gardeners for many years.

Cloning plants

Gardeners can produce new plants by taking cuttings from older plants. This is a form of artificial asexual reproduction which has been carried out for hundreds of years. How do you take a cutting? First you remove a small piece of a plant – often part of the stem or sometimes just part of the leaf. If you grow it in the right conditions, new roots and shoots will form to give you a small, complete new plant.

Using this method you can produce new plants quickly and cheaply from old plants. The cuttings will be genetically identical to the parent plants.

Many growers now use hormone rooting powders to encourage the cuttings to grow. They are most likely to develop successfully if you keep them in a moist atmosphere until their roots develop. We produce plants such as orchids and many fruit trees commercially by cloning in this way.

a) Why does a cutting look the same as its parent plant?

Figure1 Simple cloning by taking cuttings is a technique used by gardeners and nurserymen all around the world. It gives us plants like these.

Cloning tissue

Taking cuttings is a very old technique. In recent years scientists have come up with a more modern way of cloning plants called *tissue culture*. It is more expensive but it allows you to make thousands of new plants from a tiny piece of plant tissue. If you use the right mixture of plant hormones, you can make a small group of cells from the plant you want produce a big mass of identical plant cells.

Then, using a different mixture of hormones and conditions, you can stimulate each of these cells to form a tiny new plant. This type of cloning guarantees that the plants you grow will have the characteristics you want.

b) What is the advantage of tissue culture over taking cuttings?

Cloning animals

In recent years cloning technology has moved forward even further and now includes animals. In fact cloning animals is now quite common in farming, particularly cloning cattle embryos. Cows normally produce only one or two calves at a time. If you use embryo cloning, your very best cows can produce many more top-quality calves each year.

In embryo cloning, you give a top-quality cow fertility hormones to make her produce a lot of eggs. You then fertilise these eggs using sperm from a really good bull. Often this is done inside the cow, and the embryos which are produced are then gently washed out of her womb. Sometimes the eggs are collected and you add the sperm in a laboratory to produce the embryos.

SCIENCE @ WORK

Cloning cattle embryos and transferring them to host cattle is skilled and expensive work. Teams of scientists, technicians and vets are constantly working to improve the technique even more.

98

Evolution and environment

At this very early stage of development every cell of the embryo can still form all of the cells needed for a new cow. They have not become specialised.

1. Divide each embryo into several individual cells.
2. Each cell grows into an indentical embryo in the lab.
3. Transfer embryos into their host mothers, which have been given hormones to get them ready for pregnancy.
4. Identical cloned calves born. They are not biologically related to their mothers.

Early embryo (cluster of identical cells)

Figure 2 Using normal sexual reproduction, a top cow might produce 8–10 calves during her working life. Using embryo cloning she can produce more calves than that in a single year!

Cloning embryos in this way has made it possible for us to take high-quality embryos all around the world. We can take them to places where cattle with a high milk yield or lots of meat are badly needed for breeding with poor local stock. Embryo cloning is also used to make lots of identical copies of embryos that have been genetically modified to produce medically useful compounds. (See pages 102 and 103.)

GET IT RIGHT!

- Remember clones have identical genetic information.
- Make sure you are clear about the difference between a tissue and an embryo.

SUMMARY QUESTIONS

1 Match up the following definitions:

a) Cuttings	A Splitting cells apart from a developing embryo before they become specialised to produce several identical embryos.
b) Tissue cloning	B Taking a small piece of a stem or leaf and growing it on in the right conditions to produce a new plant.
c) Asexual reproduction	C Getting a few cells from a desirable plant to make a big mass of identical cells each of which can produce a tiny identical plant.
d) Embryo cloning	D Reproduction which involves only one parent, there is no joining of gametes and the offspring are genetically identical to the parent.

2 Tissue cloning and taking cuttings both give you plants which are identical to their parent. How do these two methods of plant cloning differ?

3 a) Why is the ability to clone cattle embryos so useful?
b) Draw a flow diagram to show the stages of the embryo cloning of cattle.
c) Comment on the economic and ethical issues involved in embryo cloning in cattle.

KEY POINTS

1. The genetically identical offspring produced by asexual reproduction are known as **clones**.
2. New plants can be produced quickly and cheaply by taking cuttings from older plants. The new plants are genetically identical to the older ones.
3. There are a number of more modern cloning techniques. These include tissue culture of plants and embryo cloning and transfers in animals.

99

SUMMARY ANSWERS

1 **a)** B **b)** C **c)** D **d)** A

2 **Cuttings:** simple; low technology; take small piece of plant; put it into the right conditions and it grows into a new plant genetically identical to the parent plant.
Tissue cloning: take a few cells from the plant; provide the right conditions to produce a mass of cells; separate individual cells and provide the right conditions; can form thousands of tiny plants identical to the parent.

3 **a)** It allows the production of far more calves from the best cows; can carry good breeding stock to poor areas of the world; as frozen embryos; can replicate genetically engineered animals quickly.

b) Either cow given hormones to produce large numbers of eggs → then cow inseminated with sperm → embryos collected and taken to the lab → embryos split to make more identical embryos → cells grown on again to make more identical embryos → embryos transferred to host mothers.
Or cow given hormones to produce large numbers of eggs → eggs collected and taken to lab → eggs and sperm mixed → embryos grown → embryos split up to make more identical embryos → cells grown on to make bigger embryos → embryos transferred to host mothers.

? **c)** Student shows understanding of issues involved in embryo cloning.

Answers to in-text questions

a) A cutting looks the same as its parent plant, because it is genetically identical. It has been grown from a small piece of the parent plant.

b) Tissue culture allows you to make thousands of new plants from a tiny piece of plant tissue.

KEY POINTS

The point about having identical genetical information is the most important thing to learn about clones.

B1b 6.4 New ways of cloning animals

LEARNING OBJECTIVES

Students should learn:

- The steps in the techniques of fusion cell and adult cell cloning.
- How scientists were able to clone a sheep.
- The potential benefits and risks of cloning animals.

LEARNING OUTCOMES

Most students should be able to:

- Explain the processes of fusion cell and adult cell cloning.
- Describe how Dolly the sheep was cloned.
- List some of the benefits and disadvantages of cloning animals.

Some students should also be able to:

- Evaluate the advantages and disadvantages of cloning.
- Discuss the ethical issues raised by adult cell cloning techniques.

Teaching suggestions

- **Special needs.** These students could be given pictures of Dolly the sheep and asked to make a scrapbook or poster as suggested in the Main section.
- **Gifted and talented.** Students could research ways in which animals, such as cows and sheep, can be genetically engineered and then cloned to produce useful substances in their milk. Has it been done? What substances have been produced? What substances would it be beneficial to produce?
- **Learning styles**
  Intrapersonal: Expressing personal opinions in the discussions.
  Interpersonal: Working in a group to produce a poster.
  Kinaesthetic: Making a poster.
  Visual: Viewing the PowerPoint® presentations.
  Auditory: Listening to the opinions of others.
- **ICT link-up.** There is scope here for the use of ICT in designing posters and searching the web for information. Web sites that could be useful have been given.

SPECIFICATION LINK-UP B1b.11.6

- *Modern cloning techniques include:*
  - *embryo transplants – splitting apart cells from a developing animal embryo before they become specialised, then transplanting the identical embryos into host mothers*
  - *fusion cell and adult cell cloning.*

Students should use their skills, knowledge and understanding of 'How Science Works':

- ? *to make informed judgements about economic, social and ethical issues concerning cloning . . .*

Lesson structure

STARTER

Fill in the vowels – Write a list of key words and phrases used in this chapter on the board, but leave out the vowels (e.g. CLNNG, NCLS). Ask students to write down the words and then select students to give you an answer plus a definition. (5–10 minutes)

Would you like to be a twin? – If there are twins in the class, get them to describe what it is like to be a twin. Open up the discussion and then ask the students to write down a list of advantages and disadvantages of being an identical twin. Build up a general list on the board. (10 minutes)

MAIN

The main purpose of this lesson is to ensure that the students have a sound understanding of the basic technique of fusion cell cloning and how it can be used to clone animals.

- **Fusion cell cloning** – Prepare a PowerPoint® presentation illustrating the steps of fusion cell cloning. Give students worksheets so that they can follow the sequence and write about the stages in their own words. Allow time for questions and further explanations.
- ? **Dolly the sheep** – The story of Dolly is well-documented and it would be possible to show some of the material from web sites such as the Science Museum Antenna, the BBC or the Roslin Institute, Edinburgh (where the cloning was carried out). Alternatively, if there is good access to computers, the students could be given a list of suitable sites and then carry out their own search in groups. Each group of students could then be asked to produce a poster about Dolly and her life for display in the laboratory.
- The Science Museum Antenna link is a particularly good one and covers the story of Dolly well in a sequence of topics. It combines details of the techniques involved with details of Dolly's life and demise.
- ? **Advantages and disadvantages** – Give students, working in groups, large A3 sheets of paper on which they can write their ideas about the advantages and disadvantages of this type of cloning. Ask one group to present their ideas and then discuss generally adding ideas from the other groups. At the end of the discussion, ask students to list the advantages and disadvantages in order of importance and compile a class list.

PLENARIES

The quagga . . . is fusion cell cloning the answer? – Tell the story of the quagga, a type of zebra that became extinct in the 1880s and the attempts by the Quagga Project Committee to revive it. Is this a case for the use of a cloning technique or selective breeding? Discuss cloning as an alternative to selective breeding. (10 minutes)

Why is it . . .? – Why is it easier to clone plants than it is to clone animals? Ask students the differences between plants and animals that might affect the outcome of cloning. (5 minutes)

ACTIVITY & EXTENSION IDEAS

- **Designer babies or cloned babies?** – Research the Science Museum Bionet link for the topic Design a baby? Print out or project the information. Ask students to discuss whether it would be more desirable to design a baby or clone one. What are the pros and cons of each technique if it were possible to do it?
- **How much does it all cost?** – One factor which has not so far been considered is the cost of developing these new methods for cloning animals. The initial research has to be funded and then the cost of producing cloned animals could be quite high. Humans have been producing new varieties of animals through the techniques of selective breeding for centuries, so why should money be spent on cloning? Will it make food more expensive?

Answers to in-text questions

a) Adult cell cloning or reproductive cloning.

b) It could produce medically useful proteins.

BIOLOGY VARIATION

B1b 6.4 New ways of cloning animals

LEARNING OBJECTIVES

1. How did scientists clone a sheep?
2. What are the potential benefits – and risks – of cloning animals?

True cloning of animals, without sexual reproduction involved at all, has been a major scientific breakthrough. The basic technique is known as ***fusion cell cloning***. It is the most complicated form of asexual reproduction you can find!

Fusion cell cloning

To clone a cell from an adult animal is easy. Asexual reproduction takes place all the time in your body to produce millions of identical cells. But to take a cell from an adult animal and make an embryo or even a complete identical animal is a very different thing.

Here are the steps involved:

- The nucleus is taken from an adult cell.
- At the same time the nucleus is removed from an egg cell from another animal of the same species.
- The nucleus from the original adult cell is placed in the empty egg and the new cell is given a tiny electric shock.
- This fuses the new cell together, and starts the process of cell division.
- An embryo begins to develop which is genetically identical to the original adult animal.

Adult cell cloning

Fusion cell cloning has been used to produce whole animal clones. The first large mammal ever to be cloned from another adult animal was Dolly the sheep, born in 1997. A team of scientists in Edinburgh produced Dolly from the adult cell of another sheep.

When a new animal is produced, this is known as ***adult cell*** or ***reproductive cloning***. It is still relatively rare. You still have to fuse the nucleus of one cell with the empty egg of another animal. Then you have to place the embryo which results into the womb of a third animal. It develops there until it is born.

a) What is the name of the technique which produced Dolly the sheep?

When Dolly was produced she was the only success from hundreds of attempts. The technique is still difficult and unreliable, but scientists hope that it will become easier in future.

Figure 1 Dolly the sheep was the first large mammal to be cloned from another adult animal. Her birth caused great excitement and many scientists have tried to clone other animals since.

DID YOU KNOW?

The only human clones alive at the moment are natural ones known as identical twins! But the ability to clone mammals such as Dolly the sheep has led to fears that some people will be desperate to have a clone of themselves produced – whatever the cost!

100

Evolution and environment

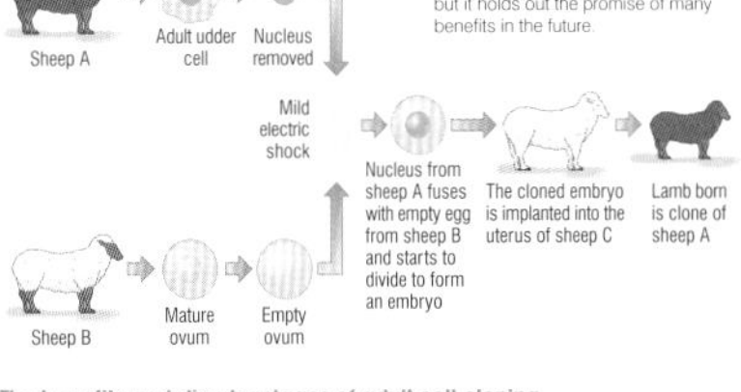
Sheep A
Adult udder cell
Nucleus removed
Mild electric shock
Nucleus from sheep A fuses with empty egg from sheep B and starts to divide to form an embryo
The cloned embryo is implanted into the uterus of sheep C
Lamb born is clone of sheep A
Sheep B
Mature ovum
Empty ovum

Figure 2 **Adult cell or reproductive cloning** is still a very difficult technique – but it holds out the promise of many benefits in the future.

The benefits and disadvantages of adult cell cloning

One big hope for adult cell cloning is that animals which have been genetically engineered to produce useful proteins in their milk can be cloned. This would give us a good way of producing large numbers of cloned, medically useful animals.

This technique could also be used to help save animals from extinction, or even bring back species of animals which died out years ago. The technique could be used to clone pets or prized animals so that they continue even after the original has died. However, some people are not happy about this idea. (See page 104.)

b) How might adult cell cloning be used to help people?

There are some disadvantages to this exciting science as well. Many people fear that the technique could lead to the cloning of human babies. At the moment this is not possible, but who knows what is in the future?

One big problem is that modern cloning produces lots of plants or animals with identical genes. In other words, cloning reduces variety in a population. This means the population is less able to survive any changes in their environment which might happen in the future. That's because if one of them does not contain a useful mutation, none of them will.

In a more natural population, at least one or two individuals can usually survive change. They go on to reproduce and restock. This could be a problem in the future for cloned crop plants, or for cloned farm animals.

SUMMARY QUESTIONS

1. Copy and complete:

   Dolly the sheep was created from the ...... cell of another sheep. She was ...... to this sheep. This technique is known as ....

2. Produce a flow chart to show how fusion cell cloning works.
3. a) List the main advantages of the development of adult cell cloning techniques and the main disadvantages.
   b) Give your own opinion about whether work on the technique should be allowed to continue. Explain your point of view.

KEY POINTS

1. Fusion cell cloning is a form of asexual reproduction
2. In adult cell cloning a whole cloned animal can result. The nucleus from a cell from an adult animal is transferred to an empty egg cell from another animal. A small electric shock fuses the cell and starts embryo development. The embryo is placed in a third animal to develop.

101

SUMMARY ANSWERS

1 Adult, identical, adult cell cloning.

2 Nucleus from adult cell
↓
Egg with nucleus removed
↓
Electric shock
↓
Fusion of cell

3 a) Advantages – production of mechanically useful proteins
– could be used to save animals from extinction
– clone pets/prized animals

Disadvantages – reduces variety in a population
– species could become extinct
– ethics of human cloning.

b) Students reasoned arguments.

KEY POINTS

The Key Points on this spread can be summarised by flow diagrams of the techniques and these diagrams would make useful revision cards.

B1b 6.5 Genetic engineering

LEARNING OBJECTIVES

Students should learn that:

- Genetic engineering involves artificially changing the genetic material of an organism.
- Genes can be transferred from one organism to another.
- Genes can be transferred into plants and animals so that they develop desired characteristics.

LEARNING OUTCOMES

Most students should be able to:

- Explain the term 'genetic engineering'.
- Describe how genes from one organism can be transferred into another organism.
- List the advantages and disadvantages of genetic engineering.
- Interpret information about cloning techniques and genetic engineering techniques.

Some students should also be able to:

- Explain the process of genetic engineering, and the difference between genetically modified organisms which produce useful proteins and organisms which are improved themselves.
- Evaluate the advantages and disadvantages of genetic engineering.

Teaching suggestions

- **Special needs.** This is a difficult area, as the concepts are so abstract. The students could look at GM labelling in supermarket products. Get the students to recognise the Soil Association symbol as meaning 'natural'.
- **Gifted and talented.** Students could be introduced to some of the terms used in genetic engineering, such as the correct names of the enzymes, plasmids, vectors, marker genes, recombinant DNA.
- **ICT link-up.** See the Word exercise in the third bullet point of the Main part of the lesson structure.

SPECIFICATION LINK-UP B1b.11.6

- *In genetic engineering, genes from the chromosomes of humans and other organisms can be 'cut out' using enzymes and transferred to cells of other organisms.*
- *Genes can also be transferred to the cells of animals or plants at an early stage in their development so that they develop with desired characteristics.*

Lesson structure

STARTER

'Glow in the dark mouse' – Search online news stories about genetic engineering such as 'glow in the dark' mice. For example, try www.bbc.co.uk. Discuss how this might come about. (5–10 minutes)

What has genetic engineering got to do with diabetes? – Ask the class if anyone knows a person with diabetes; this usually generates a response, although be careful to differentiate between Type 1 and Type 2. A short discussion on the causes and treatment will lead to the topic of insulin and how it can be made. (10 minutes)

Genetic engineering – good or bad? – Ask students to jot down three positive points about GM crops and three negative points. Give them 2 minutes to do this and then build up a list on the board. (10 minutes)

MAIN

- **What is genetic engineering?** – Take students through the sequence of selecting a gene, cutting it out, putting it into a bacterium; mention the different enzymes (not necessarily by name but according to what they are doing). A good example to choose as an illustration is the human growth hormone. The production of the hormone is under the control of one gene, so the sequence is clear. As this is a difficult concept, a worksheet for the students to fill in during the presentation will help them.
- **Make your own genetically modified bacterium** – Using a digital camera and a stop motion animation program, provide the students with the materials to make a Plasticine model of a genetically modified bacterium. They will need to remove the required gene from a chromosome model and insert it into a plasmid in a bacterium. Use a knife labelled 'enzymes'.
- **Computer exercise** – To explain genetic engineering, type a random DNA sequence into a word processor and copy/paste a part of this into a drawing of a bacterium. Now copy the two and paste as many copies of the new bacterium as quickly as you can.
- **GM circus** – Provide material from pro-GM organisations, such as Monsanto, and anti-GM organisations, such as Greenpeace and the Soil Association, together with articles from the general media coverage. Ask students to gather information from the sources during class time, and then for homework design a case for or against.

PLENARIES

Animations on genetic modification – Show this section of the *Window on Life* CD (*Sunday Times* CD). (5–10 minutes)

Designer kids quandary – Imagine you have been given the power to decide the sex, looks and personality characteristics of your own children. What would you choose and why? Discuss, asking each student to write down their choices and then pick one student to explain. That student can then pick the next and so on. (10–15 minutes)

Genetic engineering – good or bad? – Review the list of pros and cons made at the beginning. Have any ideas changed? (5–10 minutes)

Teaching suggestions continued

- **Learning styles**

  Interpersonal: Discussing the link with diabetes.

  Intrapersonal: Considering the pros and cons of genetic modification.

  Kinaesthetic: Model making to show how genetic engineering is achieved.

  Visual: Following the sequence of the presentation on genetic engineering.

  Auditory: Explaining your 'designer child' to the rest of the class.

- **ICT link-up.** Use of computer program to simulate introduction of genes into a bacterium.

- **Homework.** See last point, 'GM circus' in main part of lesson.

ACTIVITY & EXTENSION

- **Genetic diseases** – There are references on the Internet to specific genetic diseases. There is a good video on cystic fibrosis called 'Sammi's story' about Sammi Sparke who had a successful lung transplant.
- There are videos on GM crops giving a *negative* slant. These can be obtained free of charge from the organisation 'Compassion in World Farming' (Charles House, 5a Charles Street, Petersfield, Hants GU32 3EH, Tel. (01730) 264 208). However, students will be expected to present a balanced view of this topic.

B1b 6.5 Genetic engineering

LEARNING OBJECTIVES

1. What is genetic engineering?
2. How are genes transferred from one organism to another?
3. What are the issues involved in using genetic engineering?

When you clone plants and animals, you are changing the natural processes of reproduction. There is another new technology which takes the changes much further – **genetic engineering** (also known as **genetic modification**). Genetic engineering is used to change an organism and give it new characteristics which we want to see.

What is genetic engineering?

Genetic engineering involves changing the genetic material of an organism. You take a small piece of DNA – a gene – from one organism and transfer it to the genetic material of a completely different organism. So, for example, genes from the chromosomes of one of your human cells can be 'cut out' using enzymes and transferred to the cell of a bacterium. Your gene carries on making a human protein, even though it is now in a bacterium.

a) How is a gene taken out from one organism to be put into another?

If genetically engineered bacteria are cultured on a large scale they will make huge quantities of protein from other organisms. We now use them to make a number of drugs and hormones used as medicines.

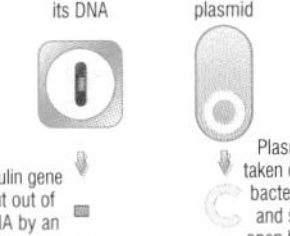

Figure 1 The principles of genetic engineering. A bacterial cell receives a gene from a human being.

Transferring genes to animal and plant cells

There is a limit to the types of proteins bacteria are capable of making. As a result, genetic engineering has moved on. Scientists have found that genes from one organism can be transferred to the cells of another type of animal or plant at an early stage of their development. As the animal or plant grows it develops with the new desired characteristics from the other organism.

b) Why are genes inserted into animals and plants as well as bacteria?

The benefits of genetic engineering

One of the biggest advantages of genetically engineered bacteria is that they can make exactly the protein needed, in exactly the amounts needed and in a very pure form. For example, people with diabetes need supplies of the hormone insulin. It used to be extracted from the pancreases of pigs and cattle but it wasn't quite the same as human insulin, and the supply was quite variable. Both of those problems have been solved by the introduction of genetically engineered human insulin. (See Figure 1.)

We can use engineered genes to improve the growth rates of plants and animals. They can be used to improve the food value of crops and to reduce the fat levels in meat. They are used to produce plants which make their own pesticide chemicals. GM food lasts longer in the supermarkets. It can also be designed to grow well in dry, hot or cold parts of the world. So it could help to solve the problems of world hunger.

A number of sheep and other mammals have also been engineered to produce life-saving human proteins in their milk. These are much more complex proteins than the ones produced by bacteria. They have the potential to save many lives.

DID YOU KNOW...

... glowing genes from jellyfish have been used to produce crop plants which give off a blue light when they are attacked by insects. Then the farmer knows when they need spraying!

Human engineering

If there is a mistake in your genetic material, you may have a genetic disease. These can be very serious. Many people hope that genetic engineering can solve the problem.

It might become possible to put 'healthy' DNA into the affected cells by genetic engineering, so they work properly. Perhaps the cells of an early embryo can be engineered so that the individual develops to be a healthy person. If these treatments become possible, many people would have new hope of a normal life for themselves or their children.

c) What do we mean by a 'genetic disease'?

Figure 2 These sheep look very normal – but they are a genetically engineered flock producing human proteins in their milk. The proteins are used in life-saving medicines.

The disadvantages of genetic engineering

Genetic engineering is still a very new science. There are many concerns about it as no-one can yet be completely sure what all of the long-term effects might be. For example, it seems possible that insects may become pesticide-resistant if they eat a constant diet of pesticide-forming plants.

Some people are concerned about the effect of eating genetically modified food on human health. Genes from genetically modified plants and animals might spread into the wildlife of the countryside. Genetically modified crops are often not fertile, which means farmers in poor countries have to buy new seed each year.

And people may want to manipulate the genes of their future children. This might be to make sure they are born healthy, but what if it is to have a child who is clever, or good-looking, or good at sport? The idea of 'designer babies' causes concern for many people. Genetic engineering raises issues for us all to think about. (See page 105.)

Figure 3 You can't tell what is genetically modified and what isn't just by looking at it! In the UK very few genetically modified foods are sold. The ones that are have to be clearly labelled. But many other countries, including the USA, are far less worried and use GM food quite widely.

SCIENCE @ WORK

There are many scientists working in genetic engineering. Some work for pharmaceutical companies developing new medicines. Others are doing medical research and some are involved in agriculture and crop breeding.

SUMMARY QUESTIONS

1. Copy and complete using the words below:

   cell engineering enzymes gene genetic transfer

   Genetic ...... involves changing the ...... material of an organism. You cut a ...... from one organism using .... Then ...... it to the ...... of a completely different organism.

2. a) Make a flow diagram that explains the stages of genetic engineering.
   b) Make two lists, one to show the possible advantages of genetic engineering and the other to show the possible disadvantages.
   c) Do you think genetic engineering is a good idea? Should it be allowed? Justify your views.

KEY POINTS

1. In genetic engineering, genes from the chromosomes of humans and other organisms can be 'cut out' using enzymes and transferred to the cells of bacteria.
2. Genes can also be transferred to the cells of animals and plants at an early stage of their development.
3. There are many potential advantages and disadvantages to the use of genetic engineering.

SUMMARY ANSWERS

1. Engineering, genetic, gene, enzymes, transfer, cell.
2. a) Suitable flow diagram based on the figure in the student book spread.

   b) **Advantages** – any suitable points such as:
   Bacteria can make human medicines and hormones.
   Improve the growth rates of plants and animals.
   Improve the food value of crops.
   Reduce the fat levels in meat.
   Produce plants that make their own pesticide chemicals.
   Crop plants can give off a blue light when attacked by insects so that the farmer knows when they need to be sprayed.
   Possible cures for genetic diseases.
   Fruit does not go bad so quickly.

   Disadvantages – any suitable points such as:
   Insects may become pesticide-resistant if they eat a constant diet of pesticide-forming plants.
   Effect on human health of eating genetically modified food.
   Genes from genetically modified plants and animals might spread into the wildlife.
   Genetically modified crops not fertile, so farmers in poor countries have to buy new seed each year.
   People may want to manipulate the genes of their future children.

   c) Student weighs pros and cons and makes a reasoned decision.

Answers to in-text questions

a) It is cut out using enzymes.

b) There is a limit to the types of protein bacteria can make.

c) A genetic disease is a disease or problem caused by a mistake in the genetic material in your cells.

KEY POINTS

Knowing the sequence of the stages of genetic engineering can help students understand the process. The advantages and disadvantages could be listed in columns and will help provide students with a balanced view.

B1b 6.6 Making choices about technology

SPECIFICATION LINK-UP

B1b.11.6

Students should use their skills, knowledge and understanding of 'How Science Works':

- *to Interpret information about cloning techniques and genetic engineering techniques*
- *to make informed judgements about the economic, social and ethical issues concerning cloning and genetic engineering, including GM crops.*

BIOLOGY VARIATION

B1b 6.6 Making choices about technology

Cc – A REAL COPYCAT!

Cc, or Copycat, was the first cloned cat to be produced. Born in 2002, she was a change of direction. Most of the research into cloning had been focused on farm and research animals – but cats are thought of first and foremost as pets.

Cc the cloned cat is unaware of the stir she has caused. Cats like this one are often well-loved pets – but should we really be cloning our old friends?

Much of the funding for cat cloning in the US comes from companies who are hoping to be able to clone people's dying or dead pets for them. It has already been shown that a succesful clone can be produced from a dead animal. Cells from beef from a slaughterhouse were used to create a live cloned calf.

But to make Cc, 188 attempts were made producing 87 cloned embryos, only one of which resulted in a kitten. Cloning your pet won't be easy or cheap. The issue is, should people be cloning their dead pets, or should they be learning to grieve, appreciate the animal they had and give a home to one of the thousands of unwanted cats already in existence? Even if a favourite pet is cloned, it may look nothing like the original because the coat colour of many cats is the result of random gene switching in the skin cells. The markings would never be the same again, even if the DNA was!

ACTIVITY

On pages 98 to 101 there is lots of information about cloning plants and animals for farming. Here you have two stories about cloning animals for money – to bring back dead pets and allow neutered racehorses to reproduce and make lots of money for their owners.

There is talk of a local company setting up a laboratory to clone animals for people across the country – cats, dogs and horses. Write a letter to your local paper either ***for*** the application or ***against*** it. Make sure you use clear sensible arguments and put the science of the situation across clearly.

THE FOAL WHO COULD CHANGE RACING – FOR GOOD! By David Turf

This little foal with her mum looks just like any other, but in fact she's made history!

The foal in this photo with her mother is no ordinary young horse. Prometea is the first cloned horse – and her surrogate mother is also her identical twin, because the foal is a clone of the mare who gave birth to her. This new technology has in turn led to a breakthrough which could change the breeding of racehorses forever.

Pieraz is a famous Arab horse who has been world endurance racing champion several times. Pieraz 2, a foal born in 2005, is his closest relative. 'So what?', I hear you say. It is very common for successful racehorses to be used for breeding. The difference here is that Peiraz was neutered when he was still a youngster – and Pieraz 2 is not his son, but his clone!

ACTIVITIES & EXTENSION IDEAS

- The activity 'Changing the genes – right or wrong' in the student text could be extended by scripting a series of interviews with the people. If the daytime TV show ran for six days, and a ten-minute slot was allotted every day, all the characters shown could be interviewed in turn. The students could work in groups and pool their ideas about each interview. This would encourage them to think of the right questions to ask and replies to give. For example, the charity worker has a point, but the decrease in use of pesticides, fertiliser and disease could mean more profit and thus the ability to buy more seed. Perhaps the charity could provide some finance (remind students of the slogan for Christian Aid Week). The seventh day could be a summary or run as a phone-in from the public!
- Students could write a letter to Gregor Mendel, explaining to him what his 'factors' are now called and how much more we know about them. Imagine what he might say about putting genes into his peas.
- There is plenty of scope in this spread for encouraging higher attaining students. Those who are really interested could investigate the qualifications needed to become a research scientist in a medical research laboratory. They need to find out: 'How long is the training? What are the job prospects?' Alternatively, they could find out how to train to be a scientific journalist.

Teaching suggestions

- **Cloning animals for money** – This is a complicated situation and before setting the students the activity, it could be beneficial to have some class discussion – a 'brainstorming' session – where they voice their thoughts about it. Ask: 'Does cloning animals for farming bring in money? What are the guarantees that Pieraz 2's offspring are going to breed another generation of successful racehorses? As for Copycat, her genes might be the same as her mother's, but will her nature be the same?' Some of the economic considerations could include the cost of training the scientists and technicians, the cost of setting up the laboratory, bringing jobs into the area and the cost to the people who want the animals. Some consideration might be given to the animal rights activists who could be against the project.
- **Videos on cloning** – There are a number of good videos: Vega produce one on cloning and stem cells (on VHS from the Open University); BBC's Horizon episode *Cloning the first human* examines the ethics of cloning.

Evolution and environment

Changing the genes – right or wrong?

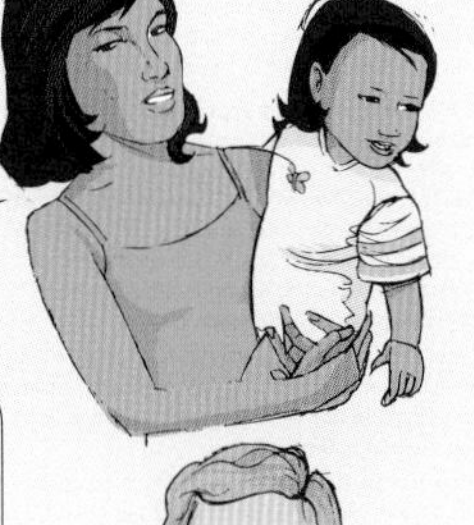

Our first daughter has been affected by a dreadful genetic disease. You don't know what it's like until it happens to you. If we had another baby, I'd want them to change the genes if they could when the embryo was really tiny. Then we'd know the baby wasn't going to be ill – and any grandchildren we had in the future would be alright as well.

One of the most exciting chances genetic engineering can give us is the development of gene therapy. Most of our research is looking at changing the genetic material in the affected cells by adding healthy genes or switching off damaged genes. But we are only treating the disease, not curing it. The affected person can pass on the faulty genes to their children, who in turn will need to be treated.
Gene therapy actually offers a way of curing genetic diseases. It would involve changing the genes of a fertilised egg or very early embryo so that the baby is born with healthy genes in all its cells. This is known as germ line gene therapy. We know it raises some major ethical issues and opinion across the world is strongly divided about it. In fact most countries, including the UK, have so far completely banned germ line gene therapy.

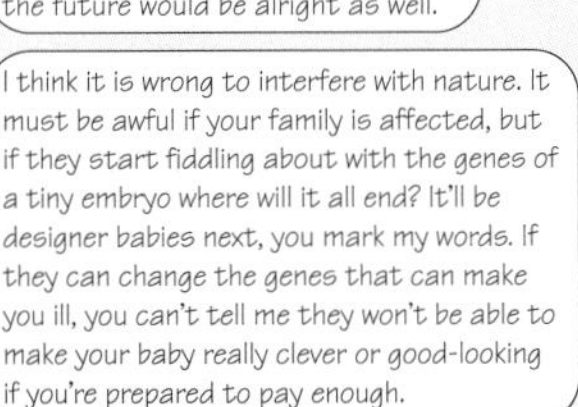

I think it is wrong to interfere with nature. It must be awful if your family is affected, but if they start fiddling about with the genes of a tiny embryo where will it all end? It'll be designer babies next, you mark my words. If they can change the genes that can make you ill, you can't tell me they won't be able to make your baby really clever or good-looking if you're prepared to pay enough.

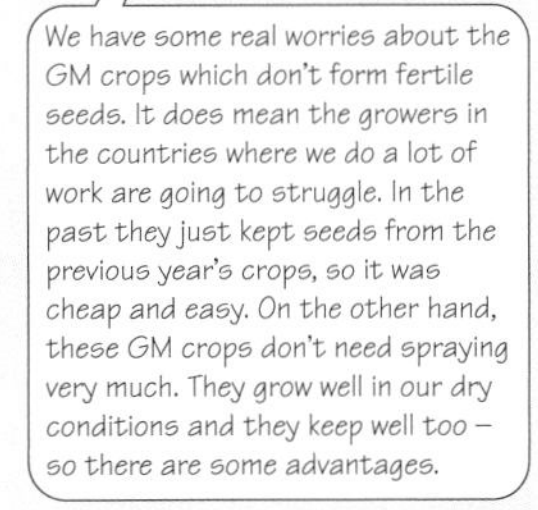

I'm more concerned about GM foods. Who knows what we're all eating nowadays. I don't want strange genes inside me, thank you very much. We've got plenty of fruit and vegetables as it is – why do we need more?
I think GM food is such a good idea. If the scientists can modify crops so they don't go off so quickly, food should get cheaper, and there will be more to go around. And what about these plants that produce pesticides – that'll stop a lot of crop spraying, so that should make our food cleaner and cheaper. It's typical of us in the UK that we moan and panic about it all.
We have some real worries about the GM crops which don't form fertile seeds. It does mean the growers in the countries where we do a lot of work are going to struggle. In the past they just kept seeds from the previous year's crops, so it was cheap and easy. On the other hand, these GM crops don't need spraying very much. They grow well in our dry conditions and they keep well too – so there are some advantages.

ACTIVITY

You are going to produce a 10-minute slot for a daytime television show entitled **'Genetic engineering – a good thing or not?'** You can ask any of the people shown here to come on your show and express their views. Using this and the information about genetic engineering on pages 102 and 103 to help you, plan out the script for your time on air. Remember that you have to inform the public about genetic engineering, entertain them and make them think about the issues involved.

105

- **Dramatisation/role play exercise** – Ask students to break into groups of four or five and to write the script for a scene where a pet is mortally wounded. The owner wants it cloned and gives reasons why. A family member or a friend is opposed to it, and each enlists a scientist to back up their respective cases. Provide prompt sheets with the summaries of the arguments for and against.
- **Alternative to the above** – Show the Monty Python 'dead parrot' sketch and get students to modify it to include the possibility of cloning a replacement.
- **Changing the genes – right or wrong?**
  - The information given in the student book can be divided into two major sections: one section concerned with genetic diseases and gene therapy and one section concerned with GM crops and GM food.
  - The germ line gene therapy is 'cutting edge' technology and needs to be considered alongside what can already be done in the way of genetic screening and gene therapy. Probably the best way of tackling these issues is to encourage students to draw up a list of advantages and disadvantages or conflicting opinions, so that they can present a balanced argument.
  - The issues about GM crops and GM foods are also emotive in a different way. There are so many misconceptions that it might be relevant to suggest that students carry out their own research (for homework) on a GM crop or a GM food, find out the facts and present the information to the rest of the class. Some good examples are the use of fungal enzymes from genetically modified yeasts, instead of bovine rennet in cheese manufacture (more than 30% of cheese is made in this way); genetically modified crops such as soya, maize and rice; and genetic modification of tomatoes.

- The web sites recommended for some of the spreads are worth investigating. Resources for Schools has some information, including animations and suggestions for activities. Related sources, such as GlaxoSmithKline, Pfizer and Industry Supports Education (ISE) are linked and all have quizzes and many interactive pages.
- **A clone poem** – Students could write a short poem summarising what they feel about cloning. Here is an example to get them started: *Clone Machine*, Mary Blackadder and Holly Tarbet (Scottish Poetry Library).

Clone Machine

What if we could make clones of ourselves?
Think of what it could do to help . . .
You wouldn't have to go to school
You could just send your clone
All you'd have to do
Is play around at home.
You could do ten things at once
Without moving a bone
Just leave the work
To your friendly old clone.
When you don't want your food
But don't want it to rot
Your clone will just eat it
And you will not!
But how would my family
Know it was me?
I'll just stick to normal
Be myself and be free!

SUMMARY ANSWERS

1 a) From a runner, which is a special stem from the parent plant with a small new identical plant on the end.

b) Asexual.

c) By sexual reproduction [this answer also needs an explanation of the production of flowers, pollination, fertilisation and the development of seeds].

d) The new plants from the packet will be similar to, but not identical to, their parents. Each one will be genetically different. The plants produced by asexual reproduction will be identical to their parents.

2 a) Traditional cuttings use parts of whole stems and roots, but tissue culture uses minute collections of cells as the starting point.

b) Embryo cloning.

c) Both allow large numbers of genetically identical individuals to be produced from good parent stock, much faster and more reliably than would be possible using traditional techniques.

d) Cloning plants uses bits of the adult plant as the raw material for cloning. Animal cloning, as it is used at the moment, involves using embryos as the raw material for the cloning (although this may change in the future).

e) There are more and more people in the world needing feeding, so techniques for reproducing high yielding plants and animals are always needed. Also, in developed countries, people demand high quality but cheap food, so techniques that reproduce valuable animals and plants more quickly are valued.

3 a) Diagram from p.102 but with growth hormone instead of insulin.

b) It is pure and free from contamination. It can be produced in large amounts relatively easily and cheaply.

Summary teaching suggestions

- **Special needs**
  These students could be shown the pictures for question 1 (or shown an actual plant and a packet of seeds) and answer the questions verbally
- **Gifted and talented**
  Students could find out more about the production of hormones using genetic engineering (question 3). They could research: 'Are the procedures always the same? How has the Human Genome Project helped this procedure?'
- **Misconceptions**
  Students need to be absolutely clear that they understand the terms 'cloning' and 'tissue culture' as applied to plants. Tissue culture involves the use of undifferentiated cells from growing points, such as buds, but cloning could apply to cuttings, such as stems, roots or buds.
- **Learning styles**
  Interpersonal: Questions 2e) and 3b) could involve class discussion and collaboration over different ideas, before students write their own answers.
  Visual: Diagrams are needed in the answers to questions 3a) and could help students learn a good answer to question 1c).
- **When to use these questions?**
  - As homework assignments after the relevant spreads have been studied.
  - As a test in preparation for the examination. Students could be warned that they need to revise the topics for homework and then you give them the questions to answer within a set time.
  - Verbally, to assess which topics need further revision and which most students know quite well.

VARIATION: B1b 6.1 – B1b 6.6

SUMMARY QUESTIONS

1

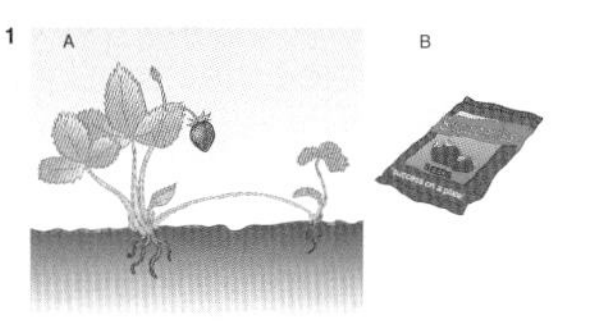

a) How has the small plant shown in diagram A been produced?
b) What sort of reproduction is this?
c) How were the seeds in B produced?
d) How are the new plants which you would grow from the packet of seeds shown in B different from the new plants shown in A?

2 Tissue culture techniques mean that 50 000 new raspberry plants can be gained from one old one instead of 2 or 3 taking cuttings. Cloning embryos from the best bred cows means that they can be genetically responsible for thirty or more calves each year instead of two or three.
a) How does tissue culture differ from taking cuttings?
b) How can one cow produce thirty or more calves in a year?
c) What are the similarities between cloning plants and cloning animals in this way?
d) What are the differences in the techniques for cloning animals and plants?
e) Why do you think there is so much interest in finding different ways to make the breeding of farm animals and plants increasingly efficient?

3 Human growth is usually controlled by the pituitary gland in your brain. If you don't make enough hormone, you don't grow properly and remain very small. This condition affects 1 in every 5000 children. Until recently the only way to get growth hormone was from the pituitary glands of dead bodies. Genetically engineered bacteria can now make plenty of pure growth hormone.
a) Draw and label a diagram to explain how a healthy human gene for making growth hormone can be taken from a human chromosome and put into a working bacterial cell.
b) What are the advantages of producing substances like growth hormone using genetic engineering?

EXAM-STYLE QUESTIONS

1 Young plants and animals resemble their parents. Th because characteristics are inherited by the young fro their parents.
Match words **A**, **B**, **C** and **D** with the spaces **1** to **4** in t sentences.

A Chromosomes **B** DNA
C Gametes **D** Genes

Genetic information is passed from parents to offspri in sex cells called ...**1**... The sex cells contain thread- structures known as ...**2**... .
The thread-like structures contain thousands of ...**3**.. that are the units of inheritance.
These units of inheritance are small sections of a dou helix called ...**4**... .

2 The table is about the production of offspring.
Match words **A**, **B**, **C** and **D** with the processes **1** to 4 the table.

A Sexual reproduction **B** Asexual reproductio
C Inheritance **D** Cloning

	Process
1	Joining of male and female sex cells (gametes) to produce young
2	Making young that are identical to both their parents and to each other, especially in agriculture
3	Producing offspring without sex cells (gametes)
4	Passing on characteristics from parents to offsprin

3 Plant tissue culture is a method used to create new plants. One method is described below:
- A small piece of tissue is removed from a plant.
- Under sterile conditions, the tissue is placed in a vessel containing nutrients.
- A mass of identical plant cells develops.
- These cells are placed in a medium containing nutrients and plant growth regulators (hormones).
- Young plants develop that are separated and grow to maturity.

(a) What type of reproduction is involved in plant tis culture?
(b) Why is a disease more likely to kill every one of a group of plants produced by tissue culture than a group grown from seeds?
(c) Suggest one advantage of growing plants from tis culture over growing them from cuttings or seeds

EXAM-STYLE ANSWERS

1 A 2
B 4
C 1
D 3 *(1 mark each*

2 A 1
B 3
C 4
D 2 *(1 mark each*

3 a) Asexual reproduction. *(1 mark*

b) The plants grown from tissue culture are genetically identical. If one member of the group is susceptible to the disease and dies, then it is highly probable that they will all die. *(1 mark*
Plants grown from seed result from sexual reproduction and therefore show genetic variety. When a disease affects the plants there is always the possibility that one or more plants may have genes that are resistant to the disease and that they will survive. *(1 mark*
*For both marks, the response must include both the reason why plants grown from tissue are more likely to all be killed **and** why some grown from seed might survive.*

c) *Either* It is more rapid/thousands of plants can be grown from a tiny piece of tissue *or* Less risk of disease initially because plants are grown in sterile conditions.
(Either answer = 1 mark
*To gain the mark, the answer must apply to **both** cuttings and seeds. Any reference to lack of variety, e.g. 'all plants will be the same/identical' is not acceptable as this only applies to cuttings and not to those produced from seeds.*

Some humans suffer from diabetes. One form of diabetes is caused by the inability to make the hormone insulin. Diabetics can lead normal lives providing they can inject insulin into their bloodstream at regular intervals. The diagram shows how insulin can be made using genetic engineering.

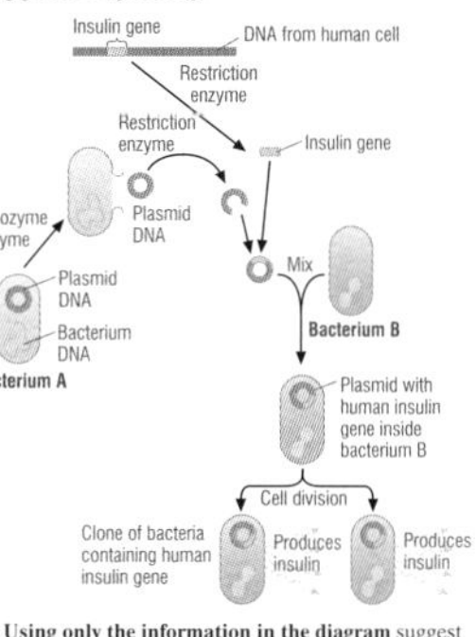

a) **Using only the information in the diagram** suggest what the following enzymes do:
(i) lysozyme enzyme (1)
(ii) restriction enzyme. (1)

b) **Using only the information in the diagram**, describe how the human insulin gene is transferred into a bacterium which then makes insulin. (6)

c) Insulin is produced from cloned bacteria that contain the human insulin gene.
(i) What is a clone? (1)
(ii) Do you think that cloning is the result of sexual or asexual reproduction? Explain your answer. (1)
(iii) A bacterium can divide every 20 minutes. Starting with 50 bacteria, how many bacteria would there be after 2 hours? Show your working. (1)

d) Insulin to treat diabetes used to be extracted from the pancreases of pigs and cattle. Give two advantages of insulin produced by genetic engineering over extracting it from animals. (2)

HOW SCIENCE WORKS QUESTIONS

1 DNA outside the nucleus

Cell membrane, Chromosomes, Nucleus, Mitochondrion, Circular chromosome, Cytoplasm

Mitochondria are sometimes referred to as the powerhouse of the cell. They carry out much of the cell's respiration. They contain their own DNA as a circular chromosome. Research has shown that this DNA is passed almost completely from the mother to the child. This is because the mitochondria are to be found in the cytoplasm. Some very rare diseases can be caused by the mitochondrial DNA mutating. These diseases can cause blindness or a disease very much like type 2 diabetes and deafness. Some children who inherit mutated mitochondrial DNA die at a very young age.

Use your scientific knowledge of cell structure and your understanding of the technology of cloning to answer the following questions.

a) Suggest how women who are at risk of passing on this mutated mitochondrial DNA might be helped. (1)
b) What are the ethical issues around this research? (2)
c) Who should be making decisions about whether or not families should be helped in this way? (3)
d) What is the role of the scientist in helping to make these decisions? (1)

2 Genetically modified (GM) crops

The first trials of GM crops were destroyed during the summer of 1999. Protesters, concerned that pollen from the crop could affect other local plants, were very pleased. The government, who started the trials, were not pleased! The government's Food Safety minister said:

'We can't operate food safety policy on a *hunch* – we have to have the science and that's why we need the trials.'

a) Explain the difference between a 'hunch' and 'science'. (2)
b) Do you think the protesters had based their ideas on a 'hunch'? Explain your answer. (1)
c) One of the purposes of the trial was to find out how far pollen from the GM crop could travel. Describe how the trial might be set up. (5)

a) (i) Breaks open/ruptures bacterial cell walls. *(1 mark)*
(ii) Breaks/cuts open strands of DNA. *(1 mark)*

b)
- Lysozyme enzyme is used to break open the wall of bacterium A.
- Restriction enzyme is used to open up the (circular) plasmid DNA released from bacterium A and to cut out the insulin gene from a strand of human DNA.
- The human insulin gene is inserted into the plasmid DNA.
- The plasmid DNA + insulin gene enters bacterium B when the two are mixed together.
- Bacterium B, with the plasmid/insulin gene, divides to form a clone.
- The cloned bacterium B produce/secrete human insulin (as instructed by the human insulin gene).

(1 mark for each point)
(6 marks in total)

c) (i) A group of genetically identical cells that are also genetically identical to the parent plant. *(1 mark)*
(ii) Asexual – as no sex cells/gametes are involved. *(1 mark)*
(iii) Start 50 bacteria
After 20 mins $50 \times 2 = 100$ bacteria
" 40 mins $100 \times 2 = 200$ "
" 60 mins $200 \times 2 = 400$ "
" 80 mins $400 \times 2 = 800$ "
" 100 mins $800 \times 2 = 1600$ "
" 120 mins $1600 \times 2 = 3200$. Answer = 3200 *(1 mark)*
To obtain the mark some relevant working must be shown as well as the correct answer. The correct answer alone earns no credit. Different forms of showing the working are possible.

d)
- Animals do not have to be killed to obtain it.
- It is cheaper to produce.
- The supply is more constant/controlled.
- The insulin produced is more effective as it is an exact copy of human insulin. *(Any 2 points for one mark each)*

Exam teaching suggestions

Question 3b) illustrates the importance of tailoring an answer to the marks allowed. It has 2 marks and so two separate points need to be made. In this case a reason why tissue cultured plants might all die **and** a reason why those from seed might not. Students frequently give one or the other explanation rather than both.

In questions such as question 4b), precisely where one marking point ends and the next begins is of less importance than listing the events in a logical order. Provided all six areas listed above are covered accurately, full marks can still be obtained even if the divisions between each point are blurred.

Question 4c) (iii) is an opportunity to impress on students the need to follow the rubric of questions. In this case to 'Show your working'. Although many students will be able to carry out the calculation in their heads they must nevertheless write it down. Students who get the answer wrong, or fail to show their working, should be encouraged to firstly list the relevant time intervals and then, starting with the initial number (50) opposite 0 minutes, simply double the numbers until they reach the end time (120 minutes).

HOW SCIENCE WORKS ANSWERS

1 a) The fertilised nucleus could be implanted into an enucleated egg from another donor woman.

b) E.g. should a child have effectively two mothers? Are the risks involved in the research low enough to match the advantages to the parents and the child?

c) Politicians should decide whether to allow the research; society should discuss the issues; and parents make the final decision as to whether to go ahead.

d) Making the information available for thorough discussion of the issues.

2 a) A 'hunch' is based on subjective feelings or opinion whereas 'science' is based on evidence collected.

b) Hunch – no experimental evidence was available.

c) Plant GM crop. After pollination, collect samples of plants at increasing distances, in all directions. Inspect the pollen collected; analyse for GM cells; plot distribution.

How science works teaching suggestions

- **Literacy guidance.** The question expecting a longer answer, where students can practise their literacy skills is: question 1c) and 2c).
- **Higher- and lower-level questions.** Question 1a) is a higher-level question. The lower-level question is 1b). Answers for both have been provided at these levels.
- **Gifted and talented.** Able students could explore the special context of mitochondrial DNA transfer from only the mother. The context could be used as the basis for a survey of attitudes.
- **How and when to use these questions.** When wishing to develop skills around the handling of modern scientific developments in an ethical context. The questions need group discussion.
- **Special needs.** Students might need some help with the context for these questions. Diagrams of the process would make these questions accessible.

B1b 7.1 The origins of life on Earth

LEARNING OBJECTIVES

Students should learn:

- The nature of fossils.
- How fossils provide evidence for the existence of prehistoric plants and animals.

LEARNING OUTCOMES

Most students should be able to:

- Explain what a fossil is.
- Suggest reasons why scientists cannot be certain about how life began on Earth.

Some students should also be able to:

- Evaluate what can be learnt from the fossil record.

Teaching suggestions

- **Special needs.** Students could make a fossil using Plasticine or plaster of Paris for a mould and stearic acid to pour in and set. (Care with hot liquids!) Alternatively, students could use a Plasticine mould and pour plaster of Paris into the impression of a shell.
- **Gifted and talented.** Set the students the problem of finding out how long a representation of a timeline would need to be if the world is 4 600 000 000 years old and 100 years was represented by a millimetre. Ask: 'How many A4 sheets of paper would be needed, allowing 1 cm overlap for gluing?' [Do not try to make one – it is over 40 km!]
- **Learning styles**

  Interpersonal: Discussing evidence from fossils.

  Intrapersonal: Making deductions about fossils from remains.

  Kinaesthetic: Researching fossils; going on a fossil hunt and looking for examples; making a fossil.

  Visual: Looking at the 'Fossil circus'.

  Auditory: Reading out answers in 'Only half the story'.

SPECIFICATION LINK-UP B1b.11.7

- *Fossils provide evidence of how much (or how little) different organisms have changed since life developed on Earth.*

Students should use their skills, knowledge and understanding of 'How Science Works':

- *to suggest reasons why scientists cannot be certain about how life began on Earth.* ?

Lesson structure

STARTER

- **Fossil circus** – Arrange a circus of numbered fossils or fossil pictures around the laboratory. Students to look and make notes, try to decide what they might be, writing answers on a numbered sheet. (10–15 minutes)
- **Can you run on your toenails?** – Sensitively find out who has got the longest fingernails in the class. Ask: 'Are they the strongest? Could you do press-ups resting on your middle finger?' Relate this to how a horse walks, and introduce equine foot development using photographs of ancient hooves. This leads to a discussion of fossil evidence. (10 minutes)

MAIN

- **Fossil formation** – Discuss the ways that fossils form such as cast formation, impressions and ice fossils. Victims of Pompeii illustrate cast formation; mammoths are found in ice; dinosaurs left footprints and animal droppings fossilise over time. For pictures, search the Internet using key words and especially look for Ardley Quarry in Oxfordshire for its dinosaur footprints.
- **Geological time scale** – This is a timeline exercise linking with the above, using a long strip of paper (till roll?), marking off the major evolutionary events for which there is evidence. An alternative to this is to use a clock face and discover that Homo sapiens evolved in the last few seconds before 12.
- **Recent fossils** – Try to get hold of some peat, look for plant remains and discuss their age (can be many thousands of years old).
- **Survivors!** – Two good examples of organisms that were thought to have been extinct and only known in the fossil record are the Maidenhair tree (*Ginkgo biloba*) and the coelacanth, both of which have been found. Pictures of ginkgo, or actual leaves could be passed around and compared with pictures of fossils from 7 million years ago. Students can spot the changes.

PLENARIES

Trapped in the ice – Show a clip from the film *Ice Age* (Twentieth Century Fox), where stages in evolution are trapped in the ice or search the web for 'ice age trailer' to see the general idea. Discuss. (5–10 minutes)

Only half the story . . . – Display a sentence with many of the letters missing, but enough left to see what it is saying. Ask the students to write their own (polite) completed version. Read out and discuss. (10–15 minutes)

What makes a good fossil? – Lead a quick discussion on why some creatures became fossilised and others did not. Ask: 'Why are there few fossils of worms and other soft-bodied creatures?' (10 minutes)

ACTIVITY & EXTENSION IDEAS

- Many fossil plants were found in the coal measures. Try finding pictures of a few, and then linking their presence in coal with why conditions were good for fossil formation. This links to the term 'fossil fuels'.
- Go on a fossil hunt. Fossils can be found in many unlikely places, such as on public buildings as fossil remains of invertebrates are found in Portland stone and other natural building materials. A trip around some municipal buildings could reveal some evidence.
- **Find out more about the fossil hunters** – Charles Lyell and Arthur Holmes are mentioned in the answer to the timeline question. Who were they? How did they become interested in fossils and dating rocks. There are good links with geology here for interested students. Why are there more fossils in some rocks than in others?

Answers to in-text questions

a) Between 3 and 4.5 billion years ago.

b) 4 billion.

c) In the ice, the whole animal is preserved so that we can see exactly what it looked like and even what it had eaten, not just its skeleton.

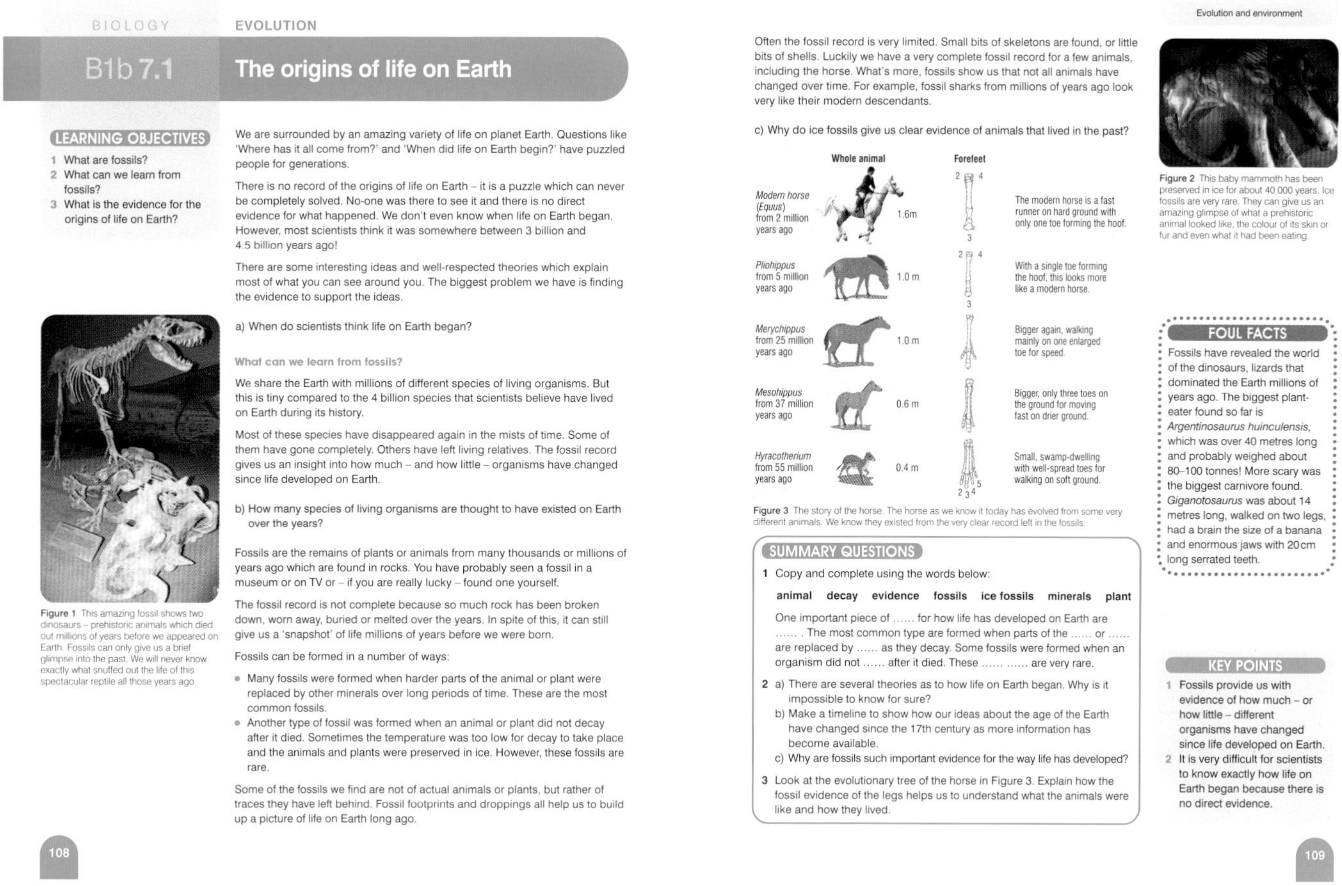

BIOLOGY EVOLUTION

B1b 7.1 The origins of life on Earth

LEARNING OBJECTIVES

1. What are fossils?
2. What can we learn from fossils?
3. What is the evidence for the origins of life on Earth?

We are surrounded by an amazing variety of life on planet Earth. Questions like 'Where has it all come from?' and 'When did life on Earth begin?' have puzzled people for generations.

There is no record of the origins of life on Earth – it is a puzzle which can never be completely solved. No-one was there to see it and there is no direct evidence for what happened. We don't even know when life on Earth began. However, most scientists think it was somewhere between 3 billion and 4.5 billion years ago!

There are some interesting ideas and well-respected theories which explain most of what you can see around you. The biggest problem we have is finding the evidence to support the ideas.

a) When do scientists think life on Earth began?

What can we learn from fossils?

We share the Earth with millions of different species of living organisms. But this is tiny compared to the 4 billion species that scientists believe have lived on Earth during its history.

Most of these species have disappeared again in the mists of time. Some of them have gone completely. Others have left living relatives. The fossil record gives us an insight into how much – and how little – organisms have changed since life developed on Earth.

b) How many species of living organisms are thought to have existed on Earth over the years?

Figure 1 This amazing fossil shows two dinosaurs – prehistoric animals which died out millions of years before we appeared on Earth. Fossils can only give us a brief glimpse into the past. We will never know exactly what snuffed out the life of this spectacular reptile all those years ago.

Fossils are the remains of plants or animals from many thousands or millions of years ago which are found in rocks. You have probably seen a fossil in a museum or on TV or – if you are really lucky – found one yourself.

The fossil record is not complete because so much rock has been broken down, worn away, buried or melted over the years. In spite of this, it can still give us a 'snapshot' of life millions of years before we were born.

Fossils can be formed in a number of ways:

- Many fossils were formed when harder parts of the animal or plant were replaced by other minerals over long periods of time. These are the most common fossils.
- Another type of fossil was formed when an animal or plant did not decay after it died. Sometimes the temperature was too low for decay to take place and the animals and plants were preserved in ice. However, these fossils are rare.

Some of the fossils we find are not of actual animals or plants, but rather of traces they have left behind. Fossil footprints and droppings all help us to build up a picture of life on Earth long ago.

108

Evolution and environment

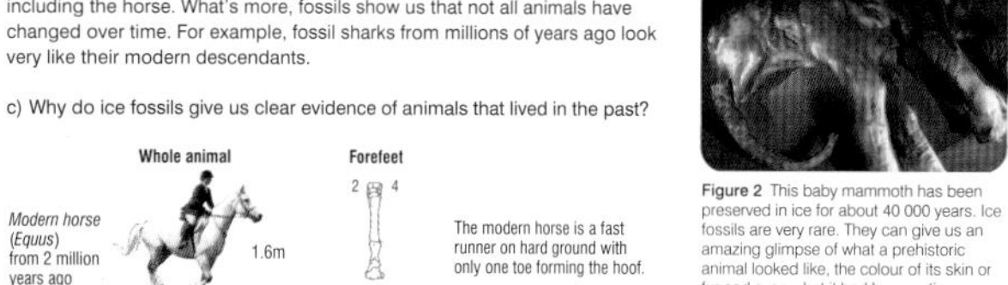

Often the fossil record is very limited. Small bits of skeletons are found, or little bits of shells. Luckily we have a very complete fossil record for a few animals, including the horse. What's more, fossils show us that not all animals have changed over time. For example, fossil sharks from millions of years ago look very like their modern descendants.

c) Why do ice fossils give us clear evidence of animals that lived in the past?

Figure 2 This baby mammoth has been preserved in ice for about 40 000 years. Ice fossils are very rare. They can give us an amazing glimpse of what a prehistoric animal looked like, the colour of its skin or fur and even what it had been eating.

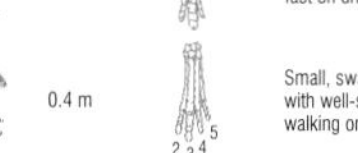

Whole animal
Forefeet
Modern horse (Equus) from 2 million years ago
1.6m
The modern horse is a fast runner on hard ground with only one toe forming the hoof.
Pliohippus from 5 million years ago
1.0 m
With a single toe forming the hoof, this looks more like a modern horse.
Merychippus from 25 million years ago
1.0 m
Bigger again, walking mainly on one enlarged toe for speed.
Mesohippus from 37 million years ago
0.6 m
Bigger, only three toes on the ground for moving fast on drier ground.
Hyracotherium from 55 million years ago
0.4 m
Small, swamp-dwelling with well-spread toes for walking on soft ground.

Figure 3 The story of the horse. The horse as we know it today has evolved from some very different animals. We know they existed from the very clear record left in the fossils.

FOUL FACTS

Fossils have revealed the world of the dinosaurs, lizards that dominated the Earth millions of years ago. The biggest plant-eater found so far is *Argentinosaurus huinculensis,* which was over 40 metres long and probably weighed about 80–100 tonnes! More scary was the biggest carnivore found. *Giganotosaurus* was about 14 metres long, walked on two legs, had a brain the size of a banana and enormous jaws with 20 cm long serrated teeth.

SUMMARY QUESTIONS

1 Copy and complete using the words below:

animal decay evidence fossils ice fossils minerals plant

One important piece of ...... for how life has developed on Earth are .... The most common type are formed when parts of the ...... or ...... are replaced by ...... as they decay. Some fossils were formed when an organism did not ...... after it died. These .... are very rare.

2 a) There are several theories as to how life on Earth began. Why is it impossible to know for sure?
b) Make a timeline to show how our ideas about the age of the Earth have changed since the 17th century as more information has become available.
c) Why are fossils such important evidence for the way life has developed?

3 Look at the evolutionary tree of the horse in Figure 3. Explain how the fossil evidence of the legs helps us to understand what the animals were like and how they lived.

KEY POINTS

1. Fossils provide us with evidence of how much – or how little – different organisms have changed since life developed on Earth.
2. It is very difficult for scientists to know exactly how life on Earth began because there is no direct evidence.

109

SUMMARY ANSWERS

1 Evidence, fossils, animal, plant, minerals, decay, ice fossils.

2 **a)** No one was there to see it and there is no direct evidence for what happened.

b) 17th century: Archbishop Usher calculated that the Earth was 6000 years old → early 19th century, Charles Lyell studied rocks and estimated that the Earth was several million years old → late 19th century, Arthur Holmes used radioactivity to date the rocks and estimated the Earth to be about 4.6 billion years old.

c) They show us how plants and animals have changed over time, and how many animals have appeared and no longer exist.

3 It shows us how tall they were, what their feet were like, what their jaws and teeth were like and the basic body shape. This in turn tells us how they might have lived, how fast they moved, what they ate; it also allows us to compare them to modern horses.

KEY POINTS

Suggest to students that they make a list of reasons why we are never likely to know how life began on Earth.

B1b 7.2 Theories of evolution

LEARNING OBJECTIVES

Students should learn:

- The theory of evolution.
- That there is evidence that evolution has taken place.

LEARNING OUTCOMES

Most students should be able to:

- State the theory of evolution.
- Describe some of the evidence that evolution has taken place.
- Suggest reasons why Darwin's theory of natural selection was only gradually accepted.
- Identify the differences between Darwin's theory of evolution and conflicting theories, e.g. Lamarck's.

Some students should also be able to:

- Suggest reasons for the different theories explaining life on Earth.

Teaching suggestions

- **Special needs**
  - Students could be presented with pictures of three or four animals with fairly obvious different characteristics (e.g. horn length in antelope, speed of response in rabbits) and asked to choose which one would survive to have offspring.
  - Alternatively, they could be asked to complete simple sentences summarising natural selection in a simple way.
- **Gifted and talented.** Students could do some more research on the attitude of the Church to evolutionary theory, in Darwin's time and also at the present time.
- **Learning styles**

  Interpersonal: Discussing theories of evolution.

  Intrapersonal: Reviewing Darwin's work; writing a comment on Darwin's work.

  Kinaesthetic: Making a cartoon summary of Darwin's voyage.

  Visual: Obtaining information about Darwin from the presentation.

  Auditory: Discussing the beaks of the finches.

SPECIFICATION LINK-UP B1b.11.7

- *The theory of evolution states that all species of living things which exist today have evolved from simple life forms which first developed more than three billion years ago.*
- *Studying the similarities and differences between species helps us to understand evolutionary and ecological relationships.*

Students should use their skills, knowledge and understanding of 'How Science Works':

- *to suggest reasons why Darwin's theory of natural selection was only gradually accepted*
- *to identify differences between Darwin's theory of evolution and conflicting theories*
- *to suggest reasons for the different theories.*

Lesson structure

STARTER

Get it in the right order – Give the first few students through the door a sheet on which there is a stage of evolution written. Place the students at the front of the class and tell them to sort themselves out into date order. The rest of the class, and then you, judge whether or not they are right. (5–10 minutes)

How much do we already know? – This is a brainstorming session on the words 'Darwin', 'Natural selection', 'Origin of Species' and 'Evolution'. Write these words on a large sheet of quartered A2 paper. Allow the sheet to be circulated around the room with students adding comments. You can circulate and observe to gain an understanding of prior knowledge. (10 minutes)

MAIN

- **Conflicting theories** – Gather together some details of as many different theories of evolution as you can: Darwin, Lamarck, the Creation, spontaneous generation, etc, giving a brief summary of each. This would be a good introduction to some of the following suggestions. Discuss the merits of each one.
- **The life and times of Charles Darwin** – Show a video (one is produced by Hawkhill Associates, Madison, WI) of Darwin's journey and discoveries, together with some details of his life (when he published his work etc.). Alternatively, search the web for 'Charles Darwin video' and you can find several online. Discuss how his trip led to his point of view.
- The showing of the video could be followed by building up a cartoon summary of Darwin's voyage. Decide on the frames and get the students to complete it.
- **What can I do with my beak?** – Search the web for 'Darwin's finches', which will clearly show the beaks, and try to match them with their function. Show a map with the location of the different finches and also a picture of the original type of finch which colonised the Galapagos Islands. Why were these islands interesting?

PLENARIES

L, D or LD? – Give students individual cards or whiteboards on which they can respond to various statements and key words from both theories, which can be projected or read out. L for Lamarck only, D for Darwin only and LD if applicable to both. (5 minutes)

We don't believe you! – Show a picture of Darwin as an ape (caricature from the *Hornet* magazine). Search the web for 'Darwin ape Hornet'. Ask the students to imagine themselves in the role of an irate Victorian and to write an indignant comment on his theory. The comment must include some of the key points. (10 minutes)

Faking the evidence – How easy would it be to fake a fossil and so provide some evidence in support of the theory? Discuss how such fakes could be made and how you could prove that they were fakes. (10 minutes)

ACTIVITY & EXTENSION IDEAS

- There are some good summaries of the issues in *A Brief History of Nearly Everything* (Bill Bryson).
- Students could find out why the Galapagos Islands are so special and research: 'Are there any other locations where similar conditions might have existed?'
- **Find out more about Darwin** – Although Darwin is probably best known for his work on the theory of evolution, he did work in other areas of biology. Use some of the information from the film and from other sources, such as libraries and the Internet, to build up an account of all his scientific work.

Answers to in-text questions

a) Useful changes or characteristics that organisms developed during their lives to help them survive are passed on to their offspring.

b) The Beagle.

c) Galapagos Islands.

d) *On the Origin of Species by Natural Selection*.

KEY POINTS

It would be helpful for students to be able to define the terms used in the key points, such as 'evolution, natural selection, extinction' and formulate for themselves a clear definition of the theory of evolution.

BIOLOGY EVOLUTION

B1b 7.2 Theories of evolution

LEARNING OBJECTIVES

1. What is the theory of evolution?
2. What is the evidence that evolution has taken place?

The theory of ***evolution*** tells us that all the species of living things alive today have evolved from the first simple life forms that existed on Earth. Most of us take these ideas for granted – but they are really quite new.

Up to the 18th century most people in Europe believed that the world had been created by God. They thought it was made, as described in the Christian Bible, a few thousand years ago. But by the beginning of the 19th century, scientists were beginning to come up with new ideas.

Lamarck's theory of evolution

Jean-Baptiste Lamarck, a French biologist, suggested that all organisms were linked by what he called a 'fountain of life'. His idea was that every type of animal evolved from primitive worms. He thought that the change from worms to other organisms was caused by the ***inheritance of acquired characteristics***.

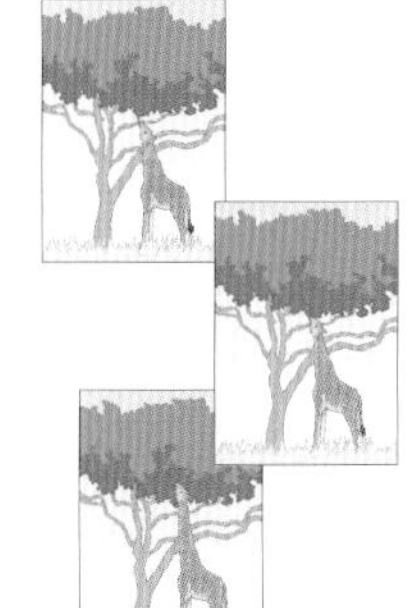

Figure 1 In Lamarck's model of evolution, giraffes have long necks because each generation stretched up to reach the highest leaves. So each new generation had a slightly longer neck!

The theory was that useful changes, developed by parents during their lives to help them survive, are passed on to their offspring. In other words, if you do lots of swimming and develop broad shoulders, your children will have broad shoulders as well!

Lamarck's theory fell down for several reasons. There was no evidence for his 'fountain of life' and people didn't like the idea of being descended from worms. People could also see quite clearly that characteristics they acquired were not passed on to their children.

a) What is meant by the phrase 'inheritance of acquired characteristics'?

Charles Darwin and the origin of species

Our modern ideas about evolution began with the work of one of the most famous scientists of all time – ***Charles Darwin***. Darwin set out in 1831 as the ship's naturalist on *HMS Beagle*. He was only 22 years old and the voyage to South America and the South Sea Islands would take five years.

Darwin planned to study geology on the trip. But as the voyage went on he became as excited by his collection of animals and plants as by his rock samples.

Figure 2 Darwin was very impressed by the giant tortoises he found on the Galapagos Islands. The tortoises on each island had different shaped shells and a slightly different way of life – and Darwin made careful drawings of them all.

b) What was the name of the ship that Darwin sailed on?

In South America, Darwin discovered a new form of the common rhea, an ostrich-like bird – but not until his party had cooked and eaten one of them! Two different types of the same bird living in slightly different areas set Darwin thinking.

On the Galapagos Islands he was amazed by the variety of species and the way they differed from island to island. Darwin found strong similarities between types of finches, iguanas and tortoises on the different islands. Yet each was different and adapted to make the most of local conditions.

110

Evolution and environment

Darwin collected huge numbers of specimens of animals and plants during the explorations of the *HMS Beagle*. He also made detailed drawings and kept written observations. The long voyage home gave him plenty of time to think about what he had seen. Charles Darwin returned home after five years with some new and different ideas forming in his mind.

c) What is the name of the famous islands where Darwin found so many interesting species?

After Darwin returned to England he spent the next 20 years working on his ideas. He knew they would meet a lot of opposition. He realised he would need lots of evidence to support his theories. He used the amazing animals and plants he had seen on his journeys as part of that evidence.

He also built up evidence from breeding experiments on pigeons to support his theory. In 1859, he published his famous book *On the Origin of Species by means of Natural Selection* (often known as *The Origin of Species*).

Darwin's central theory is that all living organisms have evolved from simpler life forms. This evolution has come about by a process of **natural selection**. Reproduction always gives more offspring than the environment can support. Only those which are most suited to their environment – the 'fittest' – will survive. When they breed, they pass on those useful characteristics to their offspring. Darwin suggested that this was how evolution takes place.

d) What was the name of Darwin's famous book?

DID YOU KNOW?

In the 21st century, we can get some of the best evidence for evolution yet by looking at the DNA of different species of animals and plants. This shows us clearly how closely related one species is to another. It backs up both Darwin's theory of evolution and his ideas of how it came about through natural selection.

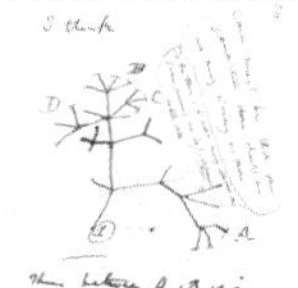

Figure 3 As Darwin studied his specimens, he began to build up a branching picture of evolution, which he tried out in different forms in his notebooks

The evidence for evolution

The fossil record shows us how species have evolved over millions of years, and how different species are related to each other. Studying the similarities and differences between species, can also help us to understand the evolutionary relationships between them. Observing how changes in the genes can be passed from one generation to another gives us more evidence still.

GET IT RIGHT!

Don't get confused between:
- the **theory of evolution** and
- the **process of natural selection**.

SUMMARY QUESTIONS

1. What is meant by the following terms:
   a) evolution? b) natural selection?
   c) inheritance of acquired characteristics?
2. Why was Darwin's theory of natural selection only accepted very gradually?
3. What was the importance of the following in the development of Darwin's ideas?
   a) South American rheas.
   b) Galapagos tortoises, iguanas and finches.
   c) The long voyage of *HMS Beagle*.
   d) The twenty years from his return to the publication of his book *The Origin of Species*.
4. Suggest reasons why Lamarck and Darwin were convinced by their theories explaining life on Earth.

KEY POINTS

1. The **theory of evolution** states that all the species which are alive today – and many more which are now extinct – evolved from simple life forms which first developed more than three billion years ago.
2. Darwin's theory is that evolution takes place through **natural selection**.
3. Studying the similarities and differences between species helps us to understand how they have evolved and how closely related they are to each other.

111

SUMMARY ANSWERS

1 **a)** All the species of living organisms which are alive today – and many more which are now extinct – have evolved from simple life forms, which first developed more than 3 billion years ago.

b) Only the animals and plants most suited to their environment – the 'fittest' – will survive to breed and so pass on their characteristics.

c) Inheritance of useful changes developed by parents during their lives to help them survive is passed on to their offspring.

2 Because, for many years, people had believed the view of the Church that God had created everything.

3 **a)** Darwin found a new species – two types of the bird living in slightly different areas made Darwin start to think about how they came about.

b) These were some of the animals in the Galapagos Islands that varied from island to island, and made Darwin wonder what had brought about the differences.

c) This gave Darwin lots of opportunities to collect specimens and time to think about his theories and ideas.

d) This gave Darwin time to work out his ideas very carefully and to collect a lot of evidence to support them.

4 Lamarck – he believed that adaptations developed gradually from characteristics acquired by parents and all species stemmed from worms (a primitive life form).
Darwin – his theory of natural selection explained the thousands of observations he had collected from the natural world.

B1b 7.3 Natural selection

LEARNING OBJECTIVES

Students should learn that:

- Individuals best suited to their environment survive to breed successfully.
- These individuals pass their genes on to the next generation.
- A mutation is a change in an existing gene.

LEARNING OUTCOMES

Most students should be able to:

- Explain what is meant by natural selection.
- State that mutation results in changes to genes.

Some students should also be able to:

- Explain how mutation can affect the evolution of an organism.

Teaching suggestions

- **Special needs.** Students could be asked to design their own mutant from an existing animal. It could be suggested that the mutant needed to survive in extreme cold, windy conditions, etc.
- **Gifted and talented.** Students could be asked to draw analogies between the evolution of animals and plants and the development of communications equipment over the last century.
- **Learning styles**

  Interpersonal: Discussing Blinky the fish. (See Starter.)

  Intrapersonal: Making deductions about the peppered moth.

  Kinaesthetic: Designing and drawing a mutant.

  Visual: Identifying favourable adaptations.

  Auditory: Explaining adaptations.

Answers to in-text questions

a) Charles Darwin.

b) Because it would be more likely to hear the approach of a fox and also to hear warning signs from other animals.

c) A change in the genes/DNA.

d) A disease of oysters where they do not grow properly; they are small, flabby and they develop pus-filled blisters and die.

SPECIFICATION LINK-UP B1b.11.7

- *Evolution occurs by natural selection:*
  - *individual organisms within a particular species may show a wide range of variation because of differences in their genes*
  - *individuals with characteristics most suited to the environment are more likely to survive and breed successfully*
  - *the genes which have enabled these individuals to survive are then passed on to the next generation.*
- *Where new forms of a gene result from mutation there may be more rapid change in a species.*

Lesson structure

STARTER

Blinky, the three-eyed mutant fish – Search the web for images of 'Blinky the fish' from The Simpsons or show sections from the video from Series 2: *Two Cars in Every Garage and Three Eyes on Every Fish*. Discuss how this mutant could have arisen and whether three eyes would be an advantage. (5–10 minutes)

Six fingers better than five? – Show a picture of a person with six fingers on each hand. Ask students to write down one advantage and one disadvantage. Discuss. (5 minutes)

Mutations – Search the web for images of 'X men' from the 2000 feature film *X-Men* (Twentieth Century Fox). Discuss for any who have not seen it. Students are asked to list the ways in which the mutations are not like real mutations and to give examples. (10 minutes)

MAIN

- **Interactive natural selection games** – There are a number of these games available on the Internet. These can be recommended to students or used in class if there is appropriate computer access. Try PBS Teacher Source at www.pbs.org or www.echalk.co.uk.
- **The peppered moth** – This moth (*Biston betularia*) with its black mutant is well-documented. Prepare a PowerPoint® presentation about the distribution of the two forms of the moth, together with some statistics on how populations have changed since the decline of industry in some areas. Students can be asked to compile a single A4 sheet summary of the evidence. This case provides support for the mechanism of survival of the fittest, natural selection and evolution.
- **The banded snail** – The activity above can be linked to another example of natural selection at work: the banded snail. Students may find evidence of this in their own gardens or in the school grounds. There are pictures of the different forms of the snail available; look for references to polymorphism. Basically, there are more variations in the banded snail and these provide camouflage in different situations. Those snails best camouflaged survive, the rest do not.
- **Mutations put to work** – The fruit fly (*Drosophila*) is used in genetics experiments to investigate the inheritance of characteristics. Show some fruit flies and photographs of some of the mutations that have been studied. Discuss how these mutations might have arisen or been induced. In order to gain an understanding of how rapidly a mutation could spread, calculate how long it would take a pair of fruit flies to produce a billion offspring if each female produces 200 offspring every two weeks.

PLENARIES

The 'mutant schoolchild' – Build up a picture of a 'mutant schoolchild', who has adapted to the school environment to an extreme extent. Suggestions from the students can be used to build up a picture on the board. Alternatively, each student could be given a sheet of paper and asked to draw or describe their mutant. Results to be displayed. (10 minutes)

Favourable adaptations? – Project or display photographs of a number of adaptations shown by plants and animals and discuss and explain how each could have arisen by natural selection. Some fairly obvious ones are thorns on stems, brightly coloured flowers, prehensile tails, swivelly eyes in chameleons etc. (10 minutes)

Card sort game – Students are given the following words or phrases: 'mutation, variation, adaptation, survival, genes passed on', and cards illustrating each of these. They are to put them in order. It can be timed to give a competitive edge. (5 minutes)

ACTIVITY & EXTENSION IDEAS

- How far is the evolution of animals linked to the evolution of plants or vice versa? A possible example of parallel evolution is that of insects and insect-pollinated plants, i.e. the evolution of flowers. This could be the basis of a discussion on links between plants and animals.
- Studies of the occurrence of heavy metal tolerant plants on spoil heaps can provide some evidence for evolution in action. Students need to be provided with some background information and then asked to summarise the process. Link to the peppered moth story.

BIOLOGY EVOLUTION

B1b 7.3 Natural selection

LEARNING OBJECTIVES

1. How does natural selection work?
2. What is mutation?

NEXT TIME YOU...

. . . look around your school playing field or the local woodland, remember you are looking at the site of a struggle for existence, where only the animals and plants best adapted for survival will go on to live and reproduce.

Scientists explain the variety of life today as the result of a process called ***natural selection***. The idea was first suggested 150 years ago by Charles Darwin.

The natural world is a harsh place, and as you saw in Chapter 5 animals and plants are always in competition with each other. Sometimes an animal or plant gains an advantage in the competition. This might be against other species or against other members of its own species. That individual is more likely to survive and breed – and this is known as ***natural selection***.

a) Who first suggested the idea of natural selection?

Survival of the fittest

Charles Darwin was the first person to describe natural selection as the 'survival of the fittest'. Reproduction is a very wasteful process. Animals and plants always produce more offspring than the environment can support.

Fruit flies can produce 200 offspring every two weeks. The yellow star thistle, an American weed, produces around 150 000 seeds per plant per year! If all those offspring survived we'd be overrun with fruit flies and yellow star thistles!

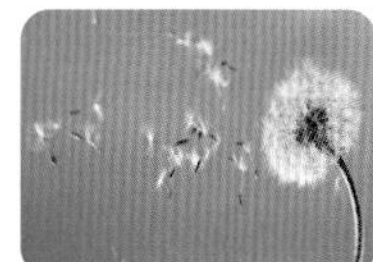

Figure 1 If all of these dandelions seeds developed to become adults and then breed themselves, there would be problems! In fact, very few of them will make it. The survivors will have a combination of genes that gives them a competitive edge over all the others.

But the individual organisms in any species show lots of variation. This is because of differences in the genes they inherit. Only the offspring with the genes best suited to their habitat manage to stay alive and breed successfully. This is natural selection at work.

Think about rabbits. The rabbits with the best all-round eyesight, the sharpest hearing and the longest legs will be the ones which are most likely to escape being eaten by a fox. They will be the ones most likely to live long enough to breed. What's more, they will pass those useful genes on to their babies. The slower, less alert rabbits will get eaten and their genes will be digested with the rest of them!

b) Why would a rabbit with sharp hearing be more likely to survive than one with less keen hearing?

Figure 2 The natural world is often brutal. Only the best adapted predators capture prey – and only the best adapted prey animals escape!

The part played by mutation

New forms of genes (new alleles) result from changes in existing genes. These changes are known as **mutations**. They are tiny changes in the long strands of DNA.

Mutations occur quite naturally through mistakes made in copying your DNA when your cells divide. Mutations introduce more variety into the genes of a species. In terms of survival, this is very important.

c) What is a mutation?

112

Evolution and environment

Many mutations have no effect on the characteristics of an organism, and some mutations are harmful. However, just occasionally a mutation has a good effect. It produces an adaptation which makes an organism better suited to its environment. This makes it more likely to survive and breed.

Whatever the adaptation, if it helps an organism survive and reproduce it will get passed on to the next generation. The mutant gene will gradually become more common in the population. It will cause the species to evolve.

When new forms of a gene arise from mutation, it may cause a more rapid change in a species. This is particularly true if circumstances change as well.

GET IT RIGHT!
Make sure you learn the main points of natural selection:
mutation → variation → adaptation → survival → genes passed on to the next generation

Natural selection in action

Have you ever eaten oysters? They are an expensive treat! They are collected from special oyster beds under the sea. Malpeque Bay in Canada has some very large oyster beds. In 1915, the oyster fishermen noticed a few oysters which were small and flabby with pus-filled blisters.

By 1922 the oyster beds were almost empty. The oysters had been wiped out by a new and devastating disease (soon known as Malpeque disease).

Fortunately a few of the shellfish carried a mutation which made them resistant to the disease. Not surprisingly, these were the only ones to survive and breed. The oyster beds filled up again and by 1940 they were producing more oysters than ever.

But the new population of oysters had evolved. As a result of natural selection, almost every oyster in Malpeque Bay now carries the allele which makes them resistant to Malpeque disease. So the disease is no longer a problem.

d) What is Malpeque disease?

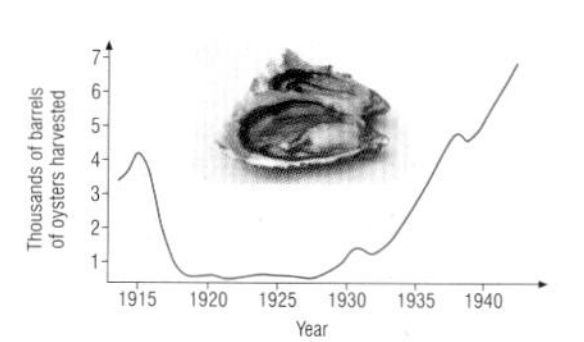

Figure 3 Oyster yields from Malpeque Bay 1915–40. As you can see, disease devastated the oyster beds. However thanks to the process of natural selection, a healthy population of oysters managed to survive and reproduce again.

SUMMARY QUESTIONS

1 Copy and complete using the words below:

adaptation breed environment generation mutation natural selection organism survive

When a ......has a good effect it produces an ...... which makes an ......better suited to its .... This makes it more likely to ...... and .... The mutation then gets passed on to the next .... This is ....

2 Give three examples from this spread of characteristics which are the result of natural selection, e.g. all-around eyesight in rabbits.

3 Explain how the following characteristics of animals and plants have come about in terms of natural selection.

a) Male peacocks have large and brightly coloured tails.
b) Cacti have spines instead of leaves.
c) Camels can tolerate their body temperature rising far higher than most other mammals.

4 Explain how mutation affects the evolution of a species.

KEY POINTS

1. New forms of genes result from changes (mutations) in existing genes.
2. Different organisms in a species show a wide range of variation because of differences in their genes.
3. The individuals with the characteristics most suited to their environment are most likely to survive and breed successfully.
4. The genes which have produced these successful characteristics are then passed on to the next generation.

SUMMARY ANSWERS

1 Mutation, adaptation, organism, environment, survive, breed, generation, natural selection.

2 [Any suitable examples from the spread.]

3 **a)** Mutation gave the male peacock a larger tail, making it more attractive to the females so they mate and pass on genes until they become normal in the population.

b) Mutation produced spines instead of leaves. Cactus loses very little water and so survives well and reproduces, passing on advantageous genes until normal in population.

c) Mutation gives temperature tolerance. These camels have an advantage, so more likely to survive and breed, passing on the mutation until it is normal in the population.

4 Mutation can result in beneficial changes to the genetic make-up of a species and help them to survive. These mutations can produce relatively rapid changes in a species.

KEY POINTS

The 'Card sort game' plenary reinforces the learning of the key points.

B1b 7.4 Extinction

LEARNING OBJECTIVES

Students should learn that:

- Environmental changes cause extinction.
- Some organisms can cause the extinction of other organisms by predation, competition or the introduction of disease.

LEARNING OUTCOMES

Most students should be able to:

- Explain what is meant by extinction.
- Explain how environmental changes cause extinction.
- Describe how new predators, new diseases and new competition from other organisms can cause the extinction of species.

Some students should also be able to:

- Evaluate the impact of introducing new organisms into an environment.

Teaching suggestions

- **Special needs.** Students to be provided with an illustrated sheet with a dead dinosaur and pictures of a thermometer, a meteorite, etc. and the key words can be transferred across or cut out and stuck on. Alternatively give them a wordsearch with the key words included.
- **Gifted and talented.** Students could prepare a brief report on: 'Does the current spate of extinctions really matter?' They could research drugs that have been discovered from wild species and speculate on what may be found in the future.
- **Learning styles**
  Interpersonal: Discussing and evaluating data on climate changes.
  Intrapersonal: Interpreting data from climate change maps.
  Kinaesthetic: Modelling a meteor impact; how much light gets through?
  Visual: Predicting trends from climate data.
  Auditory: Listening to a list of extinct animals.
- **Homework.** Ask: 'Should we try to re-create species that have become extinct?' Attempts to restore the quagga have been reported recently. This leads into a discussion of why animals become extinct at the present time. Ask: 'Are disappearing species worth preserving or should we allow nature to take its course?' Students can write their opinions for homework.

SPECIFICATION LINK-UP B1b.11.7

Extinction may be caused by:

- *changes to the environment*
- *new diseases*
- *new predators*
- *new competition.*

Lesson structure

STARTER

Dead as a dodo – Search the web for a picture of a dodo. Students are asked to make a list in rough of as many types of extinct animal as they can in 3 minutes. Check who has the largest number and get the student to read the list. Ask others in the class to check and add. Leads into a discussion of what extinction means. (5–10 minutes)

Unidentified species – Using Word, display (consecutively) 4000 letters, each to represent a million different species. Of the 4000, make two of these flash or be a different colour. These two represent the number of species alive, identified and named today. There is an unknown number (estimates range from 5 to 100 million) of unidentified species. The rest are extinct. (5 minutes)

MAIN

- **The stupidity of the dodos . . .** – Search the web for 'Ice age dodo' and mention how the film portrays them as being very stupid. Discuss how stupid the real birds were. Ask: 'What can you deduce from their appearance? Could they run fast?' A video on the extinction of the dodo is available from Channel 4 as part of the series 'Extinct' (0870 1234 344): could be shown in addition.
- **Climate changes** – Visit, for example, www.thedayaftertomorrowmovie.com to find information, and drama, about global climatic changes. Highlight changes in temperature such as ice ages. Link, if possible, with the geological time scale and the evolution of different groups of organisms. Discuss the impact of these changes on the creatures around at the time. This can link up with the ideas of competition, survival of the fittest and natural selection.
- **Taking this further** – Much has been made of global warming altering our climate and the times at which plants flower, changing patterns of migration and growing different crops in different parts of the world. Ask: 'If our climate became warmer or colder, what animals and plants would be affected in Britain?'
- **Deep impact?** – Show a video clip from the film *Deep Impact* of a simulated comet strike on Earth (search the web for 'Deep impact trailer'). Make a model, using a light sensor attached to a data logger and a glass or plastic container containing some fine dust. Shine a light through and measure intensity before and after shaking up. Discuss the effects of the lack of light on life on Earth.
- **The extinction game** – Five players have cards saying 'Climate change', 'Meteorite', 'Predators', 'Disease' and 'Competition' with appropriate pictures. Another player represents a species trying to survive. The 'species' throws a die and one of the 'extinction' players also throws a die. If the numbers match, the species is extinct and the players swap places and continue. If the numbers are different, the species has survived, collects the combined score on the dice and plays against the next 'extinction' player, until they eventually become extinct. Play to a time limit, with a small prize for the best score.

PLENARIES

Greatest impact – Using the cards from 'The extinction game', ask students to rank them in order of importance. Choose some to explain their choices. (5–10 minutes)

Quick test on definitions – Get the students to write definitions to the key words. Read some out; the rest of the class to say what it is. (5–10 minutes)

SUMMARY ANSWERS

1 Extinction, species, Earth, environment, climate, predators/competitors, competitors/predators, diseases.

2 A giant meteor hit the Earth. The impact threw up an enormous dust cloud, which spread over the entire surface of the planet. This blocked the sunlight reaching the surface and changed conditions dramatically, making everywhere colder and darker. Many types of plant became extinct, so food became short for plant eaters and many died out. Fewer plant eaters meant less food for the meat eaters and scavengers, and they gradually died out as well. Dinosaurs were reptiles, so they could not keep themselves warm as the temperatures dropped. In every way, the climate change was a disaster. By the time that things began to warm up again, survival of the fittest meant that the mammals took over.

3 a) i) The cat has kittens, the kittens breed, so soon there are lots of cats. Cats catch black-tailed mice easily, so the numbers fall and there's not enough to breed and the mice become extinct. The knock-on effect on the owls and the hawks is that part of their diet has gone, which in turn affects other prey animals.

ii) European primroses will make more food, so they have bigger leaves and set seeds which will germinate sooner. They will out-compete the English primrose and eventually it could become extinct.

b) Older species that cannot cope with changing conditions die out so only those that can compete well for food and other resources go on to evolve further.

ACTIVITY & EXTENSION IDEAS

Ask: 'Is the human race the greatest threat to endangered species?' Very little emphasis has been on endangered plants. Discussing the changing nature of agriculture could explain why some plants are becoming rare.

- **More about the quagga** – The quagga, a relative of the zebra, became extinct in the 1880s. In the 1990s, attempts were started to 'revive' the species. An Internet search can be made to find out about how this works. Link in with the homework suggestion given on page 114 and the references in the previous chapter on 'Cloning'.

BIOLOGY EVOLUTION

B1b 7.4 Extinction

LEARNING OBJECTIVES

1 Why do species become extinct?
2 What killed off the dinosaurs?

Throughout the history of life on Earth, we think a total of about 4 billion different species have existed. Yet only a few million species of living organisms are alive today. The rest have become *extinct*.

Extinction is the permanent loss of all the members of a species from the face of the Earth.

As conditions change, new species evolve which are better fitted to survive the new conditions. At the same time older species which cannot cope with the changes, and which do not compete so well for food and other resources, gradually die out. This is how evolution takes place and the balance of species on Earth gradually changes. Extinction is very important.

a) What is extinction?

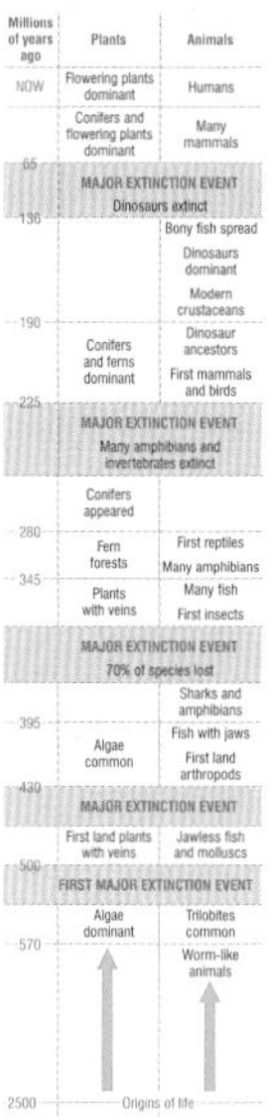

Environmental changes

Throughout history, the climate and environment of the Earth has been changing. At times the Earth has been very hot. At other times, temperatures have fallen and the Earth has been in the grip of an Ice Age.

Organisms which do well in the heat of a tropical climate won't thrive in an icy landscape. Many of them become extinct through lack of food or being too cold to breed. New species, which cope well in cold climates, evolve and thrive by natural selection.

Changes to the climate or the environment are the main cause of extinction throughout history. For example, most scientists think it was a big climate change that caused the dinosaurs to become extinct millions of years ago. This was possibly caused by a giant meteorite crashing into the Earth, creating drastic changes to the climate.

There have been five occasions during the history of the Earth when big climate changes have led to extinction on an enormous scale. Look at the major extinction events in Figure 1. These are part of the process by which evolution takes place.

Millions of years ago	Plants	Animals
NOW	Flowering plants dominant	Humans
	Conifers and flowering plants dominant	Many mammals
65	MAJOR EXTINCTION EVENT Dinosaurs extinct	
136		Bony fish spread; Dinosaurs dominant; Modern crustaceans
190	Conifers and ferns dominant	Dinosaur ancestors; First mammals and birds
225	MAJOR EXTINCTION EVENT Many amphibians and invertebrates extinct	
	Conifers appeared	
280	Fern forests	First reptiles; Many amphibians
345	Plants with veins	Many fish; First insects
	MAJOR EXTINCTION EVENT 70% of species lost	
		Sharks and amphibians
395	Algae common	Fish with jaws; First land arthropods
430	MAJOR EXTINCTION EVENT	
	First land plants with veins	Jawless fish and molluscs
500	FIRST MAJOR EXTINCTION EVENT	
	Algae dominant	Trilobites common
570		Worm-like animals
2500	Origins of life	

Figure 1 A summary of the main events in the evolution of life on Earth

Figure 2 The dinosaurs ruled the Earth for millions of years, but when the whole environment changed, they could not adapt and died out. By the time things began to warm up again, mammals, which could control their own body temperature, were becoming dominant. The age of reptiles was over.

b) How does the climate change during an Ice Age?

114

Evolution and environment

Organisms which cause extinction

The other main cause of extinction is other living things. This can take place in several different ways:

- If a new **predator** turns up in an area, it can wipe out unsuspecting prey animals very quickly. That's because the prey do not have adaptations to avoid it.
  A new predator may evolve, or an existing species might simply move into new territory. Sometimes it is our fault. The brown tree snake from Australia was brought to Guam by people following World War II. By the 1960s many bird species on Guam were becoming extinct at a rapid rate – eaten by the snakes which attacked their nests at night!
  The birds had no chance to evolve a defence to this new night-time predator. Because many of the birds of Guam are now extinct, the snakes have started eating lizards instead!
- New ***diseases*** (caused by microorganisms) can drive a species to the point of extinction. They are most likely to cause extinctions on islands, where the whole population of an animal or plant is close together.
  The Tasmanian devil in New Zealand is one example where this may happen. These rare animals are dying from a new form of cancer, which seems to attack and kill them very quickly.
- Finally, one species can drive another to extinction through successful ***competition***. You may see a new mutation which gives one type of organism a real advantage over another, or you may find people have introduced a new species by mistake.
  If the new species is really successful it may take over from the original animal or plant and make it extinct. In Australia the introduction of rabbits has been a nightmare because they eat so much and breed so fast! Other native Australian animals are dying out because they cannot compete.

DID YOU KNOW?

The big extinctions of the past took place over millions of years, but now species are becoming extinct very fast indeed! Some scientists predict that by the year 2010, a quarter of all the species on Earth will have been lost. As one species becomes extinct it affects the survival chances of another. The loss of so many different organisms so quickly could make the whole **ecosystem** of the Earth unstable.

GET IT RIGHT!

Always mention a **change** when you suggest reasons for extinction.
Watch your timescales – remember humans were not responsible for the extinction of the dinosaurs!

Figure 3 The Scottish island of North Uist has a similar problem to Guam. Someone brought a few hedgehogs onto the island to tackle garden slugs. The hedgehogs bred rapidly and are eating the eggs and chicks of the many rare sea birds which breed on the island. Now people are trying to kill the hedgehogs to save the birds!

SUMMARY QUESTIONS

1 Copy and complete using the words below:

climate competitors diseases Earth environment Extinction predators species

...... is the permanent loss of all the members of a ...... from the .... It may be caused by changes to the ...... or ......, to new ......, new ...... or possibly new ....

2 Explain how we think the dinosaurs became extinct.

3 a) Explain how each of the following situations might cause a species of animal or plant to become extinct.
i) Mouse Island has a rare species of black-tailed mice. They are preyed on by hawks and owls, but there are no mammals which eat them. A new family bring their pregnant pet cat to the island.
ii) English primroses have quite small leaves. Several people bring home packets of seeds from a European primrose which has bigger leaves and flowers very early in the spring.
b) Why is extinction an important part of evolution?

KEY POINT

1 Extinction may be caused by changes to the environment, new predators, new competitors and possibly new diseases.

115

Answers to in-text questions

a) The permanent loss of all members of a species from the face of the Earth.

b) Average temperature drops, ice caps spread across the Earth, and the sea freezes.

KEY POINTS

The 'Quick test on definitions' will help the learning of the key points.

B1b 7.5 Evolution – the debate goes on

SPECIFICATION LINK-UP

B1b.11.7

Students should use their skills, knowledge and understanding of 'How Science Works':

- *to suggest reasons why scientists cannot be certain about how life began on Earth*
- *to interpret evidence relating to evolutionary theory*
- *to suggest the reasons why Darwin's theory of natural selection was only gradually accepted*
- *to identify the differences between Darwin's theory of evolution and conflicting theories*
- *to suggest reasons for the different theories.*

BIOLOGY EVOLUTION

B1b 7.5 Evolution – the debate goes on

In some states of America fundamentalist Christians have a powerful voice. They believe that the Earth and everything in it was created exactly as it is described in the Bible. They would like to prevent the theory of evolution from being written about in school text books or taught in schools. At the very least they would like to see the Creationist view given equal emphasis.

Science versus religion?

When Charles Darwin published his book *The Origin of Species*, he knew that it would cause trouble between the scientific community and the Church. He wanted to put forward his ideas, but he did not want to unsettle faithful Christians – his beloved wife was one! Of course the book caused an uproar, and the debate still continues in some places today.

Darwin's basic principles are not in dispute among scientists, although they often like to discuss the fine details of evolution. But not everyone agrees. For some people there is no conflict between a deep faith in God and an acceptance of evolution. Others find this a problem. Religion is a system of faith and unquestioned belief. It deals with spiritual things which cannot be explained simply by using scientific methods, based on collecting evidence and data on the natural universe.

One area where the Church and science have clashed is on the age of the Earth itself. Here are some of the stages in the debate!

The age of the Earth

Fossils provide evidence that animals and plants have changed and developed over a very long time. This process is known as **evolution**. The idea of evolution suggests that the Earth itself must also have existed for billions of years. This view of the origins of life on Earth is quite recent.

In the 17th century the story of the creation of the Earth in the Christian Bible was still largely unquestioned. One famous historian, Archbishop Usher, used it to calculate that the Earth was less than 6000 years old!

During the 18th century people began to travel more. They not only discovered new lands, but amazing plants and animals as well. Our ancestors unearthed the fossil remains of massive creatures and strange plants. They began to build up evidence that the Earth had changed dramatically during its history.

By the beginning of the 19th century, the evidence for evolution was building up. Sir Charles Lyell was a British geologist. He showed that the Earth was very ancient, and was shaped by rivers and 'subterranean fires'. He estimated that the Earth was several hundred million years old and published his ideas from 1830 onwards.

Lyell's work was important to Darwin. He came up with a theory that all living organisms have arisen from evolution by natural selection. This would have taken many millions of years. Fossils also helped to confirm Darwin's ideas. They showed some of the stages at which the different animals and plants appeared and how they have changed over time. However, some people still believe that fossils were put into rocks by God to test our faith.

By the 1890s, Arthur Holmes was using radioactivity to date rocks. He established the age of the Earth as around 4.6 billion years old, giving plenty of time for evolution to have happened!

ACTIVITY

You have been asked to give evidence at the School Board meeting in the United States. The panel is trying to decide whether to allow the scientific theory of evolution to be taught in their school.

Put together a short report on why it is important that students know about the theory of evolution. Bring together some evidence of evolution occurring in the world around them. Show that it is important for students to understand the scientific view which is held by the majority of their peers in the developed world. This is true even if they choose not to accept the evidence and to hold their own beliefs.

Teaching suggestions

Many of the issues and ideas brought together in this spread have already been introduced and discussed, so material from the preceding spreads needs to be used in the activities suggested.

- **Science versus religion?** – Some background on Darwin's life and times can help us to understand why he did not publish his work sooner. Students could gather together some facts and figures and this could be presented *à la* 'This is Your Life'. The chronological sequence of events is well-charted and various characters could be introduced to talk about their part in Darwin's life. Here are some examples: the captain of *HMS Beagle*; Darwin's wife; a Galapagos turtle; a bishop; Lamarck.
- Following the above suggestion, to bring the debate up-to-date, Darwin could be put on trial for misleading the public. Arguments for and against his theory should be put forward. With a little imagination, different witnesses could give evidence and be cross-examined.
- Some information on the dating of fossils could be compiled and presented. Include dating of the rocks, remains in different strata, radioactive measurements, etc. These methods could be presented as a series of OHPs and could be used in evidence for the activity suggested.
- **Making extinction extinct?** – The Millennium Seed Bank Project (MSBP) has been described in the student book. The emphasis here is on the preservation of seeds from wild flowering plants, but consider the preservation of different varieties of crop plants. The number of different varieties of certain crop plants, such as apples, has diminished. Ask: 'Why has this happened? Why is it important to keep tissue or seeds of some of the older varieties?'
- Some of the arguments for the preservation/restoration of species could involve cloning, so this spread could be reviewed.

Making extinction extinct?

Plant species are under threat all around the world – but we're fighting back. We have a plan to protect 24 000 plant species by storing their seeds! It's a massive international project masterminded from the Royal Botanic Gardens at Kew in the UK.

In the UK we have collected seeds from all the wild flowering plants for the Millennium Seed Bank Project (MSBP). Once we've collected them they are put into storage. These UK seeds are not alone in their storage jars – seeds from around the world are sent to join them.

Sixteen countries from Jordan to Madagascar, and from Botswana to Mexico are involved in the Millennium Seed Bank project.

South Western Australia in particular has many, many plants – at least 12 000 species are known! It is one of the world's top 'hot spots' for biodiversity and seeds from all of the threatened species are being collected. Once the seeds are stored, even if the plants become extinct in the wild, we have the chance of introducing them again in the future! These plants are so important – they could be a source of new medicines in the future.

Once the seeds for the MSBP have been collected, they are checked, dried and then stored at about −20°C. Under these conditions they can survive for decades – or even centuries!

So many fascinating animals have become extinct. I mean, I'd love to see a dodo, wouldn't you? And going further back, what about a woolly mammoth walking the Earth again? These examples are a bit extreme! But there are teams around the world who are using modern technology to save species which are on the brink of extinction, or have just become extinct.

We hope cloning will be the key to the future for some species. Even if we can't clone them successfully now, by saving some of their tissue we may be able to bring the species back in the future!

For example, Banteng, the wild cattle of Java, Burma and Borneo are endangered in the wild. In 2003 a team in the US managed to produce one healthy cloned Banteng calf, and another that was abnormal. Before that, in 2000, we cloned a healthy Guar calf. Guar are very rare cattle indeed so this was quite a triumph. Sadly little Noah died from an infection just two days after he was born – but we can and will try again!

We've got some tissue from the ear of a Pyranean Ibex. The last one died in 2000. It would be fantastic to bring them back soon. And going back to mammoths . . . They are very like our modern elephants in many ways and we've found very well preserved cells in some of the ice fossils. I'm sure that one day someone will clone an extinct animal and bring the species back into existence. What's more, I'm sure prehistoric animals might even be cloned one day – and I hope I'm here to see it!

New species are being cloned all the time. In future we may be able to clone species of animals and plants that are threatened with extinction (like this banteng) and keep them going. We may even be able to clone species which became extinct some time ago using ice fossils or material found on dried specimens in museums.

ACTIVITIES

a) Extinction!
Your task is to make a poster titled 'Extinction!' for display in the school science department. You can choose what to highlight in your poster. You can use information from this chapter and you might like to use other resources as well. The following might give you some ideas:

- How extinction comes about and why it is important.
- Comparing extinction rates in the past and now – why is there so much concern?
- How can extinction be prevented/undone?

b) Interfering with nature?
Write a letter to *The Times*:

Either: express your enthusiasm for using new technologies to prevent extinction, keep species going and bring back extinct species.

Or: express your disgust at using new technologies to prevent extinction, keep species going and bring back extinct species.

Special needs

Students could make a poster about the Millennium Seed Bank Project explaining what it is and giving some reasons why we should preserve seeds from plants. Use cuttings and pre-printed statements to assist the students.

Gifted and talented

These students could research the possible use of mitochondrial DNA from preserved parts of extinct animals in attempts to restore species. Why does the mitochondrial DNA come from the female parent?

- **What would we do with a dinosaur?** Imagine what problems could arise from cloning a dinosaur. What would we do with it? Is it fair to devote money and resources to such projects?

ACTIVITIES & EXTENSIONS IDEAS

- The first activity, to give evidence for the teaching of the scientific theory of evolution, needs a balanced view. The scientific evidence could make reference to some of the many fossil finds in the USA and the evidence from the dating of fossils.
- **Extinction!** – The poster activity could be focused on extinction in general; or students might like to research the web for recently extinct species, or those that are on the *Red List of Threatened Species* (published by the International Union for the Conservation of Nature and Natural Resources (IUCN)) and choose an example to concentrate on. There are comprehensive lists of endangered plants and animals available from the web or in books such as *Plant* and *Animal* (Dorling Kindersley publications).
- **Investigation into the reasons for the recent extinction of species** – The human race is responsible for the extinction of many species. Discuss ways in which the tourist industry may be involved and how the problems could be resolved. Think about hunting, shooting, fishing, souvenirs, collectors and the building of holiday resorts. This could be extended to over-use of land, clearing for farming, over-fishing and a number of other human activities.
- **Interfering with nature?** – This approach to this exercise is similar to that in the spread on cloning. The letter could be fairly general, or it could make a case for preventing extinction of an animal such as the Giant Panda, bringing back the quagga or re-creating a woolly mammoth. Students should be encouraged to do some research if they choose to write their letter about a specific animal.

SUMMARY ANSWERS

1 **a)** Harder parts of the animal or plant are replaced by other minerals over long periods of time.

b) When the temperature is too low for decay to take place organisms are preserved in ice.

c) It tells us how much or how little organisms have changed since life developed on Earth.

d) So much rock has been crushed, buried, weathered or melted over the years.

2 **a)** **Similarities:** They both suggest evolution of living things from simpler organisms; they both suggest it took a long time; they both suggest changes passed down from parents to offspring.
Differences: Lamarck suggests primitive worms as a starting point; he suggests it is acquired characteristics which are passed on; Darwin suggests it is inherited features which are passed on, and the process of natural selection 'survival of the fittest' to decide which organisms survive and breed.

b) Any thoughtful point, e.g. it helped to pave the way for Darwin's ideas; people had already come to terms with a theory other than the Bible; debate was opened up; Darwin's made more sense.

3 Credit careful explanations that include an understanding of the basic concepts. For example, a pair of founder finches on one island with a high insect population has a mutation that results in birds with a slightly different beak shape. This makes it easier to poke its head into cracks to find insects. These birds can reach food that the others cannot, which gives them an advantage. They get more food and therefore they are more likely to survive, breed and pass on the genes for the thinner beak shape. Eventually a whole group of birds evolves with thinner beaks, which feed on insects. As they are separate from the other birds, a new species has been formed. A similar process occurs on another island where there are a lot of fruit bushes, so the birds here evolve beaks suited to eating fruit and buds etc. The birds evolve to take advantage of the available food on the islands.

4 **a)** At the same time, older species that cannot cope with the changes and that do not compete so well for food and other resources, die out. This is how evolution takes place and the balance of species on Earth gradually changes. Without extinction, less well-adapted species would struggle on and there would not be room for new species.

b) It is happening at such a fast rate.

c) Any suitable example well described.

d) Any suitable example well described.

e) Extinction is necessary for evolution to take place successfully. If you stop extinction, you could stop evolution.

Summary teaching suggestions

- **Special needs**
  In question 3, students could be given pictures of the beaks (photocopy the diagram) and cards with the different types of food written on them. The food type could then be matched with the beak type.

- **Misconceptions**
  Students should be aware that extinction is not always a bad thing and that there is always a reason for animals and plants becoming extinct.

- **Lower and higher level answers**
  There are several examples in these questions where the depth of the answer given is important. Students aiming for the higher grades should be able to write a more detailed answer to question 2b), provide a reasoned sequence of how natural selection occurred in answer to question 3, and full answers to question 4a) and e).

EVOLUTION: B1b 7.1 – B1b 7.5

SUMMARY QUESTIONS

1 a) How are rock fossils formed?
b) How are ice fossils formed?
c) What evidence do fossils give us about how life on Earth has developed?
d) Why is the fossil record not complete?

2 a) Summarise the similarities and differences between Darwin's theory of evolution and Lamarck's.
b) Why do you think Lamarck's theory was so important to the way Darwin's theory was subsequently received?

3

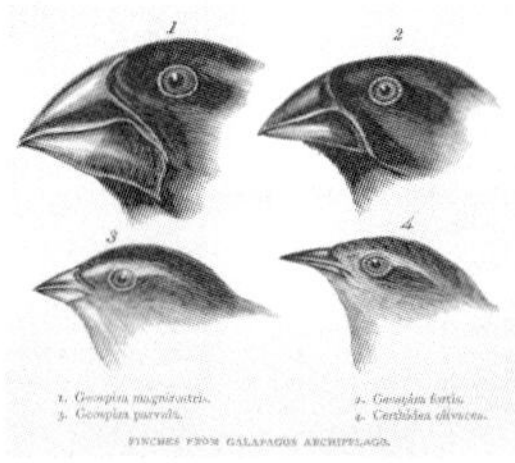

Darwin's finches – more evidence for evolution

Look at the birds in the picture. They are known as Darwin's finches. They live on the Galapagos Islands. Each one has a slightly different beak and eats a different type of food.
Explain carefully how natural selection can result in so many different beak shapes from one original type of founder finch.

4 a) Why is extinction important to the success of evolution?
b) Why are scientists so worried about the rate at which extinction is occurring now?
c) Find out and write about one animal or plant which has become extinct in the last twenty years.
d) Find out and write about one animal or plant which is close to extinction now. Explain what, if anything, is being done to help prevent it from becoming extinct.
e) Many groups are keen to prevent animals and plants becoming extinct at all costs. Why is it not necessarily a good idea to prevent extinction?

118

EXAM-STYLE QUESTIONS

1 The list contains factors that have played a part in the development of the species we see on Earth today:
A Extinction **B** Evolution
C Natural selection **D** Mutation
Match words **A**, **B**, **C** and **D** with the processes **1**, **2**, **3** and **4** in the table.

	Process
1	Change and development of organisms over a long period of time
2	Change to the amount or arrangement of genetic material within a cell
3	Permanent loss of all members of a species
4	Passing of genes to offspring by the organisms most suited to their environment

2 Which of the following does **not** play a part in evolutic by natural selection?
A Inheritance of acquired characteristics.
B Mutation of existing genes.
C Variation of individuals within a species.
D Production of offspring by individuals best suited tc their environment.

3 Which of the following would **not** normally cause the extinction of a species?
A Environmental change
B Fewer competitors
C New diseases
D More predators

4 Not all scientists agree on the exact evolutionary relationship between different primates. The diagram shows a timeline for one version of this relationship.

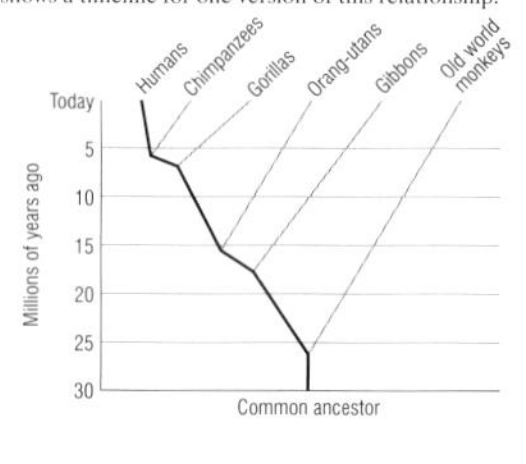

- **Homework**
  Question 4c) and d) need some research however they are tackled, so they are appropriate homework exercises where the research can be done in libraries or on the Internet.

- **When to use these questions?**
  - Question 1 is a quick way of assessing knowledge of what can be learnt from a study of fossils.
  - Question 2a) can be answered in columns and a correct answer makes a useful revision card.
  - For question 4c), students could be supplied with a list of plants and animals that have become extinct in the last 20 years, or they could find out for themselves. Some suggestions can be found using the Science News web site or other web sites. However, note that it is difficult to decide whether or not some animals and plants have *actually* become extinct in the last 20 years.
  - For question 4c), reference to the *Red List of Threatened Species* (published by the International Union for the Conservation of Nature and Natural Resources (IUCN)) will provide ideas for species to investigate.

EXAM-STYLE ANSWERS

1 **A** **3**
B **1**
C **4**
D **2** *(1 mark each)*

2 **A** *(1 mark)*

3 **B** *(1 mark)*

4 a) **D**
b) **B**
c) **D**
d) **B** *(1 mark each)*

Evolution and environment

(a) Which group is the closest relative to humans?

A Old world monkeys

B Orang-utans

C Gorillas

D Chimpanzees

(b) How many million years after the old world monkeys evolved from the common ancestor did the gorillas evolve?

A 21 B 20 C 10 D 7

(c) How many of the primate groups shown in the diagram were on Earth 20 million years ago?

A 6 B 5 C 3 D 1

(d) Many of the ancestors of the present-day primates are now extinct. How do we know these ancestors once lived?

A By studying DNA samples.

B By studying fossil records.

C By studying blood samples.

D By studying cell structure. (4)

The Galapagos Islands are a group of islands in the Pacific Ocean. The nearest country on the mainland is Ecuador, 1000 km away. By some means, a few seed-eating finches were the first birds to reach the islands. This single ancestral species has since evolved into many different species. Charles Darwin visited the islands and noted that each species had a beak adapted to the type of food it ate.

Using the theory of natural selection, explain how the ancestral species might have evolved into birds with different-shaped beaks. (6)

HOW SCIENCE WORKS QUESTIONS

It is difficult to gather data that illustrates evolution. It is possible to gather data to show natural selection, but this usually takes a long time. Simulations are useful because, while they are not factually correct, they do show how natural selection might work.

Darwin used evidence from his visit to the Galapagos Islands to show how natural selection might have worked. He used this as evidence for evolution, by natural selection. Some of the evidence he gathered was about the size and shape of the beaks of the finches on the different islands.

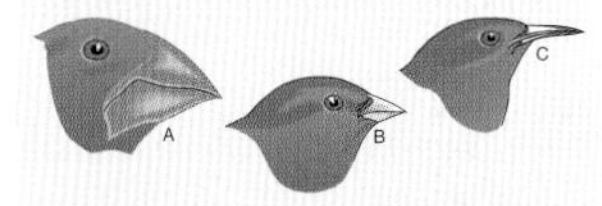

A class decided to simulate natural selection, by seeing if they could use different tools to pick up seeds. This is what they did.

Four students each chose a particular tool to pick up seeds. The teacher then scattered hundreds of seeds onto a patch of grass outside the lab. The four students were then given five minutes to pick up as many seeds as they could.

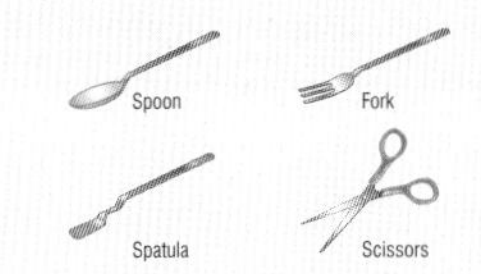

James, who was using a spoon, picked up 23 seeds, whilst Farzana, using a fork, could only pick up two. Claire managed seven seeds with the spatula, but Jenny struggled to pick up her two seeds with a pair of scissors.

a) Put the essential data into a table. (3)

b) How would this data be best presented? (1)

c) Was this a fair test? Explain your answer. (1)

d) What conclusion can you draw from this simulation? (1)

5
- When the ancestral species arrived on the islands there would be no competition from other birds.
- There would be a variety of different types of food available, e.g. vegetation, seeds, insects.
- The ancestral species would have bred rapidly and built up a large population and so created competition between its members for food.
- Sexual reproduction and/or mutations over many generations would have created variety amongst the individuals in the population.
- Some of this variety would have been in the types of beak, with different ones being suited to different food sources, e.g. a longer, thinner beak would allow its owner to capture insects in small holes or crevices that were inaccessible to birds with shorter and wider beaks.
- These birds with thinner beaks had an advantage over the other birds of an additional food supply and were therefore more likely to survive long enough to mate and raise offspring.
- These offspring would most probably possess the same genes for a long, thin beak that their parents had and so they too were more likely to survive.

(1 mark for each point up to a maximum of 6 marks)

Exam teaching suggestions

Questions 2 and 3 are examples of multiple-choice questions with a negative. Students are expected to select an option that does **not** meet the criteria set out in the first part of the question. One way to help students who find this difficult is to get them to look for the 'odd one out'. In other words, three options all meet the criteria, the fourth is the odd one out and that is the answer being sought. In this way the word 'not' can be eliminated from their thoughts and hopefully so too can the confusion it can create.

Question 4 presents students with information in a form they may not have encountered before. They need to be reassured that in this type of question the answers are all to be found in the data provided. It is testing their ability to interpret data rather than their knowledge. It is not unknown for students in examinations to simply miss out this type of question because they are frightened away by information they haven't seen previously.

Question 5 is a more taxing one and hence more suited to higher-tier students. It requires students to apply their knowledge of natural selection to a given situation. In doing so they need to marshal their information into a logical sequence and to clearly express what, for some students, are difficult ideas. Students who find this type of question a problem may find it helps them to organise their thoughts if they break the answer up into a series of bullet points. Unless the question says something to the contrary, answers can be expressed in this way. They are also less likely to become muddled if they write many short sentences/bullet points rather than fewer, longer ones.

This is a sample answer written as a continuous narrative but divided into convenient marking points. To some extent it is unimportant where a line is drawn between one marking point and another. Even the order of points may vary slightly, e.g. the first two points could be in reverse order. 7 marking points have been listed in the mark scheme although a maximum of 6 marks are allowed. This gives some flexibility in the marking, although an answer gaining all 6 marks would need to include the following 4 major points:

- New genes arising from existing ones (mutations).
- Variation amongst individuals.
- Those most suited to the environment breed successfully.
- Their genes are passed to the next generation.

One mark should be deducted for each of these points that is not adequately covered.

HOW SCIENCE WORKS ANSWERS

a)

Tool used	spoon	fork	spatula	scissors
Number of seeds	23	2	7	2

b) Data would best be presented in a bar chart, because the independent variable is categoric.

c) No, this was not a fair test. The different people could have performed differently, so it was not a test of the tools used. The results are not valid.

d) That different people using different tools can pick up different numbers of seeds. Possibly the spoon was the best tool.

How science works teaching suggestions

- **Literacy guidance.** Key terms that should be clearly understood are: table construction, validity, conclusions.
- **Higher- and lower-level questions.** Question c) is a higher-level question. The lower-level question is question a). Answers for both have been provided at these levels.
- **Gifted and talented.** Able students could consider how to adapt this style of investigation to make it more valid.
- **How and when to use these questions.** When wishing to develop fair testing. The questions could be used as a small group project.
- **Special needs.** The link between the simulation and the real situation must be made.
- **ICT link-up.** Producing the table and graph on a spreadsheet provides a good opportunity for ICT. Different presentations of the data could be considered.

B1b 8.1 The effects of the population explosion

LEARNING OBJECTIVES

Students should learn:

- The effect that the growth of the human population has had on resources, land and pollution.

LEARNING OUTCOMES

Most students should be able to:

- Explain why raw materials are rapidly being used up and more waste is being produced.
- Describe how water, air and land may become polluted with waste.

Some students should also be able to:

- Analyse and interpret scientific data concerning human population growth and environmental issues.

Teaching suggestions

- **Special needs.** Students could be given a picture of a very crowded room and asked to imagine what could happen if all the people they knew come to live in their house. Ask: 'What problems would there be?' They could fill in a list, or be shown pictures to act as visual prompts.
- **Gifted and talented.** Students could be asked to analyse why people in the developing world have such large families. They could write an article suggesting how to get round this problem.
- **Learning styles**

  Interpersonal: Discussing how many people there are in the world.

  Intrapersonal: Understanding the 'Baby counter'; report writing.

  Kinaesthetic: Research projects on effects of humans on environment.

  Visual: Making observations from pop festival pictures.

  Auditory: Composing a radio message.
- **Homework.** Use this time for research, reading or writing up reports for projects.

SPECIFICATION LINK-UP B1b.11.8

- *Rapid growth in the human population and an increase in the standard of living means that:*
  - *raw materials, including non-renewable energy resources, are rapidly being used up*
  - *increasingly more waste is produced*
  - *unless waste is properly handled more pollution will be caused.*
- *Humans reduce the amount of land available for other animals and plants by building, quarrying, farming and dumping waste.*
- *More waste is being produced which, unless properly handled, may pollute:*
  - *water – with sewage, fertiliser or toxic chemicals*
  - *air – with smoke and gases, such as sulfur dioxide which contribute to acid rain*
  - *land – with toxic chemicals, such as pesticides and herbicides, which may be washed from land into water.*

Lesson structure

STARTER

How many people in the world? – Hold a brief competition to guess how many people are alive on Earth now. The students should all stand up, holding up their estimates in front of them. Count alternately upwards and downwards towards the actual number, telling students to sit down as their answers are discounted. Award a small prize for the student/students who are nearest the correct number. (5–10 minutes)

What sort of problems do we have if we go to pop festivals? – Show a picture of masses of people at a festival like Glastonbury (search the web for 'Glastonbury'). Ask: 'What are the problems? How do we cope with the basic needs of living?' Draw comparisons with the human population on the whole of the planet. (5–10 minutes)

MAIN

- **How many babies will be born during this lesson?** – The figure is about 240 babies per minute. The point can be very pertinently made by keeping a running total on a board at the front, with a student adding to the total every 5 minutes (about 1200 every 5 minutes).
- Supply or project a partially completed graph of population numbers, and ask students to suggest what will happen over the next few hundred years if population growth continues as it is today. It might be interesting to compare world population with projected numbers for the developed world, Europe or the UK.
- **Small research projects** – This topic lends itself to the class splitting up into small groups, each group to take one aspect and prepare a report for the rest of the class. The topics could include 'pollution', 'use of resources', 'wildlife', 'quarrying', 'housing and roads', 'waste disposal', etc.
- Each group would need to be given a brief for their research project and access to suitable resources. A library box on each topic would be helpful. Probably some homework time could be allocated to this for the research, and the reports given in class, together with worksheets, so that students could make notes on the topics that other students researched.

PLENARIES

Baby counter – Look at the count started at the beginning of the lesson. Show a PowerPoint® presentation of real baby photographs very rapidly. If you give each one a name, then this emphasizes what is really happening and that it will have an impact on them and their children in the future. If you have the resources, it could be good to vary the ethnic origin and names of the children in proportion to the increases in world population. Discuss what it is that balances this rapid increase in the population of the world. (10 minutes)

Spaceship Earth analogy – Project a spaceship picture and draw out the 'Spaceship Earth analogy': we have broken into the store room, the life support system is failing and breeding is occurring quickly on finite resources. Ask the students to compose a radio message from the spaceship, asking for help or advice from anyone out there. (10 minutes)

ACTIVITY & EXTENSION IDEAS

Recycling of materials is now an important issue and many local councils have introduced separate collections of recyclable materials such as newspapers, bottles and cans. Students could investigate the nature of any such collections in their area, whether people do sort their rubbish and how popular such schemes are. The local council is a good source of information here. If schemes do exist, then students could design posters encouraging people to make use of them. If they do not, then they could decide how they would lobby their local councillors for the introduction of such a scheme. (This topic is revisited in the last spread, but introduced here: it could be a starting point.)

Answers to in-text questions

a) Over 6 billion.

b) By building houses, shops, industrial sites and roads, farming and quarrying.

c) i) Sewage, by fertilisers from farms and by toxic chemicals from industry.

ii) Smoke and poisonous gases, such as sulfur dioxide.

iii) Toxic chemicals from farming, such as pesticides and herbicides; industrial waste such as heavy metals.

BIOLOGY

HOW PEOPLE AFFECT THE PLANET

B1b 8.1 The effects of the population explosion

LEARNING OBJECTIVE

1 What effect is the growth in the number of people having on the Earth and its resources?

We have only been around on the surface of the Earth for a relatively short time – less than a million years. Yet our activity has changed the balance of nature on the planet enormously. Some of the changes we have made seem to be driving many other species to extinction. Some people worry that we may even threaten our own survival.

Human population growth

For many thousands of years people lived on the Earth in quite small numbers. There were only a few hundred millions of us! We were scattered all over the world, and the effects of our activity were usually small and local. Any changes could easily be absorbed by the environment where we lived.

But in the last 200 years or so, the human population has grown very quickly. At the end of the 20th century the human population was over 6 billion, and it is still growing

If the **population** of any other species of animal or plant had suddenly increased in this way, natural predators, lack of food, build up of waste products or diseases would have reduced it again. But we have discovered how to grow more food than we could ever gather from the wild. We can cure or prevent many killer diseases. We have no natural predators. This helps to explain why the human population has grown so fast.

Figure 1 The Earth from space. As the human population of the Earth grows, our impact on the planet gets bigger every day.

Not only have our numbers grown hugely, but in large parts of the world our standard of living has also improved enormously. In the UK we use vast amounts of electricity and fuel to heat and light our homes and places of work. We use fossil fuels like oil to produce this electricity. We also use it to move about in cars, planes, trains and boats at high speed and to make materials like plastics. We have more than enough to eat and if we are ill we can often be made better.

a) Approximately how many people are living on the Earth today?

Number of people (billions)

Year 2000

Thousands of years ago

Figure 2 This record of human population growth shows the massive increase during the last few hundred years

The effect on land and resources

The increase in the numbers of people has had a big effect on our environment. All these billions of people need land to live on. More and more land is used for the building of houses, shops, industrial sites and roads. Some of these building projects destroy the habitats of rare species of other living organisms.

We use billions of acres of land around the world for farming, to grow food and other crops for human use. Wherever people farm, the natural animal and plant population is destroyed.

In quarrying we dig up great areas of land for the resources it holds such as gravel, metal ores and diamonds. This also reduces the land available for other organisms.

b) How do people reduce the amount of land available for other animals and plants?

120

Evolution and environment

Figure 3 In the UK alone hundreds of thousands of new houses are being built, and miles of new road systems. Every time we clear land like this, the homes of countless animals and plants are destroyed.

The huge human population is an enormous drain on the resources of the Earth. Raw materials are rapidly being used up. This includes non-renewable energy resources such as oil and natural gas and metal ores which cannot be replaced.

DID YOU KNOW?

There will be 125 million births in the world this year – that's a lot of bawling babies! According to the United Nations if we keep having babies at this sort of rate, the world population will soar to 244 billion by 2150 and 134 trillion by 2300!

Pollution

The growing human population also means vastly increased amounts of waste. This is both human bodily waste and the rubbish from packaging, uneaten food and disposable goods. The dumping of this waste makes large areas of land unavailable for any other life except scavengers.

There has also been an enormous increase in manufacturing and industry to produce the goods we want. This in turn has led to industrial waste.

The waste we produce presents us with some very difficult problems. If it is not handled properly it can cause serious **pollution**. Our water may be polluted by sewage, by fertilisers from farms and by toxic chemicals from industry. The air we breathe may be polluted with smoke and poisonous gases such as sulfur dioxide. (See page 122.)

The land itself can be polluted with toxic chemicals from farming such as pesticides and herbicides, and with industrial waste such as heavy metals. These chemicals in turn can be washed from the land into the water. If the ecology of the Earth is affected by our population explosion, our use of resources and our waste, everyone will pay the price.

c) What substances commonly pollute i) water, ii) air, and iii) land?

GET IT RIGHT!

Make sure you know exactly which pollutants affect air, land and water!

SCIENCE @ WORK

Scientists working for groups as diverse as the United Nations and Greenpeace are involved both in monitoring the world population growth and in measuring and controlling levels of pollution.

Figure 4 In countries like ours, we have a very high standard of living. But a kitchen like this uses lots of resources – wood, metals, plastics and energy. This all results in pollution and the removal of resources which can never be replaced.

SUMMARY QUESTIONS

1 Copy and complete using the words below:

diseases farming food increase population predators treat two hundred

The human ...... has increased dramatically in the last ... ... years. Better ...... methods mean we have more .... We can ...... and prevent many .... We have no natural .... All this has allowed the numbers to ....

2 a) How has the standard of living increased over the last hundred years?
b) Give three ways in which people have used up different resources.

3 Write a clear paragraph explaining how the ever-increasing human population causes pollution in a number of different ways.

KEY POINTS

1 The human population is growing rapidly and the standard of living is increasing.
2 More waste is being produced. If it is not handled properly it can pollute the water, the air and the land.
3 Humans reduce the amount of land available for other animals and plants.
4 Raw materials, including non-renewable resources, are being used up rapidly.

121

SUMMARY ANSWERS

1 Population, two hundred, farming, food, treat, diseases, predators, increase.

2 a) Use of electricity for lighting, heating, TV etc.; use of fossil fuels for transport (cars, planes etc.); plastics.

b) Any three suitable suggestions, such as fossil fuels, wood, land, metals etc.

3 Look for clarity of explanation without copying the Student Book. Points covered should be:

- Growing human population increases the amount of waste: bodily waste and the rubbish from packaging, uneaten food and disposable goods.
- The dumping of waste produced by the ever-expanding human population makes large areas of land unavailable for other life.
- Driving cars etc. leads to gases from exhausts.
- Farming leads to the use of pesticides and fertiliser sprays that cause pollution.

KEY POINTS

The first of the summary questions could be used at the end of the lesson to test understanding of the key points.

B1b 8.2 Acid rain

LEARNING OBJECTIVES

Students should learn that:

- Acid rain is formed when fossil fuels are burned.
- Acid rain may damage trees.
- Plants and animals cannot survive if the water in rivers and lakes becomes too acidic.

LEARNING OUTCOMES

Most students should be able to:

- Describe how acid rain is formed.
- Describe some of the effects of acid rain on living organisms.

Some students should also be able to:

- Analyse and interpret scientific data concerning the effects of acid rain.

Teaching suggestions

- **Special needs.** Students could be asked to imagine what it would be like if the clouds rained vinegar! Give them a picture of a giant vinegar pot shaking over the roofs of a town. Ask: 'What would happen to the animals and plants? How could the rain be like vinegar?' Show them the results of 'Demonstration of combustion and the production of acid gases' and compare with the colour change if vinegar is tested.
- **Gifted and talented.** Students could research sulfur in proteins and make a link with the amino acid cysteine, as a component of proteins in plants and animals. This in turn links with the sulfur released when fossil fuels are burned.
- **Learning styles**

  Interpersonal: Discussing the acidification of lakes.

  Intrapersonal: Writing a 'Thank you' letter for the Clean Air Act.

  Kinaesthetic: Practical work on the effects of acid rain on plants or seed germination.

  Visual: Observing colour changes in pH tests.

  Auditory: Explaining the effects of acid rain.
- **Homework.** Students could carry out some research into 'cross-boundary pollution', in which one country suffers the effects of pollution originating in another. They could write a two-minute radio report on the issue for a news programme.

SPECIFICATION LINK-UP B1b.11.8

- *More waste is being produced which, unless properly handled, may pollute:*
  - *air – with smoke and gases such as sulfur dioxide which contribute to acid rain*

Students should use their skills, knowledge and understanding of 'How Science Works':

- *to analyse and interpret data concerning environmental issues.*

Lesson structure

STARTER

The acid test – If it is raining, send a student out to collect a sample of rainwater. If it is not raining, then use a sample collected earlier. Revise the pH scale (reminding students that it is a little 'p' by giving them a dried pea or getting them to draw one with a green pencil!) and then test the rainwater with pH paper. They could also test some more acid things and some alkaline liquids so that the colour change is obvious. (10 minutes)

Demonstration of combustion and the production of acid gases – Wearing eye protection, set fire to a bunch of matches inside a gas jar using a fuse (Care!). Have some hydrogen carbonate indicator (or universal indicator) in the bottom of the jar. Shake the jar after the ignition and ask the students to observe the colour change. (10 minutes)

MAIN

- **Effect of acid rain on seedlings** – This practical can either be set up as a class demonstration or the students could set up their own experiment, working in groups. For each investigation, you will need 3 or 4 trays or pots of cress seedlings, all at the same stage of growth. One set should be sprayed with water, one with 0.1 M HCl and one with 0.5 M HCl. If desired, a fourth set could be sprayed with 1.0 M HCl. The pots should be kept in the same place and observed after a few days.
- This investigation introduces 'How Science Works concepts': predictions can be made, observations recorded and conclusions drawn.
- An alternative to the experiment suggested above would be to investigate the effect of acid rain on the germination of cress seedlings. Ensure eye protection is worn. Three Petri dishes could be prepared with filter paper circles in the bottom. Moisten the filter paper of one dish with water, another with 0.1 M HCl and a third with either 0.5 or 1.0 M HCl. Sow 50 cress seeds into each Petri dish, cover and keep in the same conditions. Make daily observations and count the number of seeds that germinate in each dish. Calculate the percentage germination and display the results graphically. Again, this exercise could be used to introduce 'How Science Works' concepts.
- **Gases from fossil fuels** – Burn some coal or other sulfurous fuel on a deflagrating spoon, or capture the gas emitted by drawing it into a thistle funnel attached to a vacuum pump. If a bubble trap of indicator is included, or a pH sensor attached to a data logger can be arranged to display through a projector, then students can see the effect on pH of the gas fumes emitted. This is probably best done in a fume cupboard. Wear eye protection.
- This demonstration can be linked to a survey of the fumes emitted by generating power using fossil fuels, the introduction of 'cleaner' petrol etc. Some research on the components of the waste gases would benefit any discussions on the links between burning fossil fuels and acid rain. Students could be provided with some facts and figures available from chemistry textbooks or the Internet.
- The *People's Century* series (from PBS) has some good material on acid rain, its sources and effects.

PLENARIES

Air pollution – Search the web for images of 'smog' and show students the effects of air pollution. Read a description of what it was like to be in a 'pea souper'. Ask students to write a time travelling 'Thank you' letter to the people responsible for the Clean Air Act, making our cities more pleasant places. (10–15 minutes)

Acid lake effects – Acid rain causes the water in freshwater lakes to become acidic. Ask students to suggest what effects a change in pH might have on life in freshwater, to include the effects on insects, insect larvae, water plants and all stages of fish development. (10 minutes)

ACTIVITY & EXTENSION

- **Catalytic converters** – Students could find out how catalytic converters on cars work. Link with the concept of catalysis with reference to the platinum components and with the chemistry involved.
- **The effects of acid rain on buildings** – Show some pictures of statues or bits of buildings that have been affected by acid rain. (Quite a bit of restoration work is going on in churches and cathedrals, so there should be some good pictures available.) Give each group of students a piece of limestone to moisten, weigh and place into a beaker of dilute acid (hazard warning). The pieces of limestone need to be re-weighed after several days. Predict what will happen and explain the results. Wear eye protection.

Answers to in-text questions

a) Coal, oil, natural gas, (petrol and aviation fuel are also acceptable).

b) Sulfur dioxide and nitrogen oxides.

c) The acid dissolves and destroys the leaves, so the trees cannot make food. If it soaks into the soil it can kill the roots.

BIOLOGY — HOW PEOPLE AFFECT THE PLANET

B1b 8.2 Acid rain

LEARNING OBJECTIVES

1 How is acid rain formed?
2 What are the effects of acid rain on living organisms?

Human activities can have far-reaching effects on the environment and all the other living things which share the Earth. One of the biggest problems is the way we produce pollution.

Everybody needs air – so when the air we breathe is polluted, no-one escapes the effects. One of the major sources of air pollution is the burning of fossil fuels. We are using more and more oil, coal and natural gas. We also burn huge amounts of the fuels made from them, such as petrol, diesel and aviation fuel for planes. Fossil fuel is a non-renewable resource – there is a limited amount of it on Earth and eventually it will all be used up.

a) Name three fossil fuels.

Figure 1 Air pollution is usually invisible. Just occasionally the level is so high it can actually be seen. The brown haze you can see over this city is caused by high levels of nitrogen oxides produced from car exhausts.

The formation of acid rain

When fossil fuels are burned, carbon dioxide is released into the atmosphere as a waste product. However, carbon dioxide is not the only waste gas produced. Fossil fuels often contain sulfur impurities. When these burn they react with oxygen to form sulfur dioxide gas. At high temperatures, for example in car engines, nitrogen oxides are also released into the atmosphere.

These gases pollute the air and can cause serious breathing problems for people if the concentration gets too high. They are also involved in the formation of acid rain. This pollutes land and water over a wide area.

The sulfur dioxide and nitrogen oxides dissolve in the rain and react with oxygen in the air to form dilute sulfuric acid and nitric acid. This makes the rain more acidic – it is known as **acid rain**.

b) What are the main gases involved in the formation of acid rain?

DID YOU KNOW?

Acid rain has been measured with a pH of 2.0 – more acidic than vinegar!

The effects of acid rain

Not surprisingly, acid rain has a damaging effect on the environment. If it falls onto trees, the acid rain can cause direct damage. It may kill the leaves and, as it soaks into the soil, even the roots of the trees may be destroyed. In some parts of Europe and America, huge areas of woodland are dying as a result of acid rain.

Acid rain has an indirect effect on our environment, as well as its very direct effect on plants. As acid rain falls into lakes, rivers and streams the water in them becomes acidic. If the concentration of acid gets too high, plants and animals can no longer survive. Many lakes and streams have become dead, no longer able to support life.

c) How does acid rain kill trees?

Figure 2 These trees should be covered in leaves and full of insects, birds and other animal life. Instead they are dead and bare, killed by the action of acid rain.

Acid rain is a difficult form of air pollution to pin down and control. It is formed by pollution from factories. It also comes from the cars and other vehicles we use every day. The source of the gases is pretty widespread. Many Western countries have worked hard to stop their factories and power stations from producing these acidic gases. Unfortunately there are still many places in the world where these gases are not controlled.

122

Evolution and environment

The worst effects of acid rain are often not felt by the country which produced the pollution in the first place. The sulfur and nitrogen oxides are carried high in the air by the prevailing winds. As a result, it is often relatively 'clean' countries which get the pollution and the acid rain from their dirtier neighbours. Their own clean air goes on to benefit someone else!

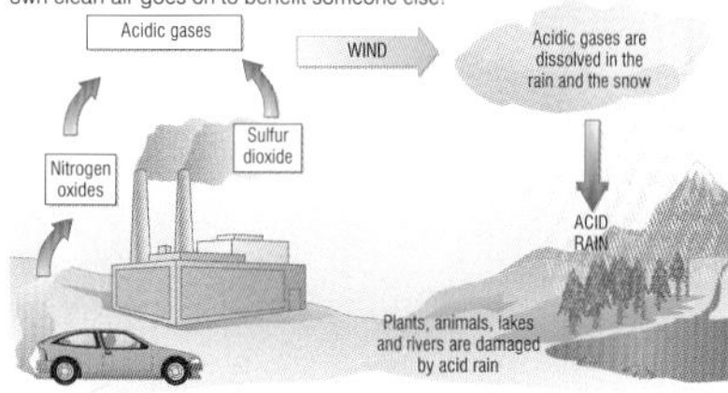

Figure 3 Air pollution in one place can cause acid rain – and serious pollution problems – somewhere else entirely. Depending on the prevailing winds, it can even be in another country!

People have become more aware of the problems caused by acid rain. The UK and other countries have introduced measures to reduce the levels of sulfur dioxide and nitrogen oxides in the air. More and more cars are fitted with catalytic converters. Once hot, these remove the acidic gases before they are released into the air. There are strict rules about the levels of sulfur dioxide and nitrogen oxides in the exhaust fumes of new cars.

Power stations are one of the main sources of acidic gases. In the UK we have introduced cleaner, low-sulfur fuels and started generating more electricity from gas and nuclear power. We have also put systems in power station chimneys to clean the flue gases before they are released into the atmosphere. As a result, the levels of sulfur dioxide in the air, and of acid rain, have fallen steadily over the last 40 years. (See Figure 4.)

SCIENCE @ WORK

Environmental Health Officers and the Environment Agency monitor the pH of our waterways, streams, lakes and soils. They also keep a careful eye on the health of the trees and the different types of animals and plants living in the water. This helps us to pick up any acid rain damage to our environment early and take action if it is needed.

GET IT RIGHT!

Be clear about the effects of the different combustion gases produced – sulphur dioxide does ***not*** affect global warming but it ***does*** cause acid rain!

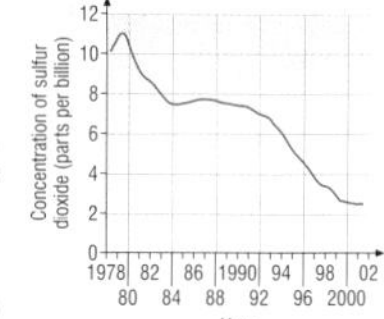

Figure 4 Graph to show levels of sulfur dioxide concentrations in the air in the UK, 1978–2002

SUMMARY QUESTIONS

1 Copy and complete using the words below:

acid rain carbon dioxide fossil nitric nitrogen oxides sulfur sulfuric

When ...... fuels are burned the pollutant gases ...., ... dioxide and .... are released into the atmosphere. The sulfur dioxide and nitrogen oxides dissolve in the rain to form dilute ...... acid and ...... acid. This is known as ....

2 Explain how pollution from cars and factories burning fossil fuels pollute:
a) the air b) the water c) the land.

3 To get rid of acid rain it is important that all the countries in an area control their production of sulfur and nitrogen oxides. If only one or two clean up their factories and cars it will not be effective. Explain why this is.

4 Use Figure 4 to help you answer this question.
a) Produce a bar chart to show the approximate levels of sulfur dioxide in the air in the UK at five-year intervals from 1980 to 2000.
b) Explain the trend you can see on your chart.

KEY POINTS

1 When we burn fossil fuels, carbon dioxide is released into the atmosphere.
2 Sulfur dioxide and nitrogen oxides can be released when fossil fuels are burnt. These gases dissolve in the rain and make it more acidic.
3 Acid rain may damage trees directly. It can make lakes and rivers too acidic so plants and animals cannot live in them.

123

SUMMARY ANSWERS

1 Fossil, carbon dioxide, sulfur, nitrogen oxides, sulfuric, nitric, acid rain.

2 **a)** Sulfur dioxide and nitrogen oxides are produced from impurities when fossil fuels are burned. These cause air pollution.

b) Sulfur dioxide and nitrogen oxides in the air dissolve in the rain and fall to the ground. The water runs into streams, rivers, etc. and lowers the pH, making them more acidic.

c) Acid rain falls on the ground, soaks in and causes it to be acidic.

3 The sulfur dioxide and nitrogen oxides produced by burning fossil fuels are carried high in the air by the prevailing winds. They can be blown hundreds of miles before they dissolve in rain and are carried to the ground. So every country needs to control emissions from the burning of fossil fuels to prevent any country being affected by acid rain.

4 **a)** The bar chart should be carefully drawn using the data provided. The axes should be labelled correctly and a suitable scale used.

b) Levels of sulfur dioxide have fallen (although they did plateau between 1984–1992) as a result of using cleaner fuels, using gas in power stations and cleaning flue gases before they are released.

KEY POINTS

A diagram of the events that cause acid rain could be built up similar to the one in the Student Book, and the students could use this as a revision card for the key points about acid rain.

B1b 8.3 Global warming

LEARNING OBJECTIVES

Students should learn that:

- Both combustion and deforestation affect the levels of carbon dioxide in the atmosphere.
- Levels of methane are increasing due to rice growing and cattle farming.
- Increasing levels of carbon dioxide and methane (greenhouse gases) may cause global warming.

LEARNING OUTCOMES

Most students should be able to:

- Explain how combustion and deforestation may cause an increase in carbon dioxide levels in the atmosphere.
- Describe how the activities of living organisms affect the composition of the atmosphere.
- Suggest some of the consequences of global warming.

Some students should also be able to:

- Explain the 'greenhouse effect'.
- Evaluate the impact of the greenhouse effect on conditions on the Earth.

Teaching suggestions

- **Special needs.** Students could visit the school greenhouse or one locally (e.g. at a garden centre). Measure the temperature inside and outside.
- **Gifted and talented.** Students could predict how a mini greenhouse would respond when full of CO_2. Get a numerical prediction and try it out.
- **Learning styles**

  Interpersonal: Discussing the deforestation problems.

  Intrapersonal: Understanding the melting of the polar ice caps.

  Kinaesthetic: Setting up a mini greenhouse.

  Visual: Following the sequence of the animation on global warming.

  Auditory: Explaining the re-radiation of heat.
- **ICT link-up.** Data loggers can be used for recording information of changes in the mini greenhouse.

SPECIFICATION LINK-UP B1b.11.8

- *Large-scale deforestation in tropical areas, for timber and to provide land for agriculture, has:*
  - *increased the release of carbon dioxide into the atmosphere (because of burning and the activities of microorganisms)*
  - *reduced the rate at which carbon dioxide is removed from the atmosphere and 'locked up' for many years as wood.*
- *Increases in the numbers of cattle and rice fields have increased the amount of methane released into the atmosphere.*
- *Carbon dioxide and methane in the atmosphere absorb most of the energy radiated by the Earth. Some of this energy is re-radiated back to the Earth and so keeps the Earth warmer than it would otherwise be. Increasing levels of these gases may be causing global warming by increasing the 'greenhouse effect'. An increase in the Earth's temperature of only a few degrees Celsius:*
  - *may cause quite big changes in the Earth's climate*
  - *may cause a rise in sea level.*

Lesson structure

STARTER

Changes in the world's climate – Find pictures of a beach and a desert on the web and show these to students. Discuss the possible complications of a change in the climate. (5–10 minutes)

Barney's burps – Show Barney from the Simpson's cartoon (search the web for 'Barney burp'). Show a can of rice pudding and ask: 'What is the link between flatulence, rice and global warming?' Discuss the link with methane as a greenhouse gas. (10 minutes)

Why is deforestation such a bad thing? – Show a picture of deforestation. Ask students for reasons why it is considered a bad thing. Compile a list on the board. Ask: 'Are there any benefits – or, more importantly, who benefits?' (5–10 minutes)

MAIN

- **Will melting the North Pole ice caps raise the sea level?** – Take a beaker $\frac{3}{4}$ full of water, mark the level on the side of the beaker. Place an ice cube in the beaker and leave it at room temperature during the lesson. By the end of an hour, the ice cube should have melted. The new level of the water in the beaker should be marked. Is there any difference in the levels?
- At the same time, a similar set of experiments can be arranged with a pile of ice cubes lined up on a chute leading into a beaker. As one melts another falls into the water. Both these demonstrations can be extrapolated to a global level and explained in terms of the effects of melting ice at both poles.
- Link the suggestion above with reference to specific places, such as holiday destinations (Maldives) and the UK. Project a map of the UK after potential global warming sea level rises. The Green Party produced a map 'The British Isle' as part of their 2005 election materials. Discuss which parts of the British Isles might disappear. Who might have to move? What would happen to East Anglia and the Fens? Would it affect where you live?
- **The mini greenhouse effect** – Set up a mini greenhouse (e.g. such as those from IKEA) with temperature sensors inside and out connected to data loggers. Ask the students to predict the differences in temperature readings with reasons and then investigate practically using an infra-red lamp.
- **Know the facts** – Many students find the concept of global warming difficult and it does get confused with other topics. An Animation, B1b 8.3 'Global warming', is available on the GCSE Biology CD ROM. Students could be given a work sheet to complete while watching this animation, so that the facts are clear.
- To see and discuss graphs showing climate change, search the BBC web site, www.bbc.co.uk, for 'climate change'.

PLENARIES

Re-radiation – Expose a piece of black paper to sunlight or an IR lamp for a few minutes. Ask a student to close their eyes and hold the paper close to their cheek. They should be able to tell you which cheek you held the paper close to. Ask students to explain and discuss the concept of re-radiation in terms of global warming. (5–10 minutes)

Key terms crossword – Supply the students with clues for the key words and terms used. You could put in initial letters for clues which students are finding difficult. (10 minutes)

Recap list – Compile a list of the causes and problems of global warming. (5 minutes)

Answers to in-text questions

a) Carbon dioxide and methane.

b) Deforestation is the cutting down of large areas of forest and burning or removing the trees. There is no replacement planting.

c) Rice growing in paddy fields and cows reared to produce cheap beef.

BIOLOGY

HOW PEOPLE AFFECT THE PLANET

B1b 8.3 Global warming

LEARNING OBJECTIVES

1. How does combustion affect the atmosphere?
2. How do living things affect the atmosphere?
3. What will be the consequences of global warming?

Many scientists are very worried that the climate of the Earth is getting warmer. This is often called **global warming**.

The greenhouse effect

Normally the Earth radiates back much of the heat energy it absorbs from the Sun. This keeps the temperature at the surface acceptable for life. Now carbon dioxide and methane are building up in the atmosphere. They are acting like a greenhouse around the Earth. The greenhouse gases absorb much of the energy which is radiated away. It can't escape out into space. As a result, the Earth and its surrounding atmosphere are warmer than they should otherwise be.

The effect is to raise the temperature of the Earth's surface. The change is very small, about 0.06°C every ten years at the moment. Not much – but an increase of only a few degrees Celsius could cause quite large changes in the Earth's climate.

Many scientists think that an increase in severe and unpredictable weather will be one of the changes we see due to global warming. Some people think the very high winds and extensive flooding seen around the world in the 21st century are early examples of the effects of global warming.

If the Earth warms up, the ice caps at the north and south poles will melt. This will cause sea levels to rise. There is evidence that this is already happening. It will mean more flooding for low-lying shores of all countries all over the world. Eventually parts of countries, or even whole countries, will disappear beneath the seas.

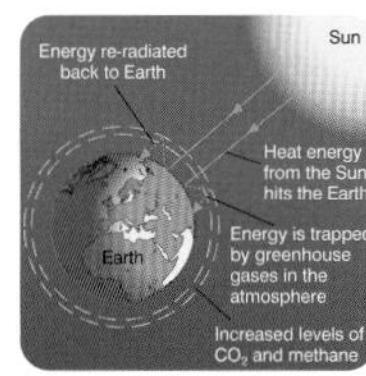

Figure 1 Many scientists believe that this simple warming effect could, if it is not controlled, change life on Earth, as we know it

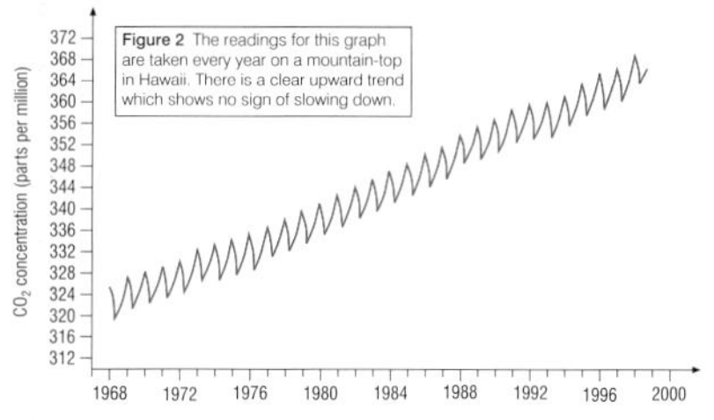

Figure 2 The readings for this graph are taken every year on a mountain-top in Hawaii. There is a clear upward trend which shows no sign of slowing down.

a) Name two greenhouse gases.

The effects of combustion

Carbon dioxide is made when we burn fossil fuels in cars, in our homes and in power stations. The number of cars and power stations around the world is steadily increasing. And respiration by all the living organisms on Earth produces carbon dioxide as well! So carbon dioxide levels are rising.

GET IT RIGHT!

Respiration and combustion **produce** carbon dioxide, photosynthesis **removes** it from the atmosphere. Greenhouse gases in the atmosphere **re-radiate** heat energy back to the surface of the Earth.

124

Evolution and environment

The effects of deforestation

All around the world large-scale deforestation is taking place. We are cutting down trees over vast areas of land for timber and to clear the land for farming. The trees are felled and burned in what is known as 'slash-and-burn' farming. The land produced is only fertile for a short time, after which more forest is destroyed. No trees are planted to replace those cut down.

Deforestation increases the amount of carbon dioxide released into the atmosphere. Burning the trees leads to an increase in carbon dioxide levels from combustion. The dead vegetation left behind decays. It is attacked by decomposing microorganisms, which releases more carbon dioxide.

Normally trees and other plants use carbon dioxide in photosynthesis. They take it from the air and it gets locked up in plant material like wood for years. So when we destroy trees we lose a vital carbon dioxide 'sink'. Dead trees don't take carbon dioxide out of the atmosphere.

For millions of years the levels of carbon dioxide released by living things into the atmosphere have been matched by the plants taking it out and the gas dissolving in the seas. As a result the levels in the air stayed about the same from year to year. But now the amount of carbon dioxide produced is increasing fast as a result of human activities. This speed means that the natural sinks cannot cope, and so the levels of carbon dioxide are building up.

b) What is deforestation?

Cows, rice and methane

Methane levels are rising too. It has two major sources. As rice grows in swampy conditions, known as paddy fields, methane is released. As the population of the world has grown so has the farming of rice, the staple diet of many countries.

The other source of methane is cattle. Cows produce methane during their digestive processes and release it at regular intervals.

In recent years the number of cattle raised to produce cheap meat for fast food, such as burgers, has grown enormously. So the levels of methane are rising. Many of these cattle are raised on farms produced by deforestation.

c) Where does the methane that is building up in the atmosphere come from?

Figure 3 Tropical rainforests are being destroyed at an alarming rate to supply the developed world with goods like mahogany toilet seats and cheap burgers

FOUL FACTS

When we lose forests, we lose biodiversity. Lots of plants and animals die out. We may well be destroying a source of new medicines or food for the future!

SUMMARY QUESTIONS

1. Define the following terms: **global warming; greenhouse gases; deforestation; a carbon sink.**
2. a) Why are the numbers of i) rice fields, and ii) cattle in the world increasing?
   b) Why is this a cause for concern?
3. Give three reasons why deforestation increases the amount of greenhouse gases in the atmosphere.
4. a) Use the data in Figure 2 to produce a bar chart showing the maximum recorded level of carbon dioxide in the atmosphere every tenth year from 1970 to the year 2000.
   b) Explain the trend you can see on your chart.
   c) Explain the greenhouse effect. How might it affect the conditions on Earth?

KEY POINTS

1. There is large-scale deforestation in tropical areas.
2. Large-scale deforestation has led to an increase of carbon dioxide into the atmosphere (from burning and the actions of microorganisms). It has also reduced the rate at which carbon dioxide is removed by plants.
3. More rice fields and cattle have led to increased levels of methane in the atmosphere.
4. Increased levels of the greenhouse gases carbon dioxide and methane may be causing global warming as a result of the greenhouse effect.

125

SUMMARY ANSWERS

1. Global warming – the increase in the average temperature of the Earth.
   Greenhouse gases – gases that absorb heat energy given off from the Earth.
   Deforestation – the cutting down of large areas of forest and burning or removing the trees without replanting.
   Carbon sink – something that removes carbon dioxide from the atmosphere, e.g. a forest.
2. **a) i)** Rice fields are increasing because there is a population explosion; rice is the staple diet for many people, so more rice is needed.
   ii) Cattle are increasing because there is a demand for cheap beef to supply burgers.
   b) Both lead to increased levels of methane in the atmosphere. Methane is a greenhouse gas.
3. Burning trees when cut down; microorganisms rotting dead plant material release carbon dioxide; loss of a major carbon sink reduces the removal of carbon dioxide.
4. **a)** Look for accurately drawn bar chart from the figures given, correctly labelled axes, neat columns etc.
   b) Levels of carbon dioxide have risen steadily since 1970 as a result of deforestation, burning fossil fuels etc.
   c) The Sun's energy heats the Earth but as it gives off heat it is re-radiated (absorbed) by greenhouse gases so it can't escape. Therefore the Earth's temperature rises. This could result in climate change such as more extreme weather events. Melting of polar ice caps will also increase sea levels and flood low-lying land.

ACTIVITY & EXTENSION

Some research could be carried out as to why cows produce so much methane. What is it in the digestive processes that results in this gas? Ask: 'Are they the only animals to produce this gas?'

KEY POINTS

The 'Key terms crossword' and the 'Recap list' in the suggestions for Plenaries will reinforce the learning of the key points about global warming.

B1b 8.4 Sustainable development

LEARNING OBJECTIVES

Students should learn that:

- Sustainable development is improving the quality of life by using natural resources wisely.
- Conservation of natural resources involves recycling waste materials and making wise use of energy.

LEARNING OUTCOMES

Most students should be able to:

- Explain the meaning of sustainable development.
- Describe ways in which families can help to conserve resources.

Some students should also be able to:

- Weigh evidence and form balanced judgements about some of the major environmental issues facing society, including the importance of sustainable development.

Teaching suggestions

- **Special needs.** Students can draw out stickers to go above light switches to remind people to turn off lights. These can be printed and distributed throughout the school.
- **Gifted and talented.** Suggest to students that they investigate how installing a solar panel on the roof of the school could help to cut the cost of heating. They would need to find out about the panels, the cost of installation and how efficient they are. Suggest the Centre for Alternative Technology web site as a reference.
- **Learning styles**

  Interpersonal: Discussing and gathering together information about how we can protect the environment.

  Intrapersonal: Writing a slogan.

  Kinaesthetic: Practical work.

  Visual: Learning to recognise lichens from the pictures.

  Auditory: Discussing the slogan.

SPECIFICATION LINK-UP B1b.11.8

- *Improving the quality of life without compromising future generations is known as 'sustainable development'. Planning is needed at local, regional and global levels to manage sustainability.*

Lesson structure

STARTER

Where does this come from? – Look at a piece of wooden furniture, a hardwood toilet seat or some advertisements for furniture. Ask: 'What kind of wood is it? Where did it come from? Where was the wood grown?' This leads into a discussion of sustainable woodland and what it means. (10 minutes)

'Think globally, act locally' – Ask: 'What does this slogan mean?' Students to discuss in small groups and then share thoughts with the rest of the class. (5–10 minutes)

Our environment – Search the web image banks for pictures of an attractive natural scene (try to include some lichen!) and contrast with a squalid litter-strewn slum or a similar picture. Ask: 'How can we affect our environment for future generations?' A PhotoPLUS, B1b 8.4 'Sustainable development' is available on the GCSE Biology CD ROM. Students to make a quick list of five bullet points each. Discuss and make a class list. (10 minutes)

MAIN

- **Sustainable energy for sustainable development?** – Discuss different sources of energy, using both renewable and non-renewable sources. Include reference to wave power, wind power and solar energy, as well as to the more conventional forms and also to the consequences of not producing enough energy. Provide students with work sheets to fill in and allow time for discussion. There are a number of excellent web sites providing information, including EarthTrends (the web site of the World Resources Institute), the dti (Department for Trade and Industry) and the World Nuclear Association. It could be useful to include some facts and figures about what proportion of the country's needs could be supplied by the different sources.
- Students could work in groups and find out more about the renewable resources. How much energy can be supplied? How reliable are they? Each group to present their findings to the rest of the class. This activity could be spread over more than one lesson in order for students to gather information. It links in with the other spreads in this chapter.
- **What is in our litter bins?** – Students are to examine the contents of the litter bins around the school (set up if necessary). Carry out a risk assessment. They find out: 'What is present in the bins? What could be recycled?' and investigate county schemes for recycling. Use the information to set up a recycling scheme in the school. If there is already one in operation, ask 'Does it work? Are the bins in the right places?'
- Set up a worm digester for fruit peel and a compost heap for all suitable organic waste in the school grounds. Discuss the limitations and problems. Information from the Henry Doubleday Foundation, Ryton Gardens, Warwickshire, could be helpful.

PLENARIES

Design a slogan – Students design a slogan for a campaign to get the school to use energy more carefully. (5–10 minutes)

What can we do? – In small groups, students discuss the question and compile a list of suggestions. Ask: 'Think about the slogan given in the starter: does anything we do have a global impact?' They then compare ideas with other groups. (10 minutes)

ACTIVITY & EXTENSION IDEAS

- Carry out a practical to demonstrate the effects of insulation. Place various thicknesses of insulating material around containers of hot water. Predictions can be made about the temperature drop over time. Measurements can be taken, graphs plotted.
- Alternatively, use pairs of digital thermometers on either side of sheets of ordinary glass and double glazed panels. If a heat source is arranged on one side of each, the heat transfer can be measured. Discuss the significance of this in terms of insulation and the conservation of energy.
- Organise a 'Get to school on public transport' day. This could involve a poster campaign, an assembly, provision of bus etc. timetables. Discuss how the use of public transport can help to conserve resources.
- **The great nuclear debate** – Should more of our energy come from nuclear power stations? This is a question which the Government will have to face as the existing power stations are decommissioned. Imagine you are in Parliament, there is a free debate: ask students to voice their opinions about this issue. (Bear in mind that students will study this in unit P1a.)

Answers to in-text question

a) Sustainable development improves the quality of our lives without risking the future generations to come.

B1b 8.4 Sustainable development

LEARNING OBJECTIVES

1. What do we mean by sustainable development?
2. How can families help to conserve resources?

As our world gets more and more crowded, we are becoming increasingly aware of the need for **sustainable development**. This combines human progress and environmental stability. It improves the quality of our lives without risking the future of generations to come.

Sustainable development

Sustainable development means looking after the environment. We need to conserve natural resources. So, for example, farmers need to look after the land. They can plough the remains of crop plants into the soil, and use animal waste instead of chemical fertilisers. They can also replant hedgerows to prevent soil erosion and avoid deforestation. These will help to make sure that growing crops will be possible for years to come.

We can see another example of sustainable development in our woodlands. We use an enormous amount of wood and paper, but we have fewer trees than almost all of our European neighbours.

However, over the last 80 years or so the Forestry Commission has developed sustainable commercial woodlands. Felling can only take place as long as replanting replaces the felled trees. Now farmed woodlands not only provide a sustainable resource but also a rich environment for a wide variety of species.

Sustainable development has to be a global idea – it is no good if it only happens in the UK. Everywhere that deforestation occurs, replanting needs to take place.

GET IT RIGHT!
Be clear about the meaning of the term 'sustainable'. Be able to give examples of how families can help conserve natural resources.

Figure 1 Sustainable woodlands have become an important and attractive part of sustainable development in the UK

Another example of the importance of sustainable development is in the management of our fishing stocks. In many areas we have taken so many fish from the sea that the populations can no longer replace themselves. The numbers of fish like cod are dropping fast.

To save the fish stocks, the numbers of fish caught ***must*** be reduced. But it needs agreement by fishermen everywhere for this sustainable development of the sea to become a reality.

a) What is sustainable development?

Conserving resources

An important part of sustainable development is using natural resources wisely. We must use only what we need, and conserve natural resources as much as possible. You and your family can help to do this in lots of different ways.

We live in a throw-away society. We use something – and then put it in the bin. This uses up resources and means land is wasted under rubbish tips. But if we recycle our waste, we save resources, use land wisely and use less energy. You can recycle your old newspapers (saving trees), glass bottles (saving energy) and aluminium cans (saving aluminium ore and energy!).

Figure 2 Our throw-away society causes problems in lots of ways. Not only do we waste resources, but a tip like this uses lots of land, and pollutes the area all around it.

Another way we can help is to make our homes more energy-efficient. Energy is one of our most important resources. Unfortunately, using electricity, gas or oil in our homes uses up some of these valuable resources. It adds to the carbon dioxide in the atmosphere as well, so the less we use the better.

We use huge amounts of energy heating our homes – and then lose it through the windows, roof and gaps around the doors. Making our homes energy-efficient helps save resources and prevent global warming. By insulating your roof spaces and the walls of your homes, and having double glazing, you will save a lot of energy. Energy-efficient boilers make the best use of your fuel. Switching things off when you have finished using them helps as well!

Finally we can look at our transport. Use your car less, and walk or cycle to places nearby. This will help you save petrol (a non-renewable resource) and also avoid adding more carbon dioxide to the atmosphere. Public transport can help as well, carrying lots of people at the same time. It may not be as convenient as using your own car, but it is certainly better for the environment.

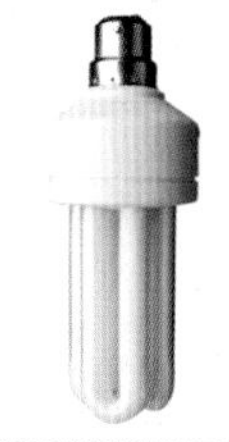

Figure 3 Just changing to energy-efficient light bulbs like these will help to conserve resources and reduce carbon dioxide levels

SUMMARY QUESTIONS

1. Copy and complete using the words below:

   cars conserving energy efficient recycle resources sustainable

   An important part of ...... development is using natural ...... wisely. This means using only what we need, and ...... natural resources as much as possible. We can ...... waste, make our homes more .... and make less use of our ....
2. List as many ways as possible in which you and your family could help to conserve natural resources.
3. Choose one aspect of sustainable development and produce a report on how it works and why it is so important.

KEY POINTS

1. Improving your quality of life without compromising future generations is known as sustainable development.
2. Sustainable development involves using natural resources wisely.
3. We can help by recycling, making our homes energy-efficient and avoiding using our cars when possible.

SUMMARY ANSWERS

1. Sustainable, resources, conserving, recycle, energy efficient, cars.
2. List of methods to conserve resources and reduce energy consumption.
3. Student report.

KEY POINTS

Each key point can be used as a starter for writing a short explanatory paragraph, using examples, thus reinforcing the general statements.

B1b 8.5 Planning for the future

LEARNING OBJECTIVES

Students should learn that:

- Living organisms can be used to detect pollution.
- Sustainable development can only be achieved by planning.
- The needs of people and the environment need to be balanced.

LEARNING OUTCOMES

Most students should be able to:

- Describe how living organisms can indicate levels of pollution.
- Explain the difference between building on green field and brown field sites.

Some students should also be able to:

- Evaluate methods of collecting environmental data and consider their reliability and value as evidence.

Teaching suggestions

- **Special needs.** These students could make a poster showing what lichens can be found locally and where they grow.
- **Gifted and talented.** Use the web site suggestion made on page 126 and find out how many wind turbines would be needed to supply the school's energy needs. Where would they be sited?
- **Learning styles**

  Interpersonal: Writing letter to local paper.

  Intrapersonal: Collaborating in group projects.

  Kinaesthetic: Practical work on lichens.

  Visual: Using keys to identify lichens.

  Auditory: Expressing opinions in Not in my back yard!
- **Homework.** There are many opportunities for Homework tasks here. For example, writing the letter to the local paper. Research on the local planning applications could involve collecting local newspapers, visiting the Library to read the Council minutes etc.

SPECIFICATION LINK-UP B1b.11.8

- *Living organisms can be used as indicators of pollution:*
  - *lichens can be used as air pollution indicators*
  - *invertebrate animals can be used as water pollution indicators.*
- *Planning is needed at local, regional and global levels to manage sustainability.*

Students should use their skills, knowledge and understanding of 'How Science Works':

- *to evaluate methods used to collect environmental data and consider their validity and reliability as evidence for environmental change.*

Lesson structure

STARTER

What is a lichen? – Collect some examples of lichens or project pictures on to the board. Ask students if they have ever seen any plants similar to the ones shown. If so, where did they see them? Lead into a discussion of where lichens grow and what type of plants they are. Most students will not know that a lichen is a combination of two plants. (10 minutes)

The importance of clean air and clean water – Setting a time limit of 3 minutes, ask students to write down as many reasons as they can think of as to why we need clean air and clean water. Go around the class collecting examples from the students and build up a list on the board. (5–10 minutes)

Brown field sites and green field sites in your area – Discuss the difference in the meaning of the terms. What are the hazards of building on brown field sites? What are the advantages? (10 minutes)

MAIN

- **Lichens as pollution monitors** – Prepare a PowerPoint® presentation on different lichens. Explain what a lichen is and that some lichens can be used to indicate levels of air pollution, because they are sensitive to sulfur dioxide in the atmosphere. Select a couple of species that are around in your area and give the students sheets with these on, to help them with identification. The yellow encrusting *Xanthoria* is quite common and tolerant, but the leafier, foliose types are more sensitive. A survey of the buildings, walls and trees in the school grounds or a convenient location can be carried out.
- The survey could just register presence or absence of the lichens, or estimates could be made of the proportion of the different types – it very much depends on your location. An area of woodland could be contrasted with a suburban street, or a cemetery (look at the gravestones) with another more populated area. Tree trunks are good places to look, so a comparison of the lichens on trees in a built-up area with trees in a woodland would yield some differences. 'How Science Works' skills related to field work investigations can be covered here.
- **Not in my back yard!** – Working in groups, the students could be given building proposals (e.g. a new runway for an airport, a major superstore, a new motorway or any project which may be of local interest) and asked to prepare a case for and a case against one of their choices. They should consider the local environment, the people affected and the economics. Choose some groups to put their cases to the rest of the class and have a vote to see which side wins their argument.

PLENARIES

Supporting the environment – Ask students to compile a list of organisations which they think help to protect the environment. Beside each one, indicate what they do to preserve sites of interest and the plants and animals. For example, the RSPB protects birds and their habitats. Do these organisations have a significant impact? (5–10 minutes)

Ecotourism – what is it and how can it help the environment? – Ask students if they have heard this term. Have any of them been to the Galapagos islands, the Everglades, diving in the Great Barrier Reef or on a visit to a National Park in this country? If they have not, ask them what they think it means. Lead into a discussion of whether or not it is a good idea to protect interesting natural ecosystems by restricting visitor access to the sites. (10 minutes)

ACTIVITY & EXTENSION IDEAS

- **The Planning Committee of your local council** – Every local council has a Planning Committee and its meetings are recorded. Find out how planning permissions are granted (or not). Are the reasons given sound, environmental ones? The local press usually publishes lists of requests for planning permission, so it would be possible to follow a particular request and find out whether permission is granted or not.
- **Find out how you can make your opinions known** – All requests for planning permission have to be made public so that people who might be affected can express an opinion.
- **My garden is an SSSI!** – Ask students to imagine that they found a rare plant or animal (perhaps a species of insect or an unusual bird) in their back garden. Write a letter to the local paper explaining this find and asking for the public to protest about the proposal to cut down a group of trees and build a new block of flats at the end of the road.

Answers to in-text questions

a) Because technology requires a lot of energy.

b) Plants or animals sensitive to varying degrees of pollution can be monitored.

B1b 8.5 Planning for the future

LEARNING OBJECTIVES

1 How do we plan for the future?
2 Can we balance the needs of people and the environment?

Most of the population of the world lives in the developing countries. Their way of life has not changed much over the centuries. There are many people, but each one uses few resources.

There are relatively few people in the developed world. However, those people are surrounded by technology, almost all of which uses a great deal of energy. Planning is needed at local, regional and global levels to make sure that the resources of the world are used in a fair and sustainable way.

a) Why does the developed world use so many more resources than the developing world?

PRACTICAL

Pollution indicators

Have a look at your local area and see how many different types of lichens you can find. How clean do you think your air is?

- Is the information we gain from lichens reliable and valid evidence of pollution?
- Compare them to using data-logging equipment to monitor pollution.

Figure 1 Lichens grow well where the air is clean. In a polluted area there would be far fewer species of lichen growing. This is why they are useful bioindicators.

Figure 2 This water looks clean and inviting – a look at some pollution indicator species would tell us if it is as clean as it looks!

Making local planning decisions

All around the UK people want to build homes, shopping centres, roads and factories. Before any building can take place, you have to apply for planning permission. For sustainable development to work, we need to consider the environment every time a planning decision is made.

To help us make the right decision we need information from field studies, carried out by scientists. They can give us information about the environment at the site of a proposed development. They can also show us the impact of a similar development elsewhere. One important tool we have in these field studies is to use living organisms as indicators of pollution.

Lichens grow on places like rocks, roofs and the bark of trees. They are very sensitive to air pollution. When the air is clean, scientists will find many different types of lichen growing. The more polluted the air, the fewer lichen species there will be. So a field check on lichen levels will give us the information we need about possible air pollution from a factory or road.

In the same way we can use invertebrate animals as water pollution indicators. The cleaner the water, the more species you will find. There are some species which are only found in the cleanest water. Others can be found even in very polluted waters.

By using living organisms as pollution indicators in this way we can reach sensible decisions about the impact a new development might have on the environment. This helps us make planning decisions which allow for the sustainable development of an area.

b) How can we use living organisms as indicators of pollution?

Brown field or green field?

As the UK population grows we need more houses. We can build them on 'green field sites' or 'brown field sites'.

Green field sites are countryside which has not been built on before. Brown field sites are usually within towns and cities, and have already been used. They are often the sites of old industrial buildings, factories, petrol stations or even rubbish tips.

Using brown field sites has many advantages for building. Water, sewage, electricity, gas and transport systems are usually already in place or easily available. No farmland or countryside is spoilt or lost.

However, brown field sites are often contaminated with chemicals and can suffer from subsidence. It is often expensive to use brown field sites because they have to be cleaned up (decontaminated) before they can be used. What's more, they can be a valuable environment in their own right.

Although it is a good idea to build on brown field sites when it is possible, the decision is not always an easy one!

Figure 3 People will soon be living in these homes – yet only a short time ago this brown field site was an old bus station

Protecting SSSIs

Around the country there are many areas which are ***SSSIs*** (***Sites of Special Scientific Interest***). These sites may have a particularly interesting or unique landscape. They may be home to rare species of plants or animals. They may be important in bird migration or as a breeding ground.

Environmentalists believe it is very important that these SSSIs are not disturbed and – most importantly – not built on. When planning decisions are made, the environmental importance of an area needs to be considered carefully.

The idea of protected areas is common now in Europe, and is spreading around the world. In terms of global planning, it is important that governments everywhere take their environmental responsibilities seriously. That is the only way we can continue to develop while sustaining the variety of life around us – and protecting the Earth from the worst effects of pollution and waste.

Figure 4 Rare plants such as these Spring Squill flowers often grow in SSSIs

DID YOU KNOW?

Protecting the environment all over the world is vital if we are to look after the variety of living things – the biodiversity – we now enjoy. We just don't know which ones will be useful in the future!

SUMMARY QUESTIONS

1 Define the following terms: pollution indicator; brown field site; green field site; SSSI.

2 Think about your local area. Choose a brown field site which you think could be used for building. Write a proposal which puts forward your plans, explaining why you have chosen the site rather than a local farmer's field.

3 a) Here is some information about four different species of living things. Explain how you would use each one as an indicator to help make planning decisions about sustainable development.
   i) Salmon are fish which are only found in clean water with lots of oxygen in it.
   ii) Lichens are very sensitive to air pollution.
   iii) Lizards are badly affected by pesticides in their insect prey.
   iv) Blood worms can survive in very polluted water.
   b) Evaluate different methods you could use to collect environmental data. Comment on their reliability and usefulness as sources of evidence.

KEY POINTS

1 We can use information gained from field studies to help us make planning decisions about sustainable development. One type of information is to use living organisms as indicators of pollution levels.
2 Brown field sites are thought to be more suitable than green field sites for new buildings.
3 Environmentalists believe it is important not to build on Sites of Special Scientific Interest (SSSIs).

SUMMARY ANSWERS

1 Pollution indicator – organism that is sensitive to effects of pollution.
Brown field site – former industrial/built up areas/sites that have been used before.
Green field site – countryside that has not been built on before.
SSSI – Sites of Special Scientific Interest which are protected against development.

2 Student proposal.

3 a) i) Salmon could be used as an indicator of very clean water and therefore any environmental effects of development near the river have been minimal.

ii) The numbers of different types of lichen can be used to monitor air quality at varying distances from a development.

iii) Any decrease in numbers of lizards could be an indicator of excess pesticides being used in an area.

iv) If only blood worms survive in a river there is significant pollution.

b) Answer should cover advantages and disadvantages of using living organisms, taking samples for analysis and continuous monitoring using sensors and data-logging equipment.

KEY POINTS

These can be reinforced by adding specific examples to the general statements made.

B1b 8.6 Environmental issues

SPECIFICATION LINK-UP

B1b.11.8

Students should use their skills, knowledge and understanding of 'How Science Works':

- *to analyse and interpret scientific data concerning environmental issues*
- *to weigh evidence and form balanced judgements about some of the major environmental issues facing society, including the importance of sustainable development.*

BIOLOGY HOW PEOPLE AFFECT THE PLANET

B1b 8.6 Environmental issues

How can we be sure?

The build-up of greenhouse gases cannot be denied, because there is hard evidence for it. However, there is still debate among scientists about whether these really cause the problems which are blamed on the 'greenhouse effect'. The majority of scientists now believe that global warming is at least partly linked to human activities such as burning of fossil fuels and deforestation. But not everyone agrees.

Some extreme weather patterns have certainly been recorded in recent years. But go back in history and it is possible to find other equally violent periods of weather long before fossil fuels were used so heavily and deforestation was happening.

The temperature of the Earth has varied greatly over millions of years. We have had both Ice Ages and times when almost the entire Earth was covered in tropical vegetation, all long before humans evolved. So some scientists argue that what we are seeing is the result of natural changes rather than a direct result of human activities.

GREENHOUSE EFFECT BLAMED FOR UNEXPECTED GALES

Environment

LOW LYING COUNTRIES SET TO DISAPPEAR AS SEA LEVELS RISE!

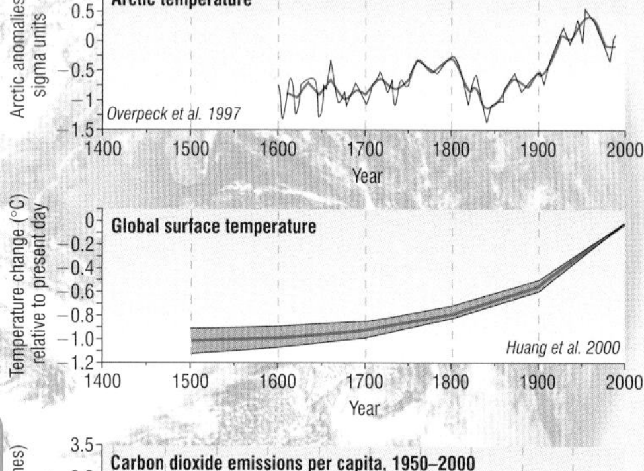

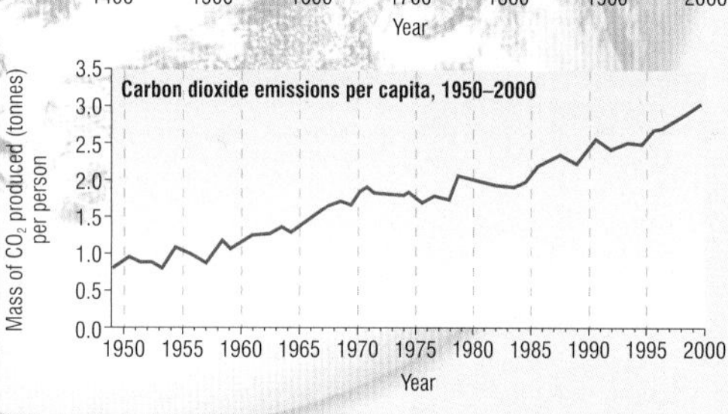

ACTIVITY

You are going to present a 3–4 minute slot on global warming for a news programme for young people. You need to explain the different scientific views and why it is so hard to be certain what is happening. You also need to present some of the scientific evidence and plan some informative and attention-grabbing graphics. Make sure you present a balanced picture, with plenty of facts – but make it interesting too or they'll all change channels!

Scientists suggest that flooding is due to global warming

Teaching suggestions

Many of the suggestions given on the previous spreads are relevant to the issues and activities given in this spread, so those not used before could be considered here to support the activities. Some additional ideas are given below.

- **How can we be sure?** – This is all about evidence and patterns of weather. The BBC News web site has facts and figures about weather patterns, carbon dioxide levels, temperatures etc., which can be useful for discussion if you have not already used them. The graphs present information and measurements over a considerable time and not just since 'records began'. (Search www.bbc.co.uk.)
- In the suggested related activity, which is to present a slot in a news programme, there is a vague indication that this is for a TV programme, rather than a radio programme. Students could consider using animations of the effect of global warming, interview a scientist who agrees and one who does not. If your school has the facilities, it would be fun for a group of students to work together and film the slot to present at an assembly or in class.
- **What can I do?** – A class discussion, stimulated by the images and captions in the Student Book spread, should result in some kind of agreement that there are areas where individuals can have some effect and other areas where they cannot. To get a discussion going, some of the following questions could be asked:
  - Can we stop deforestation by refusing to buy mahogany loo seats?
  - How do we know that the meat in cheap burgers comes from areas that have been deforested?
  - Why should we buy more expensive meat if there is cheaper meat available?
  - Where am I going to earn enough money if I cannot get employment at the logging camp?

Once some concensus of opinion has been reached, the discussion could be extended to one particular area, such as carbon dioxide emissions, energy resources etc. There are so many issues here that you could either focus on one or two, or identify a larger number and allow students to work together in groups. The suggested activity lends itself to group work, as each group can be responsible for a web page.

Evolution and environment

What can I do?

What people do in their lives really does affect our planet. More and more evidence shows that we cannot go on using the resources of the Earth in the way we do. But there are lots of different ways of looking at the problems. Here are just a few of them.

Many people are now trying to control and reduce the amount of greenhouse gases we produce. The problem is that the whole world has to agree because the Earth only has one atmosphere and this is affected by things which happen everywhere in the world. There are enormous problems with this.

One easy way to stop deforestation is to stop buying what they produce. They only cut the forests down to provide timber and beef. Don't buy mahogany loo seats, and don't eat cheap burgers! If we all stopped buying, they'd soon stop cutting down the rain forests.

It's the wealthy developed nations who have to take responsibility for most of the carbon dioxide emissions. We need to persuade our citizens to cut back a lot on using their cars. They won't do it because they're afraid if would affect their jobs and industry. I mean, America wouldn't even sign up to the Kyoto agreement, and they produce more pollution than anyone else!

I've got so little money, I need to buy things as cheaply as I can. I don't really care where the beef comes from – I can feed my kids really cheaply and give them meat nearly every day if we eat burgers and things. We're never going to see a rainforest, are we?

I just want to earn a living to support my family. We have no proper schools, hospitals and roads. I must work at the logging camp – it's the only work for miles around and I want my children to have an education.

It looks a bit different from where I live. If we can't expand our industries, what chance do we have of getting richer? We need to make more money to improve the life and health of our people. Why should we have to sacrifice our development to help solve a problem caused by other, much richer nations?

There are a lot of us who are trying to monitor and support a more responsible use of the forests. The trouble is, it's very difficult to enforce guidelines, and there are always people ready to make money fast and illegally.

ACTIVITY

Problems like greenhouse emissions, global warming and the use of non-renewable resources can seem overwhelming. But each one of us can make a difference in the choices we make in our everyday lives.
You are going to develop some web pages to be used by your school to help everyone recognise some of the problems there are. More importantly, you are going to suggest ways in which they can help to conserve resources and change attitudes.
Design a flyer to be handed out at registration time that will give everyone the web address of your site and make them want to visit it.
Think of some targets you can set for your school, or ways in which students can set targets for their own families. Make a web site – and make a difference!

131

- The activity can be centred around the school environment, and could involve:
  - recycling paper, cans and bottles
  - conserving energy (turning off lights, closing doors)
  - thinking about how you get to school.
- If the activity is extended to homes and families, then some more issues could be included:
  - re-using plastic bags
  - disposal of household items such as old refrigerators and scrap metal
  - proper insulation of houses to save energy
  - using energy-efficient light bulbs
  - considering installing solar panels
  - using public transport more.

- **Role play** – The people used in the student book photographs could be paired up and students asked to prepare speeches or a dialogue describing the problems from opposing viewpoints. For example, the Greenpeace campaigner could be matched with the poor mum. A conversation between the two people could be quite interesting! The worker at the logging camp and the woman advising against buying mahogany and eating cheap burgers might go together, as well.
- **The big nappy debate** – Disposable nappies account for a considerable proportion of the waste that the local councils have to dispose of in landfill sites. Organise a debate on the subject, getting students to put arguments for and against the use of disposable nappies.
- **Can plastic be made biodegradable?** – So much of our food nowadays comes wrapped in paper and plastic. (If you are going to have a lesson on this topic, it could be interesting to show some examples: a packet of cheese biscuits with three layers of wrapping; pre-packaged vegetables; a microwaveable meal.) Ask: 'Why is this? Is it necessary? What can we do about it?' Some years ago, there was a move to introduce plastic bags that would break down if they were buried in the ground. What has happened to this idea? A discussion on this topic reveals some of the attitudes and changes to our way of life that have taken place in the last few decades.

SUMMARY ANSWERS

1 **a)** Building houses, shops, industrial sites and roads, for farming, for waste disposal etc.

b) People need places to live, to buy food, to produce things they need, transport systems to move people and goods, to grow food, to get rid of waste and rubbish.

c) Any two sensible suggestions, e.g. recycle rubbish, build fewer houses, more flats which use up less land, use public transport so fewer roads needed etc.

2 **a)** Cars and factories produce sulfur dioxide and nitrogen oxides → up into the atmosphere → carried along by winds → dissolve in rain → fall as acid rain on land and in water.

b) i) 11 parts per billion **ii)** As appropriate. **iii)** 2.5 ppb

c) There are strict rules about the levels of sulfur dioxide and nitrogen oxides in the exhaust fumes of new cars. Cleaner, low-sulfur fuels are used in cars and power stations. More electricity is generated from gas and nuclear power. Systems are used in power station chimneys that clean the flue gases.

3 **a)** In summer months, plants photosynthesising a lot and growing fast take a lot of carbon dioxide out of the atmosphere. In winter, much less photosynthesis takes place, so carbon dioxide levels rise.

b) We can see the difference it makes when plants are actively photosynthesising in the summer, so we need plants there. If plants were not there to photosynthesise, imagine how carbon dioxide levels would build up, so it is vital to prevent deforestation.

c) More people: more cars and factories means more CO_2; more deforestation means less uptake of CO_2; more paddy fields to grow rice to feed people means more methane; more cows to produce cheap beef also producing methane. These greenhouse gases could change the Earth's climate producing more extreme weather events and altering rainfall and temperatures in different ways in different parts of the world.

4 **a)** Maintains the fertility of the soil; helps prevent soil erosion; keeps soil available for growing crops.

b) Re-uses aluminium, a non-renewable resource, and uses less energy than getting the metal from the ore.

c) This makes sure the water is not getting polluted by the sewage and that the river can cope and absorb the treated sewage.

d) This reduces the number of cars on the road, saving petrol, a non-renewable resource, and reducing pollution.

e) This makes sure houses are energy-efficient, which prevents wasting energy and reduces the use of non-renewable fuels and reduces pollution levels.

HOW PEOPLE AFFECT THE PLANET: B1b 8.1 – B1b 8.6

SUMMARY QUESTIONS

1 a) List the main ways in which humans reduce the amount of land available for other living things.
b) Explain why each of these land uses is necessary.
c) Suggest ways in which two of these different types of land use might be reduced.

2 a) Draw a flow diagram showing acid rain formation.
b) Look at Figure 4, on page 123.
i) What was the level of sulfur dioxide in the air in the UK in 1980?
ii) What was the approximate level of sulfur dioxide in the air in the UK in the year that you were born? (Make sure you give your birth year.)
iii) What was the level of sulfur dioxide in the air in 2001?
c) Explain how the levels of sulfur dioxide have been reduced in the UK since 1978.

3 In Figure 2, on page 124 you can see clearly annual variations in the levels of carbon dioxide recorded each year. These fluctuations are thought to be due to seasonal changes in the way plants are growing and photosynthesising through the year.
a) Explain how changes in plant growth and rate of photosynthesis might affect carbon dioxide levels.
b) Explain how you could use this as evidence to try and prevent the loss of plant life by deforestation.
c) How is the ever-increasing human population affecting the build-up of greenhouse gases – and what effect are they having on world climate?

4 Explain carefully how the following measures can help to support sustainable development.
a) Encouraging farmers to plough the remains of crop plants into the soil, to use animal waste as well as chemical fertilisers and to replant hedgerows.
b) Recycling all the aluminium drinks cans you use.
c) Monitoring the types of invertebrates found in a river both above and below the sewage outfall.
d) Setting up a 'walking bus' for a primary school.
e) Enforcing standards for the building trade. These include the thickness of the insulation, the materials which can be used and the fittings (e.g. boilers, light bulbs) to be put in new houses.

EXAM-STYLE QUESTIONS

1 The diagram shows a town and some of its surroundi

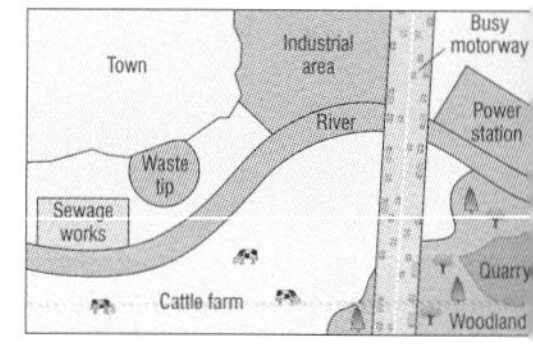

(a) A site that reduces the level of carbon dioxide in atmosphere is:
A Quarry B Waste tip
C Woodland D Town

(b) Methane is another greenhouse gas. The site tha produces the most methane is:
A Industrial area
B Cattle farm
C Power station
D Woodland

(c) Three sites produce gases that contribute to acid rain. These are:
A Town, industrial area, quarry
B Industrial area, busy motorway, power station
C Busy motorway, power station, waste tip
D Town, quarry, waste tip

(d) The two sites most likely to be damaged by acid falling on them are:
A Cattle farm and quarry
B Quarry and woodland
C Woodland and river
D River and waste tip

(e) Air pollution in the town is monitored continuou This is most likely to be done by:
A A large team of council technicians working sh
B A small team of elite scientists on call 24 hour day.
C Local people using biological indicators such lichen.
D Electronic sensors attached to data logging equipment.

- **When to use these questions?**
  - The three parts of question 1 could be used to build up a table of information about land use. One column could be causes of reduction in amount of land available, the second column for reasons and the third for ways in which the types of land use could be reduced. This would provide a valuable summary of the topic. Some discussion in class could come up with some other ideas.
  - Similarly question 3 can be used effectively to summarise fluctuations in CO_2 levels and emphasise the effects of deforestation.
  - The flow diagram in answer to question 2a) is a good revision resource.
  - Question 4 is a good question to set for homework after studying the topic of sustainable development.

Summary teaching suggestions

- **Special needs**

  For question 2a), students could be given the stages in the formation of acid rain written on cards and then asked to put them in the correct order. Similarly with question 3c), on global warming, the stages could be provided and the students put them in order.

- **Gifted and talented**

  The figures on the graph for question 2 can be manipulated and students could be asked more questions about the changes. For example, they could work out percentage changes and link the cheaper prices of low-sulfur fuels and the use of systems to remove sulfur dioxide from flue gases with changes in the levels.

EXAM-STYLE ANSWERS

1 a) **C**

b) **B**

c) **B**

d) **C**

e) **D** *(1 mark for each*

2 a) Could contribute to/affect climate change, possibly producing severe and unpredictable weather. *(1 mark*
Could produce a rise in sea level due to thermal expansion of the oceans/melting of the polar ice, leading to flooding of low-lying land. *(1 mark*

b) (i) Paddy fields/growing of rice in swampy conditions. *(1 mark*
Cattle – methane is produced during digestion. *(1 mark*
(ii) Increase in the human population has led to a greater demand for food and therefore there are more paddy fields to grow rice (rice is the staple diet of a large proportion of the world's population)/more cattle to meet the demand for milk and burgers. *(2 marks*

Evolution and environment

In 1997 the World Climate Summit took place in Kyoto. Agreement was reached to control global warming by cutting emissions of greenhouse gases.

(a) List two possible consequences of global warming. (2)

(b) One of the most important greenhouse gases is methane.
   (i) What are the two main sources of methane? (2)
   (ii) Why has the amount of methane produced increased steadily over the past 200 years? (2)

(c) Another group of greenhouse gases are the oxides of nitrogen. State the main source of these gases. (1)

(d) State two ways in which deforestation may also contribute to global warming. (2)

Lichens are plants that do not grow well where there is air pollution. The number of lichen species growing along a 15 km line from the centre of a UK city was recorded. The results were plotted on a graph.

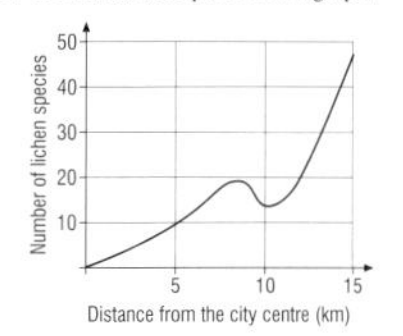

(a) (i) Name the independent variable shown above. (1)
   (ii) Is this variable categoric, ordered, discrete or continuous? (1)
   (iii) Is the dependent variable categoric, ordered, discrete or continuous? (1)

(b) Draw the headings for the table used to record the data for the graph. (1)

(c) How many species of lichen are found at a distance of 5 km from the city centre? (1)

(d) At what distance from the city centre is the least polluted air found? (1)

(e) What is the relationship between the number of lichen species and the distance from the city centre to a point 8 km from the centre? (1)

(f) Give a possible reason for the fall in the number of lichen species at a distance of 10 km from the city. (2)

HOW SCIENCE WORKS QUESTIONS

Kim needs some help to experiment on the effect of acid rain on some radish plants. She is not too certain of exactly what to do. She has the idea that if she set up some dishes with some soil and planted some radish seeds, she could then put different amounts of vinegar onto them to see what happens.

a) What advice would you give Kim as to how many different dishes she should use? (2)

b) How would you decide how many seeds to plant in each dish? (2)

c) Kim said that she wanted to keep the type of soil the same in each dish. Is this a good idea? Explain your answer. (1)

d) Kim was interested in the effects of acid rain. Do you think that vinegar was a good choice to simulate the effects of acid rain? Explain your answer. (2)

e) You could use universal indicator papers or a pH meter to measure the pH of the soil and the acid. Which would you choose . . . and why? (2)

f) Name the independent variable in Kim's investigation. (1)

g) Suggest to Kim what might be a suitable dependent variable. (1)

h) Where would you suggest Kim keeps the dishes? (2)

i) It's always a good idea to prepare a table before you start the investigation. Prepare a table for Kim to collect her results. (3)

For both marks students need to make the links: increase in human population – greater demand for food – more paddy fields/cattle. Any two of the three links earns one mark.

c) Vehicle/car exhausts. *(1 mark)*

d) • It increases the release of carbon dioxide into the atmosphere as a result of burning trees and their breakdown by microorganisms.
   • As trees absorb carbon dioxide during photosynthesis, their removal means less is taken up from the atmosphere. *(Either point – 1 mark)*

   This increases carbon dioxide concentration in the atmosphere and as carbon dioxide is a greenhouse gas this contributes to global warming. *(1 mark)*

3 a) i) Distance from the city centre. *(1 mark)*
   ii) Continuous. *(1 mark)*
   iii) Discrete. *(1 mark)*

b) *Either*

	Distance from city centre (km)				
Number of lichen species					

Or

Distance from city centre (km)	**Number of lichen species**

c) 10 *(1 mark)*

d) 15 km *(Units are required) (1 mark)*

e) The number of lichen species is directly proportional to the distance from the city centre. *(1 mark)*

f) This is likely to be the result of domestic smoke from a small town or industrial site at this point. *(1 mark)*

Exam teaching suggestions

All these questions require relatively short answers (no part of any question carries more than 2 marks). All questions therefore make useful revision exercises for use at the end of a teaching session or for homework. Questions 1 and 3 both include interpretation of pictorial/graphical information and so require some time and thought. Aspects of 'How Science Works' are covered in both these questions.

Question 2 illustrates the significance of numbering question parts. Candidates often fail to appreciate the importance of how questions are partitioned. Different questions have different themes/cover different areas of the specifications. Within a question, the main parts, e.g. a), b), c) etc., cover different aspects of the theme that are loosely connected. Sub-divisions of each part, e.g. a) (i) and a) (ii), are more closely connected. Often the second sub-division is used to move an issue on or deal with it in more depth. All this may seem obvious but students still fail to use the information to guide them in deciding what is required by a question. Take question 2b) sub-division (ii); labelled as it is means that it is closely connected with sub-division (i). The answer should therefore be related to this earlier sub-division – i.e. it should link rises in methane levels with paddy fields and cattle. However, as there are 2 marks available, this suggests that the answer also needs some explanation. A useful exercise with students not getting full marks on this part of the question is to guide them through the logic of: methane is created mostly by paddy fields and cattle – methane levels have increased – therefore there must be more paddy fields/cattle – why? – more food needed – why? – larger human population.

In question 2b), it is important that the answer says '**contributes** to climate change' or the equivalent. Answers should not say 'causes' climate change as this is a natural phenomenon that occurs cyclically in the absence of global warming.

Instead of the first point, a mark could be given for some consequence of climate change such as desertification, greater rainfall, intense storms or spread of tropical diseases.

HOW SCIENCE WORKS ANSWERS

a) She should use at least five different concentrations of acid and repeat each at least twice.

b) There should be as many as possible, without light, water, nutrients being in short supply.

c) Yes, because it is a control variable and acid might have different effects on different soils.

d) No. It doesn't normally rain vinegar! Poor validity.

e) The pH meter because it is more sensitive and it produces a continuous variable. The pH papers give an ordered or discrete variable.

f) Concentraton of vinegar added or pH.

g) E.g. height of radishes, mass of radishes, leaf area, root length.

h) In plenty of light and in conditions that do not vary between dishes.

i)

pH or vol. of vinegar (cm^3)	**Height of plant (mm) after _ days**										**Average height (mm)**

B1b Examination-Style Questions

Examiner's comments

One feature that often characterises the difference between those students who do well and those that do poorly in structured questions is the use of scientific terms and the precision of the language used. This is partly dealt with in the 'Bump the grade' section opposite, but can also be illustrated elsewhere in these questions. Take question 2 (Science B) part (a) for example.

Compare the generality of the 1 mark answer with the detail and precision of the 3 mark answer. Then again in part (g) there are similar differences between the 1 mark and 2 mark answers. In part (e) marks can again be lost through writing about reducing car use without specifying how governments might achieve this.

For success answers should be accurate, clear, precise, detailed and scientific.

Answers to Questions

Biology A

1 **A** **3**

B **4**

C **1**

D **2** *(1 mark each)*

A common error is to pair A (water storage tissue) with 4 (reducing water loss in desert plants).

2 **A** **3**

B **2**

C **4**

D **1** *(1 mark each)*

3 **1** **D** *(1 mark)*

Biology B

1 (a) • (Plant) covered/buried in mud/sediment.
- Not decayed because decay conditions/example of decay conditions absent.
- Replaced by other materials/imprints made.

(1 mark each to a maximum of 2 marks)

(b) (i) Organisms have changed/adapted *(1 mark)* over time/by natural selection. *(1 mark)*

(ii) • It shows that (some types of) organism have become extinct/no longer exist.
- It shows that (some types of) organism have become more complex.
- It shows a sequence of change/a specific example of a sequence of change.
- Rocks of different depths/ages have different fossils.

(1 mark each to a maximum of 2 marks)

continues opposite ›

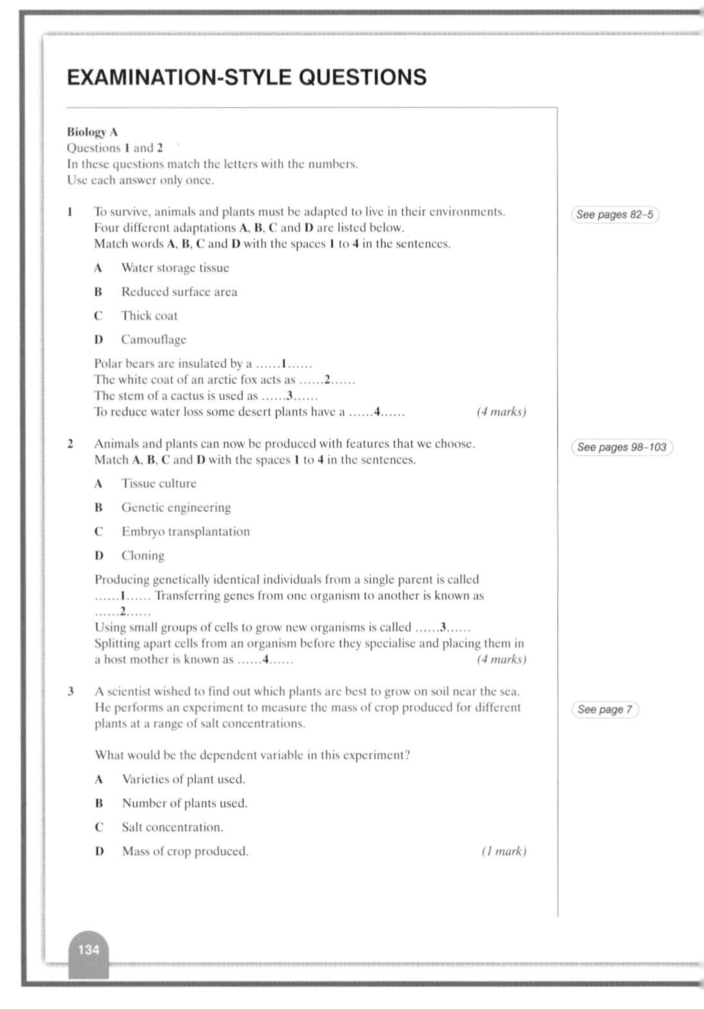

EXAMINATION-STYLE QUESTIONS

Biology A
Questions **1** and **2**
In these questions match the letters with the numbers.
Use each answer only once.

1 To survive, animals and plants must be adapted to live in their environments. Four different adaptations **A**, **B**, **C** and **D** are listed below.
Match words **A**, **B**, **C** and **D** with the spaces **1** to **4** in the sentences.

A Water storage tissue
B Reduced surface area
C Thick coat
D Camouflage

Polar bears are insulated by a ......**1**......
The white coat of an arctic fox acts as ......**2**......
The stem of a cactus is used as ......**3**......
To reduce water loss some desert plants have a ......**4**...... *(4 marks)*

See pages 82–5

2 Animals and plants can now be produced with features that we choose.
Match **A**, **B**, **C** and **D** with the spaces **1** to **4** in the sentences.

A Tissue culture
B Genetic engineering
C Embryo transplantation
D Cloning

Producing genetically identical individuals from a single parent is called ......**1**...... Transferring genes from one organism to another is known as ......**2**......
Using small groups of cells to grow new organisms is called ......**3**......
Splitting apart cells from an organism before they specialise and placing them in a host mother is known as ......**4**...... *(4 marks)*

See pages 98–103

3 A scientist wished to find out which plants are best to grow on soil near the sea. He performs an experiment to measure the mass of crop produced for different plants at a range of salt concentrations.

What would be the dependent variable in this experiment?

A Varieties of plant used.
B Number of plants used.
C Salt concentration.
D Mass of crop produced. *(1 mark)*

See page 7

134

GET IT RIGHT!

'Using the graph', 'using the information provided', 'from the data shown', etc., are common expressions on examination papers. Rather than being put off by unfamiliar information, students need to be convinced that these questions are relatively straightforward. After all, the answers are provided – they are there in the question. The skill is, of course, in finding them but at least they don't have to remember anything (not at this stage at least), just be good at interpreting and analysing information.

BUMP UP THE GRADE

Questions on examination papers sometimes include the comment: 'To gain full marks in this question you should write your ideas in good English. Put them in a sensible order and use the correct scientific words' or words to that effect. It should be impressed upon students that this is not just a request to write clearly so that the examiner finds their papers easier to mark. It is an instruction that, if not followed, will cost marks.

Ensuring that a candidate gets full marks is less about having a wide vocabulary and more about expressing themselves in a **logical** and **scientific** manner. To help those students who have difficulty with this aspect of examinations, they could attempt any of the questions in this book that carry an allowance of 4 or more marks or use the term 'explain'. These almost invariably involve the need to organise facts and express ideas. Better still, take some of the answer schemes in this book and firstly rewrite them in bad English and with imprecise scientific terms, then mix up the order of the points. Get the students to rewrite the answers in clear, logical, scientific English.

ology B

The rock shown in the diagram contains a fossil of a plant which lived millions of years ago.

(a) Describe **one** way in which this fossil might have been formed. *(2 marks)*

(b) (i) Evidence from fossils supports the theory of evolution. What is the theory of evolution? *(2 marks)*

(ii) How does fossil evidence support this theory? *(2 marks)*

See pages 108–11

The concentration of gases in glacier ice gives a measure of the composition of the air at the time the ice was formed. Because air moves rapidly around the Earth, this measure also applied to areas far away from where the measurements were taken. The graph shows the concentration, in parts per million, of carbon dioxide which has dissolved in glacier ice over the past 250 years.

Concentration of CO_2/ppm
340
320
300
280
1700 1750 1800 1850 1900 1950 2000
Year

(a) Using the graph, describe the changes in concentration of carbon dioxide in the glacier ice from 1750 to 2000. *(3 marks)*

(b) Give one piece of evidence that suggests the instruments used in this investigation had to be very sensitive. *(1 mark)*

(c) List two sources of the carbon dioxide found in the atmosphere. *(2 marks)*

(d) Explain why there has been an increase in carbon dioxide in the atmosphere since 1900. *(5 marks)*

(e) Thirty-eight nations signed the Kyoto Treaty and agreed to cut their carbon dioxide emissions by just over 5% by 2012. Suggest three different actions that governments could take to help achieve their targets. *(3 marks)*

(f) If the concentration of carbon dioxide in the atmosphere increases further, how might this affect the temperature of the air at the Earth's surface? *(1 mark)*

(g) Explain **how** carbon dioxide in the atmosphere might cause this temperature change. *(2 marks)*

GET IT RIGHT!

'Using the graph' is how question 2 part (a) begins. Another expression used by examiners is 'using only information in the diagram'. In all these cases you should only use the information provided. Do not panic if you have not met the information before. Simply work logically through the facts provided. In this case start at year 1750 and describe (say what is happening) as you work along the plotted line to the year 2000. For example, 'from 1750 to 1800 there is an increase in CO_2 . . . etc.' No explanations are required – not in this stage at least.

135

› *continues from previous page*

2 (a) *1-mark answer:*
There is a general increase over the whole period.
2-mark answer:
Mention of a 'general increase' plus some other detail, e.g. the increase is greater after 1900/there is no increase from 1800–1900.
3-mark answer:
There is an increase from 1750–1800.
No increase from 1800–1900.
Large/rapid/greater increase from 1900–2000.

(b) The overall change is only 35 ppm so the instrument has to resolve changes in concentration that are even smaller. *(1 mark)*

(c) From respiration (by living organisms).
Combustion/burning (of vegetation, fuels).
Deforestion.
(1 mark for each point to a maximum of 2)

(d)
- More industrialisation.
- **Rapid** increase in population
- leading to increased energy demands
- therefore more burning of fossil fuels.
- Increased burning of forests/rubbish.
- More transport/cars burning more fuel.
- More people/domestic animals means more respiration.
- Deforestation means less CO_2 uptake in photosynthesis.

(1 mark for each point to a maximum of 5)

(e) Reduce the burning of fossil fuels by using alternative energy sources, e.g. nuclear/wind/wave power. *(1 mark)*
Reduce car use by improving public transport/encouraging people to walk/increasing tax on petrol/diesel fuel. *(1 mark)*
Reduce energy demand (hence use of fossil fuels) by building more energy-efficient homes (grants for insulation/double glazing)/using low-energy light bulbs/more efficient heating boilers. *(1 mark)*
Any other valid point, provided it is explained, e.g.
- reduce speed limits on motorways
- recycle rubbish
- less polluting car fuels. *(maximum 3 marks)*

For a mark, the point made should involve a 'government action'. For example, 'reduce car use' is too vague – the government action to achieve this must be included, e.g. improving public transport/making public transport cheaper/increasing tax on petrol/diesel fuel.

(f) It might increase. *(1 mark)*

(g) *1-mark answer:*
CO_2 in the atmosphere acts like a greenhouse stopping some of the sun's heat escaping from the Earth.
2-mark answer:
It lets heat from the Sun into the Earth's atmosphere but stops some of it escaping because it re-radiates it back to the surface of the Earth. As a result the Earth and its atmosphere are warmer than they would otherwise be.

BIOLOGY

B2 | Additional biology

Key Stage 3 curriculum links

The following link to **'What you already know'**:

- Animal and plant cells can form tissues, and tissues can form organs; the functions of chloroplasts and cell walls in plant cells and the functions of the cell membrane, cytoplasm and nucleus in both plant and animal cells; relate cells and cell functions to life processes in a variety of organisms.
- The principles of digestion, including the role of enzymes in breaking down large molecules into smaller ones; food is used as a fuel during respiration to maintain the body's activity and as a raw material for growth and repair.
- Aerobic respiration involves a reaction in cells between oxygen and food in which glucose is broken down into carbon dioxide and water, summarised in a word equation.
- Plants need carbon dioxide, water and light for photosynthesis, and produce biomass and oxygen, summarised in a word equation. Nitrogen and other elements, in addition to carbon, hydrogen and oxygen, are required for plant growth.
- Plants carry out aerobic respiration.

QCA Scheme of work
7A Cells
8A Food and digestion
8B Respiration
9C Plants and photosynthesis
9D Plants for food

RECAP ANSWERS

1 **a)** Movement, respiration, sensitivity, growth, reproduction, excretion, nutrition.
b) Animals move whole body, plants do not; plant cells have cell walls and chloroplasts, animals do not; plants photosynthesise (make their own food), animals cannot.
c) i) Controls the cell, contains plans for making new cells and new organisms.
ii) The outer barrier of the cell, it controls substances moving into and out of the cell.
iii) Jelly (material) where all the important jobs of the cell take place.

2 **a)** For energy for our cells.
b) Big molecules need to be broken down to smaller molecules.
c) Chemical that speeds up reactions.
d) You cannot use the food for energy until it has been broken down into small molecules.

3 **a)** It makes the energy from your food available to your body.
b) glucose + oxygen → energy + carbon dioxide + water

4 **a)** It would die as it needs light to make food by photosynthesis.
b) Plants use carbon dioxide to make sugars in photosynthesis. If you talk to them you are breathing air rich in carbon dioxide on to them. They have more carbon dioxide, so they might be able to make more food and therefore grow more.
c) Minerals, such as nitrates and magnesium, are needed to make other chemicals that they need.

BIOLOGY

B2 | Additional biology

What you already know

Here is a quick reminder of previous work that you will find useful in this unit:

- Both plant and animal cells have a cell membrane, cytoplasm and a nucleu Plants cells also have cell walls and chloroplasts.
- Some cells, such as sperm, ova and root hair cells, are specially adapted tc carry out particular functions in an organism.
- Enzymes play an important part in breaking down large molecules into smaller ones during digestion.
- Food is used as a fuel during respiration to keep your body activity levels u You also need it as the raw material for growth and repair of your body cells
- Plants and animals all carry out aerobic respiration.
- Aerobic respiration involves a reaction in our cells between oxygen and food. Glucose is broken down into carbon dioxide (CO_2) and water (H_2O).
- Plants need carbon dioxide, water and light for photosynthesis. They produce biomass, in the form of new plant material, and oxygen.
- Plants also need nitrogen and other elements to grow.

RECAP QUESTIONS

1 a) Make a list of the things all living things need or do.
b) Write down three differences between animals and plants.
c) What are the jobs of:
i) the nucleus,
ii) the cell membrane,
iii) the cytoplasm,
in a cell?

2 a) Why do we need food?
b) What has to happen to the food you eat before it can be useful to your body?
c) What is an enzyme?
d) Why are enzymes so important in digestion?

3 a) Respiration takes place in all living cells. Why is it so important?
b) Write a word equation for what happens during respiration in your cells.

4 a) What would happen if you put a plant in a dark cupboard and left it for several weeks?
b) There is more carbon dioxide in the air people breathe out than in the air they breathe in. Some people claim that talking to house plants makes them grow better. What might be a scientific explanation for this claim?
c) Sunlight and water are not enough for plants to grow well.
What else do they need – and why?

136

Activity notes

This is a very open-ended activity and students would benefit from some guidelines. It can be done quite simply in a lesson, by asking students to make their individual lists and then compiling a class list, but if more detail is required then research and homework time could be allocated. The more detailed research, particularly with respect to the food that animals eat, could link with subsequent work in Chapter 3.

- **What do we mean by 'food'?** – The students could discuss the term and come to a decision about a definition. Then ask: 'What do we mean by 'different types of food'?' [It could be interpreted in many ways: fast food; energy-giving food; body-building food; baby food etc.] A brainstorming session could come up with a whole range of different types.
- **Different diets (1)** – People in different parts of the world eat different foods. In developing countries, the people rely on one food (a staple food) for a major part of their diet. For example, in SE Asia the staple food is rice and in central Africa it is roots and tubers, such as cassava and sweet potatoes. An interesting project would be to find out about the different staple foods in the world. Ask: 'Why are they the staples? What do they supply in the diet? What other foods are eaten? What can be grown successfully in a particular region?' This suggestion can be more clearly defined into the cereal crops, such as wheat, maize and

SPECIFICATION LINK-UP   Unit: Biology 2

What are animals and plants built from?

All living things are made up of cells. The structures of different types of cell are related to their functions.

How do dissolved substances get into and out of cells?

To get into and out of cells, dissolved substances need to cross cell membranes.

How do plants obtain the food they need to live and grow?

Green plants use light energy to make their own food. They obtain the raw materials they need to make this food from the air and the soil.

What happens to energy and biomass at each stage in a food chain?

By observing the numbers and sizes of the organisms in food chains, we can find out what happens to energy and biomass as it passes along the food chain.

Making connections

The plant production line!

Plants produce food for all the animals that live on Earth, including us. They do this through the process of photosynthesis. They use carbon dioxide, water, and energy from light, to make sugars and oxygen.

Feeding the world

Plants could provide enough material to feed everyone in the world. If everyone understood how pyramids of biomass work, perhaps we would all eat differently and no-one would starve!

Person / Sweetcorn — Person / Cow / Grass

Enzymes

The food you eat is made up of big molecules. They can't get out of your gut and into your bloodstream. So they can't reach the cells where they are needed. Fortunately your body makes digestive enzymes. They work in your gut to break your food down into much smaller molecules, which your body can use.

Food – vital for life!

Specialised cells

The cells in your pancreas are very specialised. Some of them (stained pink in this photo) produce enzymes needed to break down your food. Others (stained purple) make the hormone insulin which controls your blood sugar levels.

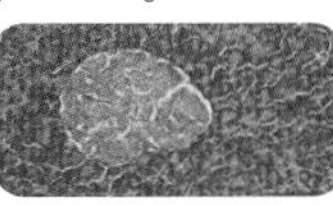

Balancing blood sugar

After you have eaten and digested a meal, the levels of sugar in your blood shoot up. You need to be able to take this sugar into your cells so they can use it. You also need to store some of the sugar to use later. The hormone insulin is vital for you to balance your blood sugar.

Inheriting problems

Most babies are born with guts that work perfectly. But some inherit genes which mean they can't feed properly. With pyloric stenosis, the baby vomits all its food back. It needs surgery to correct the fault in its gut. In cystic fibrosis the glands that make many of the digestive enzymes get clogged up with thick sticky mucus. Then they don't work at all.

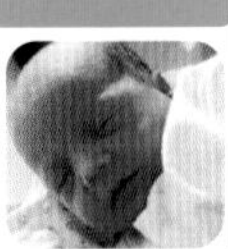

ACTIVITY

Lots of what you will learn in this unit is linked in some way to food. Every living thing needs food to survive. List, draw or find images of as many different types of food as you can.

Think about the food eaten by different types of animals and by different people around the world. There are some amazing sources of energy out there – see how many you can think of!

Chapters in this unit

Cells — How plants produce food — Energy flows — Enzymes — Homeostasis — Inheritance

137

rye, and the non-cereal staples, such as potatoes, yams and legumes. An investigation into the grass family (*Gramineae*) can yield masses of information, not only about human food but also about animal food. It would be interesting to investigate all the foods prepared from and used involving wheat.

- **Different diets (2)** – Another approach to different diets is to look at what people in different parts of Europe traditionally eat. Students could draw upon their holiday experiences to investigate the 'Mediterranean' diet, what the Greeks eat, why Scandinavian diets include more fish, etc.
- **Food eaten by animals** – There are some animals that have very specific food preferences and requirements (the Giant Panda and bamboo, for example). Some students might like to compile a fact file of such associations. Some research in a library or suitable web sites would need to be done. It could link with work done on why certain species of animals are becoming extinct (Module B1b).
- **Strange foods** – Some of the food we eat comes from strange sources. We get jelly from seaweed, food products from blue-green algae and meat substitutes from a fungus. Students could research food from strange sources. Ask: 'Can it save the world?'
- **Food in history** – Some students might like to investigate how food has changed over time. A comparison of the diet of a medieval peasant with a twenty-first century diet could be quite revealing. At the time of the Roman Conquest, the diet of a Roman soldier was different from the British diet. The Romans are thought to have been responsible for introducing rabbits and they also grew vines to make wine (ask students to think about this in relation to climate change!).

Special needs

These students could find pictures in magazines and newspapers, or draw foods and compile a scrapbook.

Teaching suggestions

- 'What you already know' and the 'Recap questions' review the extent of the students' knowledge of these topics gained from their studies at KS3. The students should understand the structure of cells, the differences between plant and animal cells and that cells are specialised to carry out particular functions. Not all students recognise individual cells and many will not appreciate the differences in size (ova and sperm cells). There could also be some confusion about exactly how much is visible at different magnifications (e.g. mitochondria and other organelles are not visible using a light microscope). The 'Recap questions' are useful to test knowledge and could be expanded by showing a series of pictures of different cells and getting students to name them.
- The roles of enzymes and aerobic respiration are fundamental concepts and necessary for an understanding of many of the processes in the cells of living organisms. At this stage, students should refer to 'glucose' being needed by cells as an energy source and not 'sugar'. The use of the terms 'fuel' and 'burn' are not recommended at this level when explaining the reaction in cells. These are useful analogies to make, but students will gain little credit for explaining respiration in such terms.
- Most students will know the outlines of the process of photosynthesis and its requirements. The misconception that plants get their food from the soil should have been corrected at KS3. The need for minerals, particularly nitrates, was introduced at KS3.
- 'The Recap questions' could be used in a quiz, similar to *Test the Nation*, where the class can be divided up into different groups so that aggregate scores and individual scores can be totalled. Prepared scoring sheets would make life easier when totalling the marks.
- **Making connections** – As indicated in the Student Book, one of the links in this module is 'food'. Using the pictures and boxes on the spread, ask the students to build up a concept map that links them all together and explains the links. Allow them 5–10 minutes to work it out on their own and then bring their ideas together on the board. Not all these pictures and boxes appear to have a direct link with the chapters in this unit – or do they? Encourage students to look ahead and link the references to the topics in the unit. For each box or picture, they could award a mark from 1 (weakest) to 5 (strongest) for its link. For example, Box 1 with its picture would be a 5, linking with Chapter 2 and Photosynthesis, but what about the others? Discuss the award of marks after allowing the students some time to look through on their own or in small groups.

Chapters in this unit

- Cells
- How plants produce food
- Energy flows
- Enzymes
- Homeostasis
- Inheritance

B2 1.1 Animal and plant cells

LEARNING OBJECTIVES

Students should learn:

- The functions of the different parts of animal cells.
- The differences between plant and animal cells.
- That the chemical reactions within cells are controlled by enzymes.

LEARNING OUTCOMES

Most students should be able to:

- Describe the structure of animal cells.
- Describe the functions of the parts of animal cells.
- List the differences between plant and animal cells and state that enzymes control the reactions inside cells.

Some students should also be able to:

- Explain how enzymes control the chemical reactions within cells.

Teaching suggestions

- **Special needs.** Students could be given an outline of an animal cell and a plant cell with labels to stick on.
- **Gifted and talented.** Students could find out more about how an electron microscope works and how it is used to look at cells. (You can find some references in more advanced Biology texts such as *Tools, Techniques and Assessment in Biology*, Adds, Larkcom, Miller and Sutton (Nelson Advanced Science Series).
- **Learning styles**

  Kinaesthetic: Examining plant and animal cells.

  Visual: Drawing and labelling cells.

  Auditory: Explaining functions of parts of cells.

  Interpersonal: Discussing what can be seen in electron micrographs.

  Intrapersonal: Reviewing and deducing functions of cell parts.

Answers to in-text questions

a) Nucleus, cytoplasm, cell membrane, mitochondria, ribosomes.

b) Plant cells have a cell wall, chloroplasts and a permanent vacuole.

c) Protein.

SPECIFICATION LINK-UP Unit: Biology 2.12.1

- *Most human cells, like most other animal cells, have the following parts:*
  - *a nucleus that controls the activities of the cell*
  - *cytoplasm in which most of the chemical reactions take place*
  - *a cell membrane which controls the passage of substances in and out of the cell*
  - *mitochondria, which is where most energy is released in respiration*
  - *ribosomes, which is where protein synthesis occurs.*
- *Plant cells have a cell wall which strengthens the cell. Plant cells often have:*
  - *chloroplasts, which absorb light energy to make food*
  - *a permanent vacuole filled with cell sap.*
- *The chemical reactions inside cells are controlled by enzymes.*

Lesson structure

STARTER

What does it do? – Write up a list of functions of parts of an animal cell on the board, splitting them up so that there is more than one function per part, e.g. 'controls activities' and 'contains chromosomes' for nucleus. Allow students to work through the list by themselves and then check. (5–10 minutes)

Plant or animal? – Show a drawing of a typical plant cell. (Search the web for 'plant cell' images.) Students say whether it is a plant or an animal cell, giving reasons. Get them to suggest labels for the parts and decide whether these are common features of cells or special to plant cells. (10 minutes)

MAIN

- **Seeing more detail** – This exercise is designed to show students that what they can see using a light microscope is limited, and that structures such as mitochondria and ribosomes are only visible using electron microscopy. The students could work in groups, each having light microscopes with slides of stained cheek cells and onion bulb inner epidermal cells and a set of electron micrographs of plant and animal cells (there are plenty in A level text books). They could identify structures in both, and make a comparison of what they can observe from the slides and from the electron micrographs.
- If the magnification of the light microscope is given and the magnification of the electron micrographs known, they can work out how much bigger the latter are. Gather together and discuss the information, particularly with respect to the structures revealed by electron microscopy. Ask: 'Why do they all appear to have membranes around them?'
- **Looking at plant cells** – Plant cells, such as rhubarb petiole epidermis or the inner epidermal cells from onion bulbs, are relatively easy to mount, stain and observe using light microscopes. In order for students to see cell structures, some staining is advisable.
- The procedure, described under 'Practical support,' could be demonstrated to the students and they can then have a go at making their own slides and drawing and labelling some cells.
- **Looking at animal cells** – Using safe, sterile procedures, students could make slides of their own cheek cells. (See 'Practical support'). Some cells could be drawn and labelled.

PLENARIES

A question of size – A typical cell is 20 μm wide (0.002 mm). You might need to talk about scales and the relationship between millimetres and micrometres. Students can then calculate how many cells will fit across the page of their textbook. Give a small prize for the first correct calculation. (5–10 minutes)

Our wonderful world – There are some excellent scanning electron micrographs (SEM) and transmission electron micrographs (TEM) of cells. Show a selection (from www.cellsalive.com) with a 'Guess what this is' attached to each one. This would help students appreciate the complexity of some structures. (10 minutes)

Practical support

Looking at cells

Equipment and materials required

Light microscopes (at least one per group of two or three students), clean microscope slides and cover slips, onion bulbs or rhubarb petiole, scalpels, scissors and mounted needles, dilute iodine solution in dropping bottles (CLEAPSS Hazcard 54), tissues, eye protection.

Details

Cut an onion in half and remove the thin inner epidermis of the leaves with forceps. This can be cut up into small squares about 5 mm square. Place a square of epidermis on a slide, trying to get it as flat as possible, and then place a drop of dilute iodine solution on top to stain the cells. Place a cover slip over the top, lowering it carefully down so that air bubbles are not trapped. Place the slide under the low power of the microscope, focusing carefully. Then switch to high power and focus using the fine adjustment.

- Underground storage organs (onion bulb) and leaf stalks (rhubarb) do not carry out photosynthesis.

For cheek cells

Equipment and materials required

Light microscopes (at least one per group of two or three students), new cotton buds, clean microscope slides and cover slips, dilute methylene blue solution, disinfectant, or another approved way, for disposal of used cotton buds and slides.

Details

The inside of the cheek is gently scraped using a sterile spatula cotton bud and the scrapings smeared on to the middle of a clean microscope slide. A drop of dilute methylene blue is added on top of the cells and covered with a cover slip. The slide can then be observed under the microscope. Some gentle pressure might be needed to spread the cells out so that they are easier to see. When finished place prepared slides and cotton buds in a container of freshly prepared sodium hypochlorite solution. (See CLEAPSS Student Safety Sheet 3 and follow Institute of Biology guidelines.)

ACTIVITY & EXTENSION IDEAS

- Using rhubarb petiole or onion bulb epidermis will not show chloroplasts, so a demonstration of some moss leaf cells or leaves of a water plant such as *Elodea* could be mounted in water and projected. The cells will be live and it is possible that the streaming of the cytoplasm can be observed.
- Showing an electron micrograph of the cellulose cell wall will demonstrate that the cellulose is in fine strands (fibrils) and that there are several layers criss-crossing, so that the cell wall acts like a sieve.

KEY POINTS

The key points about the animal and plant cells can be learnt from clear diagrams. Students should be sure that they know what structures in cells can be observed using a light microscope: ribosomes and mitochondria are only visible in electron micrographs.

B2 1.1 Animal and plant cells

LEARNING OBJECTIVES

1. What do the different parts of your cells actually do?
2. How do plant cells differ from animal cells?
3. How are all the chemical reactions which go on in your cells controlled?

The Earth is covered with a great variety of living things. The one thing all these living organisms have in common is that they are all made up of cells. Most cells are very small. You can only see them using a microscope.

The **light microscopes** you will use in school may magnify things several hundred times. Scientists have found out even more about cells using **electron microscopes** which can magnify more than a hundred thousand times!

Animal cells – structure and function

All cells have some features in common. We can see these clearly in animal cells. The cells of your body have these features, just like the cells of every other living thing!

- **A nucleus**, which controls all the activities of the cell. It also contains the instructions for making new cells or new organisms.
- The **cytoplasm**, a liquid gel in which most of the chemical reactions needed for life take place. One of the most important of these is respiration, where oxygen and sugar react to release the energy the cell needs.
- The **cell membrane**, which controls the passage of substances in and out of the cell.
- The **mitochondria**, structures in the cytoplasm where most of the energy is released during respiration.
- **Ribosomes**, where protein synthesis takes place. All the proteins needed in the cell are made here.

Ribosomes
Cell membrane
Cytoplasm
Mitochondria
Nucleus

Figure 1 A simple animal cell like this shows the features which are common to all living cells

a) What are the main features found in all living cells?

Plant cells – structure and function

Plants are very different from animals, as you may have noticed! They make their own food by photosynthesis and they do not move their whole bodies about. So while plant cells have all the features of a typical animal cell, they also contain structures which are needed for their very different way of life.

All plant cells have:

- a cell wall made of cellulose which strengthens the cell and gives it support.

Many (but not all) plant cells also have these other features:

- chloroplasts, found in all the green parts of the plant. They are green because they contain the green substance chlorophyll which gives the plant its colour. They absorb light energy to make food by photosynthesis.
- a permanent vacuole (a space in the cytoplasm filled with cell sap), which is important for keeping the cells rigid to support the plant.

b) How do plant cells differ from animal cells?

GET IT RIGHT!

Make sure you can label an animal cell and a plant cell and know the function of each of their parts.
Remember that not all plant cells have chloroplasts . . . and don't confuse chloroplasts and chlorophyll.

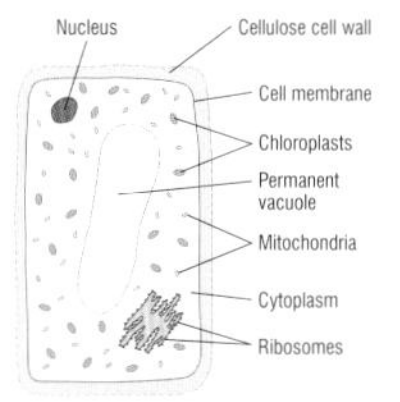

Figure 2 A plant cell has many features in common with an animal cell, but others which are unique to plants

PRACTICAL

Looking at cells

Set up a microscope to look at plant cells, e.g. from onions and rhubarb. You should see the cell wall, the cytoplasm and sometimes a vacuole but you won't see chloroplasts.

- Why won't you see any chloroplasts?

Chemical reactions in cells

Imagine 100 different reactions going on in a laboratory test tube. Chemical chaos and probably a few explosions would be the result! But this is the level of chemical activity going on all the time in your cells.

Cell chemistry works because each reaction is controlled by an enzyme. Each enzyme is a protein which controls the rate of a very specific reaction. It makes sure that the reaction takes place without becoming mixed up with any other reaction.

We find enzymes throughout the structure of a cell, but particularly in the mitochondria (and the chloroplasts in plants).

c) What are enzymes made of?

The enzymes involved in different chemical processes are usually found in different parts of the cell. So, for example, most of the enzymes controlling the reactions of:

- respiration are found in the mitochondria,
- photosynthesis are found in the chloroplasts,
- protein synthesis are found on the surface of the ribosomes.

These cell compartments help to keep your cell chemistry well under control.

SUMMARY QUESTIONS

1. a) List the main structures you would expect to find in an animal cell.
   b) You would find all of these things in a plant cell. There are three extra features which are found in plant cells. What are they?
   c) What are the main functions of these three extra structures?
2. Root cells in a plant do not have chloroplasts. Why?
3. A nucleus and mitochondria are important structures in almost all cells. Why are they so important?
4. Explain how enzymes control the chemistry of your cells.

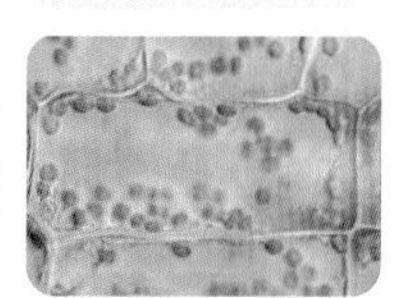

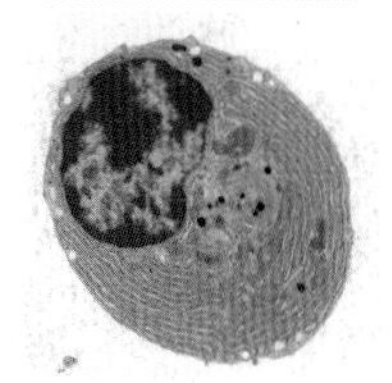

Figure 3 Diagrams of cells are much easier to understand than the real thing seen under a microscope. These pictures show a magnified plant cell and animal cell.

DID YOU KNOW?

Although most cells are so small we can only see them under the microscope, the largest cells in the world weigh 1.35 kg and are easily visible with the naked eye. The largest single cell is . . . an ostrich egg!

KEY POINTS

1. Most animal cells contain a nucleus, cytoplasm, cell membrane, mitochondria and ribosomes.
2. Plant cells contain all the structures seen in animal cells as well as a cell wall and, in many cases, chloroplasts and a permanent vacuole filled with sap.
3. Enzymes control the chemical reactions inside cells.

SUMMARY ANSWERS

1. a) Nucleus, cytoplasm, cell membrane, mitochondria, ribosomes.
   b) Cell wall, chloroplasts, permanent vacuole.
   c) Cell wall provides support and strengthening for the cell and the plant; chloroplasts for photosynthesis; permanent vacuole keeps the cells rigid to support the plant.
2. Root cells do not carry out photosynthesis. They are underground where there is no light.
3. The nucleus controls all the activities of the cell and contains the instructions for making new cells or new organisms. Mitochondria are the site of respiration, so they produce energy for the cell.
4. Each enzyme controls the rate of a very specific reaction and makes sure that it takes place without becoming mixed up with any other reaction. The enzymes involved in different chemical processes are usually found in different parts of the cell (enzymes involved in aerobic respiration are found in the mitochondria).

B2 1.2 Specialised cells

LEARNING OBJECTIVES

Students should learn that:

- Cells may be specialised to carry out particular functions.

LEARNING OUTCOMES

Most students should be able to:

- Recognise different types of cells.
- Relate the structure of given types of cells to their functions in a tissue or an organ.

Some students should also be able to:

- Relate the structure of novel cells to other functions in a tissue or organ.

Teaching suggestions

- **Special needs.** Use domino-style cards with specialised cells on one side and their special features on the other. Ask the students to play with these as dominoes. Alter the number of cards and the labelling according to ability.
- **Gifted and talented.** Suggest to students that they design a special cell found in an alien or undiscovered species. Ask them to give it an interesting, unusual or gruesome feature and make it scientifically feasible.
- **Learning styles**

  Kinaesthetic: Investigating the root hair cells using microscopes.

  Visual: Examining, drawing and labelling specialised cells.

  Auditory: Listening to the questions and answers in the plenary '20 questions'.

  Interpersonal: Discussing the differences between generalised and specialised cells.

  Intrapersonal: Interpreting specialised features of cells.

SPECIFICATION LINK-UP Unit: Biology 2.12.1

- *Cells may be specialised to carry out a particular function.*

Lesson structure

STARTER

How big can cells be? – Show a goose egg, explain that it is a single cell and break it on to a plate. They may be able to see the place where the embryo will develop (the germinal disc). Discuss with the students why it is so big and how it is specialised. If possible show an empty ostrich egg. (5–10 minutes)

Do you know what this is? – Project some images of specialised cells – do not label them but give each one a number. Students to name the ones they know and have a guess at the ones they do not. Check the answers at the end. (5–10 minutes)

Key words – Place key words and phrases from this lesson on the board. Ask students to remove and explain what each one means, leaving the rest as 'learning objectives'. (5 minutes)

MAIN

- **Root hair cells** – A few days before the lesson, sow some cress seeds on damp blotting paper or filter paper in Petri dishes. Handle the seeds by the cotyledons using forceps. When ready to use them, remove the lids and cover with cling film to keep the moisture levels high. Place the dishes under a binocular microscope and take digital photographs down the microscope. The photographs can then be stuck in the students' records. This can either be set up as a demonstration or groups of students could work together on the activity.
- If a micrometer eyepiece is inserted in the microscope, the length of some of the root hairs can be measured. The measurements can either be left as eyepiece units (eu) or converted to millimetres if the eyepiece is calibrated. This exercise will reinforce the extent to which the root hairs are specialised for the increase of the surface area available for the uptake of water.
- This exercise introduces some of the concepts of 'How Science Works', such as observation and making single measurements.
- **Sperm cells** – Video footage of sperm cell activity is readily available. There are clips available which show fertilisation, emphasising the difference in sizes of egg cells and sperm and also the relative numbers.
- Prepared slides of rat testes could be available for students to look at, observing the different stages in sperm development. Prepare a worksheet with some drawings of different stages so that students can look for specific features and make labelled drawings of their own. This activity can link with the showing of the video.
- **How structure is related to function in animal cells** – Prepare a PowerPoint presentation of a range of different animal cells, to include blood cells, neurones, muscle cells, cells from glands, fat cells and gametes. Provide each student with a worksheet with spaces for them to fill in the names of the cells, special features and how each specialised cell differs from a generalised animal cell. Allow the students to complete the worksheets individually and then go through the cells again, discussing the important points.
- **How structure is related to function in plant cells** – This could be presented in a similar manner to the above, providing a worksheet, but using a range of plant cells, such as palisade cells, guard cells, root hair cells, lignified cells (fibres) and epidermal cells. Cells from the cortex of the stem or the root could be used as generalised plant cells.
- Students are required to be able to relate the structure of different types of cell to their functions in a tissue or an organ so these exercises, with their completed worksheets, will give them a record for future reference and revision.

ACTIVITY & EXTENSION

- **Fun with colour vision** – Students could have fun with their colour vision by staring at brightly coloured cardboard and then at white paper to perceive residual false colour images. Search the web's images for 'flags' to illustrate such after-images. In the retina, there are three types of cone: sensitive to red, green or blue. When you stare at a particular colour for too long, these receptors get 'tired' or 'fatigued'. After looking at the flag with the strange colours, your receptors that are tired do not work as well. Therefore the information from all the different colour receptors is not in balance.
- **Single-celled organisms** – Show images of single-celled organisms as a contrast to specialised cells. Good examples to find on the web are *Chlamydomonas, Paramecium* and *Amoeba*, and then discuss how these organisms can carry out all the functions of life.

Practical support

Root hair cells

Equipment and materials required

Cress seedlings with root hairs, forceps, cling film, blotting paper or filter paper, digital camera, binocular microscopes, Petri dishes, micrometer eyepiece if available.

Details

See 'Main'.

KEY POINTS

The students need to be able to link the structure of specialised cells to their particular functions.

BIOLOGY CELLS

B2 1.2 Specialised cells

LEARNING OBJECTIVES

1. What different types of cells are there?
2. How is the structure of a specialised cell related to its function?

The smallest living organisms are single cells. They can carry out all of the functions of life, from feeding and respiration to excretion and reproduction. Most organisms are bigger and are made up of lots of cells. Some of those cells become **specialised** in order to carry out particular jobs.

When a cell becomes specialised its structure is adapted to suit the particular job it does. As a result, specialised cells often look very different to our 'typical' plant or animal cell. Sometimes cells become so specialised that they only have one function within the body. Good examples of this include sperm, eggs, red blood cells and nerve cells.

PRACTICAL

Observing specialised cells

Try looking at different specialised cells under a microscope.

When you look at a specialised cell there are two useful questions you can ask yourself:

- How is this cell different in structure from a generalised cell?
- How does the difference in structure help it to carry out its function?

Fat cell

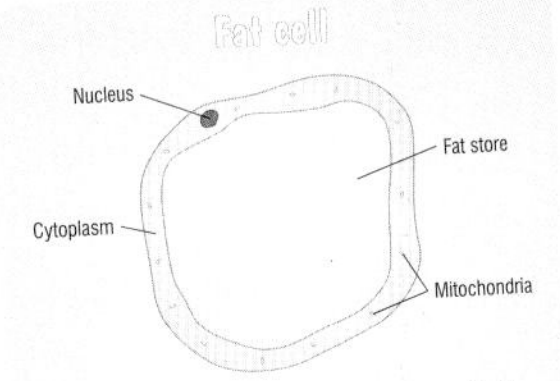

Fat cells are storage cells. If you eat more food than you need, your body makes fat and fills up the fat cells. They are important for helping animals, including us, to survive when food is in short supply. They have three main adaptations:

- **They have very little normal cytoplasm – this leaves plenty of room for large amounts of fat.**
- **They have very few mitochondria as they use very little energy.**
- **They can expand – a fat cell can end up 1000 times its original size as it fills up with fat.**

Cone cell from human eye

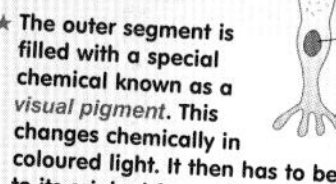

Cone cells are in the light-sensitive layer of your eye (the retina). They make it possible for you to see in colour. They have three main adaptations:

- **The outer segment is filled with a special chemical known as a *visual pigment*. This changes chemically in coloured light. It then has to be changed back to its original form. This uses up energy.**
- **The middle segment of the cell is packed full of mitochondria. They produce lots of energy. This means the visual pigment can reform and so the eye can see continually in colour.**
- **The final part of the cell is a specialised nerve ending or synapse. This connects to the optic nerve which carries impulses to your brain. When coloured light makes your visual pigment change, an impulse is triggered which crosses the synapse. This is how the response of the cone cell to coloured light passes to your brain.**

Root hair cells

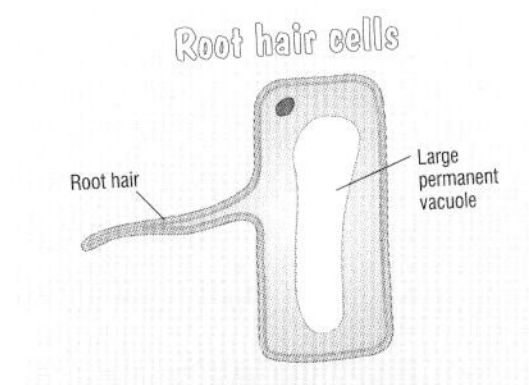

We find root hair cells close to the tips of growing roots. Their function is to enable plants to take in the water which they need. Root hair cells have three main adaptations:

- **The root hairs themselves, which increase the surface area for water to move into the cell.**
- **A large permanent vacuole, which affects the movement of water from the soil across the root hair cell.**
- **Root hair cells are always positioned close to the xylem tissue that carries water up into the rest of the plant.**

Sperm cells

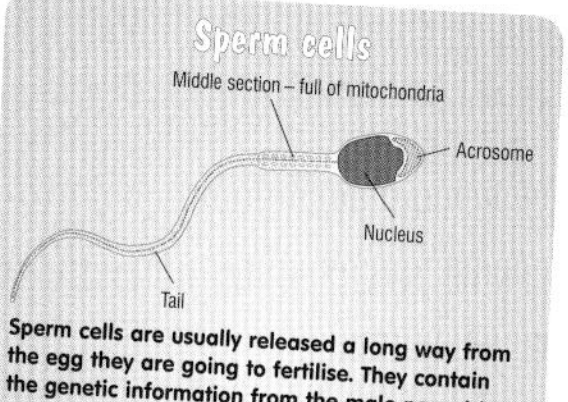

Sperm cells are usually released a long way from the egg they are going to fertilise. They contain the genetic information from the male parent to pass on to the offspring. They need to move through the female reproductive system to reach an egg. Then they have to break into the egg. They have several adaptations to make all this possible:

- **Long tails with muscle-like proteins so they can swim towards the egg.**
- **The middle section is full of mitochondria, which provide the energy for the tail to work.**
- **The acrosome, which stores digestive enzymes for breaking down the outer layers of the egg.**
- **A large nucleus, which contains the genetic information to be passed on.**

Organised cells

Specialised cells are often grouped together to form a **tissue**. Connective tissue joins bits of your body together. Nervous tissue carries information around your body and muscles move your body about.

Similarly in plants photosynthetic tissues make food by photosynthesis while storage tissues store any extra food made as starch.

In many bigger living organisms there is another level of organisation. Several different tissues work together to do particular jobs. They form an **organ** such as the heart, the kidneys or the leaf. In turn, different organs are combined in **organ systems** to carry out major functions in the body, such as transporting the blood or reproduction.

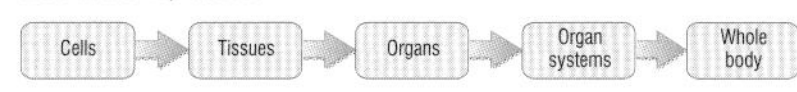

SUMMARY QUESTIONS

1. Explain how the structure of each cell on this spread is adapted to its functions.
2. Think back to two other types of specialised cells you have met in biology, e.g. motor neurones, photosynthetic cells in plants or white blood cells.
   Draw the cells you have chosen. Label them fully to show how the structures you can see are related to the function of the cells.

KEY POINTS

1. Cells may be specialised to carry out a particular function.
2. Examples of specialised cells are fat cells, cone cells, root hair cells, sperm cells.

PLENARIES

Am I colour blind? – Search a web image bank for 'colour blind' to find tests you can use to investigate colour blindness further. (5–10 minutes)

Levels of organisation – Label some Lego bricks of one colour 'Cell' and fit them together. On the back of the block formed stick a label 'Tissue'. Make some more 'Tissue' blocks from 'Cells' of other colours. Stick these together and label it 'Organ'. Have several of these and place them in a circle on the floor or the bench labelled 'Organ system'. Get the students to identify some organ systems and work back suggesting the organs, tissues and cells involved. (10 minutes)

20 Questions? – One student goes out of the room and the others decide which type of specialised cell they are. The student comes back in and asks the rest of the class questions about their specialisation to guess what they are. Repeat several, but not too many, times. (10 minutes)

SUMMARY ANSWERS

1 **Fat cell:** not much cytoplasm so room for fat storage; ability to expand to store fat; few mitochondria as do not need much energy so do not waste space.
Cone cell from human eye: outer segment containing visual pigment; middle segment packed full of mitochondria; specialised nerve ending.
Root hair cells: no chloroplasts so no photosynthesis; root hair increases surface area for water uptake; vacuole to facilitate water movement; close to xylem tissue.
Sperm cell: tail for movement to egg; mitochondria to provide energy for movement; acrosome full of digestive enzymes to break down egg; large nucleus full of genetic material.

2 [Any two cells chosen, appropriately labelled and annotated.]

B2 1.3 How do substances get in and out of cells?

LEARNING OBJECTIVES

Students should learn:

- That substances move in and out of cells by a process called diffusion.
- The factors that affect the rate of diffusion.

LEARNING OUTCOMES

Most students should be able to:

- Define diffusion in cells.
- List factors that affect the rate of diffusion.

Some students should also be able to:

- Explain the factors that affect the rate of diffusion.
- Explain how cells may be adapted to facilitate diffusion.

Teaching suggestions

- **Special needs.** Students can assemble words or cards to make a definition of diffusion. Alternatively, it can be done as an interactive exercise using Hot Potatoes JMIX or a similar interactive web-based exercise.
- **Gifted and talented.** Investigate what is meant by surface area to volume ratio. Draw out and extrapolate a SA/V graph for cutting a 2 cm × 2 cm cube in half and then in half again and so on.
- **Learning styles**

  Kinaesthetic: Dried peas activity.

  Visual: Observing and drawing colour changes in treated gel agar.

  Auditory: Discussion of net movement.

  Interpersonal: Group work in the practicals.

  Intrapersonal: Visualisation activity.

Answers to in-text questions

a) Blood diffuses through the water.

b) The difference between the numbers of particles moving in and those moving out of cells.

SPECIFICATION LINK-UP Unit: Biology 2.12.2

- *Dissolved substances can move into and out of cells by diffusion and osmosis.*
- *Diffusion is the spreading of the particles of a gas, or of any substance in solution, resulting in a net movement from a region where they are of a higher concentration. The greater the difference in concentration, the faster the rate of diffusion. Oxygen required for respiration passes through cell membranes by diffusion.*

Lesson structure

STARTER

Watching diffusion – Set up on each bench boiling tubes of water with a few crystals of potassium permanganate in the bottom. Draw them at the start of the lesson and then again at the end. (5 minutes)

Diffusion rates – Pour some perfume or clove oil on to cotton wool and place in the corner of the room. Students are to indicate when they can smell it. If each bench, or group of students, had a stopwatch or stopclock, they could time how long it took the smell to reach them. The distances could be measured and the rate of diffusion worked out. Discuss how the process could be speeded up or slowed down. (15 minutes)

Visualisation exercise – Ask the students to close their eyes and imagine a swimming pool. The pool and the room are empty. The water is completely still. You have a cup of tea in your hand. You walk up to the corner of the pool and gently pour the tea into the water, right in the corner. You turn around and leave the room in silence, locking the door behind you. No-one disturbs the room for a whole week. Then ask the students to open their eyes and discuss what would happen to the tea over the week and why. (5–10 minutes)

MAIN

- **Diffusion in gases** – It is possible to demonstrate and measure the rate of diffusion of ammonia along a glass tube. Litmus paper is used to show the progress of the gas along the tube. Glass tubes, 30 cm long and of diameter 20 mm should be marked using a felt tip pen at 2 cm intervals starting at 10 cm from one end. Each tube requires two corks, one ordinary one and one which has had a core of cork taken out the cavity plugged with cotton wool. Small squares of pink litmus paper are then dipped into distilled water, shaken and placed inside each tube with a piece of wire or a glass rod. The pieces of litmus paper should be pushed into position and lined up with the markings on the outside of the tube. Saturate the cotton wool in the cork at one end with a strong ammonia solution, then place it in the end of the tube, start the clock or stopwatch and time how long it takes each piece of litmus paper to turn from pink to completely blue.
- This experiment demonstrates that the ammonia diffuses along the tube from an area where it is in high concentration to a lower concentration. It is possible to work out the rate of diffusion. The overall time to diffuse 28 cm can be found and the individual times for the diffusion from one 2 cm mark to the next can be recorded. This gives several possibilities for discussion and calculation. It also introduces 'How Science Works' concepts.
- It is also possible to compare the rate of diffusion of the strong solution with that of a weaker solution by setting up an identical tube and timing the change in colour of the litmus squares.
- **Diffusion of liquids** – Cut wells into agar gel dyed with Universal Indicator or hydrogencarbonate indicator. Add acid at various concentrations into the wells, allow a set time and then measure the extent of the colour change around each well. Alkali could be used as well as acid for different colour changes.
- **Animated diffusion** – There are some good web-based animations illustrating diffusion. John Kyrk of Science Graphics has produced an excellent CD called Cella™. There is a good random molecule movement illustration of diffusion.

ACTIVITY & EXTENSION IDEAS

- **More diffusion in liquids** – Give students boiling tubes containing clear gelatine up to a marked level. They can then pour a thin layer of gelatine coloured with methylene blue on to the top, allow it to set and then pour on a quantity of clear gelatine equal to the volume in the bottom of the tube. (Levels could be marked for them.) The tubes should be left for a week and then the distribution of the blue colour recorded.
- **Variations** – The experiment described above can be varied by using different concentrations of methylene blue and comparing the results. It is also interesting to compare the rates of diffusion of the methylene blue with that of the ammonia. What does this tell you about diffusion in air and in liquids?
- **Two-way diffusion** – Pour some gelatine into a boiling tube and colour it with 10 drops of cresol red (it will go yellow). Mix thoroughly and allow to set. Pour a further layer of clear gelatine on top of the coloured layer. Allow this layer to set. Finally add about 5 cm^3 of ammonia solution to the top of the tube and insert a bung. Leave the tube for about 4 days. The cresol red will diffuse into the clear gelatine and the ammonia will diffuse into the gelatine. This can be shown by the cresol red turning from yellow to red.

Practical support

The details of apparatus needed for the practicals described have been given above.

The practical opportunity mentioned in the Student Book spread requires a large room or area for students to move around.

KEY POINTS

It is important to know that diffusion is a passive process: it does not require energy. It is also important to emphasise that it is the *net* movement of particles: particles move randomly all the time.

BIOLOGY CELLS

B2 1.3 How do substances get in and out of cells?

LEARNING OBJECTIVES

1 What is diffusion?
2 What affects the rate of diffusion?

Your cells need to take in substances such as oxygen and glucose. They also need to get rid of waste products and chemicals that are needed elsewhere in your body. Dissolved substances move into and out of your cells across the cell membrane. They can do this in three different ways – by **diffusion**, by **osmosis** and by **active transport**.

Diffusion

Sharks can smell their prey from a long way away – the smell reaches them by **diffusion**. Diffusion happens when the particles of a gas, or any substance in solution, spread out.

It is the net movement of particles from an area of high concentration to an area of lower concentration. It takes place because of the random movement of the particles of a gas or of a substance in solution in water. All the particles are moving and bumping into each other and this moves them all around.

Figure 1 Everyone knows that bleeding in the sea when there are sharks around is a bad idea. Sharks are sensitive to just a few particles of blood in the water. Blood from an injury spreads quickly through the sea by diffusion – and brings the sharks to investigate!

a) Why do sharks find an injured fish – or person – so easily?

Imagine a room containing a group of boys and a group of girls. If everyone closes their eyes and moves around briskly but randomly, people will bump into each other. They will scatter until the room contains a mixture of boys and girls. This gives you a good working model of diffusion.

At the moment, when the blue particles are added to the red particles they are not mixed at all

As the particles move randomly, the blue ones begin to mix with the red ones

As the particles move and spread out, they bump into each other. This helps them to keep spreading randomly

Eventually, the particles are completely mixed and diffusion is complete

Figure 2 The random movement of particles results in substances spreading out or diffusing from an area of higher concentration to an area of lower concentration

SCIENCE @ WORK

Cell biologists and biochemists work hard to discover the secrets of transport systems in your cells. They can use their findings both to understand human diseases at a cellular level and to change the nature of some organisms using genetic engineering. A salt-tolerant tomato plant which can actively move salt out of its cytoplasm into its vacuoles is just one example. This enables the plant to grow on salty ground.

Rates of diffusion

If there is a big difference in concentration between two areas, diffusion will take place quickly. However when a substance is moving from a higher concentration to one which is just a bit lower, the movement toward the less concentrated area will appear to be quite slow. This is because although some particles move into the area of lower concentration by random movement, at the same time other identical particles are leaving that area by random movement.

The overall or **net** movement = particles moving in − particles moving out

In general the bigger the difference in concentration, the faster the rate of diffusion will be. This difference between two areas of concentration is called the **concentration gradient**. The bigger the difference, the steeper the gradient will be.

b) What is meant by the net movement of particles?

142

Additional biology

Both types of particles can pass through this membrane – it is freely permeable

Steep concentration gradient

Beginning of experiment

Random movement means three blue particles have moved from left to right by diffusion

Shallow concentration gradient

Beginning of experiment

Four blue particles have moved as a result of random movement from left to right – but two have moved from right to left. There is a ***net*** movement of ***two*** particles to the right by diffusion

Figure 3 This diagram shows us how the overall movement of particles in a particular direction is more effective if there is a big difference (a steep concentration gradient) between the two areas. This is why so many body systems are adapted to maintain steep concentration gradients.

GET IT RIGHT!

Diffusion is **passive** – it takes place along a concentration gradient from high to low concentration and uses up no energy.

Concentration isn't the only thing that affects the rate of diffusion. An increase in temperature means the particles in a gas or a solution move more quickly. This in turn means diffusion will take place more rapidly as the random movement of the particles speeds up.

Diffusion in living organisms

Many important substances can move across your cell membranes by diffusion. Water is one. Simple sugars, such as glucose and amino acids from the breakdown of proteins in your gut, can also pass through cell membranes by diffusion. The oxygen you need for respiration passes from the air into your lungs and into your cells by diffusion.

Individual cells may be adapted to make diffusion easier and more rapid. The most common adaptation is to increase the surface area of the cell membrane over which diffusion occurs. Increasing the surface area means there is more room for diffusion to take place. By folding up the membrane of a cell, or the tissue lining an organ, the area over which diffusion can take place is greatly increased. So the amount of substance moved by diffusion is also greatly increased.

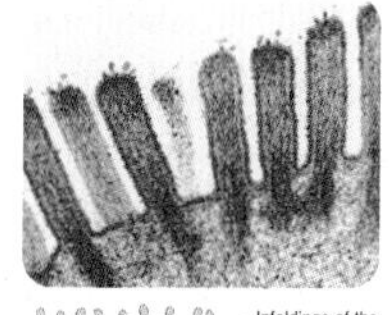

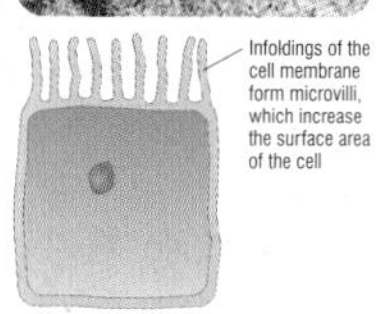
Infoldings of the cell membrane form microvilli, which increase the surface area of the cell

Figure 4 An increase in the surface area of a cell membrane means more diffusion can take place

SUMMARY QUESTIONS

Copy and complete using the words below:

Diffusion gas high low random solute

1 ...... is the net movement of particles of a ...... or a ...... from an area of ...... concentration to an area of ...... concentration as a result of the ...... movement of the particles.

2 Explain why a cut in water looks much worse than a cut on land in terms of diffusion and the movement of particles.

3 a) Explain why diffusion takes place faster as the temperature increases.
b) Explain in terms of diffusion why so many cells have folded membranes along at least one surface.

KEY POINTS

1 Dissolved substances move in and out of cells by diffusion, osmosis and active transport.
2 Diffusion is the net movement of particles from an area where they are at a high concentration to an area where they are at a lower concentration.

143

PLENARIES

Watching diffusion – Draw the tubes set up at the beginning of the lesson. (5 minutes)

Team game – Carry out a team game with students putting dried peas into a dish. One person has one pair of blunt forceps to pick up the peas and put them in to the dish, the other person has two pairs of blunt forceps to take them out using both hands. Count the peas in the dish at the start and then run the game for a minute, stop and recount the peas. Discuss the results and relate to the net movement of particles. (5–10 minutes)

Surface area to volume – Get students to fold up a sheet of A4 paper as small as they can get it without tearing. Measure the dimensions and try to fit it into a matchbox. (5–10 minutes)

SUMMARY ANSWERS

1 Diffusion, gas (solute), solute (gas), high, low, random.

2 In water the blood diffuses through the water. Its particles spread out through the water particles making the volume of blood released seem greater than the same volume of blood released on land.

3 **a)** Increase in temperature means an increase in rate of movement of particles – so random movement and collisions increase which increases the rate of diffusion.

b) Diffusion important to get substances like oxygen into or out of cells. The more surface area of membrane available, the more diffusion can take place.

B2 1.4 Osmosis

LEARNING OBJECTIVES

Students should learn:

- That water often moves across boundaries by osmosis.
- That osmosis is the diffusion of water through a partially permeable membrane from a dilute to a more concentrated solution.
- Differences in concentrations of solutes inside and outside cells cause water to move by osmosis.

LEARNING OUTCOMES

Most students should be able to:

- Define osmosis.
- Distinguish between diffusion and osmosis.
- Carry out an experiment to find out about the process of osmosis.

Some students should also be able to:

- Explain the importance of osmosis in plants and animals.
- Explain the results of experiments in terms of osmotic movement of water.

Teaching suggestions

- **Special needs.** Carry out a stop motion of a plant wilting and being re-hydrated using Intel play microscopes (the kit pot plastic ones). Explain using a football and a pump.
- **Gifted and talented.** Investigate the effect of partial drowning. What effect would it have on the water balance? Would people who have nearly drowned and had their lungs full of water for some time face any problems? If so, what might they be and how may they be overcome?
- **Learning styles**

  Kinaesthetic: Practical work on osmosis.

  Visual: Draw a warning poster for snails not to run while carrying the salt pot!

  Auditory: Participating in the 'Lower/higher' concentration activity with the juice and water.

  Interpersonal: Collecting and collating results from experiments.

  Intrapersonal: Considering what would happen if our osmotic barriers were removed.
- **Homework.** Students could use homework time to complete the writing up of the experiments or to make clear diagrams explaining diffusion and osmosis.

SPECIFICATION LINK-UP Unit: Biology 2.12.2

- *Water often moves across boundaries by osmosis. Osmosis is the diffusion of water from a dilute to a more concentrated solution through a partially permeable membrane that allows the passage of water molecules but not solute molecules.*
- *Differences in the concentrations of the solutions inside and outside a cell cause water to move into or out of the cell by osmosis.*

Lesson structure

STARTER

What happens to the chips – Show the students a bag of chips. Ask who is going to the chip shop tonight. At what time? Explain that there is always a rush on or about six o'clock, so the owners prepare the chips in advance and keep them in water. What effect does the water have on the chips? Draw out some ideas and ways of testing these ideas. A demonstration could be set up or this could lead into the experiments described in the Main section. (10 minutes)

Bouncy Castle – Show a picture of a Bouncy Castle. Has anyone got younger brothers or sisters who love these? How do they stay upright? Why don't they burst? What would happen if they were made out of flexible rubber like a thicker version of balloons? Draw out the idea of a balance of air going in, air coming out and pressure on a non-flexible skin providing support. Link with osmosis in plants providing support for plant tissues. (10 minutes)

MAIN

- **Modelling osmosis in cells** – Visking, or dialysis, tubing can be used to make model cells. Lengths of tubing, about 10 cm long, should be wetted thoroughly and one end of each tied firmly with string. Fill the tubing bags with a concentrated sugar solution (molar sucrose) and tie the open ends firmly with string. These Visking tubing bags represent cells and can be immersed in beakers of water, less concentrated and more concentrated sugar solutions.
- The results from the model cells, set up as described above, can be used to illustrate the principles of osmosis. Students can be asked to interpret each one in terms of the diffusion of water and sucrose molecules and the effect of the partially permeable membrane. Students may find it easier to understand osmosis if it is explained in terms of the diffusion of water molecules from where they are in high concentration (i.e. in a dilute solution) to where there is a lower concentration (i.e. a more concentrated solution). Diagrams help.
- **Osmosis in potato tissue** – Chips or discs of potato tissue can be immersed in different concentrations of salt or sugar solutions, left for a period of time and then their change in mass or dimensions measured. Such experiments offer opportunities for the introduction of 'How Science Works' concepts and can be used as whole investigations. The change in mass or length can be plotted against the concentration of the solution and the solution which results in the least change is considered to be equivalent to the concentration of the cell sap of the potato.
- There are variations on the above which can be tried. Some students could measure changes in dimensions (i.e. length or volume) and others could measure changes in mass. Are the results similar? Which do they consider to be the most accurate?
- **Setting up an osmometer** – A simple osmometer can be made using a length of Visking tubing, tied securely at one end, filled with a concentrated sugar solution (for quick results use syrup or treacle only slightly diluted) and then a capillary tube tied securely in the top. The whole apparatus is held in place by a clamp and stand and lowered into a beaker of water. The level of sucrose in the capillary tube is measured at the start and then again at regular intervals (5 minutes). A graph can be plotted of the distance moved by the sucrose against time.

PLENARIES

Followed up to What happens to the chips? – If the demonstrations were set up at the beginning of the experiment, they can looked at. What has happened to the chips? They can be measured, their texture assessed and the results discussed. (10 minutes)

Practical support

For the Visking tubing experiments, you will need:

- Lengths of dialysis (Visking) tubing for each group of students
- Molar sucrose solution which can be diluted according to the concentrations required
- Beakers
- String
- Small measuring cylinders or pipettes to fill the tubes.

For the potato experiments, you will need:

- Fairly large potatoes
- Cork borers to make cylinders of tissue
- Knives and tiles to cut chips
- Molar sucrose
- Boiling tubes and racks
- Rulers and balances
- Tissues or paper towels to dry potato discs or slices.

ACTIVITY & EXTENSION

- **Osmosis in animals** – Experiments on blood cells cannot be carried out, but use video footage or the diagrams in the Student Book to explain what happens to red blood cells if they are placed in solutions of different concentrations. This emphasises the importance of the correct concentration of the body fluids.
- **Wilting** – Show video footage of marathon runners (especially keeping up their fluids at drinks stations) and plants wilting and being revived.

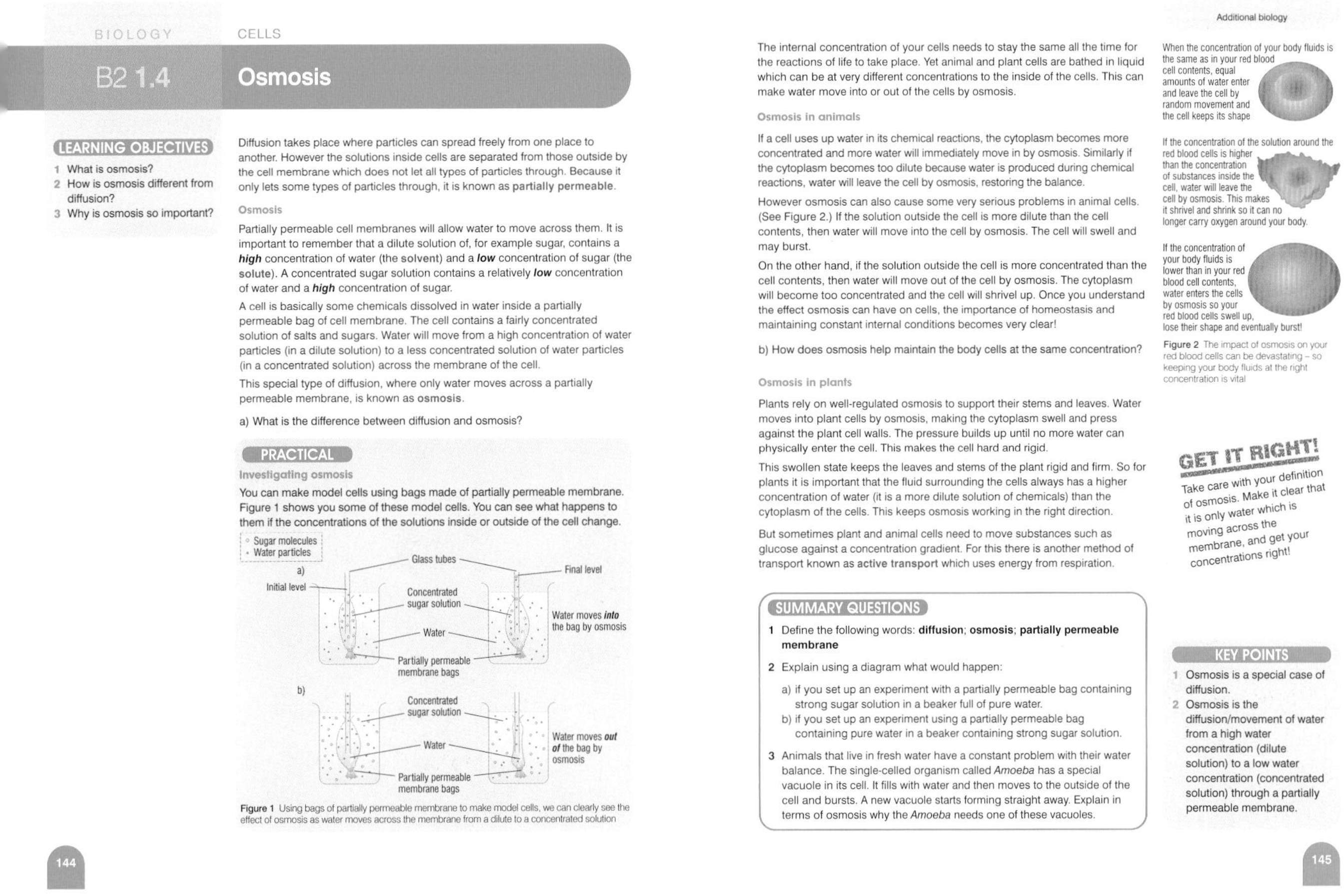

BIOLOGY CELLS

B2 1.4 Osmosis

LEARNING OBJECTIVES

1. What is osmosis?
2. How is osmosis different from diffusion?
3. Why is osmosis so important?

Diffusion takes place where particles can spread freely from one place to another. However the solutions inside cells are separated from those outside by the cell membrane which does not let all types of particles through. Because it only lets some types of particles through, it is known as **partially permeable**.

Osmosis

Partially permeable cell membranes will allow water to move across them. It is important to remember that a dilute solution of, for example sugar, contains a ***high*** concentration of water (the **solvent**) and a ***low*** concentration of sugar (the **solute**). A concentrated sugar solution contains a relatively ***low*** concentration of water and a ***high*** concentration of sugar.

A cell is basically some chemicals dissolved in water inside a partially permeable bag of cell membrane. The cell contains a fairly concentrated solution of salts and sugars. Water will move from a high concentration of water particles (in a dilute solution) to a less concentrated solution of water particles (in a concentrated solution) across the membrane of the cell.

This special type of diffusion, where only water moves across a partially permeable membrane, is known as **osmosis**.

a) What is the difference between diffusion and osmosis?

PRACTICAL

Investigating osmosis

You can make model cells using bags made of partially permeable membrane. Figure 1 shows you some of these model cells. You can see what happens to them if the concentrations of the solutions inside or outside of the cell change.

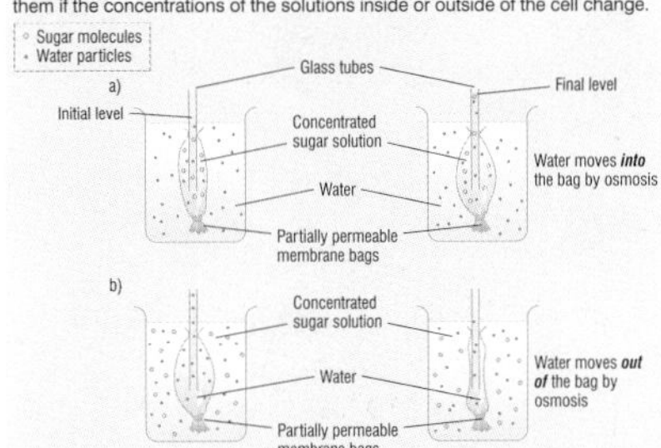

Figure 1 Using bags of partially permeable membrane to make model cells, we can clearly see the effect of osmosis as water moves across the membrane from a dilute to a concentrated solution

The internal concentration of your cells needs to stay the same all the time for the reactions of life to take place. Yet animal and plant cells are bathed in liquid which can be at very different concentrations to the inside of the cells. This can make water move into or out of the cells by osmosis.

Osmosis in animals

If a cell uses up water in its chemical reactions, the cytoplasm becomes more concentrated and more water will immediately move in by osmosis. Similarly if the cytoplasm becomes too dilute because water is produced during chemical reactions, water will leave the cell by osmosis, restoring the balance.

However osmosis can also cause some very serious problems in animal cells. (See Figure 2.) If the solution outside the cell is more dilute than the cell contents, then water will move into the cell by osmosis. The cell will swell and may burst.

On the other hand, if the solution outside the cell is more concentrated than the cell contents, then water will move out of the cell by osmosis. The cytoplasm will become too concentrated and the cell will shrivel up. Once you understand the effect osmosis can have on cells, the importance of homeostasis and maintaining constant internal conditions becomes very clear!

b) How does osmosis help maintain the body cells at the same concentration?

Osmosis in plants

Plants rely on well-regulated osmosis to support their stems and leaves. Water moves into plant cells by osmosis, making the cytoplasm swell and press against the plant cell walls. The pressure builds up until no more water can physically enter the cell. This makes the cell hard and rigid.

This swollen state keeps the leaves and stems of the plant rigid and firm. So for plants it is important that the fluid surrounding the cells always has a higher concentration of water (it is a more dilute solution of chemicals) than the cytoplasm of the cells. This keeps osmosis working in the right direction.

But sometimes plant and animal cells need to move substances such as glucose against a concentration gradient. For this there is another method of transport known as **active transport** which uses energy from respiration.

When the concentration of your body fluids is the same as in your red blood cell contents, equal amounts of water enter and leave the cell by random movement and the cell keeps its shape

If the concentration of the solution around the red blood cells is higher than the concentration of substances inside the cell, water will leave the cell by osmosis. This makes it shrivel and shrink so it can no longer carry oxygen around your body.

If the concentration of your body fluids is lower than in your red blood cell contents, water enters the cells by osmosis so your red blood cells swell up, lose their shape and eventually burst!

Figure 2 The impact of osmosis on your red blood cells can be devastating – so keeping your body fluids at the right concentration is vital

GET IT RIGHT!

Take care with your definition of osmosis. Make it clear that it is only water which is moving across the membrane, and get your concentrations right!

SUMMARY QUESTIONS

1. Define the following words: **diffusion**; **osmosis**; **partially permeable membrane**
2. Explain using a diagram what would happen:
   a) if you set up an experiment with a partially permeable bag containing strong sugar solution in a beaker full of pure water.
   b) if you set up an experiment using a partially permeable bag containing pure water in a beaker containing strong sugar solution.
3. Animals that live in fresh water have a constant problem with their water balance. The single-celled organism called *Amoeba* has a special vacuole in its cell. It fills with water and then moves to the outside of the cell and bursts. A new vacuole starts forming straight away. Explain in terms of osmosis why the *Amoeba* needs one of these vacuoles.

KEY POINTS

1. Osmosis is a special case of diffusion.
2. Osmosis is the diffusion/movement of water from a high water concentration (dilute solution) to a low water concentration (concentrated solution) through a partially permeable membrane.

SUMMARY ANSWERS

1 Diffusion – the net movement of dissolved substances or gases from an area of high concentration to an area of low concentration.
Osmosis – the net movement of water molecules from an area of high concentration to an area of low concentration through a partially permeable membrane.
Partially permeable membrane – a membrane which lets some particles through but not others.

2 a) See top part of Figure 1 in Student Book (page 144).
Water moves into the bag by osmosis. The bag becomes full and solution rises up the tube.

b) See bottom part of Figure 1 in Student Book (page 144).
Water moves out of the bag by osmosis. The bag shrinks and the level of the solution in the glass tube falls.

3 The cytoplasm of *Amoeba* contains a solution of salts and sugars so it contains a lower concentration of water particles than the water in which the organism lives. The cell membrane is partially permeable, so water constantly moves into *Amoeba* from its surroundings by osmosis. If this continued without stopping, the organism would burst. Water could be moved into the vacuole by active transport, and then the vacuole moved to the outside of the cell using energy as well. This could prevent the organism from exploding.

Answers to in-text questions

a) Osmosis is a special type of diffusion in which only water particles move across a partially permeable membrane.

b) Water moves back and forth across cell membranes to adjust concentrations.

KEY POINTS

A good definition of osmosis is essential and worth memorising by the students. It is also important to refer to *concentrations* of solutions rather than 'stronger' or 'weaker'.

B2 1.5 Cell issues

SPECIFICATION LINK-UP

Unit: Biology 2.12.1 and 12.2

This spread gives the opportunity to revisit substantive content covered in this chapter:

- *Cells may be specialised to carry out a particular function.*
- *Dissolved substances can move into and out of cells by diffusion and osmosis.*

BIOLOGY CELLS

B2 1.5 Cell issues

Discovering cells

Over the past three centuries our ideas about cells have developed as our ability to see them has improved. In 1665, the English scientist Robert Hooke designed the first working microscope and saw cells in cork.

At around the same time a Dutchman, Anton van Leeuwenhoek, also produced a microscope. It enabled him to see bacteria, microscopic animals and blood cells for the first time ever.

Almost two centuries later, by the 1840s, scientists had accepted that cells are the basic units of all living things. From then on, as optical microscopes improved, more details of the secret life inside a cell were revealed as cells were magnified up to 1000 times.

With the invention of the electron microscope in the 1930s it became possible to magnify things much more. We can now look at cells magnified up 500 000 times!

Anton van Leeuwenhoek (1632–1723)

Human cheek cells (magnified 3500 times)

Cork cells drawn by Robert Hooke

A human white blood cell at high magnification

The ability to see cells and the secret worlds inside them has developed in an amazing way since the days of the early microscopes

ACTIVITY

Produce a timeline to show how microscopes have developed since they were first invented. Annotate your timeline to show how important our discoveries about cells have been.

146

Teaching suggestions

- **Discovering cells** – Search the web for images of microscopes to show the class (try www.wikipedia.org). Start with Leeuwenhoek's and Hooke's microscopes and include some information about dates and what they observed through them. It would then be useful to include some different forms of light microscope, distinguishing between: simple, consisting of one lens; or compound, where there are separate objective and eyepiece lenses or lens systems. It is worth considering the magnification possible with these microscopes. It could be appropriate to show the same specimen viewed at the different magnifications (a simple slide of an insect leg or some plant tissue would be suitable). Other types of light microscope, such as binocular or stereoscopic, could be included as well, particularly if there are some available for the students to use.
- You might like to progress to electron microscopes, describing the differences between the transmission electron microscope (which requires smears or thin sections of material) and the scanning electron microscope (which does not involve the use of sections). There are good pictures of these available and some good images to illustrate what can be seen when specimens are viewed. Again, some idea of magnification can be given.
- If the school, or any students, have connections with a research or university department that has an electron microscope, then it could be worth trying to arrange a visit to see one being used. Many establishments are quite willing to demonstrate the preparation and viewing of specimens to small groups of interested students.
- **Beating osmosis** – The animals depicted in the Student Book spread live in different habitats and it could be helpful to students to start with the characteristics and osmotic problems associated with each habitat before considering the animals present. This can be done using a PowerPoint® presentation and providing students with work sheets for them to fill in. The suggested habitats could be freshwater, marine and terrestrial initially, but discussion could enable these to be subdivided. For example, do still freshwater habitats (ponds and lakes) differ from moving freshwater habitats (streams and rivers)? Some students might be interested in what happens in a tidal area.
- Write up a list of animals, to include invertebrates as well as vertebrates, for each major habitat. Discuss how each type of animal copes in that habitat. For example, a freshwater habitat species list could include protozoa, cnidarians, leeches, worms, insect larvae and crustaceans as well as fish and frogs. Draw up a list of different strategies for coping with osmosis.
- Students could be given the lists and sent away to do their own research using reference books or the Internet.
- **The special case of the salmon and the eels** – Salmon and eels spend part of their lives in freshwater and part of their lives in the sea, so they need to cope with the changing osmotic conditions. Students could research the life cycles and how these fish adapt.

Beating osmosis

The cells of all living organisms contain sodium chloride and other chemicals in solution. This means they can always be prone to water moving into them by osmosis. If they are immersed in a solution with a lower concentration of salts than the body cells they will tend to gain water. If in a more concentrated solution, water is lost. Either way can spell disaster. Here are just a few of the different ways in which living things attempt – largely successfully – to beat osmosis!

ACTIVITY

You have to produce some GCSE revision sheets on the exchange of materials in living organisms. Use the examples given here to help you make the sheets as lively and interesting as possible. Use any tricks which help ***you*** to remember things – and save the sheets to help you when exams are approaching!

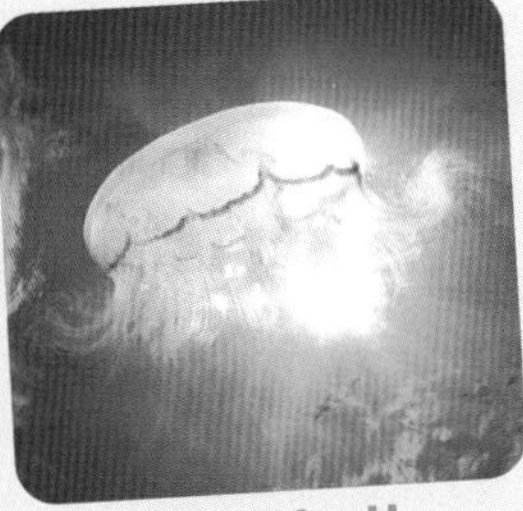

No contest!

For many marine invertebrates like this jellyfish, osmosis causes no problems because the concentration of solutes in the cells of their bodies is exactly the same as the sea water. So there is no net movement of water in or out of the cells.

Copy cats!

Living on land causes all sorts of problems for the cells, particularly if water is lost and the body fluids get concentrated. Then water will leave the body cells by osmosis fast. Many insects have taken a leaf out of the plant's book – they have a tough, waterproof outer layer which prevents water loss from the body surface. They even have breathing holes known as ***spiracles***. These can be closed up when they aren't needed – very like the stomata on the leaves of plants.

Flooding in

Fish that live in fresh water have a real problem. They need a constant flow of water over their gills to get the oxygen they need for respiration. But water moves into their gill cells and blood by osmosis at the same time. Like all vertebrates, fish have kidneys which play a big part in using osmosis to regulate their internal environment. So freshwater fish produce huge amounts of very dilute urine, which gets rid of the excess water that gets into their bodies. They also have special salt-absorbing glands. These use active transport to move salt against the concentration gradient from the water into the fish – rather like the situation in plant root cells.

The big ones

Marine vertebrates like this whale are constantly drinking salty water. The salt loading would cause water to move out of their body cells and kill them if they couldn't deal with it. Fortunately whales have extremely efficient kidneys. When a whale drinks 1000 cm^3 of sea water, it produces 670 cm^3 of very concentrated urine – and gains 330 cm^3 of pure water.

ACTIVITY & EXTENSION IDEAS

- **Discovering cells** –The activity suggested above can be started by using photographs or drawings of different types of microscope stuck to cards. On the reverse side, write the dates (approx.) when the microscopes were first invented. Ask the students to try to sequence these cards (without looking at the back, of course). It is a bit difficult to pinpoint exact dates for some, but reference to a textbook on microscopy or a suitable web site should provide the answers. Once the sequence is established, then the timeline can be established and the annotations can be discussed. Students could make individual annotated timelines, choosing their own important discoveries. Alternatively, this could be a feature of the laboratory and the class could discuss what they thought were the important discoveries to be attached to the timeline.
- It might be relevant here to remind students that Pasteur was able to show the presence of microorganisms in his broth that went bad. Without a microscope this would have been difficult. Although this is not a discovery about cells, it is an example of how microscopes can make an important contribution to research.
- **The exchange of materials in living organisms** – There is plenty of material for the suggested GCSE revision sheets activity and students could make their own or work in groups to produce a set for each major habitat. Working in groups can help to cover many different organisms and the results pooled to make a class revision resource. Before embarking on the format of the sheets, students could decide what type of approach would make the information interesting and help them to remember the facts. This activity could make use of ICT in order to find out more information and also in the production of the revision resource.
- **Individual case histories** – It could be useful to make sheets for individual species such as salmon, frogs or camels, or for groups such as marine fish, mammals and desert plants. Students should be reminded of the general nature of the activity and that it is not just about osmosis but other materials as well.
- **Don't forget the plants!** – The suggested activity refers to 'living organisms' so the introduction of some plants into the lists for each major habitat could be of benefit. Some species of thick seaweeds such as bladderwrack or *Laminaria* (broad, strap-shaped structures) could be shown to those students not familiar with seaweeds.

Special needs

Students could be given simplified lists of animals for each major habitat and asked to say whether water would go into the cells of their bodies or whether they would lose water from their cells.

Gifted and talented

Imagine a leap in resolving power, as great as that of the microscope happening, today and a large amount of new detail about the Universe becoming visible. Ask: 'What would the reactions be?' Students could write an e-mail to a friend telling them all about this fantastic new discovery. Ask: 'Will there be a finite limit to the resolving power of electron microscopes? If so, why? Could anything theoretically take their place to give even higher resolutions?'

SUMMARY ANSWERS

1 Based on labelled diagram of *Chlamydomonas*:

a) Nucleus, chloroplast, starch, cytoplasm.

b) Flagellum/flagella for moving around; eye spot for sensing light.

c) [Any sensible answer using scientific explanations.]

2 Based on a sensory nerve cell, a white blood cell and a cell with rough ER (endoplasmic reticulum) with mitochondria and ribosomes.

a) [Marks for correct labels.]

b) Sensory nerve cell: carrying nerve impulses.
White blood cell: defending the body against pathogens; part of the immune system; destroying/engulfing pathogens.
Active/secretory cell: making proteins/enzymes to be used in the body.

c) Sensory nerve cell: sensory receptor to respond to changes; long axon to carry impulses long distances around body; synapse to pass impulse; to other nerve cells; transmitter substance in the synapse to transmit impulse across gap.
White blood cell: makes antibodies on the surface; can flow and engulf bacteria etc.
Secretory cell: lots of ribosomes for protein synthesis; lots of mitochondria to provide energy.

Summary teaching suggestions

Most of the answers required are based on the diagrams or passage provided, and require facts so there is little scope for differentiating between lower and higher level answers. Higher attaining students should give full answers to question 1c), perhaps not coming to a definite conclusion but giving both animal and plant features.

- **Special needs** – Question 2 could be modified, by providing students with the unlabelled diagrams and the labels and functions on stickers that could then be attached in the right places.
- **When to use the questions?**
  - These are useful revision questions. Question 2 could be particularly useful in revising the specialisation of cells. Students could find it helpful to have a diagram of the cell and then to add annotations to the labels, thus linking structure to function.
  - Question 1 revises the differences between plant and animal cells.

CELLS: B2 1.1 – B2 1.5

SUMMARY QUESTIONS

1

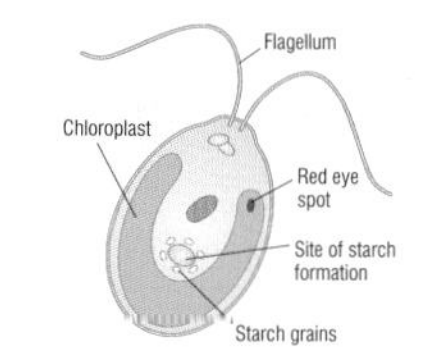

Chlamydomonas is a single-celled organism which lives under water. It can move itself to the light to photosynthesise, and stores excess food as starch.

a) What features does it have in common with most plant cells?

b) What features are not like plant cells and what are they used for?

c) Would you class *Chlamydomonas* as a plant cell or an animal cell? Explain why.

2

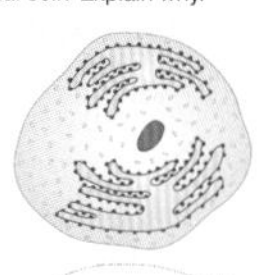

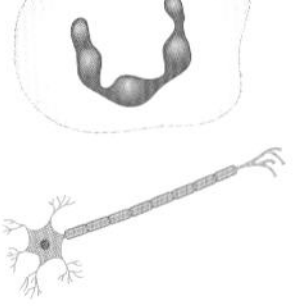

Each of these cells is specialised for a particular function in your body.

a) Copy each of these diagrams and label the cells carefully. Carry out some research if necessary.

b) Describe what you think is the function of each of these cells.

c) Explain how the structure of the cell is related to its function.

EXAM-STYLE QUESTIONS

1 The diagram is of a cell from the leaf of a plant.

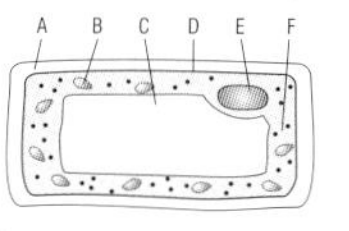

(a) Name the structures **D**, **E** and **F**.

(b) (i) What is the name of structure A?
(ii) What material is structure **A** made of?

(c) (i) What is the name of structure **C**?
(ii) What is the liquid it contains called?

(d) Structure **B** is a chloroplast. What is its function?

(e) Name two different structures that are found wit the material labelled **F**.

(f) (i) A different type of plant cell is a root hair ce What is the function of this type of cell?
(ii) State one way in which a root hair cell differs from the leaf cell shown in the diagram.

2 Copy the table below. Look at the structures listed in the first column. Fill in the empty columns by putting tick (✓) if you think it is present and a cross (✗) if you think it is absent.

Structure	Animal cell	Plant cell
Nucleus		
Cytoplasm		
Cell wall		
Cell membrane		
Chloroplast		
Permanent vacuole		

3 A student noticed that different trees give different amounts of shade on a sunny day. She decided to investigate three species of tree – oak, sycamore and ash. She thought that the more shading, the better th tree was at gathering light for photosynthesis. She wc use a light meter to record the light levels. The stude had many things to consider when deciding on a method.

(a) Should she take readings in direct sunlight as wel under the trees? Explain your answer.

EXAM-STYLE ANSWERS

1 a) **D** Cell membrane
E Nucleus
F Cytoplasm *(1 mark each*

b) i) Cell wall *(1 mark)* ii) Cellulose *(1 mark*

c) i) Vacuole *(1 mark)* ii) Cell sap *(1 mark*

d) *Either* Absorbs light energy *(1 mark*
to make/manufacture food/sugars *(1 mark*
Or (Carries out) photosynthesis *(2 marks*

The idea of light absorption and making food is needed for both marks. Either point separately earns one mark. Photosynthesis embraces both ideas and warrants both marks.

e) Ribosomes *(1 mark)* Mitochondria *(1 mark*

f) i) To absorb water/minerals (from the soil into the plant) *(1 mark*

ii) • A root hair cell has a much larger surface area (to volume ratio)/a large hair-like projection.
• A root hair cell has no chloroplasts.
(1 mark for either point

Where the leaf cell, rather than the root hair cell, is quoted the converse points are allowed. Where 'it' is used, always assume the root hair cell unless it is very clear that this is not the case.

2

Structure	Animal cell	Plant cell
Nucleus	✓	✓
Cytoplasm	✓	✓
Cell wall	✗	✓
Cell membrane	✓	✓
Chloroplast	✗	✓
Permanent vacuole	✗	✓

(1 mark for each correct pair of symbols on any one line

(b) Describe the weather that would be most appropriate when collecting the data. (1)

(c) Should the student collect data from one or more than one position? Explain your answer. (1)

(d) Explain why it would be necessary for the student to take as many readings as she could under the trees. (1)

(e) What type of independent variable has the student decided to use? (1)

(f) What type of dependent variable has she decided to use? (1)

(g) How should the student calculate the mean for each set of results? (1)

(h) Suggest how she should present her data. (1)

List **A** gives the names of different types of cells found in plants and animals. List **B** gives one special feature of each of these cells. Match each cell type with its feature by writing the relevant letter and number next to one another. (6)

List A	List B
A Fat cell	**1** Has a long tail with muscle-like proteins
B Root hair cell	**2** Can divide and change into many different types of cell
C Sperm cell	**3** Contains chloroplasts
D Leaf cell	**4** Can expand up to 1000 times its original size
E Stem cell	**5** Contains a chemical called visual pigment
F Cone cell (in eye)	**6** Has extension to increase its surface area

HOW SCIENCE WORKS QUESTIONS

Spinning cells!

It is possible to separate the different parts of a cell using a centrifuge. Your teacher might be able to show you one of these. They really are very simple. They spin around rather like a very fast spin dryer.

They are used to separate structures that might be mixed together in a liquid. One of their uses is to separate the different parts of a cell.

The cells are first broken open so that the contents spill out into the liquid. The mixture is then put into the centrifuge. The centrifuge starts to spin slowly and a pellet forms at the bottom of the tube. This is removed. The rest is put back into the centrifuge at a higher speed and the next pellet removed and so on.

Here are some results:

Centrifuge speed (rpm)	Part of cell in pellet
3000	Nuclei
10000	Mitochondria
12000	Ribosomes

(rpm = revolutions per minute)

a) From these observations can you suggest a link between the speed of the centrifuge and the size of the part of the cell found in the pellet? (1)

b) What apparatus would you need to test your suggestion? (1)

c) **i)** What was your independent variable? (1)

ii) Is your independent variable best described as categoric, discrete or continuous? (1)

d) What was your dependent variable? (1)

e) If your suggestion is correct, what results would you expect? (1)

f) What would be the easiest measurement to make to show the size of the mitochondria? (1)

g) Suggest how many mitochondria you might measure. (1)

h) How would you calculate the mean for the measurements you have taken? (2)

3 a) Yes – because she would not know how much light there was shining on the tree *(1 mark)*
OR Yes – so that she could calculate the efficiency of light gathering for photosynthesis. *(2 marks)*

b) Sunny/no clouds *(1 mark)*

c) More than one position – because the shade will vary *(1 mark)*

d) To improve the reliability of the data collected *(1 mark)*

e) Categoric *(1 mark)*

f) Continuous *(1 mark)*

g) Add up all of the results from under the tree and divide by the number of results *(1 mark)*

h) Bar chart *(1 mark)*

4 **A** 4

B 6

C 1

D 3

E 2

F 5 *(1 mark for each correct pairing)*

Exam teaching suggestions

- This test is designed primarily for foundation-level students. It tests recall and the responses needed are short with no narrative prose required. There is scope for students to pick up marks relatively easily; in question 1, part d) for example, students are asked the function of a chloroplast. The allocation of two marks for the word 'photosynthesis' might appear generous. It does however allow the not inconsiderable number of students who write 'absorbs light' to obtain one mark for a partial answer. Allowance has also been made for simple responses. In question 1, part f) for example, the words in brackets are not required for the mark, although one might expect a higher-tier candidate to include this type of detail. Part f) also illustrates the important issue of specifying in answers which of two items is being referred to when two things are compared. In this case, as the root hair cell comes first in the question (is the subject of the sentence) where a student uses 'it' they are assumed to be writing about the root hair cell. Students should, however, be discouraged from using 'it' and persuaded to use the specific name of the item and so avoid any ambiguity.
- In question 1 part a) it is very easy for students to confuse structures A (cell wall) and D (cell membrane). The need to look very carefully at what exactly the label line touches should be impressed on students.

HOW SCIENCE WORKS ANSWERS

a) The slower the centrifuge spins at the larger the cell part separated. Or reverse argument.

b) Microscope. With attachment to measure, e.g. length. Very sensitive instrument.

c) **i)** The independent variable is the centrifuge speed.
ii) Continuous variable.

d) The dependent variable is the size of part of the cell (or part of cell) in pellet.

e) For the results you would expect the mean size of part of the cell in the pellet to be larger with slower centrifuge speed.

f) Measure the length – because they are 'cigar-shaped'.

g) As many as possible! But a minimum of 10.

h) Add up all the measurements and divide by how many there are.

How science works teaching suggestions

- **Literacy guidance**
  - Key terms that should be clearly understood: how observations can be used to make hypotheses and predictions, independent, dependent variable, sensitivity.
  - Question e) requires a longer answer, where students can practise their clarity of expression skills.
- **Higher- and lower-level answers**
  - Questions b) and e) are higher level and the answers above are also at this level.
  - The lower-level questions are a), c) and d) and answers are also at this level.
- **Gifted and talented.** Able students could suggest at what speeds other cell fragments might be found.
- **How and when to use these questions.** When wishing to encourage students to use observations to formulate hypotheses, for predictions and testing.
- **Homework.** The questions could be introduced in class and then used for homework.
- **Special needs.** Students may well need to be shown a working centrifuge – safely!
- **ICT link-up.** Students could search the Internet for photographs of cell fragments and their sizes.

B2 2.1 Photosynthesis

LEARNING OBJECTIVES

Students should learn that:

- Light energy is absorbed by the chlorophyll in the chloroplasts of green plants.
- Light energy is used by converting carbon dioxide and water into sugar.
- Oxygen is released as a by-product.

LEARNING OUTCOMES

Most students should be able to:

- Summarise the process of photosynthesis in a word equation.
- Describe where the energy comes from and how it is absorbed.
- Describe experiments that show the raw materials needed and the resulting products.

Some students should also be able to:

- Explain the build up of sugars into starch during photosynthesis.

Teaching suggestions

- **Special needs.** To illustrate the point that starch is made from sugar, use Duplo or other large building blocks. Take six or eight blocks of the same colour and label each individual block 'sugar'. Stick the blocks together and place a large sticker, labelled 'starch', over the back of all of them. Cut apart into individual blocks before using. A small number on each sugar block enables you to assemble them in the correct order.
- **Gifted and talented.** Carry out the practical exercise with photosynthetic algae encapsulated in gel, based on *Science and Plants in Schools* (SAPS) student worksheet 23: 'Photosynthesis using algae wrapped in jelly balls'. Ask the students to use chemical symbols for the photosynthesis equation and to balance it.
- **Learning styles**

  Kinaesthetic: Party poppers activity; practical work testing leaves for starch.

  Visual: Making observations on leaves and describing them.

  Auditory: Listening to the song 'Feed me' from the video clip.

  Interpersonal: Discussing Audrey from *Little Shop of Horrors*.

  Intrapersonal: Individual feedback on the lesson.

SPECIFICATION LINK-UP Unit: Biology 2.12.3

- *Photosynthesis is summarised by the equation:*

  carbon dioxide + water (+ light energy) → glucose + oxygen
- *During photosynthesis:*
  - *Light energy is absorbed by a green substance called chlorophyll, which is found in chloroplasts in some plant cells.*
  - *This energy is used by converting carbon dioxide and water into sugar (glucose).*
  - *Oxygen is released as a by-product.*

Lesson structure

STARTER

Why are leaves green? – Lead a discussion based on concept cartoon-style talking head. Revise light reflection and absorbance. (5–10 minutes)

Audrey, the plant – If easily available show a video clip from the Frank Oz (1986) musical version of *Little Shop of Horrors* with Audrey, the plant, eating people. Ask: 'Was she really a plant if she eats meat?' Discuss how plants feed themselves. (10 minutes)

What will happen to my leaf? – During the growing season (or if you have plants in a greenhouse), give each student a spot label on to which they can write their initials. Allow them to choose a young leaf and stick their label on to it. They should then measure the length of the leaf and record it. Back in the laboratory, ask them to predict what will happen to the leaf and explain why. They will need to check this at a later date. (10 minutes)

MAIN

- **How leaves are adapted for photosynthesis** – In this exercise, the adaptations of leaves for the process of photosynthesis are investigated. Students should look at whole leaves and make drawings of the external appearance, labelling the features that they can see (broad lamina, green colour, veins) and annotating these labels to explain the adaptations.
- **Observing leaves** – Prepared microscope slides of transverse sections through leaves could be projected or viewed under the microscope so that students can distinguish the different tissues within the leaf. Point out the palisade tissue, the vascular tissue and the stomata. Students could draw plans of the tissues to show where photosynthesis takes place.
- **How can we show that photosynthesis has taken place?** – When carrying out experiments on photosynthesis, we can test for the products i.e. the presence of sugars or the evolution of oxygen. In most plants, the sugars are immediately converted to starch (shown by the presence of starch grains in chloroplasts). The starch test can then be used on leaves to show that photosynthesis has occurred.
- **Producing oxygen** – We can show that oxygen has been producing using water plants such as *Elodea* (Canadian pondweed) which is readily available from garden centres. The apparatus can be set up as described and kept illuminated for several hours, so that enough gas can be collected to be able to test it satisfactorily. If groups of students set up their own, it is unlikely to yield enough gas to test within a lesson.
- If students carry out the experiment, individually or in groups, several of the concepts of 'How Science Works' are introduced. They can formulate a hypothesis, make predictions, draw conclusions and evaluate the validity of experimental design.
- **Testing for starch** – To show that chlorophyll is necessary for photosynthesis, variegated plants, such as spider plant or geranium, can be used. The plants need to be kept in bright light for several hours. Keep one plant in the dark for two days to destarch it as a control. Each student can be given a leaf from an illuminated plant. A record should be made of the distribution of the green and white areas of the leaf, before testing for starch. After carrying out the test, another drawing can be made showing the areas that remain brown and those that have been stained blue/black. Comparison of the two drawings will enable a conclusion to be drawn. Testing a leaf from the control plant will show that if there is no light then no starch will be produced.

PLENARIES

Summary – Use a summary of photosynthesis with missing words. (5 minutes)

Prove it! – Write on the board, or project, a number of statements about photosynthesis. Then the students have to write out or discuss how we know each of the statements is true. (10 minutes)

ACTIVITY & EXTENSION IDEAS

- An Animation, B2 2.1 'Photosynthesis', is available on the CSE Biology CD ROM.
- **Demonstration of formation of glucose molecule** – Into one hand of each of six students, put a piece of charcoal and a balloon labelled O_2. Explain that they each represent a molecule of carbon dioxide. Ask the rest of the class to form two lines facing each other with a gap of 4 or 5 metres between them – they represent a leaf. Two students are designated as guard cells and must form an arch (barn dance style) to allow the CO_2 students to enter. In advance, remove the paper streamers from six party poppers, put a hole in the top of each and attach a large loop of string to it. Attach a similar loop of string to the pull string on the base of the popper. Get the six students to stand in a circle and put their arms through the loops so that the party poppers are suspended between them. Six more students, each carrying a bottle of water, are then to come up between the rows. They swap their bottles of water for the balloons of oxygen, so that the ring represents a molecule of glucose and the students carrying the balloons exit through the stoma. This enactment can be reversed to show respiration: the oxygen comes back in, each grabs a carbon and pulls it away, breaking the bonds between the party poppers and leaving the water behind. Students could then find out the actual structure of glucose.

Practical support

Equipment and materials required

Observing leaves

Prepared slides of sections through leaves and microscopes.

Testing for starch

Variegated plants, such as spider plant or geranium, dilute iodine solution in dropping bottles (CLEAPSS Hazcard 54), water baths to kill the leaves/make them more permeable/softer, ethanol for decolourising leaves/removing chlorophyll, white tiles or dishes to put the leaves in, forceps.

Answers to in-text questions

a) carbon dioxide + water (+ light energy) → glucose + oxygen

b) The green substance that absorbs light energy in plants.

c) Provides a large surface area for the light to fall on.

BIOLOGY — HOW PLANTS PRODUCE FOOD

B2 2.1 Photosynthesis

LEARNING OBJECTIVES

1. What is photosynthesis?
2. What are the raw materials for photosynthesis?
3. Where does the energy for photosynthesis come from and how do plants absorb it?

Like all living organisms, plants need food. It provides them with the energy for respiration, growth and reproduction. But plants aren't like us – they don't need to eat.

Plants can make their own food! They do it by **photosynthesis**. This takes place in the green parts of plants (especially the leaves) when it is light.

The process of photosynthesis

Photosynthesis can be summed up in the following equation:

carbon dioxide + water (+ light energy) → glucose + oxygen

The cells in the leaves of a plant are full of small green parts called **chloroplasts**. They contain a green substance called **chlorophyll**.

During photosynthesis, light energy is absorbed by the chlorophyll in the chloroplasts. This energy is then used to convert carbon dioxide from the air plus water from the soil into a simple sugar called **glucose**. The chemical reaction also produces oxygen gas. This is released into the air.

a) What is the word equation for photosynthesis?

Some of the glucose produced during photosynthesis is used immediately by the cells of the plant. However, a lot of the glucose made is converted into starch for storage.

Iodine solution is a yellowy-brown liquid which turns dark blue when it reacts with starch. You can use this ***iodine test for starch*** to show that photosynthesis has taken place in a plant.

PRACTICAL

Producing oxygen

You can show a plant is photosynthesising by collecting the oxygen given off as a by-product. It is very difficult to see oxygen, a colourless gas, being given off by land plants. But if you use water plants you can collect the gas which they give off when they are photosynthesising. It will relight a glowing splint, showing that it is oxygen gas.

PRACTICAL

Testing for starch

Chlorophyll is vital for photosynthesis to take place. It absorbs the light which provides the energy for the plant to make glucose and convert it into starch.

Take a leaf from a variegated plant (partly green and partly white). After treating the leaves, you use iodine solution to show how important chlorophyll is. (See Figure 1.)

- What happens in the test? Explain your observations

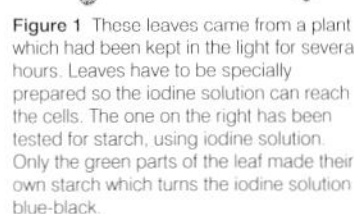

Figure 1 These leaves came from a plant which had been kept in the light for several hours. Leaves have to be specially prepared so the iodine solution can reach the cells. The one on the right has been tested for starch, using iodine solution. Only the green parts of the leaf made their own starch which turns the iodine solution blue-black.

b) What is chlorophyll?

The leaves of plants are perfectly adapted because:

- most leaves are broad, they have a big surface area for light to fall on,
- they contain chlorophyll in the chloroplasts to absorb the light energy,
- they have air spaces which allow carbon dioxide to get to the cells, and oxygen to leave them,
- they have veins, which bring plenty of water to the cells of the leaves.

150

Additional biology

All of these adaptations mean the plant can carry out as much photosynthesis as possible whenever there is light available.

c) How does the broad shape of leaves help photosynthesis to take place?

PRACTICAL

Observing leaves

Look at a whole plant leaf and then a section of a leaf under a microscope. You can see how well adapted it is.

- Compare what you can see with Figure 2 below.
- What magnification did you use?

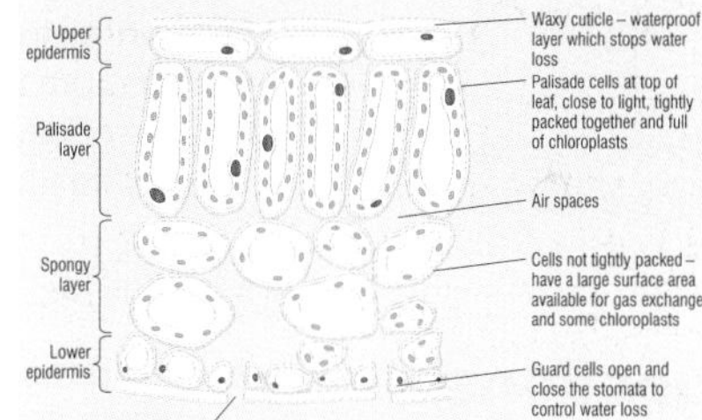

Upper epidermis
Palisade layer
Spongy layer
Lower epidermis
Waxy cuticle – waterproof layer which stops water loss
Palisade cells at top of leaf, close to light, tightly packed together and full of chloroplasts
Air spaces
Cells not tightly packed – have a large surface area available for gas exchange and some chloroplasts
Guard cells open and close the stomata to control water loss
Stomata like this allow gases to move in and out of the leaf

Figure 2 A section through a leaf

GET IT RIGHT!

Learn the equation for photosynthesis.
Be able to explain the results of experiments on photosynthesis.

NEXT TIME YOU...

. . . breathe in, remember that the oxygen in the air you are breathing was produced as a by-product of photosynthesis by plants. Luckily for us, the world's plants produce about 368 000 000 000 tonnes of oxygen every year!

SUMMARY QUESTIONS

1 Copy and complete using the words below:

carbon dioxide chlorophyll energy gas glucose light Oxygen water

During photosynthesis ...... energy is absorbed by ......, a substance found in the chloroplasts. This ...... is then used to convert .... from the air and ...... from the soil into a simple sugar called .... is also produced and released as a ......

2 a) Where does a plant get the carbon dioxide, water and light that it needs for photosynthesis?
b) Work out the path taken by a carbon atom as it moves from being part of the carbon dioxide in the air to being part of a starch molecule in a plant.

3 Design experiments to show that plants need a) carbon dioxide and b) light for photosynthesis to take place. For each experiment explain what your control would be and how you would show that photosynthesis has taken place.

KEY POINTS

1. Photosynthesis can be summed up by the equation:

   carbon dioxide + water [+ light energy] → glucose + oxygen

2. During photosynthesis light energy is absorbed by the chlorophyll in the chloroplasts. It is used to convert carbon dioxide and water into sugar (glucose). Oxygen is released as a by-product.
3. Leaves are well adapted to allow the maximum photosynthesis to take place.

151

SUMMARY ANSWERS

1 Light, chlorophyll, energy, carbon dioxide, water, glucose, oxygen, gas.

2 **a)** CO_2 comes from the air; water from the soil; light energy from the Sun/electric light.

b) From the air into the air spaces in the leaf; into plant cells; into chloroplasts; joined with water to make glucose; converted to starch for storage.

3 [In both cases, give credit for sensible suggestions, awareness of difficulties, controls (fair test).]

KEY POINTS

- It is important to be able to write out the summary equation for the process of photosynthesis.
- It is also important to be able to explain the results of experiments and to be able to describe how leaves are adapted to carry out photosynthesis.
- These key points would make good revision cards, although some of the adaptations of the leaves could be added to the last point.

B2 2.2 Limiting factors

LEARNING OBJECTIVES

Students should learn that:

- The rate of photosynthesis may be limited by low temperature and the shortage of carbon dioxide or light.
- These factors interact.
- If any of these factors are in short supply, the rate of photosynthesis is limited.

LEARNING OUTCOMES

Most students should be able to:

- List the factors that limit the rate of photosynthesis.
- Describe how the factors interact.
- Describe how the environment in which plants grow can be artificially manipulated to grow more food.

Some students should also be able to:

- Interpret data showing how the factors affect the rate of photosynthesis.
- Explain why the rate of photosynthesis is limited by low temperature, shortage of carbon dioxide or shortage of light.

SPECIFICATION LINK-UP Unit: Biology 2.12.3

- *The rate of photosynthesis may be limited by:*

  - *low temperature*
  - *shortage of carbon dioxide*
  - *shortage of light.*
- *Light, temperature and the availability of carbon dioxide interact and, in practice, any one of them may be the factor that limits photosynthesis.*

Lesson structure

STARTER

The limiting factors game – Have three sorts of counters (or small cards): one set labelled 'L' for suitable light level, another 'T' for suitable temperature and the third 'CO_2'. For each group of students, you will need a bag into which you place some of each set of counters (or cards), so that they get a mixture of 'L', 'T' and 'CO_2'. The students are to take out a counter one at a time, placing them on a base line on paper or on the desk. The aim is to make sets of three counters side by side, at which point they can start on the next layer as they have grown. If they draw out a counter which they have already got in that layer, they put it back. The objective is to grow the 'plant' to as many levels as possible during a set time. Adjust each bag's contents so that some groups run out of 'L' counters first, some of 'T' counters and some of 'CO_2' ones. Discuss the results. (10–15 minutes)

Tree rings – Examine some cross sections of tree branches using hand lenses or binocular microscopes. Ask: 'Why are there rings present?' Draw out the links between growth rate and temperature and light intensity as limiting factors. (5–10 minutes)

Oxygen production and light intensity – Show students the pondweed under the test tube practical from the previous lesson. Ask them to draw a thumbnail graph of how the rate of oxygen production would vary with light intensity. Draw on to individual ('Show Me') boards. (5 minutes)

MAIN

- **How does the intensity of light affect the rate of photosynthesis?** – This experiment is illustrated in the Student Book and is easy to set up. Students can work in groups and vary the light intensity by altering the distance of the lamp from the plant.
- This is a good experiment for developing 'How Science Works' concepts. A hypothesis can be formulated, predictions made, variables such as temperature controlled, measurements taken, results expressed as graphs and conclusions drawn. Results can be plotted as number of bubbles evolved, in a set time, against distance of the lamp from the plant. A more accurate way of plotting the results is to use light intensity, given by $1/d^2$ where d is the distance of the lamp from the plant.
- **Alternative ways of investigating variations in light intensity:**
  - A Simulation, B2 2.2 'Limiting Factors of Photosynthesis' is available on the GCSE Biology CD ROM.
  - Consider using a mercury vapour lamp light bank of the type recommended by *Science and Plants in Schools* (SAPS) for use with their 'Rapid cycling brassicas' kit. This could be an investigation into the effects of light intensity on growth rates. (Details from SAPS, Homerton College, Cambridge CB2 2PH.) Use various filters and light sensors.

Teaching suggestions

- **Special needs**
  - Students could benefit from the demonstration using the hydrogencarbonate indicator solution if it was explained that the deeper the purple colour the more photosynthesis had occurred.
  - Draw analogies with football teams. One team is held back in the league by not having enough good strikers; another may be weak in the mid-field or defence. (The students will give lots of examples!)
- **Gifted and talented.** Draw out predicted graphs for light intensity, CO_2 level and temperature during a typical day during a named month of the growing season. Explain the shapes of these graphs bearing in mind respiration. Contrast with an early spring or late autumn day.
- **Learning styles**

  Kinaesthetic: Counter activity in starter; practical work.

  Visual: Observing bubbling, colour changes etc. in practicals.

  Auditory: Listening to the explanations given.

  Intrapersonal: Analysing graph-tabulated data on limiting factors.

  Interpersonal: Discussing ideas.
- **Homework**
  - Writing up the reports of experiments could be set as a homework task.
  - Alternatively, the summary questions (particularly questions 2 and 3) could be set as homework and will test understanding of the limiting factors.
- **ICT link-up.** Use the computer simulation available on the GCSE Biology CD ROM to model the effect on plant growth of varying limiting factors.

PLENARIES

Finish off the graph – Give the students some semi-completed graphs showing limiting factors to finish off and label. (5–10 minutes)

Question loop – Students to ask and then answer questions on factors limiting photosynthesis. (5 minutes)

ACTIVITY & EXTENSION

- Try to arrange a visit to a commercial glasshouse or encourage students to find out how conditions are controlled.
- Use a computer simulation where conditions for plant growth are varied, such as the one suggested above.
- Carry out a data-logging exercise, recording the temperature and light levels as they fluctuate throughout the day.

Practical support

Equipment and materials required

How does the intensity of light affect the rate of photosynthesis?

Sprigs of *Elodea,* boiling tubes and test tube racks, bench lamps, rulers, beakers, funnels, stopwatches or stop clocks, graph paper (see centre of Figure 1 on page 152 in Student Book).

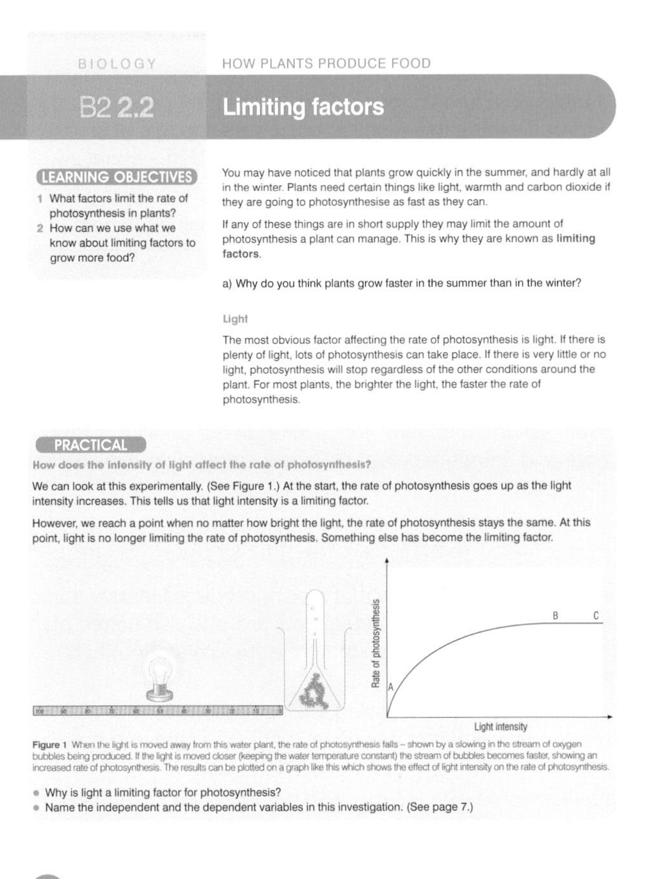

BIOLOGY HOW PLANTS PRODUCE FOOD

B2 2.2 Limiting factors

LEARNING OBJECTIVES

1. What factors limit the rate of photosynthesis in plants?
2. How can we use what we know about limiting factors to grow more food?

You may have noticed that plants grow quickly in the summer, and hardly at all in the winter. Plants need certain things like light, warmth and carbon dioxide if they are going to photosynthesise as fast as they can.

If any of these things are in short supply they may limit the amount of photosynthesis a plant can manage. This is why they are known as **limiting factors.**

a) Why do you think plants grow faster in the summer than in the winter?

Light

The most obvious factor affecting the rate of photosynthesis is light. If there is plenty of light, lots of photosynthesis can take place. If there is very little or no light, photosynthesis will stop regardless of the other conditions around the plant. For most plants, the brighter the light, the faster the rate of photosynthesis.

PRACTICAL

How does the intensity of light affect the rate of photosynthesis?

We can look at this experimentally. (See Figure 1.) At the start, the rate of photosynthesis goes up as the light intensity increases. This tells us that light intensity is a limiting factor.

However, we reach a point when no matter how bright the light, the rate of photosynthesis stays the same. At this point, light is no longer limiting the rate of photosynthesis. Something else has become the limiting factor.

Figure 1 When the light is moved away from this water plant, the rate of photosynthesis falls – shown by a slowing in the stream of oxygen bubbles being produced. If the light is moved closer (keeping the water temperature constant) the stream of bubbles becomes faster, showing an increased rate of photosynthesis. The results can be plotted on a graph like this which shows the effect of light intensity on the rate of photosynthesis.

- Why is light a limiting factor for photosynthesis?
- Name the independent and the dependent variables in this investigation. (See page 7.)

152

Additional biology

Temperature

Temperature affects all chemical reactions, including photosynthesis. As the temperature rises, the rate of photosynthesis will increase as the reaction speeds up. However, because photosynthesis takes place in living organisms it is controlled by enzymes. Enzymes are destroyed once the temperature rises to around 40 to 50°C. This means that if the temperature gets too high, the rate of photosynthesis will fall as the enzymes controlling it are denatured.

b) Why does temperature affect photosynthesis?

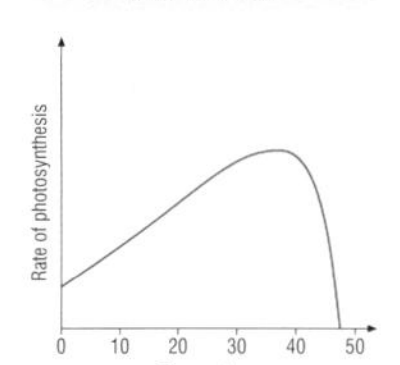

Figure 2 The rate of photosynthesis increases steadily with a rise in temperature up to a certain point. After this the enzymes are destroyed and the reaction stops completely.

Carbon dioxide levels

Plants need carbon dioxide to make glucose. The atmosphere only contains about 0.04% carbon dioxide, so carbon dioxide levels often limit the amount of photosynthesis which can take place. Increasing the carbon dioxide levels will increase the rate of photosynthesis.

For the plants you see around you on a sunny day, carbon dioxide levels are the most common limiting factor. The carbon dioxide levels around a plant tend to rise in the night as it respires but doesn't photosynthesise. Then as the light and temperature levels increase in the morning, the carbon dioxide all gets used up.

However, in a laboratory or in a greenhouse the levels of carbon dioxide can be increased artificially. This means they are no longer limiting, and the rate of photosynthesis increases with the rise in carbon dioxide.

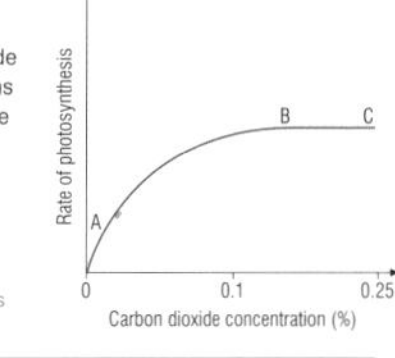

Figure 3 This graph shows the effect of increasing carbon dioxide levels on the rate of photosynthesis at a particular light level and temperature. Eventually one of the other factors becomes limiting.

DID YOU KNOW?

There are a few plants that live in very shady areas which have evolved to photosynthesise at their maximum at relatively low levels of light. For them, too much light causes the rate of photosynthesis to drop!

GET IT RIGHT!

Make sure you can explain limiting factors.
Learn to interpret graphs which show the effect of limiting factors on photosynthesis.

SUMMARY QUESTIONS

1. a) What is photosynthesis?
   b) What are the three main limiting factors that affect the rate of photosynthesis in a plant?
2. Which factors do you think would be limiting photosynthesis in the following situations? In each case, explain why the rate of photosynthesis is limited.
   a) Plants growing on a woodland floor in winter.
   b) Plants growing on a woodland floor in summer.
   c) A field of barley first thing in the morning.
   d) The same field later on in the day.
3. Look at the graph in Figure 1.
   a) Explain what is happening between points A and B on the graph.
   b) Explain what is happening between points B and C on the graph.
   c) Look at Figure 2. Explain why it is a different shape to the other two graphs on this spread.

KEY POINTS

1. There are three main factors that limit the rate of photosynthesis – light, temperature and carbon dioxide levels.
2. We can artificially change the environment in which we grow plants. We can use this to observe the effect of different factors on the rate of photosynthesis. We can also use it to control their rate of photosynthesis.

153

SUMMARY ANSWERS

1. **a)** The process by which plants use light energy trapped by chlorophyll to convert carbon dioxide and water into glucose (sugar).
   b) Carbon dioxide, light and temperature.
2. **a)** Light and temperature.
   b) Light and carbon dioxide.
   c) Temperature and light.
   d) Carbon dioxide.
3. **a)** As light intensity increases so does the rate of photosynthesis. This tells us that light intensity is a limiting factor.
   b) An increase in light intensity has no effect on the rate of photosynthesis, so it is no longer a limiting factor; something else probably is.
   c) Temperature acts as a normal limiting factor to begin with; increase in temperature increases the rate of photosynthesis. But after a certain level, the enzymes in the cells are destroyed and so no photosynthesis can take place.

Answers to in-text questions

a) It is warmer in summer and there is more light, so photosynthesis takes place more quickly making more food so plants grow.

b) Photosynthesis is a chemical reaction; temperature affects all chemical reactions. An increase in temperature speeds up the reactions as reacting particles collide more frequently and with more energy.

KEY POINTS

The key points are reinforced by the 'Get it right!' comments. Students could benefit from discussing the interaction of the factors and thinking about how the factors can be altered to control the rate of photosynthesis. Some examples of how factors are controlled in the production of glasshouse crops would be useful. For example, light, temperature and carbon dioxide levels can be controlled in order to maintain production of salad crops all year round.

B2 2.3 How plants use glucose

LEARNING OBJECTIVES

Students should learn that:

- Glucose is converted into starch for storage.
- Some of the glucose produced is used for respiration in the plant.

LEARNING OUTCOMES

Most students should be able to:

- Describe how and where carbohydrates are stored in plants.
- State that some of the carbohydrate produced is used in respiration.

Some students should also be able to:

- Explain that the energy released by plants in respiration is used to build up smaller molecules into larger molecules.
- Explain why starch is such a good storage molecule.

Teaching suggestions

- **Special needs.** Give the students samples of glucose and corn starch. Tell them to stir the powders into two beakers of water. They should observe what happens and make comments on their solubility in a simple fashion. Then ask them to filter the contents of both beakers and isolate and dry the corn starch.
- **Gifted and talented.** Students could research other ways in which plants store food. Link with question 2 of 'Summary questions'.
- **Learning styles**

  Kinaesthetic: Testing leaves and plant parts for starch.

  Visual: Making observations using microscopes.

  Auditory: Explaining the storage process to other students.

  Intrapersonal: Writing about what happens to glucose.

  Interpersonal: Discussing the results of the investigations.
- **Homework.** Students could finish off the poster suggested in a plenary or write a plant's eye view of what it does with the glucose it produces.

SPECIFICATION LINK-UP Unit: Biology 2.12.3

- *The glucose produced in photosynthesis may be converted into insoluble starch for storage. Plant cells use some of the glucose produced during photosynthesis for respiration.*

Lesson structure

STARTER

How much energy in a jelly baby? – Heat a couple of centimetres depth of sodium chlorate in a test tube until it melts. In a fume cupboard, and with great care, drop a jelly baby into the tube. A very vigorous exothermic reaction occurs issuing lots of smoke and flame and noise. Carry out a risk assessment and practise first. Use a metal-jawed clamp. (CLEAPSS Hazcard 77) (5 minutes)

Make a starch molecule – Give the students blank drawings of a chain of joined hexagons. Ask them to write 'glucose' inside each hexagon and then to turn the paper over and write 'starch' across the whole of the back. They can then stick the chains into their books as a fold-out. (5 minutes)

Showing that respiration has occurred – Issue all the students with drinking straws, and then give half of them boiling tubes containing a little fresh lime water in the bottom and the other half boiling tubes with a little hydrogencarbonate indicator solution in the bottom. Wearing safety glasses, ask them to blow gently through the drinking straws into the solutions. They should note and compare colour changes, suggesting explanations. (10 minutes)

MAIN

- **Making starch** – There are many ways to show that a plant produces starch, some of which have already been described. The most straightforward is to use potted plants: some should be kept in the dark for 48 hours (so that they are destarched) and others kept in daylight conditions.
- Students could be provided with worksheets, giving the instructions for the procedure, and then asked to test a leaf that has been kept in the light and one that has been kept in the dark.
- Leaves on plants can be half-covered with foil or initials cut out of thin card, kept in bright light for several hours and then tested for starch.
- **Where is the starch stored?** – The presence of large numbers of starch grains in potato tuber cells can be demonstrated by cutting thin slices of the tissue. Place the thin slices of tissues on microscope slides, cover with a drop of water and then with a cover slip. A drop of dilute iodine solution can be drawn through using filter paper. The starch grains will stain blue-black. In order to see the grains more clearly, it is advisable to draw some water through the slide to remove the iodine solution.
- The technique described above can be used on a variety of plant parts. Very thin sections of tissue from fruits, seeds, nuts and other plant organs can then be tested for the presence of starch grains.
- Roughly compare the starch content of fruits, such as apples, with potato tubers and some seeds.

PLENARIES

From the air to a chip – In small groups, the students could produce a series of bullet points of the stages from carbon dioxide in the air to the starch in the chips on their plates. (10 minutes)

Matching exercise – Write up a list of definitions and key words about photosynthesis and ask students to match them up. (5–10 minutes)

Formation, use and storage of glucose – Ask students to produce a poster showing how glucose is produced, what it is used for and how it is stored in the plant. They can finish this for homework. (10–15 minutes)

ACTIVITY & EXTENSION IDEAS

- Carry out the party popper demonstration reversing the process described for photosynthesis in order to demonstrate respiration. (See Activity and Extension box on page 151 in this Teacher Book.)
- Use microscopes to examine prepared slides of sections of stems, roots and leaves to show the distribution of the vascular (transporting) tissue.
- To demonstrate that plants respire, keep some *Elodea* in a boiling tube of hydrogencarbonate indicator. The boiling tube will need to have foil around it or be kept in the dark, so that photosynthesis does not occur. The cherry-red colour should turn yellow as it becomes more acid due to the evolution of carbon dioxide. Compare with the starter activity that shows respiration has occurred.

Practical support

Equipment and materials required

Making starch

Destarched and illuminated plants, water baths for killing leaves/making them permeable, ethanol to decolourise the leaves/remove chlorophyll, dilute iodine solution in dropping bottles (CLEAPSS Hazcard 54), white tiles, forceps for handling leaves, eye protection.

Where is the starch stored?

Microscopes, slides, cover slips, a variety of plant parts (potato tubers, fruits, seeds, nuts, etc.), dilute iodine solution in dropping bottles (CLEAPSS Hazcard 54), eye protection.

B2 2.3 How plants use glucose

LEARNING OBJECTIVES

1. What do plants do with the glucose they make?
2. How do plants store food?

NEXT TIME YOU…

… tuck into a plate of chips or a pile of mashed potato, remember that you are eating the winter food store of a potato plant! The starch you are enjoying was formed from glucose made in the leaves of the potato plant by photosynthesis. It was transported down from the leaves to the roots to form a tasty tuber!

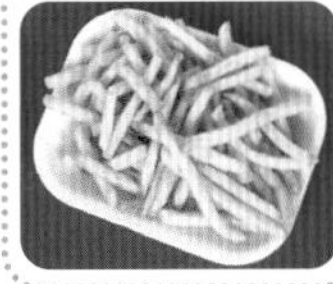
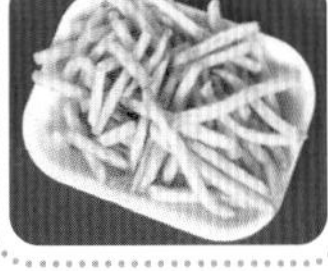

Plants make glucose when they photosynthesise. This glucose is vital for their survival. Some of the glucose produced during photosynthesis is used immediately by the cells of the plant. They use it for respiration and to provide energy for cell functions, growth and reproduction.

Respiration

Plants cells, like any other living cells, respire all the time. They break down glucose using oxygen to provide energy for their cells. Carbon dioxide and water are the waste products of the reaction.

The energy released in respiration is then used to build up smaller molecules into bigger molecules. Some of the glucose is converted into starch for storage (see below). Plants also build up sugars into more complex carbohydrates like cellulose. They use this to make new plant cell walls.

Plants use some of the energy from respiration to combine sugars with other nutrients from the soil to make amino acids. These amino acids are then built up into proteins to be used in the cells. Energy from respiration is also used to build up fats and oils to make a food store in the seeds.

a) Why do plants respire?

Transport and storage

Plants make food by photosynthesis in their leaves and other green parts. However, the food is needed all over the plant. It is moved around the plant in a special transport system.

There are two separate transport systems in plants. The **phloem** is made up of living tissue. It transports sugars made by photosynthesis from the leaves to the rest of the plant. They are carried to all the areas of the plant. These include the growing regions where the sugars are needed for making new plant material, and the storage organs where they are needed to provide a store of food for the winter.

The **xylem** is the other transport tissue. It carries water and mineral ions from the soil around the plant.

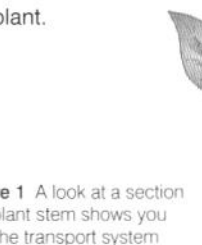
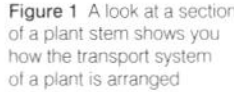
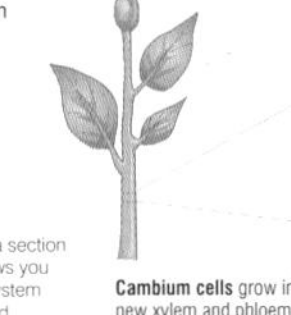
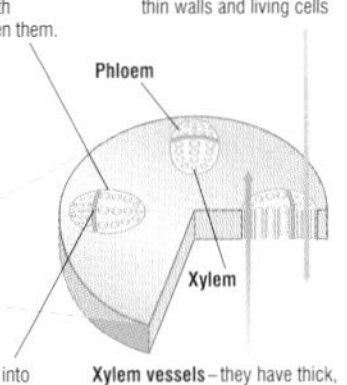
A vascular bundle. It contains **xylem** and **phloem** with **cambium cells** between them.
Phloem tubes – they have thin walls and living cells
Phloem
Xylem
Cambium cells grow into new xylem and phloem
Xylem vessels – they have thick, strong walls and are not living

Figure 1 A look at a section of a plant stem shows you how the transport system of a plant is arranged

GET IT RIGHT!

Remember:
- Plants respire 24 hours a day to release energy.
- Glucose is soluble in water, but starch is insoluble.

Plants convert some of the glucose produced in photosynthesis into starch to be stored. Glucose is **soluble** (it dissolves in water). If it was stored in plant cells it could affect the way water moves into and out of the cells. Large amounts of glucose stored in the plant cells could affect the water balance of the whole plant. Starch is **insoluble** (it doesn't dissolve in water). This means that plants can store large amounts of starch in their cells without it having any effect on the water balance of the plant.

So the main energy store in plants is starch and it is found all over a plant. It is stored in the leaves to provide an energy store for when it is dark or when light levels are low.

PRACTICAL

Making starch

You can use the presence of starch in a leaf as evidence that photosynthesis has been taking place. It is no good just adding iodine to a leaf – the waterproof cuticle and the green chlorophyll will prevent it reacting clearly with the starch. But once you have treated the leaf, adding iodine will show you clearly if the leaf has been photosynthesising or not. Look at Figure 2.

Figure 2 We use the iodine test for the presence of starch to show us that photosynthesis has taken place. The leaf on the right has been kept in the dark. It has made no glucose to turn into starch, and has used up any starch stores it had for respiration. The leaf on the left has been in the light and been able to photosynthesise. The glucose has been converted to starch which is clearly visible when it reacts with iodine and turns blue-black. The colour is removed from the leaves before testing by boiling them in ethanol.

Starch is also stored in special storage areas of a plant. Many plants produce tubers and bulbs to help them survive through the winter. These are full of stored starch. We often take advantage of these starch stores and eat them ourselves. Potatoes, carrots and onions are all full of starch to keep a plant going until spring comes again!

b) What is the main storage substance in plants?

SUMMARY QUESTIONS

1. Copy and complete using the words below:

   energy glucose growth photosynthesise respiration reproduction starch storage twenty-four

   Plants make …… when they …… . Some of the glucose produced is used by the cells of the plant for …… which goes on …… …… hours a day. It provides …… for cell functions, …… and …… . Some is converted to …… for …… .
2. List as many ways as possible in which a plant uses the glucose produced by photosynthesis.
3. a) Why is the glucose made by photosynthesis converted to starch to be stored in the plant?
   b) Where might you find starch in a plant?
   c) How could you show that a potato is a store of starch?

FOUL FACTS

Deer may look cute and cuddly – but they often kill trees. The transport tissue of a tree is found in the bark which deer love to nibble. If they nibble away a complete ring of bark, it is a disaster. Food made by photosynthesis in the leaves and water taken up by the roots cannot reach the rest of the plant – and it dies!

Figure 3 Trees like this giant redwood can be up to 30 metres tall – and then the roots spread out in all directions underground. Plants need a very effective transport system to move the food they make in their leaves distances like these.

KEY POINTS

1. Plant cells use some of the glucose they make during photosynthesis for respiration.
2. Some of the soluble glucose produced during photosynthesis is converted into insoluble starch for storage.

SUMMARY ANSWERS

1. Glucose, photosynthesise, respiration, 24, energy, growth, reproduction, starch, storage.
2. Respiration; energy for cell functions; growth; reproduction; building up smaller molecules into bigger molecules; converted into starch for storage; making cellulose; making amino acids; building up fats and oils for a food store in seeds.
3. **a)** Glucose is soluble and would affect the movement of water into and out of the plant cells. Starch is insoluble and so does not disturb the water balance of the plant.

   b) Leaves, stems, roots and storage organs.

   c) [Any sensible suggestions involving a slice of potato and dilute iodine solution.]

Answers to in-text questions

a) To provide chemical energy for their cells.

b) Starch.

KEY POINTS

The key points can be incorporated into the Plenaries 'From the air to a chip' and into the suggested poster activity.

B2 2.4 Why do plants need minerals?

LEARNING OBJECTIVES

Students should learn that:

- Plant roots absorb mineral ions (salts) needed for healthy growth.
- Nitrate is needed for protein formation and magnesium is needed for chlorophyll.
- Plants show deficiency symptoms if mineral ions are lacking.

LEARNING OUTCOMES

Most students should be able to:

- State that plants need magnesium ions to make chlorophyll and nitrate ions to make proteins.
- Describe the deficiency symptoms if nitrate and magnesium are lacking.

Teaching suggestions

- **Special needs.** Growing plants is always helpful for SEN students. Carry out on a long timescale, growing tomato plants (or sunflowers) in the early summer, with and without nitrogenous fertiliser. Each week (or more frequently if growing very quickly) hold a strip of 2–3 cm wide coloured paper next to the plant and cut the paper off at the same height as the plant. Stick the strips on to a bar chart frame, using a different colour each time. The students can predict the next and subsequent heights.
- **Gifted and talented.** Look at the structure of typical amino acids [without the structure of the R groups – this is AS level] and relate to the structure of nitrates and the need of the plant for nitrogen. They could also find out about the need for some of the other mineral ions. They could think about the differences between an etiolated plant (one grown completely in the dark) and a plant grown in a magnesium-deficient medium. Ask: 'Is there a connection?'
- **Learning styles**
  Kinaesthetic: Practical using plants.
  Visual: Checking leaves for symptoms.
  Auditory: Taking part in the Greenfingers/Blackfingers game.
  Interpersonal: Evaluation of evidence from growth experiments.
  Intrapersonal: Writing a report of findings from comparing plants.
- **ICT link-up.** Look up 'Plant mineral deficiencies' on the Internet. Reddy, T.Y. and Reddi, G.H.S. produced a good identification chart for deficiency symptoms (1997). Some good pictures of mineral-deficient tomato plants are also available.

SPECIFICATION LINK-UP Unit: Biology 2.12.3

- *Plant roots absorb mineral salts including nitrate needed for healthy growth. For healthy growth, plants need mineral ions including:*
  - *nitrate – for producing amino acids that are then used to form proteins.*
  - *magnesium – which is needed for chlorophyll production.*
- *The symptoms shown by plants growing in conditions where mineral ions are deficient include:*
  - *stunted growth if nitrate ions are deficient.*
  - *yellow leaves if magnesium ions are deficient.*

Lesson structure

STARTER

Why do we need these? – Show images of a tractor and fertiliser spreader, together with some packets of lawn fertiliser and Baby Bio or similar. (Search the web for 'fertiliser spreader' or 'lawn fertiliser'.) Ask the students to attempt to explain why these products are used. (5–10 minutes)

What is the connection? – Burn some magnesium tape (risk assess – wear eye protection and protect eyes from glare – CLEAPSS Hazcard 59). Open a bag of crisps (to release the nitrogen they are filled with to prevent oxidation). Show students a large houseplant, as big and as gaudy as possible. Ask for the connection between the three things. (10 minutes)

MAIN

- **Investigating the effect of minerals** – Provide three specimens of tomato plants per group: one plant in good condition having had all the nutrients it needs, one short of nitrogen (grown in peat rather than potting compost; add a pinch of magnesium sulphate to the water at intervals so that it does not show symptoms of magnesium deficiency) and one grown in magnesium-deficient conditions.
- If providing the plants is a problem, large coloured photographs (laminated for future use) could be provided.
- Ask students to examine and tabulate the differences. Apart from the obvious deficiency symptoms, they could measure leaves, height etc. Lead a discussion and ask the students to draw conclusions.
- **Hydroponics** – Students could then set up their own sets of plants using water culture (hydroponics). Broad bean or cereal seedlings could be used and grown in small flasks or bottles (root development can also be observed in this way). The seedlings should all be at the same stage of growth, as the plants need to grow. This is a good experiment for introducing the concepts of 'How Science Works'. Hypotheses can be formulated, predictions made, variables considered and controlled, and measurements taken. The cultures need to be aerated at intervals and the containers covered to prevent the growth of photosynthetic algae. It is possible to use duckweed in a water culture experiment. It has the advantage of growing more quickly and the growth can be assessed by the number of leaves produced. It will also show the deficiency symptoms.
- **Where does the nitrate come from?** – Draw out the sequence of events from nitrogen in the air (remind the students of the percentage) to the protein in plants. You can start with the nitrates in the soil as the centre of a spider diagram or flow chart. This can be accompanied by a modelling activity of 'Pass the N'.

PLENARIES

PowerPoint® – Show the main learning points of the lesson with blanks to complete. (5 minutes)

Greenfingers club and Blackfingers club – Give out coloured hands – in pairs, one student (Greenfingers) to give instructions as to how to get the plant to grow really well including nutrient reference, with reasons. The other (Blackfingers) is to give instructions as to how to get the plant to wither horribly and develop ghastly symptoms, with reasons including nutrient reference. (10 minutes)

ACTIVITY & EXTENSION

- **Root nodules of legumes** – Dig up some leguminous plants (peas, beans, clover) and wash the roots carefully. The root nodules should be obvious. Allow students to examine these and then discuss the relationship between the nitrogen-fixing bacteria and the crop. Link to the nitrogen cycle and crop rotation.
- **Other mineral ions** – Water culture experiments could include plants grown in solutions deficient in other mineral ions such as iron, phosfate etc. There are water culture tablets available to make up the appropriate solutions.
- **Do plants recover?** – Try giving some fertiliser, such as Baby Bio, to plants that have been grown in mineral-deficient conditions. Students can find out if they lose their symptoms and grow to become healthy plants. (They could try this activity at home.)
- **What is in the fertilisers?** – Students to investigate the components of the lawn fertiliser and Baby Bio and any other fertilisers, by looking at the boxes or containers. This will introduce the idea of commercial fertilisers not just containing one mineral ion, especially if the need for other mineral ions is demonstrated in the water culture experiments.

Practical support

Equipment and materials required

Investigating the effect of minerals

Tomato plants in pots, culture solutions lacking magnesium and nitrate.

Hydroponics

Small flasks or bottles, culture solutions lacking magnesium and nitrate, an aquarium aerator, kitchen foil to cover flasks or bottles.

Answers to in-text questions

a) Carbohydrates/glucose/sugars/starch.
b) To make proteins.
c) To make chlorophyll.

BIOLOGY — HOW PLANTS PRODUCE FOOD

Additional biology

B2 2.4 Why do plants need minerals?

LEARNING OBJECTIVES

1. What happens if plants don't get enough nitrates?
2. Why do fertilisers make your vegetables grow so well?

If you put a plant in a pot of water, and give it plenty of light and carbon dioxide, it won't survive for very long! Although plants can make their own food by photosynthesis, they cannot survive long on photosynthesis alone.

Just as you need minerals and vitamins for healthy growth, so plants need more than simply carbon dioxide, water and light to thrive. They need mineral salts from the soil to make the chemicals needed in their cells.

Figure 1 The plants on the left of this picture have been grown in a mixture containing all the minerals they need. The experimental plants on the right have been grown without nitrates. The difference in their rate of growth is clear to see.

Why do plants need nitrates?

The problem with the products of photosynthesis is that they are all carbohydrates. Carbohydrates are very important. Plants use them for energy, for storage and even for structural features like cell walls. However, a plant can't function without proteins as well. It needs proteins to act as enzymes and to make up a large part of the cytoplasm and the membranes.

a) What are the products of photosynthesis?

Glucose and starch are made up of carbon, hydrogen and oxygen. Proteins are made up of amino acids which contain carbon, hydrogen, oxygen and **nitrogen**. Plants need **nitrates** from the soil to make proteins.

These nitrates, dissolved in water, are taken up from the soil by the plant roots. If a plant is deficient in nitrates (doesn't have enough) it doesn't grow properly. It is small and stunted. So nitrates are necessary for healthy growth.

When plants die and decay the nitrates and other minerals are returned to the soil to be used by other plants.

b) Why do plants need nitrates?

Why do plants need magnesium?

It isn't only nitrates that plants need to grow well. There is a whole range of ***mineral ions*** they need. For example, plants need **magnesium** to make chlorophyll.

Chlorophyll is vital to plants. It is chlorophyll which absorbs the energy from light which makes it possible for plants to photosynthesise. So if the plant can't make chlorophyll, it can't make food and it will die. This is why magnesium ions are so important for plants – they make up part of the chlorophyll molecule.

Plants only need a tiny amount of magnesium. However, if they don't get enough, they have pale, yellowish areas on their leaves where they cannot make chlorophyll.

c) Why do plants need magnesium ions?

If any of the mineral salts that a plant needs are missing it will begin to look very sickly. This is true in the garden and for houseplants just as much as for crops in a farmer's field.

156

PRACTICAL

Investigating the effect of minerals

You can grow young plants in water containing different combinations of minerals and see the effect on their growth.

- Why are some plants grown in water with no minerals added and some with all the minerals they need?

Figure 2 The leaf on the right came from a plant that had received all the minerals it needed. The plant on the left was grown without magnesium. It is easy to see which is which – just look for the yellow patches!

If there are not enough mineral ions in the soil, your plants cannot grow properly. They will show the symptoms of mineral deficiencies. If you can pick up the symptoms soon enough and give them the mineral ions that they need, all will be well. If not, your plants will die!

Mineral ion	Why needed?	Deficiency symptoms
nitrate	making protein	stunted growth
magnesium	making chlorophyll	pale, yellow leaves

The most recent development in growing crops is **hydroponics**. You don't plant your crops in soil. Instead you plant them in water to which you add the minerals your plants need to grow as well as possible.

Hydroponic crops are usually grown in massive greenhouses where all the other factors can be controlled as well. Everything is monitored and controlled by computers 24 hours a day. The crops are very clean – no mud on the roots! And you can grow crops very quickly, and even out of their usual season.

All this means you get a good price for them. The downside is that it is an expensive way to farm, and it uses a lot of resources.

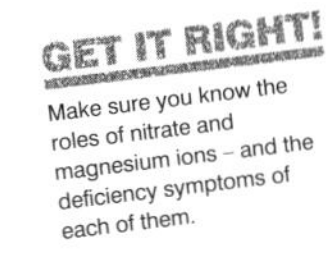

GET IT RIGHT!

Make sure you know the roles of nitrate and magnesium ions – and the deficiency symptoms of each of them.

Figure 3 If you are a farmer you harvest the crops that you grow and sell them. They are not left to die and decay naturally, returning minerals to the soil. So farmers add fertiliser to the soil to replace the minerals lost, ready for the next crop. The fertiliser may be a natural one like manure or an artificial mixture of the minerals that plants need to grow.

SUMMARY QUESTIONS

1. a) Why do plants need mineral ions?
   b) Where do they get minerals ions from?
   c) Which mineral ion is needed by plants to form proteins?
2. a) Look at the plants in Figure 1. Describe how the plants grown without nitrates differ from the plants grown with all the mineral ions they need. Why are they so different?
   b) Look at the plants in Figure 2. Describe how the plants grown without magnesium differ from the plants grown with all the mineral ions they need. Why are they so different?
3. Explain the following in terms of the mineral ions needed by plants and how they are used in the cells:
   a) Farmers spread animal manure on their fields.
   b) Gardeners recommend giving houseplants a regular mineral feed containing nitrates and magnesium.
   c) If the same type of crop is grown in the same place every year it will gradually grow less well and becomes stunted, with pale, patchy leaves.

KEY POINTS

1. Plant roots absorb mineral salts including nitrate needed for healthy growth.
2. Nitrates and magnesium are two important mineral ions needed for healthy plant growth.
3. If mineral ions are deficient, a plant develops symptoms because it cannot grow properly.

157

SUMMARY ANSWERS

1. **a)** To make the chemicals in their cells, such as proteins and chlorophyll.
   b) The water in the soil.
   c) Nitrate.
2. **a)** They are smaller and more stunted than the plants with everything. Nitrates are needed to make the amino acids that form proteins for enzymes and cell structure. Without nitrates they cannot make proteins properly so they cannot grow properly either.
   b) Their leaves have pale, yellowy patches compared to healthy plants. Magnesium is needed to make the green chemical chlorophyll. Without magnesium it cannot make enough chlorophyll so the leaves are not completely green.
3. **a)** Farmers harvest crops. They are not left to die and decay naturally, returning minerals to the soil. So farmers add fertiliser to the soil to replace the minerals lost, ready for the next crop.
   b) Houseplants are usually in pots with limited soil and so have a limited supply of minerals. Regular feeding makes sure they do not use up all the minerals they need e.g. nitrates for amino acid/protein production and magnesium as part of chlorophyll molecule.
   c) The same type of plant always takes up the same minerals so growing in the same soil each year, the minerals will become depleted. Plants show signs of mineral deficiencies, such as lack of nitrogen (stunted) and lack of magnesium (pale leaves).

KEY POINTS

Students must make sure they know why plants need nitrates and magnesium, and what the plants look like if they do not get enough of these minerals. Growing the plants or looking at good pictures of plants showing symptoms of the deficiencies will help students to remember the facts.

B2 2.5 Plant problems?

SPECIFICATION LINK-UP

Unit: Biology 2.12.3

Students should use their skills, knowledge and understanding of 'How Science Works':

- *To interpret data showing how factors affect the rate of photosynthesis and evaluate the benefits of artificially manipulating the environment in which plants are grown.*

BIOLOGY HOW PLANTS PRODUCE FOOD

B2 2.5 Plant problems?

Smallholder

'In days gone by most farms were small. Farmers fed their own families and hoped to make enough profit to survive. Different crops had to be grown each year (crop rotation) and the land was rested between crops. Fields lay fallow (no crops were grown) every few years to let the land recover. Manure from their own animals was the main fertiliser.

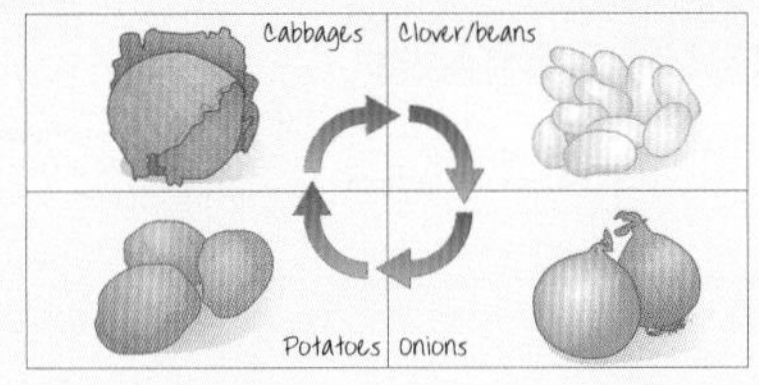

'We're trying to stick to the old ways on our little small-holding. We rotate our crops – you can see from my field plan how I do it. It helps to make sure that the minerals in the soil don't all get used up. It helps keep diseases at bay too.

'We feed our family well, and sell our extra produce to the village shop. Of course, I earn most of my money through my computer business.'

Arabic farmer

'We farm a pretty big area. I grow wheat and oil seed rape. After we have harvested, I plough the stubble back into the soil. We used to burn it off but that's not allowed now. I think it's better to plough the stuff in anyway – puts something back!

'My farm is a big business – I can't afford to have land doing nothing. So we add fertiliser to keep the mineral levels right. We need to get the best crop we can every time! Modern fertilisers mean I can plant one crop straight after the other, and I avoid fallow years altogether.

'I have to get the balance right – if I spend too much on fertiliser, I don't make enough profit. But if I don't put enough fertiliser on the fields, I don't grow enough crops! I manage to support the family pretty well with the farm, and we employ one local man as well.'

Teaching suggestions

The following suggestions are based on the material in the spread and ways in which it can be linked with the topics in the chapter. Much of this would be valuable background to gaining an understanding of present-day agriculture and could help students in carrying out the suggested activity.

- **Small-holder** – Using the material provided in the student book as a start, students could investigate the nature of small farms. Historically, many farms were worked in a similar manner with animals being raised and a variety of crops grown. In many parts of the world, farming is still carried out in this way. The methods that were, and still are, practised have a scientific foundation, but it is obvious from the passage that this type of farming does not make enough profit. Using knowledge from the chapter and by doing some research, students could find answers to the following questions:
  - What were the benefits of growing different crops every year?
  - Why were crops rotated?
  - What were the benefits of allowing the land to be rested between crops?
  - What are the pros and cons of using manure from the animals?
  - What are the disadvantages of small-holdings in the modern world?

  Hold a brainstorming session and list as many reasons as you can think of why such methods of farming are no longer economical.
- **The great organic debate** – One way in which farmers can make small farms pay is to become an 'organic' farm. Students could find out what it takes to become an organic farmer and what the advantages are. For information about organic farming and crop rotation go to www.soilassociation.org or www.rhs.org.uk.
- **Arable farming** – The modern trend in farming is for monoculture: farms on which one or possibly two crops are grown. These large farms have often arisen by the amalgamation of smaller farms. Students could consider the following consequences of monoculture:
  - the destruction of hedgerows
  - the effect of the use of artificial fertilisers on soil structure
  - the leaching of the fertilisers through the soil and into waterways
  - the problems of disease and the use of pesticides.
- **Hydroponics grower** – This system of growing plants has its advantages and its disadvantages, and raises many questions. In order to appreciate what such systems involve, students could investigate and discuss the following issues:
  - Is the system appropriate for growing all crops? They could draw up a list of crops for which this system seems suited and another list of crops which could be difficult to grow in this way.
  - Why would carbon dioxide levels need to be changed during the day?
  - What are the advantages of being able to control the temperature and light levels?
  - Why is it necessary to alter the mineral content of the water as the plants grow?
- Hydroponics systems are expensive to set up. Ask students to compile a list of ways in which such systems save money and balance it with a list of items that the grower has to budget for (apart from the initial cost of setting up the system).

Hydroponics grower

'In the laboratory you can isolate different factors and see how they limit the rate of photosynthesis. However, for most plants a mixture of these factors affects them. Early in the morning, light levels and temperature probably limit the rate of photosynthesis. Then as the level of light and the temperature rise, the carbon dioxide becomes limiting. On a bright, cold winter day, temperature probably limits the rate of the process. There is a constant interaction between the different factors.

'In commercial greenhouses we can take advantage of this knowledge of limiting factors and leave nothing to chance. We can control the temperature and the levels of light and carbon dioxide to get the fastest possible rates of photosynthesis. This makes sure our plants grow as quickly as possible. We even grow our plants in a nutritionally balanced solution rather than soil to make sure nothing limits their rate of photosynthesis and growth.'

ACTIVITY

The National Farmers Union (NFU) wants to produce a resource for schools to show how arable (crop) farming has changed over the years. Your job is to design ***either*** one large poster ***or*** a series of smaller posters that they can send out free to science departments in schools around the country.
You need to explain how plants grow, and how farmers give them what they need to grow as well as possible. Use the information on this spread and in the rest of the chapter to help you.

By controlling the temperature, light and carbon dioxide levels in a greenhouse like this we can produce the biggest possible crops – fast!

'We invested in all the computer software and control systems about two years ago. It cost us a lot of money – but we are really reaping the benefits. We can change the carbon dioxide levels in the greenhouses during the day. We control the temperature and the light levels very carefully. What's more we can change the mineral content of the water as the plants grow and get bigger.

'We sell all our stuff to one of the big supermarket chains. Our lettuces are always clean, big and crisp – and we have a really fast turnover. No more ploughing fields for us!

'Of course we don't need as many staff now. We just have lots of alarm systems in our house. Then if anything goes wrong in one of the greenhouses, day or night, we know about it straight away. The monitoring systems and computers are vital to our way of growing. As far as our plants are concerned, limiting factors are a thing of the past!'

- If possible, arrange a visit to a hydroponics system or investigate the ways in which small hydroponics systems can be set up in glasshouses.
- Discuss the advantages of hydroponics systems being set up where it would be difficult to grow crops conventionally.

pecial needs

tudents can be given, or encouraged to ollect, pictures of farm machinery from ifferent times and make a poster, showing ow cultivation of the land has changed.

ACTIVITY & EXTENSION

- The poster activity under 'Special needs' is open-ended and offers scope for a number of different approaches:
  - The cultivation of one particular crop could be charted. For example, wheat was grown in the Middle East many centuries ago and the development of the bread wheats we have today can be traced. The formation of large fields from smaller ones and the increasing mechanisation, using larger and larger tractors, could be depicted. Wheat was part of a crop rotation at one time, but development of artificial fertilisers meant that the same crop could be grown year after year. Some of the latest developments involve the breeding of shorter-stemmed varieties that are easier to harvest and produce less straw.
  - The approach could be more general and show how crop growing has changed from medieval times (strip farming and the use of manure as fertiliser) to the present day (vast fields and artificial fertilisers).
  - The emphasis could be on the types of crop grown. Ask the students: 'Have they changed over the years? Are we growing different crops from our ancestors?' They need to think about when the potato arrived in Britain. Much land is now producing crops such as oil seed rape and not the traditional root crops that were produced before. Ask: 'Why is this?'
- **Growing crops in outer space** – Is hydroponics the answer to growing crops on the Moon or on spaceships? Students might like to debate this and consider whether or not it is feasible. Ask: 'What sorts of things could be grown? Would it work?'
- **The effects of climate change** – There has been much speculation about the effects of climate change on the nature of crops that can be grown in different parts of the world. Students to consider: 'If the climate in Britain becomes warmer, what different crops might be grown? Would it be possible to grow more than one crop of tomatoes a year?'
- **How the field patterns have changed** – It may be possible to obtain aerial or other photographs of your local area that show how the fields have changed. Good sources of such information are local history museums and the archives of local newspapers.
- **Use simulation** – If not used previously in the chapter, use the Simulation, B2 2.2 'Limiting Factors of Photosynthesis' available on the GCSE Biology CD ROM.

SUMMARY ANSWERS

1 **a)** A4; B3; C2; D5; E1.

b) carbon dioxide + water (+ light energy) → glucose + oxygen

c) Starch.

2 **a)** [Credit accurately drawn graphs, correctly plotted points and correctly labelled axes, etc.]

b) Plants growing in the higher light intensity photosynthesise more and faster, so produce food and grow well. Light will not be a limiting factor, but carbon dioxide or temperature might. Light is a limiting factor in those plants grown at the lower light intensities.

3 **a)** Leaves.

b) Roots.

c) Xylem.

d) Phloem.

e) Xylem.

Summary teaching suggestions

- **Lower and higher level answers**
  There are different styles of question that test higher attaining students.
  - For example, the graph plotting exercise in question 2 and its explanation is a good discriminator. The higher attaining students should be able to include all the points given. Some students might use the figures, or some manipulation of them, to back up their answers.
- **Special needs**
  - Students could be given the words and descriptions for question 1 on sheets of paper and asked to match them up.
  - In question 2, if the graph was already plotted for them, they could answer the question verbally.
  - Question 3a) to e) could also be used if the one word answers were on cards for them to select and hold up.
- **Homework** – Question 2 would be a good homework exercise as a follow-up to the spread on limiting factors.
- **When to use the questions?** – Questions requiring short answers are useful to reinforce the content of a lesson, e.g. question 1 on photosynthesis and question 3.

HOW PLANTS PRODUCE FOOD: B2 2.1 – B2 2.5

SUMMARY QUESTIONS

1 a) Match each word related to photosynthesis to its description:

A Carbon dioxide gas	1 ...... is produced and released into the air
B Water	2 ......provides energy
C Sunlight	3 ...... from the roots moves up to the leaf through the stem
D Glucose	4 ...... is absorbed from the air
E Oxygen	5 ...... is made in the leaf and provides the plant with food

b) Write a word equation for photosynthesis.

c) Much of the glucose made in photosynthesis is turned into an insoluble storage compound. What is this compound?

2

Year	Mean height of seedlings grown in 85% full sunlight (cm)	Mean height of seedlings grown in 35% full sunlight (cm)
2000	12	10
2001	16	12.5
2002	18	14
2003	21	17
2004	28	20
2005	35	21
2006	36	23

The figures in the table show the mean growth of two sets of oak seedlings. One set was grown in 85% full sunlight, the other set in only 35% full sunlight.

a) Plot a graph to show the growth of both sets of oak seedlings.

b) Using what you know about photosynthesis and limiting factors, explain the difference in the growth of the two sets of seedlings.

3 Plants make food in one organ and take up water from the soil in another organ. But both the food and the water are needed all over the plant.

a) Where do plants make their food?

b) Where do plants take in water?

c) There are two transport tissues in a plant. One is the phloem. What is the other one?

d) Which transport tissue carries food around the plant?

e) Which transport tissue carries water around the plant?

EXAM-STYLE QUESTIONS

1 Jenny carried out an investigation to show the rate o photosynthesis in two species of plant at different lig intensities.

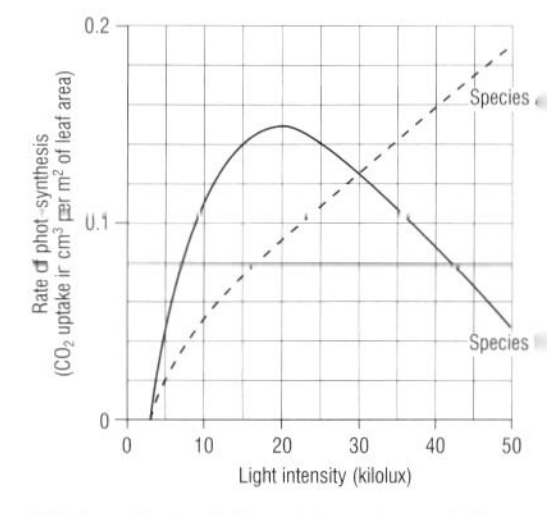

This investigation had two independent variables.

(a) Name the categoric independent variable.

(b) Name the continuous independent variable.

(c) Describe the pattern shown by species B. The results for species B were as follows:

Light intensity (kilolux)	CO_2 uptake (cm^3/m^2)
5	0.04
10	0.11
20	0.15
30	0.125
40	0.09
50	0.04

(d) Jenny was not sure where the peak of the graph should be drawn. Which extra measurements sh[e] she take to be sure of this?

(e) At what light intensity do both species photosynthesise at the same rate?

(f) If species **A** has a total leaf area of $100\,m^2$, how many cm^3 of carbon dioxide will it take up at a li intensity of 10 kilolux? Show your working.

(g) Which species shows the best adaptation to shac conditions? Using the information in the graph g reasons for your answer.

EXAM-STYLE ANSWERS

1 a) Types of species *(1 mark*

b) Light intensity *(1 mark*

c) Rapid increase in rate of photosynthesis *(1 mark*
up to a light intensity of 20 kilolux *(1 mark*
slower decrease in rate (up to 50 kilolux) *(1 mark*

d) Any reasonable spread around 20 kilolux *(1 mark*

e) 30 kilolux *(1 mark*
Both the figure and the units are required. It is important that students always give units when expressing values.

f) Reading from the graph, at 10 kilolux, there are $0.05\,cm^3$ of carbon dioxide absorbed per m^2 leaf area.
i.e. $1\,m^2$ of leaf takes up $0.05\,cm^3$ of carbon dioxide.
Therefore $100\,m^2$ of leaf will take up $0.05 \times 100\,cm^3$ of carbon dioxide.
i.e. $5\,cm^3$.
2 marks for correct answer and accurate working.
1 mark for correct answer but no working shown, or so little working shown that it is impossible to see how the answer was arrived at.
1 mark for the wrong answer but correct working, e.g. student misreads the value from the graph but answer is consistent with a correct calculation.

g) Species B shows the best adaptation to shade conditions because:
Either In the shade the light intensity is lower than out of the shade/in the sun *(1 mark*
and the rate of photosynthesis in species B is greater than that of species A at lower light intensities. *(1 mark*
Or The light intensity is less in the shade than out of it/in the sun *(1 mark*
and the maximum rate of photosynthesis of species B occurs at 20 kilolux – a much lower light intensity than the 50 kilolux that produces the maximum rate of photosynthesis for species A. *(1 mark*

h) What is the name of the sugar produced during photosynthesis? (1)

i) What is the name of the process by which this sugar is broken down to provide energy for the plant? (1)

The diagram below represents a section through a plant leaf showing the arrangement of cells as seen under a microscope.

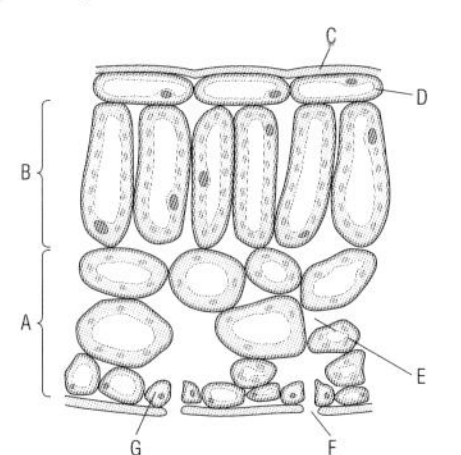

a) Name the parts labelled **E**, **F** and **G**. (3)

b) Give one function of the parts labelled
(i) **C**
(ii) **G** (2)

c) List the four letters that indicate structures that contain chloroplasts. (4)

d) The diagram shows only a small section through a leaf. State **FOUR** ways in which the **whole leaf** is adapted to carry out photosynthesis. In each case show how this feature helps the plant to carry out photosynthesis. (8)

Plants need to obtain mineral salts in order to survive.

a) Name two mineral salts that are essential to plants and in each case give a reason why they are needed. (4)

b) How do plants obtain the minerals they need? (2)

c) If crops are grown for long periods on the same piece of land, they may use up some of the minerals in the soil. State two ways in which farmers can avoid these crops dying due to lack of minerals. (2)

HOW SCIENCE WORKS QUESTIONS

Water gardens – or rather hydroponics!

Ed had seen some entrepreneurs make a fortune by growing lettuce in the middle of winter. He wanted some of the action! He knew that he would have to provide heat and light as well as the nutrients and the correct pH. He knew that the plants required water and oxygen to their roots.

None of the books told him how often he should water the lettuce. Water them too often and they would not get enough oxygen. Leave them too long without watering and they would dry out. He decided on an investigation.

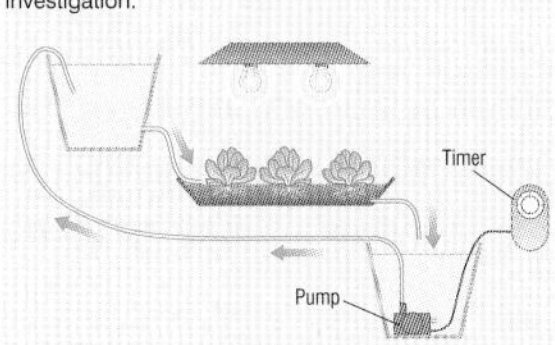

Ed set up five different trays and buckets. He set the timer differently for each tray. The lettuce would now be watered for a different number of times each day. He could therefore work out which was the best for his lettuce.

a) Suggest some time intervals for Ed to water his lettuce. (1)

b) Suggest a dependent variable he could measure. (1)

c) Explain why you have chosen this variable. (1)

d) Describe how Ed might measure this dependent variable. (1)

e) Suggest three control variables he should use. (3)

f) Explain why it would be sensible for Ed to repeat his results. (2)

Ed's first set of results showed very little difference. It did not seem to matter how often he watered them.

g) Suggest a problem he had with the design of his investigation. (1)

h) Why was it important that Ed did his own research and not ask advice from those already growing the lettuce? (2)

161

There are no marks for simply giving the correct species. With a choice of only two there is a 50% chance of arriving at this answer by guesswork alone.
To gain both marks, the student should demonstrate that light intensity is low in the shade (or conversely that it is high in the sun).

h) Glucose *(1 mark)*

i) Respiration *(1 mark)*

a) **E** air space

F stoma/stomata

G guard cell *(1 mark each)*

b) i) C – prevents water loss/reduces transpiration *(1 mark)*
ii) G – controls the movement of gases into and out of the leaf/opens and closes the stomata to allow gases in and out. *(1 mark)*

For G (guard cell) the important thing is for candidates to convey the idea of ***controlling*** *the gases going in and out. Responses such as 'open and close stomata' are acceptable as there is some semblance of control in this expression, but 'allows gases in and out' is inadequate as this is a function of the stomata rather than the guard cells.*

c) A, B, D and G. *(1 mark each)*

d)

Feature of leaf	**How it affects photosynthesis**
• Large surface area	to collect/capture as much light as possible
• Contains chloroplasts/ chlorophyll	to absorb light energy (and make sugars/glucose)
• Many air spaces	to allow CO_2 and O_2 to diffuse/move around the leaf
• Has veins	to bring water to/remove glucose from the leaf cells

1 mark for each correct feature and 1 mark for each explanation.

3 a) Nitrate *(1 mark)*
to produce amino acids/proteins *(1 mark)*
Magnesium *(1 mark)*
to make chlorophyll *(1 mark)*

b) Minerals are found in the soil solution/dissolved in water in the soil and are absorbed by the roots of the plant *(1 mark)*
by diffusion/active transport. *(1 mark)*

c) By crop rotation/growing a different crop on the land each year. *(1 mark)*
Adding fertiliser/spreading manure/muck on the land. *(1 mark)*

Exam teaching suggestions

- Reading values from graphs (question 1 part f) is a common, and relatively straightforward, exercise in examinations. One potential pitfall however is the omission of the relevant units when writing the answer. This could cost marks. If students need convincing of the need to provide units, ask them to imagine they asked someone how far it was to the nearest shop and received the reply 'twenty'.

HOW SCIENCE WORKS ANSWERS

a) Any reasonable time interval – from every 2 hours to every 12 hours.

b) Examples of dependent variable that could be measured include: height of lettuces, wet mass of lettuce.

c) Must refer to answer in b) and be related to marketable part of the plant.

d) Depends on variable chosen – describes how marketable part of plant is measured, e.g. ruler or electric balance.

e) Control variables that should be used are: volume of water, concentration of nutrients, light levels, type of lettuce, spacing.

f) More reliable than using just one lettuce but a larger sample size would further improve the reliability of the data collected.

g) e.g. poor range chosen.

h) Any advice might have been biased – they are his competitors.

How science works teaching suggestions

- **Literacy guidance**
  - Key terms that should be clearly understood: dependent variable, control variable, bias, interval measurements.
  - Question d) expects a longer answer, where students can practise their literacy skills.
- **Higher- and lower-level answers.** Questions b) and c) are higher level questions. The answers for these have been provided at this level. Question a) is a lower-level question and the answer provided is also lower level.
- **Gifted and talented.** Able students could consider the dependent variable in terms of marketable produce. They could be asked to consider all of the possible control variables and how those that cannot be controlled might be monitored.
- **How and when to use these questions.** When wishing to develop ideas about interval measurements and bias in reporting results. The questions could be small group discussion work.
- **Homework.** The questions could also be set for homework.
- **Special needs.** Photographs of hydroponic glasshouses should be used to explain the process.

B2 3.1 Pyramids of biomass

LEARNING OBJECTIVES

Students should learn that:

- Solar radiation is the source of energy for all communities of living organisms.
- Green plants capture solar energy to build up energy stores in their cells.
- The biomass at each stage in a food chain is less than it was at the previous stage.

LEARNING OUTCOMES

Most students should be able to:

- Explain where biomass comes from.
- Describe what a pyramid of biomass is and how it can be constructed.
- Interpret pyramids of biomass and construct them from appropriate information.

Some students should also be able to:

- Explain that all the biomass at one stage does not get passed on to the next stage.

Teaching suggestions

- **Special needs.** Students could be provided with the components of a food chain written on cards, which they can put in the correct order. Similarly, they can build up pyramids of biomass if provided with the components.
- **Gifted and talented.** Students to consider whether or not pyramids of biomass tell the whole story. They could write a short paragraph on what the pyramid of biomass does tell us about the relationships between the organisms.
- **Learning styles**

  Kinaesthetic: Carrying out the experiments.

  Visual: Examining the animals and plants.

  Auditory: Listening to instructions about practical exercises.

  Interpersonal: Discussing food chains and pyramids.

  Intrapersonal: Interpreting the results.
- **Homework.** Both questions 2 and 3 of the 'Summary questions' would make good homework exercises.
- **ICT link-up.** Use the Simulation, B2 3.1 'Pyramids of Biomass', available on the GCSE Biology CD ROM.

SPECIFICATION LINK-UP Unit: Biology 2.12.4

- *Radiation from the Sun is the source of energy for most communities of living organisms. Green plants capture a small part of the solar energy which reaches them. This energy is stored in the substances which make up the cells of the plants.*
- *The mass of living material (biomass) at each stage in a food chain is less than it was at the previous stage. The biomass at each stage can be drawn to scale and shown as a pyramid of biomass.*

Lesson structure

STARTER

What am I? – Prepare definitions of key words and phrases related to the topic, such as 'producer', 'consumer', 'herbivore', 'carnivore', 'detritivore', etc.; read out or project the key words and phrases only on to the board. Students write their definitions, and then you check through. (5–10 minutes)

Food chains in the school canteen – Check out the menu for lunch and get students to discuss the food chains related to items on the menu. (5–10 minutes)

MAIN

- **Investigation of leaf litter** – Using a known mass or volume of leaf litter, allows students to sort through it by hand and separate out the soil organisms into containers. It is unwise to mix organisms in case they eat each other! This sorting should remove the larger organisms, but it might be necessary to use a Tullgren funnel to find the smaller invertebrates.
- The organisms should be identified as far as possible, counted and all those of one species weighed. It should be possible to classify most families of invertebrates into different feeding types (herbivore, carnivore or detritivore). It is then easy to add up the total numbers and the total masses for the different feeding types.
- Both a pyramid of numbers and a pyramid of biomass can be constructed and compared. Students can construct these pyramids on squared paper, choosing suitable scales.
- **Pyramids of biomass for different communities** – Data can be obtained from different communities, such as a rocky shore, in woodland or open grassland.
- **A pond or freshwater habitat** – The method described in 'Practical support' can be modified to obtain a rough estimate of the biomass in a pond or stream. Sampling in water requires the use of a net and the technique needs to be standardised.
  - In flowing water, kick sampling is carried out over a certain area ($0.5\,m^2$) and the disturbed organisms are allowed to flow into a net. The net can be emptied into sampling trays and the organisms identified, grouped and their wet mass determined.
  - In still water, a sweeping technique is used. The net is swept through the water for a fixed period of time or over a fixed distance. The organisms caught are then tipped into a sampling tray and identified as before.
- Further details of the methods can be found in *Tools, Techniques and Assessment in Biology,* Adds, Larkcom, Miller and Sutton (Nelson Advanced Modular Science series). The Field Studies Council publishes excellent identification guides for plants and animals in the Aidgap series. Follow local guidelines on 'Outside Activities'.

PLENARIES

Numbers or biomass? – Students could compare a pyramid of numbers with a pyramid of biomass. The example given in the Student Book could be used or they could discuss one that they have produced from their own investigations. Ask: 'Which shows the information more accurately? Are there advantages in using numbers?' (5–10 minutes)

Anagrams with a difference – Prepare anagrams of the key words, but leave out the vowels. Show these on to the board and ask the students to work out what they are. (5–10 minutes)

Biomass is . . . – Students to complete this sentence and then to judge each other's sentences and add refinements. This could end up with a 'class' definition. (5–10 minutes)

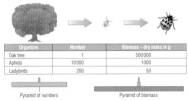

Practical support

Investigation of leaf litter

Equipment and materials required

Quadrats, sweep nets, pooters, sorting trays and small beakers, identification keys, Tullgren funnel (see *Tools, Techniques and Assessment in Biology* (Nelson Advanced Modular Science series for details), balance for weighing organisms.

Details

The method is essentially the same for the different habitats and the results can be expressed as biomass per m². (Cover any open wounds on hands and wash hands after the investigation).

- Select an area and place a 1 m² or 0.5 m² square quadrat carefully on to the ground.
- Collect the leaf litter within the quadrat or cut the plants at the base and place in a white tray.
- Search carefully and remove all the animals present. Smaller animals can be removed using a pooter, larger ones with forceps.
- Animals should be placed in suitable containers, such as plastic beakers.
- Weigh the plant material.
- Identify and sort the animals into groups.
- Weigh the groups of animals separately.
- Return the animals to their habitat.
- Construct a pyramid of biomass.

ACTIVITY & EXTENSION IDEAS

- Leaf litter from two different areas could be compared.
- Pyramids of biomass from the same area at different times of the year could be considered. For example, in the English Channel in January the biomass of animal plankton is greater than the biomass of the plant plankton (the producers). Discuss why pyramids could vary at different times of the year.
- The role of the detritivore could be discussed.

BIOLOGY ENERGY FLOWS

B2 3.1 Pyramids of biomass

LEARNING OBJECTIVES

1 Where does biomass come from?
2 What is a pyramid of biomass?

DID YOU KNOW?

Only about 1% of all the Sun's energy falling on the Earth is used by plants for photosynthesis!

Figure 1 Plants produce a huge mass of biological material in just one growing season

As you saw in the previous chapter, radiation from the Sun is the source of energy for all the groups of living things on Earth.

Light energy pours out continually onto the surface of the Earth. Green plants capture a small part of this light energy using chlorophyll. It is used in photosynthesis. So some of the energy from the Sun is stored in the substances which make up the cells of the plant. This new plant material adds to the **biomass**.

Biomass is the mass of living material in an animal or plant. Ultimately all biomass is built up using energy from the Sun. Biomass is often measured as the dry mass of biological material in grams.

a) What is the source of all the energy in the living things on Earth?

The biomass made by plants is passed on through food chains or food webs into the animals which eat the plants. It then passes on into the animals which eat other animals. No matter how long the food chain or complex the food web, the original source of all the biomass involved is the Sun.

When you look at a food chain, there are usually more producers than primary consumers, and more primary consumers than secondary consumers. If you count the number of organisms at each level you can compare them. You can show this using a **pyramid of numbers**. However, in many cases a pyramid of numbers does not accurately reflect what is happening.

b) What is a pyramid of numbers?

Pyramids of biomass

To show what is happening in food chains more accurately we can use biomass. We can draw the total amount of biomass in the living organisms at each stage of the food chain to scale and show it as a pyramid of biomass.

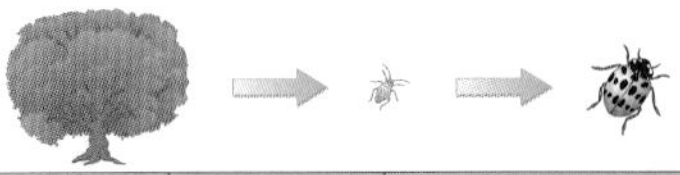

Organism	Number	Biomass – dry mass in g
Oak tree	1	500 000
Aphids	10 000	1000
Ladybirds	200	50

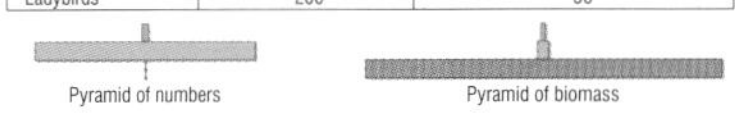

Figure 2 This food chain cannot be accurately represented using a pyramid of numbers. Using biomass shows us the amount of biological material involved at each level in a way that simple numbers cannot do.

c) What is a pyramid of biomass?

162

Additional biology

Interpreting pyramids of biomass

The biomass at each stage of a food chain is less than it was at the previous stage.

This is because:

- Not all organisms at one stage are eaten by the stage above.
- Some material taken in is passed out as waste.
- When a herbivore eats a plant, it turns some of the plant material into new herbivore. But much of the biomass from the plant is used by the herbivore in respiration to release energy for living. It does not get passed on to the carnivore when the herbivore is eaten.

So at each stage of a food chain the amount of biomass which is passed on gets less. A large amount of plant biomass supports a smaller amount of herbivore biomass. This in turn supports an even smaller amount of carnivore biomass.

In general, pyramids of biomass are drawn in proportion. Sometimes, when the biomass of one type of organism is much, much bigger than the others, this doesn't work and so the diagram can only give a rough idea.

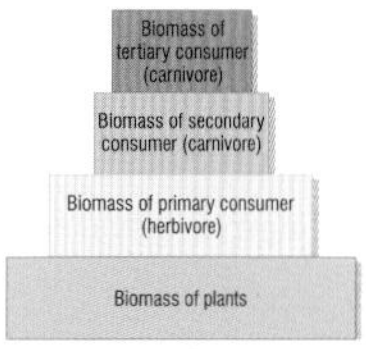

Figure 3 Any food chain can be turned into a pyramid of biomass like this

GET IT RIGHT!

Make sure you can draw pyramids of biomass when you are given the data.

GET IT RIGHT!

Make sure you can work out the proportions of the different organisms by looking at a pyramid of biomass.

SUMMARY QUESTIONS

1 a) What is biomass?
b) Why is a pyramid of biomass more useful for showing what is happening in a food chain than a pyramid of numbers?

2

Organism	Biomass, dry mass (g)
Grass	100 000
Sheep	5000
Sheep ticks	30

a) Draw a pyramid of biomass for this grassland ecosystem.
b) What would you expect the pyramid of numbers for this food chain to look like?
c) Draw the pyramids of numbers and the pyramids of biomass you would expect from the following two food chains:
i) stinging nettles → caterpillars → robin
ii) marine plants → small fish → large fish → seals → polar bear

3 a) Explain simply why the biomass from one stage of a pyramid of biomass does not all become biomass in the next stage of the pyramid.
b) Using the data in Figure 2, calculate the percentage biomass passed on from:
i) the producers to the primary consumers,
ii) the primary consumers to the secondary consumers.

KEY POINTS

1 Radiation from the Sun is the main source of energy for all living things. The Sun's energy is captured and used by plants during photosynthesis.
2 The mass of living material at each stage of a food chain is less than at the previous stage. The biomass at each stage can be drawn to scale and shown as a pyramid of biomass.

163

SUMMARY ANSWERS

1 a) The mass of living material in an animal or a plant.
b) Because it shows what is happening more accurately.

2 a) [Mark for accurately drawn pyramid.]
b) Drawn correctly it should have a large base, one sheep in the middle and more ticks at the top.
c) [Each drawn correctly as pyramids.]

3 a) Not all organisms at one stage are eaten by the stage above them. At each stage, a lot of the biomass is used in respiration to release energy for living. It does not get passed on to the next stage.
b) i) $1000/500\,000 \times 100 = 0.2\%$
ii) $50/1000 \times 100 = 5\%$

Answers to in-text questions

a) The Sun.

b) The numbers of organisms at each trophic level of a food chain drawn as a pyramid.

c) The biomass of organisms at each trophic level of a food chain drawn as a pyramid.

KEY POINTS

The practical work associated with this topic supports the learning of the key points. Students should be able to draw pyramids of biomass, working out the correct proportions from data supplied.

B2 3.2 Energy losses

LEARNING OBJECTIVES

Students should learn that:

- Materials and energy are lost in an organism's waste materials.
- Energy is used in movement.
- Energy is lost as heat to the surroundings.
- Maintaining a constant body temperature results in more heat loss and more energy use.

LEARNING OUTCOMES

Most students should be able to:

- State why materials and energy are reduced at each successive stage of a food chain.

Some students should also be able to:

- Explain in detail why energy losses are particularly large in warm-blooded organisms (mammals and birds).
- Interpret Sankey diagrams.

Teaching suggestions

- **Special needs.** Visit a local petting zoo or city farm, where the students can touch the animals and talk to the keepers about how much the animals have to eat and how they use their energy.
- **Gifted and talented.** Students could look up specific heat capacity. Ask: 'How much more energy would a 10 kg animal have at 37°C as opposed to 20°C?' Assume flesh to have approximately the same specific heat capacity as water.
- **Learning styles**

  Kinaesthetic: Practical activities.

  Visual: Viewing the PowerPoint® summary and observing the practical.

  Auditory: Listening to other students' descriptions of word meanings.

  Interpersonal: Discussing the practical exercises.

  Intrapersonal: Considering the implications of energy loss over trophic levels.
- **ICT link-up.** Students could use data loggers to monitor the temperature changes in the germinating peas experiment.

SPECIFICATION LINK-UP Unit: Biology 2.12.4

- *The amounts of material and energy contained in the biomass of organisms is reduced at each successive stage in a food chain because:*
  - *some materials and energy are always lost in the organisms' waste materials*
  - *respiration supplies all the energy needs for living processes, including movement; much of this energy is eventually lost as heat to the surroundings*
  - *these losses are especially large in mammals and birds whose bodies must be kept at a constant temperature which is usually higher than that of their surroundings.*

Lesson structure

STARTER

Sankey diagram – Show students the Sankey diagram from the Student Book available on GCSE Biology CD ROM (it shows flow of energy through a system; you could project one from the student book spread). Ask them to have a guess at what it is trying to show. (5 minutes)

Mammals vs reptiles – If possible bring in a mammal pet, such as a rabbit, rat or hamster, and a reptile pet, such as a snake or a tortoise (risk assessments needed). Students to take the skin temperatures of the two animals and comment on their necessary energy consumptions. Video footage of crocodiles or snakes eating enhances this (search at www.video.google.com). (10 minutes)

MAIN

- **The great burger race** – This is a large scale outside practical activity for a sunny day. The idea is to show energy loss through trophic levels. Arrange a course with five posts in a line about 10 m apart.
  - At the first post have a picture of the Sun, two buckets with holes in and two large barrels (fruit barrels or similar) full of water.
  - At the second post have a large picture of the Earth and two similar buckets with holes in them.
  - At the third post have a large picture of a wheat plant and two more buckets with holes in.
  - At the fourth post, have on one side a large picture of a burger, on the other side a picture of a cow. On the side with the burger, have a collecting vessel large enough to contain several buckets of water. On the side with the cow have another bucket with holes in.
  - At the fifth post have a picture of a burger and a collecting vessel on the cow side (nothing on the other side).

  Water represents the energy and it is lost through the holes in the buckets at each stage. Pairs of students start the race, collecting water from the 'Sun barrels' and passing it to the buckets of the next pair of students at the first post and so on up the course. The cow side has one more trophic level, so the student on that side should have less water in their bucket as measured with a dipstick when the time is up. Because this is hard to set up, it may be a good idea to video it for future reference!
- **Investigating the heat released by respiration** – Set up the demonstration of heat production by germinating peas. Record the temperatures and plot a graph of temperature against time.
- Discuss the results of this experiment in terms of cellular respiration. Ask: 'Why include the dead peas? Why is so much respiration taking place in the peas?'

PLENARIES

Pass the energy – In groups, start off with a large sheet of paper labelled 'Energy'. Give the paper to a student who is designated to play the role of the Sun. The paper is passed to another student representing the Earth, some being torn off for reflection. The paper is then passed along a 'food chain' with a bit torn off at each level. (5 minutes)

Jumbled answers – Give the students question 1 of the 'Summary questions' with the answers in the wrong places. Have a competition to see who can get the answers in the correct order the fastest. (5–10 minutes)

ACTIVITY & EXTENSION

- **Taking things a bit further** – 'The great burger race' activity and 'Pass the energy' plenary demonstrate how energy is passed along the food chain. You could introduce students to a consideration of how much of the solar energy intercepted by the Earth is used in photosynthesis. There are figures available, but a rough estimate is that 40% is reflected by the clouds, dust in atmosphere etc., 15% absorbed by the atmosphere (ozone layer), leaving 45%. Not all the light energy is in the right range for photosynthesis. The amount of photosynthesis that occurs depends on the season. Between 20 and 50% of the chemical energy stored by the plant is used in respiration, leaving the rest that is potentially available to the next trophic level.

Answers to in-text questions

a) Because animals do not digest everything they eat.

b) The muscles use energy to contract, and the more an animal moves about, the more energy (and biomass) it uses from its food. As the muscles contract they produce heat.

c) An animal that is able to keep its body at a constant temperature regardless of the temperature of the surroundings.

KEY POINTS

The key points are well-demonstrated by the activities and information given in the spread. They can be reinforced by testing the students with the summary questions.

Practical support

Investigating the heat released by respiration

Equipment and materials required

Two Thermos flasks, thermometers or probes for data loggers, one batch of live germinating peas, one batch of boiled, dead peas (cooled and rinsed in disinfectant), cotton wool.

Details

- Into one flask, place some live germinating seeds and a thermometer. Close the mouth of the flask with cotton wool.
- Into the second flask, place the same quantity of germinating peas that have been boiled to kill them, cooled and rinsed with disinfectant to kill microorganisms.
- Insert a thermometer and close the mouth of the flask with cotton wool.
- Keep both flasks in similar conditions and monitor the temperature in both flasks at regular intervals.
- Instead of thermometers, data loggers could be used to monitor the temperatures.

BIOLOGY ENERGY FLOWS

B2 3.2 Energy losses

LEARNING OBJECTIVES

1. How do we lose energy to the environment?
2. What is the effect of maintaining a constant body temperature?

An animal like a zebra eats grass and other small plants. It takes in a large amount of plant biomass, and converts it into a much smaller amount of zebra biomass. This is typical of a food chain.

The amounts of biomass and energy contained in living things always gets less at each stage of a food chain from plants onwards. Only a small amount of the biomass taken in gets turned into new animal material. The question is – what happens to the rest?

Figure 1 The amount of biomass in a lion is a lot less than the amount of biomass in the grass which feeds the zebra it preys on. But where does all the biomass go?

Energy loss in waste

The biomass which an animal eats is a source of energy, but not all of the energy can be used. Firstly, herbivores cannot digest all of the plant material they eat. The material they can't digest is passed out of the body in the faeces.

The meat which carnivores eat is easier to digest than plants, so they tend to need feeding less often and they produce less waste. But even carnivores often cannot digest hooves, claws, bones and teeth, so some of the biomass that they eat is always lost in their faeces.

When an animal eats more protein than it needs, the excess is broken down and passed out as urea in the urine. So biomass – and energy – are lost from the body.

a) Why is biomass lost in faeces?

Figure 2 Animals like horses eat very large amounts of biomass every day. However they also produce very large quantities of dung made up of all the biomass they couldn't actually digest!

Energy loss due to movement

Part of the biomass eaten by an animal is used for respiration in its cells. This supplies all the energy needs for the living processes taking place within the body.

Movement uses a great deal of energy. The muscles use energy to contract. So the more an animal moves about the more energy (and biomass) it uses from its food. The muscles produce heat as they contract.

b) Why do animals that move around a lot use up more of the biomass they eat than animals which don't move much?

GET IT RIGHT!

Make sure you can explain the different ways in which energy is lost between the stages of a food chain.
Check that you know how to use energy flow (Sankey) diagrams to tell if an animal is a herbivore or a carnivore, warm-blooded or cold-blooded.

164

Additional biology

Keeping a constant body temperature

Much of the energy animals produce from their food in cellular respiration is eventually lost as heat to the surroundings. Some of this heat is produced by the muscles as the animals move.

Heat losses are particularly large in mammals and birds because they are 'warm-blooded'. This means they keep their bodies at a constant temperature regardless of the temperature of the surroundings. They use up energy all the time, to keep warm when it's cold or to cool down when it's hot. Because of this, warm-blooded animals need to eat far more food than cold-blooded animals, such as fish and reptiles, to get the same increase in biomass.

c) What do we mean by a 'warm-blooded animal'?

PRACTICAL

Investigating the heat released by respiration

Even plants produce heat by cellular respiration. You can investigate this using germinating peas in a vacuum flask.

- What would be the best way to monitor the temperature continuously?
- Plan the investigation.

Figure 3 Only between 2% and10% of the biomass eaten by an animal such as this horse will get turned into new horse – the rest of the stored energy will be used or lost in other ways

Animal: herbivore – as large loss of undigested food in faeces
Energy from food — Heat, Movement, Faeces and urine

Animal: carnivore – as little loss of undigested food in faeces
Energy from food — Heat, Movement, Faeces and urine

Animal: cold-blooded – as little loss due to heat production
Energy from food — Heat, Movement, Faeces and urine

Animal: warm-blooded – as big loss due to heat production
Energy from food — Heat, Movement, Faeces and urine

Figure 4 Sankey diagrams show how energy is transferred in a system. We can use them to look at the energy which goes in to and out of an animal and predict whether it eats plants or is a carnivore. You can even tell if it is warm-blooded or cold-blooded!

SUMMARY QUESTIONS

1. Copy and complete using the words below:

   biomass body temperature energy food chain growth movement producers respiration waste

   The amounts of ...... and ...... contained in living things always gets less at each stage of a .... from ...... onwards. Biomass is lost as ...... products and used to produce energy in .... This is used for ...... and to control .... Only a small amount is used for ....
2. Explain why so much of the energy from the Sun which lands on the surface of the Earth is not turned into biomass in animals.
3. Why do warm-blooded animals need to eat more food than cold-blooded ones of the same size if they are to put on weight?

KEY POINTS

1. The amount of biomass and energy gets less at each successive stage in a food chain.
2. This is because some material is always lost in waste, and some is used for respiration to supply energy for movement and for maintaining the body temperature.

165

SUMMARY ANSWERS

1. Biomass, energy, food chain, producers, waste, respiration, movement, body temperature, growth.
2. Most of the Sun's energy is not captured by plants. Plant biomass eaten by animals cannot all be digested. Some is broken down by respiration to provide energy. Most energy is used for movement and control of body temperature. A small amount is used for growth to produce new biomass in animals.
3. Warm-blooded animals use up a lot of energy in keeping themselves warm or cooling themselves down to maintain a constant body temperature. They have to eat enough food to do this before they can gain new biomass. Cold-blooded animals do not lose energy in this way, so more of the food they take in becomes new biomass.

B2 3.3 Energy in food production

LEARNING OBJECTIVES

Students should learn that:

- The efficiency of food production can be improved by reducing the number of stages in a food chain.
- The efficiency can also be improved by limiting the movement and controlling the temperature of food animals.

LEARNING OUTCOMES

Most students should be able to:

- Explain why reducing the number of stages in a food chain increases the efficiency of food production.
- Describe the effects of limiting movement and controlling the temperature of food animals.

Some students should also be able to:

- Evaluate the positive and negative effects of managing food production.

Teaching suggestions

- **Special needs.** Students to make a collage of items to include in a balanced lunch that do not include any meat. They could use pictures, packets and wrappers to make it colourful.
- **Gifted and talented.** There are five types of vegetarians. Students could research these and suggest how each type gets the right type of proteins, vitamins and minerals without eating meat.
- **Learning styles**

  Kinaesthetic: Researching different diets.

  Visual: Viewing photographs of factory farming.

  Auditory: Reading out suggestions about diet.

  Interpersonal: Discussing in debates.

  Intrapersonal: Writing a speech for a debate.
- **Homework.** Students could write a short article describing the benefits of giving cows, in developing countries, hormones to increase their milk yield.

SPECIFICATION LINK-UP Unit: Biology 2.12.4

- *At each stage in a food chain, less material and less energy are contained in the biomass of the organisms. This means that the efficiency of food production can be improved by reducing the number of stages in food chains.*
- *The efficiency of food production can also be improved by restricting energy loss from food animals by limiting their movement and by controlling the temperature of their surroundings.*

Lesson structure

STARTER

The essentials of the diet – This is a quick review of the components of a balanced diet. Ask: 'Why do we need carbohydrates, fats and proteins? Where do we get most of the protein in our diet from?' (5–10 minutes)

The great taste test – If it is allowed, and under controlled conditions (check for allergies etc.), provide small samples of burgers/sausages made with meat and meat substitutes. Ask for volunteers to feel and smell the samples (blindfold them first) to see if they can distinguish those products made from meat and those from meat substitutes. Discuss the results and look at what the substitutes are made from. (10–15 minutes)

How many food chains in my lunch? – Show the contents of three typical lunch boxes, but vary the main component. For example, one could have a cheese or egg sandwich, another a pork pie or sausage roll and the third a portion of pasta salad. Ask the students to work out the food chains. Ask: 'Which has the fewest stages?' (10 minutes)

MAIN

- **Factory farming** – Show photographs of factory farming to include reference to battery hens, intensive rearing of pigs and veal calves and fish. There are a number of web sites with down-loadable pictures and videos, but most of them are aimed at vegetarians and some have a biased view. If possible, it would be good to stick to the facts, so that students can debate the issues later. The provision of a worksheet to accompany the presentation should encourage students to keep a balanced view.
- **Factory vs free range** – Students could write a short speech in support of the intensive rearing of chickens and one against. Each speech should include scientific facts and advantages as well as disadvantages. In the lesson, hold a debate, but students to draw lots, as to which side to support, i.e. which of their speeches to read out. In this way, they appreciate that they need to have a balanced view.
- In the suggestion above, the issue of the intensive rearing of chickens has been chosen, but you may wish to broaden the topic to include intensive rearing of any animals. Students may need homework time to prepare their speeches.
- **On being a vegetarian . . .** – Invite any vegetarians in the class to give a presentation about their diet. If there are none, then invite a member of staff or the person responsible for home economics to give a presentation. This exercise is about eating more plant food rather than the ethical issues, so the emphasis is on the types of food they eat and the variety. Try to make it as scientific as possible, students identifying where their essential nutrients are coming from. If possible, suggest they bring in recipes or samples of food. This makes a good link with food and nutrition/home economics. If the laboratory is not a suitable place for trying out foods, then perhaps the use of the home economics room can be negotiated.
- Following the exercise above, some discussion on the nature of proteins from plants could be appropriate. The concept of essential amino acids and fatty acids (without necessarily naming them) could be introduced. The different types of vegetarian (vegans, ovolacto- and pesco-vegetarians) could be researched, described and the balance of their diets discussed.
- Link with reference to staple foods, which are based mainly on vegetables, in the developing countries of the world.

PLENARIES

How yellow is my yolk – or is battery best? – It is often said that the best, most nutritious eggs have the brightest yellow yolks. Students could find out if this is true. Hard-boil an egg from each of the following sources: a battery farm, a deep litter or barn system and free range. Remove the shells, slice the eggs and compare the colour of the yolks (no tasting). Ask: 'Is there a difference? If so, can you work out why?' (5–10 minutes)

Eat less meat – Students to write down five ways in which they could change their own diets to eat less meat protein without eating less protein. Choose some to read out. (5–10 minutes)

ACTIVITY & EXTENSION IDEAS

- **The use of antibiotics in factory farming** – In order to keep the animals healthy and prevent infections, antibiotics have been routinely included in the feed given to animals. Discuss this in relation to the development of antibiotic resistance in bacteria.
- **What goes into animal feed?** – Cattle and sheep have a low protein conversion efficiency (PCE), because their natural diet is high in carbohydrates that they have to digest and convert into proteins. Pigs and poultry have a higher PCE and can be fed diets containing more protein. Students to investigate the feeding of pigs and poultry in intensive systems in relation to the number of stages in the food chain.
- **Should hormones be used to increase productivity?** – Students to investigate the use of anabolic steroids in meat and milk production. Discuss their use and consider some of the other ways in which farmers try to increase yields of meat and milk.

BIOLOGY ENERGY FLOWS

B2 3.3 Energy in food production

LEARNING OBJECTIVES

1. Why do short food chains make food production more efficient?
2. How can we manage food production to reduce energy losses?

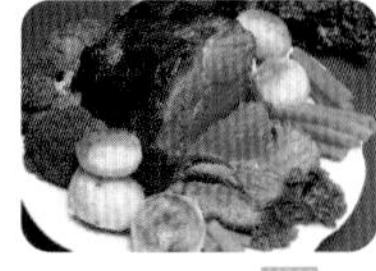

Figure 1 Reducing the number of stages in food chains could dramatically increase the efficiency of our food production. Eating less meat would mean more food for everyone.

Pyramids of biomass clearly show us that the organisms at each stage of a food chain contain less material and therefore less energy. This has some major implications for the way we human beings feed ourselves.

Food chains in food production

In the developed world much of our diet consists of meat or other animal products such as eggs, cheese and milk. The cows, goats, pigs and sheep that we use to produce our food eat plants. By the time it reaches us, much of the energy from the plant has been used up.

In some cases we even feed animals to animals. Ground up fish, for example, is often part of commercial pig and chicken feed. This means we have put another extra layer into the food chain – plant to fish, fish to pig, pig to people. What could have been biomass for us has been used as energy by other animals in the chain.

a) Name three animals which we use for food.

There is only a limited amount of the Earth's surface that we can use to grow food. The most efficient way to use this food is to grow plants and eat them directly. If we only ate plants, then in theory at least, there would be more than enough food for everyone on the Earth. As much of the biomass produced by plants as possible would be used to feed people.

But every extra stage we introduce – feeding plants to animals before we eat the food ourselves – means less biomass and energy getting to us at the end of the chain. In turn this means less food to go round the human population.

b) Why would there be more food for everyone if we all ate only plants?

Artificially managed food production

As you saw on the previous page, animals don't turn all of the food they eat into new animal. Apart from the food which can't be digested and is lost as waste, energy is used in moving around and maintaining a constant body temperature.

Farmers apply these ideas to food production. People want meat, eggs and milk – but they want them as cheaply as possible. So farmers want to get the maximum possible increase in biomass from animals without feeding them any more. There are two ways of doing this:

- Limiting the movement of food animals. Then they lose a lot less energy in moving their muscles and so will have more biomass available from their food for growth.
- Controlling the temperature of their surroundings. Then the animals will not have to use too much energy keeping warm. Again this leaves more biomass spare for growth.

GET IT RIGHT!

Make sure you can use data on food production and explain both the pros and the cons.

166

This means keeping the animals inside with restricted room to move, and a constant temperature. This is exactly what happens in the massive poultry rearing sheds where the majority of the chickens that we eat are produced.

Keeping chickens in these conditions means relatively large birds can be reared to eat in a matter of weeks. When animals are reared in this way they can appear more like factory products than farm animals. That's why these intensive methods are sometimes referred to as factory farming.

Intensive farming methods are used because there has been a steady increase in demand for cheap meat and animal products. This is the only way farmers can meet those demands from consumers.

On the other hand, these animals live very unnatural and restricted lives. More people are now aware of how our cheap meat and eggs are produced. So there has been a backlash against the conditions in which intensively reared animals live.

Many people now say they would be willing to eat meat less often and pay more if the animals they eat are raised more naturally.

Figure 2 These chickens are provided with an ideal temperature, plenty of food and very little opportunity to move. They will produce meat and lay more eggs far faster than if they were moving about and keeping themselves warm.

FOUL FACTS

Veal crates are one of the most extreme ways of rearing animals to reduce energy losses. They are narrow, solid-sided wooden boxes used for rearing calves to produce veal. They are so narrow the calves cannot turn round. They are fed an all-liquid, iron-deficient diet to produce pale, white meat. The calves are slaughtered at 4 to 6 months old having never seen the light of day. Veal crates were banned in the UK in 1990, but they are still used in Europe. They will be banned there from 2007.

Figure 3 Intensively reared pigs live in small stalls in a warm building with food delivered regularly for maximum growth. It makes life relatively easy for the farmer but costs money to run. Animals reared outside grow more slowly, but seem to have a much better quality of life. The farmer needs land, and has to cope with horrible weather – but it's cheaper as there is no artificial heating or lighting to pay for.

SUMMARY QUESTIONS

1. The world population is increasing and there are food shortages in many parts of the world. Explain, using pyramids of biomass to help you, why it would make better use of resources if people everywhere ate much less meat and more plant material.
2. Why are animals prevented from moving much and kept indoors in intensive farming?
3. a) What are the costs for a farmer of rearing animals intensively?
   b) What are the advantages of intensive rearing for a farmer?
   c) What are the advantages of less intensive rearing methods?
   d) What are the disadvantages of these more natural methods?

KEY POINTS

1. Biomass and energy are lost at each stage of a food chain. The efficiency of food production can be improved by reducing the number of stages in our food chains. It would be most efficient if we all just ate plants.
2. If you stop animals moving about and keep them warm, they lose a lot less energy. This makes food production much more efficient.

167

SUMMARY ANSWERS

1. Draw pyramids to show the bigger proportion of energy and biomass passed to people by eating plants like corn or rice than eating beef or chicken. Biomass and energy are lost at each stage of a food chain. The shorter the chain, the less resources are wasted.
2. Energy is used to move and to control body temperature. If animals do not move much and are kept warm, they waste much less energy on those things and use it to produce new biomass.
3. **a)** He has to heat animal houses, light animal houses, build animal houses and animals may be stressed.
   b) Work indoors, lower feed bills, animals grow faster so can be sold sooner and next lot started off.
   c) Animals reared more naturally (more contented?); animals healthier so lower vets bills; no heating/lighting bills.
   d) Have to deal with the weather; animals grow more slowly; need land.

Answers to in-text questions

a) Any sensible choices.

b) Because as little of the biomass produced by plants as possible would be wasted.

KEY POINTS

The key points can be reinforced by the lesson suggestions and the activities and extensions. Students do need to be advised to take a balanced approach, whatever their own views are.

B2 3.4 Decay

LEARNING OBJECTIVES

Students should learn that:

- Materials decay because they are broken down by micro-organisms.
- The decay process releases substances which plants need to grow.
- The materials are constantly cycled.

LEARNING OUTCOMES

Most students should be able to:

- Explain the role of micro-organisms in the process of decay.
- Explain why decay is important in the cycling of materials.

Some students should also be able to:

- Explain the factors which affect the rate of decay.

Teaching suggestions

- **Special needs.** Students could visit a food shop (with permission) and see the ways in which various foods are prevented from decaying. They could take photographs (with permission) and make a display.
- **Gifted and talented.** Students could find out about the human remains that have been found preserved in peat bogs. They could research: 'What are the conditions needed for peat formation? How can peat be used to provide us with information about what plants there were around thousands of years ago?'
- **Learning styles**

  Kinaesthetic: Practical work on decay.

  Visual: Making observations of rates of decay.

  Auditory: Listening to other students' answers.

  Interpersonal: Discussing experimental work.

  Intrapersonal: Writing own list of ways of preventing decay.
- **Homework.** Students could grow their own fungus on the windowsill at home. They could set up their own bread mould cultures, with damp bread, saucers and jam jars.
- **ICT link-up.** Use Animation, B2 3.4 'The Decay Process', from GCSE Biology CD ROM.

SPECIFICATION LINK-UP Unit: Biology 2.12.5

- *Living things remove materials from the environment for growth and other processes. These materials are returned to the environment either in waste materials or when living things die and decay.*
- *Materials decay because they are broken down (digested) by microorganisms. Microorganisms digest materials faster in warm, moist conditions. Many microorganisms are also more active when there is plenty of oxygen.*
- *The decay process releases substances that plants need to grow.*
- *In a stable community, the processes which remove materials are balanced by processes which return materials. The materials are constantly cycled.*

Lesson structure

STARTER

It's gone mouldy! – Have some rotting fruit and vegetables, a piece of mouldy bread and some mushrooms, suitably covered or contained in sealed dishes. Pass these around and discuss how the organisms bringing about decay get their nutrition. (5–10 minutes)

The magic pin mould – Show the students a piece of ordinary bread and a piece which you have left exposed to the air, and then covered with a beaker or a jam jar for a couple of days (it should either have a fluffy white growth on it or a mucky brown one depending on what spores are around – keep sealed under jam jar). Show a picture of pin mould or *Rhizopus* – preferably much larger than life – with sporangia full of spores (search www.images.google.com). Ask the students to make the link and explain how the bread became mouldy. (5–10 minutes)

Where do last year's leaves go? – Have some leaf litter in trays, one for each group or bench. Provide the students with forceps and containers and see how many organisms they can find in 5 minutes. A quick check should sort out the detritivores from the rest, especially if students are provided with keys. The group with the most gets a small prize. Ask: 'What would happen in a woodland if last year's leaves did not decay?' (10–15 minutes)

MAIN

- **Investigating decay** – In the Student Book, it is suggested that students plan an investigation into the effect of temperature on the rate of decay. This can be done with cubes of bread exposed to the air and then placed in different temperatures, such as in the refrigerator, classroom, etc. All other conditions, such as moisture levels, need to be kept the same. Observations will need to be made over a period of time, or allow a set time for the investigation, e.g. a week.
- This investigation introduces concepts of 'How Science Works'. Predictions can be made, variables controlled, measurements taken and conclusions drawn. The results can be assessed in a variety of ways: use digital cameras to record the appearance, assess the area of decay, etc. A time-lapse camera could be used. The Intel 'Play' digital microscopes given away to all schools as part of Science Year have this capability.
- **Investigation into how quickly different materials decay** – This investigation could take two to three weeks. Each student, or group of students, will need a plant pot containing damp soil and a selection of objects, such as a leaf, a piece of bone, a dead earthworm, an insect and a small piece of twig. (This links to 'How Science Works' – designing investigations.)
- When setting up the experiment, the students could discuss the nature of the objects and make predictions about what will happen and why. Ask: 'Why is it important not to let the soil dry out? Why use soil?' (Again, this links to 'How Science Works' concepts.)

PLENARIES

- **Stop the rot!** – How do we stop things from decaying? Students to write down five ways in which perishable foods can be treated to prevent decay. In each case, they need to explain why the treatment prevents the decay. Choose some to read out and compile a list on the board. (10 minutes)

- **Quickest rotter game** – Provide the students with a piece of paper with six empty boxes connected in a line. Arrange pairs of boxes with 'warmth', 'air' and 'moisture' written above them. In pairs, students roll the dice: 1 and 2 lets them write in the letters 'R' and 'O' in the first two boxes; 3 and 4 lets them write in 'T' and 'T' into the middle two boxes and 5 and 6 lets them write 'E' and 'N' into the last boxes. Students race to see who gets rotten first. (5–10 minutes)

ACTIVITY & EXTENSION IDEAS

- **Treatment of sewage** – Show a (simplified) diagram of a sewage treatment or photograph of the treatment of sewage (search for images at www.images.google.com). The emphasis here is to be on the microorganisms involved in the breakdown of the waste. Students need to make notes or complete worksheets outlining the main points.
- **Visit a sewage treatment works** – If possible, arrange a visit to a treatment works.
- **A longer term experiment** – If leaf discs or leaf litter are put into nylon bags with different mesh sizes and then buried in soil, the contribution made by detritivores and decomposers can be assessed. If the mesh is small, then the detritivores will be unable to gain entry and the breakdown will be brought about by the decomposers. Mesh diameter of 6 mm allows the entry of earthworms, other detritivores and decomposers. Mesh sizes of about 0.5 mm will allow entry of other detritivores, but not earthworms. A mesh diameter of 0.003 mm will only allow decomposers through. The bags should contain a known mass of leaf material and be weighed every month. (This relates to 'How Science Works' – making measurements.)

Practical support

Investigating decay

Equipment and materials required

Cubes of bread or other suitable material, Petri dishes with lids or small glass containers with lids, thermometers to register temperatures in the different locations, access to refrigerator, incubator (do not exceed 25°C), etc. to provide about three different temperatures. Fix lids with tape but do not seal.

Answers to in-text questions

a) Plants.
b) Bacteria, fungi, maggots, worms.
c) Water is needed to prevent the microorganisms from drying out/to help them absorb their soluble food; warmth is needed for the enzymes to work efficiently; oxygen is needed for aerobic respiration.

KEY POINTS

Understanding of the key points for this spread can be tested by getting students to answer the summary questions at the end of a lesson or for homework. The summary questions are also good for revision of this topic.

BIOLOGY ENERGY FLOWS

B2 3.4 Decay

LEARNING OBJECTIVES
1 Why do things decay?
2 Why is decay such an important process?

GET IT RIGHT!
You need to know the type of organisms that cause decay, the conditions needed for decay and the importance of decay in recycling nutrients.

Figure 1 These tomatoes are slowly being broken down by the action of decomposers. You can see the fungi clearly, but the bacteria are too small to be seen.

FOUL FACTS
There is a forensic research site in the USA known as the Body Farm where scientists have buried or hidden human bodies in many different conditions. They are studying every stage of human decay. The information is used by police forces all over the world when a body is found. It can help to pinpoint when a person died, and show if they were the victim of a crime.

Plants take minerals from the soil all the time. These minerals are then passed on into animals through the food chains and food webs which link all living organisms. If this was a one-way process the resources of the Earth would have been exhausted long ago!

Many trees shed their leaves each year, and most animals produce droppings at least once a day. Animals and plants eventually die as well. Fortunately all these materials are recycled and returned to the environment. We can thank a group of organisms known as the **decomposers** for this.

a) Which group of organisms take materials out of the soil?

The decay process

The decomposers are a group of microorganisms which include bacteria and fungi. They feed on waste droppings and dead organisms.

Detritus feeders, such as maggots and worms, often start the process, eating dead animals and producing waste material. The bacteria and fungi then digest everything – dead animals, plants and detritus feeders plus their waste. They use some of the nutrients to grow and reproduce. They also release waste products.

The waste products of the decomposers are carbon dioxide, water, and minerals which plants can use. When we say that things decay, they are actually being broken down and digested by microorganisms.

The recycling of materials through the process of decay makes sure that the soil remains fertile and plants can grow. It is also thanks to the decomposers that you aren't wading through the dead bodies of all the animals and plants that have ever lived!

b) Which type of organisms are the decomposers?

Conditions for decay

The speed at which things decay depends partly on the temperature. The chemical reactions in microorganisms are like those in most other living things. They work faster in warm conditions. (See Figure 3.) They slow down and even stop if conditions are too cold. Because the reactions are controlled by enzymes, they will stop altogether if the temperature gets too hot as the enzymes are denatured. You can investigate this in a simple experiment.

PRACTICAL
Investigating decay
Plan an investigation into the effect of temperature on how quickly things decay.
- Name the independent variable in this investigation. (See page 7.)

Most microorganisms also grow better in moist conditions. The moisture makes it easier to dissolve their food and also prevents them from drying out. So the decay of dead plants and animals – as well as leaves and dung – takes place far more rapidly in warm, moist conditions than it does in cold, dry ones.

168

Additional biology

Although some microbes work without oxygen, most decomposers respire like any other organism. This means they need oxygen to release energy, grow and reproduce. This is why decay takes place more rapidly when there is plenty of oxygen available.

c) Why are water, warmth and oxygen needed for the process of decay?

The importance of decay in recycling

Decomposers are vital for recycling resources in the natural world. What's more, we can take advantage of the process of decay to help us recycle our waste.

In **sewage treatment plants** we use microorganisms to break down the bodily waste we produce. This makes it safe to be released into rivers or the sea. These sewage works have been designed to provide the bacteria and other microorganisms with the conditions they need. That includes a good supply of oxygen.

Another place where the decomposers are useful is in the garden. Many gardeners have a **compost heap**. You put your grass cuttings, vegetable peelings and weeds on the compost heap. Then you leave it to let decomposing microorganisms break all the plant material down. It forms a fine, rich powdery substance known as compost. This can take up to a year.

The compost produced is full of mineral nutrients released by the decomposers. Once it is made you can dig your compost into the soil to act as a fertiliser.

Figure 2 The decomposers cannot function at low temperatures so if an organism – like this 4000 year old man – is frozen as it dies, it will be preserved with very little decay

Figure 3 Graph to show the decay rate of plant material (leaves) from two different areas of the USA. The effect of temperature can be seen clearly.

Figure 4 The decomposers are all microorganisms and so they are vulnerable to drying out. Moisture is vital for decay, along with warm temperatures and plenty of oxygen.

SUMMARY QUESTIONS

1 Copy and complete using the words below:

bacteria carbon dioxide dead decomposers digest fungi microorganisms minerals nutrients waste droppings water

The ...... are a group of ...... which includes ...... and .... They feed on .... and ...... organisms. They ...... them and use some of the .... They also release waste products which include ...., ... and ...... which plants can use.

2 The following methods are all ways of preserving foods to prevent them from decaying. Use your knowledge of the decomposing microorganisms to explain how each method works:

a) Food may be frozen.
b) Food may be cooked – cooked food keeps longer than fresh food.
c) Food may be stored in a vacuum pack – with all the air sucked out.
d) Food may be tinned – it is heated and sealed in an airtight container.

KEY POINTS
1 Living organisms remove materials from the environment as they grow. They return them when they die through the action of the decomposers.
2 Dead materials decay because they are broken down (digested) by microorganisms.
3 Decomposers work more quickly in warm, moist conditions. Many of them also need a good supply of oxygen.
4 The decay process releases substances which plants need to grow.
5 In a stable community the processes that remove materials (particularly plant growth) are balanced by the processes which return materials.

169

SUMMARY ANSWERS

1 Decomposers, microorganisms, bacteria, fungi, waste droppings, dead, digest, nutrients, carbon dioxide, water, minerals.

2 **a)** Microorganisms/decomposers do not work at very low temperatures, so when food is frozen decay processes do not take place and the food remains good.
b) Cooking destroys the microorganisms/decomposers; denatures their enzymes; no microorganisms then no decay.
c) Most decomposers require oxygen to respire; no air then no oxygen so microorganisms cannot grow.
d) Decomposers are killed; sealed in an airtight tin there is no oxygen so no action by decomposers; double protection against decay.

B2 3.5 The carbon cycle

LEARNING OBJECTIVES

Students should learn that:

- Carbon dioxide is removed from the atmosphere by photosynthesis in green plants and used to make carbohydrates, fats and proteins.
- Carbon dioxide is returned to the atmosphere when green plants, animals and decomposers respire.
- Detritus feeders and microorganisms break down the waste products and dead bodies of organisms, returning materials to the ecosystem.

LEARNING OUTCOMES

Most students should be able to:

- Describe the processes in the carbon cycle.
- Explain the importance of the activities of the detritus feeders and microorganisms in the cycling of nutrients.

Some students should also be able to:

- Explain the energy transfers in an ecosystem.

Teaching suggestions

- **Special needs.** Students to write the word 'carbon' using a pencil. Then they write it using a burned stick, a piece of burnt animal (a bone or a piece of burned beef jerky), a piece of coal and a charcoal briquette. These can be made into a poster for display along with a balloon of exhaled air.
- **Gifted and talented.** Students could speculate on whether there may be life on other planets that is not carbon-based. Ask: 'Do other elements have properties similar to carbon? What might non-carbon based life be like?'
- **Learning styles**

  Kinaesthetic: Role-playing the game 'Pass the carbon'.

  Visual: Observing the animation.

  Auditory: Writing and reading out a poem or song describing the carbon cycle.

  Interpersonal: Discussing the impact of climate change and what we should do about it.

  Intrapersonal: Writing down own opinions on the reasons for the freak weather.

SPECIFICATION LINK-UP Unit: Biology 2.12.5

- *The constant cycling of carbon is called the carbon cycle. In the carbon cycle:*
  - *carbon dioxide is removed from the environment by green plants for photosynthesis; the carbon from the carbon dioxide is used to make carbohydrates, fats and proteins that make up the body of plants*
  - *some of the carbon dioxide is returned to the atmosphere when green plants respire*
  - *when green plants are eaten by animals and these animals are eaten by other animals, some of the carbon becomes part of the fats and proteins that make up their bodies*
  - *when animals respire some of this carbon becomes carbon dioxide and is released into the atmosphere*
  - *when plants and animals die, some animals and microorganisms feed on their bodies; carbon is released into the atmosphere as carbon dioxide when these organisms respire*
  - *by the time the microorganisms and detritus feeders have broken down the waste products and dead bodies of organisms in ecosystems and cycled the materials as plant nutrients, all the energy originally captured by green plants has been transferred.*

Lesson structure

STARTER

Coal into diamonds – Show a film clip from *Superman II* where he turns a piece of coal into a diamond (or set the scene by searching the web for 'Superman II watch trailer'). Lead a discussion as to where carbon comes from and whether you can destroy it. (5–10 minutes)

Carbon cycle and climate change? – Show some pictures of freak weather or video footage of a tropical storm (search the web at www.images.google.com.) Ask: 'How is this coming about?' Get each student to write down his or her ideas and then discuss or compare with a neighbour. Collect up ideas as a class. (5–10 minutes)

Fossils in coal – Have some plant fossils in coal (real ones if possible but pictures if not). Ask students to write down how the carbon got there and what would happen to it if we burned the coal. (10 minutes)

MAIN

- **Carbon cycle** – Show Animation, B2 3.5 'Carbon Cycle', from GCSE Biology CD ROM. It is a good idea to provide the students with a worksheet and allow time for the explanation of points.
- **Role-play game 'Pass the carbon'** – In small groups, students to be labelled as parts of the carbon cycle, such as 'The atmosphere', 'Plants', 'Animals', 'Fossil fuels', etc. Have a soft ball labelled 'Carbon', which students are to pass around going from locations to other locations via the correct processes.
- **Cartoon cycle** – Students to draw, or use, pictures to make a cartoon strip illustrating how a carbon atom goes from a lion's breath, into plants, into an impala, into a lion and out again through the lion's breath. This could be done in groups and the best displayed.

PLENARIES

The carbon cycle – Label a diagram of the carbon cycle. (5 minutes)

True or false – On the board show various statements about the carbon cycle, respiration, photosynthesis and energy transfer. Students to write on 'Show me' boards whether they are true or false. (5–10 minutes)

Get it in the right order – Prepare sets of cards with stages of the carbon cycle on them. In pairs, students have to put them in a sensible order, and then compare with another pair of students and feedback. (10–15 minutes)

ACTIVITY & EXTENSION IDEAS

- **Compost heap** – If the school has a compost heap, set up a data logger to take the temperature over a period of time. If there is no compost heap, students could investigate the possibility of setting one up in a suitable position. Investigate what types of material can be composted. Why is it best to use vegetable matter only? What kinds of organisms would you expect to find in a well-established compost heap? Why does the temperature change within the compost heap?
- **Energy in decomposers** – Set fire to some dried mushrooms to show that there is energy in decomposers (risk assessment).
- **Kyoto agreement** – Students to find out more about the Kyoto agreement and why America is not signing. They should then suggest reasons why many people think that they should. Comment on progress made at Montreal conference.
- **Carbon emissions** – There has been much discussion within the European Union and worldwide about the levels of carbon released into the atmosphere. Make a collection of newspaper and magazine articles about this topic. Find out how carbon emissions are controlled and what the targets are amongst the industrialised nations.

Answers to in-text questions

a) Fossil fuels, carbonate rocks, the atmosphere, oceans and living things.

b) It removes it from the environment.

BIOLOGY ENERGY FLOWS

B2 3.5 The carbon cycle

LEARNING OBJECTIVES

1. What is the carbon cycle in nature?
2. Which processes remove carbon dioxide from the atmosphere – and which return it?

Figure 1 Within the natural cycle of life and death in the living world, mineral nutrients are cycled between living organisms and the physical environment

Imagine a stable community of plants and animals. The processes which remove materials from the environment are balanced by processes which return materials. Materials are constantly cycled through the environment. One of the most important of these is carbon.

All of the main molecules that make up our bodies (carbohydrates, proteins, fats and DNA) are based on carbon atoms combined with other elements.

The amount of carbon on the Earth is fixed. Some of the carbon is 'locked up' in fossil fuels like coal, oil and gas. It is only released when we burn them.

Huge amounts of carbon are combined with other elements in carbonate rocks like limestone and chalk. There is a pool of carbon in the form of carbon dioxide in the air. It is also found dissolved in the water of rivers, lakes and oceans. All the time a relatively small amount of available carbon is cycled between living things and the environment. We call this the **carbon cycle**.

a) What are the main sources of carbon on Earth?

Photosynthesis

Green plants use carbon dioxide from the atmosphere in photosynthesis. They use it to make carbohydrates which in turn make biomass. This is passed on to animals which eat the plants. The carbon goes on to become part of the carbohydrates, proteins and fats in their bodies.

This is how carbon is taken out of the environment. But how is it returned?

b) What effect does photosynthesis have on the distribution of carbon levels in the environment?

Respiration

Animals and plants respire all the time. They use oxygen to break down glucose, providing energy for their cells. Carbon dioxide is produced as a waste product and is returned to the atmosphere.

Also when plants and animals die their bodies are broken down by the decomposers. These decomposers release carbon dioxide into the atmosphere as they respire. All of the carbon dioxide released by the various types of living organisms is then available again. It is ready to be taken up by plants in photosynthesis.

DID YOU KNOW?

Every year about 166 gigatonnes of carbon are cycled through the living world. That's 16 000 000 000 tonnes – an awful lot of carbon!

Combustion

Fossil fuels contain carbon, which was locked away by photosynthesising plants millions of years ago. When we burn fossil fuels, we release some of that carbon back into our atmosphere:

Photosynthesis: carbon dioxide + water (+ light energy) → glucose + oxygen
Respiration: glucose + oxygen → carbon dioxide + water (+ energy)
Combustion: fossil fuel or wood + oxygen → carbon dioxide + water (+ energy)

Additional biology

The constant cycling of carbon is summarised in Figure 2.

Organic compounds in dead organisms
Death
Feeding
Death
Organic compounds in green plants
Organic compounds in consumers
Organic compounds in fossil fuels
Respiration returns CO_2 to the atmosphere
Respiration
Decay and decomposition – CO_2 released as microbes respire
Respiration
CO_2 in the air and dissolved in water, particularly oceans
Photosynthesis removes CO_2 from the environment
Burning

Figure 2 The carbon cycle in nature

For thousands of years the carbon cycle has regulated itself. However, as we burn more fossil fuels we are pouring increasing amounts of carbon dioxide into the atmosphere. Scientists fear that the carbon cycle may not cope. If the levels of carbon dioxide in our atmosphere increase it may lead to global warming.

Energy transfers

It isn't just carbon that passes through all the living organisms. The energy from the Sun also passes through all the different types of organisms. It starts with photosynthesis in plants, and is then transferred into animals. It is then transferred into the detritus feeders and decomposing microorganisms. They recycle the materials as plant nutrients.

All of the energy originally captured by green plants is eventually either:

- transferred back into the plants (in the minerals they take in),
- transferred into the decomposers, or
- transferred as heat into the environment by respiration.

GET IT RIGHT!

Make sure you can label the processes in a diagram of the carbon cycle.

Figure 3 Energy is transferred from one type of organism to another. Along the way large amounts are transferred as heat to the environment through the process of respiration.

Heat transfer to the environment from respiration
Energy ($96\ kJ/m^2$) | Mammals and birds | Energy ($24\ kJ/m^2$)
Energy ($120\ kJ/m^2$)
Energy ($1500\ kJ/m^2$) | Carnivores (animals) | Energy ($380\ kJ/m^2$)
Energy ($2000\ kJ/m^2$)
Energy ($8000\ kJ/m^2$) | Herbivores (animals) | Energy ($5000\ kJ/m^2$)
Energy ($15\,000\ kJ/m^2$)
Energy ($52\,000\ kJ/m^2$) | Plants | Energy ($23\,000\ kJ/m^2$)
Decomposers – micro-organisms and detritus feeders

This represents the energy flow through $1\ m^2$ of an ecosystem – the figures in brackets are those recorded for one particular area

SUMMARY QUESTIONS

1. a) What is the carbon cycle?
   b) What are the main processes involved in the carbon cycle?
   c) Why is the carbon cycle so important for life on Earth?
2. Explain carefully how a) carbon, and b) energy are transferred through an ecosystem.

KEY POINTS

1. The constant cycling of carbon in nature is known as the carbon cycle.
2. Carbon dioxide is removed from the atmosphere by photosynthesis. It is returned to the atmosphere through respiration and combustion.
3. The energy originally captured by green plants is eventually transferred back into plants, into decomposers or as heat into the environment.

SUMMARY ANSWERS

1 **a)** The cycling of carbon between living organisms and the environment.

b) Photosynthesis, respiration and combustion.

c) Because it prevents all the carbon from getting used up; returns carbon dioxide to the atmosphere to be available for photosynthesis again.

2 **a)** Summarise the carbon cycle – can be done by reproducing the diagram on page 171 or writing out the different stages (must cover all points in carbon cycle).

b) Energy transfer: energy from Sun used by plants for photosynthesis, energy then transferred to animals as they eat plants, into animals which eat animals, then into detritus feeders and decomposers – all energy eventually transferred back to plants in minerals, transferred into body mass of decomposers or transferred to heat in the environment.

KEY POINTS

A really clear diagram with all the organisms and processes labelled is essential in an understanding of the carbon cycle. It could be pointed out to that the cycle links many major activities of plants and animals, so a good understanding of the cycle provides them with a reference point for the processes of respiration and photosynthesis.

B2 3.6 Farming – intensive or free range?

SPECIFICATION LINK-UP

Unit: Biology 2.12.4

This spread can be used to revisit the substantive content covered in this chapter:

- *The efficiency of food production can also be improved by restricting energy loss from food animals by limiting their movement and by controlling the temperature of their surroundings.*

Students should use their skills, knowledge and understanding of 'How Science Works':

- ? *to evaluate the positive and negative effects of managing food production and distribution and to be able to recognise that practical solutions to human needs may require compromise between competing priorities.*

BIOLOGY ENERGY FLOWS
B2 3.6 Farming – intensive or free range?

Intensive farming – costs and benefits	'Free-range' farming – costs and benefits
Chickens for meat and eggs	

Benefits: • Lots of chickens in small space • Little or no food wastage • Energy wasted in movement/heat loss kept to a minimum • Maximum weight gain/number of eggs laid • Cheap eggs/chicken meat	**Benefits:** • Chickens live a more natural life • No heating/lighting costs • Less food needs supplying as they find some for themselves • Can charge more money for free-range eggs/chickens
Costs: • Chickens unable to behave naturally – may be debeaked and cannot perch • Large barns need heating and lighting • Chickens legs may break as bones unable to carry weight of rapidly growing bodies • Risk of disease with many birds closely packed together	**Costs:** • Chickens more vulnerable to weather and predators • More land needed for each bird • Eggs cannot be collected automatically • Fewer eggs laid, especially in the winter when it is cold and dark for longer periods of time

ACTIVITIES

1. Choose either cattle or chickens. Produce a leaflet to be handed out in your local shopping centre either supporting intensive farming methods or supporting free-range farming methods. In each case back up your arguments with scientific reasoning.
2. You are going take part in a debate on animal rights and farming methods. You have been chosen to speak ***either*** FOR intensive farming ***or*** AGAINST 'free-range' farming.
   You have to think carefully about the benefits to the animals of intensive methods, and the disadvantages of free-range farming.

Teaching suggestions

The material provided on cattle and chickens may provide the stimulus for either Activity 1 or Activity 2. The following suggestions could help students to focus their ideas before attempting either of the activities.

- **Comparing intensive and free-range (extensive) farming** – There are some good videos available from Compassion in World Farming. A leaflet entitled *Intensive farming and the welfare of farm animals* is freely available. This organisation also has a visiting speaker scheme. (Visit www.ciwf.org.uk.)
- **Visits to farms** – Arrange a pair of visits, one to an intensive farm and one to an extensive farm. An organisation exists to help get schoolchildren on to farms through the Access to Farms Accreditation scheme and the Growing Schools initiative from the DfES. The Federation of City Farms may well be able to help or you could write to The Greenhouse, Hereford Street. Bristol, BS3 4NA. Your local agricultural college may also be able to help you find suitable farms or show you their systems.
- **The Soil Association** – There have been references to The Soil Association on previous spreads. These are relevant for this spread as well.

For Activity 3:

- **Compost heap** – Set up a compost heap in the school grounds. Liaise with the ground staff for siting and what can be composted. This could also be part of a recycling scheme for all the waste and rubbish from the school. Free or reduced cost bins can be obtained – search the Internet for 'Recycle Now'.
- **Bin search** – Empty the contents of a school bin on to a large plastic sheet. With rubber gloves, get the students to sort out the rubbish, putting it into various piles according to how easily it will decay and what sort of recycling is possible.
- **Garden waste** – Students could investigate the local authority web sites to find out their policy on dealing with garden waste. This could be extended to finding out how much recycling is going on in your area. If the local authority is about to set up a scheme, or if there is one in operation, there is bound to be some literature available. This information could be displayed around the school and could encourage more people to make use of the scheme.

Additional biology

Intensive farming – costs and benefits	'Free-range' farming – costs and benefits
Cattle for beef 	
Benefits: • Uses the male calves produced by dairy cows • Weaning takes place by about 8 weeks and then farmers know exactly how much food each calf eats • Balance of nutrients in food changed as calf grows to maximise growth • Kept largely indoors, energy loss through movement and heat loss is kept to a minimum – can get weight gains of 1.5 kg a day! • Cheap meat **Costs:** • Feedstuff must be bought and can be expensive • Cowsheds need care and cleaning • Cowsheds have to be heated and lit	**Benefits:** • Calves are weaned naturally and stay with their mothers for up to 6 months • Feeding on grass or food grown by farmer means no contamination, such as that which led to BSE, is possible • Cattle behave and live relatively naturally **Costs:** • Animals may take slightly longer to gain weight as they are moving more actively – but they are less stressed • More land is needed to provide grazing, hay and silage

ACTIVITY

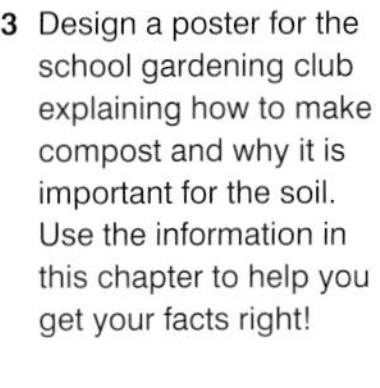

3 Design a poster for the school gardening club explaining how to make compost and why it is important for the soil. Use the information in this chapter to help you get your facts right!

173

Special needs

Students could carry out Activity 3 if they were provided with some pictures or they could use a computer to design a poster, depending on their abilities.

Homework

There is much research to be done for all these activities and ample opportunities for homework tasks. Writing a speech or doing some research in a supermarket could be done out of class time.

ICT link-up

Students can use the Internet to research their projects and these activities. ICT could also be used to design posters and leaflets.

ACTIVITY & EXTENSION

Activity 1 ?

- In the material provided, there are references to the differences in price of the food produced, but a survey in a supermarket can reveal what the costs to the consumer are. Students could carry out a survey of prices for similar products from intensive and free-range systems. They need to calculate how much more it would cost per year to eat from free-range and/or organic sources. Draw out discussion points. Ask: 'Is this an option for people on limited incomes? Should we be eating less meat anyway?'
- Also, the statements given are about costs and benefits for the farmers and producers. A leaflet to be handed out in the local shopping centre would need to contain some information for the general public. For example, the relative costs, the availability and the quality and taste of the produce. In section 3.3, a suggestion was made that students should see if they could tell the difference between a battery egg and a free-range egg.
- One further issue worth considering is the content of processed foods. Many people nowadays buy ready-prepared dishes. Ask: 'How do we know what is used to make them? Are there any products that state that free-range produce is used in the preparation of these dishes?' A supermarket survey, or even a conversation with one of the staff, might provide some information. If that fails, one of the major producers of such dishes might be able to reveal the sources of the raw materials. Some research on the web might be necessary.

Activity 2 ?

- In section 3.3, a debate on the intensive versus free-range rearing of chickens was suggested. The debate suggested here is a more general one, but students could expand their ideas from the previous debate and widen the arguments. This debate could include issues on animal rights, but students should be prepared to present and accept balanced opinions. As has been suggested before, students could prepare two speeches, one for and one against, and be prepared to use either one.

Activity 3

Information for Activity 3 is readily available and the students could design individual posters or work in groups. The activity does link well with recycling and can be part of a general scheme to reduce the amount of rubbish and encourage recycling.

SUMMARY ANSWERS

1 **a) i)** 10% **ii)** 8% **iii)** 12.5%

b) The mass of producers has to support the whole pyramid, and relatively little energy is transferred from producers to primary consumers as they are difficult to digest.

c) Relatively little energy is passed along the chain, and there is not enough by the end of the chain to support many carnivores.

d) If the animals had been birds or mammals, less energy would have been passed on between the levels. They are warm-blooded and so use a lot of energy to keep warm, which is then lost to the environment, leaving less energy to pass on along the chain.

2 **a)** [Marks for graph plotting, correct scale, labelled axes, axes correct way round, accurate points.]

b) Chickens use less energy maintaining their body temperature, so have more energy for growth.

c) To make sure that they move as little as possible. Energy is used up in movement: less movement means more energy for growth.

d) So that they grow as fast as possible to a weight when they can be eaten and another set of chickens started up. Economic reasons.

e) The line should be below the first line. Chickens outside use energy moving around and keeping warm or cool, so convert less biomass from their food for growth.

Summary teaching suggestions

- **Literary guidance** – There are several questions here that require answers written in continuous prose, giving students the opportunity to practise writing in complete sentences and organising their thoughts into a logical sequence.
- **Misconceptions** – Students should be very clear that they know and can explain the difference between a pyramid of numbers and a pyramid of biomass, together with the reasons for their use.
- **Special needs** – It is difficult to see how these questions can be adapted for students. The questions could be put verbally and the answers given verbally, particularly with respect to question 2. The students could be given strips of coloured paper equivalent to the mass of the chicken in question 2 and these could be stuck on a large grid in the correct order, so that the gain in mass is obvious.
- **When to use the questions?**
  - The whole set of questions could be used as an end-of-chapter test in a lesson.
  - They could also be used as a revision exercise to identify areas that need more explanation.
  - Certain questions can be linked to their topics. For example, question 1 can be linked to 3.2 'Energy losses' and would be a useful homework exercise following that lesson. Similarly, question 2 links with 3.3 'Energy in food production' .
  - Both questions 1 and 2 provide students with the opportunity to manipulate and interpret data, which is good practice in preparation for the examinations.

ENERGY FLOWS: B2 3.1 – B2 3.6

SUMMARY QUESTIONS

1

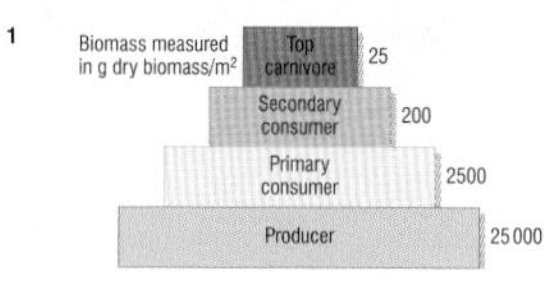

a) From this table calculate the percentage biomass passed on
i) from producers to primary consumers,
ii) from primary to secondary consumers,
iii) from secondary consumers to top carnivores.

b) In any food chain or food web, the biomass of the producers is much larger than that of any other level of the pyramid. Why is this?

c) In any food chain or food web there are only a small number of top carnivores. Use your calculations to help you explain why.

d) All of the animals in the pyramid of biomass shown here are cold-blooded. What difference would it have made to the average percentage of biomass passed on between the levels if mammals and birds had been involved? Explain the difference.

2 Chicks grown for food arrive in the broiler house as one-day-old chicks. They are slaughtered at 42 days of age when they weigh about 2 kg. The temperature, amount of food and water and light levels are carefully controlled. About 20 000 chickens are reared together in one house. The table below shows their weight gain.

Age (days)	1	7	14	21	28	35	42
Mass (g)	36	141	404	795	1180	1657	1998

a) Plot a graph to show the growth rate of one of these chickens.

b) Explain why the temperature is so carefully controlled in the broiler house.

c) Explain why so many birds are reared together in a relatively small area.

d) Why are birds for eating reared like this?

e) Draw a second line to show how you would expect a chicken reared outside in a free-range system to gain weight, and explain the difference.

174

EXAM-STYLE QUESTIONS

1 The diagram below shows a part of a food web for organisms in a lake.

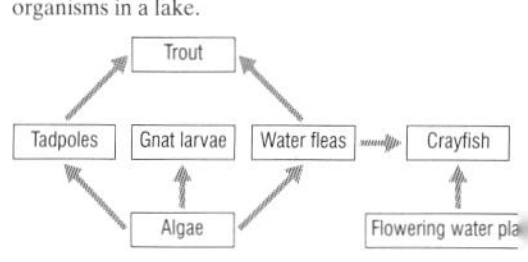

(a) Which organisms feed on algae?

(b) Which organisms are producers?

(c) Which organism is both a primary consumer and secondary consumer?

(d) Draw and label a pyramid of biomass for the food chain below:

Algae → Tadpole → Trout

(e) If a disease suddenly killed all the water fleas explain how the population of the algae might be affected.

2 The diagram below is a version of the carbon cycle.

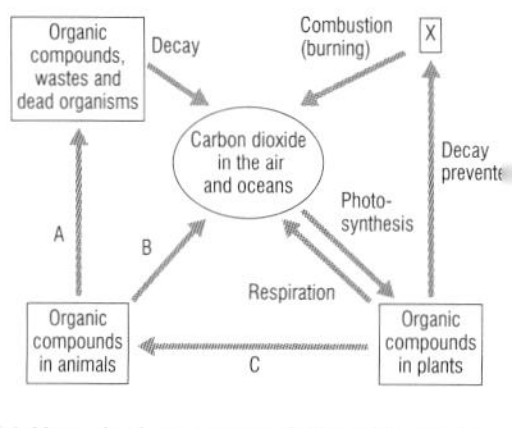

(a) Name the three processes indicated by the three arrows labelled with the letters A, B and C.

(b) In what form is the carbon in the box labelled X?

(c) The organic compounds of plants and animals are mostly in the form of three groups of substances that make up the majority of the bodies of these organisms. What are the three groups of organic compounds?

(d) The table shows the percentage of carbon cycled some of the processes involved in the carbon cyc

EXAM-STYLE ANSWERS

1 a) Tadpoles, gnat larvae and water fleas *(1 mark)*
Any one omitted/any wrong/more than three organisms = no marks

b) Algae and flowering water plants *(1 mark)*
Either omitted/either wrong/more than 2 listed = no marks

c) Crayfish *(1 mark)*

d)
Trout
Tadpole
Algae

(Accurately drawn and labelled = 1 mark)
In the absence of other information on the relative numbers of these organisms, the exact proportions are not important provided that there is a clear reduction in size as one moves up the chain.

e) The algae population would increase *(1 mark)*
because there is one less organism consuming/eating it *(1 mark)*

2 a) **A** Excretion/death
B Respiration
C Eating/feeding *(1 mark each)*

b) Fossil fuels *(1 mark)*
Allow any correct example of a fossil fuel, e.g. coal/oil/gas

c) Carbohydrates
Fats
Proteins *(1 mark each)*
DNA/vitamins or other organic compounds that make up a small proportion of the body should not be given credit as the question says 'make up the majority of the bodies'. Equally, specific examples such as glucose should not gain marks as the question states 'groups of substances'.

d) i) *Five sectors accurately plotted and labelled.* *(3 marks)*
Four sectors accurately plotted and labelled. *(2 marks)*
Three sectors accurately plotted and labelled. *(1 mark)*
Five sectors accurately plotted but no labels/wrongly labelled. *(1 mark)*

Process	Percentage of total carbon cycled
Photosynthesis	50
Respiration by animals	20
Respiration by plants	20
Respiration by microorganisms	5
Combustion/absorbed by oceans	5

(i) Draw a pie chart of these proportions. (3)

(ii) If the total amount of carbon that is cycled in one year across the Earth is 165 gigatonnes, calculate how much carbon is cycled by the respiration of plants. Show your working. (2)

(e) Respiration is an important process in recycling carbon. The word equation for respiration is shown below, with most words replaced by the letters A, B, C and D. Give the names of A, B, C and D. (4)

A + B → C + D + energy

(f) The concentration of carbon dioxide in the atmosphere has increased over the past 200 years. Suggest one human activity that might have contributed to this increase. (1)

A factory which packaged shrimps produced tonnes of waste shrimp heads. It cost money to dump these in the local tip. The managers decided to investigate the decay of shrimp heads to see if they might be used as fertiliser. They used 80 shrimp heads in 4 sealed jars. Each jar had a different amount of water. They measured the length of the shrimp heads, left them for 60 days and then measured them again:

Amount of water (cm^3)	% loss in length
40	68
50	61
60	59
70	56

(a) Explain why they decided to measure the length of shrimp heads. (1)

(b) How many shrimp heads would they have put into each jar? (1)

(c) They predicted that the more water they added the greater the breakdown of the shrimp heads. Is their prediction supported? Explain your answer. (1)

HOW SCIENCE WORKS QUESTIONS

Can this be true?

A scientist claims to have bred a featherless chicken.

The scientist says that it was the result of natural selective breeding and not genetic engineering. He claims that it will be ideal for warmer climates where the intensive breeding of chickens requires expensive air conditioning to keep the chickens at around 25°C. It will therefore be cheaper for farmers to rear these featherless chickens.

They will also be cheaper to feed as they will not need to use energy to grow feathers. Also they will cut down on the pollution caused by having to dump feathers before they are prepared for market.

'I looked to see if the date was April 1st when I read about this story,' said a geneticist.

A biologist said, 'The birds would probably find it difficult to breed without feathers.' Others claimed that it was 'ugly science' and should not be allowed.

a) Why do you think the scientist is so keen to promote his research? (1)

b) Do you think that the scientist was wrong to do the research? (1)

c) Which groups are likely to oppose such research? (1)

d) Who should make the final decision whether farmers should breed these featherless chickens or not? (1)

e) Do we know if the chickens are suffering? How could we find out? (1)

ii) Total carbon cycled = 165 gigatonnes
Plant respiration cycles 20% of the total carbon
i.e. $\frac{165 \times 20}{100} = 33$ gigatonnes

Correct working and answer	*(2 marks)*
Correct working + wrong answer (provided the answer is consequential on the error)	*(1 mark)*
Correct answer but no working shown	*(1 mark)*

e) **A** *or* **B** = glucose or oxygen
C *or* **D** = carbon dioxide or water *(1 mark each)*

f) Burning of fossil fuels/coal/oil/gas *(1 mark)*

3 a) To measure the rate of decay *(1 mark)*

b) 20 *(1 mark)*

c) No, the more water they added the less the loss of length of the shrimp heads. *(1 mark)*

Exam teaching suggestions

- All questions are relatively straightforward and require short answers. The test is pitched at foundation level with most questions predominately testing recall of knowledge. All questions could be effectively used to test learning at the end of teaching sessions on energy flow, either in class or as a homework. Question 3 is based on 'How Science Works'.
- When asked for a specific number of responses, as in question 2 part c), it is important to impress upon students the need to give only the requisite number. Some will try to hedge their bets and give extra responses in the hope that if they include the correct ones all will be well. Examiners will deduct marks for each incorrect response. The guiding principle followed in examinations is 'right + wrong = wrong'.

HOW SCIENCE WORKS ANSWERS

a) He wishes to make money from his research. He can sell the chickens for breeding.

b) This question is open for debate!

For: because he can make money; others can make more money; more people can buy cheaper chickens; less energy and food used; less pollution.

Against: birds might be in physical pain; find it difficult to breed; gives science a poor name.

c) Animal welfare groups are likely to oppose such research.

d) Society in general, the government in particular. There are some questions that science should not answer.

e) Some will argue that if they grow, then they must not be suffering. Others will argue that there is a need to understand how birds suffer and it is very difficult to produce a satisfactory dependent variable with which to judge suffering in birds. There are some questions that at the moment science cannot answer.

How science works teaching suggestions

- **Literacy guidance**
  - Key terms that should be clearly understood: the limitations of present-day science.
  - The questions expecting a longer answer, where students can practise their literacy skills are: b) and e).
- **Higher- and lower-level answers.** Questions b) and e) are higher-level questions. The answers for these have been provided at this level.Questions a) and c) are lower level and the answers provided are lower level.
- **Gifted and talented.** Able students could develop questions b) and e) into a debate on animal welfare.
- **How and when to use these questions.** When wishing to develop ideas as to the limitations of science. The questions are best discussed as a class or small, very able group. There could be links here with the English department.
- **Homework.** For homework, students could tackle another animal welfare issue using the same questions.
- **Special needs.** These students will need talking through the questions in small groups and questions rephrased.
- **ICT link-up.** To research other similar issues. Students should be made aware that Internet groups with an interest are not always factually correct in their science and they might illustrate considerable bias in support of their cause.

B2 4.1 Enzyme structure

LEARNING OBJECTIVES

Students should learn that:

- Enzymes are biological catalysts.
- Enzymes are proteins consisting of long chains of amino acids folded into a special shape.
- The special shape of an enzyme enables it to catalyse reactions.

LEARNING OUTCOMES

Most students should be able to:

- Describe how long chains of amino acids are folded and coiled into special shapes to form enzymes.
- State that enzymes speed up the rate of chemical reactions in the body.

Some students should also be able to:

- Explain the concepts of activation energy and the active site of the enzyme.

Teaching suggestions

- **Special needs.** Make blocks of Lego to represent large molecules, such as starch, proteins and fats. Label each one on one side with the name of the substrate ('starch', 'protein') and the individual bricks labelled with the name of the products ('sugars', 'amino acids'). Use plastic knives with the word 'Enzyme' on to cut up the blocks of Lego. Use a different colour-matched knife for each one. To show denaturing, dip the plastic knife in boiling water (care).
- **Gifted and talented.** Students could research the structure of proteins and use a length of Bunsen tubing to demonstrate the differences between the primary, secondary and tertiary structure. Different sequences of amino acids can be marked with a pen and the tubing can be coiled and twisted into a C shape to illustrate the active site. Record the demo on a video.
- Learning styles

  Kinaesthetic: Practical work on catalase.

  Visual: Viewing the video clips.

  Auditory: Listening to the views of others in discussions on enzyme activity.

  Interpersonal: Discussing the practicals and collaboration with peers.

  Intrapersonal: Evaluating the reliability of data obtained in practical work.

SPECIFICATION LINK-UP Unit: Biology 2.12.6

- *Catalysts increase the rate of chemical reactions. Biological catalysts are called 'enzymes'.*
- *Enzymes are protein molecules made up of long chains of amino acids. These long chains are folded to produce a special shape which enables other molecules to fit into the enzymes. This shape is vital for the enzyme's function.*

Lesson structure

STARTER

Biological stains – Bring in a cheap, clean white T-shirt and allow students to smear it with selected food and drink (tomato ketchup, mustard, egg). Show the students a box of biological washing powder (Care: if it is handed around – some people can have sensitised skin). Discuss how the stains can be removed. (5–10 minutes)

Catalysis: a rapid reaction – Demonstrate a fast-catalysed reaction, such as the breakdown of hydrogen peroxide by manganese dioxide (all should wear eye protection). Collect some of the gas produced and test it. Ask: 'How does this breakdown occur?' (5–10 minutes)

The fly – Show a clip from the film *The Fly* (David Cronenberg, 1986) where the scientist, who has been in a direct matter transfer device with a fly, starts vomiting on his food before eating it. Alternatively, show photographs of a fly's mouthparts and talk through how they function, or how a spider sucks the juice out of its victims. (For a taster, search the web for 'The Fly watch trailer'). (10 minutes)

MAIN

Enzymes in action

- **Breaking down hydrogen peroxide** – This experiment shows the action of manganese(IV) oxide, an inorganic catalyst, and a piece of liver, which contains the enzyme catalase, on hydrogen peroxide. A test tube containing hydrogen peroxide is included as a control. An additional control using a piece of boiled and cooled liver would show that the enzyme from the living tissue can be denatured.
- **Catalase in living tissue** – Catalase is present in living tissue. The more active the tissue, the greater the catalase activity. Small cubes of different tissues, such as liver, muscle, apple and potato, can be dropped into test tubes containing hydrogen peroxide (10 cm^3 to 15 cm^3 depending on the size of the tubes).
- The reactions can be described or they can be measured. (This links to: 'How Science Works' – making observations and measurements.) If the experiment is to be a qualitative one, i.e. just a simple comparison of the activity by observation, then written descriptions or comparative statements can be made.
- It is possible to make this experiment more quantitative by using the same quantities of each tissue, and then measuring the activity when placed in the same volume of hydrogen peroxide. Simple heights of froth up the tube in a given time can be measured. A more accurate measurement is given by collecting the gas evolved in a given time. (This demonstrates many of the 'How Science Works' concepts.)
- There are many variations of the catalase experiments:
  - **i)** The students could investigate the volume of gas released when different quantities of fresh liver are used in the same volume of hydrogen peroxide, i.e. varying the amount of enzyme with a fixed quantity of substrate.
  - **ii)** The converse of this is to use the same quantity of liver and vary the volumes of hydrogen peroxide used, i.e. varying the quantity of the substrate with a fixed quantity of enzyme.
- An Animation, B2 4.1 'How do enzymes work', is available on the GCSE Biology CD ROM.

PLENARIES

Word-matching exercise – On Hot Potatoes (www.halfbakedsoftware.com), or a similar Java program, students to drag and drop words from the lesson next to the correct definitions. (5 minutes)

Find the substrate for the enzyme – Using thin card, make sets of 'enzymes' of different shapes and with differently shaped 'active sites', and a corresponding set of 'substrates' that fit into the enzymes 'active sites'. (You could adapt very simple jigsaw pieces.) Students need to find the enzyme and substrate that fit together. (5–10 minutes)

ACTIVITY & EXTENSION

- **Catalase activity in plant tissues** – The different parts of a plant can be tested for catalase activity. Take equal quantities of leaf, stem and root tissue and test with hydrogen peroxide.
- Use equal quantities of germinating and non-germinating seeds to show that the more active the tissue, the greater the catalase activity.
- **Starch-digesting enzymes in maize** – This is a demonstration. Soak the maize grains for 2 days to initiate germination, and then keep them moist for a further 24 hours. Cut the maize grains in half and place cut-side down on starch/agar in Petri dishes. Place lids over the top, fixing with tape – but do not seal – and leave for 24 hours. Flood with iodine solution (CLEAPSS Hazcard 54) and the starch will stain blue-black, but there will be clear areas around the starch grains. Discuss what has happened.

Practical support

Breaking down hydrogen peroxide

Equipment and materials required

Fresh liver, potato tuber tissue, apple etc., tiles and knives for cutting, test tubes, hydrogen peroxide solution, eye protection, some method of measuring the gas given off (syringes/inverted test tubes or manometers; rulers if height of froth to be measured), stopwatches or stop clocks, water bath if liver is to be boiled and denatured.

Safety: CLEAPSS Hazcard 33. Do not dispose of organic waste down sink.

Answers to in-text questions

a) The minimum energy needed for particles to react.

b) A substance that changes the rate of a chemical reaction (usually speeds it up).

BIOLOGY

ENZYMES

B2 4.1 Enzyme structure

LEARNING OBJECTIVES

1 What is an enzyme?
2 How do enzymes speed up reactions?

DID YOU KNOW?

The lack of just one enzyme in your body can have disastrous results. If you don't make the enzyme phenylalanine hydroxylase you can't break down the amino acid phenylalanine. It builds up in your blood and causes serious brain damage. All UK babies are tested for this condition soon after birth. If they are given a special phenylalanine-free diet right from the start, the risk of brain damage can be avoided.

The cells of your body are like tiny chemical factories. Hundreds of different chemical reactions are taking place all the time. These reactions have to happen fast – you need energy for your heart to beat and to hold your body upright ***now***! They also need to be very controlled. The last thing you need is for your cells to start exploding!

Chemical reactions can only take place when different particles collide. The reacting particles don't just have to bump into each other. They need to collide with enough energy to react.

The minimum amount of energy particles must have to be able to react is known as the **activation energy**. So you will make the reaction more likely to happen if you can make it:

- more likely that reacting particles bump into each other,
- increase the energy of these collisions, or
- reduce the activation energy needed.

a) What is the activation energy of a reaction?

Controlling the rate of reactions

In everyday life we control the rates of chemical reactions all the time. When you cook food, you increase the temperature to speed up the chemical reactions. You lower the temperature to slow reactions down in your fridge or freezer. And sometimes we use special chemicals known as **catalysts** to speed up reactions for us.

A catalyst changes the rate of a chemical reaction, usually speeding it up. Catalysts are not used up in the reaction so you can use them over and over again. Different types of reactions need different catalysts. Catalysts work by bringing reacting particles together and lowering the activation energy needed for them to react.

b) What is a catalyst?

Enzymes – the biological catalysts

In your body chemical reaction rates are controlled by **enzymes**. These are special ***biological catalysts*** which speed up reactions.

Enzymes do not change the overall reaction in any way except to make it happen faster. Each one catalyses a specific type of reaction.

Enzymes are involved in:

- building large molecules from lots of smaller ones,
- changing one molecule into another, and
- breaking down large molecules into smaller ones.

Inorganic catalysts and enzymes both lower the activation energy needed for a reaction to take place.

GET IT RIGHT!

Remember that the way an enzyme works depends on the shape of the active site which allows it to bind with the substrate.

176

Additional biology

PRACTICAL

Breaking down hydrogen peroxide

Investigate the effect of i) manganese(IV) oxide, and ii) raw liver on the breakdown of hydrogen peroxide solution.

- Describe your observations and interpret the graph below.

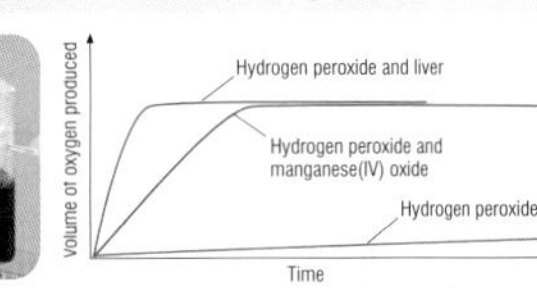

Figure 1 Hydrogen peroxide is a colourless liquid which slowly breaks down to form oxygen and water. The decomposition reaction goes much faster using manganese(IV) oxide as a catalyst. Raw liver contains an enzyme (catalase) which also speeds up the breakdown of hydrogen peroxide.

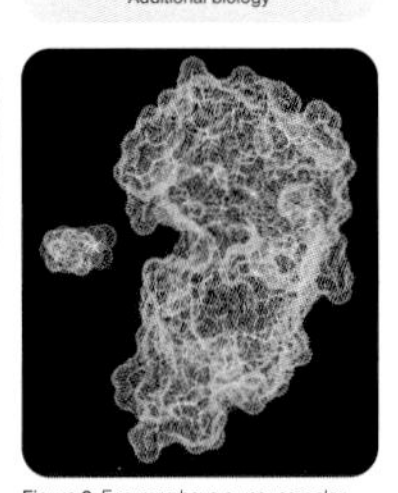

Figure 2 Enzymes have a very complex structure made up of chains of amino acids folded and coiled together. This computer-generated image shows just how complicated the structure really is!

Your enzymes are large protein molecules. They are made up of long chains of amino acids, folded and coiled to give a molecule with a very special shape. The enzyme molecule usually has a hole or indentation in it. This special shape allows other molecules to fit into the enzyme. We call this the **active site**. The shape of an enzyme is vital for the way it works.

How do enzymes work?

The substrate (reactant) of the reaction fits into the shape of the enzyme. You can think of it like a lock and key. Once it is in place the enzyme and the substrate bind together. This is called the **enzyme–substrate complex**.

Then the reaction takes place rapidly and the products are released from the surface of the enzyme. (See Figure 3.) Remember that enzymes can join together small molecules as well as breaking up large ones.

Enzymes usually work best under very specific conditions of temperature and pH. This is because anything which affects the shape of the active site also affects the ability of the enzyme to speed up a reaction.

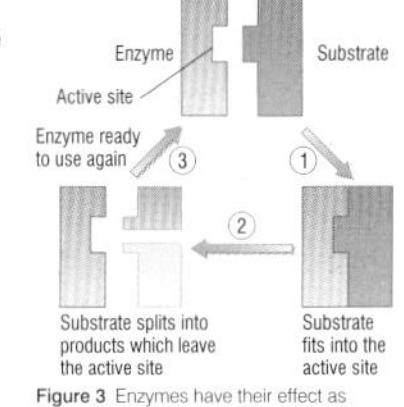

Figure 3 Enzymes have their effect as catalysts using the 'lock-and-key' mechanism shown here. You can see that anything which changes the shape of the protein molecule might change the shape of the active site and stop the enzyme from working.

SUMMARY QUESTIONS

1 Match the words and the definitions:

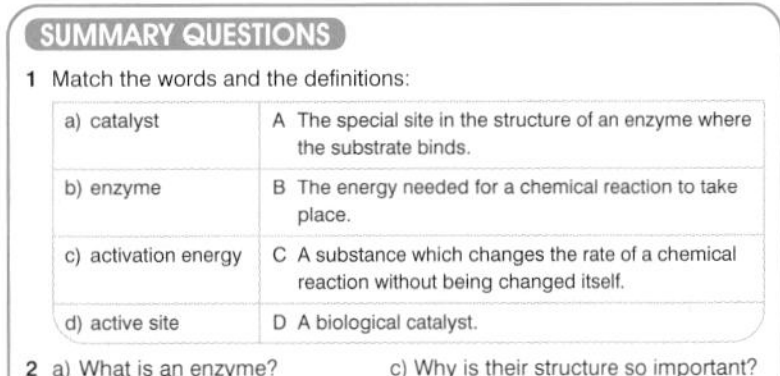

a) catalyst	A The special site in the structure of an enzyme where the substrate binds.
b) enzyme	B The energy needed for a chemical reaction to take place.
c) activation energy	C A substance which changes the rate of a chemical reaction without being changed itself.
d) active site	D A biological catalyst.

2 a) What is an enzyme?
b) What are enzymes made of?
c) Why is their structure so important?

3 a) How do enzymes act to speed up reactions in your body?
b) Why are enzymes so important in your body?

KEY POINTS

1 Catalysts increase the rate of chemical reactions. Enzymes are biological catalysts
2 Enzymes are protein molecules made up of long chains of amino acids. The chains are folded to form the active site. This is where the substrate of the reaction binds with the enzyme.

SUMMARY ANSWERS

1 a) C **b)** D **c)** B **d)** A

2 a) A biological catalyst.

b) Protein/amino acid chains.

c) The way in which the amino acid chains are folded gives them a structure that includes a special area or indentation. This is the active site where the enzyme binds.

3 a) Substrate molecule/s arrive at the active site. They fit perfectly together, like a lock and key. The activation energy is reduced once the substrate is in place. The substrate molecules react and change shape. The products leave the active site. The enzyme is left unchanged and ready to catalyse the next reaction.

b) Enzymes control reactions in cells, making them happen at the right speed. They stop reactions getting mixed up.

KEY POINTS

It is important for students to know what a catalyst is and to understand the structure of enzymes. The concept of the active site and the binding of the substrate molecules can be summarised in a series of diagrams, which students could keep for revision.

B2 4.2 Factors affecting enzyme action

LEARNING OBJECTIVES

- Enzymes are vital to all living cells.
- Changes in temperature affect the rate at which enzymes work.
- Different enzymes work best at different pH values.

LEARNING OUTCOMES

Most students should be able to:

- Describe experiments that show the effect of changes in temperature and pH on the rate of enzyme-controlled reactions.
- Describe how changes in temperature and pH affect enzyme action.

Some students should also be able to:

- Explain how changes in temperature and pH affect the active site of an enzyme.

Teaching suggestions

- **Special needs.** Students can be told that bits of milk need to be joined together by enzymes to make yoghurt. Make some yoghurt in a vacuum flask, a water bath or preferably a commercial yoghurt maker. Set up controls in the refrigerator and at room temperature. Prepare a work sheet with a results table for time taken for it to run through a funnel. Adjust the bore so that some yoghurt will very slowly flow through. Try it with boiled yoghurt (risk assessment).
- **Gifted and talented**
  - Students could research some of the organisms that live in hot springs, very cold conditions and conditions of extreme pH.
  - They could look at the role of DNA polymerase from thermophilic organisms in the Polymerase Chain Reaction for amplifying DNA.
- **Learning styles**

  Kinaesthetic: Practical work on enzymes.

  Visual: Making a diagram to show what happens when protein becomes denatured.

  Auditory: Listening to instructions about practical work.

  Interpersonal: Discussing the range and interval of values chosen for the independent variable (temperature).

  Intrapersonal: Writing a report of their investigative work.

SPECIFICATION LINK-UP Unit: Biology 2.12.6

- *Enzymes are protein molecules made up of long chain amino acids. These long chains are folded to produce a special shape that enables other molecules to fit into the enzyme. This shape is vital for the enzyme's function. High temperatures destroy this special shape. Different enzymes work best at different pH values.*
- *Enzymes inside living cells catalyse processes such as respiration, protein synthesis and photosynthesis.*

Lesson structure

STARTER

Denaturing eggs – Crack raw eggs (or get students to do this) into three beakers: one beaker at room temperature, one at a temperature where visible changes to the egg white just occur, and one at boiling point. Students to describe and explain the changes that are happening to the protein. (5–10 minutes)

What happens to milk when it goes off? – Show the students fresh and sour milk. If possible, have one that is really solid but careful risk assessment is necessary. Discuss what has happened to the milk and why putting milk in the refrigerator stops it going off. (5–10 minutes)

What happens if I get too hot? – Show a picture of a soldier or a sailor in a dry suit. Describe the sad circumstances in which a recruit died while carrying out severe exercise in a dry suit. Ask: 'Why might have he died?' Link to fever, ask: 'who has been ill with a fever?' Take some class temperatures with forehead thermometers. Ask: 'Why do your parents worry when you get very hot?' Link to denaturing enzymes. (5–10 minutes)

MAIN

Investigating the effect of temperature

- **Action of amylase on starch** – Students can use their own saliva to carry out this experiment. (See 'Practical support'.)
- A graph can be plotted of the rate of disappearance of starch (1/time taken in seconds) against the temperature.
- Many concepts of 'How Science Works' can be developed in the investigative work, e.g. hypotheses are formulated, predictions are made, variables are controlled and conclusions drawn.
- **Other enzymes** – These could be used for investigations into the effect of temperature. If the use of the students' saliva is not possible, commercial amylase could be used, but it is usually derived from fungi and can give odd results.
- If pepsin or trypsin (protein digesting enzymes) are used, the substrate to use is the white of hard-boiled eggs or an egg-white suspension made by adding 5 g of egg white to 500 cm^3 of very hot water and whisking briskly. The rate at which the egg white suspension clears can be timed at the different temperatures. More 'How Science Works' concepts are introduced here, too. The effects of varying pH can also be investigated.

PLENARIES

Effect of body temperature on digestion – Show a picture or footage of a reptile, such as a snake or a crocodile, eating a large lump of meat and show a picture of lions feeding. Ask: 'What consequences will their different body temperatures have on the rate at which they digest their meals? How often do they feed the reptiles in the Zoo?' Students to make a list and compare. (10 minutes)

The pH in the gut – Show a diagram of the human gut on to the board and write in the pH of the various regions (mouth, stomach etc.). Alongside the diagram, write a list of the major enzymes and the pH at which they work best. Students to match the enzyme with the region of the gut. (5–10 minutes)

Practical support

- **Luciferase** – Search www.images.google.com and show pictures of flashlight fish, luminous jellyfish, fungi, glow worms that all have this enzyme, which catalyses a reaction and releases energy as light. Some plants have the enzyme and glow green. Tell the true story of a pilot lost at sea from an aircraft carrier at night who navigated his way back and landed successfully after following the faint trail of light from phosphorescent algae, which glowed in the wake of the ship following its passage. Break a lightstick of the kind used by the armed forces, campers and at discos. Demonstrate the reaction of luciferase by mixing the appropriate chemicals in vitro. Students could do Internet search on luciferase to see what else they can find out about this unique enzyme.
- **Extended experiments** – With the pH experiments, the introduction of a wider range of pH values could make the experiment more reliable. The quantities of alkali and acid could be varied and the pH ascertained by testing with pH papers. Alternatively, make up solutions of known pH for use.
- **Effect of varying pH on catalase** – Potato discs can be added to hydrogen peroxide and buffer solutions and the quantity of oxygen evolved in a set time can be measured at each pH. A graph can be plotted of volume of oxygen evolved against pH and the optimum pH for catalase determined.

ACTIVITY & EXTENSION IDEAS

Investigating the effect of temperature

Equipment and materials required

Test tubes and racks, water baths for different temperatures, 2% starch solution, boiled saliva, iodine solution (CLEAPSS Hazcard 54), white tiles, glass rods, eye protection.

Details

Each student will need at least 2 cm depth of saliva in a test tube. Test tubes should be set up containing equal volumes of saliva and starch solution, shaken and then placed into water baths at different temperatures. Drops of the mixtures are then tested at 30 second intervals for the presence or absence of starch by dipping a glass rod into the mixture and then into a drop of iodine solution on a white tile. Note the colour each time and record how long it takes for the starch to disappear at each temperature. A control could be set up using boiled saliva.

BIOLOGY ENZYMES

B2 4.2 Factors affecting enzyme action

LEARNING OBJECTIVES

1. How does increasing the temperature affect your enzymes?
2. What effect does a change in pH have on your enzymes?

Leave a bottle of milk at the back of your fridge for a week or two and you'll find it is pretty disgusting. The milk will have gone off as enzymes in bacteria break down the protein structure.

Leave your milk in the Sun for a day and the same thing will happen – but much faster. Temperature affects the rate at which chemical reactions take place even when they are controlled by biological catalysts.

Biological reactions are affected by the same factors as any other chemical reactions – concentration, temperature and particle size all affect them. But in living organisms an increase in temperature only works up to a certain point.

a) Why does milk left in the Sun go off quickly?

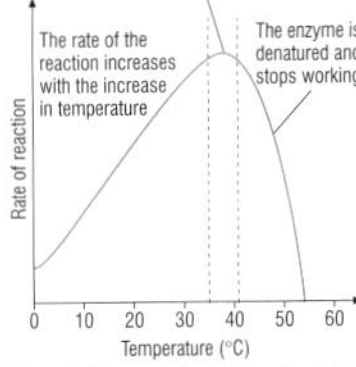

Figure 1 Like most chemical reactions, the rate of an enzyme-controlled reaction increases as the temperature rises – but only until the point where the complex protein structure of the enzyme breaks down

The effect of temperature on enzyme action

The chemical reactions which take place in living cells happen at relatively low temperatures. Like most other chemical reactions, the rate of enzyme-controlled reactions increases with an increase in temperature. The enzyme and substrate particles move faster as the temperature increases, so this makes them more likely to collide with enough energy to react.

However this is only true up to temperatures of about 40°C. After this the protein structure of the enzyme is affected by the temperature. The long amino acid chains unravel. As a result the shape of the active site changes. We say the enzyme has been **denatured**. It can no longer act as a catalyst, so the rate of the reaction drops dramatically. Most enzymes work best from about 20°C to 40°C.

b) What does it mean if an enzyme is denatured?

PRACTICAL

Investigating the effect of temperature

You can show the effect of temperature on the rate of enzyme action using simple practicals like the one shown opposite.

The enzyme amylase (found in your saliva) breaks down starch into simple sugars. You mix starch solution and amylase together and keep them at different temperatures. Then you test samples from each temperature with iodine solution at regular intervals.

In the presence of starch, iodine solution turns blue-black. But when there is no starch present, the iodine stays yellowy-brown. When the iodine solution no longer changes colour you know all the starch has been broken down.

This gives you some clear evidence of the effect of temperature on the rate of enzyme controlled reactions.

- How does iodine solution show you if starch is present?
- Why do we test starch solution without amylase added?
- What conclusion can you draw from the results?

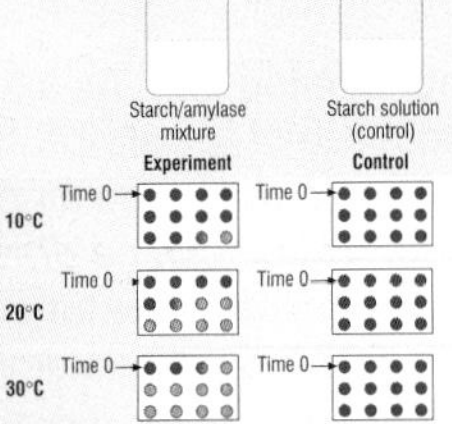

Figure 2 In each case the starch amylase mixture and the control are kept in a water bath at a given temperature. Samples are taken every five minutes and tested with iodine solution on a spotting tile.

178

Additional biology

Effect of pH on enzyme action

Enzymes have their effect by binding the reactants to a specially shaped **active site** in the protein molecule. Anything which changes the shape of this active site stops the enzyme from working. Temperature is obviously one thing which changes the shape of the protein molecule. The surrounding pH is another.

The shape of enzymes is the result of forces between the different parts of the protein molecule which hold the folded chains in place. A change in the pH affects these forces and changes the shape of the molecule. As a result, the active site is lost, so the enzyme can no longer act as a catalyst.

Different enzymes have different pH levels at which they work at their best – and a change in the pH can stop them working completely.

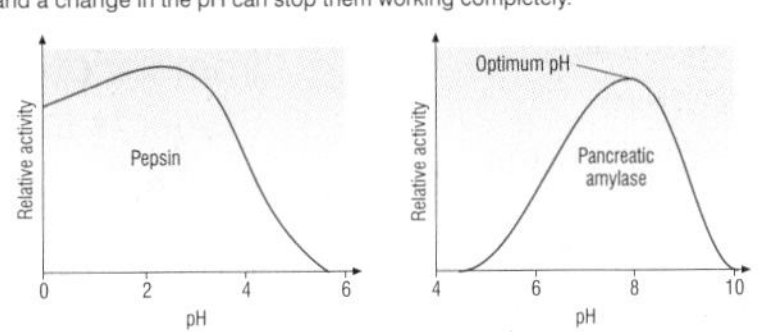

Figure 3 These two enzymes are found in quite different parts of the human gut, and they need very different conditions of pH to work at their maximum rate. Pepsin is found in the stomach, along with hydrochloric acid, while pancreatic amylase is in the small intestine along with alkaline bile.

The role of enzymes

Enzymes are vital to all living cells. They catalyse a huge range of reactions. Without them respiration, photosynthesis and protein synthesis would be impossible. This also applies to all the other reactions which take place in your cells. For the enzymes to work properly the temperature and pH must be just right. This is why it is so dangerous if your temperature goes very high when you are ill and run a fever. Once your body temperature reaches about 41°C, your enzymes start to be denatured and you will soon die.

SUMMARY QUESTIONS

1. Copy and complete using the words below:

   active site cells denatured enzyme increases protein reactions shape temperatures 40°C

   The chemical ...... which take place in living ...... happen at relatively low .... The rate of these ...... controlled reactions ...... with an increase in temperature. However this is only true up to temperatures of about .... After this the ...... structure of the enzyme is affected and the ...... of the .... is changed. The enzyme has been ....

2. Look at Figure 3.
   a) At which pH does pepsin work best?
   b) At which pH does amylase work best?
   c) What happens to the activity of the enzymes as the pH increases?
   d) Explain why this change in activity happens.

DID YOU KNOW?

Not all enzymes work best at around 40°C! Bacteria living in hot springs work at temperatures of over 80°C and higher. On the other hand, some bacteria which live in the very cold deep sea have enzymes working effectively at 0°C and below!

GET IT RIGHT!

The rate of enzyme-controlled reactions increases as the temperature goes up to about 40°C because the particles are moving faster. So substrate molecules collide with enzymes more often. Once the temperature goes much over 40°C most enzymes are denatured and no longer work as catalysts.
Enzymes aren't killed (they are molecules, not living things themselves) – use the term **denatured**.

Figure 4 The magical light display of a firefly is caused by the action of a very special enzyme called luciferase

KEY POINTS

1. Enzyme activity is affected by temperature and pH.
2. High temperatures and the wrong pH can affect the shape of the active site of an enzyme and stop it working.
3. Enzymes catalyse processes such as respiration, photosynthesis and protein synthesis in living cells.

179

SUMMARY ANSWERS

1. Reactions, cells, temperatures, enzyme, increase, 40°C, protein, shape, active site, denatured.
2. **a)** About pH 2.
   b) About pH 8.
   c) The activity levels fall fast.
   d) The increase in pH affects the shape of the active site of the enzyme so it no longer bonds to the substrate. It is denatured and no longer catalyses the reaction.

Answers to in-text questions

a) The reactions in the bacteria that make the milk go sour/bad happen faster as it is warmer.

b) The shape of the active site is lost so it does not work properly any more.

KEY POINTS

It is important to stress that each enzyme has an optimum temperature and an optimum pH at which it works. Variations in temperature and pH affect the rate at which each enzyme works. Extremes of pH and high temperature cause the enzyme to be denatured (not killed) because the shape of the enzyme is altered and the substrate no longer fits into the active site.

B2 4.3 Aerobic respiration

LEARNING OBJECTIVES

Students should learn that:

- During aerobic respiration, glucose and oxygen are used to release energy.
- Carbon dioxide and water are released as waste products.
- Most of the reactions in aerobic respiration occur inside mitochondria.

LEARNING OUTCOMES

Most students should be able to:

- Describe the raw materials and products of the process of respiration.
- Describe where the reactions take place in cells.

Some students should also be able to:

- Explain why more active cells, such as muscle cells, have greater numbers of mitochondria than less active cells.
- Design experiments to show that oxygen is taken up and carbon dioxide is released during aerobic respiration.

Teaching suggestions

- **Special needs.** Make model mitochondria from date boxes, washing liquid capsule boxes or similar, with corrugated cardboard. Fill with used batteries to indicate their role as energy carriers and display on a large poster.
- **Gifted and talented.** Give the students a sheet on the theories of Alan Templeton and Rebecca Cann with regard to 'Mitochondrial Eve' and the origins of the human species. Ask students to summarise points for and against each theory.
- **Learning styles**

  Kinaesthetic: Folding paper exercise, making a model mitochondrion.

  Visual: Observing electron microscope images of mitochondria.

  Auditory: Discussing the reasons for respiration.

  Interpersonal: Testing each other on the respiration equation.

  Intrapersonal: Appreciating the importance of the organelle.

SPECIFICATION LINK-UP Unit: Biology 2.12.6

- *During aerobic respiration (respiration which uses oxygen) chemical reactions occur which:*
  - *use glucose (a sugar) and oxygen*
  - *release energy.*
- *Most of the reactions in aerobic respiration take place inside mitochondria.*
- *Aerobic respiration is summarised by the equation:*

  glucose + oxygen → carbon dioxide + water (+ energy)

- *The energy that is released during respiration is used:*
  - *to build up larger molecules using smaller ones*
  - *in animals, to enable muscles to contract*
  - *in mammals and birds, to maintain a steady body temperature in colder surroundings*
  - *in plants, to build up sugars, nitrates and other nutrients into amino acids which are then built into proteins.*

Lesson structure

STARTER

Instant energy – Show glucose drink bottles and energy drinks. Read or show labels from various foods and read as a class (search an image bank for 'energy drink label'). Students to list the energy contents from the labels and order them in terms of sugar content and energy. (10 minutes)

Turning lime water cloudy – Draw crosses on the bottoms of test tubes with a Chinagraph pencil. Half-fill each tube with lime water. Give each student a drinking straw (use bendy straws) and a tube of lime water and tell them to blow gently through the straw into the lime water until they can no longer see the cross on the bottom from the top. Eye protection must be worn. Ask: 'How long does it take? What is happening?' Go over the reaction and introduce respiration. (10 minutes)

MAIN

Composition of inhaled and exhaled air – As a follow-up to the starter 'Turning lime water cloudy', a more refined piece of apparatus can be used to show that the air that is breathed in contains less carbon dioxide than air breathed out. Arrange two tubes of lime water, tubing and clips, so that air can be drawn in through one tube containing lime water and breathed out through another tube containing lime water. After a few breaths, it can clearly be seen that the lime water in the two tubes differs in cloudiness.

Investigating respiration

- **Using a small mammal** – A small mammal can be placed under a bell jar on a glass plate sealed with Vaseline. Air is drawn through the apparatus, first passing through a U-tube of soda lime (to remove carbon dioxide) and then through a tube of lime water (to show that carbon dioxide has been removed) before entering the bell jar. After leaving the bell jar, the air is drawn through another tube of lime water to show that carbon dioxide is given off. The small mammal does not need to be there for very long – any small mammal is usually quite active and a result is achieved fairly quickly.
- **Alternatively** – Other small animals, such as earthworms, woodlice or maggots can be used in such a demonstration/investigation.
- **Using a plant** – It is possible to substitute a potted plant for the small mammal and to show that carbon dioxide is given off during respiration in plants. The pot and soil of the plant need to be enclosed in a polythene bag and the bell jar needs to be covered in black paper to exclude light. The apparatus should be left running for a couple of days. Ask: 'Why is the plant pot covered up? Why is light excluded?'
- **Controls** – All these experiments need controls, which should be discussed with the students (this relates to 'How Science Works'– validity of experimental design). In most cases, with the bell jar experiments, the removal of the living organism should be considered as a control.

PLENARIES

How small can you get it? – Have a competition to see who can fold an A4 sheet of paper into the smallest volume. Ask: 'What is the best method?' Relate to surface area and cristae in mitochondria. (5–10 minutes)

Sort it out – Give groups of students the statements in question 1 of the summary questions on separate slips of paper. See which group can sort them out correctly in the shortest time. (10 minutes)

SUMMARY ANSWERS

1 **a)** D **b)** E **c)** A **d)** C **e)** B

2 **a)** Mitochondria are the site of energy production in the cell. They have very folded membranes inside, giving a large surface area for the enzymes that control the reactions of aerobic respiration.
b) Muscle cells are very active and need a lot of energy; therefore they need large numbers of mitochondria to supply the energy. Fat cells use very little energy so need very few mitochondria.

3 The main uses of energy in the body are for movement, for building new molecules and for heat generation. The symptoms of starvation are that people become very thin. Stored energy is used up and growth stops. New proteins are not made and there is not enough energy or raw materials. People do not want to move; they lack energy, as there is a lack of fuel for the mitochondria. People feel cold, as there is not enough fuel for the mitochondria to produce heat energy.

4 See practical box 'Investigating respiration' on page 180 in Student Book.

Practical support

Composition of inhaled and exhaled air

Equipment and materials required

Test tubes half-full of lime water in racks, test tubes with 2-hole bungs, delivery tubes (one long, one short), rubber tubing, sterile mouth pieces, Chinagraph pencil, drinking straws (bendy ones if possible) or tubing and clips, eye protection.

Investigating respiration

For these demonstrations/investigations

Lime water (to show carbon dioxide produced), soda lime in U-tube (to absorb carbon dioxide for the experiments with the small mammal and the potted plant), small mammal such as a hamster in a bell jar (for the small mammal demo.), potted plant (for the potted plant demo.), earthworms, maggots or woodlice, black paper, tubing, eye protection.

ACTIVITY & EXTENSION

- **Mitochondria** – View Internet sites on electron tomography for good pictures of mitochondria. A particularly useful site is the National Center for Microscopy and Imaging Research. (www.ncmir.ucsd.edu).
- **Aerobic exercise** – Try out some aerobics, an activity for the class in the gym to calculate energy released. Several videos will give exercises and the amount of energy used. Ask: 'Why is it called aerobic exercise?'

BIOLOGY ENZYMES

B2 4.3 Aerobic respiration

LEARNING OBJECTIVES

1 What is aerobic respiration?
2 Where in your cells does respiration take place?

One of the most important enzyme-controlled processes in living things is **aerobic respiration**.

Your digestive system, lungs and circulation all work to provide your cells with what they need for respiration to take place.

During aerobic respiration glucose (a sugar produced as a result of digestion) reacts with oxygen. This reaction releases energy which your cells can use. This energy is vital for everything else that goes on in your body.

Carbon dioxide and water are produced as waste products of the reaction.

We call the process **aerobic** respiration because it uses oxygen from the air.

Aerobic respiration can be summed up by the equation:

glucose + oxygen → carbon dioxide + water (+ **energy**)

a) Why is aerobic respiration so important?

DID YOU KNOW?

The average energy needs of a teenage boy are 11 510 kJ of energy every day – but teenage girls only need 8830 kJ a day. This is partly because on average girls are smaller than boys but also because boys have more muscle cells, which means more mitochondria demanding fuel for aerobic respiration.

PRACTICAL

Investigating respiration

Animals and plants – even bacteria – all respire. To show that cellular respiration is taking place, you can either deprive a living organism of the things it needs to respire, or show that waste products are produced from the reaction.

Depriving a living thing of food and/or oxygen would kill it – so this would be an unethical investigation. So we concentrate on the waste products of respiration. Carbon dioxide and energy in the form of heat are the easiest to identify.

Lime water goes cloudy when carbon dioxide bubbles through it. The higher the concentration of carbon dioxide, the quicker the lime water goes cloudy. This gives us an easy way of demonstrating that carbon dioxide has been produced. We can also look for a rise in temperature to show that energy is being produced during respiration.

- Plan an ethical investigation into aerobic respiration in living organisms.

Outer membrane

A mitochondrion

Folded inner membrane gives a large surface area where the enzymes which control cellular respiration are found

Figure 1 Mitochondria are the powerhouses which provide energy for all the functions of your cells

Mitochondria – the site of respiration

Aerobic respiration involves lots of chemical reactions, each one controlled by a different enzyme. Most of these reactions take place in the **mitochondria** of your cells.

Mitochondria are tiny rod-shaped bodies (**organelles**) which are found in almost all plant and animal cells. They have a folded inner membrane which provides a large surface area for the enzymes involved in aerobic respiration.

Additional biology

Cells which need a lot of energy – like muscle cells and sperm – have lots of mitochondria. Cells which use very little energy – like fat cells – have very few mitochondria.

b) Why do mitochondria have folded inner membranes?

Reasons for respiration

- Respiration releases energy from the food we eat so that the cells of the body can use it.
- Both plant and animal cells need energy to carry out the basic functions of life. They build up large molecules from smaller ones to make new cell material. Much of the energy released in respiration is used for these 'building' activities (synthesis reactions). For example in plants, the sugars, nitrates and other nutrients are built up into amino acids which are then built up into proteins.
- Another important use of the energy from respiration in animals is in making muscles contract. Muscles are working all the time in our body, whether we are aware of them or not. Even when you sleep your heart beats, you breathe and your gut churns – and these muscular activities use energy.
- Finally, mammals and birds are 'warm-blooded'. This means that our bodies are the same temperature inside almost regardless of the temperature around us. On cold days we use energy to keep our body warm, while on hot days we use energy to sweat and keep our body cool.

GET IT RIGHT!

Make sure you know the equation for respiration. Remember that aerobic respiration takes place in the mitochondria.

Figure 2 Warm-blooded animals like this bird use up some of the energy they produce by aerobic respiration just to keep a steady body temperature. When the weather is cold, they use up a lot more energy to keep warm. Giving them extra food supplies can mean the difference between life and death.

SUMMARY QUESTIONS

1 Copy and complete these sentences, matching the pairs.

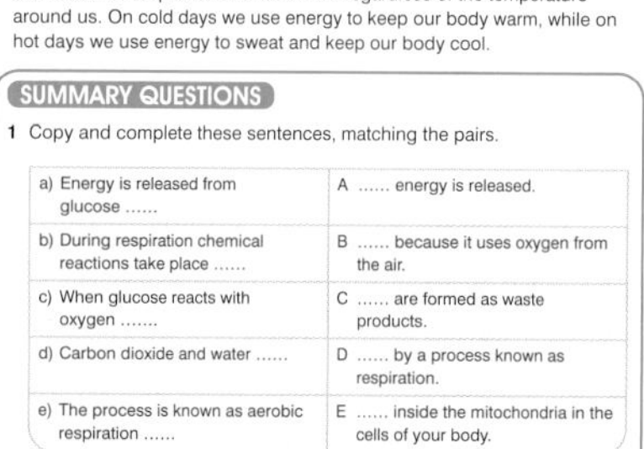

a) Energy is released from glucose	A ...... energy is released.
b) During respiration chemical reactions take place	B ...... because it uses oxygen from the air.
c) When glucose reacts with oxygen	C ...... are formed as waste products.
d) Carbon dioxide and water	D ...... by a process known as respiration.
e) The process is known as aerobic respiration	E ...... inside the mitochondria in the cells of your body.

2 Why are mitochondria so important and how is their structure adapted for the job that they do?

3 You need a regular supply of food to provide energy for your cells. If you don't get enough to eat you become thin and stop growing. You don't want to move around and you start to feel cold. There are three main uses of the energy released in your body during aerobic respiration. What are they and how does this explain the symptoms of starvation described above?

4 Suggest an experiment to show that a) oxygen is taken up, and b) carbon dioxide is released, during aerobic respiration.

KEY POINTS

1 Aerobic respiration involves chemical reactions which use oxygen and sugar and release energy. The reaction is summed up as:

glucose + oxygen → carbon dioxide + water (+ **energy**).

2 Most of the reactions in aerobic respiration take place inside the mitochondria.

Answers to in-text questions

a) It provides energy for all the functions of the cells.

b) The folded inner membranes provide a large surface for all the enzymes needed to control the reactions of respiration.

KEY POINTS

Students should learn the respiration equation. Higher attaining students could put in the chemical symbols and balance the equation. They can then test each other on the respiration equation. The site of respiration in the mitochondria is important and students should be aware that muscle tissue is more active than fatty tissue.

B2 4.4 Enzymes in digestion

LEARNING OBJECTIVES

Students should learn that:

- During digestion, the breakdown of large molecules into smaller molecules is catalysed by enzymes.
- These enzymes, which are produced by specialised cells in glands, pass out into the gut.
- The enzymes include amylases that catalyse the breakdown of starch, proteases that catalyse the breakdown of proteins and lipases that catalyse the breakdown of lipids.

LEARNING OUTCOMES

Most students should be able to:

- Explain how enzymes are involved in the digestion of our food.
- Describe the location and action of the enzymes which catalyse the breakdown of carbohydrates (starch), proteins and lipids.

Some students should also be able to:

- Explain digestion in terms of the molecules involved.

Teaching suggestions

- **Special needs.** Use flip cards with foods on one side and their components on the other. Some students might need a clue, such as starting letters or vowels. Alternatively, use an Internet version for whiteboards, or individual computers such as those created through 'Quia' (use this as a search term).
- **Gifted and talented.** Research the role of ribosomes through an interactive CD-ROM, such as the protein synthesis section of www.dnai.org.
- **Learning styles**

  Kinaesthetic: Playing floor dominoes. Carrying out practical work on model gut.

  Visual: Observing Simulation B2 4.4 'Enzymes in Digestion' from the GCSE Biology CD ROM.

  Auditory: Listening to instructions about the Visking tubing experiments.

  Interpersonal: Discussing the ethics of experimenting with the contents of another person's stomach.

  Intrapersonal: Evaluating the strength of evidence from the digestion experiments and suggesting improvements.
- **Homework.** Write the clues for the cryptic word search.

SPECIFICATION LINK-UP Unit: Biology 2.12.6

- *Some enzymes work outside the body cells. The digestive enzymes are produced by specialised cells in glands and in the lining of the gut. The enzymes pass out of the cells into the gut where they come into contact with the food molecules. They catalyse the breakdown of large molecules into smaller molecules:*
  - *the enzyme amylase is produced in the salivary glands, the pancreas and the small intestine. This enzyme catalyses the breakdown of starch into sugars in the mouth and small intestine*
  - *protease enzymes are produced by the stomach, the pancreas and the small intestine. These enzymes catalyse the breakdown of proteins into amino acids in the stomach and the small intestine*
  - *lipase enzymes are produced by the pancreas and small intestine. These enzymes catalyse the breakdown of lipids (fats and oils) into fatty acids and glycerol in the small intestine.*

Lesson structure

STARTER

Model gut – Liquidise a dinner, or similar food leaving some big bits, and spoon it into the leg cut from a pair of tights. Squeeze it so that some of the goo goes through but not the big bits. Describe how this models the situation in the intestines. Discuss limitations of this model. (5 minutes)

What we know about enzymes so far, a quick quiz – Ask ten questions on enzyme structure and the factors affecting their action. Students to swap and mark each others. (5–10 minutes)

MAIN

- **Making a model gut** – Each group of students will need two 15 cm lengths of dialysis (Visking) tubing to model the gut. (See 'Practical support'.)
- If desired, the experiments can be left for 24 hours at room temperature before testing.
- **A simplified version of the model gut** – If there is not time for the students to carry out their own experiments, then a length of dialysis tubing can be filled with a mixture of 30% glucose solution and 3% starch solutions and placed in a test tube of distilled water. If this is left for about 15 minutes, the water can be tested for starch and glucose.
- Some glucose should have diffused through the tubing into the water, but the starch should not. Tests for starch and glucose will confirm this. Note: This only demonstrates that glucose will pass through the tubing but starch will not; it does not show that the enzyme catalyses the breakdown of the starch.
- **Investigating digestion** – The model gut can be used to show the effect of changes in temperature and pH on the activity of saliva or amylase on starch. The tubing should be placed in boiling tubes, and samples of the water surrounding the tubing can be tested for starch and sugars at intervals to determine whether or not digestion has taken place.
- To investigate changes in temperature, the boiling tubes containing the enzyme-substrate mixtures in the tubing should be incubated in a range of temperatures from about 5°C to 60°C using water baths.
- To investigate the range of pH values, buffer solutions should be used, providing another opportunity to develop the investigative aspects of 'How Science Works'.

PLENARIES

Floor dominoes – Make up large 'domino' cards of food types, their components and the enzymes. Play in groups. (5–10 minutes)

Enzyme spider – Draw out a spider diagram to summarise the role of enzymes in digestion, with contributions from the class. (10 minutes)

Cryptic word search – Students to begin a word search, deciding on the words to include. Write cryptic definitions of the words as clues for homework. (10 minutes)

Practical support

Investigating digestion

Equipment and materials required

Visking or dialysis tubing, dropping pipettes, elastic bands, starch and enzyme solutions (the strengths of these solutions may need to be adjusted to give results in a lesson session), ater baths, beakers, test tubes and racks, iodine solution (CLEAPSS Hazcard 54), Benedict's solution (Harmful – CLEAPSS Hazcards 27 and 95), eye protection.

etails

Each group of students will need two 15 cm lengths of dialysis (Visking) tubing, which has been soaked in water. Each piece should be knotted securely at one end. Using a dropping pipette, fill one length of the tubing with 3% starch solution and place it in a test tube. Fold the top of the tubing over the rim of the test tube and secure with an elastic band. Remove all traces of the starch solution from the outside of the tubing by filling the test tube with water and emptying it several times. Finally, fill the test tube with water and place it in a rack.

Repeat the procedure with the second length of tubing but add 5 cm^3 saliva (safety as B2 4.2 with saliva) or amylase solution to the starch solution, and shake before filling the dialysis tubing. The test tubes should be labelled A and B and placed in a water bath at 35°C for 30 minutes. The water in the test tubes should then be tested for: starch, using iodine solution; sugars, using Benedict's solution.

SUMMARY ANSWERS

1 Food, insoluble, broken down, soluble, absorbed, cells, digestive enzymes.

2 Suitable table.

3 Large molecules cannot pass through gut membrane into blood, cannot be taken into cells or used.
glucose – absorbed, carried in blood, taken into cells, broken down by respiratory enzymes to provide energy
amino acids – absorbed, carried in blood, taken into cells and used in protein synthesis to build new proteins
fatty acids/glycerol – absorbed, carried in blood, taken into cells and used as a source of energy for respiration and for building new fats and oils.

ACTIVITY & EXTENSION IDEAS

- Search image bank for 'Alexis St Martin'. Enlarge on the story, as the students love to hear about the lumps of meat dangled inside him. Have some cubes of meat tied to pieces of string to emphasise the point.

Answers to in-text questions

a) They work outside the cells of your body.

b) Amylase.

c) Proteases.

d) Lipases.

KEY POINTS

These key points are general and students could add details of specific enzymes, their substrates and products, in order to expand their information. The production of a chart or flow diagram showing where and how food is digested in the gut could also be useful.

B2 4.4 Enzymes in digestion

LEARNING OBJECTIVES

1 How are enzymes involved in the digestion of your food?
2 What happens to the digested food?

GET IT RIGHT!

Learn the different types of digestive enzymes and the end products of the breakdown of your food. Make sure you know where the different digestive enzymes are made.

The food you eat is made up of large insoluble molecules which your body cannot absorb. They need to be broken down or ***digested*** to form smaller, soluble molecules. These can then be absorbed and used by your cells. This chemical breakdown is controlled by your digestive enzymes.

Most of your enzymes work ***inside*** the cells of your body. Your digestive enzymes are different – they work ***outside*** of your cells. They are produced by specialised cells which are found in glands (like your salivary glands and your pancreas), and in the lining of your gut.

The enzymes then pass out of these cells into the gut itself. It is here that they get mixed up with your food molecules and break them down.

Your gut is a hollow muscular tube which squeezes your food. The gut:

- helps to break up your food into small pieces with a large surface area for your enzymes to work on,
- mixes your food with your digestive juices so that the enzymes come into contact with as much of the food as possible, and
- uses its muscles to move your food along its length from one area to the next.

a) How do your digestive enzymes differ from most of your other enzymes?

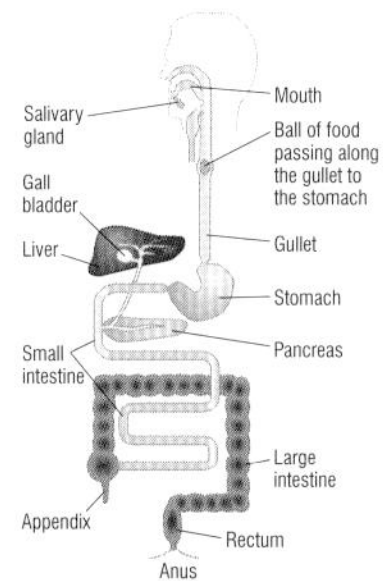

Figure 1 The human digestive system

Digesting carbohydrates

Enzymes which break down carbohydrates are known as **carbohydrases**. Starch is one of the most common carbohydrates that you eat. It is broken down into **sugars** like glucose. This reaction is catalysed by the carbohydrase called ***amylase***.

Amylase is produced in your salivary glands, so the digestion of starch starts in your mouth. Amylase is also made in your pancreas and your small intestine. No digestion takes place in the pancreas. All the enzymes made there flow into your small intestine, which is where most of the starch you eat is digested.

b) What is the name of the enzyme which breaks down starch in your gut?

Digesting proteins

The breakdown of protein food like meat, fish and cheese into amino acids is catalysed by **protease** enzymes. Proteases are produced by your stomach, your pancreas and your small intestine. The breakdown of proteins into **amino acids** takes place in your stomach and small intestine.

c) Which enzymes breaks down protein in your gut?

Digesting fats

The **lipids** (fats and oils) that you eat are broken down into **fatty acids** and **glycerol** in your small intestine. The reaction is catalysed by ***lipase*** enzymes which are made in your pancreas and your small intestine. Yet again the enzymes made in the pancreas are passed into the small intestine.

Once your food molecules have been completely digested into soluble glucose, amino acids, fatty acids and glycerol, they leave your small intestine. They pass into your blood supply to be carried around the body to the cells which need them.

d) Which enzymes break down fats in your gut?

FOUL FACTS

When Alexis St Martin suffered a terrible gunshot wound in 1822, Dr William Beaumont managed to save his life. However Alexis was left with a hole (or fistula) from his stomach to the outside world. Dr Beaumont then used this hole to find out what happened in Alexis's stomach as he digested food!

PRACTICAL

Investigating digestion

You can make a model gut using a bag of special membrane containing starch and amylase enzymes. When the enzyme has catalysed the breakdown of the starch, you can detect the presence of sugar on the outside of the 'gut'!

- How can you test for sugars?

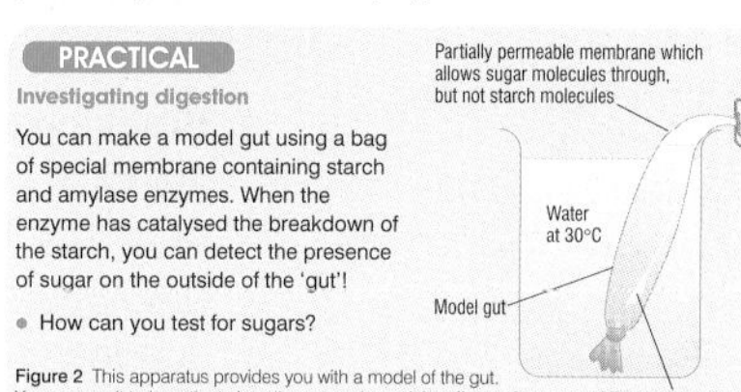

Figure 2 This apparatus provides you with a model of the gut. You can use it to investigate how the gut works and the effects of factors like temperature and pH on how the gut enzymes work.

Using the digested food

- The glucose produced by the action of amylase and other carbohydrases is used by the cells of your body in respiration.
- Fatty acids and glycerol may be used as a source of energy or to build cell membranes, make hormones and as fat stores.
- The amino acids produced when you digest protein are not used as fuel. Once inside your cells, amino acids are built up into all the proteins you need. These synthesis reactions are catalysed by enzymes. In other words, your enzymes make new enzymes as well as all the other proteins you need in your cells. This **protein synthesis** takes place in the **ribosomes**.

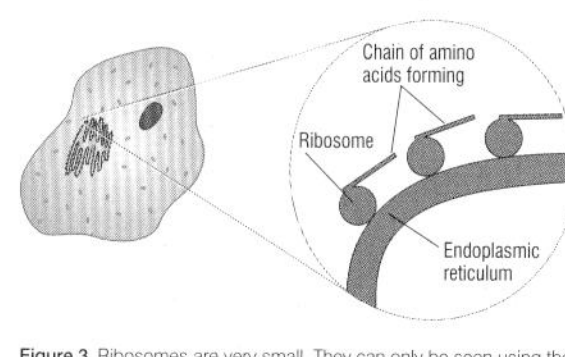

Figure 3 Ribosomes are very small. They can only be seen using the most powerful microscopes. However their role in protein synthesis means they are vital to the working of your cells and your whole body!

SUMMARY QUESTIONS

1 Copy and complete using the words below:

absorbed **broken down** **cells** **digestive enzymes** **food** **insoluble** **soluble**

The ...... you eat is made up of large ...... molecules which need to be .... to form smaller, ... molecules. These can be ...... by your body and used by your .... This chemical breakdown is controlled by your ....

2 Make a table which shows amylase, protease and lipase. For each enzyme show where it is made, which reaction it catalyses and where it works in the gut.

3 Why is digestion of your food so important? Explain in terms of the molecules involved.

KEY POINTS

1 Enzymes catalyse the breakdown of large food molecules into smaller molecules during digestion.
2 Digestive enzymes are produced inside cells but they work outside of cells in the gut.
3 Enzymes in the ribosomes catalyse the build up of proteins from amino acids.

B2 4.5 Speeding up digestion

LEARNING OBJECTIVES

Students should learn that:

- The enzymes in the stomach work most effectively in the acid conditions resulting from the production of hydrochloric acid by the stomach.
- Bile produced by the liver provides the alkaline conditions needed for the enzymes in the small intestine to work most effectively.
- Bile also emulsifies the fats increasing the surface area for the enzymes to act upon.

LEARNING OUTCOMES

Most students should be able to:

- Describe how pH affects the enzymes in the different parts of the gut.
- State that bile emulsifies fats.

Some students should also be able to:

- Explain how the emulsification of fats increases the rate of their digestion.

Teaching suggestions

- **Special needs.** Show the students a Mars bar or something similar. Ask: 'If you stuck it into your mouth whole, which bits could the saliva touch?' Get a student to chop the bar in half. Ask: 'Which bits could the saliva touch now that it could not before?' Chop again repeatedly and draw out the idea of surface area being important in digestion.
- **Gifted and talented.** Present the students with a table of results of surface area against rate of reaction. Students could draw a graph and calculate the gradient of the graph at a number of points and describe how it changes, with reasons.
- **Learning styles**

  Kinaesthetic: Carrying out practical work.

  Visual: Observing demonstrations of emulsification.

  Auditory: Discussing the activities of the lesson.

  Interpersonal: Working in groups during the practicals.

  Intrapersonal: Considering the reasons for the changes in rates of reaction or writing a letter to the doctor about gall stones.

SPECIFICATION LINK-UP Unit: Biology 2.12.6

- *Some enzymes work outside the body cells. The digestive enzymes are produced by specialised cells in glands and in the lining of the gut. The enzymes pass out of the cells into the gut where they come into contact with the food molecules. They catalyse the breakdown of large molecules into smaller molecules:*
  - *protease enzymes are produced by the stomach, the pancreas and the small intestine. These enzymes catalyse the breakdown of proteins into amino acids in the stomach and the small intestine*
  - *the stomach also produces hydrochloric acid. The enzymes in the stomach work most effectively in these acid conditions*
  - *the liver produces bile that is stored in the gall bladder before being released into the small intestine. Bile neutralises the acid that was added to the food in the stomach. This provides alkaline conditions in which enzymes in the small intestine work most effectively.*

Lesson structure

STARTER

Vomit! – Show a video clip, or still, from *Little Britain* of the woman who does the projectile vomiting over people (or discuss the sketches with the students). Search the web for 'Little Britain'. Ask: 'If we scooped up some of the sick with a spoon and tested its pH, what would it be? What does vomit taste like?' Discuss what it is and then link with bulimia and the effect of repeated vomiting on the enamel of the teeth. (10 minutes)

Emulsions – Bring in a can of emulsion paint. Ask the students what the word means (you could paint it on to a large sheet of paper). Bring in a salad, some vinegar and some olive oil. Get a student to pour some oil on top of the vinegar in a gas jar or similar vessel, shake vigorously and produce an emulsion. Students can do this themselves on a small scale in a boiling tube. Observe the globules formed and link to surface area, then to speeding up digestion. (10 minutes)

The bacon experiment – Get a volunteer to write their initials in Vaseline on a piece of bacon and immerse the bacon in protease enzyme. Students can predict what will happen and check if they're right at the end of the lesson. (5–10 minutes)

MAIN

- **Breaking down protein** – The practical described in the Student Book is easy to set up. Each group of students will require three test tubes and can set up their own experiment. An additional control tube could be added using boiled and cooled pepsin.
- **Modification to the experiment** – The experiment can be made more quantitative, and therefore extend the opportunity to teach concepts from 'How Science Works', by getting the students to formulate an hypothesis, make predictions, use stated volumes of enzyme and acid and weigh the pieces of meat used at the beginning and end of the experiment. In this case, it would be advisable to leave the experiment running for 24 hours. The pieces of meat should then be removed from the tubes, rinsed and dried on filter paper before reweighing. A bar chart can be drawn showing the percentage change in mass.
- **Comparing the action of pepsin with the action of trypsin** – Pepsin and trypsin work in different parts of the gut in different pH conditions. The experiment described in section 4.2: 'Effect of pH on enzyme action' could be used here to show that pepsin works best in acid conditions and trypsin in alkaline conditions. If specific pH values are required, then the use of buffer solutions is recommended.
- **The effect of bile salts** – The experiment described in lesson 4.4: 'Digesting fats', involves the use of bile salts and it would be appropriate to include it here.
- **Dissection of bird or mammal gut to show different regions** – Depending on the school policy and LEA regulations, a game bird or rabbit could be dissected to show the position of the liver, the gall bladder and other organs associated with the gut. If this is not allowed or advisable, then there are videos available or still pictures that can be projected. Students can be given the opportunity to see it: not for the squeamish. (Search for images on the web, looking for 'dissection gut'.) Students who object should not be forced to take part!

PLENARIES

Is the bacon done? – Check the bacon set up at the start of the lesson. If required use a web cam to display the results via a projector. (5 minutes)

Gall stones – Show some gall stones or photographs of gall stones. Discuss why gall stones occur and what might be the consequences. Get the students to write a letter to their doctor stating what problems they fear and asking advice. (10 minutes)

Colouring exercise – Give the students unlabelled diagrams of the digestive system and ask them to label them and colour in the different regions according to the different pH conditions that exist in the gut. (10 minutes)

ACTIVITY & EXTENSION

- **Increasing the surface area** – Discuss the effect of the teeth and mastication on the break up of large masses of food in the mouth. Ask: 'What is the effect on digestion in the mouth? Does chewing affect the digestion in the stomach?'
- **Design an experiment** – Students could be asked to design an experiment to find the optimum pH at which the enzyme amylase works. They need to think about where amylase is produced and then to use an appropriate range of pH values, suggest which variables need to be controlled and how to judge and evaluate the results (this is another excellent learning opportunity for 'How Science Works').

BIOLOGY ENZYMES

B2 4.5 Speeding up digestion

LEARNING OBJECTIVES

1. Why does your stomach contain hydrochloric acid?
2. What is bile and why is it so important in digestion?

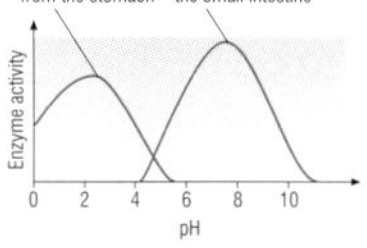

Your digestive system produces many enzymes which speed up the breakdown of the food you eat. However enzymes aren't the only important chemicals in your gut. As you saw on pages 44 and 45, enzymes are very sensitive to temperature and pH. As your body is kept at a fairly steady 37°C, your enzymes have an ideal temperature which allows them to work as fast as possible.

Keeping the pH in your gut at ideal levels isn't quite so easy. That's because different enzymes work best at different pH levels. The protease enzyme found in your stomach works best in acidic conditions.

On the other hand, the proteases made in your pancreas need alkaline conditions. Then they can catalyse protein breakdown as fast as they can. Look at the graph in Figure 1.

So your body makes a variety of different chemicals which help to give your enzymes ideal conditions all the way through your gut.

a) Why do your enzymes almost always have the right temperature to work at their best?

Pepsin – protease from the stomach
Trypsin – protease from the small intestine
Enzyme activity
0 2 4 6 8 10
pH

Figure 1 Both of these enzymes catalyse the breakdown of proteins. But as these graphs show, the enzyme found in the stomach works best at a very different pH to the one made in the pancreas and used in the small intestine.

Changing pH in the gut

You have around 35 million glands in the lining of your stomach secreting protease enzymes to digest the protein you eat. These enzymes work best in an acid pH. So your stomach also produces a concentrated solution of hydrochloric acid from the same glands. In fact your stomach produces around 3 litres of acid a day!

This acid allows your stomach protease enzymes to work very effectively. It also kills most of the bacteria which you take in with your food.

Finally, your stomach also produces a thick layer of mucus which coats your stomach walls and protects them from being digested by the acid and the enzymes!

b) How does your stomach avoid digesting itself?

PRACTICAL

Breaking down protein

You can see the effect of acid on pepsin, the protease found in the stomach, quite simply. Set up three test tubes, one containing pepsin only, one containing only hydrochloric acid and one containing a mixture of the two. Keep them at body temperature in a water bath. Add a similar sized chunk of meat to all three of them. Set up a web cam and watch for a few hours to see what happens!

- What conclusions can you make?

Figure 2 These test tubes show clearly the importance of protein-digesting enzymes and hydrochloric acid in your stomach. Meat was added to each tube at the same time.

FOUL FACTS

Pigments from your bile are largely responsible for the brown colour of your faeces. If you have a disease which stops bile getting into your gut, your faeces will be white or silvery grey!

184

Additional biology

After a few hours – depending on the size and type of the meal you have eaten – your food leaves your stomach and moves on into your small intestine. Some of the enzymes which catalyse digestion in your small intestine are made in your pancreas. Some are also made in the small intestine itself. They all work best in an alkaline environment.

The acidic liquid coming from your stomach needs to become an alkaline mix in your small intestine! So how does it happen?

Your liver carries out many important jobs in your body and one of them is producing bile. Bile is a greenish-yellow alkaline liquid which is stored in your gall bladder until it is needed.

As food comes into the small intestine from the stomach, bile is squirted onto it. The bile neutralises the acid from the stomach and then makes the semi-digested food alkaline. This provides the ideal conditions for the enzymes in the small intestine.

c) Why does the food coming into your small intestine need neutralising?

Altering the surface area

It is very important for the enzymes of the gut to have the largest possible surface area of food to work on. This is not a problem with carbohydrates and proteins. However, the fats that you eat do not mix with all the watery liquids in your gut. They stay as large globules – think of oil in water – which makes it difficult for the lipase enzymes to act.

This is the second important function of the bile. It **emulsifies** the fats in your food. This means it physically breaks up large drops of fat into smaller droplets. This provides a much bigger surface area for the lipase enzymes to act on. The larger surface area helps them chemically break down the fats more quickly into fatty acids and glycerol.

DID YOU KNOW?

Sometimes the gall bladder and bile duct get blocked by gall stones which can range in diameter from a few millimetres to several centimetres.

Figure 3 If gall stones like these block your bile duct, they not only affect your digestion, they cause absolute agony!

GET IT RIGHT!

Remember food is not digested in the liver or the pancreas.
Bile is **not** an enzyme and it does **not** break down fat molecules.
Bile emulsifies fat droplets to increase the surface area, which in turn increases the rate of fat digestion by lipase.

SUMMARY QUESTIONS

1. Copy and complete using the words below:

   alkaline emulsifies gall bladder liver neutralises small intestine

   Bile is an...... liquid produced by your .... It is stored in the .... and released onto food as it comes into the .... It ...... the acid food from the stomach and makes it alkaline. It also ...... fats.
2. Look at Figure 1.
   a) At what pH does the protease from the stomach work best?
   b) How does your body create the right pH in the stomach for this enzyme?
   c) At what pH does the protease from the intestine work best?
   d) How does your body create the right pH in the small intestine for this enzyme?
3. Draw a diagram to explain how bile produces a big surface area for lipase to work on and explain why this is important.

KEY POINTS

1. The enzymes of the stomach work best in acid conditions.
2. The enzymes made in the pancreas and the small intestine work best in alkaline conditions.
3. Bile produced by the liver neutralises acid and emulsifies fats.

185

SUMMARY ANSWERS

1. Alkaline, liver, gall bladder, small intestine, neutralises, emulsifies.
2. a) pH 2.1
   b) Hydrochloric acid is made in glands in the stomach.
   c) pH 7.4
   d) The liver produces bile that is stored in the gall bladder and released when food comes into the small intestine.
3. [Marks for a good diagram showing a large fat droplet coated in bile splitting into many small fat droplets.] This produces a larger surface area so enzymes can get to many more fat molecules and so break them down more quickly.

Answers to in-text questions

a) The body temperature is usually maintained around 37°C.

b) The stomach glands produce a thick layer of mucus.

c) The food entering the small intestine from the stomach is acidic. The enzymes of the small intestine work best in alkaline conditions.

Practical support

Breaking down protein

Equipment and materials required

For each group:

At least three test tubes and a rack, small cubes of meat, 2% pepsin solution, 0.1 M solution of hydrochloric acid, water bath at 35°C, balance, labels, filter papers, eye protection.

KEY POINTS

Students should be aware of the changes in pH through the gut and which enzymes work best in the different conditions. It is also important to make sure students understand that food is only digested in the gut and not by the liver or the pancreas. It could be helpful to refer to these organs as associated with the gut, or as accessory organs.

B2 4.6 Making use of enzymes

LEARNING OBJECTIVES

Students should learn that:

- Enzymes from microorganisms have many uses in the home and in industry.
- Proteases and lipases are used in the manufacture of biological detergents.
- Proteases, carbohydrases and isomerase are used in food manufacture.

LEARNING OUTCOMES

Most students should be able to:

- Explain how biological detergents work.
- Describe some of the ways in which enzymes are used in the food industry.

Some students should also be able to:

- Evaluate the advantages and disadvantages of using enzymes in home and in industry.

Teaching suggestions

- **Special needs.** Use name boards with a fold-over end. Write 'carbohydrate' on one and use the hinged fold-over to convert it into 'carbohydrase'. Have examples of all the enzymes required in the specification.
- **Gifted and talented.** Ask students to find out the differences in the structural formulae of glucose and fructose. They can try to work out why these sugars have different effects on the taste buds.
- **Learning styles**

  Kinaesthetic: Carrying out practical work; using structural models (like molymodels) to show interchange between glucose and fructose.

  Visual: Creating a poster.

  Auditory: Explaining to other students how temperature affects enzyme activity.

  Interpersonal: Conducting a survey of who has dishwashers in their homes and what they think of them.

  Intrapersonal: Thinking about whether or not they would eat their parent's vomit if it meant staying alive.
- **Homework.** There are some opportunities here for research at home. Students could find out how many people use biological detergents in dishwashers and washing machines, and whether or not there are differences between them. Some preparatory work for the next spread could be set as homework tasks here.

SPECIFICATION LINK-UP Unit: Biology 2.12.6

- *Some microorganisms produce enzymes that pass out of the cells. These enzymes have many uses in the home and in industry.*
- *In the home, biological detergents may contain protein-digesting and fat-digesting enzymes (proteases and lipases).*
- *In industry:*
  - *proteases are used to 'pre-digest' the protein in some baby foods*
  - *carbohydrases are used to convert starch into sugar syrup*
  - *Isomerase is used to convert glucose syrup into fructose syrup, which is much sweeter and therefore can be used in smaller quantities in slimming foods.*

Lesson structure

STARTER

Taste tests – Fructose (fruit sugar) is now available in many supermarkets. It could be interesting to make up separate solutions of fructose, sucrose and glucose of the same strength (e.g. 2 teaspoons in a beaker of water) and get the students to do a blind tasting scoring them for sweetness on a 5 point scale. (Must be done in hygienic conditions following risk assessment.) (10–15 minutes)

Baby food for lunch? – Show the students some samples of baby food. Have disposable plastic spoons and be prepared for joking. Some pelican bibs will help to create the atmosphere. Ask: 'How does this food differ from adult food? What did parents do (and some still do) before the commercially prepared baby foods were available?' (10 minutes)

MAIN

- **Investigating biological washing powder** – Using agar plates containing starch, milk and mayonnaise (or salad cream or egg yolk), the activity of enzymes in biological detergents can be demonstrated. (See 'Practical support'.)
- This demonstration can be used to:
  - Compare different biological washing powders or liquids (the advantage of liquids is that volumes can be measured and dilutions made more easily).
  - Compare dishwasher detergents with clothes washing detergents.
  - Discover whether the age of the detergent has any effect on its efficiency.
  - Discuss experimental design.
- These investigations can be used to introduce many 'How Science Works' concepts. Predictions can be made, measurements made and recorded, variables controlled and conclusions drawn. It also gives students some scope for designing their own investigations.
- **What temperature should I use?** – The experiment above can be modified to demonstrate that the proteases in a biological detergent can work at higher temperatures than trypsin from an animal source. Samples of both can be heated to temperatures of 30°C, 40°C etc. and then placed in holes in milk agar plates. Use a separate plate for each enzyme or detergent tested and the number of holes should correspond to the number of different temperatures tested. The plates should then be treated as above and the clear areas measured and recorded. A graph can then be plotted of temperature against area of clear zone.
- **Egg chunks in detergent** – Cubes of egg white immersed in a solution of biological detergent will be digested. A solution of a biological washing powder is made by dissolving 3 g of powder in 30 cm^3 of water. A cube of egg white is weighed and placed in this solution for 20 minutes, after which time it is removed, rinsed and dried. The effect of the washing powder can be assessed by reweighing.
- This investigation can be expanded to consider different variables. Comparisons can be made using different biological detergents. The strength of the detergent needed can be investigated and the optimum temperature found. Again, this fulfils many of the requirements in 'How Science Works'.

?ractical support

ıvestigating biological washing powder

quipment and materials required

ome biological detergent (either in powder or liquid form in order to make up different oncentrations if needed – avoid contact with skin), egg white in chunks/cubes (or agar lates containing starch, milk, mayonnaise, salad cream or egg yolk), test tubes and racks, ssues for drying, iodine solution (CLEAPSS Hazcard 54) balance for weighing, eye rotection, protective gloves.

etails

cork borer is used to remove cylinders of agar from the prepared plates. The number f cylinders removed depends on the number of detergents being tested. Into the holes, olutions of the detergents can be placed and the plates incubated at 25°C for about 4 hours.

dine solution is poured over the starch-agar plate and left for 5 minutes before being oured away. The diameter of clear areas around the holes can be measured and recorded. should be possible to measure clear areas around the holes on the milk-agar plates and e mayonnaise-agar plates.

PLENARY

Baby food II – Ask your students to close their eyes and envisage a situation where their parent or guardian asks them if they are hungry. They say they are starving, so the adult is sick into a bucket and gives them a spoon, saying 'OK, eat that!' If easily available, show video footage of adult seabirds regurgitating food for their young. Talk over the changes in food that make it an advantage for very young creatures to eat regurgitated food. Also discuss reasons why it is not a good idea to drink sick! (10 minutes)

BIOLOGY ENZYMES

B2 4.6 Making use of enzymes

LEARNING OBJECTIVES

1 How do biological detergents work?
2 How are enzymes used in the food industry?

Enzymes were first isolated from living cells in the 19th century, and ever since we have found more and more ways of using them in industry. Some microorganisms produce enzymes which pass out of the cells and are easy for us to use. In other cases we use the whole microorganism.

Enzymes in the home

In the past, people boiled and scrubbed their clothes to get them clean – and did it all by hand. Now we not only have washing machines to do the washing for us, we also have enzymes ready and waiting to digest the stains.

Many people use **biological detergents** to remove stains from their clothes from substances such as grass, sweat, food and blood. Biological washing powders contain proteases and lipases which break down the proteins and fats in the stains. They help provide us with a cleaner wash. We also use them at the lower temperatures that enzymes need to work best, so we use less electricity too.

Figure 1 More and more homes now have a dishwasher – and dishwasher powders contain enzymes. They digest the cooked-on proteins like eggs which are often hard to remove even in a dishwasher.

a) What is a biological washing powder?

PRACTICAL

Investigating biological washing powder

Weigh a chunk of cooked egg white and leave it in a strong solution of biological washing powder.

- What do you think will happen to the egg white?
- How can you measure just how effective the protease enzymes are?
- How could you investigate the effect of surface area in enzyme action?

Enzymes in industry

Pure enzymes have many uses in industry.

Proteases are used in the manufacture of baby foods. They 'pre-digest' some of the protein in the food. When babies first begin to eat solid foods they are not very good at it. Treating the food with protease enzymes makes it easier for a baby's digestive system to cope with. It is easier for them to get the amino acids they need from their food.

Figure 2 Learning to eat solid food isn't easy. Having some of it pre-digested by protease enzymes can make it easier to get the goodness you need to grow!

Carbohydrases (carbohydrate digesting enzymes) are used to convert starch into sugar (glucose) syrup. We use huge quantities of sugar syrup in food production – just have a look at the ingredients labels on all sorts of foods.

Starch is made by plants like corn, and it is very cheap. Using enzymes to convert this plant starch into sweet sugar provides a cheap source of sweetness for food manufacturers.

It is also important for the process of making fuel (ethanol) from plants.

b) Why does the starch need to be converted to sugar before it is used to make ethanol?

186

Additional biology

Sometimes the glucose syrup made from starch is passed into another process which uses a different set of enzymes. **Isomerase** enzyme is used to convert glucose syrup into **fructose syrup** by rearranging the atoms in the glucose molecule.

Glucose and fructose contain exactly the same amount of energy (1700 kJ or 400 kcal per 100 g) but fructose is much sweeter than glucose. This means much smaller amounts of it are needed to make food taste sweet. So fructose is widely used in 'slimming' foods. The food tastes sweet but contains fewer calories.

Figure 3 The market for slimming foods is enormous and growing all the time. Enzyme technology is being used to convert more and more glucose syrup into fructose syrup to make so-called 'slimming' foods.

GET IT RIGHT!

Remember that most enzyme names end in -ase. Some enzymes used in industry work at quite high temperatures – so don't be put off if a graph shows an optimum temperature well above 45°C!

The advantages and disadvantages of using enzymes

In an industrial process, many of the reactions need high temperatures and pressures to make them fast enough to produce the products needed. Supplying heat and building chemical plants which can stand high pressures costs a lot of money.

However, enzymes can provide the perfect answer to industrial problems like these. They catalyse reactions at relatively low temperatures and normal pressures. Enzyme-based processes are therefore often fairly cheap to run.

The main problem with enzymes is that they are very sensitive to their surroundings. For enzymes to function properly the temperature must be kept down (usually below 45°C). The pH also needs to be kept within carefully monitored limits which suit the enzyme. It costs money to control these conditions.

Whole microbes are relatively cheap, but need to be supplied with food and oxygen and their waste products removed. What's more, they use some of the substrate to grow more microbes. Pure enzymes use the substrate more efficiently, but they are also more expensive to produce.

SUMMARY QUESTIONS

1 List three enzymes and the ways in which we use them in the food industry.
2 Biological washing powders contain enzymes in tiny capsules. Explain why:
   a) they are more effective than non-biological powders at lower temperatures,
   b) they are not more effective at high temperatures.
3 Make a table to show the advantages and disadvantages of using enzymes in industry.

KEY POINTS

1 Some microorganisms produce enzymes which pass out of the cells and can be used in different ways.
2 Biological detergents may contain proteases and lipases.
3 Proteases, carbohydrases and isomerase are all used in the food industry.

187

SUMMARY ANSWERS

1 Proteases: pre-digested baby food.
Carbohydrases: convert starch to glucose syrup.
Isomerase: converts glucose syrup to fructose syrup.

2 **a)** The protease and lipase enzymes digest proteins and fats on the clothes, so the clothes get cleaner than detergent alone.
b) At temperatures above about 45°C, the enzymes may be denatured and so have no effect on cleaning.

3

Advantages	Disadvantages
Work at relatively low temperatures.	Denatured by high temperatures.
Work at relatively low pressures.	Sensitive to pH changes.
Efficient catalysts.	If whole organisms, need food, oxygen and waste products removed.

Answers to in-text questions

a) A powder that contains enzymes, usually proteases and lipases.

b) Enzymes in yeast turn sugar (glucose) into ethanol.

KEY POINTS

Students should be made aware of the use of several different enzyme groups in the manufacture of products found in the home. As these enzymes have been derived from microorganisms with optimum temperatures different from those of mammals, i.e. around 37°C, products such as washing powders may work best at higher temperatures.

B2 4.7 High-tech enzymes

SPECIFICATION LINK-UP

Unit: Biology 2.12.6

Students should use their skills, knowledge and understanding of 'How Science Works':

- *to evaluate the advantages and disadvantages of using enzymes in the home and in industry.*

Teaching suggestions

- **The washing powder debate: a question of allergies** – Show digital pictures of skin with dermatitis. Discuss the problems; ask: 'Does anyone in the class have a skin allergy or reaction to washing powders?' Show a PowerPoint® presentation on allergic skin reactions. Include information on the 'packaging' of the enzymes and also some of the instructions on the packets. Ask: 'Are the warnings clear enough?'
- **A question of temperature** – Discuss the advantages of the biological washing powders and detergents with respect to the lower temperatures needed in washing machines. Ask: 'Is this always a good thing?' Some cycles will operate at temperatures as low as 30°C, but there may be some drawbacks. Draw up a balance sheet with advantages and disadvantages of the lower operating temperatures. Ask: 'Do the advantages outweigh the disadvantages? Would you wash your baby's nappies in a low temperature wash?'
- **Dishwashers and washing machines** – Ask: 'Are the biological detergents for dishwashers and washing machines the same?' Some research into the components of biological detergents for dishwashers and those for washing clothes could be interesting. Ask: 'Are the enzymes the same?' This could be tested in one of the experiments described in the previous spread.
- **Enzymes and medicine: a simple demonstration** – Clinistix and albustix can be used to test for the presence of glucose and protein in urine. Carry out a 'Tinkle test' experiment with fake urine doctored with glucose, protein, both and neither. Discuss the benefits of such tests compared with the standard tests used in the lab (Benedict's etc.). Also discuss the value of the tests in making quick diagnoses and how they can help people with diabetes to control their condition.
- **'Sammi's story'** – Show the video *Sammi's story* (Channel 4 Television, 1995 V15.012 or animated version by Philippa

BIOLOGY ENZYMES
B2 4.7 High-tech enzymes

The washing powder debate

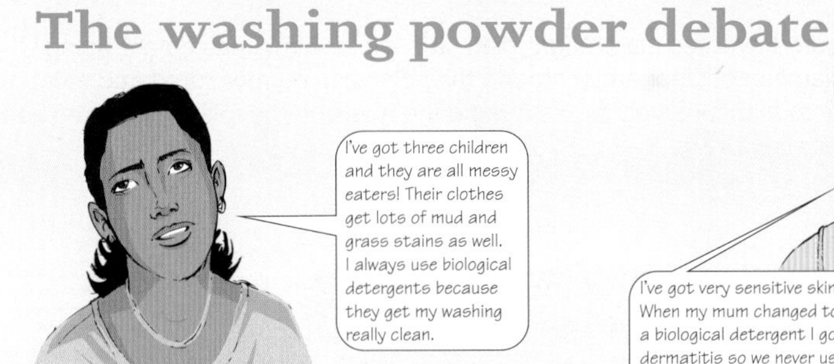

I've got three children and they are all messy eaters! Their clothes get lots of mud and grass stains as well. I always use biological detergents because they get my washing really clean.
I've got very sensitive skin. When my mum changed to a biological detergent I got dermatitis so we never use biological detergents now.

33% extra FREE
TESCO Bio
ARIEL Non-Bio
Persil

When we first started manufacturing our biological detergent we found a lot of our factory staff developed allergies. We realised they were reacting to enzyme dust in the air – proteins often trigger allergies. But once we put the enzymes in tiny capsules all the allergy problems stopped. Unfortunately it got some bad publicity and lots of people still seem to think biological detergents cause allergies.

I try to be as green as possible in my lifestyle but I'm not sure about biological detergents. Enzymes are natural, after all – but I've heard they can cause allergies. On the other hand, biological powders use a lot less electricity because they clean at lower temperatures. That's good for the environment and cheaper for me!

ACTIVITY

You are part of a team producing an article for a lifestyle magazine about biological washing powders. Create a double-page article – make it lively, interesting to look at, scientifically accurate and informative!

Allergies aren't really a problem with biological detergents. However, if the clothes aren't rinsed really thoroughly, protein-digesting enzymes can get left in the fabric. Then the enzymes may digest some of the protein in your skin and set up dermatitis. But if the detergent is used properly, there shouldn't be a problem.

188

Manning, 1994, from Health Improvement Information and Resource Centre – 02071508345 available from the cystic fibrosis charity. This goes through the necessity of taking enzymes with food due to a blocked pancreatic duct. Ask students to summarise aspects of the content of the video and explain to the rest of the class. (See www.cftrust.org.uk.)

- **Other uses** – Use the Internet to research other uses of enzymes in medicine (e.g. www.enzymestuff.com). Some of the examples given in the Student Book can be expanded as well as other examples found.
- **The raw food craze** – This is a very open-ended activity, but the two pointers given at the end can give students a starting point. Again, class discussion of the topic could be beneficial. A brainstorming session about the different aspects could draw out some themes, which could be grouped together. For example, ask: 'What does cooking do to food? How is the vitamin content affected?'
- Students could be provided with some raw food of all types (fruit, vegetables and meat). Give students the choice; ask: 'Would you eat it or not? If not, why not?' Build up a list on the board of things that most students would eat raw and things they would not. You could run a Bush Tucker trial on some vegetables that are normally eaten cooked, such as beetroot, courgettes etc. and see how many would be eaten raw.

Enzymes and medicine

Here are just some of the ways in which enzymes are used in medicine:

To diagnose disease

If your liver is damaged or diseased, some of your liver enzymes may leak out into your blood. If your symptoms suggest your liver isn't working properly, doctors can test your blood for these enzymes to find out if your liver really is damaged.

ACTIVITY

Make a poster with the title 'Enzymes in medicine' which could be used on the walls of the science department to inform pupils in KS3 and KS4. Use this material as a starting point. Also do some more research about the way enzymes can be used to help you make your poster as interesting as possible.

To diagnose and control disease

People who have diabetes have too much sugar in their blood. As a result, they also get sugar in their urine. One common test for sugar in the urine relies on a colour change on a test strip.

The test strip contains a chemical indicator and an enzyme. It is placed in a urine sample. The enzyme catalyses the breakdown of any glucose found in the urine. The products of the reaction then make the indicator change colour if glucose is present.

To cure disease

- If your pancreas is damaged or diseased it cannot make enzymes. So you have to take extra enzymes – particularly lipase – to allow you to digest your food. The enzymes are in special capsules to stop them being digested in your stomach!
- If you have a heart attack, an enzyme called streptokinase will be injected into your blood as soon as possible. It dissolves clots in the arteries of the heart wall and reduces the amount of damage done to your heart muscle.
- An enzyme from certain bacteria is being used to treat a type of leukaemia in children. The cancer cells cannot make one particular amino acid, so they need to take it from your body fluids. The enzyme catalyses the breakdown of this amino acid, so the cancer cells cannot get any and they die. Your normal cells can make the amino acid so they are not affected. Doctors hope something similar may work against other types of cancer.

Health special

IN THE RAW!

WILL YOU TRY THE NEW DIET SENSATION?

The latest food craze to sweep the US is to eat your food – including meat – completely raw. And now it's coming to the UK!

It has been reported that there are lots of health benefits to this new way of eating. It is claimed that raw food contains live enzymes which will help to give you more energy. Apparently when food is cooked these enzymes die.

One of the owners of a new raw food restaurant has been quoted as saying,

'It is an amazingly interesting way of preparing food, it is good to have live enzymes in your system and, most importantly, it is yummy.'

Dodgy science

ACTIVITY

Dodgy science is everywhere in the media. It is often used to make people think something is a really good – or really bad – idea. The way the new craze for eating raw food has been reported is a good example. Read the extract from a magazine article given here. You now have quite a lot of biological knowledge gained from this part of the course and what you have learnt earlier about things like the spread of disease. Use your knowledge to:

a) explain what is wrong with the science in this report and why it is so inaccurate,
b) explain what problems linked to eating raw meat have been ignored in this report.

189

Special needs

Carry out a washing test using biological and non-biological detergents on some white T-shirts that have been stained with grass, egg, blackcurrant juice etc. If possible make a video of the experiment as a spoof advert.

Gifted and talented

Students could do more research on the problems of eating raw food, especially with respect to the toxic substances present in some plants that are removed by cooking (oxalates in rhubarb, red kidney beans etc.). They could produce a guide leaflet to the problems.

ICT link-up

- There are numerous opportunities for the use of ICT in the production of posters, pamphlets and leaflets.
- There are also opportunities for researching on the Internet. When doing so, students should search for 'enzymes in medicine' and 'raw food diets'.

Learning styles

Kinaesthetic: Carrying out washing test or tests with Clinistix.

Visual: Displaying information on a poster.

Auditory: Discussing the pros and cons of using biological washing powders.

Interpersonal: Collaborating in the practical activities.

Intrapersonal: Reviewing knowledge gained from other spreads in completing one of the activities.

ACTIVITY & EXTENSION IDEAS

- **The biological washing powder activity** – The suggested activity provides scope for some imaginative work. Students could work in groups and provide material about different issues. The 'talking heads' in the students' book spread have most of the issues covered and there is plenty there for informative articles. Students could include some pictures of allergic reactions, a simplified diagram of the encapsulation of the enzymes in the manufacturing process, some facts and figures about energy-saving and something about the stains that can be removed. It would be sensible to include some information about the problems; any good magazine would try to give a balanced view!
- **The poster activity: enzymes and medicine** – It will be difficult to get all the information on one poster, depending on the amount of detail to be included. Students could decide to make a series of posters. Some discussion about the impact of a poster with masses of information or a poster about one condition clearly explained should help students to decide what they would like to do. The use of ICT could be encouraged here: images can be scanned in and used to create an interesting poster.
- **The raw food craze** – Once students have got the science sorted out, they need to decide how they are going to present their explanations. This 'dodgy' science was purported to be from a magazine article. Ask: 'Is the best reply to it a letter to the magazine or a counterbalancing article from 'Our Science Correspondent' or an article from someone who has tried it – and suffered?' It could be a case for reinforcing the '5-a-day' recommendation for the consumption of fruit and vegetables.

SUMMARY ANSWERS

1 a) i) 6 **ii)** 4 **iii)** 2 **iv)** 1 **v)** 3 **vi)** 5

b) Enzymes work by bringing reacting particles together and lowering the activation energy needed for them to react. Enzymes are large protein molecules with a hole or indentation known as the 'active site'. The substrate of the reaction fits into the active site of the enzyme like a lock and key. Once it is in place, the enzyme and substrate bind together. This is called the 'enzyme-substrate complex'. Then the reaction takes place rapidly and the products are released from the surface of the enzyme. Anything that affects the shape of the active site affects the ability of the enzyme to speed up a reaction. [The use of diagrams would make this explanation very clear.]

2 a) It catalyses the breakdown of starch to glucose.

b) An increase in temperature increases the rate of the enzyme-controlled reaction.

c) One tube is kept at each temperature without any enzyme to act as a control/to show what would happen if there were no enzyme present.

d) It could be improved by having an additional control of just starch solution kept at body temperature.

e) The rate of the enzyme activity would fall because it is working well in a neutral pH. Lowering the pH/making it more acid will affect the active site and reduce its effectiveness.

3 a) [Marks awarded for a good graph plot with suitable scale chosen, axes right way round, axes labelled correctly and accurately-plotted points.]

b) Alkaline.

c) This enzyme could be found in the small intestine, because it works in alkaline conditions. Other protein-digesting enzymes work in the stomach, but the conditions there are acidic.

Summary teaching suggestions

- **Literacy guidance** – The explanation required in the answer to question 1b) provides an opportunity for the students to write answers in continuous prose and practise their literary skills. Look for good grammar, clear expression and the correct spelling of biological terms.
- **Special needs**
  - Students may be able to cope with question 1a), if provided with the pre-prepared information to match up, and with question 3a) if given a prepared grid.
  - Some of the questions about the experiments (question 2) could be given verbally.
- **When to use the questions?**
  - Question 1b) is a summary of enzyme action and, together with a clear diagram, would make an excellent revision card.
  - Question 2 could be used as a stimulus to help students write up their accounts of the experiment if they carried it out. These are the kinds of questions they could be asked about any experiment on a test paper.
  - Graph plotting (question 3) is good practice for showing results. The interpretation of the curve is a more demanding skill. This could be set for homework.

ENZYMES: B2 4.1 – B2 4.7

SUMMARY QUESTIONS

1 a) Copy and complete the following sentences, matching the parts of the sentences.

i) A catalyst will speed up or slow down a reaction	1 ...... could not occur without enzymes.
ii) Living organisms make very efficient catalysts	2 ...... made of protein.
iii) All enzymes are	3 ...... binds to the active site.
iv) The reactions which keep you alive	4 ...... known as enzymes.
v) The substrate of an enzyme	5 ...... a specific type of molecule.
vi) Each type of enzyme affects	6 ...... but is not changed itself.

b) Explain how an enzyme catalyses a reaction. Use diagrams if they make your explanation clearer.

2 Use Figure 2 on page 178 to help you answer this question.

a) What effect does the enzyme amylase have on starch?

b) What do these results tell you about the effect of temperature on the action of the amylase?

c) Why is one tube of starch solution kept at each temperature without the addition of the enzyme?

d) How could you improve this investigation?

e) What do you predict would happen to the activity of the enzyme if acid from the stomach were added to the mixture?

3 The table gives some data about the relative activity levels of an enzyme at different pH levels.

pH	Relative activity
4	0
6	3
8	10
10	1

a) Plot a graph of this data.

b) Does this enzyme work best in an acid or an alkaline environment?

c) This is a protein-digesting enzyme. Where in the gut do you think it might be found? Explain your answer.

EXAM-STYLE QUESTIONS

1 (a) In the summary of aerobic respiration shown below, choose a word from each of the boxes th best completes the equation.

Glucose + BOX A → carbon dioxide + BOX B (+ er

BOX A: water / oxygen / nitrogen — BOX B: water / oxygen / nitrogen

(b) (i) State two ways in which the energy releasec during respiration is used in all animals.

(ii) How else might the energy released be usec mammals and birds only?

(iii) Give a further use of the energy released th applies to plants rather than animals.

2 A, B, C, D and E are the names of enzymes or grou of enzymes. The numbers 1, 2, 3, 4 and 5 refer to the functions or uses of each of these enzymes.

Match each letter with the appropriate number.

A	Lipase	1	Used in the manufacture baby foods
B	Amylase	2	Group of enzymes that a on carbohydrates
C	Proteases	3	Its substrate is starch
D	Isomerase	4	Used in the production c slimming foods
E	Carbohydrases	5	The products of its cataly action are glycerol and fa acids

3 Amylase is an enzyme that catalyses the conversion starch into sugar.

(a) To which of the following groups of food does starch belong?

carbohydrates fats protein vitamins

(b) Give the names of the three organs in the huma body that secrete the enzyme amylase.

The graph on the next page shows the effect of temperature on the activity of amylase.

(c) (i) At what temperature did the amylase work fastest?

(ii) Why did the amylase not work above 56°C?

(iii) State one other factor apart from temperatu that will affect the rate of reaction of amylase

EXAM-STYLE ANSWERS

1 a) Box A = oxygen *(1 mark*
Box B = water *(1 mark*

b) i) To build up larger molecules from smaller ones. *(1 mark*
To enable muscles to contract. *(1 mark*

ii) To maintain a steady/constant body temperature. *(1 mark*

iii) Used to build up sugars, nitrates and other nutrients into amino acids and then proteins. *(1 mark*

2 **A** 5
B 3
C 1
D 4
E 2 *(1 mark each*

3 a) Carbohydrates *(1 mark*

b) Salivary glands
Pancreas
Small intestine *(1 mark each*

c) i) 45°C *(1 mark*
ii) It had become denatured. *(1 mark*
iii) pH *(1 mark*

4 a)

Temperature (°C)	Time to clot for A (mins)	Time to clot for B (mins)

able to be used, with suitable format *(1 mark*
labelled *(1 mark*
labelled with units *(1 mark*

b) Range of 10 to 60°C *(1 mark*

c) At least 5 interval measurements *(1 mark*

d) e.g. shake to see if it is 'solid' *(1 mark*

Additional biology

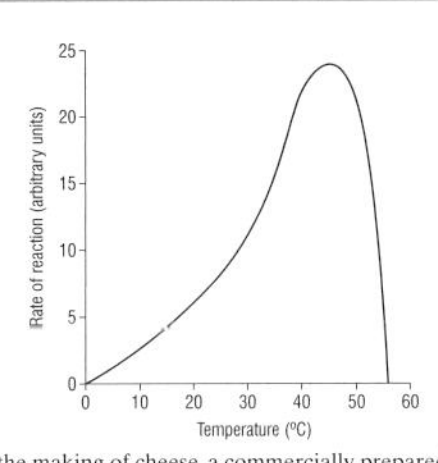

In the making of cheese, a commercially prepared form of an enzyme called rennin is used to make the protein in milk more solid. Rennin is an enzyme that is produced naturally in the stomachs of young mammals. The owner of a cheese making factory wanted to use a different source of rennin. She needed to find out the best temperature to use for the new rennin. She planned to set up 20 test tubes. All would have 20 cm^3 of milk in them: half with the rennin added (A) and half to be left without rennin (B). One tube of each type would be left in a water bath until one of them clotted. When this happened, the time taken would be recorded.

(a) Construct a table that could be used by the owner. (3)

(b) Fill in the table to show the range of temperatures she might use. (1)

(c) Fill in the table to show the interval for the independent variable. (1)

(d) Suggest how the owner might know when the milk is clotted. (1)

(e) Would you suggest that she repeats her results? Explain your answer. (1)

(f) Why do you think she used tubes A and B at each temperature? (1)

Bile is a greenish liquid that plays an important role in the digestion of food.

(a) In which organ is bile produced? (1)

(b) Where is bile stored in the body? (1)

(c) Into which region of the digestive system is bile released? (1)

(d) Describe how bile is involved in the digestion of fats. (3)

(e) What is the name of the enzyme that digests fats? (1)

(f) Name two places where this enzyme is produced in the body. (2)

HOW SCIENCE WORKS QUESTIONS

Najma had carried out a 'rates of reaction' investigation in chemistry. Her results are in Table 1 below.

Table 1 Chemistry investigation

Temperature (°C)	Time taken (secs)
20	106
30	51
40	26
50	12
60	5

When asked in biology to do a 'rates of reaction' investigation she expected to get the same results. She reasoned that in both cases she was collecting the oxygen produced from hydrogen peroxide. The only difference was that she used manganese(IV) oxide in chemistry and she was using mashed up plant cells in biology! Her results from biology are in Table 2.

Table 2 Biology investigation

Temperature (°C)	Time taken (secs)
20	114
30	96
40	80
50	120
60	No reaction

a) What was Najma's prediction for the biology investigation? (1)

b) Was her prediction supported, refuted or should she rethink the prediction? (2)

c) Najma checked her results against some results in a textbook. Why was this a good idea? (1)

d) Najma was feeling happier now that she had been supported by other scientists' results. She had also learned that enzymes had a temperature at which they worked best. How could she change her investigation so that she could find the best temperature for this enzyme? (1)

e) Najma also learned that this enzyme was called catalase and that it occurs in nearly all organisms, even those living in hot water springs. How could she change her investigation to find the best temperature for catalase in hot water spring organisms? (2)

e) Yes, because being precise about the time of clotting is very difficult. *(1 mark)*

f) B acted as a control
OR to test whether the rennin was causing the clotting. *(1 mark)*

5 a) The liver *(1 mark)*

b) The gall bladder *(1 mark)*

c) The small intestine *(1 mark)*

d)
- Neutralises the acid (added to the food in the stomach).
- Provides neutral/alkaline conditions for enzymes to function/optimum pH for enzymes.
- Emulsifies fats/breaks up fats into small droplets.
- Increases the surface area of fats (so that enzymes can work on them more effectively). *Allow any accurate description of emulsification.*

(1 mark for each point to a maximum of 3 marks)

e) Lipase *(1 mark)*

f) Pancreas *(1 mark)*
Small intestine *(1 mark)*

Exam teaching suggestions

- The first three questions and the last one (question 5) are subdivided into small bite-sized pieces, making them easy to follow and requiring short answers testing largely recall of knowledge. As such, they make suitable practice examination exercises for foundation-tier students while still providing straightforward revision tests for those at higher-tier level.
- Students need to be made aware of the significance of the preamble that accompanies certain questions. All too often they dismiss it as extraneous packaging that can be disregarded. The information is always there for a purpose. It often does more than merely set the scene; it may contain clues, even vital information, that is needed to answer the question effectively.

HOW SCIENCE WORKS ANSWERS

a) Najma's prediction was that, as the temperature was increased, the time taken for the oxygen to be collected would decrease.

b) Her prediction was supported up to about 40°C. After that, the time taken decreased until there was no reaction at all.

c) Yes, because it would check the reliability and accuracy of her results.

d) Increase the interval measurements around 40°C.

e) She would have to check the temperature of hot water springs and adjust her temperatures to match them. Collect some organisms from the spring to use instead of the plant cells she used before.

How science works teaching suggestions

- **Literacy guidance**
  - Key terms that should be clearly understood: prediction, accuracy, interval measurements.
  - The question expecting a longer answer, where students can practise their literacy skills is e).
- **Higher- and lower-level answers.** Questions b) and e) are higher-level questions. The answers for these have been provided at this level. Question c) is a lower-level question and the answer provided is also at this level.
- **Gifted and talented.** Able students could be asked to design an investigation for question e) as homework. This would be difficult if they were to use hydrogen peroxide at very high temperatures.
- **How and when to use these questions.** When wishing to develop ideas of the fallibility of predictions and the use of interval measurements. The questions are best tackled in small discussion groups.
- **Special needs.** A demonstration of the decomposition of hydrogen peroxide would bring more accessibility for some students.

B2 5.1 Controlling internal conditions

LEARNING OBJECTIVES

Students should learn that:

- Waste products, carbon dioxide and urea, have to be removed from the body.
- Carbon dioxide is removed via the lungs when we breathe out and the kidneys are involved in the removal of urea.
- Internal conditions, such as the water and ion content of the body, need to be controlled.

LEARNING OUTCOMES

Most students should be able to:

- Explain why carbon dioxide and urea need to be removed from the body.
- Describe how carbon dioxide is removed from the body via the lungs.
- Describe the role of the kidneys in removing urea and controlling the water and ion content of the body.

Some students should also be able to:

- Demonstrate that they understand the complexity of homeostatis.

Teaching suggestions

- **Special needs.** Give the students a blank diagram with a body outline on it in the centre and four arrows coming from appropriate places. The students are given the labels 'Urine', 'Faeces', 'Sweat' and 'CO_2 in breath' and asked to put them in the correct places. [Beware of additions to the body drawing!]
- **Gifted and talented.** Ask students to hold their breath for 30 seconds (risk assessment for individuals). They can feel the desire to breathe building, due to the CO_2 building up. Talk about drowning experiences. Ask: 'Does the feeling go when you breathe in? Why is this? Where might the sensors be that tell you that you have too much CO_2 in you? How can you tell?' Link to the starter: 'Can you turn the indicator yellow?'

SPECIFICATION LINK-UP Unit: Biology 2.12.7

- *Waste products which have to be removed from the body include:*
  - *carbon dioxide produced by respiration – most of this leaves the body via the lungs when we breathe out*
  - *urea produced in the liver by the breakdown of excess amino acids – this is removed by the kidneys in the urine, which is temporarily stored in the bladder.*
- *Internal conditions which are controlled include the water content of the body, the ion content of the body, temperature and blood sugar levels.*
- *If the water or ion content of the body is wrong, too much water may move into or out of the cells and damage them. Water and ions enter the body when we eat or drink.*
- *Sweating helps to cool the body. More water is lost when it is hot, and more water has to be taken as drink or in food to balance this loss.*

Lesson structure

STARTER

Keeping warm or staying cool – Show Student Book photos from the GCSE Biology CD ROM of obviously different climatic conditions. Ask: 'What will their core temperature be like?' Students to predict, respond and discuss. (5–10 minutes)

Can you turn the indicator yellow? – Show the students some hydrogencarbonate indicator. It should be cherry red when it is in equilibrium with the air. If a little dilute acid is added it turns yellow, if alkali is added it will go purple. Get a volunteer to blow into a tube of indicator through a straw. Ask: 'What does this show about the effect of CO_2 on the indicator and therefore the pH of CO_2 in solution?' Use this for students to speculate on the effect of the accumulation of CO_2 in the cells of the body. (10 minutes)

MAIN

Analysis of inspired and expired air

A sample of air is drawn into a capillary tube (called a J-tube, it consists of a syringe attached to a capillary tube bent into a square J-shape) and its volume *a* is recorded. Potassium hydroxide solution is then drawn into the tube to absorb the carbon dioxide. The volume will decrease. The new volume *b* is noted. The potassium hydroxide solution is almost all expelled. A reagent that absorbs oxygen (pyrogallol) is then drawn into the tube, causing the volume to decrease, and the new volume *c* noted. The percentage of CO_2 in the air is given by the expression: $a - b/a \times 100$

The percentage of O_2 in the air is given by the expression: $b - c/a \times 100$

- A sample of expired air can be obtained by immersing a boiling tube in a trough of water, raising it to a vertical position and then exhaling into it through a bent straw or capillary tube. The analysis of this sample of exhaled air can then be tested as above and the percentages compared.
- This is best done as a demonstration before the whole class, but you could get a volunteer to give you a sample of expired air. The samples need to be jiggled around in the J-tube so that the absorption of the gases takes place. Three samples should be measured and a mean taken. The samples need to be at room temperature before their volume is measured. (This relates to: 'How Science Works': reliability and validity of data.)

PLENARIES

Facts and figures – Produce some facts and figures from this spread (4%, 0.04%, kidneys, urea etc.) and ask students to write a sentence about the relevance of each to the current topic. Students to read some out. (10 minutes)

Spirometer – Have a spirometer for the students to look at and show how it can be used. (This relates to 'How Science Works': making measurements.) (10 minutes)

ACTIVITY & EXTENSION

- **Spirometer** – Use a volunteer student to demonstrate the use of a spirometer. Care needed and risk assessment required.
  Students should be familiar with the diagram in the Student Book and understand the principles of the use of the apparatus. They could discuss how the tracings can be used to measure lung volume and the effect of exercise on breathing.
- **pH inside cells** – Remind students of the experiment to find the optimum pH at which catalase works inside cells. If it was not done, then a demonstration emphasises the need for the elimination of carbon dioxide to keep the conditions inside cells neutral so that enzymes, such as catalase, can work most efficiently.
- **Homeostasis in the long-distance runner** – There are articles about marathon running on The Physiological Society's web site that could be read out and discussed. (See www.physoc.org.)

Teaching suggestions – continued

- **Learning styles**
  Kinaesthetic: Investigating breathing.
  Visual: Observing colour changes in indicator solutions.
  Auditory: Listening to the opinions of other students.
  Interpersonal: Collaborating with other students in kidney dissection.
  Intrapersonal: Considering the different effects of the outside environment and the maintenance of a stable internal environment.

BIOLOGY HOMEOSTASIS

B2 5.1 Controlling internal conditions

LEARNING OBJECTIVES

1. How do you keep conditions inside your body constant?
2. How do you get rid of the waste products of your cells?

For your body to work properly the conditions surrounding your millions of cells must stay as constant as possible. On the other hand, almost everything you do tends to change things.

As you move you produce heat, as you respire you produce waste, when you digest food you take millions of molecules into your body. Yet you somehow keep your internal conditions constant within a very narrow range. How do you manage this?

The answer is through **homeostasis**. As you saw on page 32, many of the functions in your body help to keep your internal environment as constant as possible. Now you are going to find out more about some of them.

a) What is homeostasis?

Removing waste products

No matter what you are doing, even sleeping, the cells of your body are constantly producing waste products as a result of the chemical reactions which are taking place. The more extreme the conditions you put yourself in, the more waste products your cells will make. There are two main poisonous waste products which would cause major problems for your body if the levels built up. These are carbon dioxide and urea.

Carbon dioxide

Carbon dioxide is produced during cellular respiration.

Every cell in your body respires, and so every cell produces carbon dioxide. It is vital that you remove this carbon dioxide. That's because if it all remained dissolved in the cytoplasm of your cells it would affect the pH. Dissolved carbon dioxide produces an acidic solution – and a lower pH would affect the working of all the enzymes in your cells!

Figure 1 Whatever you choose to do in life, the conditions inside your body will stay more-or-less exactly the same. When you think of the range of things you can do, it is amazing how the balance is maintained.

PRACTICAL

Investigating breathing

Find out the capacity of your lungs or the effect of exercise on breathing.

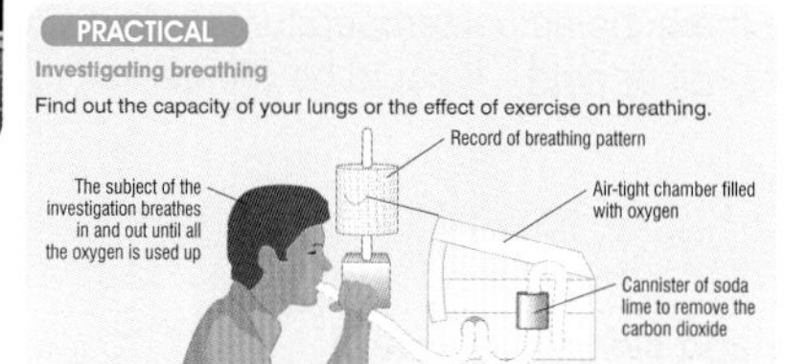

Figure 2 Because you breathe in and out of the machine all the time, you can't get rid of your waste carbon dioxide in the normal way. There has to be a special filter to remove the carbon dioxide so it doesn't poison you!

- How can we improve the reliability of investigations involving living organisms?

192

Additional biology

The carbon dioxide moves out of the cells into your blood. Your blood stream carries it back to your lungs. Almost all of the carbon dioxide you produce is removed from your body via your lungs when you breathe out. The air you breathe in contains only 0.04% carbon dioxide, but the air you breathe out contains about 4% carbon dioxide!

b) How do you remove carbon dioxide from your body?

Urea

The other main waste product of your body is **urea**.

Urea is produced in your **liver** when excess amino acids are broken down. When you eat more protein than you need, or when body tissues are worn out, the extra protein has to be broken down. Amino acids cannot be used as fuel for your body. But in your liver the amino group is removed and converted into urea.

The rest of the amino acid molecule can then be used in respiration or to make other molecules. The urea passes from the liver cells into your blood.

Urea is poisonous and if the levels build up in your blood it will cause a lot of damage. Fortunately the urea is filtered out of your blood by your **kidneys**. It is then removed in your **urine**, along with any excess water and salt.

Urine is produced all the time by your kidneys. It leaves your kidneys and is stored in your **bladder** which you then empty from time to time!

c) Where is urea made?

FOUL FACTS

The average person produces between 1.5 and 2.5 litres of urine a day – that's up to 900 litres of urine a year!

Maintaining body balance

Water and ions enter your body when you eat or drink. The water and ion content of your body are carefully controlled to prevent damage to your cells. Water is lost through breathing, through sweating and in the urine, while ions are lost in the sweat and in the urine.

If the water or ion content of your body is wrong, too much water may move into or out of your cells. That's why control is vital.

It is also very important to control your body temperature and the levels of sugar in your blood. So homeostasis plays a very important role in your body.

GET IT RIGHT!

Don't confuse urea and urine. Urea is made in the liver; urine is produced by the kidney. Urine contains urea.

SUMMARY QUESTIONS

1. Copy and complete using the words below:
   blood carbon dioxide constant controlled environment enzymes homeostasis sugar temperature urea water
   The internal ...... of your body is kept relatively ...... by a whole range of processes which make up .... Waste products such as .... and ...... have to be removed from your ...... all the time. The ...... and ion concentration of your blood are constantly ...... and so is your blood ...... level. Your body ...... is kept the same so your ...... work effectively.
2. There are two main waste products which have to be removed from the human body – carbon dioxide and urea. For each waste product, describe:
   a) how it is formed, b) why it has to be removed, c) how it is removed from the body.
3. Explain briefly a) how a period of exercise would affect the internal conditions of your body, and b) how the conditions would be returned to normal.

KEY POINTS

1. The internal conditions of your body have to be controlled to maintain a constant internal environment.
2. Poisonous waste products are made all the time and need to be removed.
3. Carbon dioxide is produced during respiration and leaves the body via the lungs when you breathe out.
4. Urea is produced by your liver as excess amino acids are broken down, and it is removed by your kidneys in the urine.

193

SUMMARY ANSWERS

1. Environment, constant, homeostasis, carbon dioxide, urea, blood, water, controlled, sugar, temperature, enzymes.
2. a) **Carbon dioxide:** formed during aerobic respiration;
   glucose + oxygen → energy + carbon dioxide + water;
   Urea: excess amino acids from protein/worn out tissues;
   amino group removed from amino acids and converted to urea in the liver.
   b) Both are poisonous to the cells/damage the body.
   c) Carbon dioxide leaves cells and is carried in the blood to the lungs; from here it is breathed out in the air.
   Urea leaves the liver in the blood and is carried to the kidneys; it is filtered out of the blood by the kidneys, forms urine and is collected into the bladder.
3. a) Period of exercise – more carbon dioxide, more heat from muscles, used up blood sugar, lost salt in sweating.
   b) Breathe faster and more deeply to get rid of excess carbon dioxide, sweat and go red to cool down, release energy from store in body to replace glucose, kidneys change concentration of urine to save salt.

Answers to in-text questions

a) The maintenance of a constant internal environment.
b) From the cells in the body it is carried in the blood to the lungs and then breathed out in the air from the lungs.
c) In the liver.

KEY POINTS

The elimination of waste materials is important for the internal environment. Students should be clear on the formation of urea in the liver, its transport in the blood and removal by the kidneys. The sequence of removal of the amino group from amino acids, its conversion to urea and its eventual elimination as urine should be emphasised and learnt.

B2 5.2 Controlling body temperature

LEARNING OBJECTIVES

Students should learn that:

- The internal temperature of the body is monitored and controlled by a thermoregulatory centre in the brain.
- The thermoregulatory centre receives information from the blood as it flows through the brain and from temperature receptors in the skin.
- If the core temperature fluctuates, responses are made so that the body is kept at the temperature at which the enzymes work best. [**HT** only]

LEARNING OUTCOMES

Most students should be able to:

- Describe how the body monitors body temperature.

Some students should also be able to:

- Explain how the body temperature is monitored and controlled by the brain in coordination with other systems.
- Describe the responses made by the body if the core temperature is too high. [**HT** only]
- Describe the responses made by the body if the core temperature drops too low. [**HT** only]

Teaching suggestions

- **Special needs.** Produce a sheet with two boxes on one side with the words 'Too hot' in one and 'Too cold' in the other. Opposite these put a number of boxes with activities (e.g. put on more clothes, stamp feet and blow on hands) and physical responses (e.g. goosebumps, sweating). Students to join the boxes with lines and add illustrations to the page.
- **Gifted and talented.** Students to consider cold-blooded animals that cannot control their body temperatures. They can write out and illustrate a product-style 'User's guide for a poikilothermic body' giving warnings and suggestions.
- **Learning styles**
  Kinaesthetic: Measuring temperatures.
  Visual: Observing photographs.
  Auditory: Listening to discussions.
  Interpersonal: Working in pairs on the heat detection exercise.
  Intrapersonal: Writing 'Dear body' messages.
- **Homework.** Design and produce a 'circuit' diagram to show how thermoregulation works in the human body.

SPECIFICATION LINK-UP Unit: Biology 2.12.7

- *Body temperature is monitored and controlled by the thermoregulatory centre in the brain. This centre has receptors sensitive to the temperature of blood flowing through the brain. Also temperature receptors in the skin send impulses to the centre giving information about skin temperature.*
- *If the core body temperature is too high:*
  - *blood vessels supplying the skin capillaries dilate so that more blood flows through the capillaries and more heat is lost.*
  - *sweat glands release more sweat which cools the body as it evaporates. [**HT** only]*
- *If the core body temperature is too low:*
  - *blood vessels supplying the skin capillaries constrict to reduce the flow of blood through the capillaries.*
  - *muscles may 'shiver' – their contraction needs respiration which releases some energy as heat. [**HT** only]*

Lesson structure

STARTER

Same temperature? – Using forehead thermometers, get the students to take their own temperatures. Collect up results and find a mean for the group. Ask: 'Why are they all about the same? How much variation is there?' (This relates to 'How Science Works': consider the reliability of data.) (5–10 minutes)

Can you tell the temperature? – Have several containers of water at different temperatures. Students are to guess the temperatures. Ask: 'How easy is it? How accurate can you be?' Give them the actual temperatures and discuss why you need to be aware of whether you are hot or cold – it could be dangerous to become too hot or too cold. (5–10 minutes)

MAIN

Differences between core and skin temperature

- Students can work in groups or this could be done as a class demonstration. Ask for volunteers (risk assessment needed) to sit with one hand in very cold water. Monitor the core temperature and skin temperature of the other hand of the volunteer using temperature sensors and data loggers. Ask: 'What happens?'
- A modification or extension of the above experiment can be done comparing heat loss from an insulated and non-insulated hand. Attach a temperature probe to each hand of a volunteer with masking tape, checking that the skin temperatures on both hands are identical. Insulate one hand fully with duvet filling, taping it and ensuring that it is of an even thickness all round. Allow time for equilibration, then record the skin temperatures of each hand and the core temperature. Change the environmental conditions (cooler temperatures, air movements) and repeat the temperature measurements.
- To cover aspects of 'How Science Works', the information gathered and the best way to present it can be discussed by the students. Ask: 'What conclusions can be drawn? Does it tell us about the thermoreceptors in the skin? Could we use this information to devise a method of measuring the heat loss from the human body?'

Comparing distributions

- Students to compare the distribution of humans with that of other mammals with respect to climatic temperatures, and include references to behavioural as well as physiological ways of controlling temperature.

Answers to in-text questions

a) If the temperature becomes too hot or too cold, it affects the action of the enzymes in the body.

b) The rate of the enzyme-controlled reactions slows down and not enough energy is made in the cells.

ACTIVITY & EXTENSION

- Use alcohol evaporating from thermometers to show the effect of evaporation on temperature. (No naked flames).
- **Does the human body temperature fluctuate much?** – Suggest to students that they monitor their own body temperature over a period of 48 hours. If the results are plotted, it can be seen that the temperature does fluctuate. Ask: 'Can it be accounted for?' If it is not possible for students to do this themselves, project a graph and get students to discuss the variations.
- **Heat detection** – In pairs, one student closes their eyes, while the other brings the palm of their hand close to the other person's cheek. The one with eyes closed says when they can feel the heat. They estimate the distance and then swap roles.
- **Animation** – Use the Animation, B2 5.2 'Controlling body temperatures', available on the GCSE Biology CD ROM.

Practical support

Body temperature

Equipment and materials required

Thermometers, temperature probes, masking tape, duvet filling, data loggers.

If clinical thermometers are used, it is advisable to use separate thermometers for each student and these need to be disinfected after use.

B2 5.2 Controlling body temperature

LEARNING OBJECTIVES

1. How does your body monitor its temperature?
2. How does your body stop you getting too hot? [Higher]
3. How does your body keep you warm? [Higher]

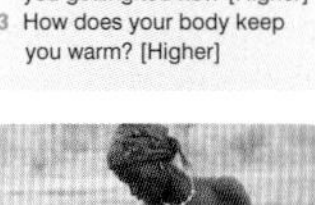
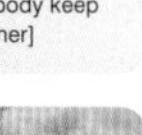

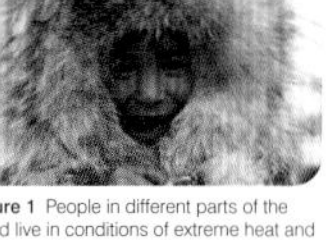
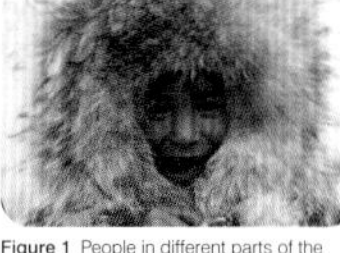

Figure 1 People in different parts of the world live in conditions of extreme heat and extreme cold and still maintain a constant internal body temperature

Wherever you go and whatever you do it is vital that your body temperature is maintained at around 37°C. This is the temperature at which your enzymes work best. Your skin temperature can vary enormously without causing harm. It is the temperature deep inside your body, known as the core body temperature, which must be kept stable.

At only a few degrees above or below normal body temperature your enzymes cannot function properly. All sorts of things can affect your internal body temperature, including:

- heat produced in your muscles during exercise,
- fevers caused by disease, and
- the external temperature rising or falling.

People can control some aspects of their own temperature. We can change our clothing, light a fire, and turn on the heating or air-conditioning. But it is our internal control mechanisms which are most important in controlling our body temperature.

a) Why is control of your body temperature so important?

Control of the temperature relies on the **thermoregulatory centre** in the brain. This centre contains receptors which are sensitive to temperature changes. They monitor the temperature of the blood flowing through the brain itself.

Extra information comes from the temperature receptors in the skin. These send impulses to the thermoregulatory centre giving information about the skin temperature. The receptors are so sensitive they can detect a difference of as little as 0.5°C!

Sweating helps to cool your body down. So the loss of salt and water when you sweat can affect your water and ion balance. If you are sweating a lot you need to take in more drink or food to replace the water and ions you have lost – just watch a marathon runner!

Cooling the body down (HIGHER)

If you get too hot, your enzymes denature and can no longer catalyse the reactions in your cells. When your core body temperature begins to rise, impulses are sent from the thermoregulatory centre to the body so more heat is lost:

- The blood vessels, which supply your skin capillaries, **dilate** (open wider). This lets more blood flow through the capillaries. Your skin flushes, so you lose more heat by radiation.
- Your rate of sweating goes up. Sweat (made up mainly of water, salt and a little protein) oozes out of your sweat glands and spreads over your skin. As the water evaporates it cools the skin, taking heat from your body. In very humid conditions, when the sweat doesn't evaporate very easily, it is very difficult to cool down.

Reducing heat loss (HIGHER)

It is just as dangerous for your core temperature to drop as it is to rise. If you get very cold, the rate of the enzyme-controlled reactions in your cells falls too low. You don't make enough energy and your cells begin to die. If your core body temperature starts to get too low, impulses are sent from your thermoregulatory centre to the body to conserve and even generate more heat.

- The blood vessels which supply your skin capillaries **constrict** (close up) to reduce the flow of blood through the capillaries. This reduces the heat lost through the surface of the skin, and makes you look pale.
- Shivering begins – your muscles contract and relax rapidly which involves lots of cellular respiration. This releases some energy as heat which you use to raise your body temperature. As you warm up, shivering stops.
- Sweat production is reduced.

b) Why is a fall in your core body temperature so dangerous?

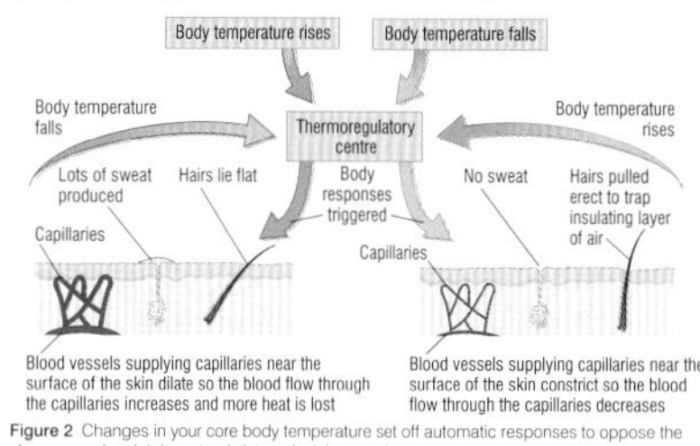

Figure 2 Changes in your core body temperature set off automatic responses to oppose the changes and maintain a steady internal environment

DID YOU KNOW?

Birds and mammals can help reduce heat loss from their bodies by pulling the hairs or feathers on their skin upright to trap an insulating layer of air. Our bodies try to do this, but we just get goose-pimples. The tiny muscles pulling on our hairs show up more than the hairs themselves!

PRACTICAL

Body temperature

Use a temperature sensor and data logger to record your skin and core body temperature on one hand as you plunge the other into icy water.

- Explain your observations.

GET IT RIGHT!

Use the terms dilate and constrict for the changes which take place in the blood vessels in your skin. Remember sweating only cools your body when the sweat actually evaporates.

SUMMARY QUESTIONS

1. Here is a jumbled list of some of the events by which your body temperature is controlled when it starts to go up. Sort them out into the right order and then copy them out.
   - A Her body temperature starts to rise.
   - B Sally takes a long, cool drink to replace the liquid she has lost through sweating.
   - C Her temperature returns to normal.
   - D Her skin goes red and her rate of sweating increases so the amount of heat lost through her skin goes up.
   - E Sally exercises hard.
2. a) Why is it so important to maintain a body temperature of about 37°C?
   b) Explain the role of i) the thermoregulatory centre in the brain and ii) the temperature sensors in the skin in maintaining a constant core body temperature.
3. Explain how the body responds to both an increase and a decrease in core temperature to return its temperature to normal levels. [Higher]

KEY POINTS

1. Your body temperature must be maintained at the level at which enzymes work best.
2. Your body temperature is monitored and controlled by the thermoregulatory centre in your brain.
3. Your body responds to cool you down if you are overheating and to warm you up if your core body temperature falls. [Higher]

SUMMARY ANSWERS

1. E, A, D, C, B.
2. **a)** This is the temperature at which enzymes work best.
   b) i) The thermoregulatory centre in the brain is sensitive to the temperature of the blood flowing through it. It also receives information about the skin temperature from receptors in the skin and coordinates the body responses to keep the core temperature at 37°C.
   ii) Temperature sensors in the skin send impulses to the thermoregulatory centre in the brain giving information about the temperature of the skin and the things it touches. This is important for maintaining the core temperature because if the external surroundings and the skin are cold, the body will tend to conserve heat to keep the core temperature up, and vice versa.
3. Look for: To lower body temperature – blood vessels supplying capillaries dilate – more blood in capillaries so more heat is lost. More sweat produced by sweat glands which cools the body as it evaporates. To raise body temperature – blood vessels supplying capillaries constrict – less blood in them. Shivering occurs by rapid muscle movement which needs respiration – releasing some heat energy. [**HT** only]

PLENARIES

Thermostat principles – Show the students a heater connected into a circuit with a bimetallic strip, or other thermostatic device, arranged so that when it drops to a given temperature the heater switches on. When the temperature rises, the heater switches off. Run through several cycles and ask the students to draw parallels with the human body. (10 minutes)

'Dear body . . .' – Students to write 'Dear body' style text messages telling the body what to do if it is too hot or too cold. (10 minutes)

Quick quiz – Ask questions on the contents of the lesson. (5–10 minutes)

KEY POINTS

The key points can be incorporated into a summary diagram of the way in which thermoregulation in the human body occurs. Students should be aware that capillaries do not contract and dilate, but it is the blood vessels that supply the capillaries that control the quantity of blood that flows through the capillaries.

B2 5.3 Controlling blood sugar

LEARNING OBJECTIVES

Students should learn that:

- The pancreas monitors and controls the level of glucose in the blood.
- The pancreas secretes insulin when the blood glucose concentration is too high, lowering the level of glucose, but when the concentration is too low, glucagon is secreted and glucose is released into the blood.
- Diabetes is caused by a lack of insulin from the pancreas.

LEARNING OUTCOMES

Most students should be able to:

- State that the pancreas monitors and controls blood sugar concentration.
- Describe the symptoms and causes of diabetes.
- Describe how diabetes can be treated.

Some students should also be able to:

- Explain how the blood sugar concentration is monitored and controlled.
- Explain the causes of diabetes and how it is treated.

Teaching suggestions

- **Special needs.** Give students a large, clear and not too complex wordsearch for the key words. The students should be provided with the definitions to the words and then cross them off as they find them.
- **Gifted and talented.** Ask the students to produce a series of Word documents showing the feedback mechanism involving insulin and glucagon, glycogen and glucose, blood glucose levels too high, too low and normal. Link these together with hyperlinks so that they form the appropriate loops.
- **Learning styles**

  Kinaesthetic: Building up feedback diagram.

  Visual: Observing animation.

  Auditory: Listening to the information gathered on the topics researched.

  Interpersonal: Taking part in role-play exercise.

  Intrapersonal: Giving feedback on research project.

SPECIFICATION LINK-UP Unit: Biology 2.12.7

- *The blood glucose concentration of the body is monitored and controlled by the pancreas. The pancreas produces the hormone insulin which allows glucose to move from the blood into the cells.*
- *Diabetes is a disease in which a person's glucose concentration may rise to a fatally high level, because the pancreas does not produce enough of the hormone insulin. Diabetes may be treated by careful attention to diet and by injecting insulin into the blood.*

Lesson structure

STARTER

Blood sugar levels – Discuss the sweet eating habits of younger brothers and sisters. Ask: 'Who eats the most at one go? Does eating a lot of sweets have an effect on their behaviour?' Talk about the blood sugar levels and speculate as to how the body copes. (5–10 minutes)

How is the disease diagnosed? – Discuss the symptoms of the disease and why they occur. Ask: 'What simple test could indicate that someone is suffering from diabetes? What is a glucose tolerance test and how can it be interpreted?' Students to analyse blood glucose graphs for normal people and for those with diabetes.

Coping with diabetes – Ask: 'Has anyone had a go at a rowing machine?' Project a picture and discuss how much energy it takes. Show footage or stills of Sir Steven Redgrave winning Olympic gold. Discuss his problem with diabetes and any other people, possibly relatives or peers, who have the condition. (10 minutes)

MAIN

- **Animation** – Show the Animation B2 5.3 from the GCSE Biology CD ROM on the control of the blood glucose levels by the pancreas. This could include diagrams/photos of pancreas tissue showing the islets of Langerhans, alpha cells and beta cells. A feedback diagram can be built up showing how the control is achieved. Give students worksheets and given time to complete these during the lesson.
- **Research the disease** – There is plenty of information about diabetes available from the doctor, from textbooks and on the Internet (e.g. www.diabetes.org.uk). Students could be given the opportunity to follow up an aspect of the condition. For example, students could investigate:
  - the causes of diabetes, including Type I and Type II
  - what happens if a diabetic has insufficient glucose in the blood, i.e. is hypoglycaemic
  - the treatment of diabetes, from the use of insulin from animals to the present-day use of genetically engineered insulin
  - the possibility of islet cell transplantation techniques
  - the research of Banting and Best and why they got the Nobel prize
  - the importance of diet for a diabetic
  - the role of the diabetic nurse in your local practice.

  For each of these suggestions, students could compile a report to be presented to the class. They could be given homework time for research and writing their reports.
- **Demonstrate Clinistix testing** – Test fake urine as suggested in section 4.7, page 188. Compare this with the way in which blood is tested now. Students could be given 'mystery' samples of fake urine to test. It could be appropriate to point out that before the invention of Clinistix, urine samples were tested with Benedict's solution. Compare a Clinistix reading with a Benedict's test on the same sample. (This relates to: 'How Science Works': aspects of sensitivity and accuracy of testing.)

PLENARIES

Why diabetes mellitus? – Discuss the origins of the word 'mellitus', with its link to honey. Tell students that many years ago doctors would taste their patient's urine to check it for sweetness. Some students may have heard of the other condition known as 'diabetes insipidus', where copious quantities of urine are produced and the patient is always thirsty. (5–10 minutes)

Spelling bee – Ask for volunteers to spell key words from this topic. Once a student has spelt a word correctly, they can choose another student to give a definition. This could reinforce the differences between 'glucose', 'glucagon' and 'glycogen', which are frequently muddled by students. (10 minutes)

ACTIVITY & EXTENSION

- **What does diabetic chocolate taste like?** – Have a range of products designed for diabetics as well as some normal ones. Ask students if they can tell the difference between a 'diabetic' one and a normal one. This can be done as a sampling exercise or as a competition to try to identify which is diabetic and which is not. (5–10 minutes)
- **Diabetic diets or get the balance right** – Students could research diabetic diets. There has been a great deal of interest in the Glycaemic Index of foods and how this can help people to lose weight and maintain a weight loss. Students could design menus suitable for people with mild forms of diabetes and research foods that are said to be for diabetics. Ask: 'What is used instead of sugar? Are these sensible recommendations for people who wish to lose weight?'
- **What can you do?** – Carry out a role-play exercise of finding someone collapsed and discovering that they had a Medic alert bracelet indicating that they are diabetic. Ask and discuss: 'What would be the sensible thing to do? Why?' If possible, get a diabetic person to talk to the class.

Answers to in-text questions

a) Blood transports glucose to the cells where it is needed for cellular respiration.

b) Insulin and glucagon.

c) Insulin is needed to enable glucose to enter the cells. If there is no insulin, the cells are deprived of fuel and therefore do not make enough energy in cellular respiration.

d) They can manage their diet by avoiding foods that are rich in carbohydrates.

KEY POINTS

The key points can be reinforced by encouraging the students to produce their own feedback diagram with all the hormones included and their effects on the glucose in the blood.

BIOLOGY HOMEOSTASIS

B2 5.3 Controlling blood sugar

LEARNING OBJECTIVES

1. How is your blood sugar level controlled?
2. What is diabetes and how is it treated?

It is very important that your cells have a constant supply of the glucose they need for cellular respiration. Glucose is transported around your body to all the cells by your blood. However you don't spend all of your time eating to keep your blood sugar levels high. Instead the level of sugar in your blood is controlled by hormones produced in your pancreas.

a) Why are the levels of glucose in your blood so important?

The pancreas and the control of blood sugar levels

When you digest a meal, large amounts of glucose pass into your blood. Without a control mechanism your blood glucose levels would vary wildly. After a meal they would soar to a point where glucose would be removed from the body in the urine. A few hours later the levels would plummet and cells would not have enough glucose to respire.

This internal chaos is prevented by your **pancreas**. The pancreas is a small pink organ found under your stomach. It constantly monitors your blood glucose concentration and controls it using two hormones known as **insulin** and **glucagon**.

When your blood glucose concentration rises above the ideal range after you have eaten a meal, insulin is released. Insulin causes your liver to remove any glucose which is not needed at the time from the blood. The soluble glucose is converted to an insoluble carbohydrate called **glycogen** which is stored in your liver.

When your blood glucose concentration falls below the ideal range, the pancreas secretes glucagon. Glucagon makes your liver break down glycogen, converting it back into glucose. In this way the stored sugar is released back into the blood.

By using these two hormones and the glycogen store in your liver, your pancreas keeps your blood glucose concentration fairly constant. Its normal concentration is usually about 90 mg glucose per 100 cm^3 of blood.

b) Which two hormones are involved in the control of your blood sugar levels?

SCIENCE @ WORK

In 2005, research scientists produced insulin-secreting cells from stem cells which cured diabetes in mice. More research is needed but the scientists hope that before long diabetes will be a disease that we can cure instead of just treating the symptoms.

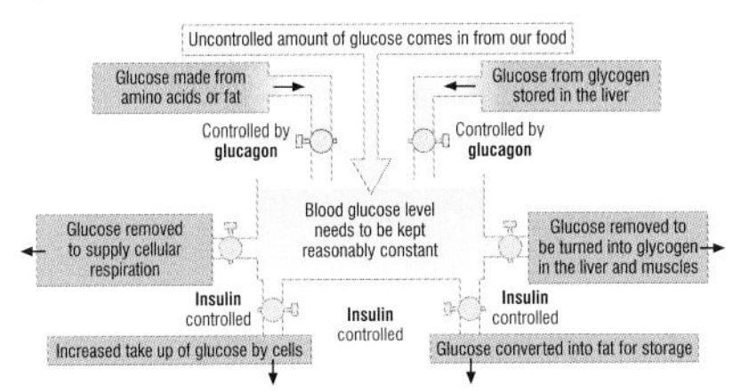

Uncontrolled amount of glucose comes in from our food
Glucose made from amino acids or fat
Glucose from glycogen stored in the liver
Controlled by glucagon
Controlled by glucagon
Blood glucose level needs to be kept reasonably constant
Glucose removed to supply cellular respiration
Glucose removed to be turned into glycogen in the liver and muscles
Insulin controlled
Insulin controlled
Insulin controlled
Increased take up of glucose by cells
Glucose converted into fat for storage

Figure 1 This model of your blood glucose control system shows the blood glucose as a tank. It has both controlled and uncontrolled inlets and outlets. In every case the control is given by the hormones insulin and glucagon.

What causes diabetes?

Most of us never think about our blood sugar levels because they are perfectly controlled by our pancreas. But for some people life isn't quite this simple. Unfortunately, their pancreas does not make enough – or any – insulin.

Without insulin your blood sugar levels get higher and higher after you eat food. Eventually your kidneys produce glucose in your urine. You produce lots of urine and feel thirsty all the time.

Without insulin, glucose cannot get into the cells of your body, so you lack energy and feel tired. You break down fat and protein to use as fuel instead, so you lose weight.

Before there was any treatment for diabetes, people would waste away. Eventually they would fall into a coma and die. Fortunately there are now some very effective ways of treating diabetes!

c) Why do people with untreated diabetes feel very tired and lack energy?

Treating diabetes

If you have a mild form of diabetes, managing your diet is enough to keep you healthy. Avoiding carbohydrate-rich foods keeps the blood sugar levels relatively low. So your reduced amount of insulin can cope with small amounts of glucose.

However, other people with diabetes need replacement insulin before meals. Insulin is a protein which would be digested in your stomach. So it is usually given as an injection to get it into your blood.

This injected insulin allows glucose to be taken into your body cells and converted into glycogen in the liver. This stops the concentration of glucose in your blood from getting too high.

Then as the blood glucose levels fall, natural glucagon makes sure glycogen is converted back to glucose. As a result your blood glucose levels are kept as stable as possible. (See graphs on page 199.)

Insulin injections treat diabetes successfully but they do not cure it. Until a cure is developed, someone with diabetes has to inject insulin several times every day of their life.

d) How can people with mild diabetes control the disease?

SUMMARY QUESTIONS

1. Define the following words: **hormone**; **insulin**; **diabetes**; **glycogen**.
2. a) Explain how your pancreas keeps the blood glucose levels of your body constant.
   b) Why is it so important to control the level of glucose in your blood?
3. What is diabetes and how can it be treated?

DID YOU KNOW?

In 2005, doctors in Japan performed a successful living transplant of pancreas tissue. Cells from a mother were given to her daughter who had severe diabetes. Within three weeks the daughter no longer needed insulin injections – her new cells were controlling her blood sugar.

Figure 2 The treatment of diabetes involves regular blood sugar tests and insulin injections. These could become a thing of the past if some of the new treatments being developed work as well as scientists hope!

KEY POINTS

1. Your blood glucose concentration is monitored and controlled by your pancreas.
2. Insulin and glucagon are the hormones involved in controlling blood sugar concentration. Insulin converts glucose to glycogen; glucagon converts glycogen to glucose.
3. In diabetes, the blood glucose may rise to fatally high levels because the pancreas does not secrete enough insulin. It can be treated by injections of insulin before meals.

SUMMARY ANSWERS

1. Hormone – A chemical message carried in the blood which causes a change in the body.
   Insulin – A hormone made in the pancreas which causes glucose to pass from the blood into the cells where it is needed for energy.
   Diabetes – A condition when the pancreas cannot make enough insulin to control the blood sugar.
   Glycogen – An insoluble carbohydrate stored in the liver.
2. **a)** Blood glucose levels go up above the ideal range. This is detected by the pancreas, which then secretes insulin. Insulin causes the liver to convert glucose to glycogen and causes glucose to move out of the blood into the cells of the body, thus lowering blood glucose levels. If the blood glucose level drops below the ideal range, this is detected by the pancreas. The pancreas secretes glucagon, which causes the liver to convert glycogen into glucose that increases the blood glucose level.
   b) Glucose needed for cells of body for respiration which provides energy for everything (could add too much glucose causes problems in kidneys, etc.)
3. Diabetes is a condition where the pancreas does not make enough or any insulin. It can be treated by injections of insulin to help balance blood sugar levels. (New possibilities are pancreas cell transplants or embryonic stem cells.)

B2 5.4 Homeostasis matters!

SPECIFICATION LINK-UP

Unit: Biology 2.12.7

Students should use their skills, knowledge and understanding of 'How Science Works':

- *to evaluate the data from the experiments by Banting and Best, which led to the discovery of insulin*
- *to evaluate modern methods of treating diabetes.*

BIOLOGY HOMEOSTASIS

B2 5.4 Homeostasis matters!

HYPOTHERMIA – THE SILENT KILLER

If your core body temperature falls too low you suffer from hypothermia. About 30 000 people die of it every year in the UK alone. Here is some more information about hypothermia:

- Hypothermia is when your body temperature drops below 35°C and the normal working of your body is affected.
- Old people, small children and people exposed in bad weather conditions are most at risk.
- Young people on outdoor expeditions are often at risk if they do not wear the right clothing. Wet weather and wind make you lose heat faster.
- The first signs of hypothermia are extreme tiredness and not wanting to move – you may not realise how cold you are.
- Up to 20% of your body heat is lost through your head.
- Warm clothing, adequate heating, regular food and warm drinks, together with exercise all help to prevent hypothermia.
- People with hypothermia have greyish-blue, puffy faces and blue lips. Their skin feels very cold to the touch. They will be drowsy, with slurred speech. As it gets worse, they will stop shivering. If the body temperature falls too low the sufferer will become unconscious and may die.
- It has been estimated that every time the temperature drops one degree Celsius below average in the winter, 8000 more elderly people will die of hypothermia.

HEAT WAVE KILLS SEVEN

The latest spell of very hot weather has led to seven deaths this week. As Britain sizzles in the latest heat-wave, with temperatures of over 33°C, people are dropping like flies.

Heat stroke and other heat-related illnesses are hitting the elderly, small babies and people with existing heart problems particularly hard.

The World Health Organisation along with the World Meteorological Organisation have suggested that a hot weather warning is added to our weather forecasts along with pollen levels, air pollution and flood warnings.

To reduce your risk of heat stroke as Britain continues to fry, stay in air-conditioned rooms where possible, drink plenty of water and take cool baths.

ACTIVITY

If more people were aware of the risks of hypothermia, fewer people would die from it. Use the information to help you design ***either*** a poster ***or*** a leaflet informing people about the dangers of hypothermia and ways to avoid it.

ACTIVITY

Climate change may well result in colder winters and hotter summers. Write an article for the lifestyle pages of a newspaper on:

- how your body copes with changes in temperature,
- the dangers to health of hot summers and cold winters, and
- the best ways to avoid any problems.

198

Teaching suggestions

- **Hypothermia: the silent killer** – Students can consider ways in which heat is lost from the body: radiation, conduction and evaporation. Project some thermographic camera pictures of people in hot and cold environments and discuss where the areas of greatest heat loss are. (Search an image bank for 'thermographic'). If the thermographs have indications of temperatures on them, it is possible to work out the difference between the core and skin temperatures in various parts of the body. Show students a silver plastic thermal blanket and ask them where they have seen these before. Ask: 'Why are they used? How do they work in reducing heat loss?' A PhotoPLUS on hypothermia is available from the GCSE Biology CD ROM.
- **Heat loss through the head** – As stated in the Student Book, up to 20% of the body heat can be lost through the head. Put a large fur hat (Russian-style with ear flaps) on a volunteer. Search for pictures of Arctic explorers, Kenny – the kid with the snorkel parka from South Park, a bald man and old photographs of people wearing hats. Discuss the benefits of hats to the young and the old. Ask: 'Why do we not wear hats so much today?'
- **The wind chill factor** – Show a PowerPoint presentation on the wind chill factor, to include a table that shows how increasing wind speed lowers the temperature on exposed skin. Students to consider the implications of this to the elderly, the very young and to people on outdoor pursuits, such as hill-walking and canoeing. Ask students to suggest why wind chill factor is made worse in wet conditions due to the evaporation of water.
- **What to take on an expedition** – Students can investigate the range of outdoor clothing available, and to recommend suitable clothing to be worn or taken on an expedition to a mountainous region. Ask: 'What properties should the clothing have?' Explain why garments are chosen. Ask: 'What should the leader have in the first aid pack in case a member of the party is hurt and needs to be left while help is sought?'
- **Evaporation and heat loss** – The differences in heat lost in wet and dry conditions can be investigated by wrapping thermometers in a cloth and using volatile substances, such as ethanol. The effect of the wind chill factor could be simulated with the use of a fan.
- **Heat wave kills seven!** – Show a PowerPoint presentation on the effects of heat stress that occurs when the body can no longer cope with excessively high temperatures. The differences between heat collapse, heat exhaustion, heat cramps and heat stroke can be illustrated. Reference to the ways in which the body temperature is controlled could be discussed. Students could explain the benefits of drinking plenty of water, keeping out of direct sunlight and replacing lost salt.
- **The heat wave forecast . . .** – We already have severe weather warnings, information about the pollen count and references to the wind chill factor in our weather reports. Suggest to students that they produce and record a radio report, to be transmitted during the weather forecasts during a prolonged heat wave, advising people on what to do.

THE DIABETES DEBATE

The treatment of diabetes has changed a great deal over the years. For centuries nothing could be done. Then in the early 1920s Frederick Banting and Charles Best realised that extracts of animal pancreas could be used to keep people with diabetes alive. For many years insulin from pigs and cows was used to treat affected people. This saved millions of lives.

In recent years, bacteria have been developed using genetic engineering which produce pure human insulin. This is now injected by the majority of people affected by diabetes.

Scientists are trying to find easier ways – like nasal sprays – to get insulin into the body. Transplanting working pancreas cells from both dead and living donors has been shown to work for some people. And for the future, scientists are hoping to use embryonic stem cells to provide people affected by diabetes with new, functioning pancreas cells which can make their own insulin.

The difference these treatments have made to the lives of people with diabetes and their families is enormous. If a cure is found, it will be even better. But most of these developments have some ethical issues linked to them.

- Banting and Best did their experiments on dogs. They made some of the dogs diabetic by removing most of their pancreas, and they extracted insulin from the pancreases of other dogs. Many dogs died in the search for a successful treatment – but the scientists found a treatment to a disease which has killed millions of people over the centuries.
- Human insulin is now mass-produced using genetically engineered bacteria. The gene for human insulin is stuck into the bacterial DNA and the bacteria make pure human protein.
- There are not enough dead donors to give pancreas transplants to the people who need them. However, in living donor transplants there is a risk to the health of the donor as they have to undergo surgery.
- Stem cell research promises a possible cure – but the stem cells come from human embryos which have been specially created for the process.

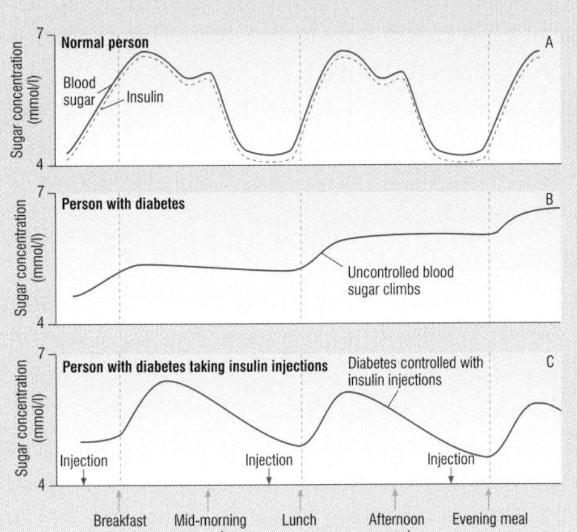

Figure 1 These graphs show the impact insulin injections have on people affected by diabetes. The injections keep the blood sugar level within safe limits. They cannot mimic the total control given by the natural production of the pancreas – but they work well enough to let people lead a full and active life.

ACTIVITY

a) You are to plan a three-minute speech for a debate. The title of the debate is:

'Ethical concerns are less important than a cure for diabetes.'

You can argue for or against the motion, but your arguments must be clear and sensible and backed up by scientific evidence.

b) Work in groups of 9 and set up a role play involving the following characters:

- Frederick Banting who first showed that animal insulin could be used to treat humans with diabetes.
- A spokesperson from a pharmaceutical company manufacturing human insulin.
- A daughter who has been cured of diabetes by receiving pancreas tissue from her mother, and her mother who donated the tissue.
- A scientist working on the development of insulin-producing cells from embryonic stem cells.
- Someone who has had diabetes since they were 10 years old.
- An animal rights activist.
- A 'pro-life' activist who is against any use of stem cells.
- A representative of a group opposed to genetic engineering.
- The chair of the discussion.

Each character must explain to the chair why research into diabetes should – or should not – continue.

The diabetes debate – Before preparing for the debate on diabetes, some further background research on Banting and Best's experiments could be done. The use of stem cells and pancreatic transplants should be researched, so that students know the advantages and the risks involved. (This relates to 'How Science Works': societal factors.)

Special needs

Use concept cartoons-style talking heads: in one a person is complaining about being too hot and receiving advice from a second, who has an empty speech balloon. Students can choose a response from a printed list. Similarly, another one can be given where a person is complaining about being too cold.

Learning styles

Kinaesthetic: Designing a poster or a leaflet.

Visual: Interpreting thermographic images.

Auditory: Listening to arguments being put forward in a debate.

Interpersonal: Role-playing in the diabetes activity.

Intrapersonal: Writing a speech for the debate.

ACTIVITY & EXTENSION IDEAS

- **Hypothermia: the silent killer** – The design of the poster or the leaflet requires some careful thought, as hypothermia can affect several groups of vulnerable people. Ask: 'Is it a good idea to make a general poster? Should there be one especially for older people who are most vulnerable? Where would you put the posters?' Students could discuss these possibilities and decide whether they think a poster or a leaflet would reach a greater number of people.
- **Heat wave kills seven** – The suggested article is very wide-ranging and gathers together some of the topics introduced in the preceding spreads, particularly with respect to the controlling of body temperature. Some other aspects of hot summers, such as the pollen, fumes and dangers of sunburn, could be incorporated. Problems associated with cold winters should include some mention of hypothermia and the elderly, as well as advice to everyone about the dangers of losing too much heat.
- The article for the newspaper could be a group activity, with contributions from different students covering the various aspects. It could be interesting to adapt the style of the article to the type of newspaper; ask: 'Would an article for a tabloid have a different approach from an article for a broadsheet?' Display the results for peer judgement.
- **The diabetes debate** – As with other debates, in order for students to maintain a balanced view, a speech for the motion and against the motion should be prepared. The ethical concerns in this case are quite complicated and students might benefit from some discussion before they plan their speeches. It could be beneficial to try the role-play activity in class, and suggest that students use some of the arguments put forward when preparing for the debate.
- **Role-play activity** – This will only work well if the students research their roles, so some homework time could be allocated for preparation.

SUMMARY ANSWERS

1 a) Diagram to show feedback – something goes up, receptor picks it up, chemical released to bring levels back down and the same idea if level drops. Annotations to show student understands principles of maintaining more-or-less constant levels.

b) For cells to work properly they need to be at the right temperature (so enzymes don't denature), they need to be surrounded by correct concentration so osmosis doesn't cause problems, they need glucose to provide energy and they need waste products to be removed as build up can change pH or poison systems. This is why the body systems must be controlled within fairly narrow limits.

2 a) Person: 37°C; lizard: 18°C.

b) The human core temperature becomes dangerously low at an external temperature of 10°C. Reactions slow down, hypothermia develops and the heart beat and breathing will stop.

c) It becomes dangerously high at 55°C. Once the body temperature goes over 40°C, enzymes start to denature and do not work properly.

d) Shivering produces heat; the blood supply to the skin and extremities is reduced to keep warmth in the core; goose pimples occur as the body tries to trap a layer of air next to the skin; behavioural changes such as moving about, putting on more clothes and turning up the heating take place.

e) Sweating occurs, so heat is lost by evaporation; blood vessels supplying the skin capillaries dilate, so more blood flows nearer the surface of the skin; heat is lost by radiation; behavioural changes occur, such as the removal of clothing or moving to a shady place.

3 a) The level of insulin rises after a meal because the levels of glucose in the blood rise. The pancreas releases insulin, allowing glucose to enter the cells and the liver to convert glucose to glycogen; the blood glucose concentration drops.

b) The level of glucose in the blood keeps rising because no insulin is produced. The glucose cannot enter the cells and no glycogen is made in the liver.

c) The insulin injections mean that glucose can enter the cells to provide them with energy. Glycogen stores are made in the liver and the levels of glucose in the blood do not get dangerously high. The pattern is quite similar to that of a person with a healthy pancreas.

d) The diet needs to be managed carefully. It is important for them to eat a diet that is relatively low in carbohydrates so that their blood glucose levels do not tend to go up too much and their pancreas can cope. It is particularly important to avoid sugary food, which makes the blood glucose level go up very rapidly. Starchy carbohydrates are broken down slowly and so the blood glucose level does not rise too quickly.

HOMEOSTASIS: B2 5.1 – B2 5.4

SUMMARY QUESTIONS

1 a) Draw and annotate a diagram explaining the basic principles of homeostasis.

b) Write a paragraph explaining why control of the conditions inside your body is so important.

2 We humans maintain our body temperature at a constant level over a wide range of environmental temperatures. Many other animals – fish, amphibians and reptiles as well as the invertebrates – cannot do this. Their body temperature is always very close to the environmental temperature.

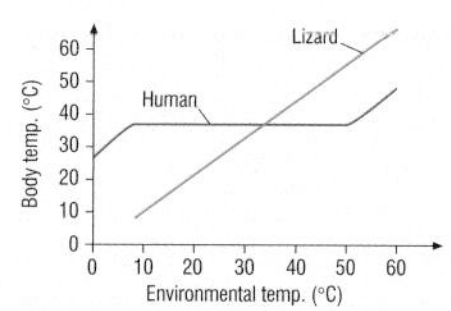

a) What is the body temperature of a person and a lizard at an atmospheric temperature of 20°C?

b) From the graph, at what external temperature does the human core temperature become dangerously low? Why is it dangerous?

c) At what external temperature does the human core temperature become dangerously high? Why is it dangerous?

d) Explain how a person maintains a constant core body temperature as the external temperature falls.

e) Explain how a person maintains a constant core body temperature as the external temperature rises. [Higher]

3 Use Figure 1 on page 199 to answer this question.

a) Look at graph A. Why does the level of insulin increase after a meal?

b) Graph B shows the blood sugar pattern of someone who has just developed diabetes and is not yet using injected insulin. What differences are there between this pattern and the one shown in A?

c) Graph C shows the effect of regular insulin injections on the blood sugar level of someone with diabetes. Why are the insulin injections so important to their health?

d) People who are mildly diabetic and those who inject insulin all have to watch the amount of carbohydrate in their diet. Explain why.

EXAM-STYLE QUESTIONS

1 Complete the passage below by choosing the corre terms from the box and matching them with the numbers in the passage.

sweating	dilate	shivering
thermoregulatory	radiation	constrict

Body temperature is controlled by the ....1.... centr the brain. On a hot day it causes blood vessels in th skin to ...2... and so lose heat by ....3.... Heat may be lost by ...4.... On a cold day the blood vessels .. to conserve heat. When cold, ...6.... may also occu create some heat. [Highe

2 The table shows the daily water loss from a typical human being.

Water lost in	Volume of water (cm^3 per
Urine	1500
X	400
Evaporation from the skin	350
Faeces	150
Sweat	100

(a) One way in which water is lost from the body ha been missed out and replaced by the letter **X**. W does **X** represent?

(b) These figures were taken on a cool day with the person at rest. State two ways in which the figur would be different if the person had been exerc on a hot day.

(c) Apart from water, what other two substances ar typically found in urine?

(d) Where is urine stored in the body?

3 (a) What is the name of the hormone that causes th liver to remove glucose from the blood?

(b) Where in the body is this hormone produced?

(c) Two people drank a solution that contained 100 glucose. The blood sugar level of each person w measured over the next three hours. The results shown in the table on the next page.

(i) On a piece of graph paper, draw a line grap the data in the table opposite.

(ii) One of the two persons is diabetic. From the graph suggest which one and give two reaso for your answer.

- When writing about control of body temperature, students get muddled about the blood flow to the skin. This problem can be remedied by a diagram showing the constriction or dilation of the arterioles supplying the capillaries. It could be pointed out that the capillary walls are only one cell thick: there is no muscle present to contract and bring about constriction or to relax causing dilation.

- **When to use the questions?**
  - Questions 2 and 3 are both graph interpretation questions and could be used as homework exercises after the relevant spreads. They could also be used for revision to practise dealing with data.
  - Question 3 uses Figure 1 on spread B2 5.4. It could be useful to discuss the first three parts of this question in class as a plenary to the lesson and then to set students the last part to do at home.

EXAM-STYLE ANSWERS

1 **1** Thermoregulatory **2** Dilate **3** Radiation **4** Sweating **5** Constrict **6** Shivering *(6 marks*

2 a) Breathing/expired air *(1 mark*

b) The figure for sweat would be higher. *(1 mark*
The figure for urine would be lower. *(1 mark*
Accept figures provided they show the appropriate increase/decrease

c) Urea *(1 mark*
Minerals/ions *(1 mark*

d) In the bladder *(1 mark*

3 a) Insulin *(1 mark*

b) In the pancreas *(1 mark*

c) i) • Time (independent variable) plotted on the x-axis/abscissa and blood sugar level on the y-axis/ordinate.
• Scale appropriate on both axes and both axes correctly labelled (including units).

Summary teaching suggestions

- **Literacy guidance** – The answers requiring explanations should be clear and unambiguous. In the answers to questions 2d) and e), students need to write a passage in continuous prose, presenting their answers in a logical sequence.
- **Lower and higher level answers**
  Higher attaining students should be able to give answers using the correct scientific terminology. References to 'glucose' rather than 'sugar' are more accurate. Students should be taught to give full answers when asked for an explanation.
- **Misconceptions**
  - It is easy to detect when students are muddled about the terms they are using – in this case, the spelling of words such as 'glucose', 'glucagon' and 'glycogen'! Some students hedge their bets and provide hybrid spellings.

Time in minutes	Blood sugar level (mg/100 cm³ blood)	
	Person X	Person Y
0 (glucose drunk)	90	90
30	160	140
60	220	90
90	200	80
20	150	70
50	130	80
80	110	90

ead the following passage about diabetes.

iabetes is a metabolic disorder in which there is an ability to control blood glucose levels due to the lack the hormone insulin. Diabetes was a fatal disease ntil in 1921 Banting and Best succeeded in isolating sulin from the pancreases of pigs and cows, having st carried out experiments on dogs. Insulin is a small otein of 51 amino acids, the sequence of which was etermined in the 1950s by Sanger. More recently the ne for human insulin has been isolated and the ormone can now be produced by bacteria as a result of netic engineering. Diabetics must test their blood gar levels regularly and inject insulin if they are to ad normal lives.

) Why do diabetics inject insulin rather than taking it by mouth? (2)

) What would happen to the blood sugar level of a diabetic who failed to inject insulin? (1)

) Suggest one other symptom of diabetes other than changes to blood sugar. (1)

) Give three advantages of using genetically engineered insulin rather than extracting the hormone from animal pancreases. (3)

) Injecting insulin only *treats* diabetes. In future it may be possible to replace the damaged pancreas by transplantation.
   (i) What would be the benefits to the person with diabetes of such treatment?
   (ii) State the drawbacks of this treatment. (4)

HOW SCIENCE WORKS QUESTIONS

You have probably heard the weatherman, during winter, tell you about the 'wind chill factor'. This is to give you a better idea of how cold your skin might feel if you were to go out whilst the temperature was low and it was windy. Remember that wind will cool the skin by evaporating moisture from it and therefore make it feel colder than the actual air temperature.

Until recently the wind chill factor was calculated by measuring temperatures of some water in a container in the Arctic. The tank of water was 10 metres above the ground.

a) Explain why this was a poor way to calculate the effect of wind chill on humans. (2)

Recently some investigations were carried out to get a better measure of the effect of wind chill on humans. The tests were carried out on humans dressed in protective clothing, except their cheeks were left exposed, so that their cheek skin temperature could be measured.

b) Why do you think the cheeks were chosen? (1)

c) The people were tested at different temperatures and wind speeds. What would have been a suitable sensitivity for the thermometer? (1)

d) How many people would have been chosen? (1)

e) How would these people have been chosen? (1)

f) Imagine you were carrying out these tests. Draw up a table that would let you fill in the results as you did the tests on just one person. (2)

g) Now fill in the table with some temperatures and some wind speeds that you think might be useful. (5)

- Points plotted accurately ±2 mm (scale may need to be varied depending upon the graph paper used).
- *Either* best fit curve *or* ruled point to point lines – accurately drawn.
- Two lines distinguished in some way (e.g. one solid/one broken) with key or labels. *(Allow the mark if both lines are drawn the same but distinguished by labels in close proximity to the relevant line.)*

(1 mark for each aspect)

ii) Person X because:
- Blood sugar level of person X rises to a much higher level (200 mg/cm³ blood) than that of person Y (140 mg/cm³ blood).
- Blood sugar level of person X continues to rise until around 60 minutes after swallowing the glucose whereas that of person Y begins to fall after 40 minutes.
- Blood sugar level of person X has not even reached the starting/normal level of 90 mg/cm³ blood after 3 hours whereas that of person Y returns to the starting/normal level after just one hour.
- Any other reasonable explanation.

(1 mark for each point to a maximum of 2 marks)

There are no marks for simply choosing the correct person as there is a 50% chance of success using only guesswork. If the student has reversed persons X and Y on the graph – allow the marks in this part of the question provided the points made are consistent with their error on the graph.

a) Because it would be broken down. *(1 mark answer)*
As insulin is a **protein**, it would be **broken down** by **enzymes** in the stomach. *(2 marks answer)*
For both marks the key words in bold (or their equivalent) must be correctly linked.

b) It would get higher/increase. *(1 mark)*

c) Glucose in the urine/increased thirst/increased hunger/urinating excessively/tiredness/weight loss/blurred vision.
(Any one for 1 mark)

d) Compared to insulin extracted from animals, genetically engineered insulin:
- is more effective (as it is an exact copy of human insulin)
- is more rapid in its action (exact copy)
- produces no immune response (exact copy)
- does not carry any risk of transmitting animal diseases
- is cheaper to produce
- does not cause patients to develop tolerance (become less sensitive) to it
- does not involve using/killing animals.

(Each point one mark to a maximum of 3 marks)

e) i) Benefits:
- Permanent cure
- No need for injections
- Cheaper in the long term (no insulin needed).

ii) Drawbacks:
- Insufficient donors
- Risk to health during operation (both to live donor and recipient)
- Some people find transplants unethical.

(1 mark for each benefit to a maximum of 2 marks)
(1 mark for each drawback to a maximum of 2 marks)
Do not allow 'nasal sprays' as this is a treatment not a cure.

HOW SCIENCE WORKS ANSWERS

a) Bucket of water does not lose heat in the same way as the human body, e.g. humans have an inner source of heat and therefore remain hotter than the surroundings. Also no humans are 10 metres tall!

b) The cheeks are usually the only part of the body left exposed in very cold conditions.

c) To at least 0.1°C

d) As many as possible (actually in one study six men and six women).

e) From a wide range of ages and sexes (actually aged 22 to 42).

f) and **g)** There will probably be many acceptable ways to record the data. These are the actual wind speeds and temperatures used. Perhaps they should have used more temperatures and wind speeds within the range!

	Air temperature (°C)		
Wind speed (km/h)	**10**	**0**	**−10**
10			
20			
30			

How science works teaching suggestions

- **Literacy guidance**
  - Key terms that should be clearly understood: range; interval measurements; sensitivity.
  - Question a) expects a longer answer, where students can practise their literacy skills.
- **How and when to use these questions.** When wishing to develop good scientific method and table construction.
The questions would be best used in small discussion groups.
- **Homework.** Questions f) and g) could be set as homework.
- **Special needs.** A demonstration of windchill could be done.
- **ICT link-up.** Data logging equipment could be used to demonstrate windchill on wet towelling.

B2 6.1 Cell division and growth

LEARNING OBJECTIVES

Students should learn that:

- Mitosis results in the production of additional cells for growth, repair and replacement.
- Before each cell division, the genetic information on the chromosomes is copied so that the new cells have the same genes as the parent cells.
- Most animal cells differentiate at an early stage but most plant cells have the ability to differentiate throughout life.

LEARNING OUTCOMES

Most students should be able to:

- Describe the process of mitosis.
- Describe how the cells produced by mitosis are genetically identical to the parent cells.

Some students should also be able to:

- Explain why plants retain the ability to grow throughout their lives whereas cell division in mature animals is involved in repair and replacement of tissues.

Teaching suggestions

- **Special needs.** Give one student two short pieces of Plasticine of one colour and two long ones of another colour. These are placed inside a ring of string representing the cell. Give the students balls of Plasticine and tell them to make copies of each and pass a set to two other students who do the same, passing on two more each until the whole class has been involved and there are many copies inside string rings on the floor. This works well in a gym.
- **Gifted and talented.** Introduce the students to the names of the stages as an introduction to AS level work.
- **Learning styles**

  Kinaesthetic: Making models with Plasticine.

  Visual: Watching animation of mitosis.

  Auditory: Discussing the properties of 'living leather'.

  Interpersonal: Compiling a small concept map for the unit and getting other students to state how the links should be labelled.

  Intrapersonal: Writing down own ideas on 'How a cell divides'.

SPECIFICATION LINK-UP Unit: Biology 2.12.8

- *In body cells, the chromosomes are normally found in pairs. Body cells divide by mitosis to produce additional cells during growth or to produce replacement cells.*
- *Most types of animal cell differentiate at an early stage, whereas many plant cells retain the ability to differentiate throughout life. In mature animals, cell division is mainly restricted to repair and replacement.*
- *The cells of the offspring produced by asexual reproduction are produced by mitosis from parental cells. They contain the same genes as the parents.*

Lesson structure

STARTER

Matching exercise – Give each student pieces of paper with 'cell', 'nucleus', 'chromosome', 'gene' and 'DNA' on them, plus definitions all muddled up. They have to join them correctly. (5 minutes)

Growth, repair or replacement – Ask if any of the students have scabs or scars. Ask: 'Who has had their hair cut most recently or cut their fingernails?' Discuss marking children's height on the back of doors. Link with Anne Frank's father marking her height when they were in hiding. Lead in to a discussion of how our 'living leather' grows and heals. (5–10 minutes)

Multiplication – Draw one cell on the board then two cells next to it. The students have to copy this down and write beneath a short paragraph entitled 'My ideas on how cells multiply'. Read some out and discuss. (10–15 minutes)

MAIN

Observing mitosis

- This can be done using prepared longitudinal sections of root tips, or the students can make their own root tip squashes. A number of different sources will give suitable root tips, although it is a good idea to choose something that does not have a large diploid number of chromosomes. Germinating broad bean or pea seeds work well, or the tips of roots produced from hyacinth or garlic bulbs suspended in water. (See 'Practical support').
- Students may require help mounting their root tips and in using microscopes. It could be helpful to give each student a worksheet with the details of the preparation on it, and then a series of diagrams or photographs showing stages so that they can have a go at identifying stages on their own slides. If a space is left beside each stage, then the students could make a sketch of what they can see on their slides. They do not need to know the names of the stages.

Cloning a cauliflower

Students could try cloning for themselves. Using sterile techniques, it is possible to grow clones of carrot, cauliflower or potato tissue on nutrient agar. Use a 3 mm tip of an 'eye' of a potato, a mini-floret from the floret of a cauliflower or a segment of carrot tap root treated with the plant growth regulator 2,4-D. The plant tissue should be sterilised in bleach, rinsed in four washes of sterilised water and then gently pressed into agar in sterilised Petri dishes. The cultures should be sealed, wrapped loosely in cling film and kept incubated in a growth cabinet at about 25°C in the light.

- Calluses should develop over the next few weeks and tiny plantlets should develop from buds. The cultures should be examined regularly and a photographic record kept. More details are available from NCBE publications.

PLENARIES

Mitosis dominoes – In groups of four, play a dominoes-style game showing the stages of mitosis and a general description. No details, i.e. named stages, are required. (10 minutes)

Growing points – Using a small potted plant, remove all the leaves, so that the growing point (main bud) and the buds in the axils of the leaves are left. Give the students a diagram of the plant with all the leaves pulled off and get them to mark where the growing points are, and to draw in what they think the plant will look like in a couple of weeks. Keep the plant in the lab and look at it when the time is up. (10 minutes)

Practical support

Observing mitosis

Equipment and materials required

Root tips of beans, peas, onions, garlic or other suitable material, dilute acetic orsein stain and dilute hydrochloric acid, watch glasses, heater/spirit lamp/hotplate, mounted needles, microscope slides and cover slips, blotting paper, microscopes.

Details of practical

mm lengths of the root tips should be cut off and placed in a watch glass containing acetic orcein stain and hydrochloric acid. This should be warmed gently for 5 minutes. The tip is then placed on a microscope slide with a few drops of the stain, teased out with a pair of mounted needles and then covered with a cover slip. Cover with blotting paper and press gently to spread out the cells. The slide can then be viewed under the microscope, remembering to focus at low power first.

ACTIVITY & EXTENSION

- **Make a mitosis flick book** – Find or make some clear diagrams of the stages of mitosis. Copy on to a sheet for each student so that they can make their own 'flick book' by cutting up the pictures and assembling them in the correct order.
- **Make your own mitosis movie** – Using Plasticine to model the chromosomes and stop motion photography with a web cam, students can make their own animation of the process of mitosis.

BIOLOGY — INHERITANCE

B2 6.1 Cell division and growth

LEARNING OBJECTIVES

1 What is mitosis?
2 Why do plants grow throughout their lives while most animals stop growing once they are adults?

New cells are needed for an organism, or part of an organism, to grow. They are also needed to replace cells which become worn out and repair damaged tissue. However the new cells must have the same genetic information in them as the originals, so they can do the same job.

Each of your cells has a nucleus containing the instructions for making whole new cells and even an entire new you! These instructions are carried in the form of genes.

A gene is a small packet of information which controls a characteristic, or part of a characteristic, of your body. The genes are grouped together on chromosomes. A chromosome may carry several hundred or even thousands of genes.

You have 46 chromosomes in the nucleus of your cells (except your gametes – sperm or ova). They come in 23 pairs. One of each pair is inherited from your father, and one from your mother.

a) Why are new cells needed?

Mitosis

Body cells divide to make new cells. The cell division which takes place in the normal body cells and produces identical daughter cells is called **mitosis**. As a result of mitosis all your body cells have the same genetic information.

In asexual reproduction, the cells of the offspring are produced by mitosis from cells of their parent. This is why they contain exactly the same genes with no variety.

How does mitosis work? Before a cell divides it produces new copies of the chromosomes in the nucleus. This means that when division takes place two genetically identical **daughter cells** are formed.

In some areas of the body of an animal or plant, cell division like this carries on rapidly all of the time. Your skin is a good example – cells are constantly being lost from the surface and new cells are constantly being formed by cell division to replace them.

b) What is mitosis?

Differentiation

In the early development of animal and plant embryos the cells are very unspecialised. Each one of them (known as **stem cells**) can become any type of cell which is needed.

In many animals, the cells become specialised very early in life. By the time a human baby is born most of its cells have become specialised for a particular job, such as liver cells, skin cells and muscle cells. They have **differentiated**. Some of their genes have been switched on and others have been switched off.

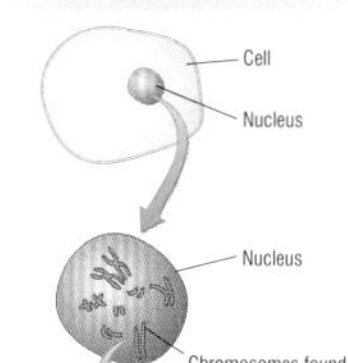

Figure 1 The nucleus of your cell contains the chromosomes that carry the genes which control the characteristics of your whole body

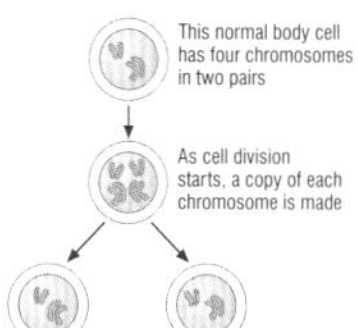

This normal body cell has four chromosomes in two pairs
As cell division starts, a copy of each chromosome is made
The cell divides in two to form two daughter cells. Each daughter cell has a nucleus containing four chromosomes identical to the ones in the original parent cell.

Figure 2 Identical daughter cells are formed by the simple division that takes place during mitosis. It supplies all the new cells needed in your body for growth, replacement and repair. Your cells really have 23 pairs of chromosomes – but for simplicity this cell is shown with only two pairs!

202

This means that when a muscle cell divides by mitosis it can only form more muscle cells. Liver cells can only produce more liver cells. So in adult animals, cell division is restricted because differentiation has occurred. Specialised cells can divide by mitosis, but this can only be used to repair damaged tissue and replace worn out cells. Each cell can only produce identical copies of itself.

In contrast, most plant cells can differentiate all through their life. Undifferentiated cells are formed at active regions of the stems and roots. In these areas mitosis takes place almost continuously.

Plants keep growing all through their lives at these 'growing points'. The plant cells produced don't differentiate until they are in their final position in the plant. What's more, the differentiation isn't permanent. If you move a plant cell from one part of a plant to another, it can re-differentiate and become a completely different type of cell. You just can't do that with animal cells – once a muscle cell, always a muscle cell!

PRACTICAL

Observing mitosis

Make a special preparation of a growing root tip to view under a microscope. Then you can see the actively dividing cells and the different stages of mitosis as it is taking place.

- Describe your observations of mitosis.

We can produce huge numbers of identical plant clones from a tiny piece of leaf tissue. Now you can see why this is possible. In the right conditions a plant cell will become unspecialised and undergo mitosis many times. In different conditions, each of these undifferentiated cells will produce more cells by mitosis. These will then differentiate to form a tiny new plant identical to the original parent.

The reason animal clones cannot be made easily is because animal cells differentiate permanently early in embryo development – and can't change back! Animal clones can only be made by cloning embryos in one way or another.

SUMMARY QUESTIONS

1 Copy and complete using the words below:

chromosomes genetic information genes growth mitosis nucleus replace

New cells are needed for ...... and to ...... worn out cells. The new cells must have the same .... in them as the originals. Each cell has a ...... containing the ...... grouped together on .... The type of cell division which produces identical cells is known as ....

2 Division of the body cells is taking place all the time in living organisms.
a) Why is it so important?
b) Explain why the chromosome number must stay the same when the cells divide to make other normal body cells.

3 The process of growth and differentiation is very different in plants and animals.
a) What is differentiation?
b) How is differentiation in animal and plant cells so different?
c) How does this difference affect the cloning of plants and animals?

Additional biology

DID YOU KNOW?

Your body cells are lost at an amazing rate – 300 million cells die every minute! No wonder mitosis takes place all the time in your body to replace them!

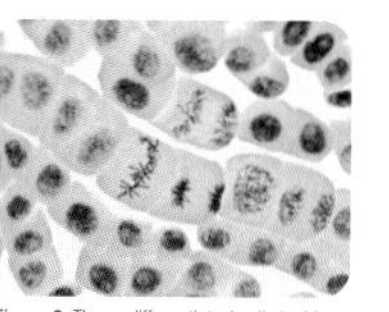

Figure 3 The undifferentiated cells in this onion root tip are dividing rapidly. You can see mitosis taking place, with the chromosomes in different positions as the cells divide.

GET IT RIGHT!

Cells produced by mitosis have identical genetic information.

KEY POINTS

1 In body cells, chromosomes are found in pairs.
2 Body cells divide by mitosis to produce more identical cells for growth, repair, replacement or in some cases asexual reproduction.
3 Most types of animal cells differentiate at an early stage of development. Many plant cells can differentiate throughout their life.

203

SUMMARY ANSWERS

1 Growth, replace, genetic information, nucleus, genes, chromosomes, mitosis.

2 a) Mitosis is important because cells die at the rate of 300 million per minute; cells are damaged; cells need to grow; in some organisms cells are needed for asexual reproduction.
b) Cells need to be replaced with identical cells to do the same job.

3 a) Differentiation is the process by which cells become specialised.
b) In animals, it occurs during embryo development and is permanent. In plants, it occurs throughout life and can be reversed or changed.
c) Plants can be cloned relatively easily. Differentiation can be reversed, mitosis is induced, conditions can be changed and more mitosis induced. The cells re-differentiate into new plant tissues. In animals, differentiation cannot be reversed, so clones cannot be made easily. In order to make clones, embryos have to be made.

Answers to in-text questions

a) New cells are needed for growth, replacement and repair.
b) Mitosis is cell division that takes place in the normal body cells and produces identical daughter cells containing exactly the same genes as their parents.

KEY POINTS

The important point to stress here is that the cells produced by mitosis are genetically identical to the parent cells. This is achieved by the copying and separation of the chromosomes during mitosis. Students could write questions to test their partner on the key points.

B2 6.2 Stem cells

LEARNING OBJECTIVES

Students should learn that:

- Stem cells are unspecialised cells found in human embryos and in some adult tissues such as the bone marrow.
- Stem cells have the potential to differentiate into different types of specialised cells.

LEARNING OUTCOMES

Most students should be able to:

- Describe the structure and location of stem cells in humans.
- Describe how stem cells have the potential to treat sick people.

Some students should also be able to:

- Explain the arguments for and against using stem cells from embryos.

Teaching suggestions

- **Special needs.** Provide the students with a pre-drawn diagram of a ball of cells and some labels of cells and organs. Students can stick the labels around the stem cells to gain an understanding of these cells giving rise to all other types of cell.
- **Gifted and talented.** Students to define for themselves when life starts. They can be given a list of criteria from contrasting organisations such as the Human Fertilisation and Embryology Authority, and Pro-Life.
- **Learning styles**

  Kinaesthetic: Researching into the topic, using web sites.

  Visual: Following a sequence of therapeutic cloning diagrams.

  Auditory: Listening to the opinions of others in the discussion on pros and cons of stem cell research.

  Interpersonal: Discussing the pros and cons of research.

  Intrapersonal: Understanding the issues involved and conversation in groups.
- **ICT link-up.** There are a number of good web sites that have useful information: *New Scientist*; *Nature*; Stem Cell Information; Christopher Reeve Foundation; Stem Cell Research Foundation. (See www.nature.com; www. stemcells.nih.gov; www.christopherreeve.org; www.newscientist.com; www.stemcellresearchfoundation.org.)

SPECIFICATION LINK-UP Unit: Biology 2.12.8

- *Most types of animal cells differentiate at an early stage whereas many plant cells retain the ability to differentiate throughout life. In mature animals, cell division is mainly restricted to repair and replacement. Cells from human embryos and adult bone marrow, called stem cells, can be made to differentiate into many different types of cells e.g. nerve cells. Treatment with these cells may help conditions such as paralysis.*

Lesson structure

STARTER

What are stem cells? – Write up the words ‘stem cells’ on the board. Students to write down, on an individual white board, a two-word phrase that it brings to mind. For lower ability students, just use the word ‘stem’ and put it into a sentence. Analyse some of the contributions. (5–10 minutes)

Growing new tissue could save lives – Search the web for some pictures of Christopher Reeve and a clip from one of his ‘Superman’ films. (Search also for Superman trailer!) Tie this in with a discussion of what happened to him and how growing new tissue, if it was possible, could have helped him. (10 minutes)

Losing limbs – Show photographs of amputees. Ask for a volunteer to wear a reverse sling for the lesson (not on their writing hand!). This can then stimulate a discussion of what it is like to lose a limb. If the class is musically orientated, play *The band played Waltzing Matilda,* by Eric Bogle – a poignant song about losing your legs. (10–15 minutes)

MAIN

- **Regenerating parts** – Some animals are able to re-grow parts of their bodies. Show photographs from search engines of lizards re-growing their tails and starfish re-growing limbs. Lead into a discussion of injuries and how they heal. Allow students to discuss injuries that they have had and what happens to them (time limit will be needed!).
- **Therapeutic cloning** – Find photographs about therapeutic cloning. There are a number of good web sites with lots of information on the use of stem cells. Students could be provided with a worksheet to be filled in. The important thing is to be aware of what is actually being done and what is hoped can be done in the future.
- **Sources of stem cells** – Show photographs of different sources of stem cells. Search the web for ‘cell division blastocyst video’ to show what happens after fertilisation. Otherwise find a photograph of a ball of stem cells. After four divisions, the cells become increasingly specialised. Discuss what would happen to the cells if they were allowed to continue development. Other sources include umbilical cord blood (rich in blood stem cells), fetal germ cells (extracted from terminated pregnancies of 5–9 weeks), frozen embryos and adult stem cells from bone marrow.
- **Pros and cons of the use of stem cells** – Both arguments need to be put forward. The cons of stem cell research could be put to the students by a visiting speaker who will argue the case. Some Catholic web sites provide a concise version of the sanctity of life (see www.justthefacts.org for some good pre-birth information). The pros can be summarised from web sites. Some useful ones are given in the ‘ICT Link-up’.
- Use a summary sheet to state the main information and hold a snowball discussion where pairs of students brainstorm the concepts, then double up as fours and continue the process. The fours then gather into groups of eight in order to compare ideas and agree on a course of action (to endorse stem cell research or not). A spokesperson from each group of eight feeds back to the whole group.

ACTIVITY & EXTENSION

- If computers are available, set up a scavenger hunt style trail of Internet sites to pull out the main bits of the pro and con arguments and details of the research carried out.
- Ask students to find out about the original research and write a report of what was discovered. (See www.stemcellresearchfoundation.org.)
- Ask students to investigate what the words 'totipotent', 'pluripotent' and 'multipotent' mean when applied to stem cells. ['Totipotent' cells are found in very early embryos (for the first three or four divisions) and can differentiate into all types of cell. 'Pluripotent' stem cells are present in later embryos and can differentiate into any cell type. 'Multipotent' stem cells are found in adults as well as embryos and will only differentiate into certain cell types.] The Stem Cell Research Foundation has illustrations on its web site.

Answers to in-text questions

a) Unspecialised cells that can differentiate to form many different types of specialised body cell.

b) Culturing human embryonic stem cells.

c) Some people think it is wrong to use a potential human being as a source of cells to help other people.

B2 6.2 Stem cells

LEARNING OBJECTIVES

1 What is special about stem cells?
2 How can we use stem cells to cure people?

Most of the cells in your body are differentiated. They are specialised and carry out particular jobs. But some of your most important cells are the completely unspecialised **stem cells**. They can differentiate (divide and change) into many different types of cell when they are needed. Human stem cells are found in human embryos and in some adult tissue including bone marrow.

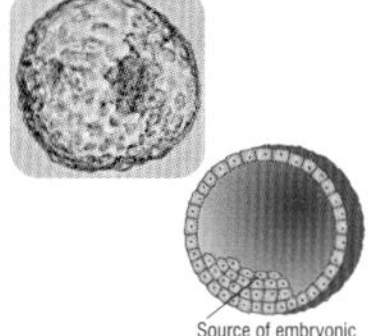

Figure 1 This ball of cells is an early human embryo. In the right conditions these few cells can form all the organs of the human body.

The function of stem cells

Stem cells divide and form the specialised cells of your body which make up your various tissues and organs. When an egg and sperm fuse to form an embryo, they form a single new cell. That cell divides and the embryo is soon a hollow ball of cells. The inner cells of this ball are the stem cells which will eventually give rise to every type of cell in your body.

Even when you are an adult some of your more specialised stem cells remain. Your bone marrow is a good source of stem cells. What's more, scientists now think there may be a tiny number of stem cells in most of the different tissues in your body. This includes your blood, brain, muscle and liver.

The stem cells can stay there for many years until your tissues are injured or affected by disease. Then they start dividing to replace the different types of damaged cells.

a) What are stem cells?

Using stem cells

Many people suffer and even die because various parts of their body stop working properly. For example, spinal injuries can cause paralysis. That's because the spinal nerves do not repair themselves. Millions of people would benefit if we could replace damaged body parts.

In 1998, there was a breakthrough. Two American scientists managed to culture human embryonic stem cells that were capable of forming other types of cells.

Scientists hope that these embryonic stem cells can be encouraged to grow into almost any different type of cell needed in the body. For example, we may be able to grow new nerve cells. If new nerves grown from stem cells could be used to reconnect the spinal nerves, people who have been paralysed could walk again.

With stem cells we might also be able to grow whole new organs which could be used in transplant surgery. These new organs would not be rejected by the body. Conditions from infertility to dementia could eventually be treated using stem cells.

Unfortunately, at the moment no-one is quite sure just how the cells in an embryo are switched on or off. We don't yet know how to form particular types of tissue. Once we know how to do this, we can really start to use stem cells effectively.

b) What was the big scientific breakthrough by American scientists in 1998?

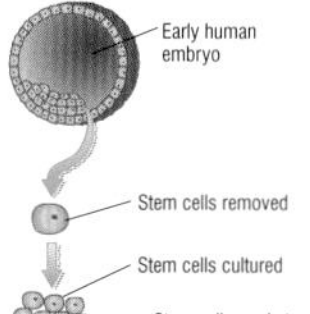

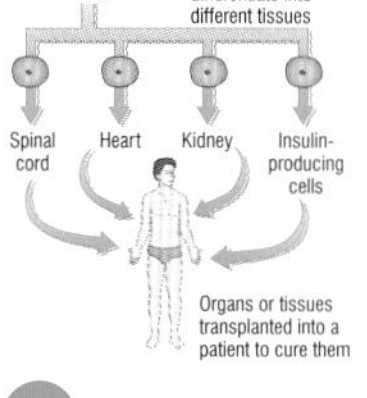

Figure 2 Some of the embryonic stem cells which scientists have produced and grown have formed into adult cells. Unfortunately no-one is quite sure how to control this process at the moment. Hopefully one day the technique shown in this diagram will be used to treat people.

Problems with stem cells

Many embryonic stem cells come from aborted embryos or from spare embryos in fertility treatment. This raises ethical problems. There are people, including many religious groups, who feel it is wrong to use a potential human being as a source of cells, even to cure others.

Some people feel that as the embryo cannot give permission, using it is a violation of its human rights. On top of this, progress with stem cells is slow. There is some concern that embryonic stem cells might cause cancer if they are used to treat sick people. This has certainly been seen in mice. Making stem cells is slow, difficult, expensive and hard to control.

c) What is the biggest ethical concern with the use of embryonic stem cells?

The future of stem cell research

We have found embryonic stem cells in the umbilical cord blood of newborn babies. These may help to overcome some of the ethical concerns.

Scientists are also finding ways of growing adult stem cells. Unfortunately the adult stem cells found so far can only develop into a limited range of cell types. However this is another possible way of avoiding the controversial use of embryonic tissue.

The area of stem cell research known as ***therapeutic cloning*** could be very useful – but it is proving very difficult.

Therapeutic cloning involves using cells from an adult person to produce a cloned early embryo of themselves as a source of perfectly matched embryonic stem cells. In theory these could then be used to heal the original donor and maybe many others as well.

Most people remain excited by the possibilities of embryonic stem cells in treating many diseases. Just how many of these early hopes will be fulfilled only time will tell!

SUMMARY QUESTIONS

1 Copy and complete using the words below:

bone marrow differentiate embryos hollow inner stem cells

Unspecialised cells known as .... can ...... (divide and change) into many different types of cell when they are needed. Human stem cells are found in human ...... and in adult .... The embryo forms a ...... ball of cells and the ...... cells of this ball are the stem cells.

2 a) Why was the work of the American scientists in 1998 such a breakthrough in stem cell research?
b) How might stem cells be used to treat patients who are paralysed after a spinal injury?

3 a) What are the advantages of using stem cells to treat a wide range of diseases?
b) What are the difficulties with stem cell research?
c) How are scientists hoping to overcome the difficulties of using embryonic stem cells in their research?

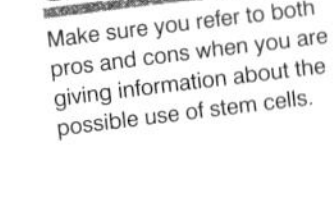

GET IT RIGHT!
Make sure you refer to both pros and cons when you are giving information about the possible use of stem cells.

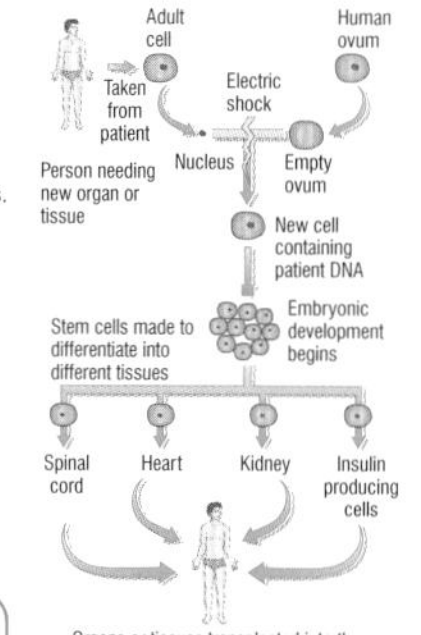

Figure 3 In 2005, a team led by Professor Woo Suk Hwang in South Korea claimed to have produced human embryos from adult cells and developed cloned stem cells from them. This seemed a huge step forward in stem cell research. But sadly, in 2006 the work was shown to be a massive scientific fraud. This was a massive blow to everyone working in stem cell research.

KEY POINT

1 Embryonic stem cells (from human embryos) and adult stem cells (from adult bone marrow) can be made to differentiate into many different types of cells.

PLENARIES

Therapeutic cloning – Draw out a sequence diagram for therapeutic cloning to illustrate how it is carried out. (10 minutes)

Poetry in motion – Give the students the key word 'stem cells'. This forms Line 1of the poem. Line 2 consists of two words describing what the key word is. Line 3 contains three words describing what the key word does. Line 4 is four words describing what the key word means to you personally. Line 5 is one word giving an alternative to the key word. Select students to read out their efforts. (10 minutes)

Anagrams – Write up or project anagrams of the key words from the lesson. Students to write them down sequentially on individual whiteboards. (5–10 minutes)

KEY POINT

This is a statement of the facts. Students should be quite clear about the unspecialised nature of stem cells and their potential to become differentiated. This could be reinforced by a diagram of an unspecialised cell surrounded by the different types of cell.

SUMMARY ANSWERS

1 Stem cells, differentiate, embryos, bone marrow, hollow, inner.

2 **a)** They cultured human stem cells for the first time.
b) They could be used to grow new nerve cells to reconnect the spinal cord.

3 **a)** They can be used to make any type of adult cell to repair or replace damaged tissues, with no rejection issues.
b) There are ethical objections and concerns over possible side effects.
c) By using stem cells from umbilical blood, adult stem cells and therapeutic cloning.

B2 6.3 Cell division in sexual reproduction

LEARNING OBJECTIVES

Students should learn that:

- Cells which divide to form gametes undergo meiosis. [**HT** only]
- Gametes have a single set of genetic information, whereas body cells have two sets.
- Fertilisation results in the formation of a cell with new pairs of chromosomes so sexual reproduction gives rise to variation.

LEARNING OUTCOMES

Most students should be able to:

- Describe what happens to the number of chromosomes during fertilisation.
- Explain how sexual reproduction gives rise to variation.

Some students should also be able to:

- Describe what happens to the chromosomes during the process of gamete formation. [**HT** only]

Teaching suggestions

- **Special needs.** Write the word 'chromosomes' twice on the board, inside a ring to represent a cell. To model meiosis, get four students to copy the word 'chromosomes' once onto a piece of A4 and put each on the board inside a ring. To model fertilisation, cut two of these 'chromosomes' words out and stick them inside a single ring. To model mitosis, take both words, stick them onto a sheet and photocopy it repeatedly. (Bear in mind that meiosis is a concept required at Higher Tier level only).
- **Gifted and talented.** Introduce the students to some of the vocabulary associated with the AS coverage of this topic (e.g. chiasmata, centrioles, centromeres). They could also look at pictures of chromosomes during Prophase I, where chiasmata formation is occurring.
- **Learning styles**

  Kinaesthetic: Plasticine modelling; using microscopes; drawing.

  Visual: Observing animations.

  Auditory: Talking over the differences between mitosis and meiosis.

  Interpersonal: Working in groups.

  Intrapersonal: Considering what would happen if reduction in chromosome numbers did not occur.

SPECIFICATION LINK-UP Unit: Biology 2.12.8

- *Body cells have two sets of genetic information; gametes have only one set.*
- *Cells in reproductive organs – testes and ovaries in humans – divide to form gametes. [**HT** only]*
- *This type of cell division in which a cell divides to form gametes is called meiosis. When a cell divides to form gametes:*
  - *copies of the chromosomes are made*
  - *then the cell divides twice to form four gametes, each with a single set of chromosomes. [**HT** only]*
- *When gametes join at fertilisation, a single body cell with new pairs of chromosomes is formed. A new individual then develops by this cell repeatedly dividing by mitosis.*
- *Sexual reproduction gives rise to variation because when gametes fuse, one of each pair of alleles comes from each parent.*

Lesson structure

STARTER

Introducing meiosis: a mnemonic for mitosis – Contrast meiosis with mitosis. Find a picture of some ghastly toes. (Search the web for 'toes'.) Get the students to copy down and remember that 'Mitosis goes on in my toes' and toes are not sexy. Also introduce meiosis as the 'reduction' division, as it reduces the number of chromosomes. (5 minutes)

Naming the sex cells – Give the students an empty grid to stick in their books. They are to complete this with the names of the sex cells from animals and plants. Have prompts ready if needed. For the higher attaining students, you could introduce the correct spellings i.e. 'spermatozoa' etc. (5–10 minutes)

Internet – Show an Internet-based flash animation summary of mitosis if easily available. (5 minutes)

MAIN

- **Simulation** – Use the Simulation B2 6.3 'Meiosis' available from the GCSE Biology CD ROM to introduce the topic.
- A flow diagram of the events of meiosis can be built up, showing that there are similarities in that the chromosomes are copied, but that there are two divisions rather than one. It is probably best to concentrate on the formation of sperm to begin with (because four observable cells result from the division) and follow up with slides of testis showing stages in sperm development.
- **Microscopic examination of testis slides** – The best prepared slides are of rat or grasshopper testis squashes. Provide the students with a sheet showing stages in the development of sperm that they are likely to be able to see on their slides, and a flow diagram to show the different stages of division. They may need help with their microscopes, as they will need to use high power if they are to see any chromosomes.
- Alternatively, sections of testis could be projected on to the board and students could identify the different cells with reasons for their choices.
- **Microscopic examination of ovary slides** – Slides could be projected and viewed by the class, or slides could be viewed using a microscope. There will be obvious differences in size of the sperms and the eggs. For higher attaining students it would be possible to explain what happens during the meiotic divisions that produce ova, e.g. the formation of the polar bodies.
- **Modelling meiosis and the need for reduction** – Using model chromosomes, firstly without a reduction in number of the chromosomes, show how the number of chromosomes would go on increasing. Follow this with the reduction part of the division, so that gametes have half the number and the correct number is restored on fertilisation.

- **Introducing variation** – Using Plasticine of different colours, it is possible to show how variation can occur during the process of meiosis. Students, in groups, could make models showing how the chromosomes separate, perhaps showing some exchange of genes, and then matching one set of gametes with another set, to represent the sperm and the ovum at fertilisation.

PLENARIES

True or false? – Present students with statements about mitosis and meiosis. They are to write 'True' or 'False' on 'Show me' boards, e.g.

'Mitosis is necessary for growth, repair and replacement of tissues.' [True]
'In meiosis, the number of chromosomes stays the same.' [False]
'Meiosis takes place in the testes.' [True]
'Mitosis involves two divisions of the chromosomes.' [False]
'Mitosis results in genetically identical cells.' [True] (5–10 minutes)

Use summary questions – Give students 5 minutes to read and answer the summary questions given in the Student Book . Check answers verbally to test the students' understanding of the topic. (10 minutes)

ACTIVITY & EXTENSION IDEAS

- **Differences between mitosis and meiosis** – Students to make a leaflet or poster summarising the differences between mitosis and meiosis. They should make it memorable, perhaps using the 'non-sexy toes' statement.
- **Meiosis in plants** – Preparations of squashes of immature anthers from developing buds of lily show the stages of meiosis and chromosomes very clearly.

BIOLOGY INHERITANCE

B2 6.3 Cell division in sexual reproduction

LEARNING OBJECTIVES

1. What happens to your chromosomes when your gametes are formed? [Higher]
2. How does sexual reproduction give rise to variation?

Mitosis is taking place all the time, in tissues all over your body. But mitosis is not the only type of cell division. There is another type which takes place only in the reproductive organs of animals and plants. **Meiosis** results in sex cells with only half the original number of chromosomes.

DID YOU KNOW?
About 80% of fertilised eggs never make it to become a live baby – in fact about 50% never even implant into the lining of the womb.

Meiosis

HIGHER

The reproductive organs in people, like most animals, are the **ovaries** and the **testes**. This is where the sex cells (the gametes) are made. The female gametes or ova are made in the ovaries. The male gametes or **sperm** are made in the testes.

The gametes are formed by meiosis, which is a special form of cell division where the chromosome number is reduced by half. When a cell divides to form gametes, the first stage is very similar to normal body cell division. The chromosomes are copied so there are four sets of chromosomes. The cell then divides twice in quick succession to form four gametes, each with a single set of chromosomes.

GET IT RIGHT!
Be careful with the spelling of mitosis and meiosis. Make sure you know the differences between the two processes.

Why is meiosis so important?

Your normal body cells have 46 chromosomes in two matching sets – 23 come from your mother and 23 from your father. If two 'normal' body cells joined together in sexual reproduction, the new cell would have 92 chromosomes, which simply wouldn't work!

Fortunately, as a result of meiosis, your sex cells contain only one set of chromosomes, exactly half of the full chromosome number. So when the gametes join together at fertilisation, the new cell formed contains the right number of 46 chromosomes.

a) What are the names of the male and female gametes and how do they differ from normal body cells?

A cell in the reproductive organs looks just like a normal body cell before it starts to divide and form gametes

As in normal cell division, the first step is that the chromosomes are copied

The cell divides in two, and these new cells immediately divide again

This gives four sex cells, each with a single set of chromosomes – in this case two instead of the original four

Figure 1 The formation of sex cells in the ovaries and testes involves a special kind of cell division to halve the chromosome number. The original cell is shown with only two pairs of chromosomes to make it easier to follow what is happening.

DID YOU KNOW?
One testis can produce over 200 million sperm each day. As most boys and men have two working testes, that gives a total of 400 million sperm produced by meiosis every 24 hours! Only one sperm is needed to fertilise an egg. However as each tiny sperm needs to travel 100 000 times its own length to reach the ovum, less than one in a million ever completes the journey – so it's a good thing that plenty are made!

206

Additional biology

HIGHER

In girls, the first stage of meiosis is completed before they are even born. The tiny ovaries of a baby girl contain all the ova she will ever have.

In boys, meiosis doesn't start until puberty when the testes start to produce sperm. It then carries on for the rest of their lives.

Each gamete you produce is slightly different from all the others. The combination of chromosomes will be different. What's more, there is some exchange of genes between the chromosomes during the process of meiosis. This means that no two eggs or sperm are the same. This introduces lots of variety into the genetic mix of the offspring.

b) What type of cell division is needed to produce the gametes?

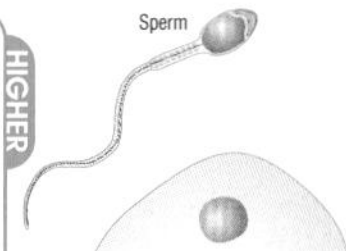

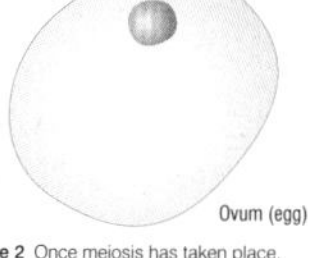

Figure 2 Once meiosis has taken place, the male and female gametes develop very differently – they are adapted for very different jobs

Fertilisation

More variety is added when fertilisation takes place. Each sex cell has a single set of chromosomes. When two sex cells join during fertilisation the new cell formed has a full set of chromosomes. In humans, the egg cell has 23 chromosomes and so does the sperm. When they join together they produce a new normal cell with the full human complement of 46 chromosomes.

The combination of genes on the chromosomes of every newly fertilised ovum is completely unique. Once fertilisation is complete, the unique new cell begins to divide by mitosis. This will continue long after the fetus is fully developed and the baby is born.

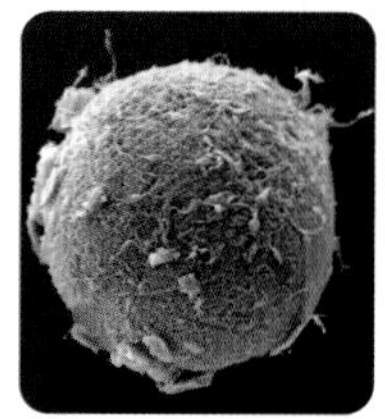

Figure 3 At the moment of fertilisation the chromosomes in the two gametes are combined so the new cell has a complete set, like any other body cell. This cell will then grow and reproduce by mitosis to form a new individual.

Variation

The differences between asexual and sexual reproduction are a reflection of the different types of cell division involved in the two processes.

In asexual reproduction the offspring are produced as a result of mitosis from the parent cells. (See the start of this chapter.) So they contain exactly the same chromosomes and the same genes as their parents. There is no variation in the genetic material.

In sexual reproduction the gametes are produced by meiosis in the sex organs of the parents. This introduces variety as each gamete is different. Then when the gametes fuse, one of each pair of chromosomes, and so one of each pair of genes, comes from each parent.

The combination of genes in the new pair will contain **alleles** (different forms of the gene) from each parent. This also helps to produce different characteristics in the offspring.

SUMMARY QUESTIONS

1. a) How many pairs of chromosomes are there in a normal human body cell?
   b) How many chromosomes are there in a human egg cell?
   c) How many chromosomes are there in a fertilised human egg cell?
2. Sexual reproduction results in variety. Explain how.
3. a) What is the name of the special type of cell division which produces gametes from ordinary body cells? Describe what happens to the chromosomes in this process.
   b) Where in your body would this type of cell division take place?
   c) Why is this type of cell division so important in sexual reproduction? [Higher]

KEY POINTS

1. Cells in the reproductive organs divide to form the gametes (sex cells).
2. Body cells have two sets of chromosomes; gametes have only one set.
3. Gametes are formed from body cells by meiosis. [Higher]
4. Sexual reproduction gives rise to variety because genetic information from two parents is combined.

207

SUMMARY ANSWERS

1. **a)** 23 pairs **b)** 23 **c)** 46
2. As the gametes are formed, each gamete has a different combination of chromosomes and there is some exchange of genes. This introduces variation, as each gamete is different. In sexual reproduction, two unique gametes from two different people join together, so the combination of chromosomes and the mix of alleles on the chromosomes will be unique.
3. **a)** Meiosis. After the chromosomes are copied, the cell divides twice quickly resulting in sex cells each with half the number of chromosomes.
   b) In the reproductive organs/in the ovary or the testes.
   c) Sexual reproduction involves the joining of gametes from mother and father. The chromosome number of the body cells needs to be halved to make the gametes, otherwise the number of chromosomes in the cell would just get bigger and bigger when gametes joined at fertilisation. Meiosis halves the chromosome number. [**HT** only]

Answers to in-text questions

a) Sperm, ova, half the number of chromosomes.

b) Meiosis.

KEY POINTS

The differences between mitosis and meiosis are important and it is also important that the students spell the words correctly. They could test each other. Many of the suggested activities aim to show students how to remember the differences.

B2 6.4 From Mendel to DNA

LEARNING OBJECTIVES

Students should learn:

- About the work of Mendel and why its importance was not recognised until after his death.
- Why DNA fingerprinting is possible.
- How specific proteins are made.

LEARNING OUTCOMES

Most students should be able to:

- Describe Mendel's discoveries.
- Recognise why Mendel's ideas were not accepted in his time.
- Describe how DNA fingerprinting is used to identify individuals.

Some students should also be able to:

- Explain simply the structure of DNA.
- Explain how a gene codes for a specific protein. [**HT** only]

Teaching suggestions

- **Special needs.** Using different coloured (yellow and green) dried peas, glue and a large sheet of paper, ask the students to make a large poster showing Mendel's experiment as depicted in the Student Book.
- **Gifted and talented.** Ask students to draw up a plan for one of Mendel's experiments and calculate how long it took him to get his results. What precautions would he have to take? Did he use controls? Apply some of the criteria needed for 'How Science Works' to his experiments. Would you do it the same way as he did it? What different techniques might you use?
- **Learning styles**
  Kinaesthetic: Growing plants and counting seedlings.
  Visual: Video or PowerPoint® on Mendel.
  Auditory: Discussion of Mendel's work.
  Interpersonal: Working with partners in practical session.
  Intrapersonal: Writing a report or a letter to Mendel.
- **Homework.** Follow up the plenary interview with Mendel by writing a contemporary newspaper report about Mendel's work, along the lines of 'Mad monk sits and counts peas . . .'.
  Or Write a letter to Mendel explaining about chromosomes and how his work provided the foundation for modern genetics.

SPECIFICATION LINK-UP Unit: Biology 2.12.8

- *Some characteristics are controlled by a single gene. Each gene may have different forms called alleles.*
- *An allele which controls the development of a characteristic when it is present on only one of the chromosomes is a dominant allele.*
- *An allele which controls the development of a characteristic only if the dominant allele is not present is a recessive allele.*
- *Chromosomes are made up of large molecules of DNA (deoxyribose nucleic acid). A gene is a small section of DNA.*
- *Each gene codes for a particular combination of amino acids which make a specific protein. [**HT** only]*
- *Each person (apart from identical twins) has unique DNA. This can be used to identify individuals in a process known as DNA fingerprinting.*

Students should use their skills, knowledge and understanding of 'How Science Works':

- *to explain why Mendel proposed the idea of separately inherited factors; and why the importance of this discovery was not recognised until after his death.*

Lesson structure

STARTER

How did Mendel start? – Give the students a collection of dried peas to sort out into groups. Include smooth and wrinkled skins, yellow and green if possible. Ask them to predict what would happen if the peas were planted. Would you get peas identical to the ones you planted? Discuss. (10 minutes)

A model of DNA – If possible have a model of DNA showing its structure with the different bases, the deoxyribose and the phosphate groups. Get the students to identify the component parts and discuss the coding on a simple level. (10 minutes)

MAIN

- **The life and work of Mendel** – Video or PowerPoint® presentation on Mendel's life and work. There is plenty of information available and scope for introducing students to the demands of research (think about all those plants he grew and seeds he counted). Consider the characteristics that he investigated, introduce some of the easier terms, such as 'pure-breeding' and some of the simple ratios. Discuss his technique. (This relates to 'How Science Works': acceptance of new ideas.) (See www.mendelweb.org. A web image bank will yield many illustrations on a search for 'Gregor Mendel'.)
- **Grow your own genetics experiment** – It is possible to obtain sets of seeds from monohybrid genetic crosses from suppliers. When the seeds grow, it is possible to observe differences between the seedlings and make predictions about the genetic constitution of the parent plants. Tobacco (colour of cotyledons, hairiness of stem, colour of stem and leaf shape), tomato (leaf shape, hairiness of stem, colour of stem) and cucumber (bitterness of leaves) are all suitable for class use. The seeds are sown in seed trays, kept in light, airy conditions and watered every two or three days. They will be ready for scoring the characteristics after about 15 to 20 days. These seeds usually come provided with instructions and an explanation of the parental cross which produced them.
- **Genetic fingerprints** – Prepare a PowerPoint® presentation on genetic fingerprinting. The technique can be fairly simply explained (see NCBE publications or web site for details) and then the implications discussed. Some examples of different uses can be given e.g. forensic evidence, paternity issues. There are some good images available on the Internet.

PLENARIES

Press conference – Select a student who is prepared to be Mendel. Other students are to interview him about his work and why he did not get recognition at the time. The student can choose other members of the class to represent workers who followed up his discoveries. (10–15 minutes)

Anagram quiz – Give the students anagrams of the key words in this topic and a small prize for the student who solves the puzzles in the fastest time. (5–10 minutes)

Be a detective! – Present the students with some genetic evidence (some DNA from a murder weapon and three sets of genetic fingerprints). Let them work out who did the crime. (5–10 minutes)

ACTIVITY & EXTENSION IDEAS

- **Did Mendel fiddle his results?** – There are several ways in which you could consider Mendel to be lucky. His choice of plants to work on, the characteristics he chose, the numbers he obtained – all these worked out well for him. Give students some of his results and let them work out the ratios. If there are any budding mathematicians in the group, ask if they can work out the probability of getting such good results. There are suggestions that he knew what he wanted to prove before he set up his experiments. Ask: 'What do you think?'
- **How to get enough DNA for a fingerprint** – Sometimes the quantity of DNA left at a crime scene is very small, but using PCR (the polymerase chain reaction) this can be increased. Find out how this works. Use a search engine and key in the words or go to the NCBE web site for more information.

BIOLOGY INHERITANCE

B2 6.4 From Mendel to DNA

LEARNING OBJECTIVES

1. What did Mendel's experiments teach us about inheritance?
2. What are DNA fingerprints?
3. How are specific proteins made in the body? [Higher]

GET IT RIGHT!

Mendel knew nothing of chromosomes and genes. Make sure you don't confuse modern knowledge with what Mendel knew when he did his experiments.

For hundreds of years people had no idea about how information moved from one generation to the next. Yet now we can identify people by the genetic information in their cells!

Mendel's discoveries

Gregor Mendel was born in 1822 in Brunn, Czechoslovakia. Clever but poor, he became a monk to get an education.

He worked in the monastery gardens and became fascinated by the peas growing there. He decided to carry out some breeding experiments, using pure strains of round peas, wrinkled peas, green peas and yellow peas for his work. Mendel cross-bred the peas and counted the different offspring carefully. He found that characteristics were inherited in clear and predictable patterns.

Mendel explained his results by suggesting there were separate units of inherited material. He realised some characteristics were dominant over others and that they never mixed together. This was an amazing idea for the time.

a) Why did Gregor Mendel become a monk?

Mendel kept records of everything he did, and analysed his results. This was almost unheard of in those days! Finally in 1866, when he was 44 years old, Mendel published his findings.

He never saw chromosomes and never heard of genes. Yet he explained some of the basic laws of genetics in a way we still use today.

Sadly Mendel's genius was ahead of his time. As no-one knew about genes or chromosomes, people simply didn't understand his theories. He died twenty years later with his ideas still ignored – but convinced that he was right!

b) What was unusual about Mendel's scientific technique at the time?

Sixteen years after his death, Gregor Mendel's work was finally recognised. By 1900, people had seen chromosomes through a microscope. Three scientists, discovered Mendel's papers and repeated his experiments. When they published their results, they gave Mendel the credit for what they observed! From then on ideas about genetics developed fast. It was suggested that Mendel's units of inheritance might be carried on the chromosomes seen beneath the microscope. And so the science of genetics as we know it today was born.

DNA – the molecule of inheritance

The work of Gregor Mendel was just the start of our understanding of inheritance. Today, we know that our features are inherited on genes carried on our chromosomes. We also know what those chromosomes are made of.

Your chromosomes are made up of long molecules of a chemical known as DNA (**d**eoxyribose **n**ucleic **a**cid). Your genes are small sections of this DNA. The DNA carries the instructions to make the proteins which form most of your cell structures. These proteins also include the enzymes which control your cell chemistry.

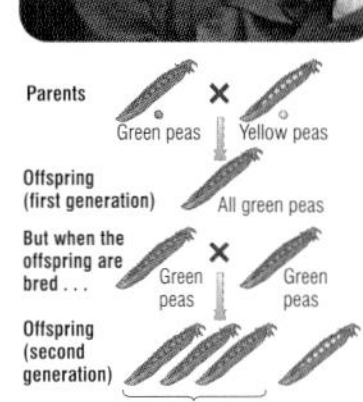

Figure 1 Gregor Mendel, the father of modern genetics. When he died in 1884 he was still hoping that eventually other people would acknowledge his discoveries. In the 21st century, we know just how right he was!

Additional biology

HIGHER

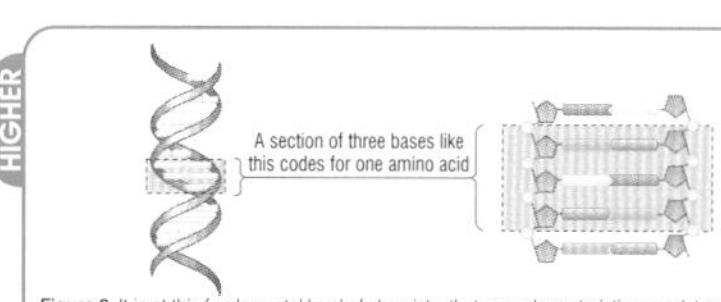

Figure 2 It is at this fundamental level of chemistry that your characteristics are determined. A small quirk of chemistry would have resulted in a very different you – a very strange thought.

The long strands of your DNA are made up of combinations of four different chemical bases. (See Figure 2.) These are grouped into threes and each group of three codes form an amino acid.

Each gene is made up of hundreds or thousands of these bases. The order of the bases controls the order in which the amino acids are put together so that they make a particular protein for use in your body cells. Each gene codes for a particular combination of amino acids which make a specific protein.

A change or mutation in a single group of bases can be enough to change or disrupt the whole protein structure and the way it works.

DID YOU KNOW?

The first time DNA fingerprinting was used to solve a crime it not only identified Colin Pitchfork as the murderer of two teenage girls, it also cleared another innocent man of the same crimes!

DNA fingerprinting

Unless you have an identical twin, your DNA is unique to you. Other members of your family will have strong similarities in their DNA, but each individual has their own unique blueprint. Only identical twins have the same DNA. That's because they have both developed from the same original cell.

The unique patterns in your DNA can be used to identify you. A technique known as 'DNA fingerprinting' can be applied.

Certain areas of your DNA produce very variable patterns under the microscope. These patterns are more similar between people who are related than between total strangers. The patterns are known as ***DNA fingerprints***. They can be produced from very tiny samples of DNA from body fluids such as blood, saliva and semen.

The likelihood of two identical samples coming from different people (apart from identical twins) is millions to one. As a result DNA fingerprinting is enormously useful in solving crimes. It is also used to show who is the biological father of a child when there is doubt.

Figure 3 DNA fingerprints like these can be used to identify the guilty – and the innocent – in a crime investigation

SUMMARY QUESTIONS

1. a) How did Mendel's experiments with peas convince him that there were distinct 'units of inheritance' which were not blended together in offspring?
   b) Why didn't people accept his ideas?
   c) The development of the microscope played an important part in helping to convince people that Mendel was right. How?
2. Two men claim to be the father of the same child. Explain how DNA fingerprinting could be used to find out which one is the real father.
3. Explain the saying 'One gene, one protein'. [Higher]

KEY POINTS

1. Gregor Mendel was the first person to suggest separately inherited factors which we now call genes.
2. Chromosomes are made up of large molecules of DNA.
3. A gene is a small section of DNA which codes for a particular combination of amino acids which make a specific protein. [Higher]
4. Everyone (except identical twins) has unique DNA which can be used to identify them using DNA fingerprinting.

SUMMARY ANSWERS

1. a) He found that characteristics were inherited in clear and predictable patterns. He realised some characteristics were dominant over others and that they never mixed together.
   b) No-one could see the units of inheritance so there was no proof of their existence. People were not used to studying careful records of results.
   c) Once people could see chromosomes, a mechanism for Mendel's ideas of inheritance became possible.
2. The DNA fingerprint of the real father would have similarities to the DNA fingerprint of the child whereas that of the other man would not.
3. A gene is made up of groups of three base pairs. Each group of three base pairs codes for a single amino acid. The order of the base pairs in the gene determines the sequence of the amino acids which are joined together to make a protein – so each gene codes for a unique protein.

Answers to in-text questions

a) Mendel became a monk because he was clever but poor and the only way to get an education if you were poor was to join the Church.

b) He kept records and analysed his results.

KEY POINTS

These key points are straightforward and can be reinforced by being made into revision cards. The higher ability students would benefit from adding their own annotations to expand the basic concepts.

B2 6.5 Inheritance in action

LEARNING OBJECTIVES

Students should learn that:

- Characteristics are controlled by genes which have different forms called alleles.
- Dominant alleles control the development of a characteristic when they are present on only one of the chromosomes, but recessive alleles must be present on both chromosomes for the characteristic to develop.
- In humans, the sex chromosomes determine whether you are female (XX) or male (XY).

LEARNING OUTCOMES

Most students should be able to:

- Explain how the inheritance of characteristics is controlled by dominant and recessive alleles.
- Explain how sex is determined in humans.

Some students should also be able to:

- Predict and/or explain the outcomes of crosses between individuals for each possible combination of dominant and recessive alleles of the same gene. [**HT** only]
- Construct genetic diagrams. [**HT** only]

Teaching suggestions

- **Special needs.** Play a card game using dominant and recessive cards for lobed ears, dimples and tongue rolling. Some students might be able to cope with the sex determination game in its simplest form.
- **Gifted and talented.**
  - If the school has the facilities, students could try scoring the *Drosophila* crosses, particularly if the flies have been used for sixth-form classes.
  - Alternatively, they could research haemophilia and trace it through the offspring of Queen Victoria. They could produce a PowerPoint® report to be shown to the rest of the group.

SPECIFICATION LINK-UP Unit: Biology 2.12.8

- *An allele that controls the development of a characteristic when it is present on only one of the chromosomes is a dominant allele.*
- *An allele that controls the development of characteristics only if the dominant allele is not present is a recessive allele.*
- *In human body cells, one of the 23 pairs of chromosomes carries the genes which determine sex. In females, the sex chromosomes are the same (XX), in males the sex chromosomes are different (XY).*

Lesson structure

STARTER

Get the words right – Put up at the front of the room word cards with the important terms from the spread on them (e.g. allele, chromosome, dominant, recessive, Mendel, inheritance etc.). Ask the students to write down sentences containing any one of these words. Select from responses, noting key ideas. (10 minutes)

Can you? – Ask some of these: 'Can you roll your tongue? Can you taste quinine? Do you have dimples? Do you have dangly ear lobes? Do you have straight thumbs or bendy thumbs?' Discuss some of these characteristics. (The invitation to poke one's tongue out at the teacher usually goes down well!) (5–10 minutes)

MAIN

- **Grow your own genetics experiment** – If students did not carry out this exercise when studying the previous spread, it could be done here as an illustration of inheritance in action.
- **The Human Genome Project** – Inherited conditions in humans are due to mutations of the DNA. The Human Genome Project has mapped all the human chromosomes. Prepare a PowerPoint® presentation on this project, including references to why it was done, how it was done and how long it took. Give the students prepared work sheets and allow time for them to make their own notes. Discuss the implications. Information is available on The Wellcome Trust web site and the Human Genome Project Information web site.
- **Sex determination game** – Prepare sets of sperm cards with either an X or a Y on the back and egg cards, all with X on the back. Working in pairs, the students are to turn one sperm card and one egg card over at a time. In a table, they note the sperm chromosome, the egg chromosome, the combination, the gender and give the baby a name. Run for about 5–7 minutes and then see who has the biggest family. Ask: 'Are there more boys than girls? What does this tell us about the ratio of the sexes?'

PLENARIES

Punnett square – Draw a Punnett square on the board with a number of different dominant and recessive alleles. Use some of the examples mentioned in the 'Can you?' starter. Get the students to fill in pre-printed frames and to describe the crosses. (10–15 minutes)

Human karyotypes – Show students some human karyotypes where the chromosomes have not been matched into pairs. Can they identify the sex chromosomes and decide where the karotype is from a male of a female? Compare with karyotypes where the chromosomes are in pairs. When is this type of information helpful? (5–10 minutes)

ACTIVITY & EXTENSION IDEAS

- **Modify the sex determination game** – Using a symbol to represent a characteristic, such as tongue rolling, which can be stuck on the cards, the inheritance of a human characteristic (not a sex-linked one) can be investigated at the same time. Just add another column to the table. It would be possible to investigate sex-linkage using this game, but it is beyond the specification.

Teaching suggestions – continued

- **Learning styles**

  Kinaesthetic: Growing plants and counting seedlings.

  Visual: Viewing PowerPoint® presentation on the Human Genome Project.

  Auditory: Discussing the Human Genome Project.

  Interpersonal: Working with partners in sex determination game.

  Intrapersonal: Completing own work sheet.

- **ICT link-up.** There are some excellent web sites with genetics games that can be played on-line, e.g. the Canadian Museum of Nature web site (nature.ca/genome).

B2 6.5 Inheritance in action

LEARNING OBJECTIVES

1. How is sex determined in humans?
2. Can you predict what features a child might inherit? [Higher]

Ideas about genetics, chromosomes and genes are everywhere in the 21st century. We read about them in the papers, see them on TV and learn about them in science lessons. The way features are passed from one generation to another follow some clear patterns. We can use these to predict what may be passed on.

How inheritance works

Scientiest have built on the work of Gregor Mendel. We now understand how genetic information is passed from parent to offspring.

Human beings have 23 pairs of chromosomes. In 22 cases, each chromosome in the pair is a similar shape and has genes carrying information about the same things. But one pair of chromosomes may be different – these are the **sex chromosomes**. Two X chromosomes mean you are female. However, one X chromosome and a much smaller one, known as the Y chromosome, give a male.

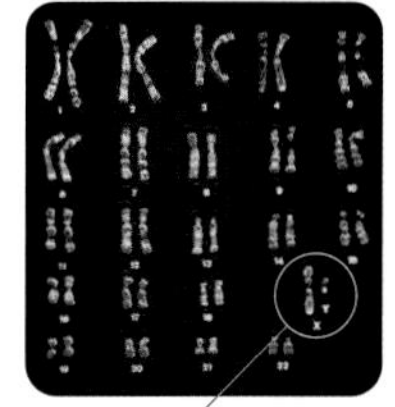

Figure 1 The chromosomes of the human male. The X chromosome carries genes controlling lots of different features. The Y chromosome is much smaller than the X chromosome and carries information mainly about maleness!

a) Twins are born. Twin A is XY and twin B is XX. What sex are the two babies?

The chromosomes we inherit carry our genetic information in the form of genes. Many of these genes have different forms, known as alleles. (See page 207.) A gene can be pictured as a position on a chromosome. An allele is the particular form of information in that position on an individual chromosome. For example, the gene for dimples may have the dimple or the no-dimple allele in place.

HIGHER

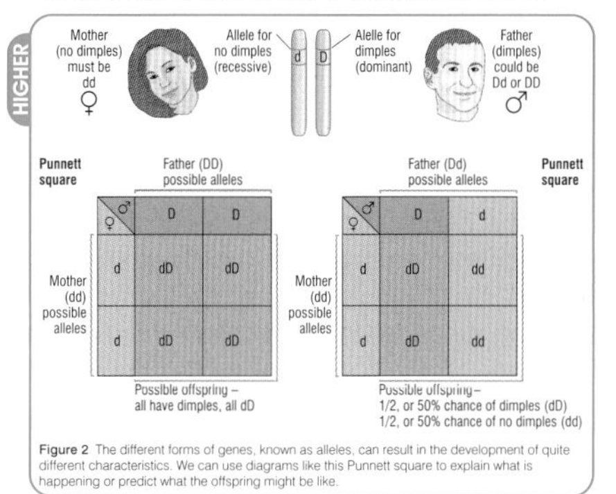

Figure 2 The different forms of genes, known as alleles, can result in the development of quite different characteristics. We can use diagrams like this Punnett square to explain what is happening or predict what the offspring might be like.

Most of your characteristics, like your eye colour and nose shape, are controlled by a number of genes. However, some characteristics, like dimples or having attached earlobes, are controlled by a single gene. Often there are only two possible alleles for a particular feature. However, sometimes you can inherit one from a number of different possibilities.

Some alleles control the development of a characteristic even when they are only present on one of your chromosomes. These alleles are **dominant**, e.g. dimples and dangly earlobes.

Some alleles only control the development of a characteristic if they are present on both alleles – in other words, no dominant allele is present. These alleles are **recessive**, e.g. no dimples and attached earlobes.

NEXT TIME YOU...

... look at a group of people, just think of the millions of genes – and the different alleles of those genes – which are controlling what they look like. Once you start counting dimples and dangly earlobes, it's difficult to stop

HIGHER

How does inheritance work?

We can use a simple model to help us understand how inheritance works. It explains how different features are passed on from one generation to another.

Imagine a bag containing marbles. If you put your hand in and – without looking – picked out two marbles at a time, what pairs might you get? If the bag contained only red marbles or only blue marbles, the pairs would all be the same. But if the bag held a mixture of red and blue marbles you could end up with three possible pairs – two blue marbles, two red marbles or one of each.

This is what happens when you inherit genes from your parents, depending on the different alleles they have. For example, if both of your parents have two alleles for dimples (like the red marbles) you will definitely inherit two dimple alleles – and you will have dimples! If both of your parents have two alleles for no dimples, you will inherit alleles for no dimples and you will be dimple free.

But if your parents both have one allele for dimples and one for no dimples, you could end up with two dimple alleles, two no dimple alleles – or one of each.

DID YOU KNOW?

Sex determination varies from species to species. In birds, females are XY and the males are XX. In some species of reptiles the males are XY and the females XX. In others the males are XX and the females XY. Some reptiles including alligators and tortoises don't have any sex chromosomes at all. The sex of the babies is decided by the temperature of the eggs as they incubate. And there are some species of fish and snails which change sex at different stages of their lives!

SUMMARY QUESTIONS

1 Copy and complete:

male   sex chromosomes   23   22   X   XX   Y

Human beings have ...... pairs of chromosomes. In ...... pairs the chromosomes are always the same. The final pair are known as .... If you inherit ...... you will be female, while an ...... and a ...... make you ....

2 a) What is meant by the term 'dominant allele'?
b) What is meant by the term 'recessive allele'?
c) Try and discover as many human characteristics as you can which are inherited on a single gene. Which alleles are dominant and which are recessive?

3 Use a Punnett square like the one in Figure 2 to show the possible offspring from a cross between two people who both have dimples and the genotype Dd. [Higher]

KEY POINTS

1. In human body cells the sex chromosomes determine whether you are female (XX) or male (XY).
2. Some features are controlled by a single gene.
3. Genes can have different forms called alleles.
4. Some alleles are dominant and some are recessive.
5. We can construct genetic diagrams to predict features. [Higher]

SUMMARY ANSWERS

1 23, 22, sex chromosomes, XX, X, Y, male.

2 a) Dominant allele – an allele which controls the development of a characteristic even when it is present on only one of the chromosomes.

b) Recessive allele – an allele which only controls the development of a characteristic if it is present on both chromosomes.

c) 2 marks for each case where students identify correctly the single gene characteristic and the dominant and recessive alleles.

3 [Marks awarded for drawing a Punnett square correctly with the appropriate gametes.] DD, Dd, dD, dd; 3 with dimples: 1 with no dimples. [**HT** only]

Answer to in-text question

a) A is male and B is female.

KEY POINTS

It is important to use the correct terms and not to muddle genes and alleles. Plenty of practice in working out crosses will help students become familiar with these terms and with the idea of characteristics having dominant and recessive alleles.

B2 6.6 Inherited conditions in humans

LEARNING OBJECTIVES

Students should learn that:

- Some human disorders are inherited.
- Some disorders are the result of the inheritance of a dominant allele (Huntington's disease), but others are the result of the inheritance of two recessive alleles (cystic fibrosis).
- Embryos can be screened for genetic disorders.

LEARNING OUTCOMES

Most students should be able to:

- State that some human disorders may be inherited.
- Describe how genetic disorders caused by a dominant allele are inherited.
- Explain how a genetic disorder caused by a recessive allele must be inherited from both parents.
- List some issues concerning embryo screening.

Some students should also be able to:

- Draw genetic diagrams to show how genetic disorders are passed on. [**HT** only]
- Make informed judgements about the economic, social and ethical issues concerning embryo screening that they have studied or from information that is presented to them.

Teaching suggestions

- **Special needs.** Students could work out genetic crosses using large printed grids and cards with the alleles on. They could work out the ratios and show them underneath.
- **Gifted and talented.** Students could do some research on the frequency of genes in populations. We are told that 1 person in 25 carries the allele for cystic fibrosis. Ask: 'How has this been calculated?' They could find out about the Hardy-Weinberg law and how it works. The law itself is fairly straightforward, students could work out how they can use it to inform people that the incidence of the alleles for certain conditions is quite high.

SPECIFICATION LINK-UP Unit: Biology 2.12.8

- *Some disorders are inherited:*
  - *Huntington's disease – a disorder of the nervous system – is caused by a dominant allele of a gene and can therefore be passed on by only one parent who has the disorder.*
  - *Cystic fibrosis – a disorder of cell membranes – must be inherited from both parents. The parents may be carriers of the disorder without actually having the disorder themselves. It is caused by a recessive allele of a gene and can therefore be passed on by parents, neither of whom has the disorder.*

  (Attention is drawn to the potential sensitivity needed in teaching about inherited disorders.)
- *Embryos can be screened for the alleles that cause these and other genetic disorders.*

Lesson structure

STARTER

Infectious or genetic or . . .? – Read students a list of diseases, including some infectious and some genetic. Students to respond by writing on 'Show me' boards whether a disease is infectious (writing I) or genetic (writing G). If they do not know then they should write a question mark. Draw up a list on the board in two columns. (5–10 minutes)

Interpreting a pedigree diagram – Introduce pedigree diagrams with their conventions: circles for females, squares for males, ways in which affected and carriers are shown. Illustrate with an invented 'condition' and get students to work out some of the offspring. (5–10 minutes)

MAIN

Huntington's disease

- Find and show images on Huntington's disease. The basic facts, symptoms and inheritability can be presented, together with some of the diagnostic tests. (See www.hdfoundation.org.)
- This could lead to a discussion on the disease and whether or not genetic testing is a good idea. It is a distressing condition, so it is advisable to find out whether anyone is affected before the presentation.

Cystic fibrosis

- Useful web sites can be found, including 'The Cystic Fibrosis Trust', the 'Cystic Fibrosis Foundation' (see www.cftrust.org.uk and www.cff.org), and the students could be asked to research different aspects of the disease in groups and put together a lesson on the condition.
- One group could describe the disease and its symptoms, another the genetics of how it is inherited and a further group could review the different treatments. Ask: 'According to the statistics, 1 in 25 people carries the allele, so is it worth being screened for it?'

Inherited or not?

It is difficult to know whether a particular disease or condition is inherited or not. The only way to find out is to carry out pedigree analysis and go back through the generations if possible. Suggest to students that they think of a particular family trait and see if they can draw up a pedigree within their own family. It is probably better to choose a characteristic, such as dangly ear lobes or dimples, rather than a disease unless a student has a particular interest.

PLENARIES

Play the inheritance game – The sex determination game, from the previous spread, could be modified by adding a dominant or recessive genetic disease sticker to some of the cards and to see what happens to the offspring. Allow 5 minutes for the game and then add up how many are affected offspring, how many are carriers and how many are unaffected by the disease. (10–15 minutes)

Statistics or chance? – Much emphasis is put on the ratios of incidence of the condition, but it does not necessarily follow that it works like that. Ask: 'Why are there some families where there are no boys or no girls?' Every child of a sufferer from Huntington's disease could inherit the disease. Students to try tossing a coin to see if they get equal numbers of heads and tails or if they get a run of heads. Discuss the implications if it was your family. (10 minutes)

ACTIVITY & EXTENSION

- **Other inherited conditions** – There are an estimated 6000 different conditions due to the inheritance of a single gene in humans. Students could research some of the suggested web sites (that of the Wellcome Trust is particularly useful – www.wellcome.ac.uk) for other conditions. Some suggestions could include certain forms of diabetes, phenylketonuria and sickle cell anaemia.
- **Sex-linked genetic diseases** – The best-known sex-linked genetic diseases are haemophilia and colour blindness. Discuss the inheritance of conditions with genes located on the X chromosome. Draw up Punnett squares to show how the alleles are inherited and the probability of the disease occurring. It was suggested on the last spread that gifted and talented students did some research on haemophilia. This could be a good opportunity for them to present their findings.

GET IT RIGHT!

It is sensible for students to be able to use Punnett diagrams: it makes the interpretation of any genetic cross much easier. [Higher Tier only] It could be a good idea to stress to students that they are dealing with ratios and probabilities. What happens in real life is not always the same!

BIOLOGY INHERITANCE

B2 6.6 Inherited conditions in humans

LEARNING OBJECTIVES

1. How are human genetic disorders inherited?
2. Can you predict if a child will inherit a genetic disorder? [Higher]

Not all diseases are infectious. Sometimes diseases are the result of a problem in your genes and can be passed on from parent to child. They are known as **genetic diseases** or **genetic disorders**.

We can use our knowledge of dominant and recessive alleles to work out the risk of inheriting a genetic disease.

a) How is a genetic disease different from an infectious disease?

DID YOU KNOW?

Around 7500 babies, children and young adults are affected by cystic fibrosis in the UK alone. Between 6800 and 8000 people are affected by Huntington's disease.

Huntington's disease

One example of a very serious, although very rare, genetic disorder is Huntington's disease. This is a disorder of the nervous system. It is caused by a dominant allele and so it can be inherited from one parent who has the disease. If one of your parents is affected by Huntington's you have a 50% chance of inheriting the disease. That's because half of their gametes will contain the faulty allele.

The symptoms of this inherited disease usually appear when you are between 30 and 50 years old. Sadly, the condition is fatal. Because the disease does not appear until middle-age, many people have already had children and passed on the faulty allele before they realise they are affected.

b) You may inherit Huntington's disease even if only one of your parents is affected. Why?

Figure 1 Modern medicine and determination mean that many sufferers from cystic fibrosis manage to lead full and active lives. However, the cells in their bodies are still carrying the faulty alleles and cannot function properly.

Cystic fibrosis

Another genetic disease which has been studied in great detail is **cystic fibrosis**. This is a disorder which affects many organs of the body, particularly the lungs and the pancreas.

The organs become clogged up by a very thick sticky mucus which stops them working properly. The reproductive system is affected so most people with cystic fibrosis are infertile.

Treatment for cystic fibrosis includes physiotherapy and antibiotics to help keep the lungs clear of mucus and infections. Enzymes are used to replace the ones the pancreas cannot produce and to thin the mucus.

However, although treatments are getting better all the time, there is still no cure.

Cystic fibrosis is caused by a recessive allele so it must be inherited from both parents. Children affected by cystic fibrosis are born to parents who do not suffer from the disease. They have a dominant healthy allele which means their bodies work normally but they carry the cystic fibrosis allele. Because it gives them no symptoms, they have no idea it is there.

People who have a silent disease-causing allele like this are known as **carriers**. In the UK, one person in 25 carries the cystic fibrosis allele. Most of them will never be aware of it, unless they happen to have children with a partner who also carries the allele. Then there is a 25% (one in four) chance that any child they have will be affected.

c) You will only inherit cystic fibrosis if you get the allele from both parents. Why?

DID YOU KNOW?

Some genetic disorders cause such chaos in the cells that the fetus does not develop properly and miscarries. Yet other genetic disorders – such as colour-blindness – have little effect on people's lives!

212

Additional biology

The genetic lottery

When the genes from parents are combined, it is called a genetic cross. We can show this using a genetic diagram (see Figures 2 and 3). A genetic diagram shows us:

- the alleles for a characteristic carried by the parents,
- the possible gametes which can be formed from these, and
- how these could combine to form the characteristic in their offspring.

When looking at the possibility of inheriting genetic diseases, it is important to remember that every time an egg and a sperm meet it is down to chance which alleles combine. So if two parents who both carry the cystic fibrosis allele have four children, there is a 25% chance (one in four) that each child might have the disease.

But in fact all four children could have cystic fibrosis, or none of them might be affected. They might all be carriers, or none of them might inherit the faulty alleles at all.
It's all down to chance!

Parent with Huntington's disease Hh
Normal parent hh

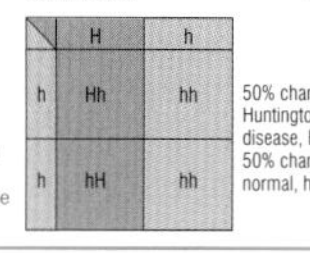

	H	h
h	Hh	hh
h	hH	hh

50% chance Huntington's disease, Hh or hH
50% chance normal, hh

Figure 2 A genetic diagram for Huntington's disease shows us how a dominant allele can affect offspring. It is important to realise that this shows that the chance of passing on the disease allele is 50%, but it cannot tell us which, if any, of the children will actually inherit the allele.

HIGHER

Both parents are carriers, so Cc

	C	c
C	CC	Cc
c	cC	cc

25% normal (CC)
50% carriers (Cc)
25% affected by cystic fibrosis (cc)

3/4, or 75% chance normal
1/4, or 25% chance cystic fibrosis

Figure 3 The arrival of a child with cystic fibrosis in a family often comes as a complete shock. The faulty alleles can be covered up by normal alleles for generations until two carriers have a child and by chance both of the cystic fibrosis alleles are passed on.

Curing genetic diseases

So far we have no way of curing genetic diseases. Scientists hope that genetic engineering will enable them to cut out faulty alleles and replace them with healthy ones. They have tried this in people affected by cystic fibrosis. But so far they have not managed to cure anyone.

There are genetic tests which can show people in affected families if they carry the faulty allele. This allows them to make choices such as whether to have a family. It is also possible to screen embryos for the alleles which cause these and other genetic disorders. These tests are very useful but raise many ethical issues. (See page 215.)

SUMMARY QUESTIONS

1. a) What is Huntington's disease?
   b) Why can one parent carrying the allele for Huntington's disease pass it on to their children even though the other parent is not affected?
2. At the moment, only people who have genetic diseases in their family are given genetic screening. What would be the pros and cons of screening everyone for diseases like cystic fibrosis and Huntington's disease?
3. a) Why are carriers of cystic fibrosis not affected by the disease themselves?
   b) A couple have a baby who has cystic fibrosis. Neither of the couple, nor their parents, have any signs of the disease.
   Draw genetic diagrams of the grandparents and the parents to show how this could happen. [Higher]

GET IT RIGHT!

If one parent has a characteristic caused by a single dominant allele (e.g. Huntington's disease, dangly earlobes) you have a 50% chance of inheriting it.
If one parent has two dominant alleles (e.g. for Huntington's disease, dangly earlobes) you have 100% chance of inheriting it.
If both parents have a recessive allele for a characteristic (e.g. cystic fibrosis, attached earlobes) you have a 25% chance of inheriting that characteristic.

KEY POINTS

1. Some disorders are inherited.
2. Huntington's disease is caused by a dominant allele of a gene and can be inherited from only one parent.
3. Cystic fibrosis is caused by a recessive allele of a gene and so must be inherited from both parents.

213

SUMMARY ANSWERS

1. **a)** A genetic disease; a disorder of the nervous system which is fatal.
   b) The faulty allele is dominant so the offspring will have the disease with only one allele.
2. [Marks to be awarded for any sensible relevant points.] Can plan whether or not to have children; can choose partner without damaged allele; may worry people unnecessarily; have embryo screening if there is a risk; might not be able to get insurance if it is known that you have a genetic disease.
3. **a)** Carriers have a normal dominant allele, so their body works normally.
   b) Genetic diagrams based on Figure 3, page 213 in Student Book, showing how the cc arises. [**HT** only]

Teaching suggestions – continued

- **Learning styles**
  Kinaesthetic: Playing the inheritance game.
  Visual: Viewing a presentation on Huntington's disease.
  Auditory: Listening to the reports of others on their investigations.
  Interpersonal: Discussing the research on cystic fibrosis.
  Intrapersonal: Working out an individual family pedigree and researching the Human Genome Project.
- **Homework.** There is plenty of scope for homework exercises associated with this topic.
- **ICT link-up.** A great deal of research can be carried out by students, either using libraries or the Internet.

Answers to in-text questions

a) A genetic disease – inherited from parents.
Infectious disease – caught from other people.
b) Huntington's disease is caused by a dominant allele.
c) Cystic fibrosis is caused by a recessive allele.

KEY POINTS

The key points for this spread are covered in the summary questions.

B2 6.7 Stem cells and embryos – an ethical minefield

SPECIFICATION LINK-UP

Unit: Biology 2.12.8

Students should use their skills, knowledge and understanding of 'How Science Works':

- *to make informed judgements about the social and ethical issues concerning the use of stem cells from embryos in medical research and treatments.*
- *to make informed judgements about the economic, social and ethical issues concerning embryo screening that they have studied or from information that is presented to them.*

BIOLOGY INHERITANCE

B2 6.7 Stem cells and embryos – an ethical minefield

The stem cell dilemma

Doctors have treated people with adult stem cells for many years by giving bone marrow transplants. Now scientists are moving ever closer to treating very ill people using embryonic stem cells. This area of medicine raises many issues. Here are just a few different opinions:

I think it is absolutely wrong to use human embryos in this way. Each life is precious to God. They may only be tiny balls of cells – but they could become people.

The accident happened so quickly. Now I'm stuck in this wheelchair for the rest of my life. I can't walk or even control when I go to the loo. It would be wonderful if they could develop cell stem therapy. I want them to heal my spinal cord so I can walk again!

It may become possible to take stem cells from the umbilical cord of every newborn baby. They could be frozen and stored ready for when the person might need them later in their life.
The embryos we use would all be destroyed anyway. Now we are even making our own embryos from adult cells. We could do so much good for people that we all feel it is very important for the research to continue.

It was terrible to see my husband suffer. By the time he died he didn't know who I was or any of the children. If these stem cells can cure Alzheimer's disease then we should do the research as fast as possible.
We need to be careful. There are some real problems with these stem cell treatments. We don't want to solve one problem and cause another.
I am going to volunteer to let them use some of my cells for therapeutic cloning. It is too late to help me now, but I'd like to think I could help other people.

ACTIVITY

Here is an opportunity to make your voice heard. Your class is going to produce a large wall display covered with articles both for and against stem cell research. Your display is aimed at students in Years 10 and 11, so make sure the level of content is right for your target group.
Try and carry out a survey or vote of your target group before the display is put up. Find out:

- how many people support stem cell research,
- what proportion are completely against it, and
- how many haven't made up their minds.

Record your findings

Work on your own or in a small group. Each group is to produce one piece of display material. Make sure that some of you give information in favour of stem cell research and others against. Use a variety of resources to help you – the material in this chapter is a good starting point. Make sure that your ideas are backed up with as much scientific evidence as possible.
Once the material has been displayed for a week or two, repeat your initial survey or vote. Analyse the data to see if easy access to information has changed people's views!

214

Teaching suggestions

- **The stem cell dilemma** – Several web sites have already been suggested as good sources of information about stem cell research, and it was suggested on a previous spread that students considered the pros and cons of the research programme and the use of stem cells. All this material should be useful when carrying out the suggested activity.
- **Can we know too much?** – Students could discuss the pros and cons of genetic screening using the material given in the Student Book and from any other information they have. This discussion could result in the need for more information so the web sites given on the previous spread could be investigated.
- If anyone knows a genetic counsellor, it might be a good opportunity to invite them in to the school to give a short talk and answer questions about the topic.

Special needs

Students can be asked simple questions, such as 'If you or someone you loved was ill, would you donate bone marrow even if it hurt?'

Learning styles

Kinaesthetic: Role play in the interviews.

Visual: Producing material for the display about stem cell research.

Auditory: Listening to the interviews.

Interpersonal: Discussing the presentation of interviews.

Intrapersonal: Writing a postcard from hospital to the person who has just donated bone marrow to you.

Additional biology

Can we know too much?

Today we not only understand the causes of many genetic disorders, we can also test for them. But being able to test for a genetic disorder doesn't necessarily mean we should always do it.

- People in families affected by Huntington's disease can take a genetic test which tells them if they have inherited the faulty gene. If they have, they know that they will develop the fatal disease as they get older and may pass on the gene to their children. Some people in affected families take the test and use it to help them decide whether to marry or have a family. Others prefer not to know.
- If a couple have a genetic disease in their family or already have a child with a genetic disorder, they can have a developing embryo tested during pregnancy. Cells from the embryo are checked. If it is affected, the parents have a choice. They may decide to keep the baby, knowing that it will have a genetic disorder when it is born. On the other hand, they may decide to have an abortion. This prevents the birth of a child with serious problems and allows them to try again to have a healthy baby.
- Some couples who have a genetic disease in the family or who already have a child affected by a genetic disease have their embryos screened before they are implanted in the mother. Embryos are produced by IVF (*in vitro* fertilisation). Doctors remove a single cell from each embryo and screen it for genetic diseases. Only healthy embryos free from genetic disease are implanted back into their mother. Using this method, only healthy babies are born.

ACTIVITY

Many couples who have a genetic disease in the family spend time with a genetic counsellor to help them understand what is happening and the choices they have. Plan a role-play of an interview with a genetic counsellor.

Either: Plan the role of the counsellor. Make sure you have all the information you need to talk to a couple who have already got one child with cystic fibrosis who would like to have another child. You need to be able to explain the chances of another child being affected and the choices that are open to them.

Or: Plan the role of a parent who already has one child with cystic fibrosis and who wants to have another child. Work in pairs to give the views of a couple if you like. Think carefully about the factors which will affect your decision such as: Can you cope with another sick child? Are you prepared to have an abortion? Do you have religious views on the matter? What is fairest to the unborn child – and the child you already have? Is it ethical to choose embryos to implant?

215

ACTIVITY & EXTENSION

- **The stem cell dilemma** – Suitable material for this activity has already been discussed in previous spreads. One of the ways in which the views of the target group can be discovered is by means of a questionnaire. The questions need to be phrased correctly to elicit the information: students must not just use the three bullet points given on the Student Book spread. It might be worth asking whether people have heard of some of the arguments for and against before asking their opinion. Carefully designed questions should yield this information. The results of such a questionnaire could form part of the display.
- **Analysis of results** – The students can display the results of the original survey in a variety of different ways. They should try to be bold and innovative, making use of ICT. The subsequent survey and any changes of opinion can reveal how good their material was. Again, bold display of the data is needed. They might like to find out if their target group needed more information or whether it was pitched at the right level.
- **Bone marrow transplants** – Information about bone marrow transplants can be obtained from the Anthony Nolan Trust. It would be interesting to know whether any of the staff are members or have any interest in it.
- **Continuum** – Students place themselves, or a card with their name on, along a line from two opposing viewpoints: for and against using embryonic stem cells. They could be asked to justify their position when challenged by their peers. This could also be used to canvas opinion amongst Years 10 and 11.
- **Remember!** – There are different aspects of stem cell research: they do not all involve the use of embryos or fetal tissue.
- **Can we know too much?** – This activity is a difficult one for some students to do without carrying out research. It also requires a degree of confidence. Give students time to discuss the activities in small groups. Perhaps they could plan it like a film script or scenes for a TV documentary. The questions suggested could be expanded and some about the lives of the couple included. Ask: 'How demanding are their jobs? How have they coped with the child they already have?' If each group approached the activity from a slightly different perspective, then these role-play exercises could prove interesting to the class and give them a balanced view. For example, one interview could feature parents who would not consider an abortion, another could be with parents who are prepared to use screening and IVF.

SUMMARY ANSWERS

1 a) Mitosis is cell division that takes place in the normal body cells and produces genetically identical daughter cells.

b) [Marks awarded for correct sequence of diagrams with suitable annotations.]

c) All the divisions from the fertilised egg to the baby are mitosis. After birth, all the divisions for growth are mitosis, together with all the divisions involved in repair and replacement of damaged tissues.

2 a) Stem cells are unspecialised cells which can differentiate (divide and change) into many different types of cell when they are needed.

b) They may be used to repair damaged body parts, e.g. grow new spinal nerves to cure paralysis; grow new organs for transplants; repair brains in demented patients. [Accept any other sensible suggestions.]

c) **For:** They offer tremendous hope of new treatments; they remove the need for donors in transplants; they could cure paralysis, heart disease, dementia etc.; can grow tissues to order.
Against: They use tissue from human embryos; it's wrong to use embryos, as these could become people; embryos cannot give permission; stem cells could develop into cancers.
[Accept any other valid points on either side of the debate.]

3 a) Meiosis is a special form of cell division where the chromosome number is reduced by half. It takes place in the reproductive organs (the ovaries and testes).

b) [Marks awarded for correct sequence of diagrams with appropriate annotations.]

c) Meiosis is important because it halves the chromosome number of the cells so that when two gametes fuse at fertilisation, the normal chromosome number is restored. It also allows variety to be introduced as the chromosomes separate in cell divisions and when the two gametes join.
[**HT** only]

4 [Give credit for valid points made and the way in which the letter is written.]

5 a) Sami's alleles are **ss**. We know this because she has curved thumbs and the recessive allele is curved thumbs. She must have inherited two recessive alleles to have inherited the characteristic.

b) If the baby has curved thumbs, then Josh is **Ss**. The baby has inherited a recessive allele from each parent, so Josh must have a recessive allele. We know he also has a dominant allele as he has straight thumbs.

		Sami	
		s	s
Josh	S	Ss	Ss
	s	ss	ss

c) If the baby has straight thumbs, then Josh could be either **Ss** or **SS**. We know that the baby has inherited one recessive allele from mother, and we know that Josh has one dominant allele but we do not know if he has two dominant alleles. If he has two dominant alleles then:

		Sami		
		s	s	
Josh	S	Ss	Ss	
	S	Ss	Ss	

[**HT** only]

INHERITANCE: B2 6.1 – B2 6.7

SUMMARY QUESTIONS

1 a) What is mitosis?
b) Explain, using diagrams, what takes place when a cell divides by mitosis.
c) Mitosis is very important during the development of a baby from a fertilised egg. It is also important all through life. Why?

2 a) What are stem cells?
b) It is hoped that many different medical problems may be cured using stem cells. Explain how this might work.
c) There are some ethical issues about the use of embryonic stem cells. Explain the arguments both for and against their use.

3 a) What is meiosis and where does it take place?
b) Explain, using labelled diagrams, what takes place when a cell divides by meiosis.
c) Why is meiosis so important?
[Higher]

4 Hugo de Vries is one of the scientists who made the same discoveries as Mendel several years after his death. Write a letter from Hugo to one of his friends after he has found Mendel's writings. Explain what Mendel did, why no-one took any notice of him and how the situation is so different now for you if you were doing the same sort of experiments.

5 Whether you have a straight thumb or a curved one is decided by a single gene with two alleles. The straight allele **S** is dominant to the curved allele **s**. Use this information to help you answer these questions.
Josh has straight thumbs but Sami has curved thumbs. They are expecting a baby.
a) We know exactly what Sami's thumb alleles are. What are they and how do you know?
b) If the baby has curved thumbs, what does this tell you about Josh's thumb alleles? Fill in a Punnett square to show the genetics of your explanation.
c) If the baby has straight thumbs, what does this tell us about Josh's thumb alleles? Fill in a Punnett square to show the genetics of your explanation.
[Higher]

EXAM-STYLE QUESTIONS

1 The diagram below is of stages in sexual reproductic a mammal.

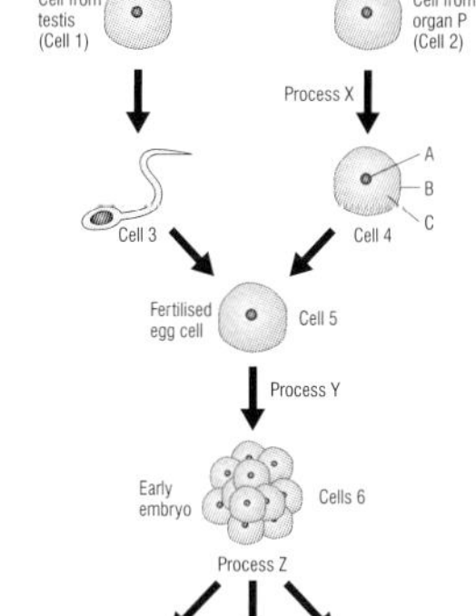

(a) What is the name of organ **P**?
(b) Give the names of parts **A**, **B** and **C** in cell 4.
(c) What is the name of cell 3?
(d) What type of cell division takes place in processe **Y** and **Z**?
(e) Which two of the cells labelled 1–6:
(i) are genetically identical to one another?
(ii) are known as gametes?
(f) Cells 6 will in due course change into a range of different cell types.
(i) What name is given to the type of cell labelle as cells 6?
(ii) What is the process called by which these ce change into different cell types?

Summary teaching suggestions

- **Lower and higher level answers** – Question 5 has been flaggec as suitable for higher ability students. The answers to section b) and c) of this question should have the Punnett squares in order to qualify for full marks.
- **Homework** – All the questions here are suitable as homework exercises and many would be useful for revision of the topic.
- **When to use the questions?**
  - The questions do link to particular topics within the chapter and can be used to test students' understanding at the end of a topic if the summary questions have been used in the lessor
  - Question 2 could be set as an exercise after holding a debate about the ethics of using stem cells. Students are then encouraged to present a balanced view of the issues involved.
  - Question 4 could be used when teaching about Mendel and the letter-writing suggestion has already been made as an activity.
- **Special needs** – Students could answer some of the questions verbally, especially parts of questions 1 and 2. In the question on the processes of mitosis, students could be given diagrams, or pictures, of the stages that they could put into the correct order and label or annotate with pre-printed labels.

EXAM-STYLE ANSWERS

1 a) Ovary *(1 mark*

b) **A** Nucleus
B Cell membrane
C Cytoplasm *(1 mark each*

c) Sperm *(1 mark*

d) Mitosis *(1 mark*

e) i) Cell 5 and cells 6 *(1 mark*
ii) Cell 3 and cell 4 *(1 mark*

f) i) Stem cells *(1 mark*
ii) Differentiation *(1 mark*

ystic fibrosis is a condition in which people suffer from
e accumulation of thick and sticky mucus in their
ngs. The chart shows part of a family tree in which
me members have cystic fibrosis.

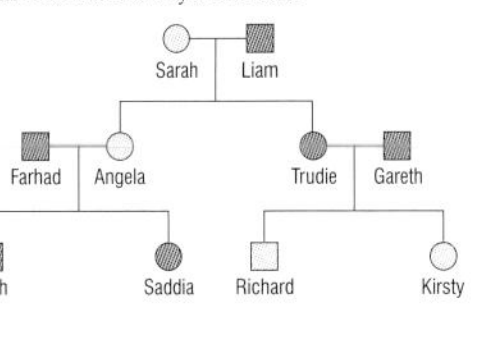

) Using two pieces of evidence from the family tree, explain why cystic fibrosis appears to be controlled by a recessive gene. (2)

) Trudie and Gareth want to have another child. What is the chance that this child will inherit cystic fibrosis? Explain, with the aid of a genetic diagram, how you reached your answer. (4)

) The letters **A**, **B** and **C** show the three different possible combinations of alleles possessed by the members of this family tree

A dominant and dominant

B dominant and recessive

C recessive and recessive

For each of the individuals below, give the letter that represents the alleles they possess.

(i) Liam (1)
(ii) Angela (1)
(iii) Saddia (1)

) Explain how it is possible that Farhad and Angela could have a child with cystic fibrosis. (3)

[Higher}

HOW SCIENCE WORKS QUESTIONS

Amjid grew some purple flowering pea plants he had bought at the garden centre.

Here are his results.

Seeds planted	247
Purple-flowered plants	242
White-flowered plants	1
Seeds not growing	4

a) Is the white flowered plant an anomaly? (1)

b) Are the seeds that did not grow anomalies? (1)

c) What might Amjid do with the white-flowered plant? (1)

Amjid was interested in these plants, so he collected the seed from some of the purple-flowered plants and used them in the garden the following year. He made a careful note of what happened.

Here are his results:

Seeds planted	406
Purple-flowered plants	295
White-flowered plants	102
Seeds not growing	6

Amjid was slightly surprised. He did expect to find that a third of his flowers would be white.

d) Suggest how Amjid could display his results. (1)

e) Check the accuracy of Amjid's results. How accurate were they? (3)

f) How could Amjid have improved his method of growing the peas to make his results more valid? (1)

a) *Any two accurate pieces of evidence, e.g.:*

- Sarah has cystic fibrosis but her daughter Trudie does not.
- Angela has cystic fibrosis but neither of her children does.
- Neither Trudie nor Gareth has cystic fibrosis but both their children do.

(Any two points, 1 mark each)

The first two points show the condition cannot be dominant and therefore it must be recessive. The third point shows it to be recessive.

b) Trudie and Gareth already have two children with cystic fibrosis and therefore both must carry the allele for the condition. *(1 mark)*

As neither Trudie nor Gareth suffer from cystic fibrosis themselves, they must both possess one dominant and one recessive allele. *(1 mark)*

Let F = normal allele and f = cystic fibrosis allele
Trudie = Ff and Gareth = Ff
Sex cells (gametes) for both = F and f

		Gareth	
		F	f
Trudie	F	FF	Ff
	f	Ff	ff

(1 mark)

As cystic fibrosis is a recessive condition, only a child with both recessive alleles (ff) will be affected. Only one in four (25%) of the possible offspring have the ff alleles and therefore the chance of the next child having cystic fibrosis is one in four (25%). *(1 mark)*

c) i) Liam **B** *(1 mark)*
ii) Angela **C** *(1 mark)*
iii) Saddia **B** *(1 mark)*

d)
- To have cystic fibrosis the child would need to have two recessive alleles (ff).
- The mother/Angela has cystic fibrosis and so is ff and must therefore pass one recessive allele to each of her children.
- If the father/Farhad has one dominant and one recessive allele (Ff), there is a 50% chance that a child might inherit the second recessive allele from him.

(1 mark for each point)

There are a number of different ways to explain why a child with cystic fibrosis is possible. Provided the reasoning is logical and genetically accurate, even if longwinded, full credit should be given. [**HT** only]

Exam teaching suggestions

- While question 1 tests recall in places, it also requires students to demonstrate an understanding of cell division and associated material.
- Question 2 is designed for higher-tier students. It involves predicting and explaining the outcome of genetic crosses between individuals with different combinations of alleles and also the construction of genetic diagrams. Both these skills are much more likely to be tested on Higher Tier papers.
- Question 2, part b). The use of letters other than F and f are acceptable – indeed C and c are much more likely to be used. While it makes sense to use the first letter of cystic fibrosis, it can also lead to marks being lost because the higher and lower case versions of the letter are the same in form and differ only in size. Unfortunately candidates often fail to distinguish the size adequately and so either get mixed up themselves or make it impossible for the examiner to follow what is happening.

HOW SCIENCE WORKS ANSWERS

a) Yes, the white flowered plant is an anomaly because it is not as expected.

b) Yes, the seeds that did not grow are also anomalies probably due to the way they were grown or some genetic problem.

c) As an anomaly, the white flowered plant should be investigated, e.g. to see if the colour was a result of a mutation or because of the particular conditions in which it was grown. He could breed from it, plant it in a different soil, etc.

d) The results are best presented in a bar graph.

e) Response should hypothesise that the parents were heterozygous and draw a Punnett square to predict a 3:1 ratio and show how accurate his actual results were by comparing with his predicted results.

f) To improve his method of growing the peas to make his results more valid Amjid could have grown them under control conditions.

How science works teaching suggestions

- **Higher- and lower-level answers.** Question e) is a higher-level question and the answers provided above are also at this level. Question f) is lower level and the answer provided is also lower level.
- **How and when to use these questions.** When wishing to develop the importance of checking on the cause of anomalies and developing ideas of scientific method.
  The questions are best done as small group work which will need some support for those struggling with the genetics.
- **Homework.** Question c) could be developed into experimental design for homework.
- **ICT link-up.** There are many programs which allow interactive work in genetics, for example, Drosophilia Genetics at www.newbyte.co.uk.

B2 Examination-Style Questions

Answers to Questions

1 (a) *Quality of written communication* *(1 mark)*
The mark should be given where correct scientific terms are used and the ideas are given in a sensible order. The mark can be awarded for a scientific and logical answer, even if it is inaccurate; it cannot be given if the answer is non-scientific or nonsensical.

- Microorganisms/bacteria/fungi/saprotrophs/saprophytes/saprobionts
- digest/break down organic matter/leaves/decompose (reference to decomposers)/decay/rot
- use of enzymes/correctly named example
- absorption by diffusion/active transport
- respiration/combustion
- carbon dioxide can be used (by trees) in photosynthesis.

(1 mark for any point to a maximum of 3)

(b)
- warmth/suitable temperature *(heat/hot weather are not acceptable)*
- damp/water/rain/humid/moisture
- oxygen
- suitable pH.

(1 mark for any point to a maximum of 2)

2 (a) The concentration of fructose increases *(1 mark)*
then levels off/rate of increase slows *(1 mark)*

(b) (i) They acted as controls. *(1 mark)*

(ii) Exactly the same as tube A. *(1 mark)*

(c) (i) less sugar is used/cheaper than using glucose *(1 mark)*

(ii) food is just as sweet/fructose is sweeter *(1 mark)*
there is less sugar to convert to fat/less surplus energy *(1 mark)*

3 (a) Mouth temperature was used in both investigations for all those tested. *(1 mark)*

(b) He carried out the largest survey. *(1 mark)*

(c) E.g. tests carried out several times on the same people; used a digital thermometer which is less easy to misread; more recent thermometers are more likely to be more accurate. *(1 mark)*

(d) E.g. more accurate diagnosis of disease and therefore more appropriate treatment. *(1 mark)*

continues opposite ›

EXAMINATION-STYLE QUESTIONS

1 Each autumn, many trees lose their leaves.

See pages 168–71

(a) Describe how carbon compounds in the leaves can be recycled so that they can be used again by the trees. *(4 marks)*

To gain full marks in this question you should write your ideas in good English. Put them into a sensible order and use the correct scientific words.

(b) Give **two** environmental conditions that speed up the processes that you have described in part (a). *(2 marks)*

GET IT RIG
When giving a change environmental condit remember to say in w direction the change place. Use terms su 'higher', 'lower', 'mo 'less'.

2 In an investigation, an enzyme was added to glucose syrup in test tube A. In another test tube (B) glucose was left without the enzyme. In a third test tube (C) the enzyme was left without the glucose. The concentrations of glucose, fructose and the enzyme were measured for thirty minutes. The results for test tube A are shown in the graph.

See pages 136–9

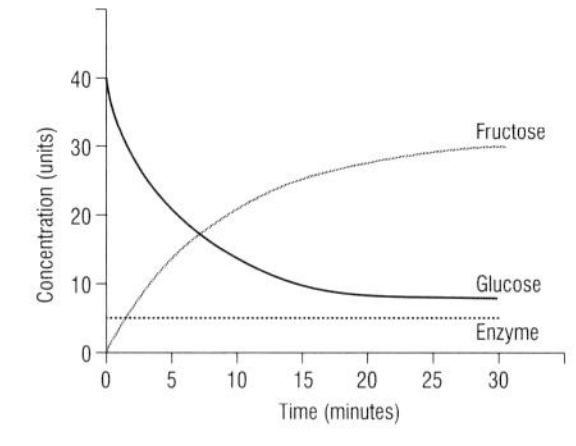

(a) Describe the changes in the concentration of fructose. *(2 marks)*

(b) (i) Explain why test tubes B and C were used. *(1 mark)*

(ii) How should tubes B and C have been treated. *(1 mark)*

(c) Fructose is often added to foods used by people on a slimming diet.

See pages 186–7

(i) Give **one** advantage of this for the company making the slimming food. *(1 mark)*

(ii) Explain **one** advantage of this for a person on a slimming diet. *(2 marks)*

218

GET IT RIGHT!

iven that even a single mark can improve a final grade, it is nportant to teach students the significance of being precise specially in the use of scientific terms. Another common failure to ccurately distinguish related terms occurs in the use of 'breathing' nd 'respiration'. A useful exercise to help students avoid these rrors is to compile a list of commonly confused pairs of terms when narking students' work. These could then be used as an exercise in vhich students are asked to state at least one way in which the two erms in each pair can be clearly distinguished.

n question 4, part b) i), the importance of reading every word of a uestion is stressed. Here the key word is 'system' indicating that nervous system' is required rather than a specific part of that ystem such as the brain.

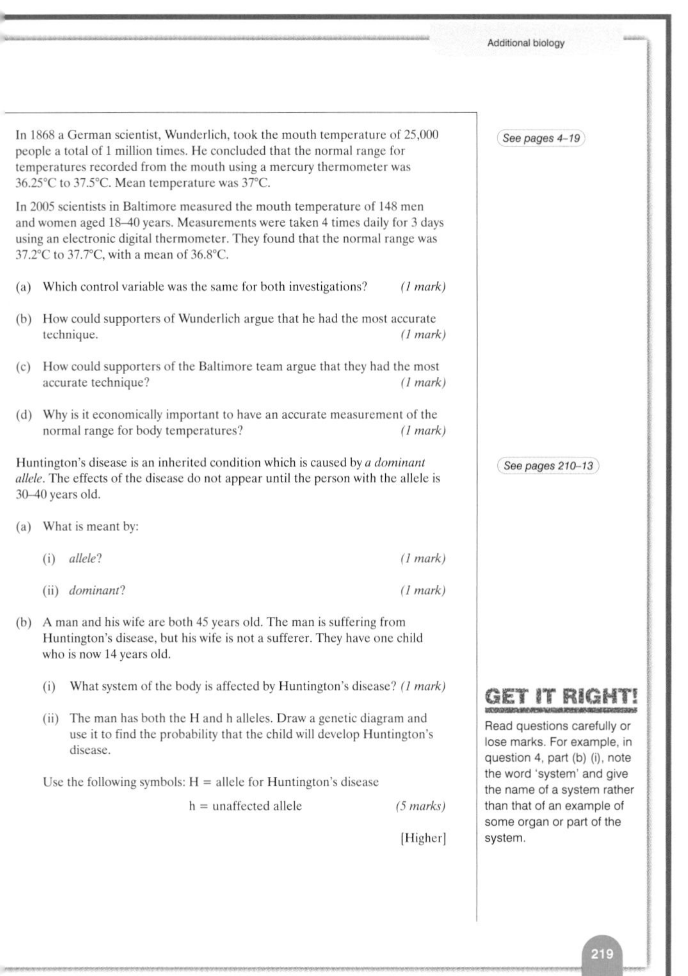

Additional biology

In 1868 a German scientist, Wunderlich, took the mouth temperature of 25,000 people a total of 1 million times. He concluded that the normal range for temperatures recorded from the mouth using a mercury thermometer was 36.25°C to 37.5°C. Mean temperature was 37°C.

In 2005 scientists in Baltimore measured the mouth temperature of 148 men and women aged 18–40 years. Measurements were taken 4 times daily for 3 days using an electronic digital thermometer. They found that the normal range was 37.2°C to 37.7°C, with a mean of 36.8°C.

See pages 4–19

(a) Which control variable was the same for both investigations? *(1 mark)*

(b) How could supporters of Wunderlich argue that he had the most accurate technique. *(1 mark)*

(c) How could supporters of the Baltimore team argue that they had the most accurate technique? *(1 mark)*

(d) Why is it economically important to have an accurate measurement of the normal range for body temperatures? *(1 mark)*

Huntington's disease is an inherited condition which is caused by *a dominant allele*. The effects of the disease do not appear until the person with the allele is 30–40 years old.

See pages 210–13

(a) What is meant by:

(i) *allele?* *(1 mark)*

(ii) *dominant?* *(1 mark)*

(b) A man and his wife are both 45 years old. The man is suffering from Huntington's disease, but his wife is not a sufferer. They have one child who is now 14 years old.

(i) What system of the body is affected by Huntington's disease? *(1 mark)*

(ii) The man has both the H and h alleles. Draw a genetic diagram and use it to find the probability that the child will develop Huntington's disease.

Use the following symbols: H = allele for Huntington's disease
h = unaffected allele *(5 marks)*

[Higher]

GET IT RIGHT!
Read questions carefully or lose marks. For example, in question 4, part (b) (i), note the word 'system' and give the name of a system rather than that of an example of some organ or part of the system.

219

BUMP UP THE GRADE

Some students fail to pace themselves adequately during an examination, either finishing too early or running out of time. Students usually notice how many marks are available for each portion of a question, but cannot translate this into the time they should spend on answering it. One way of overcoming the problem is to train students to divide the total mark allocation into the time available in minutes. Both pieces of information appear on the front cover of the examination paper. In general there is around one minute for each available mark. It is not practical to time each separate mark but the total for each question or double page is usually given in the bottom corner of each page. A glance at this when turning the page can help students pace themselves through the examination and so avoid careless loss of marks as a result of rushing parts of the paper.

› *continues from previous page*

4 (a) (i) *Either* one of two (/of several) forms of a gene
Or (a variant) form of a gene *(1 mark)*

(ii) *Either* expressed even if only one copy is inherited
Or expressed/seen in heterozygote *(1 mark)*

(b) (i) nervous (*'brain' is not 'a system' and therefore not allowed*) *(1 mark)*

(ii) Man/affected = **Hh**, and wife unaffected = **hh** *(1 mark)*
correct gametes from parental genotypes *(1 mark)*
F_1 genotypes correctly derived from parental gametes *(1 mark)*
Identification of **Hh** in F_1 as having Huntington's disease *(1 mark)*
Correct probability from F_1 genotypes, e.g. $\frac{1}{2}$/0.5/50%/1 in 2/1:1/50:50 *(1 mark)*
Care should be taken not to allow '1:2' or '50/50'.
As the question specifically asks for 'a genetic diagram', a mark must be deducted if one is omitted, even though the answer itself is correct. Provided the chain of logic can be picked up from the previous statement, the following mark can be given even if the previous statement was wrong. In other words, an error should only be penalised once as long as the rest that follows is logical and genetically accurate. [**HT** only]

B3 | Further biology

Key Stage 3 curriculum links

The following link to **'What you already know'**:

- Cells and functions of cells related to life processes in a variety of organisms.
- Products of digestion are absorbed into the bloodstream and transported throughout the body.
- The role of lung structure in gas exchange, including the effects of smoking.
- Aerobic respiration involves a reaction in cells between oxygen and food, in which glucose is broken down into carbon dioxide and water. The reactants and products of respiration are transported throughout the body in the bloodstream.
- Plants need carbon dioxide, water and light for photosynthesis, and produce biomass and oxygen. The role of root hairs in absorbing water and minerals from the soil.
- The growth and reproduction of bacteria and viruses can affect health.

QCA Scheme of work

7A Cells
8A Food and digestion
8B Respiration
8C Microbes and disease
9C Plants and photosynthesis

Links with Units B1 and B2

The following link to units in the specification that should have been studied earlier.

From Unit B1, the students should have studied:

- The effect of smoking on the lungs.
- How the growth and reproduction of bacteria and viruses can affect health.

From Unit B2, the students should have studied:

- Cells and cell functions.
- The need for carbon dioxide, water and light for photosynthesis in plants.
- The absorption of water and mineral ions by the root hairs of plants.
- Aerobic respiration.
- Enzymes and digestion.
- The need to remove waste products in order to maintain a constant internal environment.

RECAP ANSWERS

1 a) The exchange of gases – oxygen from the air into the blood when you breathe in, carbon dioxide from the blood into the air to be removed from the body when you breathe out.

b) The cilia are inactivated so mucus builds up in the lungs; tar coats the surface of the lungs making gas exchange less effective and can cause asthma, bronchitis, emphysema and lung cancer.

2 a) A transport system is needed to carry the substances you need around the body and to remove waste products from your cells.

b) The soluble products of digestion pass out of your gut into the blood to be carried around your body.

3 a) Plants need carbon dioxide, water and light to make food in the process of photosynthesis.

b) They get the carbon dioxide that they need by diffusion from the air through the leaves.

c) Plant roots are important because they take up water and minerals from the soil; they anchor the plants in the soil.

4 a) Infectious diseases are caused by microorganisms, which can be passed from one person to another.

b) Bacteria, viruses, fungi, protists.

c) Bacteria consist of cells; reproduce by splitting in two; killed by antibiotics. Viruses only survive by taking over the cell of another organism and using it for replication.

Activity notes

- A class discussion on the care of pets in hot weather could be beneficial. Apart from the obvious need for the provision of wate[r] during hot weather, students could make a list of the situations where pets are stressed by the heat. For example, dogs left in unventilated cars get very hot.
- A leaflet can contain more information than a poster; these coul[d] be made in groups. The students could allocate tasks within the[ir] group and the leaflet could be produced using ICT. If there are connections with a local vet, then advice as to what to include could be sought. You never know, the local vet might be prepared to display the leaflets or distribute some if they achieve a good standard!

BIOLOGY

B3 | Further biology

What you already know

Here is a quick reminder of previous work that you will find useful in this unit:

- You can relate cells and cell functions to life processes in a variety of organisms.
- The products of digestion are absorbed into your bloodstream and transported throughout the body.
- The structure of your lungs plays a role in gas exchange.
- Smoking affects your lungs.
- Aerobic respiration involves a reaction in your cells between oxygen and glucose. The glucose is broken down into carbon dioxide and water.
- The reactants and products of respiration are transported throughout your body in the bloodstream.
- Plants need carbon dioxide, water and light for photosynthesis.
- The root hairs of plants play a role in absorbing water and minerals from th[e] soil.
- The growth and reproduction of bacteria and viruses can affect your healt[h]

The lungs play a vital part in gas exchange

RECAP QUESTIONS

1 a) What are the main functions of your lungs?
b) If you smoke, your lungs often do not work as well. Explain how smoking can affect your lungs.
2 a) Why is a transport system so important in your body?
b) What happens to the food you eat once it has been digested in your gut?
3 a) Explain why plants need carbon dioxide, water and light.
b) How do they get the carbon dioxide that they need?
c) Why are plant roots so important?
4 a) What is an infectious disease?
b) What types of microorganisms cause infectious diseases?
c) What are the differences between bacteria and viruses?

SPECIFICATION LINK-UP

Unit: Biology 3

This unit covers the following:

How do dissolved substances get into and out of animals and plants?

The cells in animals and plants all need oxygen to be able to release energy for the jobs they do. They all produce carbon dioxide as a waste product.

How are dissolved materials transported around the body?

Substances are transported around the body by the circulation system (the heart, the blood vessels and the blood). They are transported from where they are taken into the body to the cells, or from the cells to where they are removed from the body.

How does exercise affect the exchanges taking place within the body?

The human body needs to react to the increased demand for energy during exercise.

How do exchanges in the kidney help us to maintain the internal environment in mammals and how has biology helped us to treat kidney disease?

People whose kidneys do not function properly die because toxic substances accumulate in their blood. Their lives can be saved by using dialysis machines or having a healthy kidney transplanted.

How are microorganisms used to make food and drink?

People from different cultures have known for thousands of years how to use microorganisms to make various types of food and drink, such as bread, beer, wine and yoghurt.

What other useful substances can we make using microorganisms?

Microorganisms are used on a large scale to make many useful substances, including antibiotics such as penicillin, foods such as mycoprotein and fuels such as biogas and ethanol.

How can we be sure we are using microorganisms safely?

If the microorganisms that we want to use are contaminated, the other microorganisms that are present may produce harmful substances. So it is only safe to use microorganisms if we have a pure culture containing only one particular species of microorganism.

Chapters in this unit

- **Exchange of materials**
- **Transporting substances around the body**
- **Microbiology**

Making connections

ll living organisms need water for the reactions to take lace in their cells. Moving water into and out of the ells, using it for photosynthesis, to get rid of waste or o cool down. These are all important processes for nimals and plants. In fact, almost all biology eventually epends on water!

Plants need water for photosynthesis. They have to take the water in from the soil through their roots. When water is in short supply, plants have some spectacular adaptations to help them survive!

When you breathe out, vater evaporates from the urfaces of your lungs into he air and is lost. You can ee this clearly in the vinter when it condenses n the cold air as you reathe out. This loss happens all the time, day and ight, summer and winter. So you have to take in nough water to replace it.

Fish take the oxygen they need from water. Their gills are specially adapted to make this possible. But whether fish live in fresh or salt water, they have to deal with the problem of water moving in to or out of the cells of their gills as it flows over them. Fish have several different adaptations to help them balance the water levels in their body.

The amount of water you take into your body varies enormously from one day to the next. Your body has to cope with whatever it is given. The concentration of your body fluids has to stay more or less the same whatever you drink and whatever you do. You use water to cool down, to remove waste and to carry materials around your body. All the chemical reactions of your cells take place dissolved in water. It's not surprising that you need your kidneys to maintain the water balance of your body!

Microorganisms are no different to other living things – water is vital for the chemical reactions in their cells to take place. That's why drying food preserves it. The microbes which cause decay simply can't grow.

ACTIVITY

Pet animals depend on their owners to keep them safe and well. Many types of pets, such as cats and dogs, cannot lose heat by sweating over most of their bodies. That's because of their thick layer of fur. They sweat through their feet and pant, cooling down as water evaporates from the surface of their mouths.

Design a poster or a leaflet for your local vet's surgery explaining why water is so important for animals. In particular, describe how to care for dogs and cats in hot weather.

Chapters in this unit

Exchange of materials – Transporting substances around the body – Microbiology

Teaching suggestions

The theme of the unit opener is the need for water in plants and animals.

- Students could be presented with 'What you already know' statements and asked to write down the links with the previous units of the specification, i.e. what they have already studied at KS4 for GCSE. This could indicate some areas where they may need to look back at their notes before embarking on the new material.
- A knowledge of cells and cell functions is fundamental to an understanding of how dissolved substances get into and out of cells and the transport of substances around the bodies of plants and animals. The role of cell membranes is particularly important when considering the processes of osmosis and active transport.
- Before considering the transport of dissolved materials, students need to have a clear idea of why transport systems are needed and some understanding of surface area to volume ratios.
- **Recap questions** – These questions could be used as starters for the relevant topics. In this unit, it is not suggested that they all be used together as they cover a wider range of topics than in previous unit openers.
  - Question 1 about the lungs could be used before any new material about the lungs is studied.
  - Question 2a) is more open-ended and could be used as the basis of a class discussion. Allow students 5 minutes to jot down their own thoughts and then build up a list on the board, or use a blank outline diagram of the human body and put in the various transport systems.
- **Making connections** – The picture references in this spread involve water and its uses in plants and animals. The information could be used to build up a list of the uses of water in the bodies of plants and animals. Students, working in groups, could compile lists and then pool their ideas with the whole class. Ask: 'How many different functions are shown? Are there any more to add to the list that are not shown? Is there a difference between how land and aquatic organisms use water?'
- **Importance of water** – Is it possible to decide which property or function of water is the most important?
- The activity box could involve the students in some research into the physiology of their pets, particularly cats and dogs. Many students do not realise that animals with thick fur cannot lose heat by sweating in the same way that humans can. More information can be obtained from the Internet, from libraries and from the local veterinary surgery.

B3 1.1 Active transport

LEARNING OBJECTIVES

Students should learn:

- That active transport is the absorption of substances against a concentration gradient. (**HT** only)
- That active transport enables cells to take up ions from very dilute solutions. (**HT** only)
- That sugars and ions can pass through cell membranes. (**HT** only)

LEARNING OUTCOMES

Most students should be able to:

- Describe how active transport differs from diffusion and osmosis. (**HT** only)
- Describe examples of active transport in plants and animals. (**HT** only)

Some students should also be able to:

- Explain the importance of active transport to plants and animals. (**HT** only)
- Explain how active transport across a cell membrane takes place. (**HT** only)
- Explain the action of cyanide. (**HT** only)

Teaching suggestions

- **Special needs**
  - Use the jelly babies game if it has not been used as a starter.
  - Alternatively, try using a piece of hose-pipe with a perforated section where it passes a large card labelled 'kidney'. Show, or give, the students some salty and sweet food. Say that we like it because we need salt and sugar in our blood (but not too much!). Pour a mixture of salt and sugar into the tube, arranging a dish to catch it as it comes out at the kidney. Have another funnel in the upper surface of the tube below the perforations but still inside the kidney diagram. Get the students to pour the salt and sugar back in. Explain that we need to reabsorb some substances even when there are more of them on the inside than on the outside and that this takes energy.
- **Gifted and talented.** Students could do Internet research on the number of ATP molecules produced during respiration of a glucose molecule. Relate this to the energy required to take in a molecule of glucose by active transport. If needed, set up a 'Scavenger Hunt' style series of URLs with the data needed on them.

SPECIFICATION LINK-UP Unit: Biology 3.13.1

- *Substances are sometimes absorbed against a concentration gradient. This requires the use of energy from respiration. The process is called active transport. It enables cells to absorb ions from very dilute solutions. Other substances, such as sugar and ions, can also pass through cell membranes. [**HT** only]*

Lesson structure

STARTER

Quick quiz – Students to answer ten short questions on slips of paper on osmosis and diffusion to recap work done so far. Their peers mark the answers. (5–10 minutes)

'Hungry hippos' game – Remind students of this game, where they have to grab marbles from a central arena using hippo-shaped scoops. The marbles caught end up in the traps. Use this as an analogy to describe taking molecules from an area of low concentration to an area of high concentration. (5 minutes)

Jelly babies game – This is similar to the last starter, where students take jelly babies from a plate in turns until each of them has more in their hands than are on the plate. (They should not eat them in class.) Stop the game at this point and discuss what would happen due to diffusion (or osmosis if putting the sweets into a paper bag). Ask: 'Is it worth expending the energy to get and keep the sugar?' (5–10 minutes)

MAIN

- Use Simulation B3 1.1 'Active transport' From the GCSE Biology CD ROM. Give students a work sheet with questions and allow time for completion.
- If easily available, show some animations on active transport. Note: It is very difficult to explain how active transport works without referring to carrier proteins and pumps in the membranes.
- **Active transport in plants** – Show a PowerPoint® presentation on the need for certain mineral ions for healthy growth, the presence of these ions in the soil solution and the cell sap, and the way in which plants accumulate ions against the concentration gradient.
- A useful example of the need for energy in respiration is to describe a hydroponics system, where solutions are aerated to provide oxygen for the respiration of the roots.
- **Salt glands in marine birds and reptiles** – Show photographs of marine vertebrates and discuss the problems of the salt in their diets and how they get rid of it. There is some information on the Internet, especially from the RSPB web site.
- **Energy needed** – To get across the idea of energy being needed, use a revolving door analogy. If something valuable is on the far side (students to decide what it is!), then it is worthwhile keeping on giving the door a good hard shove, even if you have got plenty inside already.
  See www.rsob.org.uk

PLENARIES

Cyanide pills – Show video footage of someone taking a cyanide pill (WW2 films) or a photograph of a real cyanide capsule as used in WW2 by spies. How would this work to kill you so quickly? Seach for video at www.britishoath.com (5 minutes)

Diffusion, osmosis and active transport – If not used already, show the students the animations (or make your own illustrative diagrams), call them 'a', 'b' and 'c' and get them to say which is which and why. (5–10 minutes)

Where in the body? – Give the students a blank body diagram each and get them to draw on where active transport will take place. Annotate with reasons. Examine in pairs. (10 minutes)

ACTIVITY & EXTENSION

- **Cystic fibrosis and membrane structure** – Students could investigate the cause of cystic fibrosis, finding out about the mutation that involves a protein involved in active transport.
- **Energy needed** – Discuss the energy requirements of active transport. Link to respiration, and discuss the rate of uptake of ions linked to the rate of respiration using the graph in the students' book.
- **Lots of mitochondria!** – Get students to suggest cells and tissues that are 'active' and contain many mitochondria. (Liver, kidney, muscle cells are examples and can be linked with active uptake. Link this with the plenary 'Where in the body?').

Answers to in-text questions

a) Active transport moves things against a concentration gradient, using energy.

b) Mitochondria are the sites of cellular respiration that produce the energy needed for active transport. So cells where lots of active transport take place need lots of mitochondria.

Teaching suggestions continued

- **Learning styles**

  Kinaesthetic: Modelling of active transport using Plasticine or filling in diagrams of the body with the regions where active transport is used.

  Visual: Looking at the PowerPoint® presentations.

  Auditory: Taking part in class discussions.

  Interpersonal: Working in groups and setting short crosswords for each other on the three topics covered so far in this chapter.

  Intrapersonal: Considering the circumstances that would necessitate the use of a cyanide pill.

B3 1.1 Active transport

LEARNING OBJECTIVES

1. What is active transport? [Higher]
2. Why is active transport so important? [Higher]

There are two main ways in which dissolved substances are moved into and out of cells. Substances move by diffusion, along a concentration gradient which must be in the right direction to be useful to the cells. Osmosis depends on a concentration gradient of water and a partially permeable membrane. Only water moves in osmosis. However, sometimes substances needed by your body have to be moved against a concentration gradient, or across a partially permeable membrane. The process is known as **active transport**.

HIGHER

GET IT RIGHT!

Remember active transport takes place **against** a concentration gradient from low to high concentration and it requires energy from respiration.

Active transport

Active transport allows cells to move substances from an area of low concentration to an area of high concentration. So substances move against the concentration gradient. As a result, cells can absorb ions from very dilute solutions. It also makes it possible for them to move substances like sugars and ions from one place to another through the cell membranes.

a) How does active transport differ from diffusion and osmosis?

It takes energy for the active transport system to carry a molecule across the membrane and then return it to its original position. (See Figure 1.) That energy comes from cellular respiration. Scientists have shown in a number of different cells that the rate of respiration and the rate of active transport are closely linked. (See Figure 2.)

In other words, if a cell is making lots of energy, it can carry out lots of active transport. These cells include root hair cells and your gut lining cells. Cells involved in a lot of active transport usually have lots of **mitochondria** to provide the energy they need.

Rate of active transport
0
Rate of respiration

Figure 2 The rate of active transport depends on the rate of respiration

b) Why do cells which carry out a lot of active transport have lots of mitochondria?

Cell membrane
Solute molecule to be transported
Transport protein
Solute recognised and grabbed by transport protein
Outside cell
Inside cell
Protein rotates in membrane and releases solute inside the cell (using energy)
Protein rotates back again (often using energy)

Figure 1 Sometimes it is worth using up energy when a resource is particularly valuable and its transport is really important!

The importance of active transport

Active transport is widely used in cells. There are some situations where it is particularly important. For example, the mineral ions in the soil are usually found in very dilute solutions. These solutions are more dilute than the solution within the plant cells. By using active transport, plants can absorb these mineral ions, even though it is against a concentration gradient. (See Figure 3.)

Glucose is always moved out of your gut and kidney tubules into your blood, even when it is against a large concentration gradient.

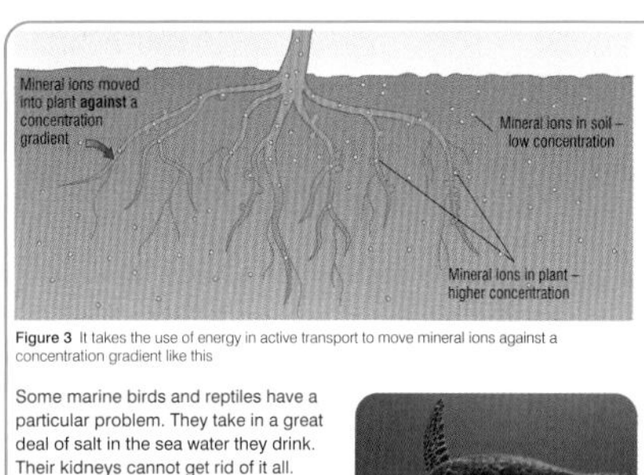

Figure 3 It takes the use of energy in active transport to move mineral ions against a concentration gradient like this

Some marine birds and reptiles have a particular problem. They take in a great deal of salt in the sea water they drink. Their kidneys cannot get rid of it all.

The answer is special **salt glands**, which are usually found near the eyes and nostrils. Sodium ions are moved out of the body into the salt glands. The glands then produce a very strong salt solution – up to six times more salty than their urine!

Figure 4 Marine animals like this turtle live in very extreme conditions. The salt glands which some animals have as an adaptation for survival depend on active transport to move salt out of their bodies.

The sodium ions have to be moved against a very big concentration gradient. So active transport is vital to the survival of these marine creatures.

SUMMARY QUESTIONS

1 Copy and complete using the words below:

concentration active transport osmosis against mitochondria diffusion energy

...... and ...... depend on a ...... gradient in the right direction to work. Substances are moved ...... a gradient by .... which uses ...... produced by ....

2 a) Explain how active transport works in a cell.
b) Give some examples of a situation when a substance cannot be moved into a cell by osmosis or diffusion, and how active transport solves the problem.

3 The processes of diffusion and osmosis do not need energy to take place. Why does an organism have to provide energy for active transport and where does it come from?

4 a) Explain why cyanide is such an effective poison.
b) Why is active transport so important for animals which live in the sea?

HIGHER

FOUL FACTS

Cyanide is a deadly poison. It smells faintly of almonds and once you have taken it, you quickly die. Cyanide kills you because it stops the reactions of respiration in your mitochondria. If you give individual cells cyanide, all active transport stops as their energy supply dries up. But if you supply the cells with energy, even though the mitochondria are still poisoned, active transport starts up again!

DID YOU KNOW?

People affected by the genetic disease cystic fibrosis (see page 212) produce thick, sticky mucus in their lungs, gut and reproductive systems. This fatal condition is the result of a mutation which affects a protein involved in the active transport system of the mucus-producing cells.

KEY POINTS

1. Substances are sometimes absorbed against a concentration gradient by active transport.
2. Active transport uses energy from respiration.
3. Cells can absorb ions from very dilute solutions and move molecules through cell membranes using active transport.

SUMMARY ANSWERS

1 Diffusion, osmosis, concentration, against, active transport, energy, respiration.

2 **a)** A transport protein or system in the membrane is usually used. The substrate molecule binds to the transport protein in the membrane. This then moves across the membrane carrying the substance to the other side. The substrate is then released and the carrier molecule returns to its original position. This all uses energy. [A diagram similar to the one in the students' book could be used to explain this.]

b) Active transport is needed when the substance required is at a lower concentration outside the cell than inside (e.g. mineral ions in the soil for plants, glucose in kidneys, etc.), when an unwanted substance is constantly moving in to a cell along a concentration gradient (e.g. water into Amoeba), and when a cell needs to move an unwanted substance against a concentration gradient (e.g. salt glands). [**HT** only]

3 When substances needed by your body have to be moved against a concentration gradient, or across a partially permeable membrane, the cell uses a transport system across the membrane and this needs energy to move the desired substrate. Energy produced by respiration in the mitochondria is used. [**HT** only]

4 **a)** Cyanide stops the reactions in the mitochondria, so there is no energy available for active transport and other energy-dependent reactions in the body.

b) They need to get rid of the excess salt and they use active transport to secrete the salt into special salt glands that remove it from the body. [**HT** only]

KEY POINTS

The important points are that active transport is *active*; diffusion and osmosis are *passive*. A link with the word 'activity' should suggest the use of energy and respiration.

B3 1.2 Exchange of gases in the lungs

LEARNING OBJECTIVES

Students should learn:

- That the lungs are located in the thorax, protected by the ribcage and separated from the abdomen by the diaphragm.
- That oxygen from the air taken in to the lungs diffuses into the bloodstream and carbon dioxide diffuses out of the bloodstream into the air in the lungs.
- That the alveoli provide a large surface area over which gases can readily diffuse into and out of the blood.

LEARNING OUTCOMES

Most students should be able to:

- Describe the structure of the lungs and how breathing is brought about.
- Describe the function of the alveoli.
- Explain how the alveoli are adapted for the efficient exchange of gases.

Some students should also be able to:

- Explain how movements of the ribcage and the diaphragm bring about changes in the volume of the thorax during breathing.
- Evaluate the importance of adaptations which give increased s.a. to the effectiveness of gas exchange, e.g. by recognition of the problems of COPD.

Teaching suggestions

- **Special needs.** Students could make a large cut-out-and-stick model of the thorax.
- **Gifted and talented.** Students could carry out a short research activity on the gases used in scuba diving. Ask: 'Why are helium and nitrogen used with oxygen as well as compressed air?'
- **Learning styles**

  Kinaesthetic: Dissecting the sheep or pig's trachea and lungs.

  Visual: Observing of interactive ICT.

  Auditory: Listening to students describing what is happening at each stage of gas exchange.

  Interpersonal: Taking part in the interactive modelling of gas exchange.

  Intrapersonal: Considering how you can breathe for other people if they are unconscious.

SPECIFICATION LINK-UP Unit: Biology 3.13.1

- *Many organ systems are specialised for exchanging materials.*
- *In humans:*
  - *The surface area of the lungs is increased by the alveoli.*
- *The lungs are in the upper part of the body (thorax) protected by the ribcage and separated from the lower part of the body (abdomen) by the diaphragm.*
- *The breathing system takes air into and out of the body so that oxygen from the air can diffuse into the bloodstream and carbon dioxide can diffuse out of the bloodstream into the air.*
- *The alveoli provide a very large, moist surface, richly supplied with blood capillaries so that gases can readily diffuse into and out of the blood. [**HT** only]*

Lesson structure

STARTER

Looking at blood vessels – Students to get a partner to pull down their lower lip as far as they comfortably can (care with body fluids). They then have a close look at the blood vessels. Ask: 'Can you see two different colours? Which and why?' They then reciprocate with the partner. (**Safety**: wash hands before and after.) (5 minutes)
Blood samples – Ask: 'Has any one in the class had a blood sample taken? What colour is it in the syringe?' Show some images of blood in bags from the transfusion service. If you spilled it on to the floor, what colour would it go? (5–10 minutes)

MAIN

- **Breathing model** – Using a bell jar, a sheet of rubber and two balloons, it is possible to make a working model of the thorax. The bell jar represents the thorax; the sheet of rubber is tied firmly around the base and represents the diaphragm; the two balloons, representing the lungs, are attached to the two branches of a Y-shaped glass tube, which is inserted through the cork at the top of the bell jar.
- When the 'diaphragm' is pulled downwards, the 'lungs' should inflate.
- Get the students to describe and explain what is happening. Ask: 'In what ways does this differ from a human thorax?'
- **Looking at lungs** – Obtain the lungs and trachea of a sheep or pig from a local butcher (find one where they slaughter their own or can get them for you when given notice).
- Provide the students with a work sheet, so that they can fill in details of colour, texture, size and what happens when air is introduced into the trachea via a hose. When attaching a hose (or bellows) to the trachea, make sure the joint is airtight and avoid blowback if blowing into it. **Safety**: Wash hands. Use a cycle pump or foot pump to inflate the lungs. Keep the lungs in a large plastic bag to contain aerosols. CLEAPSS leaflet P52.
- **Spare ribs!** – As a sequel to the exercise above, show the students some (uncooked) spare ribs – a rack of ribs if possible. Observe the muscles and the cartilage and link with how they are arranged in the thorax. This introduces the idea that there are muscles controlling the movements of the ribcage.
- **Analysis of inhaled and exhaled air** – The experiment shown in the Student Book simply indicates that the air breathed in contains less carbon dioxide than the air breathed out. **Safety**: Eye protection. Lime water CLEAPSS Hazcard 18.
- A more sophisticated experiment using a J-tube to analyse the oxygen and carbon dioxide content of inhaled and exhaled air was described in lesson B2.5.1, 'Controlling internal conditions'. If it was not done earlier, it would fit in well here. If it was done, then students could be reminded of the experiment and the results compared with the table given in the Student Book.

PLENARIES

Simulations of gas exchange at the alveoli – Use a digital teaching tool such as Multi Media Science School to display and interact with simulations of gas exchange at the alveoli. (10 minutes)
O_2 in, CO_2 out – Give students cards to hold (or pin on badges on hats) that represent the parts of the respiratory system. Let one student represent 'oxygen' and get them to pass down the system and eventually into the blood and to the cells, where they join with a student labelled 'carbon'. They both come back up and out as 'CO_2'. Get the students involved so that they describe what is happening at each stage. (10 minutes)

ACTIVITY & EXTENSION

- **Alveoli under the microscope** – Either mount some lung tissue from the sheep or pig's lungs looked at earlier, or show students prepared slides of lung tissue. Get them to comment on the blood supply and the proximity of the capillaries to the air sacs. If there are red blood cells on the slides, then it is possible to emphasise the thin nature of the alveolar wall and the short diffusion paths for the gases.
- **CPR** – Get the students to think about how you can breathe for someone who is unconscious and not breathing. Link to PSHE. Show clips from *Casualty* or a similar programme. Get someone in to demonstrate on the model used when learning Cardiopulmonary resuscitation (do not attempt to try it out on a real person!).

Answers to in-text questions

a) The chest, where the lungs and heart are found; the upper part of the body.
b) Gas exchange – oxygen into the blood, carbon dioxide out of the blood.

Practical support

Comparing air breathed out and air breathed in

Equipment and materials required

See diagram in Student Book, page 225.

Details

For the experiment illustrated in the Student Book, you will need sets of apparatus made up as shown. If students are using this apparatus, they should be supervised. As an alternative to lime water, hydrogencarbonate indicator could be used. It will be cherry red when in equilibrium with atmospheric air, but turns yellow as carbon dioxide is bubbled through it.

Safety: Eye protection. Lime water CLEAPSS Hazcard 18. Hydrogencarbonate indicator CLEAPSS Recipe Card 34

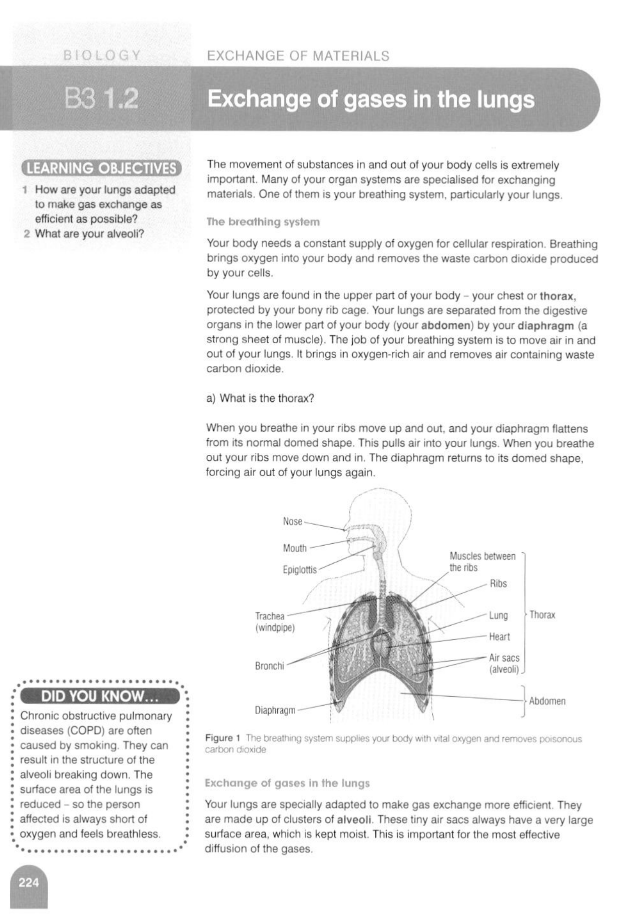
BIOLOGY EXCHANGE OF MATERIALS

B3 1.2 Exchange of gases in the lungs

LEARNING OBJECTIVES
1 How are your lungs adapted to make gas exchange as efficient as possible?
2 What are your alveoli?

The movement of substances in and out of your body cells is extremely important. Many of your organ systems are specialised for exchanging materials. One of them is your breathing system, particularly your lungs.

The breathing system

Your body needs a constant supply of oxygen for cellular respiration. Breathing brings oxygen into your body and removes the waste carbon dioxide produced by your cells.

Your lungs are found in the upper part of your body – your chest or **thorax**, protected by your bony rib cage. Your lungs are separated from the digestive organs in the lower part of your body (your **abdomen**) by your **diaphragm** (a strong sheet of muscle). The job of your breathing system is to move air in and out of your lungs. It brings in oxygen-rich air and removes air containing waste carbon dioxide.

a) What is the thorax?

When you breathe in your ribs move up and out, and your diaphragm flattens from its normal domed shape. This pulls air into your lungs. When you breathe out your ribs move down and in. The diaphragm returns to its domed shape, forcing air out of your lungs again.

Figure 1 The breathing system supplies your body with vital oxygen and removes poisonous carbon dioxide

Exchange of gases in the lungs

Your lungs are specially adapted to make gas exchange more efficient. They are made up of clusters of **alveoli**. These tiny air sacs always have a very large surface area, which is kept moist. This is important for the most effective diffusion of the gases.

DID YOU KNOW...
Chronic obstructive pulmonary diseases (COPD) are often caused by smoking. They can result in the structure of the alveoli breaking down. The surface area of the lungs is reduced – so the person affected is always short of oxygen and feels breathless.

224

Further biology

b) What is the function of the alveoli?

The alveoli also have a rich blood supply. This maintains a concentration gradient in both directions. Oxygen is constantly removed into the blood and more carbon dioxide is constantly delivered to the lungs. As a result, gas exchange takes place along the steepest concentration gradients possible, making it rapid and effective. The layer of cells between the air in the lungs and the blood in the capillaries is also very thin. This lets diffusion take place over the shortest possible distance.

An analysis of the gases in inhaled and exhaled air shows clearly the differences in the quantities of some of the main gases.

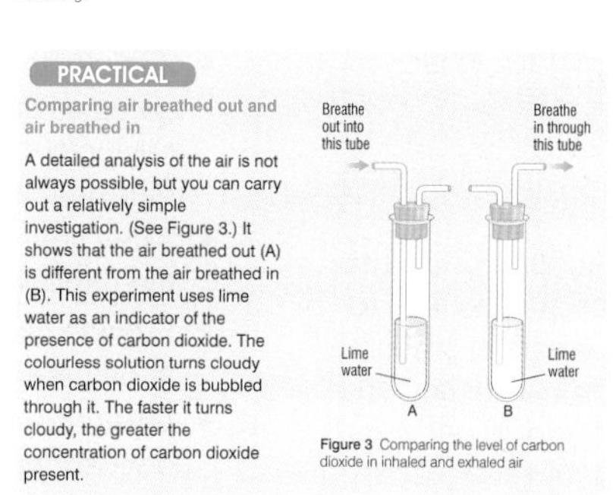

Figure 2 The alveoli are adapted so that gas exchange can take place as efficiently as possible in the lungs

FOUL FACTS
If all the alveoli in your lungs were spread out flat, they would have a surface area about the size of a tennis court!

Atmospheric gas	Air breathed in	Air breathed out
nitrogen	About 80%	About 80%
oxygen	20%	About 16%
carbon dioxide	0.04%	About 4%

KEY POINTS
1 Your breathing system takes air into and out of your body.
2 Oxygen from the air diffuses into your bloodstream and carbon dioxide diffuses out.
3 The alveoli of the lungs provide a very large, moist surface area with a rich blood supply and thin walls to make diffusion as effective as possible.

PRACTICAL
Comparing air breathed out and air breathed in

A detailed analysis of the air is not always possible, but you can carry out a relatively simple investigation. (See Figure 3.) It shows that the air breathed out (A) is different from the air breathed in (B). This experiment uses lime water as an indicator of the presence of carbon dioxide. The colourless solution turns cloudy when carbon dioxide is bubbled through it. The faster it turns cloudy, the greater the concentration of carbon dioxide present.

Breathe out into this tube
Breathe in through this tube
Lime water
Lime water
A
B

Figure 3 Comparing the level of carbon dioxide in inhaled and exhaled air

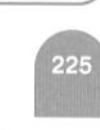

SUMMARY QUESTIONS
1 What is meant by the term gaseous exchange and why is it so important in your body?
2 How are the lungs adapted to allow gas exchange to take place as effectively as possible?
3 Draw a bar chart to show the difference in composition between inhaled and exhaled air. (Use the table above.)

225

SUMMARY ANSWERS

1 Gaseous exchange is the exchange of the gases oxygen and carbon dioxide in the lungs. This is vital because oxygen is needed by the cells for cellular respiration to provide energy while carbon dioxide is a poisonous waste product which must not be allowed to build up.

2 The lung tissue is arranged into clusters of alveoli (tiny air sacs), which give ideal conditions for rapid gas exchange by diffusion. They have a very large surface area that is kept moist, and a rich blood supply so that a concentration gradient is maintained in both directions. Oxygen is constantly removed into the blood and more carbon dioxide is constantly delivered to the lungs. This makes sure that gas exchange can take place along the steepest concentration gradients possible, so that it occurs rapidly and effectively. The walls between the air in the lungs and the blood in the capillaries are also very thin, so that diffusion takes place over the shortest possible distance. [Students could answer with an annotated diagram.]

3 Student bar chart.

KEY POINTS

The key points here would make very good revision cards, particularly if accompanied by diagrams summarising breathing movements and the exchange of gases at the alveoli. The exchange of gases at the alveoli could be annotated with reference to the relevant diffusion/concentration gradients of oxygen and carbon dioxide.

B3 1.3 Exchange in the gut

LEARNING OBJECTIVES

Students should learn:

- That the villi increase the surface area of the small intestine.
- That the villi have an extensive network of capillaries to absorb the products of digestion.
- That the products of digestion are absorbed by diffusion and active transport.

LEARNING OUTCOMES

Most students should be able to:

- Describe the adaptations of the small intestine that increase the efficiency of absorption.
- Describe the structure of a villus.

Some students should also be able to:

- Explain how exchange surfaces are adapted to maximise efficiency.
- Explain how food is moved from the gut into the blood by active transport as well as diffusion.

Teaching suggestions

- **Special needs.** Students can make a model ileum by sticking towelling to the inside of a wide-bore, flexible, plastic pipe (or pink rain jacket sleeve) and then inverting it. Place wicks into the model and lead them to the pipes, symbolising the blood supply.
- **Gifted and talented.** Students could use geometry to work out the surface area to volume ratio of a 10 cm length of smooth tube, one with a hundred villi per cm^2 and one with one hundred microvilli per villus. Each villus is 2 mm in length and 0.2 mm in diameter. Each microvillus is 100 microns in length and 10 microns in diameter. For ease of calculation, they can assume perfect cylinders. They then use the formula $\pi r^2 D$ for surface area, where $\pi = 3.14$, $r =$ radius of villus or microvillus and $D =$ length of villus or microvillus.
- **Learning styles**

  Kinaesthetic: Playing card games.

  Visual: Observing villi under microscopes.

  Auditory: Talking through bacon sandwich journey with a peer.

  Interpersonal: Modelling exercise with tokens, as described in the plenary 'Importance of a good blood supply'.

  Intrapersonal: Carrying out surface area calculations.
- **Homework.** Students to finish off the story of the bacon sandwich.

SPECIFICATION LINK-UP Unit: Biology 3.13.1

- *Many organ systems are specialised for exchanging materials.*
- *The villi of the small intestine provide a large surface area, with an extensive network of capillaries, to absorb the products of digestion by diffusion and active transport.*

Lesson structure

STARTER

Efficient absorption – Spill a drink on purpose next to a student (avoiding them and any of their possessions). Give them a piece of cloth with poor absorbent qualities (e.g. a piece of synthetic trouser material) and ask them to clean it up. Do the same next to another student, but give them a fluffy towel to dry it up with. Draw out a discussion as to why the towel is so much better than the trouser material. Link this to the digestive system. (5–10 minutes)

Looking at the ileum – Examine a piece of inside-out cow's ileum. This is easily and cheaply available in pet stores as dog chews (sometimes labelled as tripe, but avoid honeycomb pieces). Keep it sealed inside its plastic bag as it can be smelly. Get the students to describe the inside-out surface and suggest reasons why it is like that. **Safety:** Wash hands after handling. (5–10 minutes)

Getting through the gut wall – Make some large molecules, such as starch and proteins, out of Duplo or similar building bricks and use plastic knives, representing enzymes to cut them up. A mixture of the large molecules and the smaller ones is placed into a Christmas tree net, or similar large mesh bag. Ask: 'Which ones go through?' (5–10 minutes)

MAIN

- Use Simulation B3 1.3 'Exchange in the gut' from the GCSE Biology CD ROM.
- There are some good scanning electron micrographs, but other prepared sections may be difficult to interpret unless accompanied by a diagram.
- If students are provided with a work sheet they can make their own sketches and notes as the presentation proceeds. It is possible to present the small intestine as having two important functions: it provides a large surface area for the completion of digestion as well as for the absorption of the products of digestion.
- Allow time for students to view sections of small intestine for themselves and to note the capillary network.
- Provide diagrams or use the students' book to help the students identify the structures in both the ileum and the villi.
- **Endoscopy** – There are several web sites where it is possible to download endoscope pictures of the small intestine. A video sequence could be shown to students separately. Search the Internet for 'video endoscopy'.
- **Small molecules** – The importance of the digestion of large, insoluble molecules into smaller, soluble ones can be demonstrated by using Visking, or dialysis, tubing to model the gut.
- These experiments have already been described in B2.4.4 'Enzymes in digestion'. A number of different experiments were described and it could be appropriate here to set up some that were not done.
- **Alternatively . . .** – Students could be asked to design an experiment to show the need for the digestion of large molecules, such as starch, into smaller insoluble sugars that could pass through the gut wall. They could then use their previous knowledge to help them.

PLENARIES

How big are your intestines? – Go to the gym or a large outdoor space and mark out an area of 2000 m^2. Describe this as being the surface area of your intestines when fully spread out. Ask: 'How can this be?' Back in the lab, give each group of students a ball of string and a small match box. Run a competition to see who can get the longest piece of string inside the matchbox. Link this with the length of the small intestine in the abdominal cavity. (10–15 minutes)

A bacon sandwich: my story – Describe the fate of a bacon sandwich from eating it to the defecation of the remains. Draw out what happens to all the parts, the bread, the butter and the bacon. Use writing frames and support material if needed. Discuss and start in class and students to finish off for homework. (10 minutes)

Importance of a good blood supply – Carry out a modelling exercise where some students are designated as being villi. They give soluble food tokens to students representing the blood. These students take the food tokens to the liver (another student) who counts them per minute. Carry this out in single file slowly, then multiples as fast as possible. Draw out the importance of a good blood supply to the villi. (10 minutes)

EXCHANGE OF MATERIALS

B3 1.3 Exchange in the gut

LEARNING OBJECTIVES

1. What are the adaptations in your small intestine which allow you to absorb food efficiently?
2. Why are your villi so important?

Your gut is an area of your body where the exchange of materials is extremely important. The food you eat is broken down in your gut. It forms simple sugars such as glucose, amino acids, fatty acids and glycerol. But these products of digestion are of no use if they stay in your gut. They would simply be passed out of your body in your faeces.

Absorption in the small intestine

The molecules from food need to be made available to your body cells. In cells they provide fuel for respiration and the building blocks of all the tissues of your body. For this to happen they must move from the inside of your small intestine into your bloodstream. They do this by a combination of diffusion and active transport.

a) Why must the products of digestion get into your bloodstream?

This explains one reason why it is so important that your food is broken down into a soluble form during digestion. Only when the molecules are dissolved in water can diffusion take place.

The digested food molecules are then small enough to pass freely through the walls of the small intestine into the blood vessels. They move in this direction because there is a very high concentration of food molecules in the gut and a much lower concentration in the blood. In other words, they move into the blood along a steep concentration gradient.

b) Why is it so important that your food is broken down into smaller molecules?

The lining of the small intestine is folded into thousands of tiny finger-like projections known as **villi**. These greatly increase the uptake of digested food by diffusion. (See Figure 1.) Only a certain number of digested food molecules can diffuse over a given surface area of gut lining at any one time.

DID YOU KNOW...

The villi aren't the only way in which the surface area of the small intestine is increased. Each individual villus is itself covered in many microscopic microvilli. This increases the surface area available for diffusion many times over! Although your gut is only around 7 metres long, the way it is folded into your body along with the villi and microvilli give you a surface area of 2000 m^2!

Figure 1 Without the villi of the small intestine we would be unable to absorb enough digested food to survive. They increase the surface area available for diffusion many times.

226

Further biology

So increasing the surface area means there is more room for diffusion to take place.

Like the lungs, the lining of the small intestine has an excellent blood supply. This carries away the digested food molecules as soon as they have diffused from one side to the other. So a steep concentration gradient is maintained all the time, from the inside of the intestine to the blood. This in turn makes sure diffusion is as rapid and efficient as possible.

Diffusion isn't the only way in which dissolved food substances move from the gut into the blood. What happens as the food moves down the small intestine and the time since the last meal gets longer?

Glucose and other dissolved food molecules are moved from the small intestine into the blood by active transport, against the concentration gradient. (See page 222.) This makes sure that none of the digested food is wasted and lost in the faeces.

c) Why is it so important that the villi have a rich blood supply?

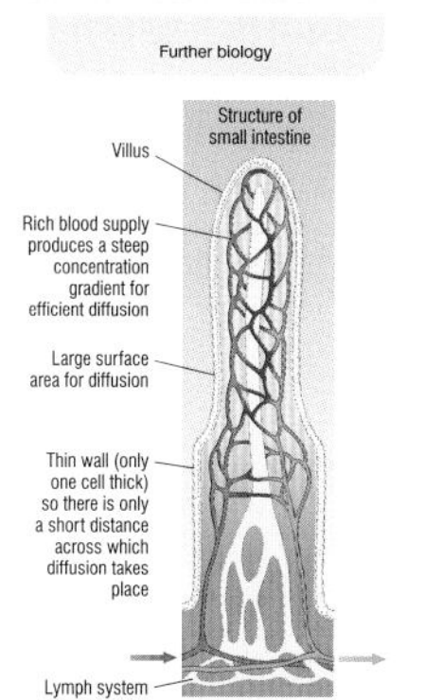

Figure 2 Thousands of finger-like projections in the wall of the small intestine – the villi – make it possible for all the digested food molecules to be transferred from your small intestine into your blood by diffusion

Exchange of materials in other organisms

Human beings are not the only organisms where an exchange of materials is important. Whether it is the gills of a fish or the kidneys of a desert rat, certain adaptations will always be seen:

- a large surface area to give plenty of opportunity for substances to diffuse,
- a rich blood supply to remove the substances, maintaining a steep concentration gradient and carrying them to where they are needed,
- moist surfaces for substances to dissolve,
- a short distance between the two areas so diffusion happens effectively. You can read more about this on the next two pages.

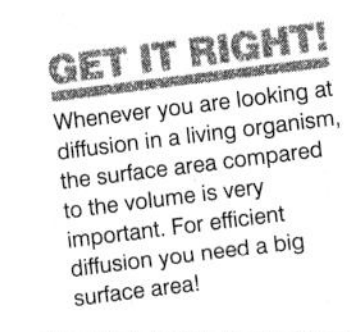

GET IT RIGHT!
Whenever you are looking at diffusion in a living organism, the surface area compared to the volume is very important. For efficient diffusion you need a big surface area!

SUMMARY QUESTIONS

1 Match A, B, C or D to the correct ending (1 to 4).

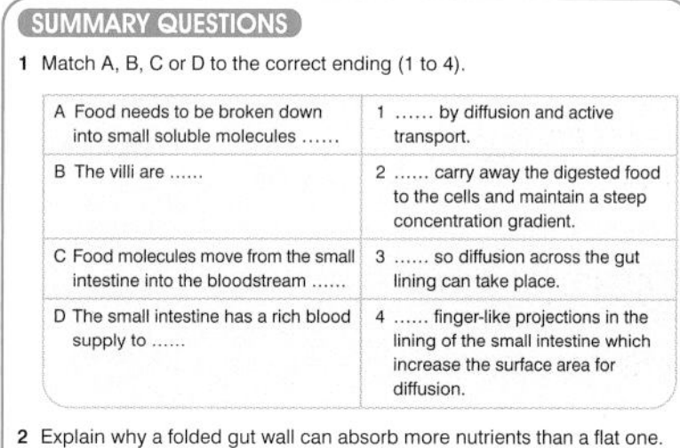

A Food needs to be broken down into small soluble molecules	1 ...... by diffusion and active transport.
B The villi are	2 ...... carry away the digested food to the cells and maintain a steep concentration gradient.
C Food molecules move from the small intestine into the bloodstream	3 ...... so diffusion across the gut lining can take place.
D The small intestine has a rich blood supply to	4 ...... finger-like projections in the lining of the small intestine which increase the surface area for diffusion.

2 Explain why a folded gut wall can absorb more nutrients than a flat one.

3 Places where materials are exchanged in the body always have a large, moist surface area, short distances across boundaries and a rich blood supply. Why is each of these features important for successful diffusion?

KEY POINTS

1. The villi in the small intestine provide a large surface area with an extensive network of capillaries. This makes them well adapted to absorb the products of digestion by diffusion and active transport.
2. In material exchanges, the surface area : volume ratio is always important – a big surface area is vital for successful diffusion.

227

SUMMARY ANSWERS

1. Food needs to be broken down into soluble molecules so diffusion across the gut lining can take place.
   The villi are finger-like projections in the lining of the small intestine, which increase the surface area for diffusion.
   Food molecules move from the small intestine into the bloodstream by diffusion and active transport.
   The small intestine has a rich blood supply to carry away the digested food to the cells and maintain a steep concentration gradient.
2. A folded gut wall has a much larger surface area over which nutrients can be absorbed.
3. An exchange surface has:
   - a large surface area to give plenty of opportunity for substances to diffuse
   - a rich blood supply to remove the substances, maintaining a steep concentration gradient and carrying them to where they are needed
   - moist surfaces for substances to dissolve
   - thin walls between the two areas as diffusion occurs most effectively over a short distance.

Answers to in-text questions

a) So that the soluble products can be carried to the cells of the body where they will be used.

b) So they are soluble and diffusion can take place.

c) To carry dissolved food molecules to the cells of the body; to maintain a concentration gradient.

KEY POINTS

The key points of this spread can be summarised by a series of drawings of a section of the small intestine, a diagram of a villus and a close-up of an epithelial cell. Each diagram should be fully annotated with the appropriate adaptations.

B3 1.4 Exchange of materials in other organisms

LEARNING OBJECTIVES

Students should learn:

- That the gas and solute exchange surfaces of many organisms are adapted to be as effective as possible.
- That the gills of fish and amphibians are adapted for gas exchange in water.
- That the respiratory system of insects is adapted for gas exchange in air.
- That the respiratory system of adult amphibians is often adapted for gas exchange in air and water.

LEARNING OUTCOMES

Most students should be able to:

- Describe the respiratory systems of a fish, an amphibian and an insect.
- List the main adaptations of a successful respiratory surface.

Some students should also be able to:

- Describe how a fish gill works and explain why it is ineffective in air.
- Explain the differences between the respiratory adaptations of larval and adult amphibians.
- Relate the respiratory adaptations of insects to general principles of gas exchange organs.

Teaching suggestions

- **Special needs.** These students could be provided with an aquarium containing small fish, such as guppies or goldfish, and describe the movements the fish make. They could try counting opercular movements and working out a rate.
- **Gifted and talented.** The solubility of oxygen in water decreases as the temperature of the water increases. Design an experiment to investigate the effect of the temperature of the water on the breathing movements of fish in an aquarium.
- **Learning styles**

  Kinaesthetic: Investigate the external features and gills of a fish.

  Visual: Watching the video of the frog life cycle.

  Auditory: Listening to descriptions of what the phrase 'A fish out of water' means.

  Interpersonal: Discussing the problems of amphibians.

  Intrapersonal: Working out the breathing rate of a fish or a locust.

SPECIFICATION LINK-UP Unit: Biology 3.1.4

Students should use their skills, knowledge and understanding of how science works:

- *to explain how other gas and solute exchange surfaces in humans and other organisms are adapted to maximum effectiveness.*

Lesson structure

STARTER

Exchange surfaces – Show a series of pictures of a variety of living organisms (suggest protoctistans, seaweeds, worms, molluscs, etc., need some obvious ones and some more obsure) and ask the students to write down where gas exchange takes place. Check answers. (10 minutes)

Like a fish out of water! – Ask a student to describe what this phrase means to them. Discuss with the class. Show a video of a fish floundering around on land or ask any student who goes fishing to describe what happens and why. (10 minutes)

MAIN

- **Investigation of the gills of a fish** – For this exercise, it would be good for the students to work in groups and you will need one whole fish per group. Herrings or mackerel would be suitable, but sprats could be substituted. The fish should be as fresh as possible. Students can be provided with a work sheet on which they can record observations about the external appearance of the fish, the location of the gills, the connection between the mouth and the gills and the appearance of the gills. The operculum (gill cover) can be cut away and the gills observed *in situ*. Small pieces of gill tissue can then be removed and observed with a hand lens or under the low power of a microscope.
- Alternatively, the exercise can be done as a demonstration. The observations of the gill structures can be projected on to a screen.
- **Frogs and tadpoles** – Show a video of the life cycle of the frog. There are several available: try the BBC web site, Science & Nature: Animals. Ask students to write down how gas exchange occurs at each stage of the life cycle. Discuss the problems of animals, such as amphibians, that spend some stage of their life cycle in water and some on land.
- **A plague of locusts!** – Show the students some live locusts (obtainable from suppliers). They will need to be kept in a special vivarium and provided with light and food (privet leaves, grass, etc.). These insects are large enough for the spiracles to be visible on the thorax and the abdomen. It is possible to set up a demonstration of the action of the thoracic and abdominal spiracles by fixing a live locust to a cork mat and using a binocular microscope.
- The students could be given a locust in a beaker (with a lid on it to prevent escape) or a large boiling tube, and a hand lens. They could make observations of the opening and closing of the spiracles. It is possible to count the movements and to work out the rate of breathing. If the temperature increases slightly, does the rate change?
- Link the previous exercise with a large diagram of the tracheal system of an insect, such as the locust, projected on to the board. Discuss the pathway of the air and the way in which the tracheal system supplies oxygen to the muscles and the organs of the body.
- Use Simulation B3 1.4 'Exchange of material' from the GCSE Biology CD ROM.

PLENARIES

Quick quiz – Prepare ten short questions on the topic of the lesson (to include the properties of exchange surfaces). Students to write down answers. (5 minutes)

Breathing on land vs breathing in water – Build up a table comparing the two processes, to include sources of oxygen, where gas exchange occurs, structures involved. Discuss which appears to be most efficient. (10 minutes)

ACTIVITY & EXTENSION

- **Exchange in land invertebrates** – Make a list of all the soft-bodied creatures that are found in gardens (such as earthworms, slugs, snails, spiders, millipedes, etc.). and find out how exchange of gases occurs in each group. Do they have gills? Are they restricted in their habitats? How fast do they move?
- **Cockles and mussels!** – Investigate the exchange surfaces of marine molluscs. Where are their gills? What other functions do their gills have in addition to exchange of gases?
- **A marine aquarium** – Set up a seawater aquarium in the laboratory. A visit to the seashore will enable you to stock it with anemones, molluscs and other marine invertebrates, which can then be observed in their natural habitat.
- **Looking at fish gills** – Show students a fresh herring and remove the operculum to show the gills. Mount a small piece of gill and project so that students can see the structure. Students can check the adaptations shown against the list given in the Student Book.

Safety: Take care with sharp instruments.

Practical support

Hand lenses or binocular microscopes, microscope slides and cover slips will be needed if the fish gills are to be viewed.
If the students are to remove the gill covers, then they will need dissecting boards and scalpels.

KEY POINT

The key point is reinforced by the Quick Quiz plenary. The students should be able to explain why each of the features of exchange surfaces is necessary for maximum efficiency.

BIOLOGY

EXCHANGE OF MATERIALS

B3 1.4 Exchange of materials in other organisms

LEARNING OBJECTIVE

1 How does gas and solute exchange take place in other organisms?

DID YOU KNOW?

There is about 20 times as much oxygen in the air as there is in water. Perhaps that is why most warm blooded animals, which need a lot of oxygen to maintain their body temperature, live on the land and breathe air!

Gas exchange in a fish

Fish cannot get oxygen directly from the water they live in because their bodies are covered in protective scales. Fortunately fish have evolved a very effective respiratory system which works really well in water.

Gills are made up of many thin layers of tissue with a rich blood supply. (See Figure 1 below.) The gills are thin so there is only a short distance for the gases to diffuse across. The surfaces are always moist as they work in water!

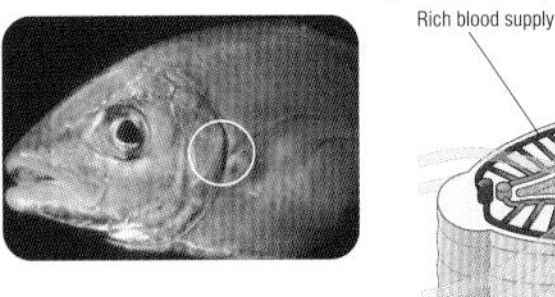

Rich blood supply
Very thin tissue – short distance for gases to diffuse across
Constant flow of water
Large surface area
Gill stacks

Figure 1 The gills of a fish – another example of an organ adapted for efficient gas exchange

In bony fish the gills are contained in a special gill cavity. Water is pumped over them constantly to maintain a concentration gradient. Fish such as sharks have to keep swimming all the time to keep water moving over their gills!

a) Why do fish need gills?

Unfortunately gills can't work in air – a fish out of water 'suffocates'. Without water surrounding them, the gill stacks all stick together. There simply isn't a big enough surface area available for the fish to get the oxygen it needs to survive from the air.

b) Why do fish die when they are taken out of water?

Tadpoles and frogs

Frogs are amphibians and have a very strange life history. The eggs hatch into tadpoles which spend all their time in water. Young tadpoles have frilly external gills with a large surface area and a rich blood supply. The tadpoles get all their oxygen by diffusion from the water through these gills. In the same way carbon dioxide diffuses out along a concentration gradient into the water.

When the tadpoles turn into frogs they spend a lot of time on the land – but they can still breathe in water!

c) Why would external gills be no use for an adult frog?

228

Further biology

The external gills disappear and are reabsorbed into the body of the developing frog. We say that the tadpole undergoes ***metamorphosis***. An adult frog has very moist skin with a rich blood supply, and under normal conditions most of its gas exchange takes place through the skin.
The mouth – which is very large and thin skinned – is also important for gas exchange. If it gets hot or the frog is being very active on land, it also has a pair of very simple lungs. These can be used to increase the surface area available for gas exchange to take place. When the frog is in the water, all the gas exchange takes place through the skin.

Figure 2 Tadpoles get all the oxygen they need from the water around them, while adult frogs can exchange gases with either water or air. They both have different adaptations for gas exchange. But the respiratory surfaces of both tadpoles and frogs have a big surface area, a rich blood supply, short diffusion routes and need to be moist.

The respiratory system in insects

Many insects are very active so their muscles need a lot of oxygen. However, little or no gaseous exchange can take place through the tough outer covering of insects. To supply their needs, insects have an internal respiratory system which supplies oxygen directly to their cells and removes carbon dioxide. (See Figure 3.)

If you look along the side of an insect you can see the **spiracles**. These can open when the insect needs plenty of oxygen but close when they don't. This prevents water loss, rather like the guard cells of plants.

The spiracles lead into a system of tubes which run right into the cells of the tissues themselves. Most of the gas exchange takes place in the **tracheoles**. These tiny tubes are freely permeable to gases. They are very moist and air is pumped in and out of them by the insect to maintain a concentration gradient.

There is no blood supply in an insect. However, the tracheoles have a very large surface area and come into close contact with individual cells in the body of the insect. So they are very effective at gas exchange.

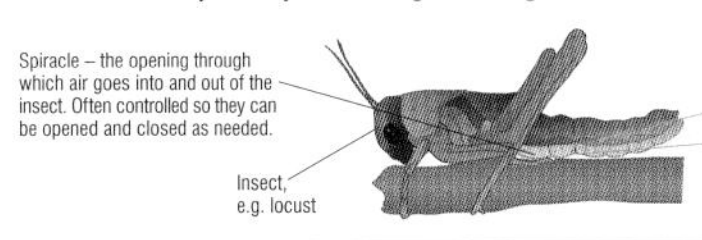

Spiracle – the opening through which air goes into and out of the insect. Often controlled so they can be opened and closed as needed.
Insect, e.g. locust

Tracheoles are tiny tubes with a large surface area and moist lining. They are freely permeable to gases and pass right into the tissue of the insect, between the cells. Most gas exchange takes place here.
Trachea – largest tubes carrying air into the insect's body. Lined with rings of chitin, they are quite impermeable to gases, so little gaseous exchange takes place here.

Figure 3 The respiratory system of an insect has to do the same job as your breathing system. In spite of the very different design, there are many similar features which make gas exchange successful.

SUMMARY QUESTIONS

1 Copy and complete using the words below:

adaptations surfaces solute short surface area blood supply

Wherever gas or …… exchange is important, certain …… will always be seen. These include a large …… ……, a rich …… ……, moist …… and …… diffusion distances.

2 a) Why do fish need a constant flow of water over their gills?
b) Why are fish gills arranged in stacks?
c) Why do you think the human breathing system does not work in water?

3 Draw and label a diagram of a tadpole and a frog, showing clearly the different ways in which gas exchange takes place.

4 What are the main features of the respiratory system of an insect and how are they important in successful gas exchange?

KEY POINT

1 Whatever the organism, gas and solute exchange depends on a large surface area, moist surfaces, short diffusion distances and a large concentration gradient.

229

SUMMARY ANSWERS

1 solute, adaptations, surface area, blood supply, surfaces, short

2 **a)** To maintain a concentration gradient so oxygen diffuses from the water into the blood in the gills and carbon dioxide diffuses in the opposite direction from the blood into the water.
b) To give a large surface area of diffusion in a relatively small space.
c) Not enough surface area to get the oxygen needed as there is less oxygen dissolved in water than in the air, difficulty of getting rid of enough carbon dioxide as not enough surface area and difficulty of moving water into and out of the lungs to maintain diffusion gradient.

3 Mark for recognisable diagrams, clear labelling and explanation.

4 Main features: spiracles – holes which control the movement of gases in and out of the body and also water loss. They enable the insect to maintain a concentration gradient between the tissues and the air, and also allow more air in when the energy needs are greater while reducing water loss when they are low. Important to keep water levels up in the insect to keep surfaces moist for diffusion. Internal respiratory system of narrow tubes which supplies oxygen to tissues and removes carbon dioxide. These carry the respiratory gases deep into the tissues where they are needed so gaseous exchange can take place. Very tiny tubes which are moist, freely permeable to gases and have a very large surface area that enable effective gas exchange yet protects from drying out.

Answers to in-text questions

a) To take oxygen from the water for cellular respiration and to get rid of waste carbon dioxide.
b) The gill stacks stick together so there isn't enough surface area for the fish to get the oxygen it needs from the air, or to get rid of waste carbon dioxide and so it dies.
c) Because they spend a lot of time out of water.

B3 1.5 Exchange in plants

LEARNING OBJECTIVES

Students should learn:

- That carbon dioxide enters the leaf cells by diffusion through stomata.
- That water and mineral ions are absorbed by root hair cells.
- That the root hairs increase the surface area of the roots and the flattened shape and internal air spaces increase the surface area of the leaves.

LEARNING OUTCOMES

Most students should be able to:

- Describe how leaves are adapted for gaseous exchange.
- Describe how roots are adapted for the efficient uptake of water and mineral ions.

Some students should be able to:

- Explain why plants do not need carbon dioxide from the air under any circumstances.
- Apply the principles of exchange surfaces to exchange mechanisms in plants.

Teaching suggestions

- **Special needs.** Use a modification of the starter 'Round leaves v. flat leaves'. Give each student a block of Plasticine and see who can make the largest, thinnest leaf. Give them a jumbled sentence to complete on why a large surface area is an adaptation.
- **Gifted and talented.** Students could investigate the root systems of plants growing in different environments to see how they are adapted for the efficient uptake of water and mineral ions. They can compare root systems of some desert plants and dune plants with typical flowering plants.
- **Learning styles**

  Kinaesthetic: Preparing epidermal peels.

  Visual: Observing the root hairs.

  Auditory: Listening to the 'conversations between the molecules'.

  Interpersonal: Discussing adaptations of leaves.

  Intrapersonal: Completing homework exercise on the pathways of the gases.

SPECIFICATION LINK-UP Unit: Biology 3.13.1

- *In plants:*
  - *Carbon dioxide enters the leaf cells by diffusion.*
  - *Most of the water and mineral ions are absorbed by root hair cells.*
- *The surface area of the roots is increased by root hairs and the surface area of leaves by the flattened shape and internal air spaces.*
- *Plants have stomata to obtain carbon dioxide from the atmosphere.*

Lesson structure

STARTER

Round leaves vs flat leaves – Give each student a cube of Plasticine and ask them measure its volume and then to make a round thick leaf shape with it. Measure the surface area by placing on graph paper and drawing round it. Then ask them to flatten the leaf and make it as thin as possible. Measure the new surface area and work out the surface area to volume ratios of both leaves. Relate this increase in surface area to the greater efficiency of gas exchange. (10 minutes)

Revising leaf structure – Give each student a blank diagram of the external structure of a leaf, and a diagram of a transverse section through a leaf with the different tissues drawn in but not labelled. Allow 5 minutes for students to label all the parts they can. Students can then check each other's diagrams. (10 minutes)

MAIN

- **Gas exchange in plants** – Give each student a leaf (could be the one they will use to make a nail varnish peel) and project a transverse section through a leaf showing all the cells. **Safety:** Nail varnish is flammable and the vapour is harmful. You will need a good section that shows a distinct palisade layer and a definitely spongy mesophyll with large air spaces. Get the students to write down all the features that they think are adaptations enabling efficient gaseous exchange, both externally and internally. Gather the information together and make a list on the board.
- **Looking at stomata** – Use fresh leaves (privet is good).
- **Investigating root systems** – Use binocular microscopes to observe the root hairs on young cress seedlings. If cress seeds are sown on damp filter paper in Petri dishes, they will germinate and the roots will grow in a few days. Provided that the atmosphere in the dish is kept moist, it should be possible to see the root hairs with a microscope.
- Use prepared slides of longitudinal sections through young roots to show root hairs and, if possible, carry out measurements. Find out how far the root hair region extends. Ask: 'Can you see young root hairs developing or older root hairs breaking down?'
- Use Simulation B3 1.5 'Exchange in plants'.

PLENARIES

What was your journey like? – In small groups, students should write a conversation between a water molecule, a carbon dioxide molecule and a mineral ion as they meet in a leaf. They should describe their journeys to get there (as people go on about roads and journeys as small talk at parties) and ponder their fate. (10 minutes)

Transplant – Get students to explain why it is important to keep a ball of soil around seedlings or bedding plants when you plant them out. Ask: 'Why do young trees come from the nursery with their roots in a ball of soil?' (5 minutes)

Answers to in-text questions

a) The flattened shape of a leaf increases the surface area for diffusion; thin leaves have shorter distances for the carbon dioxide to diffuse from the outside air to the active photosynthesising cells; the many air spaces allow carbon dioxide to come into contact with lots of cells speeding up diffusion.

b) The leaves only photosynthesise in the light.

SUMMARY ANSWERS

1 The flattened shape of the plant leaves increases the surface area for diffusion. Most plants have thin leaves, which means the distance for the carbon dioxide to diffuse from the outside air to the active photosynthesising cells is kept as short as possible. The many air spaces inside the leaf allow carbon dioxide to come into contact with lots of cells, providing a large surface area for diffusion.

2 **a)** Stomata are small openings all over the leaf surface.

b) The stomata open during daylight allowing air into the leaves so that carbon dioxide enters the cells for photosynthesis, but they close the rest of the time to control the loss of water.

c) The opening and closing of the stomata is controlled by the guard cells.

3 **a)** Plant roots are thin, divided structures with a large surface area. The cells on the outside of the roots near the growing tips also have extensions, called root hairs, which increase the surface area for the uptake of substances from the soil. These tiny projections from the cells push out between the soil particles. The membranes of the root hair cells also have microvilli that increase the surface area for diffusion and osmosis even more. The water then has only a short distance to move across the root to the xylem, where it is moved up and around the plant.
Plant roots are also adapted to take in mineral ions using active transport. They have plenty of mitochondria to supply the energy they need, as well as all the advantages of the large surface area and short pathways needed for the movement of water as well.

b) The adaptations are very similar: large surface area, moist surfaces, and small distances to travel. Plants are not always as effective as animals at maintaining concentration gradients through active circulation, but they have plenty of active transport systems to help them out.

KEY POINTS

The emphasis here is on surface area available for the exchange of gases and the uptake of water and mineral ions. Both of these processes have strong links to photosynthesis and respiration in plants, so there is the opportunity to make cross-links in the specification.

Practical support 

Looking at stomata

Equipment and materials required

Fresh privet leaves, clear nail varnish, paintbrushes, forceps, microscope slides and cover slips, microscopes.

Details

Apply a thin layer of clear nail varnish to the lower surface of the leaf. Allow the nail varnish to dry and then carefully peel it off using forceps. Place the 'peel' in a drop of water on a microscope slide and cover it with a cover slip. Look at the slide using the low power of the microscope. The stomata should be visible, but use the high power of the microscope to see the more detailed structure, including the guard cells. Having made 'peels' of the lower epidermis of the leaf, the students could investigate the upper epidermis, comparing the numbers of stomata on each side. Ask: 'Are they the same? Which surface has the greater number?'
The density of the stomata can be determined. The area of the leaf can be found by drawing around it on graph paper and counting the number of squares. Using a calibrated eye piece graticule in the eyepiece of the microscope, the number of stomata in a field of view of known area can be counted and hence the total number of stomata on the leaf can be calculated, or the number per cm^2.

B3 1.5 Exchange in plants

LEARNING OBJECTIVES

1 How are the leaves of plants adapted for gaseous exchange?
2 How are roots adapted for the efficient uptake of water and mineral ions?

Animals aren't the only living organisms that need to exchange materials. Plants rely heavily on diffusion to get the carbon dioxide they need for photosynthesis. Also important are osmosis to take water from the soil and active transport to obtain minerals from the soil. Plants have adaptations which make these exchanges as efficient as possible.

Gas exchange in plants

Plants need carbon dioxide and water for photosynthesis to take place. They get the carbon dioxide they need by diffusion through their leaves. The flattened shape of the leaves increases the surface area for diffusion. Most plants have thin leaves. This means the distance for the carbon dioxide to diffuse from the outside air to the photosynthesising cells is kept as short as possible.

What's more, the many air spaces inside the leaf allow carbon dioxide to come into contact with lots of cells. This provides lots of surface area for diffusion.

a) How are leaves adapted for efficient diffusion of carbon dioxide?

However, there is a problem. Leaf cells constantly lose water by evaporation. If carbon dioxide could diffuse freely in and out of the leaves, water vapour would also be lost very quickly. Then the leaves – and the plant – would die.

The leaf cells do not need carbon dioxide all the time. When it is dark, they don't need carbon dioxide because they are not photosynthesising. When light is a limiting factor on the rate of photosynthesis, the carbon dioxide produced by respiration can be used for photosynthesis. But on bright, warm, sunny days a lot of carbon dioxide needs to come into the leaves by diffusion.

So leaves are adapted to allow carbon dioxide in only when it is needed. They are covered with a waxy cuticle. This is a waterproof and gas-proof layer. Then all over the leaf surface there are small openings known as **stomata**. The stomata can be opened when the plant needs to allow air into the leaves so that carbon dioxide enters the cells. But they can be closed the rest of the time to control the loss of water. The opening and closing of the stomata is controlled by the guard cells.

b) Why don't leaves need carbon dioxide all the time?

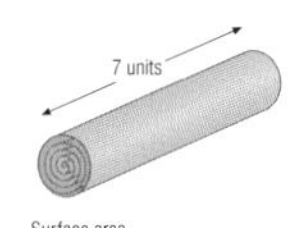

Surface area = 22 units²

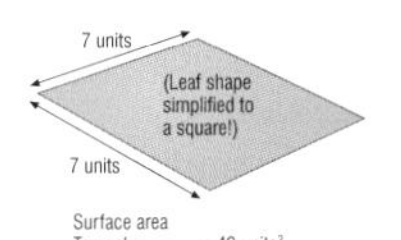

Surface area
Top only = 49 units²
Top and bottom = 98 units²

Figure 1 The wide, flat shape of most leaves greatly increases the surface area for collecting light and exchanging gases, compared with more cylindrical leaves

PRACTICAL

Looking at stomata

You can look at stomata by coating the surface of a leaf with nail varnish. Allow the varnish to dry, peel off the layer of varnish and look at it under a microscope – the stomata will be revealed!

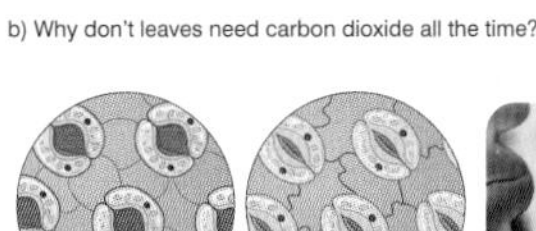

Figure 2 Guards cells open and close the stomata to control the carbon dioxide going into the leaf

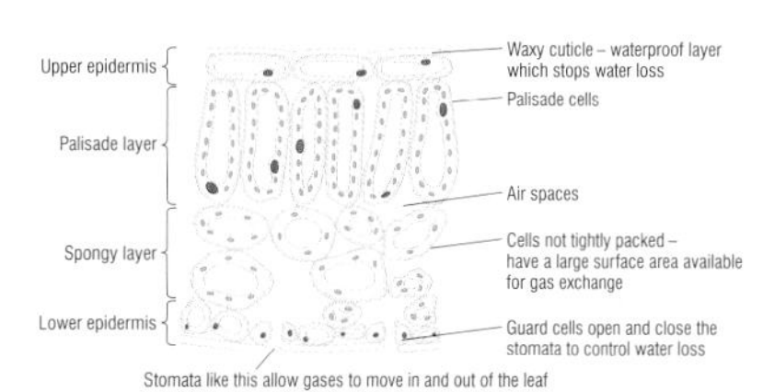

Figure 3 The arrangement of the cells inside a leaf, with plenty of air spaces and short diffusion distances, means that the carbon dioxide needed for photosynthesis reaches the cells as efficiently as possible

Uptake of water and mineral ions in plants

If you pull up a plant you will see a mesh of tiny white roots. These are adapted to enable plants to take water and minerals from the soil as efficiently as possible. Water is vital for plants. They need it to maintain the shape of their cells and for photosynthesis. They also need minerals to make proteins and other chemicals.

The roots themselves are thin, divided tubes with a large surface area. The cells on the outside of the roots near the growing tips also have their own adaptations. They increase the surface area for the uptake of substances from the soil. Known as root hair cells, they have tiny projections out from the cells which push out between the soil particles.

The membranes of the root hair cells also have microvilli. These increase the surface area for diffusion and osmosis even more. The water then has only a short distance to move across the root to the **xylem**, where it is moved up and around the plant.

Plant roots are also adapted to take in mineral ions using active transport. (See pages 222–3.) They have plenty of mitochondria to supply the energy they need. They also have all the advantages of a large surface area and the short pathways needed for the movement of water.

SUMMARY QUESTIONS

1 How is a plant leaf adapted for the diffusion of carbon dioxide from the air?
2 a) What are stomata?
b) What is their role in the plant?
c) How are they controlled?
3 a) How are plant roots adapted for the absorption of water and minerals?
b) How do the adaptations of plants for the exchange of materials compare with human adaptations in the lungs and the gut?

DID YOU KNOW...

The root hairs and microvilli of plants have an amazing effect – a 1 m² area of lawn grass has 350 m² of root surface area!

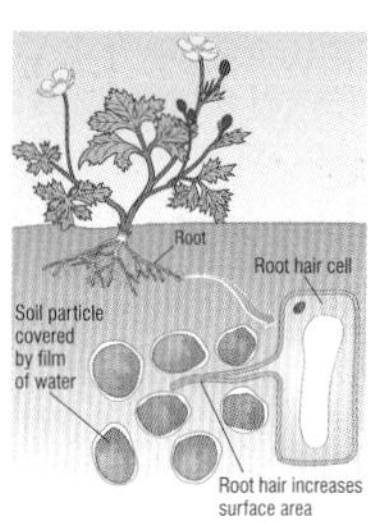

Figure 4 Many small roots, and the presence of microscopic root hairs on the individual root cells all increase diffusion of substances from the soil into the plant

KEY POINTS

1 Plants have stomata which allow them to obtain carbon dioxide from the atmosphere.
2 Carbon dioxide enters the leaf by diffusion. Leaves have a flat thin shape and internal air spaces to increase the surface area available for diffusion.
3 Most of the water and mineral ions needed by a plant are absorbed by the root hair cells which increase the surface area of the roots.

B3 1.6 Transpiration

LEARNING OBJECTIVES

Students should learn:

- That transpiration is the loss of water vapour through the stomata on the leaves of a plant.
- That the rate of transpiration is more rapid in hot, dry and windy conditions.
- That when plants lose water faster than it is replaced, the stomata can close to prevent further wilting.

LEARNING OUTCOMES

Most students should be able to:

- Explain why transpiration occurs.
- Describe the effect of environmental conditions on transpiration.
- Explain how water loss may be controlled.

Some students should be able to:

- Explain how to compromise between the need for carbon dioxide and water loss.

Teaching suggestions

- **Special needs.** Using pre-printed tables and graphs with the axes already drawn, carry out a mass–loss experiment with two plants, one exposed to moving warm air e.g. from a hair dryer. This can be done on a large scale using two spring balances at the front or individually with assistance if available. Remember to cover the pots of the plants with polythene bags, or to make sure there is no water evaporating from anywhere except the aerial parts of the plants.
- **Gifted and talented.** Here are things for students to think about:
  - What would limit the height to which water can travel up a tree trunk?
  - How would you estimate the leaf surface area of a whole tree?
  - How could you investigate the effect of humidity on the transpiration rate?
- **Learning styles**
  Kinaesthetic: Completing card sorting exercise.
  Visual: Producing a cartoon summary of the transpiration process.
  Auditory: Listening to discussions.
  Interpersonal: Working in groups on potometer experiments.
  Intrapersonal: Thinking about the graph interpretation.
- **Homework.** Give students graphs of water uptake (absorption) and water loss (transpiration) over a 24-hour period. (The peaks for transpiration and absorption are at different times.) They are to interpret these.

SPECIFICATION LINK-UP Unit: Biology 3.13.1

- *Plants lose water vapour from the surface of their leaves. This loss of water vapour is called transpiration. Transpiration is more rapid in hot, dry and windy conditions. Most of the transpiration is through stomata. The size of stomata is controlled by guard cells that surround them. If plants lose water faster than it is replaced by the roots, the stomata can close to prevent wilting.*

Lesson structure

STARTER

A clean shirt – Say that you have to go to a party tonight and your favourite shirt is dirty. Get a T-shirt and dip it into a bucket of water. Ask the students to suggest ways in which you can dry it the fastest. Relate this to evaporation from plant leaves. (5 minutes)

Brainstorm session: how does water get to the top of trees? – Working in groups, students to write down their ideas on large sheets of paper. Share results. (10 minutes)

Transpiration process – Use Animation B1 5.2 'Stomata' from the GCSE Biology CD ROM. (5 minutes)

MAIN

- **Modelling transpiration** – Place a wick through a drinking straw and clip the wick to a small piece of card with blotting paper attached to it, to imitate a leaf. Place the imitation leaf into a boiling tube containing dyed water. The water will travel up the wick and evaporate from the blotting paper.
- To cover investigative aspects of 'How Science Works', the variables, such as leaf size, temperature and wind speed, can be altered and the rate of transpiration can be ascertained by weight loss under these different conditions.
- This exercise lends itself to group work. One group could investigate leaf size, another the effects of temperature and so on. They need to report back at the end of the practical session.
- **Using potometers** – Potometers measure the rate of water uptake, which is linked with the rate at which water evaporates from the leaf surface. The best plant material to use is a woody twig, which can be firmly attached to the tubing. It is important that there are no bubbles in the system and that the whole apparatus is watertight.
- Once set up, it needs to be left to allow the plant to settle down after the handling. Introduction of an air bubble enables measurements of the water uptake to be made. Either the distance travelled by the bubble in a set time or the time taken for the bubble to travel a set distance can be measured.
- Discuss reliability and precision of measurements ('How Science Works'). Repetitions are necessary to calculate a mean rate under each set of conditions.
- This exercise can be used to develop other 'How Science Works' concepts: predictions can be made, hypotheses formulated, measurements taken and the results plotted.

PLENARIES

Computer simulation of potometer experiment – This can be done as a plenary if not done as part of the main session. Discuss the conditions. (5 minutes)

Sequencing session – Make cards of the stages in the process of transpiration from water uptake in the soil to evaporation from the leaf cells. Get students to put these into the correct order. This makes an excellent summary for a revision card. (5–10 minutes)

Graph interpretation – Give students a graph of the transpiration rate of a plant over 24 hours. Break it into sections labelled with letters. The students have to explain why the rate changes at different times of day. There could be different explanations for some sections. Combine thoughts at the end of the session. (10 minutes)

SUMMARY ANSWERS

1 Water evaporates from the surface of the leaves. As this water evaporates, water is pulled up through the xylem to take its place. This constant moving of water molecules through the xylem from the roots to the leaves is known as the 'transpiration stream'.

2 **a)** Transpiration is the loss of water vapour from the surface of plant leaves through the stomata.

b) The waxy cuticle and the guard cells prevent the loss of water vapour.

c) Petroleum jelly on the upper surface reduces water loss, as it would cover the stomata so less evaporation occurs. It also reduces the uptake of water, but not greatly as most stomata are on the underside of the leaves.

d) Petroleum jelly on the lower surface would greatly reduce the loss of water from the leaf, as most of the stomata would be covered and therefore no evaporation would take place. In turn, the rate of water uptake would be very much reduced.

e) The rate of transpiration would increase because the rapid air movement across the leaf would increase the rate of evaporation of water and so increase the uptake of water as well.

f) The uptake of water from the cut end of the plant stem.

3 [Any appropriate description that shows awareness of the need to keep all other conditions constant, etc.]

Answers to in-text questions

a) The transpiration stream is constant movement of water molecules through the xylem, from the roots to the leaves, where it evaporates out through the stomata.

b) Sunny, hot, dry, windy.

KEY POINTS

Students need to be sure that they understand that transpiration is a consequence of gaseous exchange in land plants. The conditions which cause a rapid rate of transpiration are those which favour the evaporation of water, i.e. increase in temperature, decrease in humidity and increase in air movements. Students might find it helpful to consider the diffusion gradients involved.

Practical support

Evidence for transpiration

Equipment and materials required

One potometer per group, preferably set up with the shoot inserted, electric fan to create air movements, bench lamps to provide light, hair dryer to provide warmer temperature (but it will create air movement as well), petroleum jelly to block stomata.

Details

It is important that there are no bubbles in the system and that the whole apparatus is watertight. Once set up, it needs to be left to allow the plant to settle down after the handling. Introduction of an air bubble enables measurements of the water uptake to be made. Either the distance travelled by the bubble in a set time or the time taken for the bubble to travel a set distance can be measured.

The electric fan will **increase the air movements**, the bench lamps will allow **the effect of light** to be investigated and the hair dryer allows the **effect of temperature** to be investigated. The **effect of changing the leaf area** can also be investigated, either by removing some of the leaves or by blocking the stomata with petroleum jelly. It is possible to calculate the uptake per unit area by measuring the total area of the leaves.

Safety: Take care with electrical equipment.

B3 1.6 Transpiration

LEARNING OBJECTIVES

1 What is transpiration?
2 When do plants transpire fastest?
3 How does wilting aid survival?

Figure 1 The transpiration stream can pull many litres of water up to 30 metres above the surface of the earth in a giant redwood tree like this one!

The top of a tree may be many metres from the ground, yet the leaves at the top need water just as much as the lower branches. So how do they get the water they need?

Water loss from the leaves

Plants have holes called stomata on the surfaces of their leaves. These stomata are opened to allow carbon dioxide into the plant for photosynthesis. But all the time that the stomata are open to allow carbon dioxide in, plants lose water vapour from the surface of their leaves. This loss of water vapour is what we call **transpiration**.

Stomata can be opened and closed by the guard cells which surround them. Losing water through the stomata is a side effect of opening them to let carbon dioxide in.

As water evaporates from the surface of the leaves, water is pulled up through the xylem to take its place. This constant movement of water molecules through the xylem from the roots to the leaves is known as the **transpiration stream**. It is driven purely by the evaporation of water from the leaves. So anything which affects the rate of evaporation will affect transpiration.

a) What is the transpiration stream?

The effect of the environment on transpiration

Firstly, conditions which increase the rate of photosynthesis will increase the rate of transpiration. Increased rates of photosynthesis mean more stomata are opened up to let carbon dioxide in. In turn, more water is lost by evaporation through the open stomata. So warm, sunny conditions increase the rate of transpiration.

Conditions which increase the rate of evaporation of water when the stomata are open will also make transpiration happen more rapidly. Hot, dry, windy conditions increase the rate of transpiration.

Figure 2 The transpiration stream

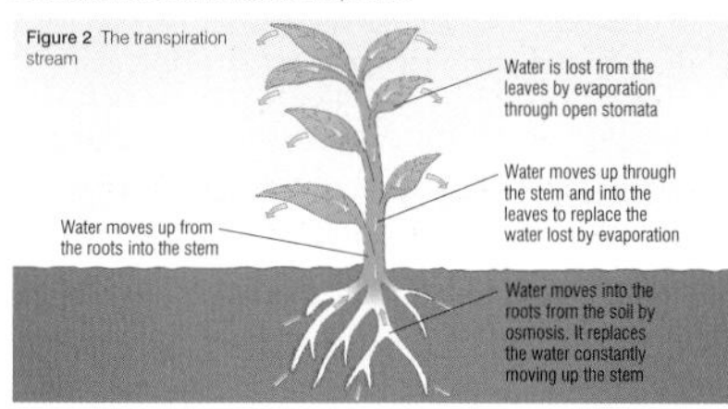
Water is lost from the leaves by evaporation through open stomata
Water moves up through the stem and into the leaves to replace the water lost by evaporation
Water moves up from the roots into the stem
Water moves into the roots from the soil by osmosis. It replaces the water constantly moving up the stem

b) Give three conditions that will increase the rate of transpiration from a leaf.

Controlling water loss

Plants have a constant problem – they need to open their stomata to photosynthesise but this means they are going to lose water. Most plants have a variety of adaptations which help them to photosynthesise as much as possible, while at the same time losing as little water as possible!

Most leaves have a waxy, waterproof layer (known as the **cuticle**) to prevent uncontrolled water loss. In very hot environments the cuticle may be very thick and shiny. Most of the stomata are found on the underside of the leaves. This means that they are not as exposed to the light and heat of the Sun, and reduces the time they are open.

If a plant begins to lose water faster than it is replaced by the roots, it can take some drastic measures. The whole plant may wilt. **Wilting** is a protection mechanism against further water loss. The leaves all collapse and hang down so the surface area available for water loss by evaporation is greatly reduced.

The stomata close, which stops photosynthesis and risks overheating. But this prevents most water loss and any further wilting. The plant will remain wilted until the temperature drops, the Sun goes in or it rains!

DID YOU KNOW...

Aquatic plants like water lilies have their stomata on the upper surface, so openings are not under water. On the other hand, they are never short of water so lots of transpiration isn't a problem for them!

GET IT RIGHT!

Remember that plants have to make a compromise – when their stomata are open to get carbon dioxide for photosynthesis they have to pay the price through water loss by transpiration.

PRACTICAL

Evidence for transpiration

There are a number of experiments which can be done to investigate the movement of water in plants by transpiration. Many of them use a piece of apparatus known as a potometer.

A potometer can be used to show how the uptake of water by the plant changes with different conditions. This gives you a good idea of the amount of water lost by the plant in transpiration.

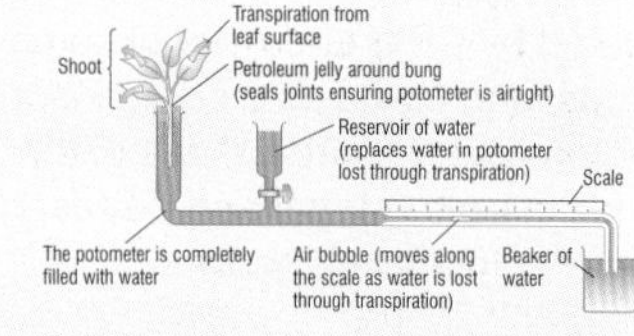
Transpiration from leaf surface
Shoot
Petroleum jelly around bung (seals joints ensuring potometer is airtight)
Reservoir of water (replaces water in potometer lost through transpiration)
Scale
The potometer is completely filled with water
Air bubble (moves along the scale as water is lost through transpiration)
Beaker of water

Figure 3 A potometer is used to show the water uptake of a plant under different conditions

SUMMARY QUESTIONS

1 Describe how water moves up a plant in the transpiration stream.

2 a) What is transpiration?
b) What part of the leaves helps them to prevent losing water under normal conditions?
c) If the top surfaces of the leaves were coated in petroleum jelly, how do you think it would affect the rate at which a plant takes up and loses water?
d) If the bottom surfaces of the leaves were coated in petroleum jelly, how do you think it would affect the rate at which a plant takes up and loses water?
e) What do you think would happen to the rate of transpiration if you turned a fan onto the leaves of the plant? Explain your answer.
f) What does a potometer actually measure?

3 Suggest an investigation using a potometer to show the effect of either light or temperature on the rate of transpiration from a leafy stem.

KEY POINTS

1 The loss of water vapour from the surface of plant leaves is known as transpiration.
2 Water is lost through the stomata which are opened and closed by guard cells to let in carbon dioxide for photosynthesis.
3 Water is pulled up through the xylem from the roots to replace the water lost from the leaves in the transpiration stream.
4 Transpiration is more rapid in hot, dry, windy or light conditions.

B3 1.7 Transport problems

SPECIFICATION LINK-UP

Unit: Biology 3.13.1

Students should use their skills, knowledge and understanding of How Science Works:

- *To explain how gas and solute exchange surfaces in humans and other organisms are adapted to maximise effectiveness.*

BIOLOGY EXCHANGE OF MATERIALS

B3 1.7 Transport problems

Black lung!

OBITUARIES

Jack Higgins was well-known for his sense of humour and his skill in organising the village fete. For many years he was captain of the cricket team until the lung disease, which finally killed him, made him hang up his boots four years ago. Jack started working down the mine when he was only 16 years old, following his father and older brothers. The family have been mining coal for 150 years and he was proud of his work. His enthusiasm for life will be sadly missed in the village. He leaves a wife and three children.

The threat of 'black lung'

The disease 'black lung' is as threatening as its name. Black lung, or as it is properly known, coal worker's pneumoconiosis (CWP), is a form of chronic obstructive pulmonary disease (COPD) found only in mine workers.

The black dust produced as coal is mined gets breathed into the lungs. Normal amounts of dust are removed by the mucus produced in your breathing system. However, in a mine there is so much coal dust that your body cannot cope. Once the dust gets into the alveoli, an immune reaction is triggered. In the end the structure of the alveoli is lost. Instead of pink tissue containing millions of tiny air sacs, the lungs are blackened and have relatively few large air sacs.

The sufferer cannot get enough oxygen into their body, so they become tired, weak and less able to do things. They have to breathe pure oxygen to help them, until eventually even that isn't enough and they die.

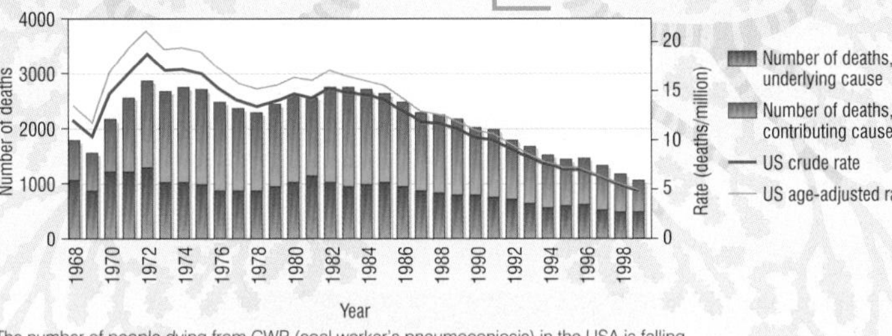

The number of people dying from CWP (coal worker's pneumoconiosis) in the USA is falling steadily. This is partly due to the fact that fewer people are working in the coal industry. Also, as people realised that coal dust in the mines caused the fatal disease, the level of dust in the mines has been controlled and miners are provided with masks to filter out the deadly particles.

ACTIVITY

Work in a group to put together a Powerpoint® presentation on coal worker's pneumoconiosis (black lung). Using this material and what you have learnt in this chapter as your starting point, make sure you cover the lungs, how they work, how CWP affects the lungs and prevents effective gaseous exchange and the social impact of diseases like this. You might include some history of the disease, and ways in which it can be prevented.

234

Teaching suggestions

Activities

- **Black lung!** – In order to find out more, students could be encouraged to search the Internet. There are a number of sites that are good starting points (Wellcome Library, Miner's Advice, US Department of Labor) and the causes and symptoms of the disease are easily available. Most of these sites give references and so further searches can be made.
- **Looking at lungs** – It would be useful to show students pictures of lung tissue affected by CPW and have pictures of normal lung tissue as a comparison. Diagrams or prepared slides of the alveoli of both normal and diseased lungs could also be shown.
- **Health and Safety regulations** – The fall in the number of deaths, as shown by the graph, is due to several factors. It could be useful to check Health and Safety regulations in the UK to find out what is required when working in dusty situations.
- **Beating osmosis** – As an introduction, use a clip from a video showing sea creatures ('Finding Nemo', 'Sponge Bob Square Pants' or similar). Ask: 'What problems would the creatures face due to osmosis?' This could be a group discussion with the leaders reporting back.
- Show a picture of a life-raft adrift at sea or a clip from the Tom Hanks' film *Castaway*. (Search the Internet for 'Castaway movie trailer'.) Read the 'Water, water everywhere nor any drop to drink' bit from Coleridge's *Rime of the Ancient Mariner*. Ask: 'What would happen to you if you drank seawater?' Discuss. Try to find out what is recommended in such situations. Ask: 'Should one drink one's own urine? Or is a seawater enema the answer?' You could mention the 'Survival' programme where a family was adrift for days in a small boat and were kept alive by means of seawater enemas.
- **No contest!** – Discuss the situation in marine invertebrates. Students should be clear that there is movement of water in to and out of the organisms, but no *net* movement in either direction. Contrast this with the situation in freshwater invertebrates, where there is a net movement of water into the organisms. These organisms have evolved mechanisms to get rid of excess water [no need for details, but it could be interesting to introduce students to the idea of putting a freshwater organism like *Amoeba* into increasing concentrations of seawater and noting how the regulating mechanism is affected]. The freshwater organisms have to expend energy in order to get rid of excess water.
- **Copy cats!** – Project images of different beetles and other insects with obvious tough outer layers. If possible allow students close-up views of live locusts where it is

ACTIVITY & EXTENSION IDEAS

- For the black lung activity, it is recommended that students work in groups to put together the PowerPoint® presentation. There are several areas to investigate, so the work can be divided up easily.
- The historical aspects of this activity could involve some investigation into the changes that have been made in the way in which coal is mined. Ask: 'Were miners more susceptible to the disease a century ago? Is there any link with diet or living conditions? Has improved medical treatment been able to help?'
- For the activity associated with beating osmosis, there is scope to make a number of GCSE revision sheets. 'Exchange of materials' is a broad topic and the sheets can cover many aspects of the material in this chapter. Students can decide whether to use a cartoon approach, diagrams, bullet points or any other means. If some ideas are needed to trigger the students, then they could refer to web sites such as BBC 'bite-size' (www.bbc.co.uk).
- As an alternative to the sheets, students might like to consider using tapes and recording the information so that they can use a Walkman or MP3 player. Some students could find this method of learning the material useful.
- An extension to the information on the freshwater fish would be to consider what happens to fish such as salmon that start life in freshwater, migrate to the sea and then return to freshwater to spawn.

Further biology

Cellular exchanges

The exchange of materials is vital to life as we know it. Diffusion, osmosis and active transport are necessary to move substances we need from one place to another – and to get rid of substances which would cause problems. Here are a few more of the different ways in which living organisms manage to transport substances into and out of their cells!

Keep it moving!

Bony fish, like cod and the goldfish you may keep as a pet, pump water over their gills. This helps to maintain a steep concentration gradient between the water and the blood flowing through the gill stacks. This in turn means that oxygen moves by diffusion into the blood and carbon dioxide moves out.

But most of the cartilaginous fish – like sharks – don't have a pumping mechanism. They also lack the swim bladders which make other fish buoyant. So if a shark stops swimming, it will start to sink to the bottom. Once it lands on the bottom, water is no longer flowing over the gills and it cannot get the oxygen it needs to live. So most sharks need to keep swimming to stay alive.

Paramecium and other single-celled freshwater organisms have a real problem because the concentration of solutes in their cells is greater than concentration in the water they live in. So water tends to flow into the cells by osmosis, which could make then swell up and burst – not good news! *Paramecium* deal with the problem using a contractile vacuole. Water is moved out of the cytoplasm into the vacuole against a concentration gradient by active transport. Once the vacuole is getting full, it is moved – again using energy from cellular respiration – to the surface of the cell where the water is released.

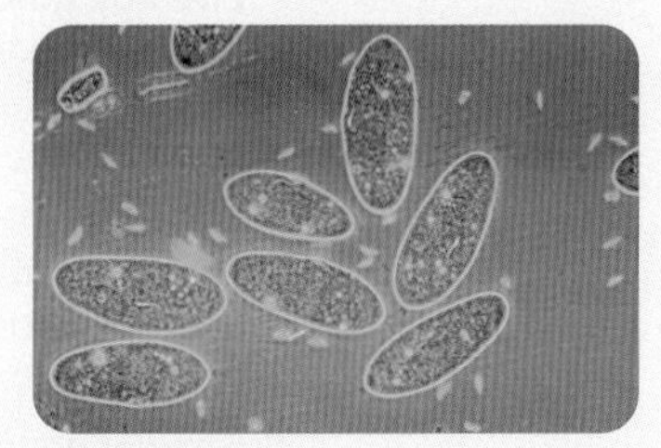

ACTIVITY

Design a wall poster – or three posters – on exchange mechanisms in animals and plants. Cover diffusion, osmosis and active transport. For each one find as many interesting examples as possible to show why the process is so important to living organisms.

235

possible to see the spiracles and the ventilation movements. Draw parallels between leaves in dry situations (thick cuticle, protected stomata, etc.) and the insects' tough outer layers.

- **Other strategies on land** – Consider how other land animals, both invertebrates and vertebrates, cope with the drying conditions on land and the evaporation of water from the surface of the body. Get students to think about soft-bodied animals such as earthworms, slugs and snails, and to compare them with the beetles and vertebrates.
- **Flooding in** – The text in the Student Book is self-explanatory. Animals living in freshwater do have to get rid of excess water. Fish have scales on their bodies which does prevent water uptake over the external body surface, but the gills are exposed to the water. They have a different problem in that they need to take up ions (particularly chloride ions) against a concentration gradient. To help students understand this, it could be useful to make a diagram of a freshwater fish and annotate it for the uptake and loss of water and ions with the relevant arrows to indicate direction.
- **The big ones** – Students could research the physiology of marine vertebrates, such as the whales. Ask: 'Do they have special mechanisms for getting rid of the excess salts that they ingest with the seawater?'

Special needs

A game can be played with coloured counters. Draw a jellyfish on a large sheet of paper and place a number of coloured counters inside it and the same number outside. In pairs, the students can move the counters one at a time into and out of the jellyfish, as long as there is always the same number going in as coming out. Repeat with a (freshwater) fish sheet, but this time have a pot marked 'urine' into which the counters can be placed and emptied to the outside from time to time.

Gifted and talented

Students could investigate what happens in marine fish where there is a tendency to lose water from the gills.

SUMMARY ANSWERS

1

Diffusion	Osmosis	Active transport
Diffusion is the net movement of particles from an area of high concentration to an area of lower concentration	Osmosis is the net movement of water from a high concentration of water molecules (a dilute solution) to a lower concentration of water molecules (a more concentrated solution) across a partially permeable membrane	Active transport is the movement of a substance from a low concentration to a higher concentration, or across a partially permeable membrane
Diffusion takes place because of the random movements of the particles of a gas or of a substance in solution in water	Although all the particles are moving randomly, only the water molecules can pass through the partially permeable membrane	Active transport involves transport or carrier proteins which carry specific substances across a membrane
Diffusion takes place along a concentration gradient	Osmosis takes place along a concentration gradient of water molecules	Active transport takes place against a concentration gradient
No energy from the cell is involved	No energy from the cell is involved	Active transport involves the use of energy from cellular respiration

Diffusion is very useful but relatively uncontrolled so can cause problems for the cell. **Osmosis** is vital for many processes, including keeping plants upright, but it can cause problems if the concentrations inside and outside a cell are different. Water may enter or leave the cell in an uncontrolled way.

Active transport is vital for moving substances around a cell or an organism against concentration gradients, but it can be poisoned. If active transport fails, the cell is in trouble.

2 Large surface area of gill stacks; Very thin tissue so a short distance for gases to diffuse; Rich blood supply.

3 **a)** The alveoli provide a very large surface area with thin walls and a rich blood supply.

b) There is more carbon dioxide in exhaled air as the lime water goes milky/cloudy far more quickly than inhaled air drawn through the solution.

4 **Plant leaf:** large surface area for diffusion; thin so small distances for diffusion; stomata allowing gases in and out; guard cells to control size of stomata so controlling diffusion; large internal surface area; moist internal surfaces for diffusion. **Plant root:** root hair cells and microvilli give huge surface area for osmosis and diffusion; small distances for diffusion; moist internal surfaces for diffusion. All these are similar to the leaf, but active transport needed for uptake of mineral ions against the concentration gradient.

5 The main points to cover are: water in the soil; moves by osmosis into root hair cell; across root cells by osmosis [and diffusion if given, but not diffusion on its own] into xylem; up through roots and stems into leaves; out into air spaces in leaves by evaporation; out through stomata by diffusion along a concentration gradient.
[This can be answered by means of a well-annotated diagram.]

EXCHANGE OF MATERIALS: B3 1.1 – B3 1.7

SUMMARY QUESTIONS

1 Produce a table to compare diffusion, osmosis and active transport. Write a brief explanation of the advantages and disadvantages of all three processes in cells.

2 Look at the diagram of a fish gill below:

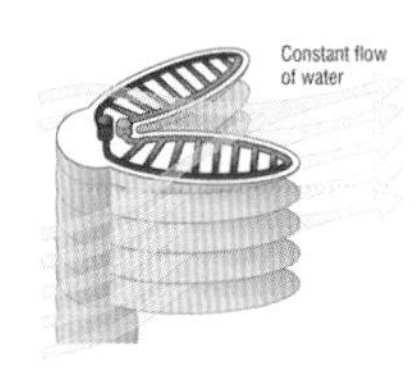

Explain carefully how it is adapted for the exchange of gases in water.

3 a) How are the lungs adapted to allow the exchange of oxygen and carbon dioxide between the air and the blood?

b) Explain what the experiment shown below tells us about inhaled and exhaled air.

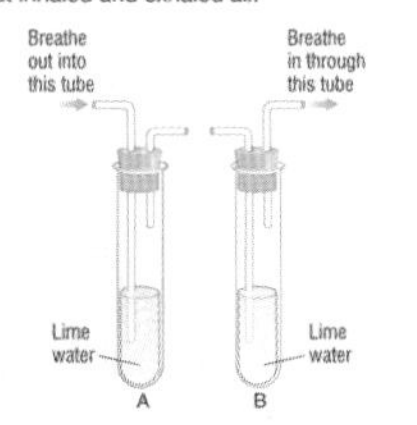

4 Compare the adaptations of a plant leaf for the exchange of carbon dioxide, oxygen and water vapour with the adaptations of the roots for the absorption of water and minerals.

5 Tell the story of water particles as they travel from the soil around the roots of a plant to the point when they reach the air surrounding the leaves on a sunny, windy day.

EXAM-STYLE QUESTIONS

1 The diagram shows two plant cells that have been immersed for the same time in different solutions.

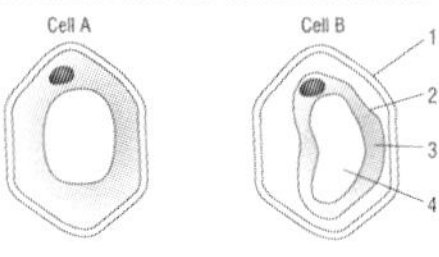

(a) Label the parts numbered **1**, **2**, **3** and **4** in cell **B**.

(b) The contents of cell **B** have shrunk because water moved out of the structure labelled **4**. What proc is most likely to have caused this water to move o

(c) One cell has been immersed in distilled water an the other in a solution of sugar. Which cell has be immersed in distilled water?

(d) The structure labelled **2** in cell **B** allows water to across it, but not other substances such as sugar. What name is given to structures that have this property?

(e) Explain how cells in the same state as cell **A** can to support the leaves of plants.

2 (a) Copy out the table below and complete the boxes writing 'Yes' or 'No' as appropriate.

	Diffusion	Osmosis	Active transpo
Occurs against a concentration gradient			
Needs energy			
Method by which oxygen is absorbed in the lungs			
Method by which digestive products are absorbed in the gut			
Method by which water is absorbed by plant roots			

(b) For the efficient transfer of materials by diffusion the exchange surface needs to have certain featu such as a large surface area, that increase the rat diffusion. Name two other features other than a l surface area that increase the rate of diffusion by exchange surface.

Summary teaching suggestions

- **Literacy guidance** – Many of the answers require extended prose, including the use of technical terms and detailed instructions, so there are opportunities for students to practise these skills.
- **Misconceptions** – The distinctions between diffusion, osmosis and active transport need to be clearly established. Question 1 enables students to write clear definitions and distinctions. The answers to this question provide a good revision resource.
- **Special needs**
  - Question 1 could be modified by providing the students with statements, written on paper or card, and a blank table into which the statements could be placed.
- **When to use these questions?**
  - **As homework tasks** – Designing experiments, models and demonstrations could be set for homework and then discussed in class to decide on the most appropriate solutions. Homework time gives students an opportunity to spend some time considering alternatives – time that might no be available during a lesson.
  - **As starters** – Questions could be used at the start of specific lessons on the relevant topics. Again, the experimental design questions are open-ended and could provide the stimulus or introduction to a practical session.
  - **As revision aids** – The answers to all questions could be usefully converted to revision cards, providing concise summaries of the topics.

The table in question 1 summarises the differences between ways in which molecules and ions move into and out of cells. Both questions 2 and 4 can be answered by means of annotatec diagrams and the information needed to answer question 3 can be given in the form of a table.

Further biology

(c) In each of the following situations, name the particular structure that is adapted to produce a large surface area for the exchange of materials.

(i) Absorption of oxygen into the blood in the lungs. (1)

(ii) Absorption of glucose from the gut into the bloodstream. (1)

(iii) Absorption of water by the roots of plants. (1)

[Higher]

The diagram shows a small portion of the surface of a leaf as seen under a microscope.

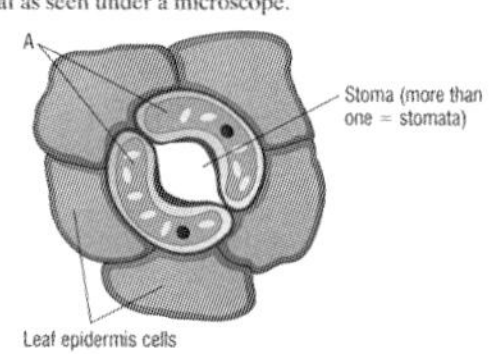

(a) Name the structures labelled **A**. (1)

(b) Stomata are small pores in the leaf through which carbon dioxide and other gases can be exchanged between the inside of the leaf and the outside.

(i) What process in the plant requires this carbon dioxide? (1)

(ii) From what other process occurring within the plant cells might this carbon dioxide also be obtained? (1)

(iii) By what process does this carbon dioxide enter the plant through stomata? (1)

(c) State two adaptations of leaves that produce a large surface area for the absorption of carbon dioxide. (2)

(d) Water vapour can be lost from the plant through stomata.

(i) What name is given to this loss of water vapour? (1)

(ii) Explain how this loss of water vapour can be controlled by the plant. (2)

(iii) The loss of water vapour is affected by environmental conditions. State three changes in environmental conditions that would **decrease** the rate of water loss in plants. (3)

HOW SCIENCE WORKS

Alana was fascinated by the idea that she could get a measure of the concentration of the cytoplasm of some plant cells by a simple osmosis investigation. She prepared a range of sodium chloride solutions. She cut out some potato chips to more or less the same shape and size. She dried the outside of the chips and weighed them. Alana then put them into the different solutions and left them for half an hour. She then dried and weighed them again.

Alana's results are in the table below. L and R are the repeats

Concentration (mol/dm³)	Weight before (g)	Weight after (g)	Difference (g)	Mean difference (g)
L 0.0	8.584	8.873	+0.289	0.268
R 0.0	8.454	8.701	+0.247	
L 0.1	7.223	7.361	+0.138	0.144
R 0.1	8.048	8.198	+0.150	
L 0.2	8.157	8.059	−0.098	
R 0.2	8.236	8.198	−0.038	
L 0.3	7.720	7.433	−0.287	−0.266
R 0.3	8.590	8.344	−0.246	
L 0.4	7.032	6.616	−0.416	−0.467
R 0.4	7.798	7.361	−0.437	
L 0.5	8.286	7.789	−0.497	−0.487
R 0.5	7.399	6.922	−0.477	

a) Which important variable did Alana not control? (1)

b) Calculate the mean difference for the 0.2 (mol/dm³) concentration. (1)

c) Draw a graph for the mean differences shown in the different concentrations of sodium chloride solutions. Draw a line of best fit. (3)

d) Are there any anomalies to be found from the graph? If so, which results would you consider anomalous? (1)

e) What should be done with any anomalies? (1)

f) Which concentration could be considered to be similar to that of the cytoplasm in the potato cells? (1)

g) How accurate do you think this concentration is? (1)

h) How could you increase the accuracy of this concentration? (1)

EXAM-STYLE ANSWERS

a) **1** (Cellulose) cell wall **2** Cell membrane
3 Cytoplasm **4** Vacuole/cell sap *(1 mark each)*

b) Osmosis *(1 mark)*

c) Cell A *(1 mark)*

d) Partially permeable *(1 mark)*

e)
- Water enters the vacuoles of leaf cells making them swell.
- Causing them to press their cytoplasm/cell contents against the (rigid) cell wall.
- The pressure builds up until there is no net entry of water.
- This swollen state (turgor) causes leaf cells to press firmly against each other supporting the leaf and making it rigid.

(1 mark for each point to a maximum of 3)

a)

	Diffusion	Osmosis	Active transport
Occurs against a concentration gradient	No	No	Yes
Needs energy	No	No	Yes
Method by which oxygen is absorbed in the lungs	Yes	No	No
Method by which digestive products are absorbed in the gut	Yes	No	Yes
Method by which water is absorbed by plant roots	No	Yes	No

(1 mark for each set of three correct answers in any row)

b)
- A rich blood supply (to maintain a steep concentration gradient).
- A moist surface (for substances to dissolve).
- Thin walls/membranes between the two areas.

(1 mark each to a maximum of 2)

c) i) Alveoli ii) Villi iii) Root hairs *(1 mark each)*

3 a) Guard cells *(1 mark)*

b) i) Photosynthesis ii) Respiration iii) Diffusion *(1 mark each)*

c) They have a flattened shape. *(1 mark)*
They have internal air spaces. *(1 mark)*

d) i) Transpiration *(1 mark)*
ii) By the stomata closing *(1 mark)*
due to the action of the guard cells. *(1 mark)*
iii)
- Decrease in/lower temperature (colder).
- More water in the air/higher humidity.
- Less air movement/less windy/more still air.

(1 mark each)

Exam teaching suggestions

All three questions require short, often just one-word, answers: they almost entirely test recall of factual information found in chapter B3.1. Any one question will make a short end of topic test that can be simply and easily marked in a relatively short time. All questions are suitable for foundation-level students.

Question 1, part e) and question 3, part d) illustrate the need to qualify answers with all important adjectives like more/less, higher/lower, increase/decrease, etc. Whenever a 'change' is asked for in a question, students should be conditioned into immediately thinking – I need to state the **direction** of change.

HOW SCIENCE WORKS ANSWERS

a) Students might suggest many variables such as how the chips were dried, the volume of water used. Perhaps the most crucial though is the weight of the potatoes before they were put into the solutions.

b) The mean difference is −0.068 g.

c) The graph should show the mean difference in weight on the *y* axis and concentration on the *x* axis. The plots should be accurately plotted. A line of best fit should be drawn.

d) and **e)** The plot at 0.0 M should not really be considered as an anomaly. Some interesting discussion could develop around when you have natural variation and when it counts as an anomaly. Also it could well be that the plant cells are fully turgid and therefore the graph would level off at these molarities.
Other plots are also close enough to the line of best fit, with the exception of 0.5 M. Does the graph start to level off or is this an anomaly? Further investigation is needed. There is room for much open debate with these results.

f) Around 0.6 M.

g) Concentration is reasonably accurate, but there are some doubts about the accuracy of the readings and also where the line of best fit is actually drawn.

h) Repeating the investigation using smaller-interval readings between 0.1 M and 0.2 M concentrations would increase the accuracy of the results.

How science works teaching suggestions

- **How and when to use these questions.** When wishing to develop a detailed appreciation of the need for designing accuracy into an investigation. Also the use and abuse of anomalies.
The questions are best tackled in group discussions.

B3 2.1 The circulatory system

LEARNING OBJECTIVES

Students should learn:

- That there is a double circulation: one circulation carrying blood to the lungs and back to the heart, the other carrying blood to the rest of the body and back to the heart.
- That the heart pumps blood into the arteries, which carry blood to the organs; the blood returns to the heart in veins and in the organs, the blood flows through capillaries.
- That substances needed by the cells in the body tissues pass out of the blood, and substances produced by the cells pass into the blood through the walls of the capillaries.

LEARNING OUTCOMES

Most students should be able to:

- Explain what is meant by 'a double circulation'.
- Describe the action of the heart and the functions of the different blood vessels.
- State that substances enter and leave the blood in the capillaries.

Some students should also be able to:

- Describe a double circulation.
- Explain how substances enter and leave the blood in the capillaries.

Teaching suggestions

- **Special needs.** Make models of the blood vessels from thin and thick-walled tubing. Capillaries can be made from plasticine discs rolled out to make very small tubes. Match the models with the names of the vessels.
- **Gifted and talented.** Research the ideas of Galen (try the BBC history web site). What might lead the ancients to these ideas? Also research William Harvey. How were his ideas different? How well were they accepted? What should we conclude about the permanence of scientific knowledge?

SPECIFICATION LINK-UP Unit: Biology 3.13.2

- *The heart pumps blood around the body. Blood flows from the heart to the organs through arteries and returns through veins. In the organs, blood flows through capillaries. Substances needed by cells in the body tissues pass out of the blood, and substances produced by the cells pass into the blood through the walls of the capillaries.*
- *There are two separate circulation systems, one to the lungs and one to all the other organs of the body.*
- *Blood plasma transports:*
  - *carbon dioxide from the organs to the lungs*
  - *soluble products of digestion from the small intestine to other organs*
  - *urea from the liver to the kidneys.*

Lesson structure

STARTER

Interactive heart animation – Show the students an interactive heart animation. Answer questions and talk through the workings of the heart. (10–15 minutes)

Capillary loops – Smear clove oil on the cuticle of a finger or a thumb and then observe through a microscope with top illumination. Alternatively, look at photographs or videos of developing chick embryos. Ask: 'How does the blood circulate?' (10 minutes)

MAIN

- **Examining a sheep's heart and blood vessels** – Obtain complete sheep's hearts (or pigs') from a butcher (you will need to order these as they are usually trimmed before being sold). Ideally, have one heart per group of students and provide the students with a work sheet, listing things to look for and suggesting that they make sketches of the different valves. It could be useful to include on the work sheet a drawing of the external appearance of the heart, to help the students locate the different blood vessels as real hearts look very different from the diagrams in the Student Book.
- If it is possible to have at least one heart with associated lungs, students can see the links with the lungs and trace the pulmonary circulation. It can also be used in the plenary later to feel the vessels.
- It is also useful to have a model heart (the sort that comes apart to expose the internal structure) available for reference.
- **Blood vessels** –Link structure to function and emphasise the links between the blood vessels: arteries → arterioles → capillaries → venules → veins. Refer to Simulation B3 2.1 'Circulatory system'.
- You could link the above with 'What goes where?' and describe what is taken up from the plasma at the capillary level. There are videos that show exchange of materials at the capillaries.
- **Blood flow** – Working in pairs, students can find each other's pulse at the wrist (radial pulse) and on the side of the neck (carotid pulse). Ask: 'Is it easy to find? Why is the middle finger of the hand used on the wrist and not the thumb?'
- Discuss what is happening and why the pulse can be felt. Students could be told of the other points (front ankle, posterior-tibial and femoral) and possibly try out the ankle ones. Ask: 'What information does the pulse give?'
- **Interactive heart animation** – If the animation was not used as a starter, you can use it as part of the main lesson. There is much information on the animation and it could be beneficial to students to spend more time on it.

PLENARIES

Feel the blood vessels – From the dissection of the heart and lungs of a sheep or a pig, cut out a section of aorta and a section of vena cava. Allow students to place their fingers inside to feel the elasticity of the tissues (latex gloves recommended or very thorough washing afterwards). Students will also appreciate the thickness of the walls and the difference in diameter of the lumen of each. (10 minutes)

Taboo – Play a game of 'Taboo', where a student has to get another to guess a certain word (chosen from the vocabulary of this spread) without using specified words or phrases. (10–15 minutes)

- **Efficiency of the double circulation** – A useful comparison can be made between the single circulation in a fish (mentioned in the previous chapter) and the double circulation in mammals. This can be illustrated with simple diagrams. Students can discuss the advantages of keeping oxygenated and deoxygenated blood separate, the differences in pressure of blood going to lungs and that needed to pump blood to the rest of the body.
- **'Artificial' blood** – Students can research blood substitutes. There has been on-going research in this field for some time. Ask: 'Does it seem a good idea? When could artificial blood be used?'
- **What is in the plasma?** – The plasma of the blood carries a large number of solutes, some of which are needed by cells and others that are waste products. Build up a list of these solutes and, using a large diagram of the human body with the major organs drawn in, consider what is removed and added to the plasma as it passes through the various organs.
- **Foul fact illustration** – For the blood-thirsty students, use the Black Knight sketch from the film *Monty Python and the The Holy Grail,* to illustrate the spurting blood.

Teaching suggestions continued

- **Learning styles**

  Kinaesthetic: Carrying out heart dissection and pulse finding exercise.

  Visual: Watching interactive heart animation.

  Auditory: Playing taboo game.

  Interpersonal: Collaborating in the investigation of the heart and blood vessels.

  Intrapersonal: Memorising the parts of the circulatory system and their function.

- **Homework.** There are some good opportunities here for students to make a set of Revision cards about the blood vessels and circulatory system.

B3 2.1 The circulatory system

LEARNING OBJECTIVES

1. How does your circulatory system work?
2. Where do substances enter and leave your blood?

You are made up of billions of cells and most of them are a long way from a direct source of food or oxygen. A transport system is absolutely vital to supply the needs of your body cells and remove the waste material they produce. This is the function of your blood circulation system. It has three elements – the pipes (**blood vessels**), the pump (the **heart**) and the medium (the **blood**).

A double circulation

You actually have not one transport system but two. Humans have a **double circulation**. One carries blood from your heart to your lungs and back again to exchange oxygen and carbon dioxide with the air. The other carries blood all around the rest of your body and back again.

A double circulation like this is very important in warm-blooded, active animals like ourselves. It makes our circulatory system very efficient. Fully oxygenated blood returns to the heart from the lungs and can then be sent off to the different parts of the body. This means more areas of your body can receive fully oxygenated blood quickly.

a) Why do we need a blood circulation system?

Body

Heart

Lungs

Figure 1 The two separate circulation systems supply the lungs and the rest of the body

The blood vessels

You have three main types of blood vessels. They are adapted to carry out particular functions within your body, although they are all carrying the same blood.

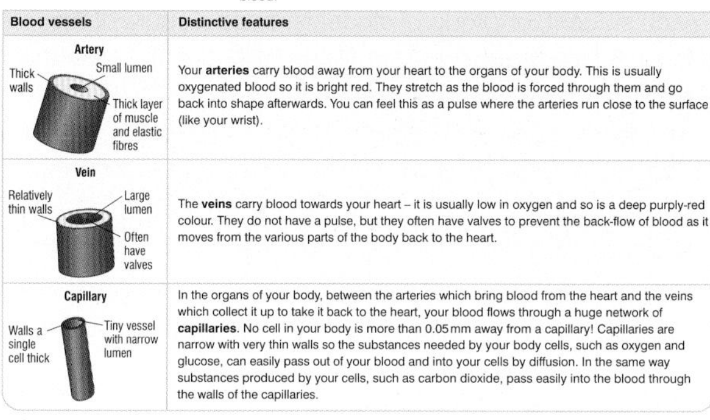

Blood vessels	Distinctive features
Artery (Thick walls; Small lumen; Thick layer of muscle and elastic fibres)	Your **arteries** carry blood away from your heart to the organs of your body. This is usually oxygenated blood so it is bright red. They stretch as the blood is forced through them and go back into shape afterwards. You can feel this as a pulse where the arteries run close to the surface (like your wrist).
Vein (Relatively thin walls; Large lumen; Often have valves)	The **veins** carry blood towards your heart – it is usually low in oxygen and so is a deep purply-red colour. They do not have a pulse, but they often have valves to prevent the back-flow of blood as it moves from the various parts of the body back to the heart.
Capillary (Walls a single cell thick; Tiny vessel with narrow lumen)	In the organs of your body, between the arteries which bring blood from the heart and the veins which collect it up to take it back to the heart, your blood flows through a huge network of **capillaries**. No cell in your body is more than 0.05 mm away from a capillary! Capillaries are narrow with very thin walls so the substances needed by your body cells, such as oxygen and glucose, can easily pass out of your blood and into your cells by diffusion. In the same way substances produced by your cells, such as carbon dioxide, pass easily into the blood through the walls of the capillaries.

b) Substances can only enter and leave the blood in the capillaries. Why is this?

The heart as a pump

Your heart is made up of two pumps (for the double circulation) which beat together about 70 times each minute. The walls of your heart are almost entirely muscle, supplied with oxygen by the coronary blood vessels.

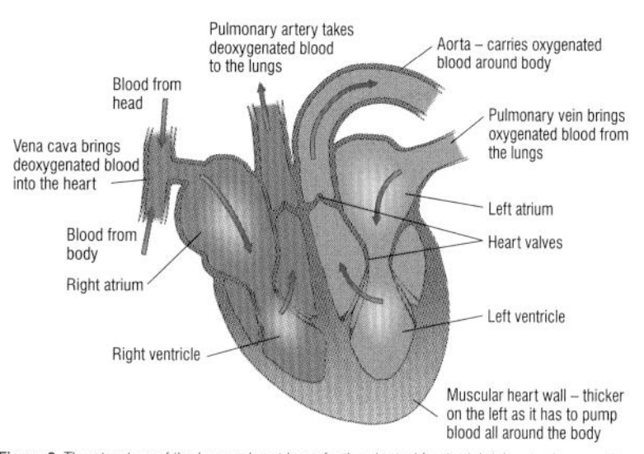

Figure 2 The structure of the human heart is perfectly adapted for the job it has to do, pumping blood to your lungs and your body. The two sides of the heart fill and empty at the same time to give a strong, co-ordinated heart beat.

PRACTICAL

Blood flow

You can practise finding your pulse and look for the valves in the veins in your hands and wrist.

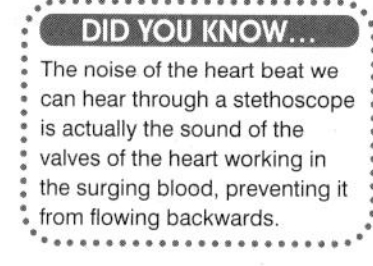

DID YOU KNOW...

The noise of the heart beat we can hear through a stethoscope is actually the sound of the valves of the heart working in the surging blood, preventing it from flowing backwards.

FOUL FACTS

Because the blood in the arteries is under pressure, it is very dangerous if an artery is cut. The blood spurts out rapidly every time the heart beats. If a large artery is cut the blood may spurt a couple of metres or more.

SUMMARY QUESTIONS

1. a) Draw a diagram which explains the way the arteries, veins and capillaries are linked to each other and to the heart.
   b) Label the diagram and explain what is happening in the capillaries.
2. Blood in the arteries is usually bright red because it is full of oxygen. This is not true of the arteries leaving the heart for the lungs. Why not?

KEY POINTS

1. The body transport system consists of the blood vessels, the heart and the blood.
2. Human beings have a double circulation.
3. The heart works as a pump, moving blood around the body.
4. The three main types of blood vessels are the arteries, veins and capillaries.

SUMMARY ANSWERS

1. a) Make sure students show the capillary network between arteries and veins.
   b) [These should be labelled: heart, lungs, artery to lungs, capillaries in lungs, vein to heart, artery to body, capillaries in organs of the body, vein to heart.]
2. Blood carried from heart to lungs is deoxygenated blood from the body, so it is dark (purply) red until it picks up oxygen again in the lungs.

Answers to in-text questions

a) A blood circulation system transports substances like oxygen, glucose and carbon dioxide around our bodies.

b) Only the capillaries have the thin walls needed to enable substances to diffuse across them into and out of the blood.

Practical support

Blood flow

Equipment and materials required

Stopwatches or stop clocks if the pulse rate is to be measured.

If dissection is carried out:

Safety: Wash hands if handling animal material. Take care with sharp instruments.

KEY POINTS

The key points state the basic facts about the components of the circulatory system. These are easily summarised on a simple diagram. Students could expand the statements with annotations of their own in order to make a good revision resource.

B3 2.2 Transport in the blood

LEARNING OBJECTIVES

Students should learn:

- That red blood cells contain a red pigment called haemoglobin, which transports oxygen from the lungs to the body organs.
- That in the lungs, oxygen combines with haemoglobin to form oxyhaemoglobin.
- That in the other body organs, the oxyhaemoglobin splits into oxygen and haemoglobin.
- That the blood plasma transports useful substances, such as glucose, to the cells and removes waste products, such as carbon dioxide, as the blood passes through the capillaries of the organs.

LEARNING OUTCOMES

Most students should be able to:

- Describe the structure of red blood cells and their role in the blood.
- Describe how oxygen is transported from the lungs to the other body organs by haemoglobin.

Some students should also be able to:

- Explain how red blood cells are adapted to carry out their function.
- Describe the reversible reaction between haemoglobin and oxygen.

Teaching suggestions

- **Special needs**
  - Students could make models of red blood cells as suggested in the first plenary.
  - Play a game of floor dominoes with A4 cards, each with a part of the blood and a function.
- **Gifted and talented.** They could do a calculation exercise, using the information given in the Student Book. They could work out approximately how many red blood cells there are in the room. This could be continued to calculate the number of molecules of haemoglobin and hence the number of molecules of oxygen capable of being taken up.

SPECIFICATION LINK-UP Unit: Biology 3.13.2

- *Red blood cells transport oxygen from the lungs to the organs. Red blood cells have no nucleus. They are packed with a red pigment called haemoglobin. In the lungs, haemoglobin combines with oxygen to form oxyhaemoglobin. In other organs, oxyhaemoglobin splits up into haemoglobin and oxygen.*

Lesson structure

STARTER

Iron in the blood – Show the students a clip from the film *X-Men 2,* where Magneto is in a plastic prison and he escapes by getting an accomplice to inject one of the guards with extra iron, which he extracts from his blood and turns into bullets. Discuss the science behind this. (10 minutes)

Black pudding – Let the students look at a piece of black pudding (under hygienic conditions if regulations allow – also be sensitive to religious and racial issues). Ask: 'What is black pudding made of? Why would it be good for you if you were anaemic?' Discuss how, during WW2, there were more blood donors than were needed and the government seriously considered making black pudding with it as the population were very short of protein! (10 minutes)

Blood composition – Refer to Photo PLUS B3 2.2 'Blood cells' on GCSE Biology CD ROM. Discuss what students already know about the composition of the blood. (10 minutes)

MAIN

- **Microscopic examination of blood** – Using prepared slides of blood films, get students to identify the cells. They could be provided with a work sheet showing the different types of cell and then asked to find and draw as many as they can find on their slides.
- It could be interesting for each student, or group of students, to do a count of the numbers of the different cells in a field of view. Individual counts could be collated and some idea of the relative numbers/proportions of the different types could be achieved.
- If calibrated eyepiece graticules are available, students could measure the diameter of the different cells. (This relates to: 'How Science Works': making measurements.)
- Link the measurements with the diameter of capillaries and review the structure of lung tissue, to highlight the close proximity of the red blood cells to the alveoli in the lungs for gas exchange.
- **Oxygenating the blood** – Obtain some pig's blood from an abattoir and use sodium citrate, or other anti-coagulant, to prevent clotting.
- Bubble some oxygen through this blood. Observe the colour changes and ask students to explain what is happening.
- Link this with a video clip of blood entering and leaving a heart/lung machine during open-heart surgery. Again, discuss the colour changes.
- Discuss the affinity of haemoglobin for oxygen. Remind students of what happens when they cut themselves [the blood is always bright red].

PLENARIES

Biconcavity – Use sweets of an appropriate shape to demonstrate what is meant by a biconcave disc. Give students small blocks of red Plasticine to make models of red blood cells. Draw out the main adaptations – lack of nucleus; surface area to volume ratio. (10–15 minutes)

A loop game – Use this to summarise the work covered so far on the circulation and the composition of the blood. (5 minutes)

'Ten questions' – Play a game of 'ten questions' to identify yourself as either a part of the blood, part of the heart, a blood vessel or a substance carried. This can be played from the front or in groups. Keep scores and hold a competition. (10 minutes)

ACTIVITY & EXTENSION

- **Blood clotting** – Video yourself pricking a finger and then running a sterile needle through the drop of blood until the fibres of fibrin start to be drawn out. This shows the role of the platelets. *Do not carry this out in front of the class.* Mention self-sealing tanks on fighter aircraft and racing cars.
- **Other vertebrate blood** – Mammal blood is different from other vertebrate blood. Show slides of bird, fish, amphibian and reptile blood and get students to list the differences. Get them to think about how bloodstains may be identified and the forensic implications of this.
- **Effects of smoking** – This could be a good opportunity to review the effects of smoking on the carriage of oxygen in the blood and carbon monoxide poisoning.

Practical support

Safety: Handle pig's blood carefully and hygienically. Wash hands after use.

Teaching suggestions continued

- **Learning styles**

  Kinaesthetic: Modelling a red blood cell.

  Visual: Examining blood slides and observation of colour changes.

  Auditory: Discussing black pudding.

  Interpersonal: Playing 'ten questions' game.

  Intrapersonal: Writing down feelings about some groups refusing to allow blood transfusions.

- **Homework.** Students could write a short paragraph about the uses of artificial blood, bringing in a discussion about people refusing to allow blood transfusions. Encourage students not to be judgemental about this issue. If necessary, they could also carry out some research to find out what the religious objections are.

BIOLOGY TRANSPORTING SUBSTANCES AROUND THE BODY

B3 2.2 Transport in the blood

LEARNING OBJECTIVES

1. What is blood made up of?
2. How are red blood cells adapted to carry oxygen around your body?

Your blood is a complex mixture of cells and liquid which carries a huge range of substances around your body. The liquid part of your blood is called the plasma. It carries red blood cells, white blood cells and **platelets**.

The white blood cells are part of your immune system which is your defence against disease. The platelets are involved in the clotting of your blood. But it is the blood plasma and the red blood cells which are involved in the transport of materials around your body.

a) What are the roles of the white blood cells and the platelets in your blood?

Figure 1 Blood plasma is a yellow liquid which transports everything you need – and need to get rid of – around your body to the right places

The blood plasma as a transport medium

Your **blood plasma** is a yellow liquid – the red colour of whole blood comes from your red blood cells. The plasma transports all of your blood cells and a number of other things around your body. Carbon dioxide produced in the organs of the body is carried in the plasma back to the lungs.

Similarly **urea**, a waste product formed in your liver from the breakdown of proteins, is carried in the plasma to your kidneys. In the kidneys the urea is removed from your blood to form urine. (See page 246).

All the small, soluble products of digestion pass into the blood from your gut. They are carried in the plasma around your body to the organs and individual cells which need them.

b) What is transported in your blood plasma?

Red blood cells

There are more **red blood cells** than any other type of blood cell in your body. You have about 5 million red blood cells in each 1 mm^3 of your blood – and the average person has between 4.7–5.0 litres of blood in their body! The red blood cells pick up oxygen from your lungs and carry it to the tissues and cells where it is needed. Your red blood cells have a number of adaptations which make them very efficient at their job:

- they have a very unusual shape – they are ***biconcave discs***. This means they are concave – pushed in – on both sides. This gives them an increased surface area : volume ratio over which the diffusion of oxygen can take place,
- they are packed full of a special red pigment called **haemoglobin**, which can carry oxygen,
- they do not have a nucleus – this makes more space to pack in molecules of haemoglobin.

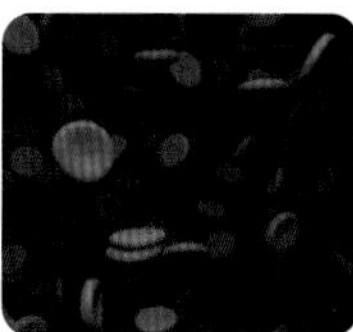

Figure 2 Red blood cells have a unique shape which helps them pick up and carry as much oxygen as possible

240

Further biology

The formation and breakdown of oxyhaemoglobin

Haemoglobin is a large protein molecule folded around four iron atoms. In a high concentration of oxygen, such as in the lungs, the haemoglobin reacts with oxygen to form **oxyhaemoglobin**. This is bright scarlet in colour, which is why the blood in most of your arteries is bright red.

In areas where the concentration of oxygen is lower, such as the cells and organs of the body, the reaction reverses. The oxyhaemoglobin splits to give haemoglobin and oxygen. The oxygen then diffuses into the cells where it is needed. Haemoglobin is purply-red – the colour of the blood in your veins.

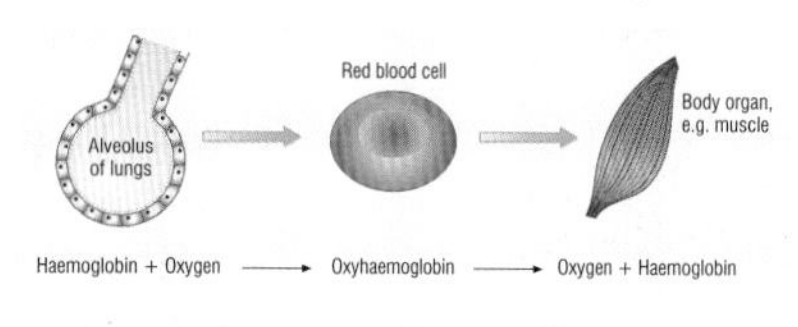

Figure 3 This reversible reaction makes active life as we know it possible by carrying oxygen to all the places where it is really needed

Because haemoglobin is based on iron, if your diet lacks iron, your body cannot make enough red blood cells and you suffer from anaemia. People who are anaemic are pale and lack energy. That's because they cannot carry enough oxygen around the body for their needs

DID YOU KNOW...

One red blood cell contains about 250 million molecules of haemoglobin, which allow it to carry 1000 million molecules of oxygen.

SCIENCE @ WORK

Scientists get involved all sorts of different jobs. For example, some religious groups don't allow blood transfusions even to save lives. So scientists have developed an artificial blood which can be used during surgery and after blood is lost in an accident.

SUMMARY QUESTIONS

1 Copy and complete using the words below:

transported glucose red blood cells urea blood lungs plasma oxygen

Substances are ...... around your body in your .... Dissolved food molecules such as ...... and waste substances such as ...... are carried in the ......, while ...... is carried from the ...... to the cells by your ....

2 a) Why is it not accurate to describe the blood as a red liquid?
b) What actually makes the blood red?
c) Give three important functions of the blood plasma.

KEY POINTS

1. Your blood is the main transport medium of your body.
2. Your blood plasma transports dissolved food molecules, carbon dioxide and urea.
3. Your red blood cells are adapted to transport oxygen from your lungs to the organs of your body.
4. Red blood cells are biconcave discs which have no nucleus and are packed with the red pigment haemoglobin.
5. Oxygen is carried by haemoglobin which becomes oxyhaemoglobin in a reversible reaction.

241

SUMMARY ANSWERS

1. Transported, blood. glucose, urea, plasma, oxygen, lungs, red blood cells.
2. a) Blood plasma is a yellow liquid with cells suspended in it.
   b) Red blood cells.
   c) Three of: transports waste products, digested food, carbon dioxide, blood cells, hormones.

Answers to in-text questions

a) White blood cells form part of the immune system, a defence against invading organisms. Platelets are involved with the clotting of the blood.

b) All of your blood cells, oxygen (in red blood cells), carbon dioxide, urea, soluble products of digestion are transported in the blood plasma.

KEY POINTS

These summarise the main points made in this spread and their understanding can be tested using the in-text and summary questions as a plenary, or playing the loop game or 'ten questions' plenaries.

B3 2.3 The effect of exercise on the body

LEARNING OBJECTIVES

Students should learn:

- That muscles need energy from respiration in order to contract.
- That, during exercise, there is an increase in the blood flow to the muscles so more sugar and oxygen is supplied and carbon dioxide removed.
- That glycogen provides a store of energy in the muscles.

LEARNING OUTCOMES

- Describe how the body responds to the demands of exercise.
- Describe how glycogen is used in the body.

Some students should also be able to:

- Interpret data on the use of oxygen/heart rate increase during exercise.
- Relate the responses of the body to exercise and the ability of the muscles to contract efficiently.

Teaching suggestions

- **Special needs.** Create and use jig-saw sheets of the human body and the changes which happen due to exercise. Blank jig-saw sheets can be bought or you can make your own. Dependent on the difficulties faced, use a variety of clues if needed.
- **Gifted and talented.** A young woman has been found dead in a field. No one knows her identity. Students could write a letter of advice from a pathologist telling the police what they can find out about her exercise habits and lifestyle from her corpse, to aid with her identification.
- **Learning styles**

  Kinaesthetic: Carrying out practical investigations on pulse rates.

  Visual: Drawing and labelling diagrams.

  Auditory: Discussing with peers to predict results of investigations.

  Interpersonal: Working in pairs on pulse rate and breathing rate experiments.

  Intrapersonal: Writing weekly exercise chart.
- **Homework.** Students could be given question 3 of the summary questions to do. Some class discussion beforehand would be of benefit, in ensuring the validity and reliability of any data that they might think of collecting.

SPECIFICATION LINK-UP Unit: Biology 3.13.3

- *The energy that is released during respiration is used to enable muscles to contract.*
- *During exercise a number of changes take place:*
  - *the heart rate increases*
  - *the rate and depth of breathing increases*
  - *the arteries supplying the muscles dilate.*
- *These changes increase the blood flow to the muscles and so increase the supply of sugar and oxygen and increase the rate of removal of carbon dioxide.*
- *Glycogen stores in the muscle are used during exercise.*

Lesson structure

STARTER

- **Exercise – how much do you get?** – Show a clip from an old exercise video such as Mr Motivator, with some good 1980s clothes to have a laugh at. Also, show some footage of modern gym clubs. Carry out a quick survey to find out the health and exercise activities of the class members and their families. Use Simulation B3 2.3 'Exercise'. (10 minutes)
- **Cardiac muscle contraction** – Show an MPEG (Moving Picture Experts Group) or a video clip of contracting heart muscle cell. Discuss what the energy source for this movement will be and the reaction involved. (5–10 minutes)
- **Breathing out a Mars bar** – Ask if anyone has ever breathed out a Mars bar (or similar confectionery). Ask if anyone has ever eaten one. Get a student to read out the contents from a wrapper. Ask: 'What happens to the glucose?' Get a student to breathe out through a tube of lime water and another to exhale on to a cold glass (a round-bottomed flask full of cold water). Test the condensation with cobalt chloride paper. Link the two to the glucose and remind students of the equation for aerobic respiration. Use a molecular model to show the respiration process. (10–15 minutes)

MAIN

- **Effect of exercise on heart rate** – This investigation is best carried out in pairs so that students can record each other's pulse rates. Before starting, the students should decide on the level and period of exercise. The simplest investigation could concentrate on one level of exercise, such as walking on the spot for a set time.
- The resting pulse rate in beats per minute should be determined. Ideally, this should be done three times and a mean taken. The student then undertakes the exercise and the pulse rate recorded immediately and at set intervals, such as every minute afterwards, until the rate returns to normal.
- A graph can be plotted of heart/pulse rate against time. The students exchange roles.
- There are many variations on this investigation that can be carried out:
  - The intensity of the exercise can be varied.
  - The increase in breathing rate (number of breaths per minutes) can be investigated either separately from the pulse rate or in conjunction with it.
  - In addition to the performance of individuals, comparisons could be made between members of the class who exercise regularly and those who do not.
- This investigation can be used for teaching/assessing 'How Science Works': hypotheses can be formulated and predictions made, measurements taken and results tabulated, fulfilling many of the investigative requirements.
- Digital pulse monitors can be used and it is possible to use data loggers and get live read-out graphs that can be displayed through a projector.

PLENARIES

Breathing out on the Moon – Get the students to imagine being on the dark side of the Moon, removing their space helmet and breathing out. Before their head exploded, they might catch a glimpse of carbon dioxide from their last breath falling gently as carbon dioxide snow. Get students to draw a picture to illustrate this. (5–10 minutes)

'Bloaty Slobs' club – Imagine a 'Bloaty Slobs' club, where the membership rules included the banning of exercise. Write down instructions for the club's Unfitness Enforcers, giving a list of tell-tale signs that would indicate the person being investigated has been exercising. Use imagination and illustration, being careful regarding the size sensitivity of some students. (10 minutes)

SUMMARY ANSWERS

1 **Heart rate:** increases as soon as the exercise starts as a result of anticipation. It climbs rapidly, followed by a steady rise and then falls off quite sharply as the exercise finishes. Increased heart rate supplies muscles with the extra blood they need to bring food and oxygen to the muscle fibres, and to remove the waste products which rapidly build up. **Breathing rate:** increases more slowly and evenly than the heart rate, but remains high for some time after exercise. To begin with, increased heart rate supplies enough oxygen, then breathing rate needs to increase to meet demand. When exercise stops, breathing rate remains high until the oxygen debt is paid off.

2 [Mark depending on ideas presented. Look for clear, sensible ideas, safe investigation, realistic expectations, appropriate methods of recording and analysing, awareness of weakness in investigation.]

Practical support

The practicals suggested in this spread do not require complex apparatus. The pulse and breathing rate investigations simply require stopwatches or stop-clocks. If a spirometer is used, then follow the instructions supplied with it.

Safety: Lime water: wear eye protection – CLEAPSS Hazcard 18. Avoid prolonged skin contact with cobalt chloride paper. Wash hands.

KEY POINTS

Testing fitness

The need for an increased supply of glucose and oxygen and the more rapid removal of carbon dioxide should be linked to respiration.

BIOLOGY TRANSPORTING SUBSTANCES AROUND THE BODY

B3 2.3 The effect of exercise on the body

LEARNING OBJECTIVES

1 How does your body respond to the increased demands for oxygen during exercise?
2 What is glycogen and how is it used in the body?

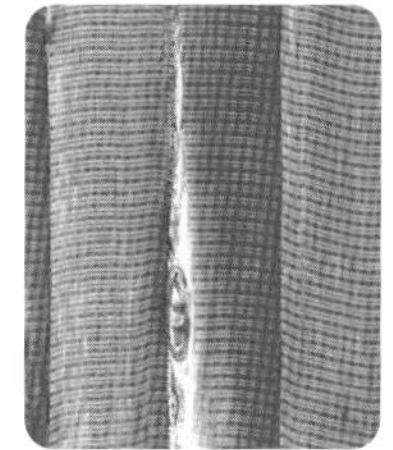

Figure 1 All the work done by the muscles is based on these very special protein fibres which contract when they work and relax afterwards

Your muscles use a lot of energy. They move you about and help support your body against gravity. Your heart is made of muscle, and the movement of food along your gut depends on muscles too.

Muscle tissue is made up of protein fibres which contract when they are supplied with energy from respiration. (See Figure 1.) Muscle fibres need a lot of energy to contract. They contain many mitochondria to supply the energy they need by aerobic respiration.

Your muscles also contain **glycogen** stores. Glycogen is a carbohydrate which can be converted rapidly to glucose. This supplies the fuel needed to provide the energy for cellular respiration when your muscles contract.

glucose + oxygen → carbon dioxide + water (+ energy)

Muscles fibres usually occur in big blocks or groups known as muscles. Your muscles contract to cause movement. They relax when their role is finished which allows other muscles to work.

a) What is aerobic respiration?

The response to exercise

Even when you are resting your muscles use up a certain amount of oxygen and glucose. This is because some of your muscles fibres are constantly contracting to keep you in position against the pull of gravity. Muscles are also involved in your life processes such as breathing and circulation of the blood.

But when you begin to exercise, your muscles start contracting harder and faster. As a result they need more glucose and oxygen to supply their energy needs. During exercise the muscles also produce increased amounts of carbon dioxide. This needs to be removed for them to keep working effectively.

b) Why do you need more energy when you exercise?

So during exercise, when muscular activity increases, several changes take place in your body:

- Your heart rate increases and the arteries supplying blood to your muscles dilate. These changes increase the blood flow to your exercising muscles. This in turn increases the supply of oxygen and glucose and increases the removal of carbon dioxide.
- Your breathing rate increases and you breathe more deeply. These changes mean that not only do you breathe more often, but you also bring more air into your lungs each time you breathe. This increases the amount of oxygen brought into your body and picked up by your red blood cells. The oxygen is carried to the exercising muscles. It also means that more carbon dioxide can be removed from the blood in the lungs and breathed out.

c) Why do you produce more carbon dioxide when you are exercising hard?

DID YOU KNOW?

The maximum rate to which you should push your heart is usually calculated as approximately 220 beats/minute minus your age. When you exercise, you should ideally get your heart rate into the range between 60% to 90% of your maximum!

The benefits of exercise

Your heart and lungs benefit from regular exercise. Both the heart and the lungs become larger. They both develop a bigger and very efficient blood supply. This means they function as effectively as possible at all times, whether you are exercising or not.

	Before getting fit	After getting fit
Amount of blood pumped out of the heart during each beat (cm^3)	64	80
Heart volume (cm^3)	120	140
Breathing rate (no. of breaths per minute)	14	12
Pulse rate	72	63

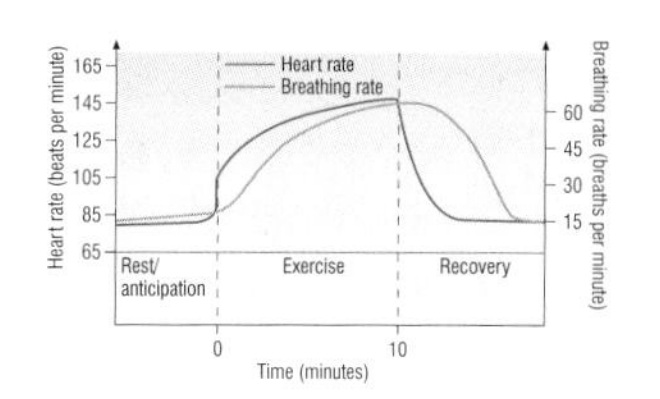

Figure 2 During exercise the heart rate and breathing rate increase to supply the muscles with what they need and remove the extra waste produced

PRACTICAL

Testing fitness

A good way of telling how fit you are is to measure your resting heart rate and breathing rate. The fitter you are, the lower they will be! Then see what happens when you exercise. The increase in your heart rate and breathing rate and how fast they return to normal is another way of finding out how fit you are – or aren't!

SUMMARY QUESTIONS

1 Using Figure 2, describe the effect of exercise on the heart rate and the breathing rate of a fit person and explain why these changes happen.
2 Plan an investigation into the fitness levels of your class mates. Describe what you would do and how you would record and analyse your results. What pattern would you expect to see?

Further biology

Figure 3 Not everyone can be as fit as a top-class athlete – but doing exercise helps keep your heart and lungs healthy at whatever level you take part

GET IT RIGHT!

Exercise results in an increased rate of delivery of oxygen to the muscles – not just 'more oxygen'.

Be clear about the difference between the rate and the depth of breathing.

KEY POINTS

1 The energy that is released during respiration is used to enable muscles to contract.
2 When you use your muscles you need glucose and oxygen to be supplied at a faster rate. The rate at which carbon dioxide is removed from muscle tissues needs to increase too.
3 Body responses to exercise include an increase in heart rate, and increase in breathing rate and depth of breathing. The arteries supplying blood to the muscles dilate and the glycogen stores in the muscle are converted to glucose to use as fuel for respiration.
4 Regular exercise benefits the muscles, heart and lungs.

242

243

ACTIVITY & EXTENSION IDEAS

- **Glycogen and its importance** – Introduce the structure of glycogen and its importance as a storage carbohydrate in the liver and the muscles. Link with its role in the maintenance of a steady level of glucose in the blood. Students can be reminded of the conversion of glucose to glycogen, stimulated by the release of insulin and the conversion of glycogen to glucose when the glucose levels in the blood decrease.
- **Changes in depth and rate of breathing** – If a spirometer is available, then it could be used to demonstrate changes in depth and rate of breathing after exercise. The resting rates should be measured followed by a period of exercise and then more measurements taken. If the apparatus is not available, show students' video clips of the apparatus in use and project spirometer traces for the students to interpret. It is possible to calculate depth and rate of breathing from the spirometer tracings.
- **Can I improve my fitness?** – Encourage students to write a weekly exercise chart for themselves. They could suggest improvements and try to improve their fitness over a period of time.

Answers to in-text questions

a) Aerobic respiration is the complete breakdown of glucose using oxygen to produce energy with carbon dioxide and water as waste products:

Glucose + oxygen → energy + carbon dioxide + water

b) Your muscles are contracting harder for longer so need energy.

c) The muscles are contracting more therefore using more energy, so there is more aerobic respiration and more carbon dioxide produced which is a waste product of the process.

B3 2.4 Anaerobic respiration

LEARNING OBJECTIVES

Students should learn:

- That during long periods of vigorous activity, muscles respire anaerobically in order to obtain energy.
- That less energy is released by anaerobic respiration than aerobic respiration. (**HT** only)
- That during anaerobic respiration, incomplete breakdown of glucose results in the formation of lactic acid and the building up of an oxygen debt. (**HT** only)

LEARNING OUTCOMES

Most students should be able to:

- Explain why muscles respire anaerobically during vigorous exercise.
- Explain why less energy is released by anaerobic respiration. (**HT** only)
- Describe the oxygen debt and how it is repaid. (**HT** only)

Some students should also be able to:

- Interpret data relating to the effects of exercise on the human body. (**HT** only)
- Explain the principle of oxygen debt and why speed of recovery from exercise is a measure of physical fitness. (**HT** only)

Teaching suggestions

- **Special needs.** Make cards with the relevant words and symbols for students to compose equations for aerobic and anaerobic respiration.
- **Gifted and talented.** Research the differences between the different energy systems used by muscles: the aerobic system, the alactic anaerobic system and the lactic acid system.
- **Learning styles**

  Kinaesthetic: Carrying out practical work on breathing and muscle exercises.

  Visual: Watching the videos.

  Auditory: Discussing a standardised muscle test.

  Interpersonal: Working together on the experiments.

  Intrapersonal: Writing your own fitness programme.
- **Homework.** Students could design a training programme to prepare themselves for competing in the London Marathon. This should include reference to diet as well as exercise regimes.

SPECIFICATION LINK-UP Unit: Biology 3.13.3

- *If muscles are subjected to long periods of vigorous activity, they become fatigued, i.e. they stop contracting efficiently. If insufficient oxygen is reaching the muscles, they use anaerobic respiration to obtain energy.*
- *Anaerobic respiration is the incomplete breakdown of glucose and produces lactic acid. As the breakdown of glucose is incomplete, much less energy is released than during aerobic respiration. Anaerobic respiration results in an oxygen debt that has to be repaid in order to oxidise lactic acid to carbon dioxide and water. [**HT** only]*

Lesson structure

STARTER

Wile E Coyote and Road Runner – Show the students a video clip of Wile E Coyote and Road Runner at the start of the episode 'Lickety Splat', where Wile E runs very hard and gets out of breath. Draw out a thumbnail sketch of the graph you would expect of his breathing rate against time, labelling what is happening in each section. (10 minutes)

Sprinting! – Show a video of a 100 m sprint (one from the Olympics or the World Championships so that there could be some well-known people running), preferably one where the athletes are shown immediately before and afterwards. (Search video.google.com for sports.) Get students to observe the behaviour of the athletes. Comment on breathing, whether they collapse, etc. Ask: 'Are they breathing deeply? Can they talk?' Use Simulation B3 2.4 'Breathing and respiration'. (10 minutes)

MAIN

- **Build up of lactic acid in the muscles** – Students to work in pairs and devise a simple repetitive action, such as stepping up and down on to a low bench, lifting a book from the bench to shoulder height or raising one arm and clenching and unclenching the fist twice a second. One student to perform the action as many times as they can before tiring, while the other student keeps a record of the number of actions and the time.
- A period of recovery time is allowed – the student to decide when they are ready to resume the activity, but record the time. Ask: 'Can they do the same number of actions before tiring again? Are they performing the action at the same speed as before? Why does the student slow down?'
- There are variations on this investigation that can be discussed and students could be asked to design a standard test that everyone could do, and that could be used to determine whether muscle fatigue varied from person to person. For example, the same action could be carried out for a set time and a set recovery time allowed. Students could find out if they could continue longer doing that, than carrying out the investigation as first suggested.
- **Breathing rates** – If the variation in breathing rate with activity was not used in the previous spread, it could be investigated here. (This relates to 'How Science Works'.)
- Again, it would be sensible for students to work in pairs, so that the record keeping is done by the partner, and then the roles can be reversed. In this case, it could be appropriate to vary the intensity of the exercise, starting with walking on the spot, then running on the spot and so on. Carry out the exercise for a set time and record breathing rates until they return to normal, before starting on a more vigorous exercise.

PLENARIES

The long distance runner – Show some video footage of a marathon race, at the beginning, during and at the end. Students to observe the behaviour of the athletes and compare with the sprint shown as a starter (or show both here if preferred). Ask: 'Do the athletes seem so out of breath? Or are they breathing as deeply?' Discuss why there are differences in behaviour. (10 minutes)

Energy yields – Anaerobic respiration in yeasts produces alcohol. Show students that there is energy locked up in alcohol by igniting some in controlled conditions (could use it in a spirit lamp or similar). Link to the energy still in lactic acid. For some students, it could be appropriate to compare the structures of glucose, lactic acid and carbon dioxide, in terms of the numbers of C, H and O atoms in the molecules. (10 minutes)

SUMMARY ANSWERS

1 **Aerobic respiration:** cellular respiration using oxygen.
Anaerobic respiration: cellular respiration that does not use oxygen and produces less energy.
Oxygen debt: the amount of oxygen needed to break down the lactic acid built up in your muscles during a period of anaerobic respiration.
Blood doping: the illegal practice used by some athletes of removing some of their own blood, storing it and then transfusing it back just before a competition to increase the oxygen carrying capacity of the blood.

2 **a)** The muscles become fatigued. After a long period of exercise, your muscles become short of oxygen and switch from aerobic to anaerobic respiration, which is less efficient. The glucose molecules are not broken down completely, so that less energy is released than during aerobic respiration. The end products of anaerobic respiration are lactic acid and water. The lower energy levels, combined with the build up of lactic acid in your muscles, makes them ache and work much less effectively.

b) The waste lactic acid you produce during exercise as a result of anaerobic respiration has to be broken down to produce carbon dioxide and water. This needs oxygen, and the amount of oxygen needed to break down the lactic acid is known as the oxygen debt. Even though your leg muscles have stopped, your heart rate and breathing rate stay high to supply extra oxygen, until you have broken down all the lactic acid and paid off the oxygen debt. [**HT** only]

3 **a)** Both people are exercising hard. The fit person's breathing rate goes up slower, doesn't go as high and comes down faster than the unfit person.

b) The breathing of the fit person doesn't need to increase immediately as their fit heart will simply pump more blood to the muscles. Their lungs will be larger than those of an unfit person so they will not need to breathe as quickly, and because they can keep their muscles better supplied with blood and oxygen they will not fatigue as quickly. They won't build up such a large oxygen debt, so their breathing will return to normal faster.

c) They could exercise more regularly and build up their own levels of fitness. Then their heart and lung capacity would increase and they would not get as breathless when they exercised. [**HT** only]

Practical support

Making lactic acid

- No specialised apparatus is required to carry out the suggested practical activity, apart from stopwatches or stop-clocks. If a spirometer is used, follow the instructions given. CLEAPSS Handbook CD-ROM 1436 and section 11.8.2.
- Note: no student should feel under pressure to take part in any of the activities, particularly if they have any medical condition.

Answers to in-text questions

a) Anaerobic respiration does not use oxygen; incomplete breakdown of glucose/sugars; lactic acid end product instead of carbon dioxide.

b) The amount of oxygen needed to break down the lactic acid built up during a period of anaerobic respiration.

KEY POINTS

Some of the definitions asked for in question 1 of the summary questions could be used to reinforce the key points of this topic. This could be a good time to review aerobic and anaerobic respiration, drawing up a table of differences between them.

B3 2.4 Anaerobic respiration

LEARNING OBJECTIVES

1 Why do muscles use anaerobic respiration to obtain energy?
2 Why is less energy released by anaerobic respiration than aerobic respiration? [Higher]
3 What is an oxygen debt? [Higher]

Your everyday muscle movements are made possible with energy produced by aerobic respiration. However, during vigorous exercise your muscle cells may become short of oxygen. In spite of your increased heart and breathing rates, the blood simply cannot supply oxygen fast enough. When this happens the muscle cells can still obtain energy from glucose. But they have to do it by a type of respiration which does not use oxygen – **anaerobic respiration**.

Muscle fatigue

When you have been using your muscle fibres for vigorous exercise for a long time they can become fatigued. This means they stop contracting efficiently. At this stage they are usually very short of oxygen, so they switch to anaerobic respiration.

However, anaerobic respiration is not as efficient as aerobic respiration. In anaerobic respiration the glucose molecules are not broken down completely. So far less energy is released than during aerobic respiration.

The end products of anaerobic respiration are **lactic acid** and water, instead of the carbon dioxide and water produced by aerobic respiration.

Anaerobic respiration:

glucose → lactic acid (+ energy)

a) How does anaerobic respiration differ from aerobic respiration?

DID YOU KNOW? (HIGHER)

Training at altitude so your blood carries more oxygen is legal. There are other ways of increasing your red blood cell count to avoid oxygen debt (see next page) which are not. Sometimes athletes remove some of their own blood, store it, and then, just before a competition, transfuse it back again (blood doping). Others use hormones to stimulate the growth of more red blood cells. Both of these methods give an athlete extra red blood cells to carry more oxygen to the working muscles so they can run faster or compete better. Both of them are illegal.

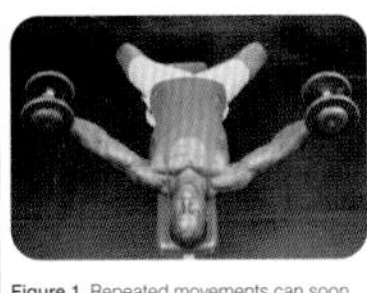

Figure 1 Repeated movements can soon lead to anaerobic respiration in your muscles – particularly if you're not used to it

PRACTICAL

Making lactic acid

Carry out a single repetitive action such as stepping up and down or lifting a weight or a book from the bench to your shoulder time after time. You will soon feel the effect of a build up of lactic acid in your muscles.

- How can you tell when your muscles have started to respire anaerobically?

Oxygen debt

If you have been exercising hard, you often carry on puffing and panting for some time after you stop. The length of time you remain out of breath depends on how fit you are. But why do you keeping breathing faster and more deeply when you have stopped using your muscles?

The waste lactic acid you produce during anaerobic respiration presents your body with a problem. You cannot simply get rid of lactic acid by breathing it out as you can with carbon dioxide. As a result, when the exercise is over lactic acid has to be broken down to produce carbon dioxide and water. This needs oxygen, and the amount of oxygen needed to break down the lactic acid is known as the ***oxygen debt***.

Even though your leg muscles have stopped, your heart rate and breathing rate stay high to supply extra oxygen until you have paid off the oxygen debt. The bigger the debt, the longer you will puff and pant!

Oxygen debt repayment:

lactic acid + oxygen → carbon dioxide and water

b) What is an oxygen debt?

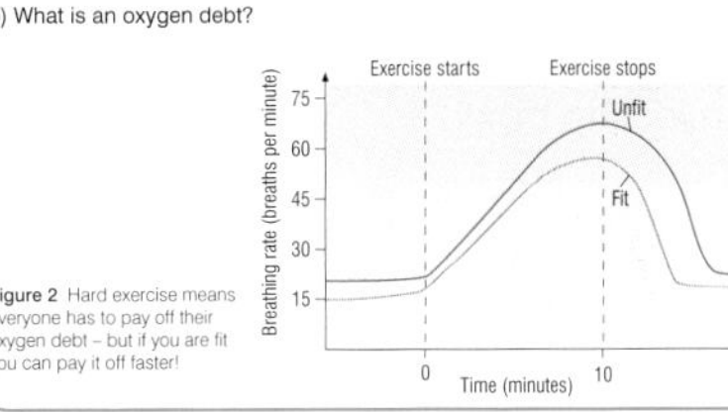

Figure 2 Hard exercise means everyone has to pay off their oxygen debt – but if you are fit you can pay it off faster!

DID YOU KNOW? (HIGHER)

In a 100 m sprint race some athletes do not breathe at all. This means that the muscles use the oxygen taken in the start of the race and then don't get any more oxygen until the race is over. Although the race only takes a few seconds, a tremendous amount of energy is used up so a big oxygen debt can develop.

Figure 3 Training hard is the simplest way to avoid anaerobic respiration as it enables your body to get more oxygen to your muscles and to remove carbon dioxide more efficiently

SUMMARY QUESTIONS

1 Define the following words:

aerobic respiration; anaerobic respiration; oxygen debt; blood doping [Higher]

2 a) If you exercise very hard or for a long time your muscles begin to ache and do not work so effectively. Explain why.
b) If you exercise very hard you often puff and pant for some time after you stop. Explain what is happening. [Higher]

3 Look at Figure 2.
a) Explain what is happening to both people.
b) Why is the graph for the unfit person different from the graph for the fit person?
c) What could the unfit person do to change their body reactions to be more like those of the fit person? [Higher]

KEY POINTS

1 If muscles work hard for a long time they become fatigued and don't contract properly. If they don't get enough oxygen they will respire anaerobically.
2 Anaerobic respiration is respiration without oxygen. Glucose is broken down to form lactic acid, water and a small amount of energy. [Higher]
3 After exercise, oxygen is still needed to break down the lactic acid which has built up. The amount of oxygen needed is known as the oxygen debt. [Higher]

B3 2.5 The human kidney

LEARNING OBJECTIVES

Students should learn:

- That urine, containing urea, excess mineral ions and water, is removed from the body by the kidneys.
- That sugar, mineral ions and water needed by the body are reabsorbed into the blood as it passes through the kidneys.

LEARNING OUTCOMES

Most students should be able to:

- Describe how the kidneys produce urine.
- Explain the importance of the kidneys in the removal of urea and the regulation of the water content of the body.

Some students should also be able to:

- Explain the role of the kidney in homeostasis through different scenarios of water loading and water stress.
- Explain that sugar and dissolved ions may be actively absorbed against a concentration gradient in the kidney tubules (**HT** only)

Teaching suggestions

- **Special needs.** Make a model kidney out of chicken wire. Pass coloured beads into it – blue ones representing water, yellow ones representing urea, white ones representing mineral ions and small sweets representing glucose. Include some big red and white balls, which cannot get out, representing blood cells. Arrange a collecting bowl underneath and get students to put the beads and sweets that escape into two tubes, a red one labelled 'back into the blood', the other labelled 'to the bladder and out' with a picture of a toilet on it.
- **Gifted and talented.** Students could look at kidney structure in more detail, examining prepared slides and trying to identify the parts of the nephron, such as glomeruli, loop of Henle and the cells of the tubules.
- **Learning styles**

  Kinaesthetic: Dissecting kidneys.

  Visual: Observing the video clips.

  Auditory: Listening to the opinions of others in discussions.

  Interpersonal: Taking part in dissection practical.

  Intrapersonal: Calculating volumes of urine produced.

SPECIFICATION LINK-UP Unit: Biology 3.13.4

- *A healthy kidney produces urine by:*
  - *first filtering the blood*
  - *reabsorbing all the sugar*
  - *reabsorbing the dissolved ions needed by the body*
  - *reabsorbing as much water as the body needs*
  - *releasing urea, excess ions and water as urine.*
- *Sugar and dissolved ions may be actively absorbed against a concentration gradient. [**HT** only]*

Lesson structure

STARTER

'Kidney trouble' – Show a clip from *The Simpsons* ('Kidney Trouble', 1008 AABFO4, Series 10) where Homer doesn't let Abe get out of the car to urinate and his kidneys explode! (Or search the Internet for 'Simpson's Kidney Trouble' for background information.) Discuss the anatomy behind this. (10 minutes)

Do you know where your kidneys are? – Ask students to place there hands on the outside of their bodies to indicate where they think their kidneys are. Inspect to see who can get it right. [The correct location can be shown by resting one's hands on the top of the hips with the thumbs pointing forwards. Where the hands go at the back is roughly in line with where the kidneys are – much higher up than most people think.] Discuss the protection they have and link with why boxers wear wide belts. (5–10 minutes)

MAIN

- **Kidney function** – Use Simulation B3 2.5 'Human kidney'. Include some statistics of the volume of blood filtered each day, the reabsorption of glucose and amino acids and the way in which the kidney controls the water and ion content of the blood.
- **Kidney dissection** – Obtain some fresh lamb's or pig's kidneys with the fat and vessels attached. Students to work in pairs or small groups and either provide a work sheet or talk them through the observations and dissection.
- It is worth looking at the outside to see the blood vessels and to point out that the fat surrounding the kidney is all the protection they have. Using a scalpel, the kidney should be sliced horizontally, so that the cortex, medulla and the ureters can be seen. Students to identify the renal artery, the renal vein, the ureter and the collecting area (pelvis) for the urine. The cortex and the medulla can be distinguished by their difference in colour. Ask: 'Why is the outer part (the cortex) darker red than the inner part (the medulla)?'
- **The effect of drinking on urine production** – This investigation can either be presented to the students as collected data, or they could carry it out themselves at home. The idea is to find out what effect drinking a large volume of water has on the volume and colour of the urine.
- The person carrying out the investigation should empty their bladder as completely as possible. After 15 minutes, they should urinate again into a measuring cylinder, record the volume produced and retain a small sample in a sealed specimen tube. A litre of water should then be drunk.
- After 15 minutes, the person should urinate, record the volume produced and retain a further sample in a specimen tube. The volume of urine produced at 15 minute intervals should be recorded and a sample taken for as long as possible.
- No extra liquid should be drunk during the experiment. The volume of urine produced can be plotted against time and the colour recorded at the different intervals.
- A slight variation of the above could be to suggest to the students that they design an experiment to investigate the effect of drinking a large quantity of water on the production of urine. They could then be given some figures to plot and colours to comment on. (This relates to: 'How Science Works': relationships between variables.) Ask: 'How could the samples be tested to see if there were any glucose or amino acids present?'

PLENARIES

Why do peanuts make you thirsty? – Get the students to write a note for the back of a packet of peanuts explaining why they can make you thirsty. To fit on to the back of the packet, they need to make it either exactly 20 words long; or let the students do it on a computer giving them a text box of fixed size that they must fill. (5–10 minutes)

Rate of flow – To observe what the flow rate through the kidneys looks like, set a flow rate of 1200 cm^3 per minute (20 cm^3 per second) through a hose from a container by adjusting a clamp. (5–10 minutes)

ACTIVITY & EXTENSION IDEAS

- **The effect of different activities on the volume and content of the urine** – Predict and explain how each of the following activities might alter the volume and the composition of the urine: eating a Mars bar, running a marathon, drinking two pints of lager, having a bath, eating two packets of crisps.
- **Where does urea come from?** – Show a flow chart with the formation of urea from excess amino acids from protein in the diet in the liver, and its removal from the blood by the kidneys and elimination in the urine. This could be the basis of a very useful revision aid.

Practical support

Kidney dissection

ake care with sharp scalpels. Wash hands fter the experiment. Students should wash nd control their own glassware. Urine hould be disposed of in the toilet, and ontaminated equipment should be placed n a container of disinfectant (fresh sodium hlorate(I) solution). Wear eye protection.

Answers to in-text questions

a) Urea is a waste product from the breakdown of amino acids/proteins in the liver.

b) Processed food often contains a lot of salt. Excess salt is removed by the kidneys and excreted in the urine. If you eat a lot of salty food, the kidneys have to remove the extra salt from the blood.

BIOLOGY

TRANSPORTING SUBSTANCES AROUND THE BODY

B3 2.5 The human kidney

LEARNING OBJECTIVES

1 Why are your kidneys so important?
2 How does your kidney work?
3 How are sugar and dissolved ions moved back into the blood? [Higher]

Your kidneys are one of the main organs which help to maintain homeostasis, keeping the conditions inside your body as constant as possible.

What are the functions of your kidney?

Your kidneys are very important in your body for homeostasis. For example, you produce urea in your liver when you break down excess amino acids. These excess amino acids come from protein in the food you eat and from the breakdown of your worn out body tissues. Urea is poisonous, but your kidneys filter it out of your blood and get rid of it in your urine.

a) What is urea?

GET IT RIGHT!

Make sure you don't confuse **urea** and **urine**.

Remember that the amount of water and ions reabsorbed by your kidneys varies.

Your kidneys are also vital in the water balance of your body. If the concentration of your body fluids change, water will move in to or out of your cells by osmosis. (See page 144.) This could damage or destroy the cells. You gain water when you drink and eat. You lose water constantly from your lungs because water evaporates into the air in your lungs and is breathed out. Whenever you exercise or get hot you sweat and lose more water.

So how do your kidneys balance all these changes? They remove any excess water and it leaves your body as urine. If you are short of water your kidneys conserve it. You produce very little urine and most water is saved for use in your body. If you drink too much water then your kidneys produce lots of urine to get rid of the excess.

The ion concentration of your body is very important. You take in mineral ions with your food. The amount you take in varies. Sometimes you eat very little.

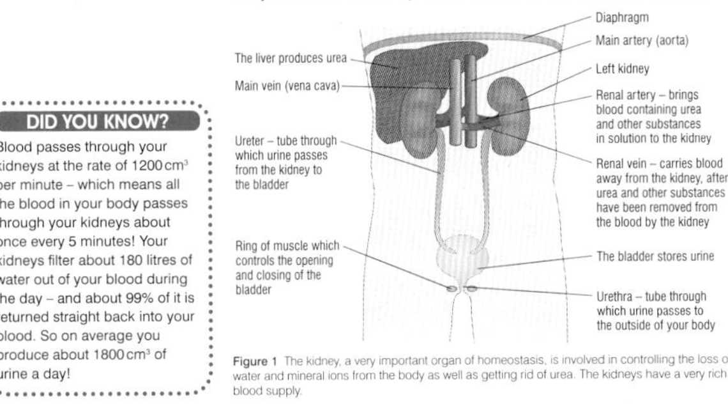

Diaphragm
Main artery (aorta)
The liver produces urea
Left kidney
Main vein (vena cava)
Renal artery – brings blood containing urea and other substances in solution to the kidney
Ureter – tube through which urine passes from the kidney to the bladder
Renal vein – carries blood away from the kidney, after urea and other substances have been removed from the blood by the kidney
Ring of muscle which controls the opening and closing of the bladder
The bladder stores urine
Urethra – tube through which urine passes to the outside of your body

Figure 1 The kidney, a very important organ of homeostasis, is involved in controlling the loss of water and mineral ions from the body as well as getting rid of urea. The kidneys have a very rich blood supply.

DID YOU KNOW?

Blood passes through your kidneys at the rate of 1200 cm^3 per minute – which means all the blood in your body passes through your kidneys about once every 5 minutes! Your kidneys filter about 180 litres of water out of your blood during the day – and about 99% of it is returned straight back into your blood. So on average you produce about 1800 cm^3 of urine a day!

246

Further biology

But if you eat processed food which is high in salt, you take in a big load of minerals. Some are lost through your skin when you sweat. Again your kidneys are most important in keeping a mineral ion balance. They remove excess mineral ions (particularly salt) which are lost in the urine.

b) Why do your kidneys work hard after you eat a lot of processed food?

How do your kidneys work?

Your kidneys filter your blood and then take back (re-absorb) everything your body needs. So sugar (glucose), amino acids, mineral salts and urea all move out of your blood into the kidney tubules by diffusion along a concentration gradient. The blood cells are left behind – they are too big to pass through the membrane of the tubule.

Then ***all*** of the sugar is reabsorbed back into the blood by active transport. But the amount of water and the dissolved mineral ions which are reabsorbed varies. It depends on what is needed by your body. This is known as ***selective reabsorption***.

HIGHER

Both sugar and dissolved mineral ions move back into the blood both by diffusion along a concentration gradient and by active transport. Active transport is used to move them against a concentration gradient. This makes sure no sugar is left in the urine and the right quantity of dissolved mineral ions is reabsorbed.

The amount of water reabsorbed depends on what your body needs. It is controlled by a very sensitive feedback mechanism. Urea is lost in your urine. However, some of it leaves the kidney tubules and moves back into your blood by diffusion along a concentration gradient.

Your kidneys have a very rich blood supply and they produce urine all the time. It trickles into your bladder where it is stored until the bladder is full and you choose to empty it!

What does urine contain?

Your urine contains the waste urea along with excess mineral ions and water not needed by your body. The exact quantities vary depending on what you have taken in and given out. For example, on a hot day if you drink little and exercise a lot you will produce very little urine. This will be concentrated and relatively dark yellow. On a cool day if you drink a lot of liquid and do very little you will produce a lot of dilute, almost colourless urine.

DID YOU KNOW?

Water, glucose, urea and salt are all colourless. So why is urine yellow? **Urobilins** are yellow pigments which come from the breakdown of haemoglobin in your liver. They are another waste product excreted by your kidneys in the urine along with everything else – and make it yellow!

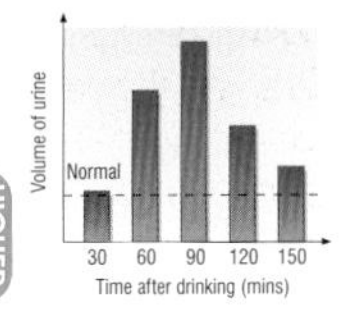

Volume of urine
Normal
30
60
90
120
150
Time after drinking (mins)

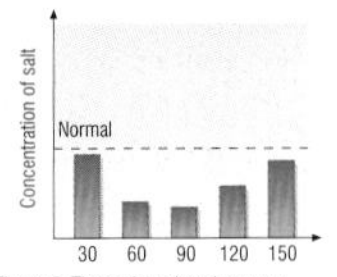

Concentration of salt
Normal
30
60
90
120
150

Figure 3 These data show how your kidneys respond when you drink a lot. They show volume of urine produced and the concentration of salt in the urine after a student drank a large volume of water.

SUMMARY QUESTIONS

1 Urea is one of the main waste products of your body. Describe carefully
 a) how it is formed,
 b) why it has to be removed,
 c) how it is removed from your body.

2 Explain how your kidneys would maintain the water and mineral balance of your blood on
 a) a cool day when you stayed inside and drank lots of cups of tea,
 b) a hot sports day when you ran three races and had forgotten your drink bottle.

KEY POINTS

1 A healthy kidney produces urine by filtering the blood. It then reabsorbs ***all*** of the sugar, and the mineral ions and water needed by your body.
2 Excess mineral ions and water along with urea are removed in the urine.
3 Sugar and dissolved ions can be actively reabsorbed against a concentration gradient. [Higher]

247

SUMMARY ANSWERS

1 **a)** Urea is produced in your liver when you break down excess amino acids from protein in the food you eat, and from the breakdown of your worn out body cells and tissues.

b) Urea is poisonous.

c) Your kidneys filter it out of your blood and get rid of it in your urine.

2 **a)** Your blood would become diluted. The kidneys would retain all but the excess salt and lose a lot of water, so you would produce a lot of very dilute urine.

b) Your kidneys would conserve both salt (because you are losing it in sweat) and water, so you would produce small quantities of very concentrated urine.

KEY POINTS

The functions of the kidney in regulating the water and ion content of the body and the removal of urea are important areas for the students to understand. The difference between urea and urine should be emphasised. A simple diagram of a kidney tubule could be annotated to show where filtration and reabsorption of substances takes place.

B3 2.6 Dialysis – an artificial kidney

LEARNING OBJECTIVES

Students should learn:

- That kidney failure can be treated by dialysis.
- That dialysis removes the urea from the blood.
- That dialysis restores the concentrations of dissolved substances in the blood to normal levels.

LEARNING OUTCOMES

Most students should be able to:

- Describe how dialysis is used to treat kidney failure.
- Explain why dialysis needs to be carried out at regular intervals.
- Describe what happens during kidney dialysis.
- List some of the advantages and disadvantages of kidney dialysis.

Some students should also be able to:

- Give a detailed explanation of kidney dialysis in terms of diffusion and concentration gradients.
- Evaluate the pros and cons of dialysis treatment.

Teaching suggestions

- **Special needs**
  - Students would benefit from a visit from a suitably briefed visiting speaker, such as a nurse or a dialysis patient.
  - As an alternative, use a loop of cellulose acetate tubing filled with a mixture of red particles which will not pass through (fine sand to represent red blood cells), and some yellow dye such as fluorescein (representing waste products) which will pass through. Put a spoonful of salt and a spoonful of sugar into the mixture in the tube representing the blood and also into the water in the beaker outside which represents the dialysis fluid. Get the students to see that you are doing this so that the concentrations of sugar and salt are the same and there is no concentration gradient. (This could work for the class as a whole, as well.)
- **Gifted and talented.** Follow a scavenger hunt through a series of linked web sites (on a school intranet, through Word document links or using software such as Quia) to complete a table summarising the differences and similarities between haemo-dialysis and peritoneal dialysis.

SPECIFICATION LINK-UP Unit: Biology 3.13.4

- *People who suffer from kidney failure may be treated either by using a kidney dialysis machine or by having a healthy kidney transplanted.*
- *In a dialysis machine, a person's blood flows between partially permeable membranes. The dialysis fluid contains the same concentration of useful substances as the blood. This ensures that glucose and useful mineral ions are not lost. Urea passes out from the blood into the dialysis fluid. Treatment by dialysis restores the concentrations of dissolved substances in the blood to normal levels and has to be carried out at regular intervals.*

Lesson structure

STARTER

Dialysis – what we know so far – Show a PowerPoint® review of dialysis. Students could complete a set of questions following viewing and discussion. (10 minutes)

The importance of kidneys – Discuss how you would feel if you had the wrong kidney removed. Talk over the importance of kidneys and what could go wrong. (10 minutes)

Receiving dialysis – Show a short video or project pictures of a kidney patient undergoing dialysis. Discuss what is happening. Search the Internet for 'Video kidney dialysis'.) (10–15 minutes)

MAIN

- **Practical work on dialysis** – If students have not done any practical work using dialysis tubing, then it could be helpful to set up a demonstration or allow them to carry out some simple experiments to show that small molecules, such as glucose, pass through the tubing while larger molecules do not. Experiments such as the model gut (see lesson 1.3 'Exchange in the gut') or the use of tubing to show water uptake and loss in cells could be used.
- **Dialysis and kidney machines** – Use the Animation B3 2.6 'Dialysis' on the GCSE Biology CD ROM. There is some good information on the School Science web site and the Nephron Information Centre. The information on both these web sites is comprehensive and includes images, animations and questions.
- Student worksheets would help to focus on the issues and provide the students with the facts. This is a good opportunity to point out the differences between dialysis and the normal functioning of the kidney.
- **Visiting speaker** – A brief talk from a nurse experienced in dialysis, or from a person who undergoes regular dialysis, followed by a question and answer session, could be of benefit.
- There are links here with the question of transplants and also some careers information. You could combine a talk on dialysis with some discussion of transplants (see next spread).

PLENARIES

Leaflet on dialysis – In groups, students can produce a leaflet for kidney patients explaining what dialysis is and how it works. Use word processing or desktop publishing software if available. (10–15 minutes)

Virtual dialysis – If not used as a starter or in the main part of the lesson, show the virtual dialysis clip from the Nephron Information Centre (www.nephron.com). (10 minutes)

Role play – Students can take the roles of patient, doctor and family member in making decisions about the treatment of kidney failure. In particular, they should discuss the advantages and disadvantages of dialysis. Open up the discussion to the other students. (10–15 minutes)

Teaching suggestions continued

- **Learning styles**

  Kinaesthetic: Carrying out dialysis practical.

  Visual: Watching video of kidney patient during dialysis.

  Auditory: Taking part in role play; listening to the views of others.

  Interpersonal: Taking part in role play with doctor, patient and relative.

  Intrapersonal: Creating a monologue or video diary script.

ACTIVITY & EXTENSION IDEAS

- **Role play** – Students to write a monologue or script for a video diary presentation of what it feels like to be a kidney patient and have to undergo dialysis on a regular basis. This could be set as a homework exercise and then some selected to be read out in class.
- **What should I eat?** – Patients undergoing dialysis have to be careful about their diet. Ask: 'What recommendations would you make to a patient? Why does the quantity of protein eaten need to be controlled? And why should salt and fluid intake make a difference? Suppose you were diagnosed with renal failure, what foods would you miss most?'
- **What stays in and what comes out?** – It could be helpful in understanding the way in which haemo-dialysis works to build up a diagram of the concentrations of substances either side of the dialysing membrane. The concentration gradients in the dialysate are adjusted so that they are the same as in blood. Students can then consider the differences in concentration and be very clear about what comes out and what stays.

B3 2.6 Dialysis – an artificial kidney

LEARNING OBJECTIVES

1. Why do kidneys fail?
2. What is dialysis and how does it work?

Your kidneys can be damaged and destroyed by infections. Some people have a genetic problem which means their kidneys fail, and sometimes the kidneys are damaged during an accident. Whatever the cause, untreated failure of both your kidneys can lead to death. Toxins, such as urea, build up in the body and the salt and water balance of your body is lost.

Dialysis

For centuries kidney failure meant certain death, but now we have two effective methods of treating kidney failure. We can carry out the function of the kidney artificially using **dialysis** or we can replace the failed kidneys with a healthy one in a **kidney transplant**. (See pages 250–1.)

a) Why is kidney failure such a threat to life?

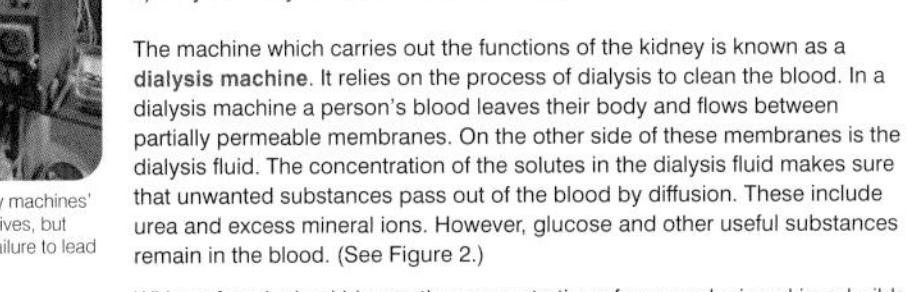

Figure 1 These 'artificial kidney machines' have not only saved countless lives, but allowed sufferers from kidney failure to lead full, active lives

The machine which carries out the functions of the kidney is known as a **dialysis machine**. It relies on the process of dialysis to clean the blood. In a dialysis machine a person's blood leaves their body and flows between partially permeable membranes. On the other side of these membranes is the dialysis fluid. The concentration of the solutes in the dialysis fluid makes sure that unwanted substances pass out of the blood by diffusion. These include urea and excess mineral ions. However, glucose and other useful substances remain in the blood. (See Figure 2.)

Without functioning kidneys, the concentration of urea and mineral ions builds up in the blood. Treatment by dialysis restores the concentrations of these dissolved substances to normal levels. Then as the patient carries on with normal life, urea and other substances build up again. So the dialysis has to be repeated at regular intervals.

It takes around eight hours for dialysis to be complete. So people with kidney failure have to remain attached to a dialysis machine for many hours several times a week. They also have to manage their diets carefully. This helps keep their blood chemistry as stable as possible.

b) What process does dialysis depend on?

SCIENCE @ WORK

Scientists have developed a way of carrying out dialysis within the body of the patient. The dialysis fluid is put into the body cavity and removed hours later. People can do this for themselves at home without a machine. They rely on a huge support team. This includes the scientists who developed the technique, the doctors who treat the patients, the nurses who train and manage the patients and the technical support staff who deliver the bags of dialysis fluid and fetch the patients into hospital fast if things go wrong.

How dialysis works

During dialysis, it is vital that patients lose the excess urea and mineral ions which have built up in their systems. It is equally important that they do not lose useful substances such as glucose and useful mineral ions from their blood.

The loss of these substances is prevented by the careful control of the dialysis fluid. The dialysis fluid contains the same concentration of glucose and mineral ions as normal blood plasma so there is ***no*** net movement of glucose and useful mineral ions out of the blood. It also contains normal plasma levels of mineral ions so excess ions are lost from the blood along a concentration gradient, but no more.

In contrast the dialysis fluid contains no urea, so there is a strong concentration gradient from the blood to the fluid for this substance and all the urea leaves the blood. The whole process of dialysis depends on diffusion along concentration gradients which have to be maintained by the flow of fluid. There is no active transport.

Many people go to hospital to receive dialysis, but in 1964 home dialysis machines were made available for the first time. This meant that at least some people with kidney failure could set up their dialysis in their own home. Even with all our modern technology, dialysis machines are still quite large – certainly much bigger than the kidneys they replace!

Dialysis has some disadvantages. You have to follow a very carefully controlled diet and there are regular, long sessions connected to a dialysis machine. Over many years, the balance of substances in the blood can become more difficult to control however careful the dialysis. But for many people with kidney failure, dialysis means life rather than death, because we have successfully copied the action of the living kidney in the body.

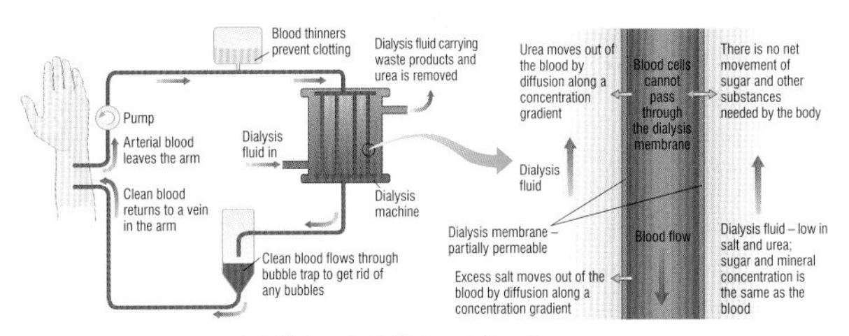

Figure 2 A dialysis machine relies on simple diffusion to clean the blood, removing the waste products which would damage the body as they built up. Changing the concentration of substances in the dialysis fluid allows unwanted substances to be removed from the blood while the concentration of useful substances remains unchanged.

SUMMARY QUESTIONS

1. Why is kidney failure potentially life threatening?
2. Produce a flow chart to explain how a dialysis machine works.
3. a) Why do people with kidney failure have to control their intake of protein and salt?
   b) Why can patients with kidney failure eat and drink what they like during the first few hours of dialysis?
4. a) Explain the importance of dialysis fluid containing no urea and normal plasma levels of salt, glucose and minerals.
   b) Both blood and dialysis fluid are constantly circulated through the dialysis machine. Explain why the constant circulation of dialysis fluid is so important.

KEY POINTS

1. People suffering from kidney failure may be treated by regular sessions on a kidney dialysis machine or by having a kidney transplant.
2. In a dialysis machine, the concentration of dissolved substances in the blood is restored to normal levels.
3. The levels of useful substances in the blood are maintained, while urea and excess salt pass out from the blood into the dialysis fluid.

SUMMARY ANSWERS

1. There is a build-up of toxins (e.g. urea) in the body and the salt and water balance is disrupted.
2. Blood out of artery → through pump → blood thinners added to prevent clotting → blood passes through dialysis membranes and excess salt and urea are removed → clean blood into bubble trap → blood returns to vein in an arm
3. **a)** People with kidney failure cannot remove excess salt or get rid of the urea produced by the breakdown of excess amino acids.
   b) The excess salt and urea are removed during the process of dialysis.
4. **a)** There is no urea, so a steep concentration gradient exists from the blood to the dialysis fluid.
   Normal plasma levels of salt, glucose etc. are present, so there is no net loss or gain due to diffusion.
   b) To help maintain concentration gradients for diffusion.

Answers to in-text questions

a) If the kidneys fail, there is a build-up of toxins such as urea and salt in the blood, which can change the internal environment and can cause death due to water moving in or out of cells or poisoning due to urea.

b) Diffusion.

KEY POINTS

A clear understanding of diffusion and concentration gradients is needed to understand and explain the basis of kidney dialysis. Simple diagrams help. Students should be aware of the differences between dialysis and the normal working of the kidney. There is no active transport involved in dialysis.

B3 2.7 Kidney transplants

LEARNING OBJECTIVES

Students should learn:

- That a kidney transplant involves the replacement of a diseased kidney by a healthy one from a donor.
- That precautions need to be taken to prevent the rejection of the transplanted kidney by the immune system.
- List the advantages and disadvantages of kidney transplants.

LEARNING OUTCOMES

Most students should be able to:

- Describe how a diseased kidney is replaced by a healthy one.
- Explain the problems of rejection by the immune system.
- List the ways in which rejection is prevented.
- List the advantages and disadvantages of having a kidney transplant.

Some students should also be able to:

- Evaluate the advantages and disadvantages of treating kidney failure by dialysis or kidney transplant.
- Explain the issues of rejection and the advantages of close tissue matches for success.

Teaching suggestions

- **Special needs.** Students could complete a concept map of the pros and cons of transplants and dialysis. Use prompts if needed, such as initial letters, colour coding of key words and the places they go on the map.
- **Gifted and talented.** Students could draw up arguments for and against the use of xeno-transplants. If necessary, do a preliminary trawl of the web for information on xeno-transplants.
- **Learning styles**

  Kinaesthetic: Making the snakes and ladders game.

  Visual: Designing the poster.

  Auditory: Listening to the arguments in the debate.

  Interpersonal: Debating organ donation.

  Intrapersonal: Writing the 'Thank you' letter.

SPECIFICATION LINK-UP Unit: Biology 3.13.4

- *A kidney transplant enables a diseased kidney to be replaced with a healthy one from a donor. However, the donor kidney may be rejected by the immune system unless precautions are taken.*
- *To prevent rejection of the transplanted kidney:*
  - *A donor kidney with a 'tissue type' similar to that of the recipient is used.*
  - *The recipient is treated with drugs that suppress the immune system.*

Lesson structure

STARTER

'Kidney trouble' – Show a clip from *The Simpson's,* 'Kidney trouble' (1008 *AABF04 Series 10*), where Homer first runs away from giving Grandpa a kidney transplant then is conned into doing so but plans to get one off Bart. (10 minutes)

'Liver donor' – Show the 'Liver donor' sketch from the Monty Python film *The Meaning of Life* (warn the students that it is rather gory!). Discuss donor cards. Issue some to have a look at. Get students to write down their feelings about organ donation. (10 minutes)

Immuno-suppressant drugs – Write the words on the board and get the students to break the word 'immuno-suppressant' into bits. What does 'immuno' mean? What does 'suppress' mean? Why are these drugs needed? Draw into a discussion of how our bodies recognise foreign protein and link to kidney transplants. (10 minutes)

MAIN

- **Kidney transplants** – Refer to Electronic Resource 'Transplant v. dialysis'. Pictures and more detailed information are available from web sites. The Nephron Information Centre has information that could be helpful.

Students could be given a work sheet and encouraged to keep a tally of pros and cons as the presentation proceeds.

- **Video of surgery** – Show a video of a kidney transplant operation (Search the Internet for 'Video kidney transplant.).
- **Pros and cons** – Draw up a table of the pros and cons of kidney transplants. Expand this to compare transplants with dialysis.
- Using the information given in the Student Book about the relative costs of the two procedures, students could work out the difference in cost to the NHS of a patient on dialysis for 20 years and a patient who has two kidney transplants in 20 years. (This relates to: 'How Science Works': making decisions.)

PLENARIES

Kidney 'Snakes and Ladders' – Draw up a series of statements for a 'Snakes and Ladders' game of kidney problems. Include the benefits and difficulties associated with both transplants and dialysis. Discuss in class and make a game for homework. (10–15 minutes)

Design a poster – Students could design a poster to encourage people to carry donor cards or encouraging kidney donation. Display some of the results. (10 minutes)

A bit of me once belonged to someone else – Discuss how it would feel to have an organ donated to you from someone else. Ask: 'Would it matter which organ it was?' Watch excerpts or give a plot summary and discuss the 1946 film *The Beast with Five Fingers,* directed by Robert Florey. For a real life situation, discuss Chris Hallam, the first person to have an arm transplant. He eventually asked for it to be removed. BBC News web site has the story and pictures. (10–15 minutes)

- **Homework**
  - The poster and 'Thank you' letter activities could be used for homework tasks.
  - In addition, homework time could be used for the preparation of the speeches for the debate on organ donation.

ACTIVITY & EXTENSION

- **Kidneys for sale** – Debate whether or not people should be able to sell their kidneys if they want to. Students could research the scale of this and the legal issues involved. The *Sunday Mirror* web site has some good material on this, and there are sites where information can be gathered.
- **Organ donation** – Organise a debate on the topic of organ donation: 'Should organ donation be assumed unless you actively opt out?' The preparation for this could be a homework task, to write a speech in favour of the motion and one against the motion. [Note: this activity has been suggested for the 'Issues' spread at the end of the chapter.]
- **Finding kidney donors** – Draw up a list of sources of kidneys for transplantation. Ask: 'Why is there such a shortage? How would you set about campaigning for more donors?' This exercise could draw together some of the issues raised in other suggested activities.
- **Thank you letter** – Students could write an imaginary letter to the relative of the person from whom they received a kidney. In their letter, they could describe what it has meant in their life and what they can now do that was not possible before. Alternatively, they could write a letter to a close relative who has donated one of their kidneys to them.

BIOLOGY TRANSPORTING SUBSTANCES AROUND THE BODY

B3 2.7 Kidney transplants

LEARNING OBJECTIVES

1. Why are kidney transplants sometimes rejected?
2. Which is better – dialysis or a kidney transplant?

If your kidneys fail, your life can be saved using a dialysis machine. However once your body has been returned to health and your blood is kept balanced by regular dialysis, another treatment becomes possible – a **kidney transplant**.

What is a kidney transplant?

If your kidneys have failed they may be replaced in a transplant operation by a single healthy kidney from a **donor**. The donor kidney is joined to the normal blood vessels in the groin of the person getting the new kidney (the **recipient**). If all goes well, it will function normally to clean and balance the blood. One kidney is quite capable of keeping your blood chemistry in balance and removing your waste urea for a lifetime.

However, there are some difficulties to overcome in transplanting a kidney from one person to another. The main problem is that your new kidney comes from a different person, so the antigens on the cell surfaces will be different to yours. This means there is a risk that the donor kidney will be rejected by your immune system. When this happens your immune system destroys the new organ. Everything is done to make sure the new kidney is not rejected but it is always a risk.

a) There is just one situation where there is no risk of a new kidney being rejected. What do you think that might be?

There are a number of ways of reducing the risk of rejection. The match between the donor and the recipient is made as close as it can be. Whenever possible donor kidneys with a 'tissue type' similar to the recipient will be used so their antigens are very similar. For example, they will come from people with the same blood group. Also the recipient of the new kidney will be given drugs that suppress their immune response and stop it working (**immunosuppressant drugs**) – for the rest of their lives. As immunosuppressant drugs get better, the need for a really close tissue match is getting less important.

The down-side of these drugs is that the patients cannot deal with infectious diseases very well. They have to take great care if they become ill in any way. Most people feel this is a small price to pay for a new, working kidney and the immunosuppressant drugs are improving all the time.

Transplanted organs don't last forever – the average transplanted kidney works for around nine years. Once the organ starts to fail the patient has to return to dialysis until another suitable kidney is found.

SCIENCE @ WORK

Scientists are working hard to solve the problem of the lack of organ donors. Some are working on **xenotransplantation**, producing genetically engineered pigs and other animals with hearts and kidneys which can be used for human transplants. Other scientists are hoping **stem cell** research (see page 204) will enable them to grow new kidneys on demand, so that no-one dies waiting for a suitable organ to become available.

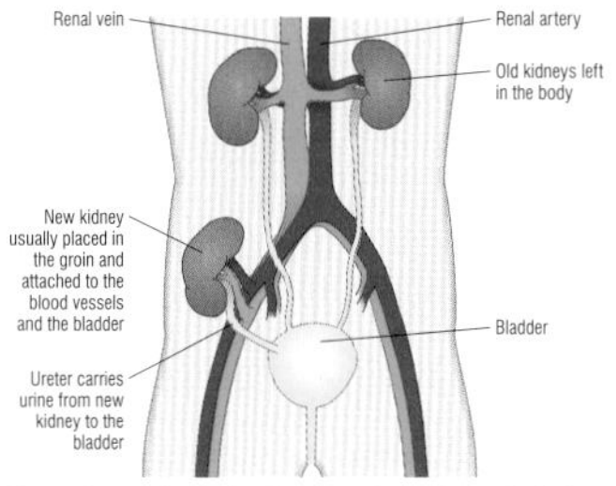

Figure 1 A donor kidney is placed in the body where it takes over the functions of the organs which have failed

250

Further biology

Dialysis v transplants

The great advantage of receiving a kidney transplant is that you no longer have to live like someone with kidney failure. You can eat what you like and are free from the restrictions which come with regular dialysis sessions. An almost completely normal life is the dream of everyone waiting for a kidney transplant.

The disadvantages are mainly to do with the risk of rejection. You have to take medicine every day of your life in case the kidney is rejected. You also need regular check-ups to see if your body has started to reject the new organ. And of course, the biggest disadvantage is that you may never get the chance of a transplant at all.

Dialysis is much more readily available than donor organs, so it is there when your kidneys fail. It enables you to lead a relatively normal life, although you are tied to a special diet and regular sessions on the machine.

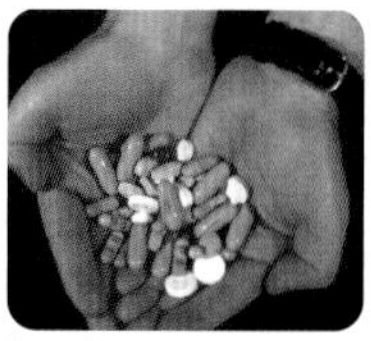

Figure 2 These are some of the immunosuppressant drugs taken by the recipients of kidney transplants. They can have unpleasant long-term side effects. However, most people feel that it is a relatively small price to pay for the hope of a normal life.

Finding the donors

One of the biggest problems in the kidney transplant programme is the lack of donor kidneys. The main source of kidneys is from people who die suddenly and unexpectedly either from road accidents or from strokes and heart attacks. In the UK, organs can only be taken from people if they carry an organ donor card or are on the on-line register giving permission for their organs to be used in this way – or if the bereaved relatives give consent.

Figure 3 This young man has been given a new lease of life by a kidney transplant. Not everyone who suffers from kidney failure is so lucky, due to a lack of donors.

Because many of us do not carry donor cards, there are never enough donor kidneys to go around. What's more, as cars become safer, fewer people die in traffic accidents and become potential donors. At any one time there are thousands of people having kidney dialysis who would love to have a kidney transplant but who never get the opportunity.

DID YOU KNOW?

Dialysis is more expensive than a transplant. It costs an average of £30 000 a year to treat a patient with dialysis. In comparison a kidney transplant costs £20 000 in the first year (including the surgery and hospital stay) and £6500 a year afterwards for the anti-rejection drugs.

SUMMARY QUESTIONS

1. How does a kidney transplant overcome the problems of kidney failure?
2. Sometimes – in very rare cases – a healthy kidney is taken from a live donor to be given to someone with kidney failure. Usually the live donor is a close family member – the parent, brother or sister of the recipient. Live donor kidney transplants have a higher rate of success than normal transplants from dead, unrelated donors.
   a) Suggest two reasons why live transplants have a higher success rate than normal transplants.
   b) Why do you think that live donor transplants are relatively rare?
3. Produce a table to compare the advantages and disadvantages of treating kidney failure with dialysis or with a kidney transplant. Which treatment do you think is preferable and why?

KEY POINTS

1. In a kidney transplant diseased or damaged kidneys are replaced with a healthy kidney from a donor.
2. The donor kidney may be rejected by the recipient's immune system. To try and prevent rejection the tissue types of the donor and the recipient are matched as closely as possible and immunosuppressant drugs are used.

251

SUMMARY ANSWERS

1 The transplanted kidney takes on the functions of the failed kidneys and balances the blood chemistry.

2 a) Live organs have no tissue damage; family donors are a close tissue match.
  b) Taking organs from a living healthy person can threaten their health. It is a big step to take.

3

Dialysis	Transplant
Machines available	Need donor, often not available
No problem with tissue matching	Need tissue match
Twice a week at least, for life	Surgery every ten years or more
Expensive long term	After surgery, relatively low cost of medicine
Always have to watch diet etc.	Can lead relatively normal life etc.

[Preferable treatment personal choice, but must be justified by rational argument.]

Answers to in-text question

a) If the donor and the recipient are identical twins they will have the same antigens.

KEY POINTS

The key points of this spread are clear and covered well by the material. The importance of the immuno-suppressant drugs should be emphasised. Students should be aware that they are expected to be able to evaluate the advantages and disadvantages of transplants in treating kidney failure.

B3 2.8 Transporting substances around the body – the past and the future

SPECIFICATION LINK-UP

Unit: Biology 3.13.4

Students should use their skills, knowledge and understanding of 'How Science Works'.

- *to evaluate the advantages and disadvantages of treating kidney failure by dialysis or kidney transplant.*

Teaching suggestions

Understanding the circulatory system – a complicated story!

This section of the spread concentrates on the historical aspects. If it is not proposed to set the students the activity, then a review of the ideas of the various scientists provides an interesting opportunity to explore scientific ideas.

- Students could divide into groups, each group taking one of the scientists and working out how to explain their theories to the rest of the class. They should be prepared to answer questions and defend their ideas. (This relates to: 'How Science Works': how scientific knowledge develops and changes.)
- Students could produce a newspaper headline for each belief or theory. They could try to think of a really catchy title (one for a tabloid and one for a broadsheet) followed by a short paragraph explaining the ideas in non-scientific terms.
- Students could prepare a three-line conversation based on one of the characters in the spread. Their peers are to guess who they are and then reciprocate. This could be repeated several times with different characters, if time allows.
- Consider the findings of all the scientists. Ask: 'How would they find out about circulation of the blood? How many of them used animals? How many references are there to vivisection?' Students to debate how far our knowledge would have progressed without using animals.
- Did William Harvey find out all there was to know about the circulation of the blood? Students could investigate whether any other scientist since his time has made any significant discoveries.
- William Harvey was a court physician. Students could find out the difference between a physician and a surgeon. Ask: 'Why might a surgeon be more likely to have a knowledge of blood circulation than a physician?'

BIOLOGY TRANSPORTING SUBSTANCES AROUND THE BODY
B3 2.8 Transporting substances around the body – the past and the future

Understanding the circulatory system
A complicated story!

ACTIVITY

The History of the Heart

Choose one of the scientists mentioned here and plan a display for a Museum of Medicine on him. Decide what objects you might like to display, and write out information cards or produce a PowerPoint presentation which could accompany your display. Remember that to make it interesting you need little details about the person as well as the story of their scientific discovery. Use web resources such as www.timelinescience.org to help you.

1 The Ancient Chinese model

The circulation of the blood was understood in China by the second century BC at the latest – about two thousand years before it was accepted in Europe!

The ancient Chinese thought there were two separate circulations of fluids in the body with the heart acting as a pump. They even made a model heart with bellows and bamboo tubes!

2 Galen of Pergamum ca 130–ca 200 AD

Galen was a Greek doctor who lived in the second century AD who taught his students that there were two distinct types of blood. 'Nutritive blood' was made by the liver and consumed by the organs and 'vital blood' was made by the heart and moved through the arteries to carry the 'vital spirits'. Galen believed that the heart acted not to pump blood, but to suck it in from the veins.

Galen carried out lots of dissections and vivisections on animals. He was the first person to show that experiments were important in medicine. In spite of this he got the circulation completely wrong – but his ideas were accepted for over a thousand years!

3 Ibn-El-Nafis 1208–88

Ala'El-Deen Ibn-El-Nafis was a brilliant Arab doctor. He worked out the correct anatomy of the heart and lungs and the way the blood flowed through them. He was also the first person known to record the blood supply to the heart itself.

Around 300 years after his death, some of his writings were translated into Latin and became available in Europe. Soon afterwards some European scientists and doctors began to make the same discoveries!

4 Michael Servetus 1511–53

Michael Servetus was a Spanish doctor who was trained in Paris. He accurately described the circulation of the blood from the heart to the lungs and back again. Unfortunately he wrote about it as part of some revolutionary religious arguments which upset Protestants and Catholics alike, so his findings were ignored.

Servetus met a grisly fate. First he was burned to death as a heretic by the Protestants. A few months later he was executed again by the Catholic Inquisition. As he was already dead, they executed an effigy of him!

5 William Harvey 1578–1657

William Harvey was court physician to both King James I and King Charles I. At the same time he carried out lots of research on the blood flow in the human body. Harvey questioned the teachings of Galen and investigated them scientifically.

He carried out lots of experiments. For example, when Harvey removed the beating heart from a living animal, it continued to beat, acting as a pump, not a sucking organ.

Harvey's notes show that he developed his ideas about the circulation as early as 1615 but he didn't publish his findings until 1628. Why did he wait so long? To rebel against the ideas of Galen could quickly end the career of any doctor, and this is what Harvey risked. After his work was published, many doctors and scientists rejected him and his findings. Some of his patients left him. Controversy raged for twenty years – yet today Harvey's work is considered to be one of the most important contributions in the history of medicine.

- More information about each of the scientists, their life and times, can be found on the Internet. Recommended web sites are those of the BBC History web site, www.timelinescience.org, and Medicina Antiqua, a UCL web site (for Galen and Harvey particularly).
- Refer to Electronic Resource B3 2.8 'Blood circulation' on the GCSE Biology CD ROM.

Special needs

- Students could be given pictures, names and dates to make a timeline or to compile a scrapbook of the scientists mentioned in the spread.
- Show *The Simpsons* 'Hell Toupe' from Halloween Special Treehouse of Horror IX (AABF01), where a villain called 'Snake' is sentenced to death for smoking in the Kwik-E-Mart for the third time. Homer gets Snake's hair as a transplant and it intermittently takes over his character, leading to mayhem. Watch and discuss.

Transplant surgery – points of view

Some aspects of this topic have already been covered in previous spreads, so this could provide an opportunity to take the debate further.

I sold one of my kidneys last year. The money has paid for my children to go to school and my wife to have some medicine.

Our faiths all support organ donation – there is no greater gift to give someone than life itself.

Our only comfort in those awful days after the accident was the thought that although we had lost our daughter, other people would get the chance of life. We've been told that two different people had successful kidney transplants and they used her liver and heart as well…

We are desperate for our son to get a new kidney – he really struggles with dialysis. The awful thing is that someone else has to die for him to live a full life. If someone offered to sell us a kidney, I really think we would buy it

It's not that I have anything against kidney transplants, but I don't think stem cell research or transplanting parts of other animals into people should be allowed.

I don't want to think about dying. I think all this business about donor cards is just morbid!

ACTIVITY

Most people agree that kidney transplants are a good thing. However there are still some ethical issues involved. In some parts of the world organs are bought and sold. Some of the new technologies which could increase the number of organs available raise their own ethical questions. Some people simply don't want to think about it.

a) Write two short speeches, one in favour of organ donation and the other raising issues about it.
b) At the moment you have to sign up to be an organ donor. In some countries everyone is automatically an organ donor unless they opt out. This provides many more organs for transplantation. It is suggested that this method is used in the UK. What do you think – should things stay as they are or should they change?

Write a brief paragraph explaining your personal point of view.
Then design a poster or leaflet supporting kidney transplants to be handed out at blood donation sessions, in hospitals and at doctor's surgeries.

- Points which come from the pictures in the Student Book could be discussed in class. The pros and cons of selling kidneys is a big issue. Use role play to highlight the problems of selling kidneys. One aspect that could be expanded is the reasons for buying a kidney. Ask: 'What are your thoughts about it? Is it fair? If people are willing to sell kidneys, there must be some people willing to accept them. How much is it worth?' There is some information about this on some of the newspaper web sites.
- **Consider the statistics** – it could be interesting to do some research on the numbers of people who suffer renal failure, the number of people waiting for transplants and the number of people on dialysis. This could be made more general by considering all transplants, leading to a discussion on donor cards.
- **Religious and ethical aspects** – As shown in the Student Book, the leaders of some of the major faiths do not appear to consider that organ donation is wrong. Ask: ' Is there any religion which has objections? What are the ethical considerations? Is it right to take a kidney from a dead person but not to take one from a living person?'
- **Growing new organs** – Students could investigate the research being done on the growth of human organs on other animals (xeno transplants). Ask: 'Is this the way to go in the future?'

ACTIVITY & EXTENSION IDEAS

Understanding the circulatory system

- **Activity 1, the timeline** – This could be a class project to be displayed in the laboratory. For each of the ideas, a brief description is required. It could also be interesting to show how long certain ideas persisted. For example, Galen's teachings and influence on theories of the circulation persisted from around 200 AD until Harvey's time. Some method of indicating this could be devised. The brief descriptions could be in the form of diagrams or cartoon representations.
- **Activity 2** – Each student should choose a scientist; time will be needed for research, using web sites or books. Some sources of information have been given in the 'Teaching suggestions', but if the exhibit is to be interesting some personal details could be included. It is interesting to know what the people looked like. Ask: 'Were they married? Where did they live? Are there any pages of their notebooks or research findings still in existence? What sort of conditions did they do their research in? Did they work in a laboratory?' PowerPoint® presentations are useful ways of showing information and can include animations or models.

Transplant surgery – points of view

The activities outlined here have already been referred to in the previous spread.

- **Speeches** – It is suggested that the students write two short speeches about organ donation, one in favour and one against. As suggested earlier, this could then form the basis of a class debate on organ donation.
- **The donor card issue** – The design of a poster or a leaflet can be discussed. As has been suggested previously for such an activity as this, the simpler and more direct the message, the more likely it is to get across to people. Students could consider campaigns that have used TV advertising, such as the anti-smoking campaign and the appeal for more blood donors. If facilities are available, a group of students could consider making a video to appeal for more people to carry donor cards. (A work sheet is available.)
- **Lobby your MP** – Ask: 'Does it need a change in the legislation to have to opt out of being an organ donor? If so, then where do you start?' Students could write a letter to their MP in support of a change in the law. Aspiring politicians could try drafting a Bill to change the law!

SUMMARY ANSWERS

1 a) If there is a 'hole in the heart', then oxygenated and deoxygenated blood is no longer kept separate. The oxygenated and deoxygenated blood mix and so the level of oxygen in the blood going around the body is not as high as it should be. This explains the blue colour and the lack of energy. If surgeons close up the hole, then the heart works perfectly normally and the two types of blood no longer mix.

b) The heart muscle itself becomes starved of oxygen and so cannot work properly. This is why the chest pain develops. If the damaged blood vessels are replaced by healthy ones, then the blood flow to the heart muscle is restored. It is no longer starved of oxygen and can work properly again.

2 a) The shape of the red blood cells gives maximum surface area for oxygen to enter by diffusion. The red blood cells contain lots of haemoglobin to maximise the carriage of oxygen.

b) i) After giving blood, the athlete will not perform as well. There is less blood available to carry oxygen and food to the muscles as they exercise.

ii) Iron is needed for the formation of haemoglobin, the red pigment in red blood cells that carries oxygen around the body. So if there is a lack of iron in the diet, there will be less haemoglobin, a lack of oxygen, resulting in a lack of energy and a tired feeling.

3 a) [Credit will be given in the subsequent answers for extracting and using the information on the bar charts.]

b) i) Increased fitness means that the heart has a greater volume and pumps more blood at each beat. The heart beats more slowly at rest.

ii) Increased fitness affects the lungs by lowering the breathing rate.

4 a) This is to ensure the right concentration gradients are present so that as much urea as possible leaves the blood in the dialysis machine and that the right levels of salt, glucose and minerals are maintained.

b) The constant circulation of the dialysis fluid is to maintain the right concentration gradient. As urea, salt etc. move from the blood into the dialysis fluid by diffusion, the fluid is removed and more 'clean' fluid arrives to maintain the maximum concentration gradient.

c) Look for well thought out similarities and differences, e.g. similarities: both remove urea from the blood, both remove excess salt and water from the blood, both maintain blood glucose levels, both balance levels of mineral ions in the blood etc. differences: dialysis several times a week, kidneys working constantly, need to manage diet on dialysis, eat what you like with working kidneys; set levels in dialysis fluid, kidneys adjust constantly to the intake and outputs of the body; produces used dialysis fluid while kidney produces urine etc.

Summary teaching suggestions

- **Literacy guidance** – In the answers, the students should be encouraged to use the technical terms in their correct contexts. Most answers require some extended explanations, written in complete sentences with attention to spelling, punctuation and correct grammar.
- **Lower and higher level answers**
  - Questions 1, 2 and 3 require straightforward answers.
  - The answers to the rest of the questions require more detailed explanations and the higher level answers should include all the relevant details.
- **Special needs**
  - Students may be able to make bar charts in answer to question 3 and, given the alternatives to choose from, could complete sentences to answer the questions.
  - The answers to some of the other questions could be adapted so that students fill in the appropriate words in sentences.

TRANSPORTING SUBSTANCES AROUND THE BODY: B3 2.1 – B3 2.8

SUMMARY QUESTIONS

1 Here are descriptions of two heart problems and how they may be overcome. In each case use what you know about the heart and the circulatory system to explain the problems caused by the condition and how the treatment helps.

a) Sometimes babies are born with a 'hole in the heart' – there is a gap in the central dividing wall of the heart. They may look blue in colour and have very little energy. Surgeons can close up the hole.

b) The blood vessels supplying blood to the heart muscle itself may become clogged with fatty material. The person affected may get chest pain when they exercise or even have a heart attack. Doctors may be able to replace the clogged up blood vessels with bits of healthy blood vessels taken from other parts of the patient's body.

2 a) Explain how red blood cells are adapted for carrying oxygen around your body.

b) In each of the following examples, explain the effect on the blood and what this will mean to the person involved:

i) an athlete running a race after acting as a blood donor and giving blood,

ii) someone who eats a diet low in iron.

3 It is often said that taking regular exercise and getting fit is good for your heart and your lungs.

a) Make bar charts to show the data given on page 243.

b) Use the information on your bar charts to explain exactly what effect increased fitness has on

i) your heart, and

ii) your lungs.

4 a) Explain the importance of dialysis fluid containing no urea and normal plasma levels of salt, glucose and minerals for the successful treatment of the blood.

b) Both blood and dialysis fluid are constantly circulated through the dialysis machine. Explain why the constant circulation of dialysis fluid is so important.

c) What are the main similarities and differences between a working kidney and a dialysis machine?

EXAM-STYLE QUESTIONS

1 The table below shows which parts of the blood transp different substances from one part of the body to another. Some words have been replaced by the letters A–G. State the word or words represented by these letters.

Part of the blood	Substance transported	Transported from	Transported to
A	Carbon dioxide	Various organs	B
Blood plasma	Soluble products of digestion	C	Various organs
Blood plasma	D	E	Kidneys
F	G	Lungs	Various organs

2 The diagram shows the rate of blood flow to various parts of the body when a person is at rest and when th are exercising.

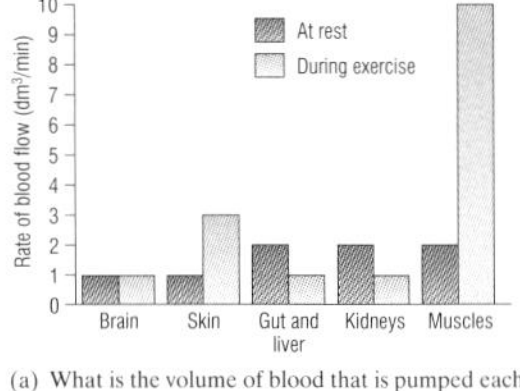

(a) What is the volume of blood that is pumped each minute to the muscles when:

(i) at rest?

(ii) during exercise?

(b) (i) What is the total volume of blood that is pumped each minute to **all** of the organs whe the person is exercising?

(ii) What is the percentage increase in blood pump to all the organs when the person is exercising compared to the volume pumped when at rest?

(c) Suggest reasons for the changes in the rate of blo flow to the following organs during exercise compared to the flow when at rest.

(i) muscles

(ii) skin

(iii) gut and liver

- **When to use the questions?**
  - Some of the questions could be used in conjunction with the in-text and summary questions to assess the students' understanding of the topics, either in **a** plenary session or as **a** starter before going on to study a further, related topic.
  - Question 2b) could be extended and students could be asked to suggest other factors or situations that could have an affect on the blood and the transport of oxygen. For example, the effects of disease could be discussed.

EXAM-STYLE ANSWERS

1 **A** Blood plasma **B** Lungs

C Small intestine **D** Urea

E Liver **F** Red blood cells

G Oxygen *(1 mark each*

2 a) i) 2 dm^3 *(1 mark)* ii) 10 dm^3 *(1 mark*

b) i) 16 dm^3 *(1 mark)* ii) 100 (%) *(1 mark*

c) i) *2 mark answer*

The rate of blood flow increases because the blood brings the oxygen and glucose needed for respiration to produce the energy to make the muscles work and also removes the carbon dioxide produced.

For both marks the two ideas required are:

- To work muscles need energy from respiration.
- Blood supplies glucose/oxygen (removes carbon dioxide) for respiration.

There are a wide range of responses worth one mark. Examples include:

- Blood brings oxygen/glucose so that muscles can respire more.
- Blood removes carbon dioxide from muscles.
- Blood provides the nutrients/materials that enable muscles to release energy/respire.

ii) • During exercise muscles produce much more heat. *(1 mark*

(d) Variation in the distribution of blood to organs is one change that occurs during exercise. State two other changes that take place within the body when exercise is taking place. (2)

The passage below about anaerobic respiration has some words replaced by the numbers **1**, **2**, **3** and **4**. Match the words **A**, **B**, **C** and **D** with the appropriate numbers in each sentence of the passage.

A energy **B** glucose
C oxygen **D** water

Anaerobic respiration is the partial breakdown of ....**1**.... to produce lactic acid.

Compared to aerobic respiration much less ....**2**.... is released.

Anaerobic respiration results in an ....**3**.... debt.

This debt must be repaid in order to convert the lactic acid to carbon dioxide and ....**4**.... [Higher]

The kidneys remove wastes from the liquid portion of the blood.

(a) Name the solution of waste that is stored in the bladder. (1)

(b) Describe how a healthy kidney produces this solution of waste. (5)

(c) If a person's kidney fails to function properly they may die because toxic substances cannot be removed from the blood. One way in which they may be kept alive is by using a dialysis machine. During dialysis, blood passes through branched or coiled tubes on the other side of which is dialysis fluid.

(i) Suggest a reason why the tubes containing blood are branched or coiled. (1)

(ii) What special property does the membrane of the tubes possess? (1)

(iii) Name one substance that diffuses from the blood into the dialysis fluid. (1)

(iv) Why do useful substances such as glucose and some mineral ions not also pass out of the blood? (1)

(d) Another form of treatment is to replace a diseased kidney with one transplanted from a donor. The new kidney may however be rejected by the immune system of the person receiving it. State three ways in which rejection of the transplanted kidney may be prevented. (3)

HOW SCIENCE WORKS

Kidney dialysis machines are designed to remove harmful substances such as urea and excess salts and water from the body. They also maintain a healthy level of potassium and sodium ions. People who need to use a dialysis machine will visit a treatment centre about three times a week for about 3 to 5 hours.

Their blood supply is connected to the machine. Every month the patient's blood is checked to ensure that the machine is working properly. The blood is tested for its urea content (URR test). The results have to be above 64%. Another test checks the amount of blood being filtered compared with the amount of fluid in the body (Kt/V test). This needs to be more than 1.1.

Look at this chart for a patient and answer the questions that follow.

Test	Target	Jan	Feb	Mar	April	May	June	July	Aug
Kt/V	≥ 1.2	1.1	1.15	1.2	1.23	1.24	1.2	1.2	1.2
URR	≥ 65	60	62	64	65	66	65	65	65
Weight (kg)	80	81	80	82	81	79	80	81	81

a) What was the range for the Kt/V test? (1)

b) What was the pattern for the Kt/V test? (3)

c) How does the Kt/V test results compare to those for the URR test? (2)

d) Can you say that there is a causal link between the two sets of test results? (1)

e) For how many months were the patient's tests both satisfactory? (1)

f) The doctors were concerned at the beginning of the year that the test results were a little low for this patient. What should they have done to get a better idea of what was happening? (1)

g) Urologists say that the two tests really measure the same thing. Why then is it a good idea to do both tests? (1)

h) What are the economic issues related to kidney dialysis? (2)

i) What are the social issues related to kidney dialysis? (2)

- This heat is carried by the blood to the skin where it is lost to the outside. *(1 mark)*

iii) • There is only a fixed volume of blood in the body/blood is needed elsewhere/needed by muscles/needed by the skin. *(1 mark)*

- The functions of the gut and liver/the digestion of food is not urgent and so the blood can be temporarily diverted to the muscles/skin (where it is more urgently required). *(1 mark)*

d) • The rate/depth of breathing increases. *(1 mark)*
- The heart beat/rate increases. *(1 mark)*

3 **A** 2 **B** 1 **C** 3 **D** 4 *(1 mark each)*

4 a) Urine *(1 mark)*

b) • The kidney filters out water, sugar (glucose), amino acids, mineral ions and urea from the blood
- leaving behind blood cells and proteins.
- All the sugar (glucose) is then reabsorbed
- by active transport.
- As much water as the body needs is (selectively) reabsorbed.
- The mineral ions that are needed by the body are (selectively) reabsorbed (by active transport).
- All urea, excess ions and excess water are released from the kidney and pass to the bladder (as urine).

(1 mark for each point to a maximum of 5)

The active transport mark can be given for the reabsorption of either glucose or mineral ions but not for both.

c) i) To provide a large surface area (for exchange of materials). *(1 mark)*

ii) They are partially permeable/any accurate description of partially permeable. *(1 mark)*

iii) Urea. *(1 mark)*

iv) Because the concentration of glucose and some mineral ions/these substances in the dialysis fluid is the same as that of the blood. *(1 mark)*

d) • The recipient may be treated with drugs that suppress their immune system.
- The bone marrow of the recipient may be treated with radiation to reduce or stop white cell production.
- The recipient should be kept in sterile conditions for a period of time after the transplant operation.
- The donor kidney should be tissue matched with the kidney of the recipient to ensure that the two are as similar as possible.

(1 mark for each point to a maximum of 3)

Exam teaching suggestions

Questions 1 and 3 are most suited to students at foundation level. They require short responses and therefore make useful quick revision tests at the end of a class session. Questions 2 and 4 mostly require students to describe and explain situations and to express their ideas in a logical, scientific way. As such, these questions are more suited to higher-tier students and best used as a homework exercise or a formal test in the classroom. The marking of questions 2 and 4 requires some interpretation in places and so is only suitable for assessment by the teacher.

HOW SCIENCE WORKS ANSWERS

a) The range for the Kt/V test was 1.1 to 1.24 Kt/V.

b) Pattern shows a gradual increase until May then the pattern dropped to a constant level.

c) They both show a similar pattern. Some students might notice that the URR remains below an acceptable level for one extra month.

d) No. It is most likely to be by association.

e) 5 months.

f) Taken the tests more regularly. The patient is dialysed 12 times a month.

g) Increases the reliability of the results.

h) Economic issues related to kidney dialysis: cost of machines, cost of staffing, cost of loss of lifestyle.

i) Social issues related to kidney dialysis: when money is limited, who will be allowed to benefit from kidney dialysis and who will be left to die?

How science works teaching suggestions

- **Literacy guidance**
  - Key terms that should be understood: range, causal links and those by association, reliability, patterns, interval measurements, social and economic issues.
  - Question c) expects a longer answer, where students can practise their literacy skills.
- **Higher- and lower-level answers.** Questions d) and f) are higher-level questions. The answers for these have been provided at this level. Question a) is lower level and the answers provided are lower level.
- **Gifted and talented.** Able students could review the data more closely. Kt/V, where K is the flow of blood through the machine; t = time in minutes that the patient is receiving dialysis; V = the body fluid content estimated as 60% of the body mass. Body mass is therefore included in the data chart. URR is calculated as the reduction in urea content during dialysis as a percentage of the content pre-dialysis.
- **How and when to use these questions.** When wishing to develop an awareness of the need for careful collection of data and its careful analysis. The questions relating to social and economic issues are best tackled by individuals and then discussed in groups or the whole class.

B3 3.1 Growing microbes

LEARNING OBJECTIVES

Students should learn:

- That microorganisms can be cultured in the laboratory in a medium containing the nutrients they require.
- That the sterilisation of apparatus and media is essential when making uncontaminated cultures.
- That the risk of growing dangerous pathogens is reduced by incubating the cultures at temperatures below 25°C.

LEARNING OUTCOMES

Most students should be able to:

- Describe the conditions needed for microorganisms to grow well in culture.
- Describe the safe procedure for the production of uncontaminated cultures of microorganisms.
- Explain why incubation temperatures in the laboratory should be kept below 25°C.

Some students should also be able to:

- Explain how pathogens might arise on plates inoculated by harmless bacteria.
- Explain the exponential growth of bacteria in ideal conditions and how we try to provide those in a culture.

Teaching suggestions

- **Special needs.** The students could do a 'spot the mistakes' exercise. Carry out a version of the practical demonstration but do a number of things wrong which the students have to identify.
- **Gifted and talented.** Give the students a table of data on times for reproduction of a bacterium at different temperatures. Get them to draw out a graph and work out a formula to describe the relationship between time and temperature. They can extrapolate the graph, labelling any change points (e.g. the point at which bacterial enzymes denature).
- **Learning styles**
  Interpersonal: Taking part in the 'Which disease?' plenary.
  Intrapersonal: Writing down things to do at home to reduce bacterial growth.
  Kinaesthetic: Doing the practical.
  Visual: Observing the PowerPoint.
  Auditory: Listening to instructions and safety regulations.

SPECIFICATION LINK-UP Unit: Biology 3.13.7

- *Microorganisms can be grown in a culture medium containing carbohydrates as an energy source, mineral ions, and in some cases supplementary protein and vitamins. These nutrients are often contained in an agar medium which can be poured into a Petri dish.*
- *In order to prepare useful products, uncontaminated cultures of microorganisms are required. For this:*
  - *Petri dishes and culture media must be sterilised before use to kill unwanted microorganisms.*
  - *Inoculating loops used to transfer microorganisms to the media must be sterilised by passing them through a flame.*
  - *The lid of the Petri dish should be sealed with adhesive tape to prevent microorganisms from the air contaminating the culture.*
- *In school and college laboratories, cultures should be incubated at a maximum temperature of 25°C to prevent the growth of pathogens that might be harmful to humans. In industrial conditions, higher temperatures can produce more rapid growth.*

Lesson structure

STARTER

Who would like to eat some microbes? – Show students some blue cheese and some live yoghurt. If permitted, have some samples for tasting. Discuss how these products are made and what organisms are involved. (5–10 minutes)

Biohazard! – Show the students the biohazard safety sign (or give them a copy to stick into their books). Ask them to guess what it means. Discuss and list the conditions under which this warning must be displayed. Talk about safety and general precautions when handling microorganisms. (10 minutes)

MAIN

- **Demonstration of aseptic techniques** – The students could observe the process of setting up a sterile culture. This should include swabbing down the bench, the use of sterile dishes, the pouring of agar plates and inoculating them from a culture of bacteria. Show the use of a special inoculating needle or loop which is sterilised both before and after the inoculation process. The correct method of just raising the lid of the dish and then sealing after inoculation should also be shown. Labelling the dishes is also important.
- **Have a go yourself** – Following the demonstration, students could set up some cultures of their own. It is easier if they are provided with sterile agar plates already poured and set. A culture of a harmless bacterium, such as *Bacillus subtilis*, could be used.
- The students could be shown how to make streak plates and spread plates (using sterile glass spreaders) of bacteria. The lids on Petri dishes should be secured with sticky tape and then incubated at 25°C for two days before inspection.
- **Growth in broth** – Microorganisms can be grown in nutrient broth. Get some students to inoculate small flasks of nutrient broth as well as making the plates.
- Discuss the advantages and disadvantages of using agar plates or nutrient broth in which to grow the microorganisms.
- **Microbes everywhere!** – Expose some sterile plates very briefly in different situations, such as in the laboratory and outside in the open air. On to the surface of sterile agar plates, put a small quantity of dust from the floor, a drop of tap water and place a finger gently on the surface. Seal with tape, label clearly on the bottom of the dish and incubate. Observe after a couple of days.

PLENARIES

The calculator game – Give the students a range of multiplication rates for microorganisms and get them to calculate, for each, how long it would take one organism to become a million. (5–10 minutes)

How fast can they grow? – Show the students a film clip or MPEG of bacteria reproducing quickly. Suppose that these are growing in someone's glass of milk (or a footballer's half-time drink – pick a team that people either love or hate). Half of the class like the person, and the other half do not and would like them to have a nasty surprise. Get half the class to write down three ways in which the process of bacterial growth could be speeded up, and the other half to write down ways in which the process could be slowed down. Read out and collect on the board. Remind about the dangers of food safety in real life. (10 minutes)

ACTIVITY & EXTENSION

- **Preventing the growth of microbes** – Ask the students to write down a list of things you can do at home to prevent the growth of microorganisms.
- **Growth rates** – It is difficult to measure the growth rate of a bacterium grown on a solid agar medium. However, it is possible to show the growth rate of a microorganism by inoculating agar plates in the centre with a mould such as *Mucor hiemalis* (bread mould). This fungus will grow in the lab and does not need incubating. The growth rate can be determined by measuring the diameter at set intervals of time. The results can be plotted as increase in diameter of colony against time.

Practical support

Growing microbes safely

Equipment and materials required

Sterile Petri dishes, sterile nutrient agar, inoculating loop, Bunsen burner, culture of *Bacillus subtilis* or other suitable bacterium, Chinagraph pencil or marker pen, sealing tape, incubator at 25°C.

Recipes for agar and more general information can be obtained from standard microbiology textbooks and from the National Centre for Biotechnology Education (NCBE) publications. They have a useful web site.

Prepare sterile bench tops with disinfectant or alcohol well before the experiment. No naked flames during this procedure.

Details

See 'Main' section in lesson structure and Student Book, page 257.

Answers to in-text questions

a) Bacteria, viruses, fungi and protoctistans.

b) Carbohydrate, mineral ions, proteins/fats, water, warmth, sometimes oxygen.

SUMMARY ANSWERS

1. Microorganisms are cultured in order to learn about them, the nutrients they need, the effect of antibiotics, etc.
2. They cannot make their own food.
3. This is the temperature of the human body and we cannot risk growing pathogens that might be dangerous to people.
4. **a)** Growth is incredibly fast; it is exponential as it doubles every 20 minutes.
   b) Change in temperature, lack of food or oxygen, build up of waste products, acid conditions, and any other sensible point.

KEY POINTS

The key points of this topic are well demonstrated by the practical work. Students should be recommended to include safety precautions when writing up the accounts of their investigations.

BIOLOGY MICROBIOLOGY

B3 3.1 Growing microbes

LEARNING OBJECTIVES

1. What conditions do microbes need to grow well?
2. How can you grow microorganisms safely in the laboratory?

Microbiology is the study of microorganisms. These are tiny living organisms such as bacteria, viruses, fungi and protoctista which are usually too small to be seen with the naked eye. To see and understand them properly you need to use a microscope. There are many different reasons why we study microorganisms. They play a vital role in decay and the recycling of nutrients in the environment. (See pages 168–9.) They can cause disease and they are enormously useful to people.

Many microorganisms can be grown in the laboratory. This allows us to learn a lot more about them. We can find out which nutrients they need to survive and which chemicals will kill them. We can also discover which microbes can be useful to us and which cause deadly disease.

a) What are the four main types of microorganisms?

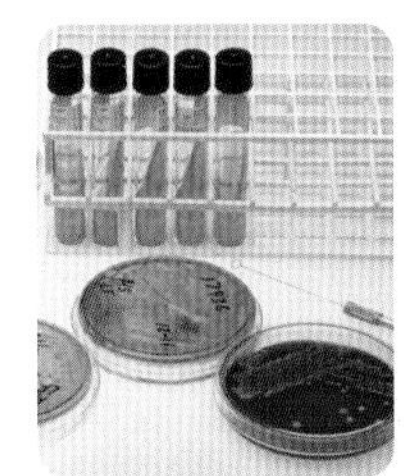

Figure 1 Culturing microorganisms like bacteria makes it possible for us to see what they are like and what they need to grow

What do microbes need to grow?

If you want to find out more about microorganisms you need to **culture** them. In other words, grow very large numbers of them so that you can see the colony as a whole.

To culture microorganisms you must provide them with everything they need. This usually involves providing a culture medium containing carbohydrate to act as an energy source. Along with this, various mineral ions and in some cases extra, supplementary protein and vitamins are included.

The nutrients are often contained in an **agar** medium. Agar is a substance which dissolves in hot water and sets to form a jelly. You pour hot agar containing all the necessary nutrients into a **Petri dish**. Then leave it to cool and set before you add any microorganisms. The other way to provide nutrients to grow microorganisms is as a broth in a culture flask. Whichever way you do it, you usually need to provide warmth and oxygen as well.

b) What do bacteria need to grow?

Safety precautions in the lab

You have to take great care when you culture microorganisms. This applies even when the microbes you want to grow are completely harmless. There is always the risk that a mutation may take place resulting in a new and dangerous pathogen. Also your culture may be contaminated by disease-causing pathogens which are present in the air, soil or water around you.

Not only do you want to avoid contamination by any dangerous miroorganisms, you also need to keep any pure strains of bacteria that you are growing free from other microorganisms. There are always millions of microbes – some harmful and some not – in the air, on the lab surfaces and on your own skin. So whenever you are culturing microorganisms, you must carry out very strict health and safety procedures.

SCIENCE @ WORK

Some scientists work with the most dangerous pathogens – microbes which can cause deadly diseases like ebola fever. So they have to take extreme safety measures. They work in labs with negative pressure gradients, so air moves in not out when doors are opened. They change their clothes before entering and leaving the lab. They also spend much of their time with their arms inside special sealed safety cabinets.

PRACTICAL

Growing microbes safely

Sterilise the Petri dishes and the nutrient agar before they are used. This kills unwanted microorganisms. You sterilise them using heat, often in a special oven known as an autoclave. The autoclave uses steam at high pressure.

Then inoculate the sterile nutrient agar with the microorganism you want to grow following the steps shown below:

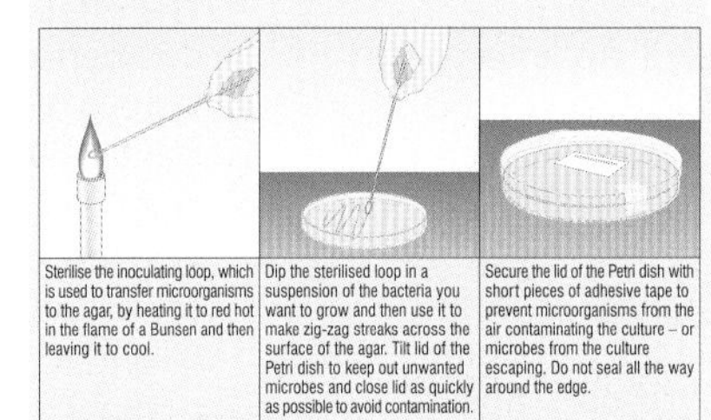

Sterilise the inoculating loop, which is used to transfer microorganisms to the agar, by heating it to red hot in the flame of a Bunsen and then leaving it to cool.
Dip the sterilised loop in a suspension of the bacteria you want to grow and then use it to make zig-zag streaks across the surface of the agar. Tilt lid of the Petri dish to keep out unwanted microbes and close lid as quickly as possible to avoid contamination.
Secure the lid of the Petri dish with short pieces of adhesive tape to prevent microorganisms from the air contaminating the culture – or microbes from the culture escaping. Do not seal all the way around the edge.

Figure 4 Culturing microorganisms in the lab

Once you have taken these steps, you need to incubate the sealed Petri dishes to allow the microorganisms to grow. In schools and college labs the maximum temperature at which cultures should be incubated is 25°C. This relatively low temperature greatly reduces the likelihood of pathogens growing which might be harmful to people. In industrial conditions, bacterial cultures are often grown at higher temperatures to promote rapid growth of the microbes. Once you have observed the microbes you have grown, you need to re-sterilise your Petri dishes complete with their cultures by heating them to 100°C and then throwing them away.

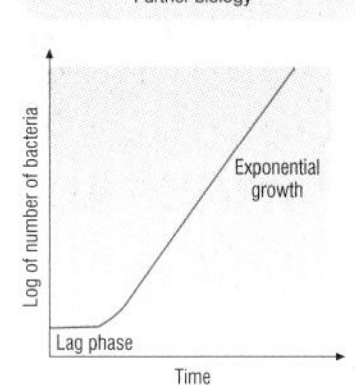

Figure 3 If you give bacteria the right conditions they can grow and divide very rapidly – which is why it is relatively easy to culture them in the lab

FOUL FACTS

You are surrounded by disease-causing bacteria all the time. If you cultured bacteria at 37°C – human body temperature – there would be a very high risk of growing some very nasty pathogens indeed.

GET IT RIGHT!

Make sure you can explain why we culture bacteria at 25°C rather than 37°C in the lab at school.

SUMMARY QUESTIONS

1. Why do we culture microorganisms in the lab?
2. Why do you think that all bacteria need a carbohydrate energy source as part of their nutrient medium?
3. Explain why cultures are not incubated at 37°C in a school lab.
4. Look at the graph in Figure 3.
   a) What does this tell you about the growth of bacteria in ideal conditions?
   b) A Petri dish provides ideal conditions for bacteria as they start to grow, but the ideal conditions do not last forever. What might limit the growth of the bacteria in a culture on a Petri dish?

KEY POINTS

1. Microorganisms can be grown in an agar culture medium with a carbohydrate energy source and various minerals, vitamins and proteins.
2. You need to take careful safety measures and use sterilised equipment to grow uncontaminated cultures of microorganisms and to avoid the growth of harmful pathogens.

B3 3.2 Food production using yeast

LEARNING OBJECTIVES

Students should learn:

- That yeast is a single-celled organism used in making bread and alcoholic drinks.
- That yeast can respire aerobically (producing carbon dioxide and water) and anaerobically (producing carbon dioxide and alcohol in a process called fermentation).
- That in brewing beer, the yeast uses the sugars produced by germinating barley grains but in wine-making, the yeast uses the natural sugars in the grapes.

LEARNING OUTCOMES

Most students should be able to:

- Describe the structure of yeast.
- Explain how the respiratory activities of yeast are linked to its use in making bread and alcoholic drinks.
- Describe how yeast is involved in the processes of making bread and alcoholic drinks.

Some students should also be able to:

- Explain the role of aerobic and anaerobic respiration in the way we use yeasts.

Teaching suggestions

- **Special needs.** Using Microsoft Word, students could make a poster advertising the wonderful properties of yeast. Provide them with a page of images to choose from and text to include, for them to drag and drop into a framework.
- **Gifted and talented.** Set up an Sc1 investigation to find out the maximum alcohol percentage achievable. Special strains of yeast are available from home brewing shops and the claim is made that these are able to produce wines with an alcohol content over 20%. Introduce the students to hydrometers and the concept of specific gravity.
- **Learning styles**

  Kinaesthetic: Carrying out practical work with yeast.

  Visual: Observing budding yeast.

  Auditory: Listening to discussion on uses of yeast.

  Interpersonal: Discussing what life without yeast would be like.

  Intrapersonal: Turning the concept map into a revision card.

SPECIFICATION LINK-UP Unit: Biology 3.13.5

- *Microorganisms are used to make food and drink:*
  - *yeast is used in making bread and alcoholic drinks.*
  - *bacteria are used in yoghurt and cheese manufacture.*
- *Yeast is a single-celled organism. Each cell has a nucleus, cytoplasm and a membrane surrounded by a cell wall.*
- *Yeast can respire without oxygen (anaerobic respiration) producing carbon dioxide and ethanol (alcohol). This is called fermentation. In the presence of oxygen yeast carries out aerobic respiration and produces carbon dioxide and water. Aerobic respiration provides more energy and is necessary for the yeast to grow and reproduce.*
- *In brewing beer and wine-making, carbohydrates are used as an energy source for yeast to respire. For making beer:*
  - *the starch in barley grains is broken down into a sugary solution by enzymes in the germinating grains, in a process called malting*
  - *the sugary solution is extracted then fermented*
  - *hops are then added to give beer flavour.*
- *In wine-making the yeast uses the natural sugars in the grapes as its energy source.*

Lesson structure

STARTER

Connections – Show the students a jar of Marmite, a loaf of bread, a bottle of beer and a bottle of wine. Ask: 'What is the connection?' Discuss the origins and history of these. (5–10 minutes)

Budding yeast cells – Set up a slide under a microscope attached to a projector. Show yeast cells budding. If Intel Play microscope is used or video capture software, you can take stop motion shots on time delay and sequence. Ask students what they think this is. Return to it at the end of the lesson. (5 minutes)

MAIN

- **Is yeast needed to make bread dough rise?** – The change in height of the dough can be plotted against time.
- **The effect of temperature on the rising of bread dough** – Dough in measuring cylinders can be placed in water baths at different temperatures and changes in the volume determined by measuring the height/volume of the dough. The changes in volume should be expressed as percentage changes.
- This is another investigative opportunity to cover aspects of 'How Science Works', as students can formulate hypotheses, take measurements and draw conclusions. There are a number of variables that need to be controlled and precautions taken to ensure reliability and comparability of results.

PLENARIES

Yeast budding – Return to the 'Budding yeast cells' starter set up at the beginning of the lesson. Observe any changes. (5 minutes)

Concept map – Students could complete a concept map on yeast, linking together the key words from the section and labelling the links. (10 minutes)

- **Homework.** There are lots of opportunities here for projects to carry out at home.
  - Students could expand their concept maps into useful revision resources.
  - They could also research and write an account of the structure of yeast cells and how they reproduce.

ACTIVITY & EXTENSION

- **Making ginger beer** – Ginger beer is easy to make from simple ingredients. If it is carried out in the laboratory, then it cannot be consumed. If this activity is to be undertaken, then try negotiating use of the Food technology room, but the method can be discussed in class and the recipe tried out at home. There are many recipes available. Try The Foody web site, BBC Woman's Hour web site; or to order a Ginger Beer 'plant', try The Hamstead Brewing Centre (they have other brewing kits as well).
- **Visit a brewery** – Many local breweries encourage visits and often have interesting displays. Local vineyards can also be visited. If there is not an opportunity to do this as a class activity, then encourage students to visit such establishments on family trips or holidays. They could report back to the class.
- **Low alcohol beers** – Find out how low alcohol beers are made. Ask: 'What does the 'lite' mean? How much alcohol do they contain?' Link this with the drink/driving problem and review work done on alcohol abuse earlier in the course.

Answers to in-text questions

a) **Aerobic:** plenty of energy for growth and reproduction.
Anaerobic: can survive a long time in low oxygen concentrations.

b) Carbon dioxide.

SUMMARY ANSWERS

1 **Aerobic respiration:** Sugar + oxygen → water + carbon dioxide
Anaerobic respiration: Sugar → ethanol (alcohol) + carbon dioxide

2 **a)** Sugar is added to give it a substrate (food) for respiration.
b) The yeast grows and divides as fast as possible in the warm temperature.
c) Bubbles of gas expand in the heat of the oven.

3 **a)** Barley grains malted → sugary solution from malting mixed with yeast and fermented → hops added for flavour and bitterness → beer clears → put into barrels, bottles or cans
b) **Similarities:** It depends on the fermentation of yeast; produces alcohol.
Differences: It uses sugar from fruit e.g. grapes as substrate for the yeast; yeast removed by filtering; the wine needs time, sometimes years, to mature in the bottle before it is ready for drinking

Practical support

Using yeast to make bread

Equipment and materials required

A suspension of yeast mixed with sugar and a little water, flour, beakers and suitable containers.

It is not usually feasible to bake bread in the laboratory, but the rising of the dough can be observed. If the effect of temperature is to be investigated, then suitable water baths will be needed.

Details

Get students to make small quantities of dough using 50 g of flour mixed with a little water (not too much). Divide the dough in half and add some yeast suspension made with a little yeast, water and 10 g of sugar to one half. Grease the insides of two small measuring cylinders and roll each portion of dough out into a sausage shape, so that it will fit into the measuring cylinders. Measure the height of the dough in each measuring cylinder at the start and then at set times through the lesson. It may be necessary to put the cylinders into a warm water bath or to use a yeast suspension that is active in order to get results during the lesson time.

KEY POINTS

The key points are reinforced by the practical work, which covers all the aspects of the uses of yeast. Students should know about the necessity for yeast to respire anaerobically when alcoholic drinks are produced.

B3 3.2 Food production using yeast

LEARNING OBJECTIVES

1 Why is yeast used to make bread?
2 Why is oxygen kept out when you make wine or beer?

One of the microorganisms which is most useful to people is yeast. The yeasts are single celled organisms. Each yeast cell has a nucleus, cytoplasm, and a membrane surrounded by a cell wall. The main way in which yeasts reproduce is by asexual budding – splitting in two to form new yeast cells.

When yeast cells have plenty of oxygen they respire aerobically. They break down sugar to provide energy for the cells, producing water and carbon dioxide as waste products.

However yeast can also respire anaerobically. When yeast cells break down sugar in the absence of oxygen, they produce ethanol and carbon dioxide.

Ethanol is commonly referred to as alcohol. The anaerobic respiration of yeast is sometimes called ***fermentation***.

The yeast cells need aerobic respiration because it provides more energy than anaerobic. This allows them to grow and reproduce. However, once there are large numbers of yeast cells, they can survive for a long time in low oxygen conditions. They will break down all the available sugar to produce ethanol.

We have used yeast for making bread and alcoholic drinks almost as far back as human records go. We know yeast was used to make bread in Egypt 6000 years ago. Not only that, some ancient wine found in Iran is over 7000 years old.

a) What are the advantages to yeast of being able to use both aerobic and anaerobic respiration?

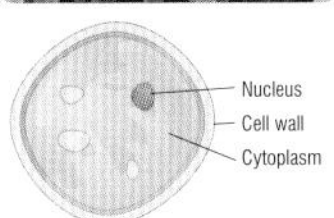

Figure 1 Yeast cells – these microscopic organisms have been useful to us for centuries

PRACTICAL

Using yeast to make bread

When you make basic bread, you mix yeast with sugar to provide it with an energy source for respiration. Just 1 gram of yeast contains about 25 billion cells! Then you mix the yeast and sugar with water and flour. Kneading the mixture makes sure the yeast is evenly spread throughout the dough and improves the texture. Then you leave the mixture somewhere warm.

As the yeast grows and respires it produces carbon dioxide, making the bread rise. When you bake the bread the bubbles of gas expand in the high temperature, giving the cooked bread a light texture.
The yeast cells are killed during the cooking process.

b) What by-product of yeast respiration makes bread rise?

Figure 2 When you make bread the dough is usually left to rise at least twice, to give the yeast a chance to make as much carbon dioxide as possible

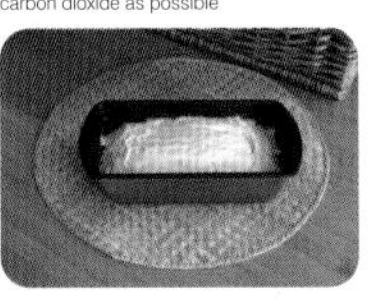

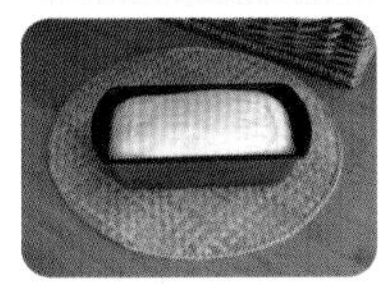

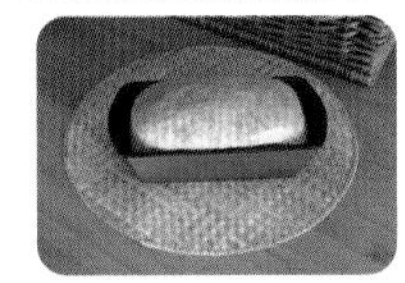

Making alcoholic drinks

When fruit fall to the ground and begin to decay, wild yeasts on the skin break down the fruit sugar. They form ethanol and carbon dioxide. These fermented fruits can cause animals to become drunk when they eat them. This is probably how our ancestors discovered alcohol!

We now use this same reaction in a controlled way to make beers and wine. In both cases the yeast must be supplied with carbohydrates to act as an energy source for respiration.

Beer making depends on a process called **malting**. You soak barley grains in water and keep them warm. As germination begins, enzymes break down the starch in the barley grains into a sugary solution. You then extract the sugary solution produced by malting and use it as an energy source for the yeast. The yeast and sugar solution mixture is then fermented to produce alcohol. You can add hops at this stage to give flavour. Finally you allow the beer to clear and develop its flavour fully before putting it in barrels, bottles or tins to be sold.

DID YOU KNOW?

Beer is a traditional drink in the UK and there are still many smaller independent brewers making high-quality local beers. Ringwood Brewery in Hampshire may be local – but they still manage to produce 30 000 brewer's barrels each year. That's around 8.5 ***million*** pints – a lot of beer!

DID YOU KNOW?

In one pint of beer there are 500 000 000 yeast cells!

Figure 3 Brewing at Ringwood Brewery in Hampshire. In tanks like these billions of yeast cells respire anaerobically, turning sugar into alcohol every day of the year. Who would guess, looking at the finished product, where it has all come from?!

In contrast, ***wine making*** uses the natural sugar found in fruit such as grapes as the energy source for the yeast. You press the grapes and mix the juice with yeast and water. You then let the yeast respire anaerobically until most of the sugar has been used up. At this stage you filter the wine to remove the yeast and put it in bottles. It will be stored for some time to mature before it is sold. Most wine sold commercially is made from grapes. However, wine can actually be made from almost any fruit or vegetable. The yeast isn't at all fussy about where the sugar it uses comes from!

Interestingly, alcohol in large amounts is poisonous to yeast as well as to people. This is why the alcohol content of wine is rarely more than 14%. Once it gets much higher, it kills all the yeast and stops the fermentation!

GET IT RIGHT!

Remember: yeast can respire aerobically in bread making but ***must*** respire anaerobically to make alcoholic drinks.

SUMMARY QUESTIONS

1 Produce word equations to show the difference between aerobic and anaerobic respiration in yeast.

2 a) In breadmaking, why is the yeast mixed with sugar before it is added to the flour and water?
b) Why is the bread dough kept warm while it is rising?
c) Why are cooked rolls or loaves bigger than the risen dough which is placed in the oven?

3 a) Make a flow chart to explain the process of making beer.
b) Explain how wine making is similar to and different from brewing beer.

KEY POINTS

1 Yeast is a single-celled organism which can respire aerobically producing carbon dioxide and water. This reaction is used in breadmaking to make the dough rise.
2 Yeast can also respire anaerobically producing ethanol and carbon dioxide in a process known as fermentation.
3 The fermentation reaction of yeast is used to produce ethanol in the production of beer and wine.

B3 3.3 Food production using bacteria

LEARNING OBJECTIVES

Students should learn:

- That bacteria are used in the production of yoghurt and cheese from milk.
- That both yoghurt and cheese require starter cultures of lactic acid bacteria.
- That in yoghurt production all the milk is used, but in cheese manufacture the curds are separated from the liquid whey.

LEARNING OUTCOMES

Most students should be able to:

- Describe how bacteria are involved in the production of yoghurt.
- Describe which steps in cheese manufacture require the action of bacteria.

Some students should also be able to:

- Compare and contrast the processes of cheese and yoghurt production.

Teaching suggestions

- **Special needs.** Give the students the stages in yoghurt production to arrange in the correct sequence.
- **Gifted and talented.** Students can investigate butter manufacture, where bacteria are also used. They can draw out a flow chart and compare with that for cheese manufacture.
- **Learning styles**

  Kinaesthetic: Carrying out practical work on yoghurt making.

  Visual: Drawing flow charts of the processes.

  Auditory: Listening to explanations of how yoghurt and cheese are made.

  Interpersonal: Working in groups for practical work.

  Intrapersonal: Completing homework exercise on cost of yoghurt.
- **Homework.** Students could research the cost difference between making yoghurt at home, using live yoghurt as a starter, and buying yoghurt from the supermarket. They could consider the variety you can achieve at home, by adding your own flavours and fruit. Ask: 'What are the advantages and disadvantages?'

SPECIFICATION LINK-UP Unit: Biology 3.13.5

- *Bacteria are used in yoghurt and cheese manufacture.*
- *In the production of yoghurt:*
  - *a starter of bacteria is added to warm milk.*
  - *the bacteria ferment the milk sugar (lactose) producing lactic acid.*
  - *the lactic acid causes the milk to clot and solidify into yoghurt.*

Lesson structure

STARTER

Bacterial food – Show some cheese, live bacteria drinks such as Yakult and live yoghurts, if not used as a starter for a previous lesson. (Do not taste these in the laboratory.) Discuss the use of bacteria in food and the popularity of yoghurt. Ask: 'Why do you think that products such as cheese and yoghurt were first made?' (5–10 minutes)

What happens to milk? – Set up some milk samples to keep at various temperatures. Students to predict what will happen to them. (5–10 minutes)

MAIN

- **Making yoghurt** – Natural live yoghurt is used as the starter culture in this experiment. See 'Practical support' for details.
- There are several further investigations that can be carried out using this method of making yoghurt, many of which introduce the concepts of 'How Science Works'.
- For example, the effect of temperature can be investigated, using a range of temperatures between 20°C and 50°C. The quality of the final product can be judged by its viscosity. The relative viscosity of different samples can be measured by the time it takes for a known volume to pass through a filter funnel.
- **Using different milk** – Investigate how different types of milk affect the quality of the yoghurt. Use skimmed, semi-skimmed and whole milk for a comparison. Also try making yoghurt with milk from sheep or goats (available in many of the larger supermarkets). The pH can be monitored as before, or the viscosity measured as described.
- **Cheese-making** – Search the Internet for pictures of cheese-making. There are several web sites with pictures and details of cheese manufacture. The students need to know the role of bacteria in the process, but the other steps could be included as well. Refer to electronic resource B3 3.3 'Bacteria' on the GCSE Biology CD ROM.

PLENARIES

How many cheeses can you name? – Give students 5 minutes to write down the names of as many different types of cheese as they can. Then compile a class list, with a small prize for the person who can name the greatest number. (10 minutes)

Lactose intolerance – Ask students if they have heard of lactose intolerance. If they have not, explain what it is and what effects it has. Ask why yoghurt is acceptable to lactose-intolerant people. Discuss the possibilities of making yoghurt with lactose-reduced milk. Ask: 'What problems might there be with using this type of milk?' (10 minutes)

KEY POINTS

Students need to understand that it is the lactose in the milk that is the substrate for the bacteria. This can be reinforced by referring to lactic acid fermentation and distinguishing the reactions from the fermentation which yeast undergo in anaerobic respiration. It could be helpful for students to compare the two equations, one for the fermentation of lactose and the other for the fermentation of glucose.

ACTIVITY & EXTENSION

- **Different names for yoghurt-type preparations** – All over the world, milk is used to make different types of yoghurt. Students can find out where the following are made and what type of milk is involved: kumiss, lassi, raita, filmjolk, kefir and amaas. They can add others to the list if they find them. One research method is to investigate the uses of different types of milk, i.e. sheep, goat, buffalo, camel, etc.
- **Make your own flow charts** – Students to design their own flow charts of the processes of cheese manufacture and yoghurt production. These can be used as revision aids. They should highlight where bacteria are involved and include the names of the bacteria.
- **'Low fat' yoghurt and others** – If not done earlier, students could carry out a survey of the different types of yoghurt available. Ask: 'Do the 'low fat' ones have significantly lower fat content than others? How could this be arranged during manufacture?

SUMMARY ANSWERS

1 Starter culture of bacteria added to warm milk → mixture kept warm → lactic fermentation takes place → yoghurt produced

2 Starter culture of bacteria added to warm milk → kept warm → lactic fermentation takes place → enzymes added to increase separation of solid part → curds cooked in whey → curds separated from whey → curds cut and salted; bacteria and fungi may be added to give different flavours and textures → curds packed and pressed → cheese left to dry and ripen

3 The initial stages of both processes are very similar. In yoghurt production, there is no complete separation of curds and whey. The bacteria simply thicken the milk. It only takes hours and the milk is preserved for a few days. Cheese making is much more complex, with many different steps. It can take years for the cheese to ripen, but the milk is preserved for a very long time.

Practical support

Making yoghurt

Equipment and materials required

UHT milk (or other milk which will need to be heat-treated), natural live yoghurt to use as a starter culture, boiling tubes, measuring cylinders, cling film, pH meter, narrow range pH test papers or probe and data logger, water bath at 43°C.

These are the basic requirements and if some of the suggested variations are used, then different types of milk and water baths at different temperatures will be needed.

Details

Place 10 cm^3 UHT milk in a boiling tube and add 1 cm^3 of natural live yoghurt. Mix the two together, record the pH, cover the tube with cling film and incubate in a water bath at 43°C. Record the pH and changes in the appearance of the yogurt at intervals of 30 minutes for up to 5 hours. If UHT milk is not used, then the milk should be heated to about 90°C for 15 minutes to kill the bacteria present in the milk. The pH can be measured using a pH meter, pH papers or using a probe and a data logger.

Answers to in-text questions

a) Cows, sheep, goats, horse, camel, [any other sensible suggestion], all provide milk for people.

b) The lactic acid produced by the bacteria makes the milk clot and solidify.

BIOLOGY MICROBIOLOGY

B3 3.3 Food production using bacteria

LEARNING OBJECTIVES

1 Why are bacteria used in yoghurt making?
2 What part do bacteria play in making cheese?

People began to domesticate animals quite early in history. They soon realised that the milk that female animals made for their babies could be used as food for us too! However, there is one big drawback in using milk as part of the diet. It very rapidly goes off, starts to smell and tastes disgusting! However, it didn't take people long to find ways of changing the milk into foods which lasted much longer than milk itself. These changes depended on the action of microorganisms.

Yoghurt has long been a staple part of the diet in the Middle East but has only become popular in the UK quite recently. Cheese, on the other hand, has been around for a very long time almost all over the world.

Figure 1 In the UK we get most of our milk from cows, but around the world a number of different animals including camels, horses, sheep and goats are used for milking

a) How many different animals can you think of that are used to provide milk for people?

Making yoghurt

Traditionally, yoghurt is fermented whole milk but now we can make it from semi-skimmed milk, skimmed milk and even soya milk. Yoghurt is formed by the action of bacteria on the **lactose** (milk sugar) in the milk.

GET IT RIGHT!

Don't confuse **lactose** and **lactic acid**!

PRACTICAL

To make yoghurt you add a starter culture of the right kind of bacteria to warm milk. You then keep the mixture warm so the bacteria begin to grow, reproduce and ferment. As the bacteria break down the lactose in the milk they produce **lactic acid** which gives the yoghurt its sharp, tangy taste. This is known as **lactic fermentation**. The lactic acid produced by the bacteria causes the milk to clot and solidify into yoghurt. The action of the bacteria also gives the yoghurt a smooth, thick texture.

Once the yoghurt-forming bacteria have worked on the milk, they also help prevent the growth of the bacteria which normally send the milk bad. Yoghurt, if it is kept cool, will last almost three weeks before it goes off. Ordinary milk only lasts a few days – and then only if it's kept really cold.

Once you have made your basic yoghurt you can mix in all sorts of colourings, flavourings or fruit purees. This is also done industrially to give us the wide variety of different types and flavours of yoghurts we find on the supermarket shelves.

FOUL FACTS

No-one really knows when yoghurt making first started. It seems to have come from Turkey. Legend has it that travellers took some milk on a journey in a bag made of a sheep's stomach. The heat of the Sun and the bacteria in the stomach worked on the milk, and in the cool of the desert night they discovered the bag was full of yoghurt!

b) Why is yoghurt so much thicker than milk?

SCIENCE @ WORK

The baking, brewing and dairy industries employ thousands of people. Some work directly with the microorganisms and some with all the other aspects of the products from development to marketing.

260

Figure 2 In just a few years yoghurts have become one of the most widely eaten dairy products in the UK!

Cheese-making

Like yoghurt-making, cheese-making depends on the reactions of bacteria with milk. These change the texture and taste and also preserve the milk. Cheese-making is very successful at preserving milk, and some cheeses can survive for years without decay.

Just as in yoghurt-making, you add a starter culture of bacteria to warm milk. The difference is in the type of bacteria added. The bacteria in cheese making also convert lactose to lactic acid, but they make much more of the lactic acid. As a result the solid part (the curds) are much more solid than the yoghurt ones.

Enzymes are also added to increase the separation of the milk. When it has completely curdled you can separate the curds from the liquid whey. Then you can use the curds for cheese-making. The whey is often used as animal feed.

Next you cut and mix the curds with salt along with other bacteria or moulds before you press them and leave them to dry out. The bacteria and moulds which you add at this stage of the process are very important. They affect the development of the final flavour and texture of the cheese as it ripens. The ripening may take months or years depending on the cheese being made.

DID YOU KNOW?

Almost one third of all the cow's milk produced in the USA is turned into cheese!

Figure 3 Curds are formed by the action of one set of bacteria on the milk. Then the final flavour and texture of the cheese may well depend on other bacteria added at this stage of the process.

SUMMARY QUESTIONS

1 Produce a flow chart to summarise the production of yoghurt from milk.
2 Produce a flow chart to summarise the production of cheese from milk.
3 Compare the processes of yoghurt-making and cheese-making and summarise the differences between them.

KEY POINTS

1 Bacteria are used in making both yoghurt and cheese.
2 In the production of yoghurt a starter culture of bacteria acts on warm milk. Lactose is converted to lactic acid in a lactic fermentation reaction. This changes the texture and taste of the milk to make yoghurt.
3 In cheese-making a different starter culture is added to warm milk giving a lactic fermentation which results in solid curds and liquid whey. The curds are often mixed with other bacteria or moulds before they are left to ripen into cheese.

261

B3 3.4 Large-scale microbe production

LEARNING OBJECTIVES

Students should learn:

- That large-scale culture of microorganisms takes place in vessels called fermenters.
- That fermenters provide the conditions needed to ensure optimum growth.
- That the fungus *Fusarium* is used to produce a protein-rich food called mycoprotein.

LEARNING OUTCOMES

Most students should be able to:

- Describe how microorganisms can be cultured on a large scale in fermenters.
- Explain how conditions are controlled in fermenters for optimum growth of the microorganisms.
- Describe the production of mycoprotein.

Some students should also be able to:

- Explain why conditions in a fermenter need to be controlled.

Teaching suggestions

- **Special needs.** Students to play 'Pin the label on the fermenter'. Project an unlabelled diagram of a fermenter (of a type likely to come up in the exam). Blindfold a student, help them to the board and give them a large label, with a piece of Blu-Tack on the back, for one of the parts of the fermenter. The other students have to direct the blindfolded student to the correct spot to stick the label by saying 'up', 'down', 'right' or 'left'. You can give clues as needed, e.g. by whispering to some students. The blindfolded student then gets to choose who should be blindfolded next.
- **Gifted and talented.** Ask students to produce an article for a newspaper, summarising in an engaging way, the development of mycoprotein and speculating on its future.
- **Learning styles**

  Kinaesthetic: Carrying out fermentation practical.

  Visual: Observing presentations.

  Auditory: Reading out the limerick.

  Interpersonal: Working in groups on taste surveys.

  Intrapersonal: Writing about the consequences of banning the eating of animals.

SPECIFICATION LINK-UP Unit: Biology 3.13.6

- *Microorganisms can be grown in large vessels called fermenters to produce useful products such as antibiotics. Industrial fermenters usually have:*
  - *An air supply, to provide oxygen for respiration of the microorganisms.*
  - *A stirrer to keep the microorganisms in suspension and maintain an even temperature.*
  - *A water-cooled jacket to remove heat produced by the respiring microorganisms.*
  - *Instruments to monitor factors such as pH and temperature.*
- *The fungus* Fusarium *is used to make mycoprotein, a protein-rich food suitable for vegetarians. The fungus is grown on starch in aerobic conditions and the biomass is harvested and purified.*

Lesson structure

STARTER

Beer – Beer is made in large containers in breweries. Show students some pictures of the large vessels used for the fermentation process in beer-brewing. Pour a 500 cm^3 glass of fake beer. Ask: 'How many of these would fit into the fermenting tank? Think of some of the problems which might occur when fermentation occurs in such large volumes and write them down.' (5–10 minutes)

Is it meat or is it fungus? – If possible (perhaps using the Food technology room), provide small samples of some meat products and their corresponding Quorn substitutes. Ask students to decide whether they have been given meat or fungus. Review the Quorn products and mark them on a scale of 1 to 10 for taste and preparation. Alternatively hand around but don't taste. (15 minutes)

MAIN

- **Fermenters** – Use Photo PLUS B3 3.4 'Yoghurt' from the GCSE Biology CD ROM.
- Discuss a diagram of an industrial fermenter, with all the inlets and outlets labelled, to include: the air supply, the cooling jacket and the probes which monitor pH and temperature.
- **A simple fermenter** – Having described large fermenters, it would be appropriate to show students a simple fermenter that can be used in the laboratory and to set it up to grow an organism, such as yeast, under controlled conditions. Simple fermenters are available from the NCBE (www.ncbe.reading.ac.uk), together with the instructions on how to set up and use them.
- The progress of the growth of the yeast can be shown by taking samples at regular intervals and counting the number of cells. In this way, a growth curve could be plotted. (This relates to 'How Science Works': relationships between variables.) If enough fermenters are available, students can work in groups and set up their own.
- **The production of mycoprotein** – There is a web site, Mycoprotein Education (www.mycoproteineducation.com), which gives a great deal of information about the production of Quorn. There are diagrams and flow charts showing the steps in production, together with information about the nutritional content.
- Students could be provided with a work sheet, so that they can build up a resource for themselves, which encompasses the manufacturing process and the nutritional benefits.

PLENARIES

Compose a limerick . . . – Students to produce a limerick by completing the following 'A man who ate nothing but Quorn . . . '. Students to work in groups and then read out the results. (10 minutes)

True or false? – Give students a series of statements about the differences between mycoprotein and meat (include some nutritional ones as well as structural). Students to hold up 'True' or 'False' cards. (5–10 minutes)

ACTIVITY & EXTENSION IDEAS

- **Batch culture or continuous culture**? – Find out what can be produced by batch culture and what is produced using continuous culture.
- **Can you tell the difference?** – Using meat sausages and Quorn sausages on cocktail sticks, students could carry out a survey amongst the students of a different year group to determine whether they can distinguish between the meat and the Quorn. Announce the results on posters around the school.
- **Single cell protein (SCP)** – Quorn is not the only example of a food stuff produced by microorganisms. Other examples include 'Pruteen', 'Pekilo', *Spirulina* and 'Toprina G'. Students could research one of these products and find out the organism used, the substrates it grows on, the product and what it is used for. Ask: 'What are the major disadvantages of single cell protein? Why is it not more widely used?'
- **No more animals to be eaten!** – Students could consider the consequences of a government decision that the eating of animals should be banned and that the nation should eat more protein from microorganisms. Ask: 'What would it mean to the farming industry? How much would it cost? What would be the long term consequences?'

Answers to in-text questions

a) Antibiotics.

b) pH affects the shape of enzymes so they do not work any more. If the pH in a fermenter changes, the microbial enzymes will not work and the culture stops growing or dies.

BIOLOGY MICROBIOLOGY

Further biology

B3 3.4 Large-scale microbe production

LEARNING OBJECTIVES

1. How are microbes cultured on a large scale?
2. How can fungi be used as a meat substitute?

It is one thing to grow microorganisms in the laboratory on a small scale in a Petri dish. In those conditions it is relatively easy for you to provide the food, oxygen and warmth the microorganisms need to grow. The cultures are not usually kept for long enough for the build-up of waste products to be a problem.

But it becomes a very different story when we need to grow microorganisms on an industrial scale. Increasingly we want to make use of materials made by the microbes. These may be drugs, like antibiotics, or food. This means we need to keep a very large fermentation going for a long time so we can harvest the products.

Scaling up brings its own problems. Imagine moving a culture of microorganism from an agar plate to vessels which might hold 1 000 000 dm^3 of medium. This is a very big step to manage successfully.

a) Which drugs are commonly made by microbes?

Fermenters

In ideal conditions the numbers of bacteria can double every 20 minutes. However, ideal conditions are rare and what usually happens in a bacterial culture is shown in Figure 1.

As the numbers of microorganisms begin to rise, conditions change. The food is used up. The metabolism of all the millions of microorganisms causes the temperature to rise. Oxygen levels fall as it is used up in respiration.

The carbon dioxide waste from respiration can also alter the pH of the culture. If the pH changes, the activity of the enzymes in the culture can be affected so it stops growing or dies. Other waste products may begin to build up and poison the culture. In industrial fermentations, very large microbe cultures are involved. So problems like these can develop very rapidly.

Lag phase as the bacteria start to grow | Exponential phase – bacteria divide every twenty minutes | Stationary phase of plateau – growth slows and stops | Death phase – bacteria are dying off faster than they are dividing

Theoretical growth is exponential!

A

B

Log of number of bacteria

Time

Figure 1 In real life rather than ideal conditions, all sorts of factors make bacterial growth slow down

When we grow microbes on an industrial scale we use large vessels known as **fermenters**. These industrial fermenters are designed to overcome the problems which stop a culture from growing well. They react to changes, keeping the conditions as stable as possible. This in turn means we can get the maximum yield. Industrial fermenters usually have:

- an oxygen supply to provide oxygen for respiration of the microorganisms,
- a stirrer to keep the microorganisms in suspension. This maintains an even temperature and makes sure that oxygen and food are evenly spread throughout the culture,
- a water-cooled jacket which removes the excess heat produced by the respiring microorganisms. Any rise in temperature is used to heat the water which is constantly removed and replaced with more cold water,
- measuring instruments which constantly monitor factors such as the pH and temperature so that changes can be made if necessary.

262

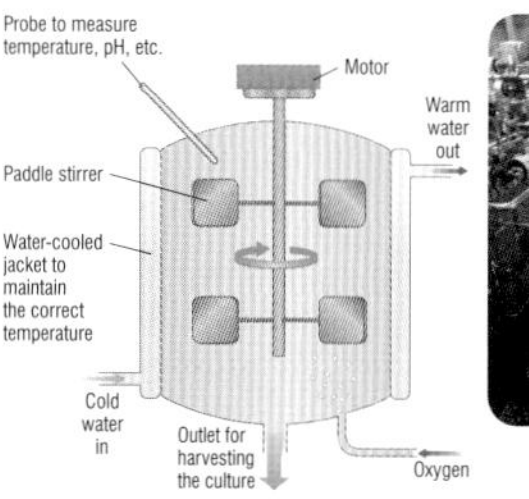

Figure 2 The design of fermenters is improving all the time as new ways are developed to keep conditions inside the fermenter as stable as possible

b) Why does a change in pH in a fermenter have such a serious effect on the microorganisms?

Mycoprotein production

Recently a completely new food based on fungi has been developed. It is known as **mycoprotein**, which means 'protein from fungus'. It is produced using the fungus *Fusarium*. This grows and reproduces rapidly on a relatively cheap sugar syrup (made from waste carbohydrate) in large specialised fermenters. It needs aerobic conditions to grow successfully. Then it can double its mass every five hours!

The fungal biomass is harvested and purified. Then it is dried and processed to make mycoprotein. This is a pale yellow solid with a faint taste of mushrooms. On its own it has very little flavour.

However, mycoprotein can be given a range of tastes and flavours to make it similar to many familiar foods. It is a high-protein, low-fat meat substitute. So it is used by vegetarians and people who want to reduce the fat in their diet plus people who just like the taste!

When mycoprotein was first developed people thought a world food shortage was on its way. They were looking for new ways to make protein cheaply and efficiently. The food shortage never happened, but the fungus based food continued. It is versatile, high in protein and fibre, low in fat and calories and so is widely used in the developed world.

Figure 3 Mycoprotein can be made to look like meat, chicken and fish. You can have mycoprotein pies and burgers. It is also sold relatively unflavoured for people to use in their own way. Mycoprotein can even be ground into flour that can be used to make snacks and a variety of sweet things.

DID YOU KNOW?

The protein content of mycoprotein is similar to that of prime beef!

SUMMARY QUESTIONS

1. a) Describe how curve B differs from curve A in Figure 1.
   b) What causes the difference between the growth pattern in ideal conditions and the growth pattern in real conditions?
2. How might an industrial fermenter be adapted to ensure the best possible conditions for the growth of a particular microorganism?
3. Why do the following factors tend to change during a fermentation process?
   a) Temperature. b) Oxygen levels. c) pH levels.

KEY POINTS

1. Microorganisms can be grown on a large scale in vessels known as fermenters to make useful products such as antibiotics and mycoprotein food.
2. Industrial fermenters usually have a range of features to make sure the fermentation takes place in the best possible conditions for a maximum yield of the product.
3. The fungus *Fusarium* is grown on sugar syrup in aerobic conditions to produce mycoprotein foods.

263

SUMMARY ANSWERS

1. **a)** In curve A, the growth of bacteria continues unchecked. In curve B, the bacteria start to grow exponentially but the growth levels out and finally the population numbers fall.
   b) In ideal conditions, there are no limiting factors on bacterial growth. In a real culture, food and oxygen supplies run out, waste products build up, pH levels change and the bacteria become overcrowded. These limitations mean the population growth steadies and finally more bacteria die than are produced, so numbers start to fall.
2. The fermenter needs regular additions of food and oxygen, stirrers to ensure everything is mixed together evenly, measures to remove excess waste and/or neutralise acid produced.
3. **a)** The effect of the metabolic reactions of the millions of bacteria produces heat (think compost heap!).
   b) Oxygen levels fall as oxygen is used up in cellular respiration.
   c) Carbon dioxide levels increase and cause a fall in pH.

KEY POINTS

Students should have a good understanding of the conditions for the optimum working of large-scale fermenters. Labelled and annotated diagrams of fermenters would be appropriate revision aids. The summary questions could be used to test students' understanding of this lesson.

B3 3.5 Antibiotic production

LEARNING OBJECTIVES

Students should learn:

- That the antibiotic penicillin can be produced on a large scale in a fermenter, by growing the mould *Penicillium* on a medium containing sugar and other nutrients.
- That the penicillin is produced by the fungus when most of the nutrients needed for growth have been used up.

LEARNING OUTCOMES

Most students should be able to:

- Explain the circumstances in which penicillin is produced.
- Describe the conditions needed for the production of penicillin.

Some students should also be able to:

- Explain the requirement for a source of nitrogen in the fermentation broth.
- Explain why penicillin is not produced for the first 40 hours of the process.

Teaching suggestions

- **Special needs.** Get students to make a large poster of 'Things you should know about antibiotics'. Supply the students with the statements to arrange on the poster, or they could make them. Use statements from the starter 'Antibiotics: what you should know'.
- **Gifted and talented**
  - Students could research the meaning of the terms 'narrow-spectrum' and 'broad-spectrum' when applied to antibiotics.
  - Ask: 'What do the terms 'microbicidal' and 'microbistatic' mean?
  - Students could also find out how antibiotics affect the structure and physiology of bacteria.
- **Learning styles**

  Interpersonal: Participating in the games.

  Intrapersonal: Writing letter to Sir Alexander Fleming.

  Kinaesthetic: Experiment on effect of antibiotics on bacteria.

  Visual: Watching presentations.

  Auditory: Listening to the experiences of others being treated with antibiotics.

SPECIFICATION LINK-UP Unit: Biology 3.13.6

The antibiotic penicillin is made by growing the mould Penicillium *in a fermenter. The medium contains sugar and other nutrients, e.g. a source of nitrogen. The* Penicillium *only starts to make penicillin after using up most of the nutrients for growth.*

Lesson structure

STARTER

Who might have died? – Ask the students to put their hands up if they have ever taken antibiotics. Ask: 'What for?' From their responses, draw out those who would have been likely to have died before the advent of penicillin. Get them to rest face down on the desk. Explain how serious some infections were and that they really would have died if antibiotics had not been available. Get them to ask their grandparents or older relatives about a time without antibiotics and collect true stories. (5–10 minutes)

Antibiotics: what you should know – As the students enter the room, give them a sheet of paper headed 'Antibiotics – what you should know' and some ruled lines. Give students a time limit of 3–4 minutes to complete. Pick one student to read out one fact from their list. That student can then pick on someone else and so on. Vary this by specifying gender or instructing them to pick on someone they would not normally ask. Collect suggestions at the front and compile a list that can be printed off and distributed later; this will be a useful revision aid. (10 minutes)

MAIN

- **Production of antibiotics** – Use PhotoPLUS B3 3.5 'Penicillin'. Expand the ideas to include other antibiotics. Students to have worksheets to complete and make their own notes. This presentation should focus on the production of antibiotics in general.
- **The story of penicillin** – Students can research the discovery of penicillin and the involvement of Fleming, Florey and Chain. For this approach, they need to be told to find out as much as they can about the discovery of penicillin and about the contributions made by these three scientists. This could be a homework exercise. With contributions from the students, build up a flow chart of events in the discovery of penicillin, annotating the stages with extra information.
- **Sir Alexander Fleming** – Build up a biography of Sir Alexander Fleming. There is a great deal of information available about his life from web sites, such as timelinescience and the Wellcome Museum. There is a museum at St Mary's Hospital devoted to the discovery of penicillin.
- **Effect of antibiotics on bacteria** – It is possible to set up agar plates inoculated with a bacterium, such as *Bacillus subtilis*, and place discs impregnated with different antibiotics (Oxoid Multodiscs) on to the agar. The effect of the antibiotic on the bacterium can be judged by the diameter of the clear zone around each disc, but this will depend on the concentration of each antibiotic on the disc so it is better not to compare the sizes of the zones for each antibiotic. A variety of different bacteria can be used, but check for safety of use. For more information, refer to NCBE (www.ncbe.reading.ac.uk) or a microbiology text. **Safety:** Follow aseptic techniques described previously.

PLENARIES

Antibiotic acrostic – Get the students to suggest words associated with this topic and then compile an acrostic. Students to start writing clues and definitions. Complete for homework. (10 minutes)

Matching pairs – Have the names of the people involved, their roles in the discovery and development, the details of the fermentation process etc. in a pair of mixed up lists. Give these to students on pieces of paper and then see who can successfully match the pairs correctly in the fastest time. (10 minutes)

ACTIVITY & EXTENSION IDEAS

- **Secondary metabolites** – Penicillin and other antibiotics are anti-microbial compounds produced by living organisms. These substances are made by the organism late in its cycle of growth. Project a graph showing the yield of fungal mycelium and penicillin production against time during a batch fermentation. Mark in various key points, such as the time when the nutrients are beginning to run out. Ask: 'Why do you think the organisms produce these substances?' Discuss what is happening and how this affects the way in which the antibiotic is extracted from the process.
- **Other antibiotics** – Patients are not prescribed 'penicillin' as such. Get the students to say what they have been prescribed. Ask: 'Did they have injections, capsules or was it in syrup?' Discuss the different forms of penicillin and why these have been produced. List reasons for slightly altering the structure of an antibiotic.
- **Other uses of antibiotics** – Antibiotics are not just used in the treatment of human diseases. List the other ways in which they are used [in agriculture, animal feeds, household products etc.]. Ask: 'How does this affect their use in treating human diseases?'
- **Why are antibiotics so effective?** – Get students to recall what a good medicine should do, from work done in a previous unit. Then discuss how antibiotics fit into this ideal. Introduce the idea that antibiotics are toxic to bacterial cells but not to human cells. Remind students of the structure of a bacterium and then outline how the antibiotics affect the bacterial cells.

Teaching suggestions continued

- **Homework.**
  - Several suggestions for homework activities have already been made (research on the discovery of penicillin and completion of the acrostic).
  - Students could write an imaginary letter to Alexander Fleming thanking him for his discovery and telling him how it had saved their life.

Answers to in-text questions

a) Leaving plates open, leaving plates on windowsill, or forgetting to look at plates would not be allowed!

b) It was so difficult to extract the antibiotic from the mould and unstable once it was extracted.

BIOLOGY MICROBIOLOGY

B3 3.5 Antibiotic production

LEARNING OBJECTIVES

1. How was penicillin discovered and developed?
2. What does the *Penicillium* fungus need to grow?

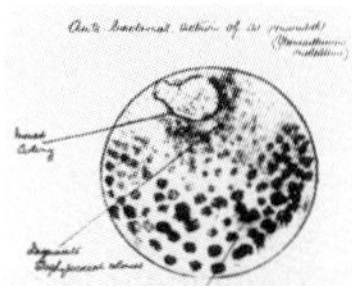

Figure 1 The keen eyes of Alexander Fleming noticed the clear areas on his plates. He realised he had made a discovery of enormous potential.

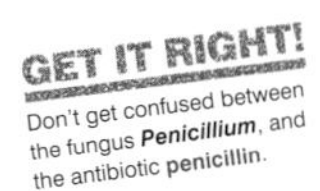

GET IT RIGHT!
Don't get confused between the fungus Penicillium, and the antibiotic penicillin.

Figure 2 Alexander Fleming, Howard Florey and Ernst Chain each received a well-deserved Nobel Prize for their work on the miracle drug penicillin. This reflected the fact that it wasn't just the original discovery that was important. It wasn't until penicillin could be made on a large scale at plants like this early production unit in the UK that the impact of the drug on our health could really be seen.

Penicillin is one of the best known medicines in the world. It has revolutionised medicine since it was first manufactured.

The discovery of penicillin

In 1928, Alexander Fleming was a young researcher at St Mary's Medical School, London. He left some of the plates on which he was culturing bacteria uncovered near an open window. When he remembered to look at them he found bacteria were growing on the surface of his dishes as he expected. But Fleming also noticed spots of mould growing, and around them were clear areas of agar. The bacteria were no longer growing. Whatever had blown in and started growing on his plates was producing a chemical which killed the bacteria.

a) Which parts of Fleming's experimental technique would not be allowed in your school labs?

Fleming found that the microorganism which had invaded his Petri dishes was a common mould called *Penicillium notatum*. He set about trying to extract the substance which killed bacteria but found it almost impossible with the technology available at that time.

He managed to get a tiny amount and used it to treat an infected wound. He called his extract **penicillin**. But it was very difficult to extract, and very unstable once he had got it. So Fleming decided he wouldn't be able to get useful amounts of penicillin from his mould. He saved his cultures and returned to other areas of research.

b) Why did Fleming give up on penicillin?

During the Second World War the need for a drug to kill bacterial infections became ever more urgent. Howard Florey and Ernst Chain were working at Oxford University in a desperate search to find an antibacterial drug. They turned to Fleming's mould and finally managed to extract enough penicillin to show what it could do.

After successful animal trials, they tried to save the life of a London policeman dying of a blood infection. The dying man made an amazing recovery. However, the supply of penicillin ran out, the infection regained its hold and he died. More months of work produced enough penicillin to save the life of a boy dying of a bacterial infection. But what was needed was enough of the drug to treat thousands of wounded and sick soldiers. British factories were dedicated to the war effort, so Chain and Florey turned to the American pharmaceutical industry for help in developing a manufacturing process.

Fleming's original mould was incredibly difficult to grow in large cultures. Then a mould growing on a melon in a market was found to yield 200 times more penicillin than the original. What's more it grew relatively easily in deep tanks, making large-scale production possible. By 1945, we were producing enough penicillin each year to treat 7 million people.

264

Further biology

Modern penicillin production

The production of penicillin did not stop with the end of the war. We now use modern strains of *Penicillium* mould which give even higher yields. We grow the mould in a sterilised medium. It contains sugar, amino acids, mineral salts and other nutrients. It is made from soaking corn in water (corn steep liquor mould is grown in huge 10 000 dm^3 fermenters).

We use huge 10 000 dm^3 fermenters which have strong paddles to keep stirring the broth. That's because the penicillin mould needs lots of oxygen to thrive. Sterile air is blown in to provide the oxygen. We control the temperature by a cooling jacket which surrounds the fermenter.

During the first 40 hours of the fermentation the mould grows rapidly, using up most of the nutrients in the broth. It is only after most of the nutrients have gone that the mould begin to make penicillin. This is why there is a 40-hour lag period from the start of the fermentation to the start of the production of penicillin. We have to provide enough food to allow lots of mould to grow. Then limit the supplies so that it produces the penicillin we need!

Over a period of about 140 hours broth is removed regularly and small amounts of nutrients are added. This allows us to get the maximum yield of the drug, which is then extracted from the broth. It is purified and turned into medicines – almost 30 000 tonnes a year! Although we now have many other antibiotics, penicillin is still used all over the world.

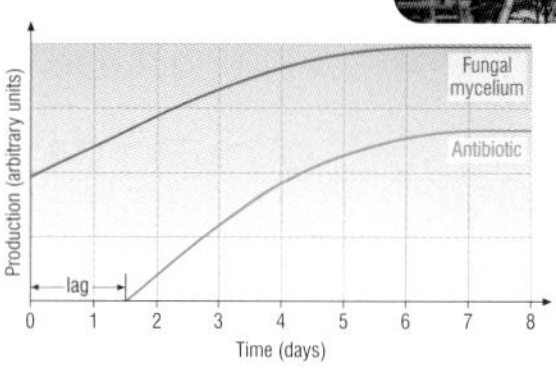

Figure 3 The production of penicillin is geared to the growth of the Penicillium mould, which doesn't start making the chemical we need until it starts to run out of food!

FOUL FACTS

The bacterium Fleming was working on when he discovered penicillin was a type of *Staphylococcus aureus*, which causes appalling boils and skin infections. In spite of all our progress, one of the same family is involved in the MRSA (Methycillin Resistant *Staphylococcus aureus*) problems we have in hospitals today!

DID YOU KNOW?

Fleming's original mould, *Penicillium notatum*, yielded only one part penicillin for every million parts of fermentation broth!

SUMMARY QUESTIONS

1. Produce a timeline to show the stages in the discovery and development of penicillin.
2. Why has the development of penicillin had such an impact on the quality of human life?
3. What have been the difficulties of producing penicillin on a commercial scale?

KEY POINTS

1. The antibiotic penicillin was discovered by Alexander Fleming. The method of mass production was the work of Howard Florey and Ernst Chain.
2. Penicillin is made by growing the mould *Penicillium* in a fermenter.
3. The medium contains sugars and other nutrients and has a good supply of oxygen.
4. The mould only starts making penicillin after most of the nutrients are used up.

265

SUMMARY ANSWERS

1. **1928** Fleming discovers *Penicillium notatum* and its effect on bacteria.
   1934 Fleming gave up work on penicillin.
   1938 Florey and Chain begin work on penicillin, using it on people 1939–45.
   Second World War Florey and Chain worked with pharmaceutical companies in US to manufacture penicillin to treat soldiers.
2. It enables doctors to treat and cure bacterial diseases that previously killed millions of people. It also led to the discovery and development of many other antibiotics.
3. Biggest difficulty – *penicillium* first discovered by Fleming had minute yield and compound very unstable and difficult to extract. Newer strains used by Florey and Chain etc. had a higher yield but still needed to grow huge amounts of mould to get the medicine.
   Have to manage lag period – time of growth before drug starts to be made – and also maintain conditions not too hot or cold. Have to oxygenate large tanks. Extracting and purifying the penicillin difficult.

KEY POINTS

The key points in this spread are concerned with the discovery and production of penicillin and other antibiotics. Students should be careful to distinguish between the antibiotic penicillin and the fungus *Penicillium*. This could be emphasised by showing them a picture of the fungus, or better still a culture of it on an agar plate.

B3 3.6 Biogas

LEARNING OBJECTIVES

Students should learn:

- That biogas, mainly consisting of methane, can be produced by the anaerobic fermentation of waste materials containing carbohydrates.
- That when biogas is burnt it can provide energy for domestic and industrial use.
- That different designs of biogas generators are suitable for different circumstances.

LEARNING OUTCOMES

Most students should be able to:

- Describe the nature of biogas.
- Explain how biogas is produced in a generator.

Some students should also be able to:

- Evaluate the advantages and disadvantages of given designs of biogas generator.

Teaching suggestions

- **Special needs.** Make cards showing dung coming from a cow, dung being placed in a fermenter, bubbles of gas coming from the mixture and a flame coming from the end of a gas pipe attached to the fermenter. Make caption cards to go with these, either complete, with initial letters or vowels depending on difficulty level.
- **Gifted and talented.** Students could investigate in more detail the process through which faeces and urine can be broken down to methane. They could draw up a flow chart showing the three stages, putting in the names of the different groups of bacteria, their substrates and their products.
- **Learning styles**

  Kinaesthetic: Making a model generator.

  Visual: Watching the film or video clips.

  Auditory: Listening to the opinions of others.

  Interpersonal: Discussing the advantages of biogas generators.

  Intrapersonal: Writing paragraph about biogas generator in their home.

Answers to in-text questions

a) Methane is the main component of biogas.

b) One in which heat is produced.

SPECIFICATION LINK-UP Unit: Biology 3.13.6

- *Fuels can be made from natural products by fermentation. Biogas, mainly methane, can be produced by anaerobic fermentation of a wide range of plant products or waste material containing carbohydrates.*
- *On a large scale, waste from, for example, sugar factories or sewage works can be used. On a small scale, biogas generators can be used to supply the energy needs of individual families or farms. Many different microorganisms are involved in the breakdown of materials in biogas production.*

Students should use their skill, knowledge and understanding of 'How Science Works':

- *to evaluate the advantages and disadvantages of given designs of biogas generators.*

Lesson structure

STARTER

'Sprouts of Doom' – Show the students excerpts from the 'Sprouts of Doom' episode on the video/DVD *Bottom – Mindless violence,* Adrian Edmondson and Rik Mayall, which shows ignited flatulence resulting from over-indulgence in brassicas near sources of ignition [view first and judge suitability]. Discuss why this is possible. (10 minutes)

Will o' the Wisp – Discuss the folk tales associated with Will o' the Wisp. Give students a sheet of some of the tales (see web site) and some DARTs (Directed Activities Related to Text), such as sequencing a mixed set of paragraphs, highlighting the real explanation sections from the folk tale sections. Try to create a Will o' the Wisp effect by bubbling methane through water and igniting the bubbles as they surface. A tiny amount of detergent helps this process. [Risk assess and try out first.] **Safety:** use eye protection. (10–15 minutes)

Gas power – Show the students a photograph of a gas-powered bus as used during the Second World War. Show a video clip from *Dad's Army*, Episode 1 series 3 ,'The Armoured Might of Lance Corporal Jones', where Corporal Jones's van has been converted to run off gas. Explain that sometimes the gas came from fermented chicken dung. Lead into a discussion of fermenters. (10 minutes)

MAIN

- **A biogas generator** – Project a diagram of a biogas generator on to the board and describe the parts. Use PhotoPLUS B3 3.6 'Biogas' from the GCSE Biology CD ROM.
- The fixed-dome biogas generator and the domestic generator are fairly simple, but the biogas plant designed for use on farms can be used to power an engine that generates electricity. A flow chart would help students to understand the two stages of the process.
- Discuss the types of material that can be used and what happens to them.
- Additional information from alternative technology web sites could be researched.
- **Biogas** – The composition of biogas is given in the Student Book on page 266, ask: 'but where do the components come from?' List the materials that can be used, such as plant material and faeces, and what they are made up of. Most of the material contains carbohydrate.
- Give the students a simplified version (flow chart) of the stages so that they can see how the breakdown occurs.
- The first stage of the breakdown is the breakdown of the carbohydrates, lipids and proteins into simple sugars, amino acids, fatty acids and glycerol by aerobic bacteria.
- The second stage involves bacteria that convert the sugars and other compounds into acetic and other acids, with some carbon dioxide and hydrogen produced. This stage occurs as the oxygen levels are decreasing.
- The third stage occurs only in anaerobic conditions and bacteria convert the acids to methane.
- Annotate the flow chart to give the temperature conditions needed.

PLENARIES

Make a model digester – Provide the students with card, washing-up liquid or other plastic bottles that fit inside each other, and other scrap materials, and ask them to make a model digester. Display the results and offer a small prize for the most realistic. (10 minutes)

The right generator for the job – Project pictures of different types of biogas generator (to include domestic, farm and industrial if possible) on to the board and get students to say where they could be used and give their reasons why. (5–10 minutes)

Word search – Students to design a word search using words from this topic and the previous ones. They write definitions for the words and, if necessary, finish for homework. These can be swapped around between the students at a later date and used for revision purposes. (5–10 minutes plus homework time)

ACTIVITY & EXTENSION IDEAS

- **Use of biogas in the UK** – Students to do some Internet research. Set up a scavenger hunt style trail of URLs for the students to follow and give them a Word framework to fill in. Investigate landfill sites and the use of the gases.
- **No more fossil fuels!** – Get students to imagine a time in the relatively near future when all the fossil fuels are used up. Discuss how biogas could replace fossil fuels. Students could compose a leaflet from their local council to householders, telling them about the organic refuse collection service and how to exchange your full waste containers for bottles of compressed methane. Ask: 'What modifications would need to be made in the home.

SUMMARY ANSWERS

1 Many different bacteria are needed to digest the wide range of carbohydrates and other compounds in the waste mixture.

2 a) More than one generator is needed. As one is being cleaned out, another is producing gas

b) A batch type digester will use plant material rather than the waste needed for the other types shown. In some places animal waste may be in short supply, or there may be religious or hygiene objections to using waste, yet the batch digester can still be used to produce biogas.

3 a) A floating drum digester can be fed continuously, so that you don't need a large amount of plant material at the beginning as you do for the batch digester. It can use just animal waste or animal waste mixed with vegetation so more flexible. Continuous supply of gas, although the pressure can fluctuate depending on what has been fed in. Only need one.

b) A fixed dome digester – again uses animal waste and vegetable matter – more expensive to buy but only need one as works continuously and provides steady gas pressure.

KEY POINTS

The key points can be covered by ensuring that students have a clear understanding of the processes involved in the production of biogas. This can be achieved by means of a simple annotated flow chart.

B3 3.6 Biogas

LEARNING OBJECTIVES

1 What is biogas?
2 How is biogas produced?

Everyone needs fuel of some sort but there is only a finite amount of fossil fuels like coal, oil and gas to use. Even wood and peat are getting scarcer. Around the world, we all need other, renewable forms of fuel. The generation of **biogas** from human and animal waste is becoming increasingly important in both the developing and the developed world.

Figure 1 Biogas generators like this have made an enormous difference to many families by producing cheap and readily available fuel

What is biogas?

Biogas is a flammable mixture of gases formed when bacteria break down plant material or the waste products of animals in anaerobic conditions. It is mainly **methane** but the composition of the mixture varies. It depends on what is put into the generator and which bacteria are present:

Type of gas	Percentage in the mixture by volume
methane	50–80
carbon dioxide	15–45
water	5
other gases including hydrogen	0–1
hydrogen sulfide	0–3

a) What is the main component of biogas?

FOUL FACTS

Biogas is produced naturally in sewers and rubbish dumps. In the days before electricity, biogas was taken from the London sewers and used as fuel for the gas lamps which lit the steets.

Biogas generators

Around the world millions of tonnes of faeces and urine are made by animals like cows, pigs, sheep and chickens. We produce our fair share of waste materials too! Also, in many places, plant material grows very rapidly. Both the plant material and the animal waste contain carbohydrates. They make up a potentially enormous energy resource – but how can we use it?

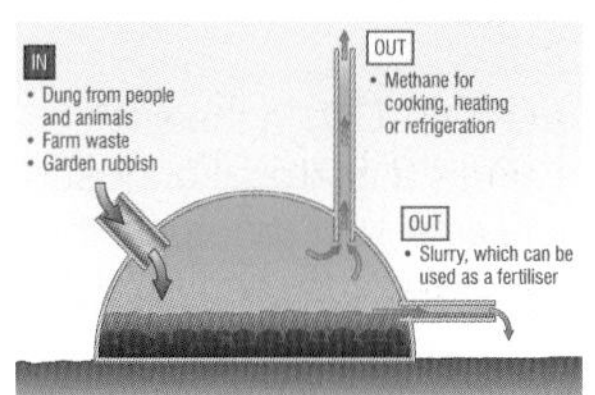
IN
• Dung from people and animals
• Farm waste
• Garden rubbish
OUT
• Methane for cooking, heating or refrigeration
OUT
• Slurry, which can be used as a fertiliser

Figure 2 Biogas generators take in waste material or plants and biogas and useful fertilisers come out the other end

The bacteria involved in biogas production work best at a temperature of around 30°C. So biogas generators tend to work best in hot countries. However, the process generates heat (the reactions are ***exothermic***). This means that if you put some heat energy in at the beginning to start things off, and have your generator well insulated to prevent heat loss, biogas generators will work anywhere.

Under ideal conditions, 10 kg of dry dung can produce 3 m^3 of biogas. That will give you 3 hours of cooking, 3 hours of lighting or 24 hours of running a refrigerator. Not only that, but you can use the waste from your generator as a fertiliser!

b) What is an exothermic reaction?

Scaling up the process

At the moment most biogas generators around the world operate on a relatively small scale. They supply the energy needs of one family, a farm or at most a whole village.

What you put into your small generator has a big effect on what comes out. Biogas units are widely used in China. There are well over 7 million biogas units, which produce as much energy as 22 million tonnes of coal. Waste vegetables, animal dung and human waste are the main raw materials. These Chinese digesters produce excellent fertiliser but relatively low-quality biogas.

In India, there are religious and social taboos against using human waste in biodigesters. As a result only cattle and buffalo dung is put into the biogas generators. This produces very high quality gas, but much less fertiliser.

There are also different sizes and designs of biogas generators. The type chosen will depend on local conditions. For example, many fermenters are sunk into the ground, which provides very good insulation. Others are built above ground, which may be easier and cheaper but offers less insulation. If the night-time temperatures fall a long way it could cause problems.

Many countries are now looking at biogas generators and experimenting with using them on a larger scale. The waste material we produce from sugar factories, sewage farms and rubbish tips could be used to produce biogas. We have some problems to overcome with scaling the process up, but early progress looks promising.

Biogas could well be an important fuel for the future for all of us. It would help us to get rid of much of the waste we produce as well as providing a clean and renewable energy supply.

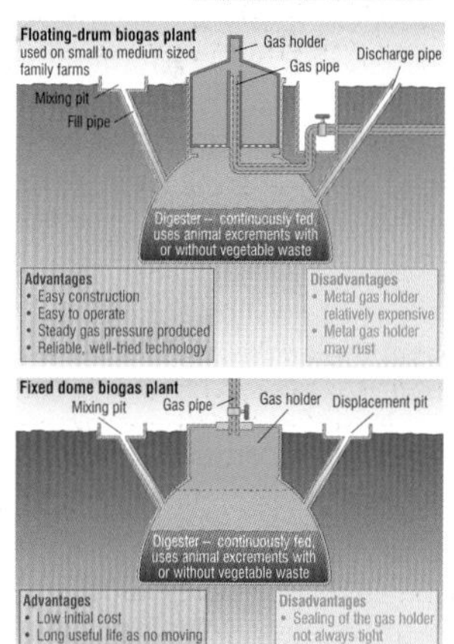
Floating-drum biogas plant
used on small to medium sized family farms
Gas holder
Gas pipe
Discharge pipe
Mixing pit
Fill pipe
Digester – continuously fed, uses animal excrements with or without vegetable waste
Advantages
• Easy construction
• Easy to operate
• Steady gas pressure produced
• Reliable, well-tried technology
Disadvantages
• Metal gas holder relatively expensive
• Metal gas holder may rust
Fixed dome biogas plant
Mixing pit
Gas pipe
Gas holder
Displacement pit
Digester – continuously fed, uses animal excrements with or without vegetable waste
Advantages
• Low initial cost
• Long useful life as no moving or rusting parts involved
• Well-insulated
Disadvantages
• Sealing of the gas holder not always tight
• Gas pressure fluctuates

Figure 3 Different types of biogas generators have different advantages and disadvantages – it is a case of choosing the right design for the conditions available

SUMMARY QUESTIONS

1 Why is it important that there are many different bacteria present in a biogas generator rather than a culture containing just one or two microbes?

2 Some types of biogas generators are set up with a large amount of plant material like straw and a starter mixture of bacteria, and left to produce gas. These ***batch digesters*** are very effective. Once gas generation begins to drop, the generator is emptied and cleaned out and the process starts again.

a) Using a generator like this, how could you be sure of a continuous supply of gas for cooking?
b) What are the advantages of a batch type digester over the types shown in Figure 3?

3 What are the advantages of a) a floating drum, and b), fixed dome biogas plant, over a batch type digester?

KEY POINTS

1 Biogas – mainly methane – can be produced by anaerobic fermentation of a wide range of plant products and waste materials that contain carbohydrates.
2 Many different organisms are involved in the breakdown of material in biogas production.

B3 3.7 More biofuels

LEARNING OBJECTIVES

Students should learn:

- That sugar cane juices and glucose from maize starch can be fermented anaerobically to produce fuels.
- That the ethanol is distilled from the products of fermentation and can be used in motor vehicle fuels.

LEARNING OUTCOMES

Most students should be able to:

- Describe how ethanol-based fuels are produced by the fermentation of sugars.

Some students should also be able to:

- Interpret economic and environmental data relating to the production of fuels by fermentation and their uses.

Teaching suggestions

- **Special needs.** Play a simplified form of the 'Pass the briquette' game. Form groups of three and pass a charcoal briquette around. One student to be labelled 'CO_2 in the air', another 'sugar in plants' and a third 'alcohol in fuel'. Students have to pass the briquette from one to another stating what happens at each exchange (photosynthesis, distillation, combustion). Have prompt cards ready if they are needed, and repeat the cycling until they are familiar with it.
- **Gifted and talented.** Students could research the possibility of obtaining fuel from fast growing trees and plants. Ask: 'Which species of plants are already planted as 'fast growing'? What would be needed for the plant material to be quickly and economically converted into biofuel?'
- **Learning styles**

  Kinaesthetic: Passing the briquette activity.

  Visual: Observing the PowerPoint® presentation.

  Auditory: Discussing the pros and cons of biofuels.

  Interpersonal: Taking part in the production of the TV advert.

  Intrapersonal: Writing part of the script for the advert.

SPECIFICATION LINK-UP Unit: Biology 3.13.6

- *Ethanol-based fuels can be produced by the anaerobic fermentation of sugar cane juices and from glucose derived from maize starch by the action of carbohydrase. The ethanol is distilled from the products of the fermentation and can be used in motor vehicle fuels.*

Students should use their skill, knowledge and understanding of 'How Science Works':

- *to interpret economic and environmental data relating to production of fuels by fermentation and their use.*

Lesson structure

STARTER

Biofuels – Show the students a clip from *The Simpsons,* Series 4 (9F14) 'Duffless', where Homer has given up drinking for a month and is thinking about drinking Biofuel, filling up his car on a 'one for you, one for me' basis. Discuss biofuels. (5–10 minutes)

Energy in alcohol – Set fire to some alcohol [a small amount – risk assess] in a (glass!) Petri dish. Pour a small amount into a film can and ignite when it has evaporated using a piezo spark gun. Care where you point the lid! Try it out first. A version of the gas-powered rocket from RSC '100 classic chemistry demonstrations' using alcohol instead of gas will also do. Care to ensure all the alcohol has evaporated. Try it out beforehand. (5–10 minutes)

MAIN

- **Exposition** – Remind students of the process of the alcoholic fermentation of glucose by yeasts and write up the equation.
- List the sources of the sugars (sugar cane juice, bagasse and molasses in warmer countries; sugar beet; maize) and the processes involved.
- Discuss gasohol and what it consists of. Ask: 'Is gasohol more environmentally friendly?' Draw out from the students the economic implications of producing alcohol for fuel. Link with their knowledge of the countries involved. Ask; ' Why is gasohol successful in Brazil? Why has it not been produced in the UK?'
- **Biofuels** – Prepare a PowerPoint® presentation on biofuels, to include reference to fermentation processes, production of gasohol and the use of vegetable oils, such as coconut oil, palm and castor oil, sunflower oil and rape-seed oil.
- Present the students with a series of questions. Ask: 'Would the design of cars need to change? What sort of quantities would need to be produced to satisfy demands? Would agriculture need to change? How much do you think it would cost? Would it be the same for all countries?'
- **How green are biofuels?** – Lead a discussion on the environmental considerations. Put this into the context of carbon emissions and compare with the combustion of fossil fuels.
- If students have access to computers, they could research this aspect and present their findings. For the contrasting points of view, there are several web sites you and they could try, such as World Land Trust, Power Plants and West Wales Eco centre.

PLENARIES

True or false? – Give students 10 statements about biofuels and their origin. They hold up 'True' or 'False' cards for each one. (5 minutes)

Pass the briquette – In groups of three, designate one student as 'CO_2 in the air', another as 'sugar in plants' and the third as 'alcohol in fuel'. Pass a charcoal briquette around from one to another stating what happens at each passing over (photosynthesis, distillation, combustion). Students to pass the briquette in the right order and repeat until they are familiar with the sequence. Run a competition to see which group is fastest and can keep it up without making a mistake for the longest time. (10 minutes)

ACTIVITY & EXTENSION IDEAS

- **Use gasohol!** – In groups, write a TV advert for gasohol. Make sure the environmental benefits are flagged up, as well as some 'science'. If facilities are available, take the best script, get the students to perform it, record it digitally and show the rest of the class.
- **The carbon-neutral car?** – As a variation on the suggestion above, students to script a presentation about ethanol-based fuel on a *Top Gear* programme. This could take the form of a contest between using ethanol-based fuel and conventional petrol or diesel. Use the format of the programme to generate some interest as well as presenting the facts.
- **Digesting cellulose** – There are microorganisms that can digest cellulose present in soil and in the ruminant gut (the cud chewers). It is possible to set up a demonstration to show the presence of some of these moulds in soil. Add some finely shredded filter paper to liquid agar and pour into Petri dishes. Obtain a sample of fresh soil, divide it in half and sterilise one half by baking it. Place fresh soil and sterilised soil in separate bottles with some sterile water. Shake it up thoroughly and then inoculate the plates with 1 cm^3 samples from the sterile soil and from the fresh soil. Tape up the plates and incubate. Make observations at intervals and record the differences between the two sets of plates. If desired, a separate set of plates can be set up that have been inoculated with a cellulose solution. This experiment can be carried out using a solution of cellulose and the results judged by changes in viscosity of the cellulose with time.
- **Debate the issue** – Students could hold a debate on the subject of whether the government should fund the development of engines that run on ethanol-based fuels. Each student could write a short speech in favour of the proposal and one arguing the case against it. (This relates to: 'How Science Works': societal issues.)

BIOLOGY MICROBIOLOGY

B3 3.7 More biofuels

LEARNING OBJECTIVES

1. How can yeast produce fuel for your car?
2. What is the environmental impact of biofuels?

In tropical countries plants grow fast. Sugar cane grows about 4 to 5 metres in a year. It has a juice which is very high in carbohydrates, particularly sucrose. Maize (which we often refer to as sweet corn) is another fast grower. We can break the starch in the maize kernels down into glucose using the enzyme **carbohydrase**. (See page 186.) But can we convert the carbohydrates we grow into clean and efficient fuels?

Figure 1 The Sun and the rain in areas like the Caribbean allow plants like this sugar cane to photosynthesise and grow very rapidly. The next step is to convert them into usable fuel.

Ethanol-based fuels

If sugar-rich products from cane and maize are fermented anaerobically with yeast, the sugars are broken down incompletely to give **ethanol** and water. You can extract the ethanol from the products of fermentation by **distillation**, and you can then use it in cars as a fuel.

Car engines need special modification to be able to use pure ethanol as a fuel, but it is not a major job. Many cars can run on a mixture of petrol and ethanol without any problems at all.

a) Why is it cheaper to produce ethanol-based fuels from sugar cane than from maize?

GET IT RIGHT!

At the moment the advantages and disadvantages of using biofuels depends on the space, climate and economy of a country – but from an environmental point of view everyone benefits.

The advantages and disadvantages of ethanol as a fuel

In many ways ethanol is an ideal fuel. It is efficient and it does not produce toxic gases when you burn it. It is much less polluting than conventional fuels which produce carbon monoxide, sulfur dioxide and nitrogen oxides. In addition, you can mix ethanol with conventional petrol to make a fuel known as gasohol. This is being done increasingly in the USA and it reduces pollution levels considerably. However, there is still some pollution from the petrol part of the mix.

Using ethanol as a fuel is known as **carbon neutral**. This means there is no overall increase in carbon dioxide in the atmosphere when you burn ethanol. The original plants removed carbon dioxide from the air during photosynthesis. When you burn the ethanol, you simply return it.

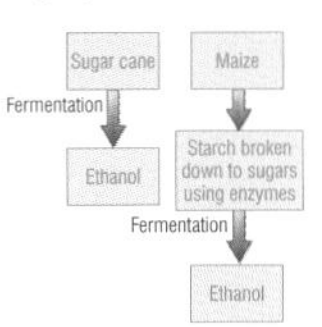

Figure 2 The starch in maize needs to be broken down by enzymes before yeast can use it as fuel for anaerobic respiration. Although it takes more steps to produce ethanol from maize than from sugar cane, maize can be grown in many more countries around the world.

The biggest difficulty with using plant-based fuels for our cars is that it takes a lot of plant material to produce the ethanol. As a result, the use of ethanol as a fuel has largely been limited to countries with enough space and a suitable climate to grow lots of plant material as fast as possible.

Brazil was the trailblazer for ethanol as a fuel when oil prices shot up in the 1970s. The Brazilians grew their own 'green petrol' and slashed the money paid out on oil imports which were crippling the economy of the country. They were very successful, and in the 1980s 90% of the cars produced in Brazil had ethanol-powered engines.

However, when oil prices dropped again the Brazilian government couldn't afford to subsidise ethanol as a fuel. As a result they began to move back to petrol driven cars.

Now people all over the world are worried about the environmental problems caused by the burning of fossil fuels. Interest in clean alternatives such as ethanol is soaring. Brazil is again taking a lead, supporting countries such as India with advice on technology for producing ethanol from plants.

b) What is meant by the term 'carbon neutral'?

SCIENCE @ WORK

Scientists are attempting to find ways of producing economically viable quantities of ethanol from plants which grow fast and well in Europe. They have tried pine trees and beet but have not yet been very successful. Now they are looking at fast growing grasses!

In America, the use of gasohol – a mixture of 90% petrol and 10% ethanol – is increasing all the time. A lot of the ethanol is fermented and distilled from maize grown within the USA itself. However as the use of ethanol grows the Americans are also importing ethanol from places like Brazil and the Caribbean. In 2004, the USA imported 160 million gallons of ethanol fuel on top of its own production!

The main problem is finding enough ethanol. If we Europeans added 5% ethanol to our fuel it would reduce carbon dioxide emissions – but we would need 7.5 billion litres of ethanol a year. That's more than half of the total production level in Brazil!

The methods of ethanol production we use at the moment leave large quantities of unused cellulose from the plant material. To make ethanol production work financially in the long term, we need to find a way to use this cellulose. We might develop biogas generators which can break down the excess cellulose into methane, another useful fuel.

Genetically engineered bacteria or enzymes may be able to break down the cellulose in straw and hay and make it available for yeast to make more ethanol. We don't know exactly what the future will hold, but it seems likely that ethanol-based fuel mixe s will be part of it!

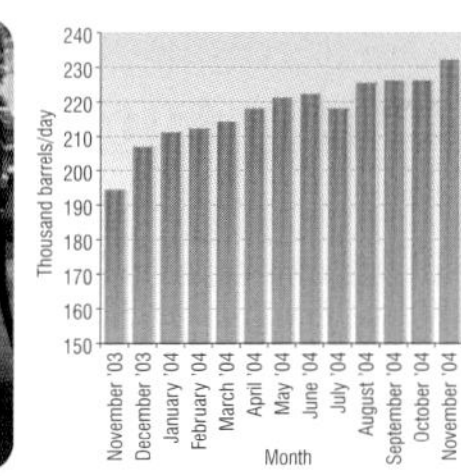

Figure 3 Increasing demand for gasohol in the USA has lead to increasing production of ethanol from maize, as the data clearly shows

SUMMARY QUESTIONS

1. Make a table to summarise the advantages and disadvantages of ethanol as a fuel for cars.
2. Use the data in Figure 3 to help you answer the following questions:
   a) What was the increase in ethanol production in the USA between
      i) November and December 2003,
      ii) October and November 2004?
   b) What was the percentage increase in USA ethanol production between November 2003 and November 2004?
   c) Much of the ethanol made is mixed with normal petrol to make a more ecologically sound motor fuel. If each barrel of gasohol sold contains 10% ethanol, how many barrels of gasohol could have been produced in January 2004?

KEY POINTS

1. Ethanol based fuels can be produced by the anaerobic fermentation of sugar cane juices and from glucose derived from maize starch by the action of the enzyme carbohydrase.
2. Ethanol is distilled from the fermentation products and can be used as a fuel in motor vehicles on its own or mixed with petrol to produce gasohol.

SUMMARY ANSWERS

1

Advantages	Disadvantages
Efficient	Expensive to produce
Produces little pollution when it burns	Needs huge amount of plant material
Carbon dioxide neutral	Leaves large amounts of cellulose waste
Can be mixed with conventional fuel	Pure ethanol – cars need to be adapted but mixtures (gasohol) can be used in normal cars

2 **a)** **i)** 13–14 000 barrels/day; **ii)** 7 000 barrels/day
b) increase is 38 000 barrels/day, 38 ÷ 195 = 19%
c) 211 000 barrels of ethanol is 10%; ×10 to get 100%; 2 110 000 barrels

Answers to in-text questions

a) There is one less step: the yeast can use sugar from sugar cane. In the other method, it is necessary to convert starch from maize to glucose using the enzyme carbohydrase.

b) There is no overall increase in the levels of carbon dioxide in the air as a result of a fuel being burnt.

KEY POINTS

The key points of this topic extend the knowledge of the uses of microorganisms and fermentations. The students would benefit from practice in answering questions which ask for the interpretation of data relating to the use of fuels produced by fermentations.

B3 3.8 The microbe revolution – yes or no?

SPECIFICATION LINK-UP

Unit: Biology 3.13.5 and 6

Students should use their skill, knowledge and understanding of 'How Science Works':

- *to explain how scientists such as Spallanzani, Schwann and Pasteur were involved in the development of the theory of biogenesis.*
- *to evaluate the advantages and disadvantages of given designs of biogas generator.*

BIOLOGY MICROBIOLOGY

B3 3.8 The microbe revolution – yes or no?

An historical perspective on microbes

Lazzaro Spallanzani

Theodor Schwann

Louis Pasteur

For centuries people believed in ***spontaneous generation***. They thought that living things appeared from nothing in water or on food as it decayed.

In 1765, Lazzaro Spallanzani set up two sets of containers full of broth. One set was left open to the air. The other was boiled to kill anything living and then sealed. Only the second set remained sterile. Spallanzani claimed this showed spontaneous generation didn't happen. His opponents said boiling killed a 'vital principle' in the air so nothing could live there!

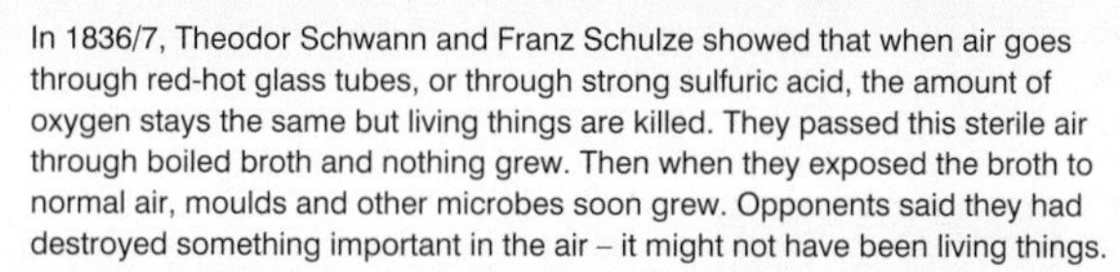

In 1836/7, Theodor Schwann and Franz Schulze showed that when air goes through red-hot glass tubes, or through strong sulfuric acid, the amount of oxygen stays the same but living things are killed. They passed this sterile air through boiled broth and nothing grew. Then when they exposed the broth to normal air, moulds and other microbes soon grew. Opponents said they had destroyed something important in the air – it might not have been living things.

In the 1850s Louis Pasteur first proved that microorganisms (yeast) are responsible for fermentation. Later he carried out a series of experiments with swan-necked flasks. These showed clearly that the moulds and bacteria which appeared in liquids were tiny organisms that had come from the air. Spontaneous generation did not exist.

ACTIVITY

1 This account gives you a taste of some of the controversy which surrounded the scientists who started working with microorganisms in the past. Louis Pasteur is the most famous of them. Find out as much as you can about him and produce plans for a brief biography of Pasteur to be part of a science programme aimed at young people. Make sure you explain what he did and why it was so important. Try and find something about the sort of person he was as well, to make it even more interesting!

Teaching suggestions

This spread has four themes: historical, safety in handling microorganisms, a new type of micro-food and biogas.

An historical perspective on microbes

Some of the aspects of this historical approach have been raised before, particularly with respect to the work of Pasteur, but Spallanzani and Schwann are less well-known.

- **Spontaneous generation** – Students could research spontaneous generation and write a short paragraph about it, explaining what people believed and for how long the ideas persisted. This would be important background to any further work on the scientists, as it is just as well to know what they were trying to disprove. Ask: 'Could the theories account for some of the experiments they tried to do?'
- **'The genesis tub'** – Show the students a clip from *The Simpsons* (Treehouse of Horror, VII 4F02), where Lisa dissolves her tooth in a cup of soda and it spontaneously gives rise to an evolving ecosystem of microscopic creatures, eventually finishing up with a mini-race who think she is God. Link with the theories of spontaneous generation.
- **Schwann and Schulze** – Further information on the work of Schwann and Schulze, whose experiments were carried out about two decades before Pasteur, who showed that yeast was responsible for fermentations, could also be useful in setting the scene for the work of Pasteur.
- **Pasteur** – Review the work already done on **Pasteur** and use different sources to find out more about him. If not covered previously, remind the students about pasteurisation, how it is done and what happens in the milk. Also review Pasteur's broth experiment. If not carried out previously, then a demonstration could be set up (see page 256).

Grow your own!

The activity suggested depends on getting across the safety rules of working with microorganisms in a school or college laboratory.

- Suggest to students that they write down all the rules they can remember about handling microorganisms. Discuss the list. Ask: 'Are there any that could be added? Are there some more important than others? Could you put them into an order of importance, the top ten rules, for example?'
- Give the students a banner headline along the lines 'School student in hospital after science lesson'. Ask them to write the article, stating the cause, the student's story, the mother's reaction and a 'spokesperson' from the school.

Further biology

Grow your own!

ACTIVITY

2 When students work with microorganisms in the school labs for the first time they have some funny ideas …! Make a poster or a safety leaflet which could be given to students before they start working with microorganisms for the first time. You need to make sure they understand the need for care and good hygiene, both to get good results and to keep everyone safe.

A microbe miracle?

Imagine you work for a food company. A research team has come to see you. They have been working on a new form of food which can be manufactured safely, quickly and cheaply. The new foodstuff can be used in different ways, and it is high in vitamins, fibre and protein but low in fat. The production of this new food helps to solve some environmental problems by getting rid of waste from human activities. Your task is to look at the details of the food and its production, and decide how you might market it.

This food is based on a microorganism which grows on human sewage, supplied very cheaply by local authorities. The microorganism turns sewage into protein with 30% efficiency. It also acts as an effective treatment for the sewage. Once the microbes have been harvested the liquid which remains in the production tanks can be pumped safely into rivers or the sea. The microorganism product is a creamy colour with very little taste or smell. It can be processed to give it a wide range of textures and tastes, both savoury and sweet.

ACTIVITY

3 a) Make a list of the points in favour of the new food and the points against it.

b) Discuss these points and use them to decide how you will overcome any problems.

c) Think of a name for your new food. Then make an outline plan of the ways you will try and advertise it. Produce a draft advert to go in a magazine.

Fuel for school?

You go to school all through the year. The school is heated and lit; there is hot water to wash your hands and the whole system uses an awful lot of energy. Yet your school also produces an enormous amount of waste of various sorts.

ACTIVITY

4 Put together a presentation to explain the benefits of installing a biogas generator at your school. Your presentation will be seen by the parents, the governors and the local education authority. You need to explain how your idea will work – where the fuel supply for the generator might come from and how the biogas produced might be used – and all the benefits it could bring. Be prepared to answer questions and push your ideas as hard as possible.

271

A microbe miracle?

This 'food from sewage' idea is an open-ended activity and provides a great deal of scope for imaginative students. Again, it is probably a group activity rather than a project for an individual.

Some ideas which could be worth following up are given below:

- Students could research into the composition of sewage. The activity suggested in the spread is based on a microorganism which grows on human sewage. A large proportion of the sewage which reaches treatment works is made up of detergents and industrial wastes, so it needs to be made clear whether or not these need to be removed before the food is grown, or whether human sewage is to be kept separate from the waste from domestic appliances, industrial waste and water from the roads.
- Students could research into the organisms present in human sewage. Human sewage can contain pathogenic organisms, such as *Salmonella* and *Escherichia coli*. Students need to consider how these organisms can be removed.
- Research into the organisms involved and treatment of sewage in different areas, such as septic tanks and sewage lagoons.
- Students need to find out: 'What kind of microorganism? Is it a fungus or a bacterium? Does this make a difference?' They also need to refer to the work on mycoprotein.
- A food which can be given both sweet and savoury flavours could be difficult to market, so students might need to consider how the product might be used: 'Is there more demand for sweet foods? Or is it more likely that the new food could be used as a meat substitute like mycoprotein?'

Fuel for school?

This topic has been covered extensively in the previous spread, so students should have plenty of information about biogas generators of different types.

Students need to think about:

- The different types of waste: not all the waste is suitable for use in a generator.
- Who will sort it? Is it in the job description of the school caretaker? Will the caretaker need to be paid extra? Or will someone else have to be employed?
- What happens to the material that is not suitable?
- What size of generator? A web search will indicate types and sizes, which the authorities will need to be made aware of if they are to be persuaded.
- Where will it go? Would planning permission be needed?
- How much will it cost to install? Will there be enough fuel generated?

Some research on the web indicates that there are schools and other institutions, such as prisons, where biogas generators have been installed and used to provide fuel. Perhaps the students could use an example, already installed and working, as part of their presentations.

Special needs

Many of the suggested activities can be modified for the students. The leaflet and poster activities can be adapted, and detailed writing frames with suggestion sheets and prompts can be supplied.

SUMMARY ANSWERS

1 a) 1 Collect a sample of soil in a sterile container and mix with sterile water.
2 Collect a Petri dish containing sterile agar.
3 Sterilise an inoculating loop by heating it to red hot in the flame of a Bunsen burner and then leaving it to cool.
4 Using the right hand, dip the sterilised loop into the water in which the soil has been mixed.
5 Carefully lift the lid of the Petri dish with the left hand and transfer a drop of the water from the loop on to the surface of the agar, making zig-zag streaks across the surface of the agar.
6 Replace the lid of the Petri dish as quickly as possible to avoid contamination.
7 Secure the lid of the Petri dish with adhesive tape to prevent microorganisms from the air contaminating the culture or microorganisms from your culture escaping. Label the Petri dish clearly.
(These instructions could be modified according to your routine.)

b) Safety precautions are necessary because you do not know what microorganisms might appear in your culture. Some microorganisms can mutate to produce harmful forms. Some disease-causing microorganisms from your skin or breath might have landed on the agar. You need to be sure that you are only growing bacteria from the soil sample.

2 a) Bread dough rises due to the carbon dioxide produced by the yeast as it respires. Yeast is a living organism and its respiration is controlled by enzymes. Enzymes work faster as the temperature increases up to about 40°C. Changing the temperature at which the bread dough is kept changes the rate at which the yeast cells are growing and respiring. In turn, this affects the amount of carbon dioxide produced and thus the rate at which the bread dough rises.

b) Yeast cells are living organisms and fermentation depends on enzyme-controlled reactions. If it is too cold, then little fermentation takes place and little ethanol is produced. If it is too hot, the enzymes become denatured, so there is no fermentation at all. At the ideal temperature, the yeast cells grow fast and ethanol is produced.

c) If some yeast is left in the bottle, fermentation continues after the wine is bottled and more carbon dioxide is produced.

3 [Students should use what they know about yoghurt and cheese making, and find out about the growing market in 'probiotic' yoghurt and drinks. The reports should cover the main issues and show some evidence of careful research into what is meant by 'friendly' bacteria and how beneficial they might be. Credit the inclusion of non-biased information rather than the claims made by the manufacturers.]

4 Advantages and disadvantages of using microorganisms in the food and drugs industries:

Advantages	Disadvantages
Easy to handle; no animal rights issues	Need suitable conditions; temperature must be controlled; culture supplied with nutrients and oxygen
Reproduce themselves, so relatively cheap to set up	If things go wrong, then whole culture may die
Do not need high temperatures and pressures to work	Limit to the size of fermenters, so limited production
Can be genetically modified to produce specific products	Limit to substances that can be made by microorganisms

[There may be other valid points which can be credited.]

MICROBIOLOGY: B3 3.1 – B3 3.8

SUMMARY QUESTIONS

1 a) Produce a set of instructions for setting up and growing a culture of soil bacteria in the school lab.
b) Explain why all the safety precautions are necessary.

2 a) You can leave bread dough in a fridge for hours without it rising. Put it somewhere warm and it starts to rise again. Bread usually rises fully in an hour or two. In a cool room it may take as long as twenty four hours. Explain how these differences come about.
b) Temperature is vital for successful beer and wine making. Why is it so important?
c) To make sparkling wine or champagne, a small amount of yeast is left in the bottle. What effect does this have and what is the gas that makes the bubbles in the drink?

3 Write a brief report on 'Bacteria in the dairy industry'. Use what you know about yoghurt and cheese-making, and find out about the growing market in 'probiotic' yoghurts and drinks. Try to answer the question: What are 'friendly bacteria' and does eating them really do you any good?

4 Why are microorganisms so useful in industrial processes? Make a table to summarise the advantages and disadvantages of using these living organisms in the food and drugs industries.

5 Write a letter from a mother to her friend in the late 1940s, describing the recovery of her child from a serious throat infection thanks to the use of penicillin.

6 a) Suggest ways in which people might improve the quality of the biogas produced in their fermenters.
b) Suggest reasons for and against the use of fermenters and biogas in the UK.

7 Write a letter to your local authority explaining why you think they should look into the idea of running all their vehicles – buses, emergency vehicles, etc. – on ethanol or gasohol. Explain the potential value of ethanol in helping to prevent the greenhouse effect and global warming.

EXAM-STYLE QUESTIONS

1 The passage below describes how a student grows a culture of microorganisms on an agar plate. Petri di[shes] and a culture medium containing agar, carbohydrat[e,] protein and mineral ions are heated to 120°C for 15 minutes. The culture medium is poured into the Pet[ri] dishes and left to set. A wire inoculating loop is pass[ed] through a flame until red hot, allowed to cool and th[en] dipped into a container of microorganisms. The loo[p is] then streaked across the medium in the Petri dish. T[he] Petri dish is sealed with adhesive tape and incubate[d at] a temperature not exceeding 25°C.

In each case, give one reason why the following procedures were carried out:
(a) Carbohydrate was included in the culture mediu[m].
(b) The culture medium and Petri dishes were heat[ed] to 120°C for 15 minutes.
(c) The wire inoculating loop was cooled before bei[ng] used to transfer microorganisms.
(d) The Petri dish was sealed with adhesive tape.
(e) The temperature at which the microorganisms a[re] grown was not allowed to exceed 25°C.

2 In countries such as China, India and Nepal, biogas generators, similar to the one shown below, are common. Although they may cost over £300 to set u[p] they are an energy-efficient way of producing biogas [in] the long term.

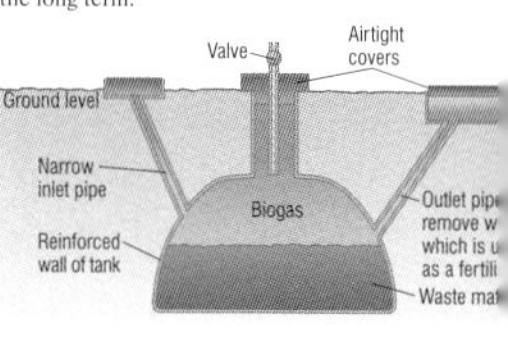

(a) Biogas is a mixture of different gases. Which gas makes up the largest proportion of this mixture?
(b) Give one use of biogas.
(c) Suggest two types of waste material that could b[e] added to the container in order to generate biog[as].

272

5 [In this letter, credit as many scientific points as possible. In addition, give credit for the assessment of the situation in the context of what might have happened without the use of penicillin.]

6 a) The quality of biogas produced could be improved by changing the mix of bacteria in the fermenter and by changing the mix of waste put into the fermenter.

b) **Reasons for could include:** could be done on a small scale; at a local level in schools or hospitals which produce a lot of waste; would be ecologically sound; could save money; could be done in villages and by farmers as well; bigger fermenters could be used at municipal tips.

Reasons against might include: smell; public health and hygiene issues; difficulty in collecting vegetable and food waste; many areas do not have animals; objections to the use of human waste.

7 This letter would need to show a clear understanding of the science issues both in the production of ethanol and gasohol and also with respect to the greenhouse effect and global warming.

Summary teaching suggestions

- **Literacy guidance**
  - Several of these questions (questions 3, 5 and 7) require the students to write answers in extended prose. These provide opportunities to check spelling, punctuation and the use of good grammar.
  - In all the answers, students should be made aware of the importance of the correct use of technical terms. In question 1, the selection and order of the instructions is a good test of the ability of a student to organise their material into a logical sequence.

d) Explain how the waste material is converted into biogas. (3)

e) Suggest a reason why
(i) airtight covers are used. (1)
(ii) the wall of the tank is reinforced. (1)

f) Biogas generators have both advantages and disadvantages over other forms of energy generation.
(i) Suggest two advantages of a biogas generator. (2)
(ii) Suggest one disadvantage of a biogas generator. (1)

The diagram shows a fermenter that can be used to produce the antibiotic penicillin using the mould *Penicillium*.

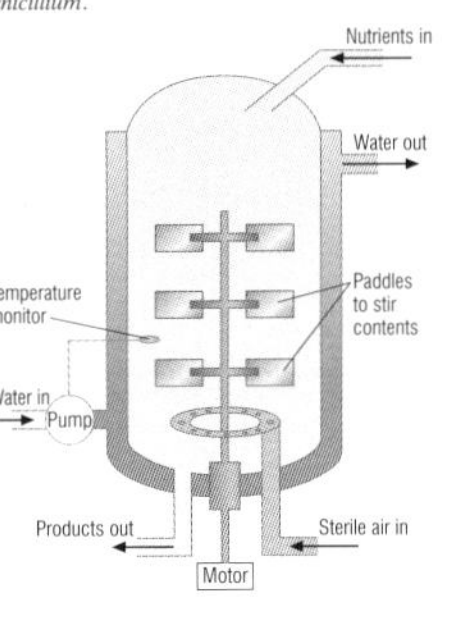

a) Name two nutrients that a mould like *Penicillium* requires. (2)

b) (i) Suggest a reason why air is bubbled through the fermenter. (1)
(ii) Why must this air be sterile? (1)

c) (i) Explain why it is necessary to pump water around the fermenter. (2)
(ii) Suggest a reason why the temperature monitor is attached to the pump circulating the water around the fermenter. (1)

d) Suggest **two** reasons why the contents of the fermenter are continually stirred. (2)

HOW SCIENCE WORKS

A class of students were set the task of finding the best temperature at which to manufacture yoghurt. They checked out the recipe for yoghurt. All it said was that after preparing the milk and the culture they should be left in a warm cupboard until the yoghurt is set.

Each group followed the same recipe. They were each assigned a temperature at which to keep their yoghurt.

a) Suggest a suitable range and number of different temperatures to use. (2)

b) If you were not at all sure about the answer to question a), what could you do to find out? (1)

c) What would be a suitable end point? That is, how will everybody know when the yoghurt is ready? (1)

d) What would be a suitable number of repeats? Give reasons for your answer. (1)

e) Construct a suitable table for the results. (3)

f) Suggest how the results might be displayed. (1)

g) This information could be very useful to anyone wanting to be involved in the yoghurt industry. Suggest some of the ways in which this information could be exploited by different companies. (2)

h) Explain why you think that you might have some difficulty convincing an industrialist of the reliability of your data. (1)

EXAM-STYLE ANSWERS

1 a) To provide an energy source *(1 mark)*

b) To sterilise them/kill unwanted organisms. *(1 mark)*

c) To prevent the microorganisms being killed by heat. *(1 mark)*

d) To prevent other microorganisms from the air contaminating the culture of microorganisms being grown. *(1 mark)*

e) To reduce the chance of the growth of harmful/pathogenic microorganisms that might contaminate the culture medium. *(1 mark)*

2 a) Methane/CH_4. *(1 mark)*

b) It can be burnt as a source of heat/light/refrigeration. *(1 mark)*

c) • Human sewage.
• Animal/agricultural waste/faeces/manure.
• (Organic) domestic waste (e.g. vegetable/food waste).
• Farm waste (e.g. stalks/leaves left after crop harvesting).
• Industrial waste *(1 mark for each point to a maximum of 2)*

d) • The waste material contains a high proportion of carbohydrate.
• A mixture of populations of different bacteria/microorganisms is added/already present.
• The bacteria/microorganisms breakdown/digest the carbohydrate in the waste.
• By anaerobic respiration/fermentation.
• Producing methane/biogas (as a waste product).
(1 mark for each point to a maximum of 3)

e) i) To maintain anaerobic conditions in the generator/to prevent oxygen from entering. *(1 mark)*
ii) To prevent it bursting/exploding. *(1 mark)*

f) i) • Less reliance on/reduced use of fossil fuels.
• Little smoke/sulphur dioxide produced.
• Renewable energy source/self-sufficiency.
• Fertiliser is a useful by-product.
• Wastes are efficiently disposed of.
• Cheaper/economic in the long term.
(1 mark for each point to a maximum of 2)

ii) • Methane is explosive/explosion risk/fire risk.
• Initial costs are high.
• Some training on use of generator is required.
(1 mark for any one point)

3 a) • Carbohydrate/sugar.
• A source of nitrogen/protein. *(1 mark each)*

b) i) To provide oxygen (for aerobic respiration/fermentation by *Penicillium*). *(1 mark)*
ii) To prevent the culture medium being contaminated with other microorganisms. *(1 mark)*

c) i) • To cool the contents of the fermenter.
• which increases in temperature as a result of respiration by *Penicillium*.
• A high temperature could kill *Penicillium*.
(1 mark for each point to a maximum of 2)

ii) So that when the fermenter contents have cooled to the desired/optimum temperature it can switch off the pump that circulates the cooling water. *(1 mark)*
Allow the reverse argument – it switches the pump on when the temperature rises.

d) • To keep the *Penicillium* in suspension/stop the *Penicillium* from settling out.
• To maintain the same temperature throughout.
• To ensure that the *Penicillium* and the nutrients are always in contact with one another.
(1 mark for each point to a maximum of 2)

HOW SCIENCE WORKS ANSWERS

a) Optimum temperature for one type of yoghurt is 43°C. However they are most likely to suggest any temperature from 37°C to 60°C. These temperatures should be discussed. They must also suggest at least five different temperatures within a sensible range.

b) They might suggest a preliminary test.

c) They should suggest assessing how runny it is and some suitable measure, e.g. tip the yoghurt to 45° until it no longer moves. They might also suggest a pH below 7.

d) At least two, but suggest three because of the inaccuracy built into the end point.

e)

Temperature (°C)	Time (min)	Time (min)	Time (min)	Mean Time (min)

f) Graph, temperature on X axis and mean time on Y axis.

g) E.g. home-made yoghurt makers; manufacturers of yoghurt; magazines with yoghurt recipes.

h) A school student does not have the status of a research scientist or access to data recording which is accurate and sensitive.

B3 Examination-Style Questions

Examiner's comments

These questions would be most useful as a class exercise completed under examination conditions, especially after a homework has been set to revise the whole unit. In this way it would introduce students to working without assistance and under time constraints.

Alternatively the questions could be given as a homework with students being encouraged to use the Student Book and their own notes to research their answers. The expectation would be that most responses would be accurate. If students do not have access to textbooks at home, the questions could be tried as a class exercise. The purpose of using the questions in this manner is to reinforce what has been taught rather than to test what has been learnt. When used in this way, it is still important that students' responses are checked thoroughly as some students will rely on their knowledge alone and not bother to check their answers. As a result there may be errors and if these are not rectified the inaccuracies are simply reinforced in their minds.

Answers to questions

1 **A** Transpiration **B** Breathing
C Diffusion **D** Active transport
E Osmosis *(1 mark each)*

2 (a) arteries (b) lungs
(c) plasma (d) haemoglobin
(e) liver *(1 mark each)*

3 (a) (i) More blood is pumped at each beat. *(1 mark)*

(ii) There are fewer heart beats. *(1 mark)*
Allow correctly quoted figures that illustrate the same general conclusions.

(iii) Then you can see how many people do not fit the pattern or have produced anomolous results. *(1 mark)*

(b) (i) Cardiac output can be calculated by multiplying the volume of blood at each beat by the number of beats per minute.
i.e. $100 \times 210 = 21\,000\ cm^3$
*Correct answer **and** appropriate working.* *(2 marks)*
Correct figures (100 × 210) but answer calculated inaccurately. *(1 mark)*
If 'trained' individuals or 'at rest' chosen but the figures 'volume × number of beats' are correct and the answer correctly calculated. *(1 mark)*

(ii) Untrained individual during exercise = 210 beats per min.
Untrained individual at rest = 70 beats per min.
Increase 210 − 70 = 140 beats per min.
% increase = $140/70 \times 100$
= $2/1 \times 100 =$ **200%**
Correct working and answer *(2 marks)*
Correct figures but calculation error *(1 mark)*

(c) • During exercise the muscles require more energy as they are working faster.
• The energy is produced from respiration that requires oxygen and glucose and produces carbon dioxide.

continues opposite ›

EXAMINATION-STYLE QUESTIONS

1 Complete the table by choosing the word from the list in the box that matches the description in the table. Write the chosen word next to the relevant letter in the 'process' column in the table. *(5 marks)*

See pages 222–33

active transport **breathing** **dialysis** **diffusion** **digestion**
osmosis **respiration** **transpiration**

Process	Description
A	Loss of water from plant leaves
B	Movement of air in and out of the lungs
C	Movement of particles of a substance from high to low concentration
D	Movement of substances from low to high concentration using energy from respiration
E	Movement of water from dilute to concentrated solution through a partially permeable membrane

2 In the following questions, write down one answer from the list that follows the question.

See pages 238–49

(a) Through which vessels does blood flow from the heart to the organs? *(1 mark)*
arteries **capillaries** **veins**

(b) There are two separate circulations in the body, one goes to the organs. Where does the other go to? *(1 mark)*
intestines **kidneys** **lungs**

(c) In which part of the blood is carbon dioxide transported? *(1 mark)*
plasma **red blood cells** **white cells**

(d) What is the name of the pigment in red blood cells that transports oxygen? *(1 mark)*
glycogen **haemoglobin** **lactic acid**

(e) Blood plasma transports urea to the kidneys from which organ? *(1 mark)*
liver **lungs** **muscles**

GET IT RIG
Never look for a patte responses where you choose from a list. It as likely that it is alw first word or item in any other combinati Often, as in this cas choices are simply presented in alpha order.

3 The table compares the volume of blood pumped and the number of heartbeats of individuals at rest and during strenuous exercise. The individuals were of two types – trained athletes and untrained non-athletes.

See pages 242–5

State	Training level	Volume of blood pumped out of the heart at each beat /cm^3	Number of heartbeats per minute
At rest	Untrained	65	70
	Trained	90	52
During strenuous exercise	Untrained	100	210
	Trained	155	190

274

BUMP UP THE GRADE

The surest way to ensure examination success is to prepare adequately. The key to this is revision. There are a few basic principles that candidates can usefully adopt.

- Work in a place with the least likelihood of distraction.
- Keep revision sessions short – say 30–60 minutes.
- Take a break of at least 15 minutes between sessions; have a complete change – leave the room, take a walk, watch TV.
- Vary the revision by changing topic or subject – variety helps relieve boredom.
- Test learning regularly by practising past papers or trying exercises such as those throughout this book.
- Organise revision by producing a written timetable and keeping to it.

GET IT RIGHT!

In the first two questions, candidates need to make choices from lists of items. Multiple choice questions of this, and other, types are common on examination papers. Candidates sometimes look for a pattern of answers or simply think, 'My last 4 answers were option B, so it can't be option B again this time.' It is normally the case that the list of choices is arranged alphabetically. It is worth impressing on students that this is the rationale behind the sequences given and that examiners do not spend time trying to ensure that the first, second, third, etc. (or each letter) on the list are always used as an answer.

In question 5 there is some flexibility allowed in the use of terms like microorganisms, bacteria, fungi and microbes. Depending on the question asked, this may not always the case. It is therefore beneficial for students to understand the difference between viruses, bacteria and fungi and also that microorganisms (preferable) and microbes are general terms encompassing not just bacteria and fungi but also protoctists and other groups. The term 'germs' should never be used as it is imprecise and not scientific.

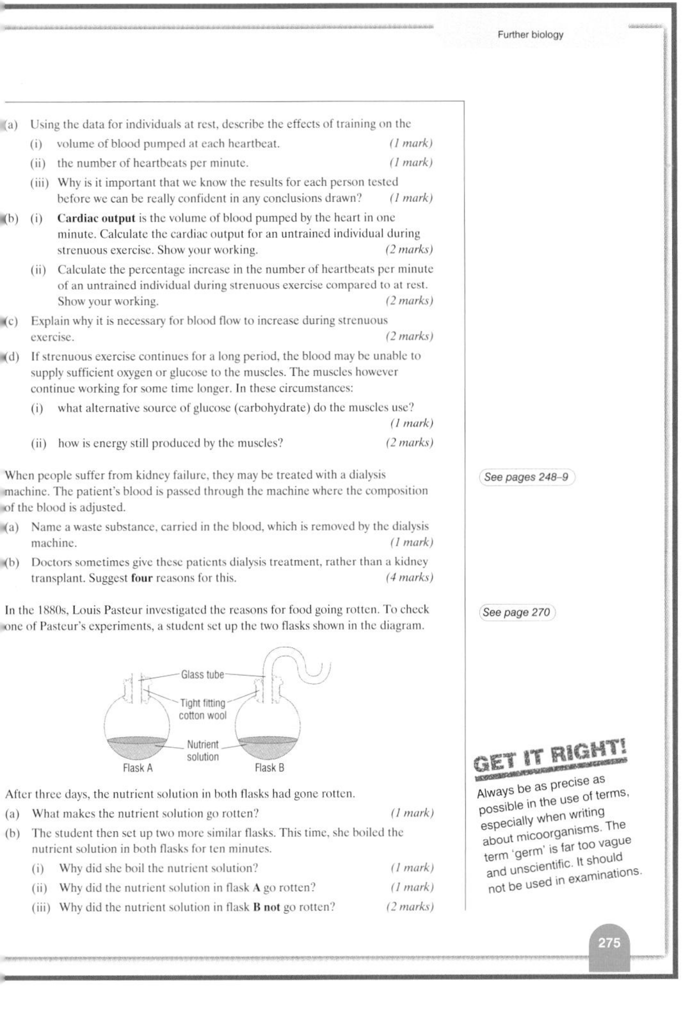
Further biology

(a) Using the data for individuals at rest, describe the effects of training on the
  (i) volume of blood pumped at each heartbeat. *(1 mark)*
  (ii) the number of heartbeats per minute. *(1 mark)*
  (iii) Why is it important that we know the results for each person tested before we can be really confident in any conclusions drawn? *(1 mark)*

(b) (i) **Cardiac output** is the volume of blood pumped by the heart in one minute. Calculate the cardiac output for an untrained individual during strenuous exercise. Show your working. *(2 marks)*
  (ii) Calculate the percentage increase in the number of heartbeats per minute of an untrained individual during strenuous exercise compared to at rest. Show your working. *(2 marks)*

(c) Explain why it is necessary for blood flow to increase during strenuous exercise. *(2 marks)*

(d) If strenuous exercise continues for a long period, the blood may be unable to supply sufficient oxygen or glucose to the muscles. The muscles however continue working for some time longer. In these circumstances:
  (i) what alternative source of glucose (carbohydrate) do the muscles use? *(1 mark)*
  (ii) how is energy still produced by the muscles? *(2 marks)*

When people suffer from kidney failure, they may be treated with a dialysis machine. The patient's blood is passed through the machine where the composition of the blood is adjusted.

See pages 248–9

(a) Name a waste substance, carried in the blood, which is removed by the dialysis machine. *(1 mark)*

(b) Doctors sometimes give these patients dialysis treatment, rather than a kidney transplant. Suggest **four** reasons for this. *(4 marks)*

In the 1880s, Louis Pasteur investigated the reasons for food going rotten. To check one of Pasteur's experiments, a student set up the two flasks shown in the diagram.

See page 270

Glass tube
Tight fitting cotton wool
Nutrient solution
Flask A
Flask B

After three days, the nutrient solution in both flasks had gone rotten.

(a) What makes the nutrient solution go rotten? *(1 mark)*

(b) The student then set up two more similar flasks. This time, she boiled the nutrient solution in both flasks for ten minutes.
  (i) Why did she boil the nutrient solution? *(1 mark)*
  (ii) Why did the nutrient solution in flask **A** go rotten? *(1 mark)*
  (iii) Why did the nutrient solution in flask **B** **not** go rotten? *(2 marks)*

GET IT RIGHT!
Always be as precise as possible in the use of terms, especially when writing about micoorganisms. The term 'germ' is far too vague and unscientific. It should not be used in examinations.

275

BUMP UP THE GRADE

There are a couple of other useful things that are worth discussing with students.

Use the spare minutes – revision can be less onerous if short periods of time that are otherwise wasted can be utilised. Try to persuade students to write down useful words or phrases when revising and to keep these with them. These can then be looked over, or used as short test, when they have a spare few minutes such as when on the bus/train, when waiting for dinner/someone to call/to go out. The cumulative effect of these sessions can be considerable and help bump the grade.

› *continues from previous page*

- Increased blood flow is needed to supply the oxygen and glucose and to remove the carbon dioxide.

(1 mark each point to a maximum of 2)

(d) (i) Glycogen (stored in the muscles). *(1 mark)*

(ii) • By the incomplete breakdown of glucose.
• During anaerobic respiration. *(1 mark each)*

4 (a) Urea *(1 mark)*

(b)
- Suitable for short-term treatment.
- No long-term drug treatment is required.
- No chance of rejection.
- No risk/less risk than during surgery (e.g. from anaesthetic).
- Operations unsuitable for certain individuals/operations risky for elderly or those in poor health.
- Risk of infection during/after transplant.
- Suitable kidneys are not always available for transplant/there is a long waiting list for kidneys.
- Dialysis is less painful than transplant surgery.

(1 mark for each point to a maximum of 4)

Accept any reverse points that correctly argue against transplant surgery.

5 (a) (The action of) bacteria/fungi/microorganisms/microbes. *(1 mark)*

(b) (i) To kill the bacteria/fungi/microorganisms/microbes (already present in the solution). *(1 mark)*

'To sterilise it' is an acceptable alternative.

(ii) Bacteria/fungi/microorganisms/microbes can get in (as they can enter down the straight tube). *(1 mark)*

(iii) • Bacteria/fungi/microorganisms/microbes cannot get in/get stuck
• In the curved part of the tube/because of the shape of the tube.

(1 mark each)

Notes